AF863385

# Lecture Notes in Mechanical Engineering

**Series Editors**

Francisco Cavas-Martínez, Departamento de Estructuras, Universidad Politécnica de Cartagena, Cartagena, Murcia, Spain

Fakher Chaari, National School of Engineers, University of Sfax, Sfax, Tunisia

Francesco Gherardini, Dipartimento di Ingegneria, Università di Modena e Reggio Emilia, Modena, Italy

Mohamed Haddar, National School of Engineers of Sfax (ENIS), Sfax, Tunisia

Vitalii Ivanov, Department of Manufacturing Engineering Machine and Tools, Sumy State University, Sumy, Ukraine

Young W. Kwon, Department of Manufacturing Engineering and Aerospace Engineering, Graduate School of Engineering and Applied Science, Monterey, CA, USA

Justyna Trojanowska, Poznan University of Technology, Poznan, Poland

**Lecture Notes in Mechanical Engineering (LNME)** publishes the latest developments in Mechanical Engineering—quickly, informally and with high quality. Original research reported in proceedings and post-proceedings represents the core of LNME. Volumes published in LNME embrace all aspects, subfields and new challenges of mechanical engineering. Topics in the series include:

- Engineering Design
- Machinery and Machine Elements
- Mechanical Structures and Stress Analysis
- Automotive Engineering
- Engine Technology
- Aerospace Technology and Astronautics
- Nanotechnology and Microengineering
- Control, Robotics, Mechatronics
- MEMS
- Theoretical and Applied Mechanics
- Dynamical Systems, Control
- Fluid Mechanics
- Engineering Thermodynamics, Heat and Mass Transfer
- Manufacturing
- Precision Engineering, Instrumentation, Measurement
- Materials Engineering
- Tribology and Surface Technology

To submit a proposal or request further information, please contact the Springer Editor of your location:

**China:** Dr. Mengchu Huang at mengchu.huang@springer.com
**India:** Priya Vyas at priya.vyas@springer.com
**Rest of Asia, Australia, New Zealand:** Swati Meherishi at swati.meherishi@springer.com
**All other countries:** Dr. Leontina Di Cecco at Leontina.dicecco@springer.com

To submit a proposal for a monograph, please check our Springer Tracts in Mechanical Engineering at http://www.springer.com/series/11693 or contact Leontina.dicecco@springer.com

**Indexed by SCOPUS. All books published in the series are submitted for consideration in Web of Science.**

More information about this series at http://www.springer.com/series/11236

Santhakumar Mohan · S. Shankar ·
G. Rajeshkumar
Editors

# Materials, Design, and Manufacturing for Sustainable Environment

## Select Proceedings of ICMDMSE 2020

Springer

*Editors*
Santhakumar Mohan
Department of Mechanical Engineering
Indian Institute of Technology Palakkad
Kozhippara, Kerala, India

S. Shankar
Department of Mechatronics Engineering
Kongu Engineering College
Erode, Tamil Nadu, India

G. Rajeshkumar
Department of Mechanical Engineering
PSG Institute of Technology and Applied Research
Coimbatore, Tamil Nadu, India

ISSN 2195-4356 ISSN 2195-4364 (electronic)
Lecture Notes in Mechanical Engineering
ISBN 978-981-15-9808-1 ISBN 978-981-15-9809-8 (eBook)
https://doi.org/10.1007/978-981-15-9809-8

© The Editor(s) (if applicable) and The Author(s), under exclusive license to Springer Nature Singapore Pte Ltd. 2021
This work is subject to copyright. All rights are solely and exclusively licensed by the Publisher, whether the whole or part of the material is concerned, specifically the rights of translation, reprinting, reuse of illustrations, recitation, broadcasting, reproduction on microfilms or in any other physical way, and transmission or information storage and retrieval, electronic adaptation, computer software, or by similar or dissimilar methodology now known or hereafter developed.
The use of general descriptive names, registered names, trademarks, service marks, etc. in this publication does not imply, even in the absence of a specific statement, that such names are exempt from the relevant protective laws and regulations and therefore free for general use.
The publisher, the authors and the editors are safe to assume that the advice and information in this book are believed to be true and accurate at the date of publication. Neither the publisher nor the authors or the editors give a warranty, expressed or implied, with respect to the material contained herein or for any errors or omissions that may have been made. The publisher remains neutral with regard to jurisdictional claims in published maps and institutional affiliations.

This Springer imprint is published by the registered company Springer Nature Singapore Pte Ltd.
The registered company address is: 152 Beach Road, #21-01/04 Gateway East, Singapore 189721, Singapore

# ICMDMSE 2020—Organizing Committee

## Chief Patron

Shri. L. Gopalakrishnan, Managing Trustee, PSG & Sons' Charities

## Patrons

Dr. P. V. Mohanram, Principal, PSG iTech
Dr. G. Chandramohan, Vice-Principal, PSG iTech
Dr. Jacob Chandapillai, Director, FCRI, Palakkad

## Chairs

Dr. N. Saravanakumar, Professor and Head, Mech, PSG iTech
Dr. S. Ram Mohan, Deputy Director, FCRI, Palakkad

## Co-chair

Dr. R. Ramesh, PSG iTech

## Organizing Secretaries

Dr. D. Elangovan, PSG iTech
Dr. P. Manojkumar, PSG iTech
Dr. G. Rajeshkumar, PSG iTech

## Publications

Dr. V. Rajkumar, PSG iTech
Dr. K. Sivananda Devi, PSG iTech
Mr. S. Nanthakumar, PSG iTech

## Public Relations

Mr. K. Anantharaman, PSG iTech
Mr. C. Gopalakrishnan, PSG iTech
Mr. J. Nagarjun, PSG iTech

## Publicity and Media

Mr. T. Premkumar, PSG iTech
Mr. M. Senthilvel, PSG iTech
Mr. R. Avinashkumar, PSG iTech

# Preface

The First International Conference on Materials, Design and Manufacturing for Sustainable Environment (ICMDMSE 2020) by the Department of Mechanical Engineering, PSG Institute of Technology and Applied Research, Coimbatore, India, was conducted on March 13 and 14, 2020, in association with Fluid Control Research Institute (FCRI), Palakkad. The scope of the conference includes original research and the latest advances in the field, focusing on advanced materials, vibration, tribology, finite element method (FEM), computational fluid dynamics (CFD), heat transfer, fluid mechanics, energy engineering, manufacturing technology, additive manufacturing, robotics and automation, automobile engineering, industrial and production engineering, Industry 4.0, MEMS and nanotechnology, sustainable environment, optimization techniques, condition monitoring system and new paradigms in technology management.

Over 162 papers were received from India and other countries. Based on the comments of reviewers and the scientific merits of the submitted manuscripts, 110 articles were accepted for presentation in the conference, out of which 65 papers have been selected for the publication in the Lecture Notes in Mechanical Engineering. The deliberations of ICMDMSE 2020 were categorized under (i) materials and design, (ii) manufacturing and industrial engineering, (iii) thermal and energy engineering, (iv) digital solutions and (v) allied applications. All registered authors discussed their ideas and professionally interacted with other delegates.

We want to express our thanks to the reviewers who contributed to the review process with their experience and scientific background. Only through their effort, it was possible to complete the review process in a short span of time. Furthermore, we thank the keynote speakers Dr. A. Senthilkumar, National University of Singapore; Dr. RamaGopal V Sarepaka, Senior Vice President, SPDT & IR Optics; Dr. C. Balaji, IIT Madras; Dr. Rajesh Ranganathan, CIT Coimbatore; and Dr. N. Siva Shanmugam, NIT Trichy, for sharing their in-depth research knowledge in their fields with the participants.

We would like to thank PSG Management, FCRI—Palakkad, Indian Institute of Technology Palakkad, all the participants, session chairs, session co-chairs, committee members, reviewers, international and national advisory committee members, co-sponsors and all the people who have directly or indirectly contributed to the success of this conference. Our special thanks to Dr. P. V. Mohanram, Dr. G. Chandramohan, Dr. N. Saravanakumar, Dr. R. Ramesh, Dr. D. Elangovan and Dr. P. Manojkumar, PSG iTech; and Dr. Jacob Chandapillai and Dr. S. Ram Mohan, FCRI, for their immense contributions. The editors would also like to thank Springer Editorial Team for their support and for publishing the papers as part of the *Lecture Notes in Mechanical Engineering* series.

Kozhippara, India
Erode, India
Coimbatore, India
May 2020

Dr. Santhakumar Mohan
Dr. S. Shankar
Dr. G. Rajeshkumar

# Contents

# About the Editors

**Dr. Santhakumar Mohan** got his Ph.D. (Robotics and Control) from the Indian Institute of Technology Madras, Chennai (India) in 2010. From June 2010 to March 2011, he worked as an assistant professor in the Department of Mechanical Engineering at National Institute of Technology Calicut (NITC), Kerala (India). He then worked as world-class university (WCU) postdoctoral fellow at the Korean Advanced Institute of Science and Technology (KAIST), Daejeon (Republic of Korea), in addition to this, he received another prestigious Brain Korea 21 (BK21) Post-doctoral fellowship with the same institute from September 2011 to March 2012. In 2012, he joined the faculty of Mechanical Engineering at the Indian Institute of Technology Indore. He is holding visiting faculty positions at IISc Bangalore, India, RWTH Aachen, Germany and ECN, France. His active research areas include underwater vehicle and underwater manipulator design and control, parallel robotic platforms, assistive robots, field and service robots, intelligent motion control, and, dynamic modeling and control of dynamic systems. Furthermore, he has received the outstanding young Scientist for the year 2014 from Korea Robotics Society, European Master on Advanced Robotics Plus (EMARO+) fellowship (2018-2019) and Alexander von Humboldt (AvH) Fellowship (2016-2017). He has published more than 100 articles in various journals and conference proceedings.

**Dr. S. Shankar** is currently working as a Professor at the Department of Mechatronics Engineering, Kongu Engineering College, Erode. He obtained his B.E. (Mechanical Engineering) and M.E. (Engineering Design) from Bharathiar University, Coimbatore, and Ph.D. from the Indian Institute of Technology Madras, Chennai. His major areas of research interests include computational mechanics, tribology, biomechanics, composite materials and human factors in design. He has published 84 papers in referred international journals and more than 50 conference publications. Dr. Shankar received the IE(I) Young Engineering Award in 2012-2013 and AICTE Career Award for Young Teachers in 2014-2015.

**Dr. G. Rajeshkumar** is currently an Associate Professor at the Department of Mechanical Engineering, PSG Institute of Technology and Applied Research, Coimbatore. He obtained his Bachelor of Engineering degree from Sri Ramakrishna Institute of Technology, Coimbatore, in 2010 and his Master of Engineering at Kongu Engineering College, Erode, in 2012 with a Gold Medal. He received his Ph.D. in Mechanical Engineering from Anna University, Chennai, in 2017. His major areas of research interests include polymer matrix composites, mechanism of needle insertion in soft materials, tribology and vibrations. He has published about 52 articles in peer-reviewed international journals and conference proceedings, and a textbook.

# Experimental Analysis of Tribological Behaviour of Jute Fiber-Reinforced Nanoclay Filled Epoxy Composites

**S. Ramakrishnan, K. Krishnamurthy, and G. Rajeshkumar**

**Abstract** The current study is aimed at studying the effect of reinforcing fiber length, fiber content (wt.%), surface modification and nanoclay dispersion on the tribological properties of jute fiber-reinforced epoxy composites (JFRECs). The epoxy matrix composites were developed by compression moulding technique for both untreated and NaOH treated, randomly oriented short jute fibers in the presence and absence of Cloisite 20A nanoclay. Furthermore, the fiber length (10–40 mm), the weight fraction of fiber (0–25%), NaOH molar concentration for surface modification (5 and 10%) and nanoclay dispersion (1–7 wt.%) were varied to obtain the optimum mechanical properties. From the obtained data, it was noted that mercerized (5% of NaOH) JFRECs along with 5 wt% of nanoclay showed superior mechanical behaviour as a result of augmented interfacial bonding amid the fiber and matrix. However, the properties got reduced for 7 wt.% of nanoclay dispersed composites due to its wide agglomeration. Furthermore, better tribological properties were noted for the epoxy composites reinforced with alkali (5% molar concentration of NaOH) treated jute fibers having 20 mm length and weight fraction of 20% along with 5 wt.% of nanoclay loading.

**Keywords** Jute fiber · Nanoclay · Tribological behaviour · Epoxy matrix composites

---

S. Ramakrishnan (✉)
Department of Mechanical Engineering, Kongu Engineering College, Erode, Tamil Nadu, India
e-mail: ramkidesign@gmail.com

K. Krishnamurthy
Department of Mechatronics Engineering, Kongu Engineering College, Erode, Tamil Nadu, India
e-mail: krish9842722881@gmail.com

G. Rajeshkumar
Department of Mechanical Engineering, PSG Institute of Technology and Applied Research, Coimbatore, Tamil Nadu, India
e-mail: grajeshkumar.me@gmail.com

© The Editor(s) (if applicable) and The Author(s), under exclusive license to Springer Nature Singapore Pte Ltd. 2021

S. Mohan et al. (eds.), *Materials, Design, and Manufacturing for Sustainable Environment*, Lecture Notes in Mechanical Engineering,
https://doi.org/10.1007/978-981-15-9809-8_1

# 1 Introduction

The study related to friction and wear properties of JFRECs is very essential to estimate the reliability of machine components. The scientific community is striving to improve the tribological (wear and friction) behaviours of lignocellulosic fiber composites to minimize the failure of components under tribological loads. Under tribological loading, the material deterioration occurs due to friction and induced energy dissipation between the surfaces in contact. Majority of the engineering components are subjected to frictional losses and the tribological study becomes imperative [1–4]. Barari et al. [5] investigated the curing kinetics of cellulose-based nano-fiber-reinforced epoxy composites and found that the composites reinforced with surface modified cellulose nano-fibers had better tribological properties. Friedrich et al. [6] have reported that the PMCs reinforced with traditional fillers hybridized with inorganic nanoparticles (1–3 vol.%) showed better tribological behaviour (minimum friction and minimum wear) against steel counterparts. Chand and Dwivedi [7] revealed that adding up the coupling agent at the melt mixing stage imparted improved resistance to wear for the jute fiber-reinforced polypropylene composites. El-Tayeb [8] found that polyester composites reinforced with short sugarcane fiber exhibited improved resistance to wear along with friction coefficient equivalent to that of polyester composite reinforced with glass fiber. Yousif et al. [9] observed that polyester composites reinforced with betelnut fiber had enhanced tribological (friction and wear) behaviours under the wet contact state with reference to the dried out contact state. Mylsamy and Rajendran [10] performed an empirical analysis to determine the tribological behaviour of epoxy composites incorporated with agave fiber. From the observations, it was revealed that epoxy composites reinforced with short agave fiber exhibited better wear behaviour when sliding against stainless steel. Shalwan and Yousif [11] observed that the tribological behaviour of composites was dependent on the volume/weight fraction, fiber surface modification, orientations and physical characteristics of the natural fibers. Prabhakar et al. [12] have reported that incorporation of controlled quantity of nanofillers improved the wear behaviour of nanocomposites considerably. Mahesha et al. noticed that the wear loss of basalt fiber-reinforced composites decreased with addition of $TiO_2$ and nanoclay. This phenomenon stems from the augmented adhesion between fiber and matrix along with better stress transfer between them [13]. Based on their extensive review, Devnani and Sinha [14] have affirmed that the mechanical behaviour, in particular wear resistance, can be augmented significantly by adding optimal concentration nanofillers to natural fiber-reinforced epoxy composites. Mohanty and Srivastava [15] observed that alumina nanoparticles enhanced the bonding that exists between the matrix and reinforcement and successively, improved the wear performance of polymer composites.

From the above studies, it is quite evident that the synergistic impact of fiber addition and nanofiller loading on the tribological behaviours of the polymer composites need to be explored deeply. Hence the present study is aimed at enhancing the tribological behaviours of epoxy matrix composites by reinforcing them with untreated and mercerized jute fibers along with nanoclay.

# 2 Materials and Methods

## 2.1 *Materials*

In the current analysis, the epoxy resin (LY556) and its own amine-based hardener (HY951) were effectively used for the matrix scheme. Jute fibers at various proportions were used as reinforcement and nanoclay (Cloisite 20A) was utilized as additional reinforcement. The resin and hardener were acquired from the Covai Seenu & Company at Coimbatore, Tamil Nadu, South India. The jute fiber was purchased from Raju Traders at Kolkata, West Bengal, India. Moreover, Cloisite 20A nanoclay (modified organically) was acquired from Southern Clay Products (Texas, North America). The densities of the jute fiber, resin, hardener and nanoclay are 1.3–1.46 g/cm$^3$, 1.15–1.20 g/cm$^3$, 0.968 g/cm$^3$ and 1.77 g/cm$^3$, respectively.

## 2.2 *Surface Modification of the Jute Fiber*

Jute fibers were first rinsed thoroughly with the distilled water and then dried for nearly 3–4 h. The dried jute fibers were then separately soaked in two different molar concentrations of NaOH solutions (5 and 10%) at ambient temperature of 28 °C for 4 h. The fiber to solution proportion of 1:20 was utilized in the current study. The NaOH residues present on the fiber surface were washed away by running water and distilled water. Furthermore, the jute fibers were taken out, normalized with acetic acid (diluted) and dried out at room temperature for about 24 h.

## 2.3 *Fabrication of the Composites*

In the current work, the compression moulding technique was used for fabricating the composite panels at various fiber weight fractions (5, 10, 15, 20 and 25%), different fiber lengths (10, 20, 30 and 40 mm) and weight quantity of nanoclay (1, 3, 5 and 7%), respectively. Specially prepared steel dies were used to prepare the composite specimen with randomly oriented short jute fibers. The composite plates were prepared in two categories: (i) composites using the untreated and mercerized fibers and (ii) composites using the mercerized fibers along with nanoclay. The steel rollers were used to ensure uniform fiber distribution and orientation. The resin was mixed with its amine hardener in the ratio of 10: to prepare the matrix system. For the first category of composites, the matrix system was poured onto the jute fibers, which were first spread in the female die. Following this process, the mould was closed and subjected to a compressive force of 4.5 bar for about 6 h. For the second category of composites, the nanoclay was uniformly dispersed in the epoxy resin with the aid of magnetic stirring (2 h) coupled with high-intensity ultrasonication (1 h). Composite panels of size 300 × 300 × 3 $mm^3$ were exposed to a post-curing procedure at 60 °C in the oven for 4 h before preparing the required test specimen.

### 2.4 *Wear Analysis*

The wear properties of the composite specimens were determined by using the pin-on-disk device as per the ASTM G99 standard. The wear test under dry sliding condition was performed on the specimens, fabricated by normal hand layup method using MS mould (chrome-plated) at ambient temperature. Three identical specimens having size 75 × 75 × 10 $mm^3$ were tested for each case with different normal loads (10, 20 and 30 N), sliding speeds (1, 2 and 3 m/s) and sliding distances (1, 2 and 3 km) and the average of those values was recorded for further studies.

## 3 Results and Discussions

### 3.1 *Wear Analysis*

#### 3.1.1 Effect of Fiber Length and Weight Fraction (wt.%) on Specific Wear Rate (SWR)

Figure 1a–c present the effects of the fiber length and weight fraction on Specific Wear Rate (SWR) of untreated JFRECs at different values of normal load $F_N$ (10, 20 and 30 N). From the obtained results, the epoxy composites reinforced with 10 mm length fibers were found to exhibit higher SWR for all weight fractions, followed by 30 and 40 mm. The minimum wear rate was exhibited by the epoxy composites reinforced with 20 mm jute fibers. This could be due to the influence of the critical fiber length along with better fiber-resin interface adhesion. The epoxy composites reinforced with 10 mm fibers suffered higher pull-outs compared to other samples and therefore, the SWR was high. However, JFRECs reinforced with 40 and 30 mm length fibers showed slightly higher SWR than 20 mm length fibers due to fiber entanglement effect.

It was noted that SWR values decreased gradually with the fiber weight fraction upto 20% and then slightly increased with a further increase in wt.%. This trend confirms the general reinforcing ability of jute fibers. At all values of $F_N$, the

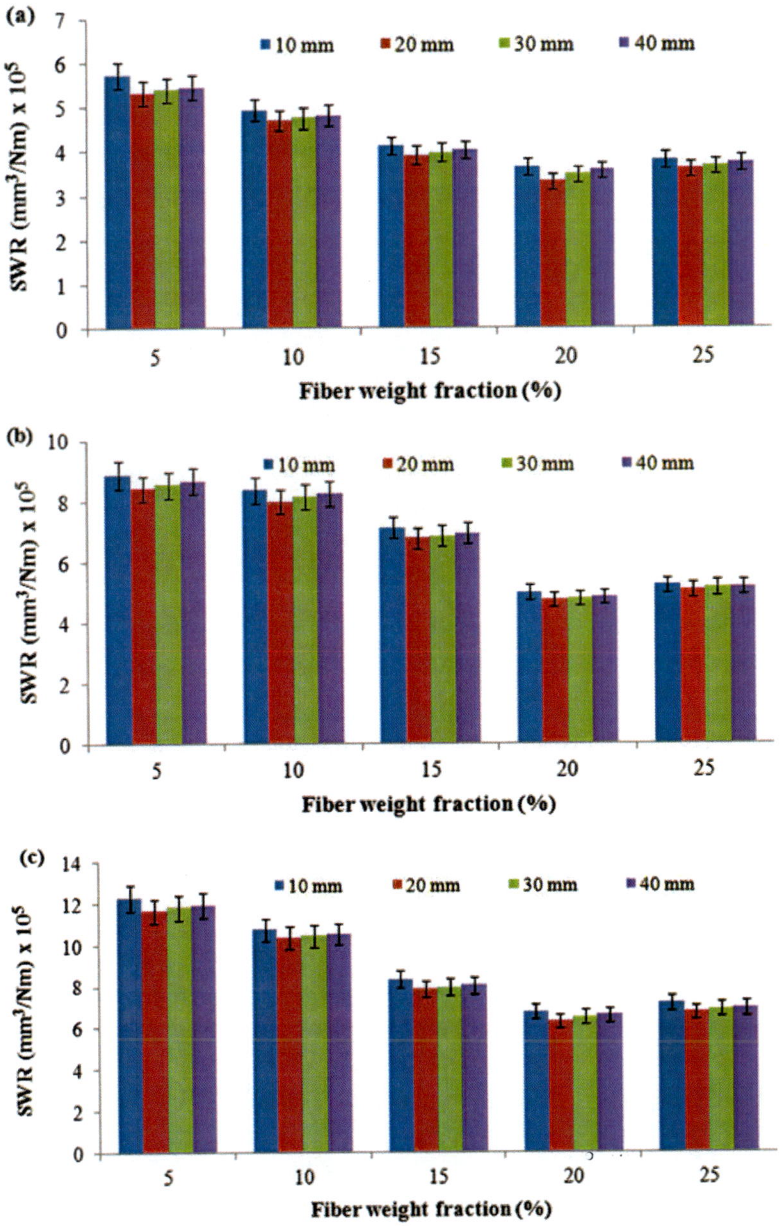


**Fig. 1** Effect of fiber length and weight fraction (%) on SWR at **a** 10 N, **b** 20 N, and **c** 30 N

composites reinforced with 20 wt.% of jute fibers exhibited minimum SWR followed by 25 wt.% of jute fibers. The incorporation of jute fiber in the matrix protects it from abrasion due to sliding. However, the epoxy composites that are reinforced with 25 wt.% of jute fibers had slightly higher SWR owing to poor adhesion bonding existing between the fiber and matrix.

#### 3.1.2 Effect of NaOH Treatment on SWR

The tribological behaviour of the NFRPCs is normally influenced by fiber-matrix adhesion bonding. The hydrophilic nature of ligno-cellulosic fibers inhibits interfacial bonding between them and the polymer matrix. However, the alkali treatment of jute fibers improves the interfacial bonding with matrix, which in turn augments the tribological behaviour (SWR) of JFRECs. From the preceding section, it is evident that the epoxy composite samples prepared with 20 mm length jute fibers exhibited better tribological behaviour (lower SWR) at 20 wt.%.

Therefore, the consequence of NaOH treatment on the SWR of the above-mentioned composite sample is discussed for the sake of conciseness. Figure 2a–c present the effect of mercerization on SWR of the epoxy composites at various loads of 10, 20 and 30 N, respectively. It was noted that the SWR of epoxy composites decreased with NaOH treatment. This behaviour is credited to the superior adhesion between alkali-treated fibers and the matrix. However, it slightly increases for 10% of NaOH treated fibers with respect to 5% of NaOH treated fibers. This decrease is owing to the reduced fiber-matrix adhesion due to fiber damage at higher concentration of NaOH (>5%).

The untreated fibers cannot provide sustained support to the matrix against rubbing for a longer distance of sliding due to weak adhesion and succumb to fiber pull-outs. However, the alkali-treated fibers provide sustained support to the matrix due to enhanced adhesion bonding and fiber pull-outs are also reduced.

#### 3.1.3 Effect of Nanoclay Addition on SWR

From the previous section results, it is evident that epoxy composites that are reinforced with 5% treated jute fibers of length 20 mm demonstrated minimum SWR at 20 wt.%. However, literature indicate that the mechanical behaviour of NFRPCs can be further improved by incorporating measured quantity of nanoclay. Hence, an attempt was carried out to further boost the tribological property of the above JFRECs by incorporating nanoclay at different weight fractions (1, 3, 5 and 7%). The effect of nanoclay incorporation on SWR is shown in Fig. 3a–c at $F_N$ values of 10, 20 and 30 N, respectively. It can be noted that SWR of the composites decreased gradually with nanoclay addition upto 5 wt.%. The probable basis behind the drop in SWR is the augmented adhesion bonding existing between matrix and fiber owing to the presence of nanoclay platelets at the interface [16]. However, at elevated nanoclay loading of 7 wt.% the SWR slightly increases owing to the wide agglomeration effect.

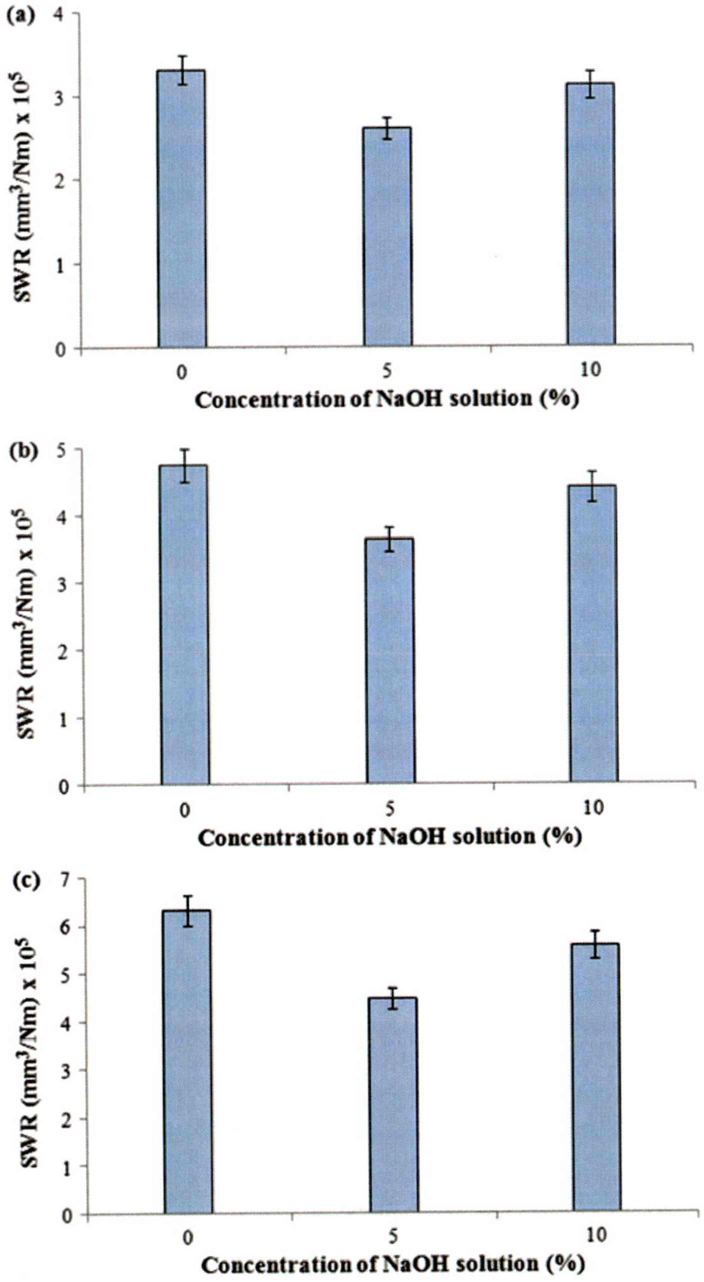


**Fig. 2** Effect of NaOH treatment on SWR at **a** 10 N, **b** 20 N and **c** 30 N

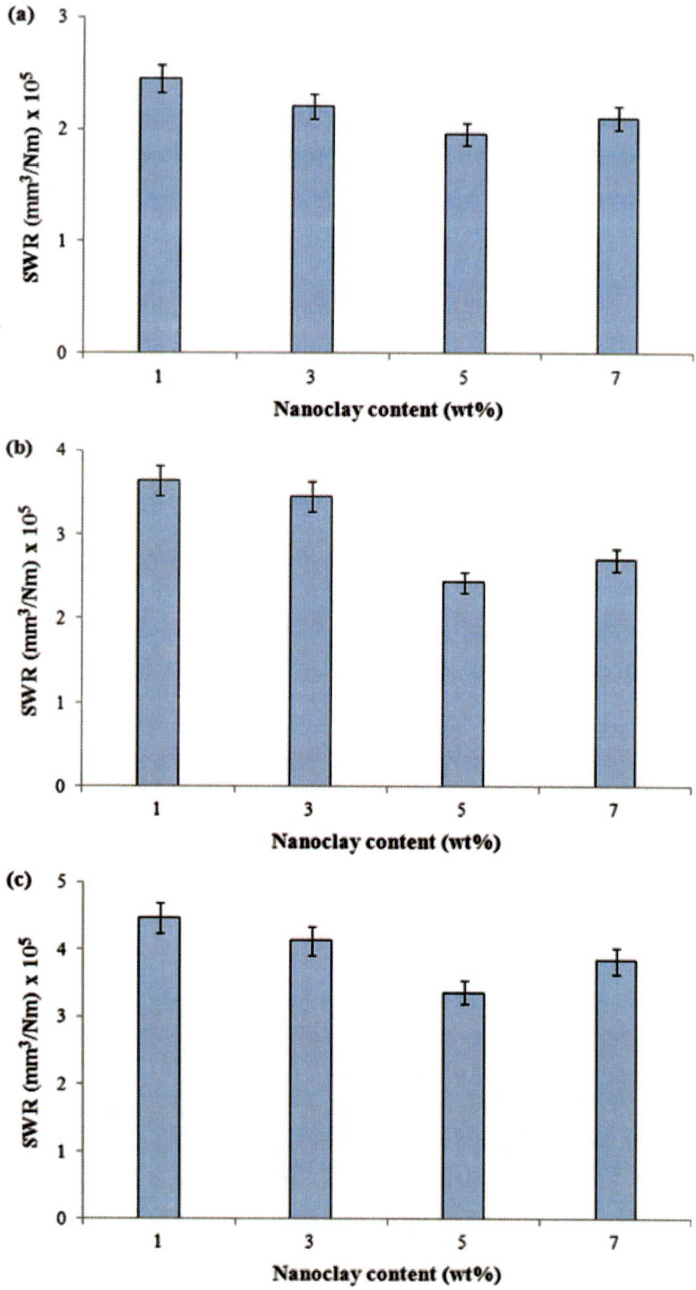


**Fig. 3** Effect of nanoclay addition on SWR at **a** 10 N, **b** 20 N and **c** 30 N

### 3.1.4 Effect of Fiber Length and Weight Fraction on Coefficient of Friction (μ)

The dual effects of fiber length and wt.% (untreated fiber condition) on $\mu$ values of JFRECs are revealed in Fig. 4a–c. It was noted that coefficient of friction ($\mu$) is higher for 10 mm fiber-reinforced composites followed by 40, 30 and 20 mm fiber incorporated epoxy composites.

The 20 mm jute fiber-reinforced composites showed the lowest $\mu$ value owing to better embedding of jute fibers in the epoxy matrix at critical length. The $\mu$ values were found to decline with raise in normal load ($F_N$) and sliding speed ($V_s$). However, they increase when the sliding distance ($D_s$) increases.

The drop in $\mu$ value with increase in load can be credited to the influence of load-induced temperature rise between the contact surfaces [17]. When the temperature increases, thermal stresses are developed due to non-uniform thermal gradients. As a result, the fiber-matrix interfacial bonding is weakened and fiber pull-outs occur. The fiber detachment from the matrix leads to low values of $\mu$ under dry sliding.

Furthermore, at low value of $F_N$ the coefficient of friction value is higher owing to the mechanical oriented inter-locking of irregularities between contact surfaces [18]. From the graphs, it is evident that $\mu$ value is lower for composites prepared with 20 mm length of untreated fibers at all weight fractions.

### 3.1.5 Effect of NaOH Treatment on μ

The jute fibers are subjected to alkali treatment at two different concentrations of NaOH (5 and 10%) to improve the wear behaviour of JFRECs and the acquired results are presented in Fig. 5a–c at 10, 20 and 30 N, respectively.

From the preceding section, it is observed that the JFRECs reinforced using 20 wt% of 20 mm length jute fibers exhibited better tribological behaviour (lower $\mu$). Therefore, the effect of NaOH treatment on the $\mu$ value of the above-mentioned epoxy composite sample is discussed for the sake of simplicity and brevity.

It can be noted that the coefficient of friction was lower for 5% of NaOH JFRECs with reference to that of 10% of NaOH treated and untreated JFRECs, respectively. This confirms the fact that the alkali-treated jute fibers shows superior interfacial adhesion bond with epoxy matrix. Due to better adhesion bonding, the $\mu$ value decreases for treated fiber composites. The same trend is noted for all weight fractions.

### 3.1.6 Effect of Nanoclay Addition on μ

Similar to SWR, the coefficient of friction ($\mu$) was observed to be lower for JFRECs reinforced with 5% treated jute fibers of 20 mm length at 20 wt.%. Furthermore, an endeavour was carried out to augment the tribological behaviour of the JFRECs by adding different quantities of nanoclay (1–7 wt.%) and the results are plotted in

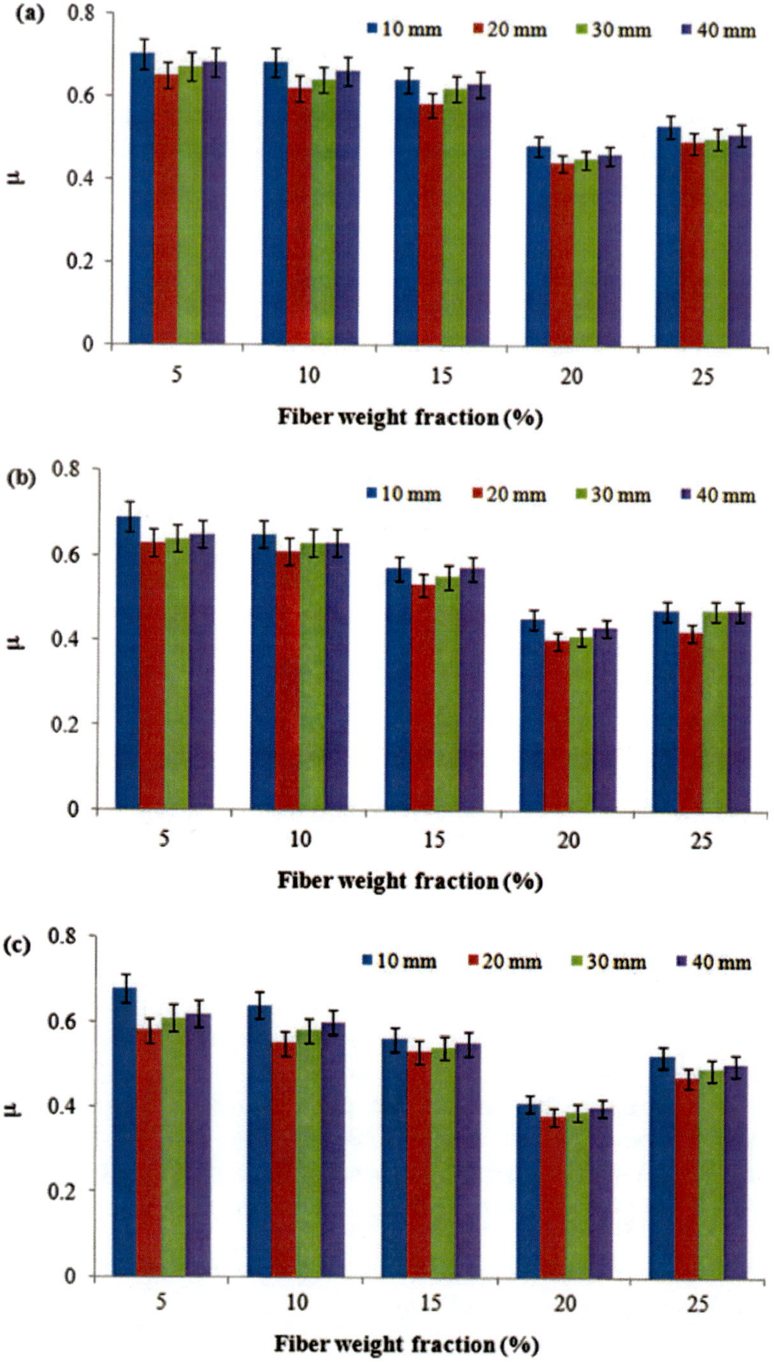


**Fig. 4** Effect of fiber length and weight fraction (%) on $\mu$ at **a** 10 N, **b** 20 N and **c** 30 N

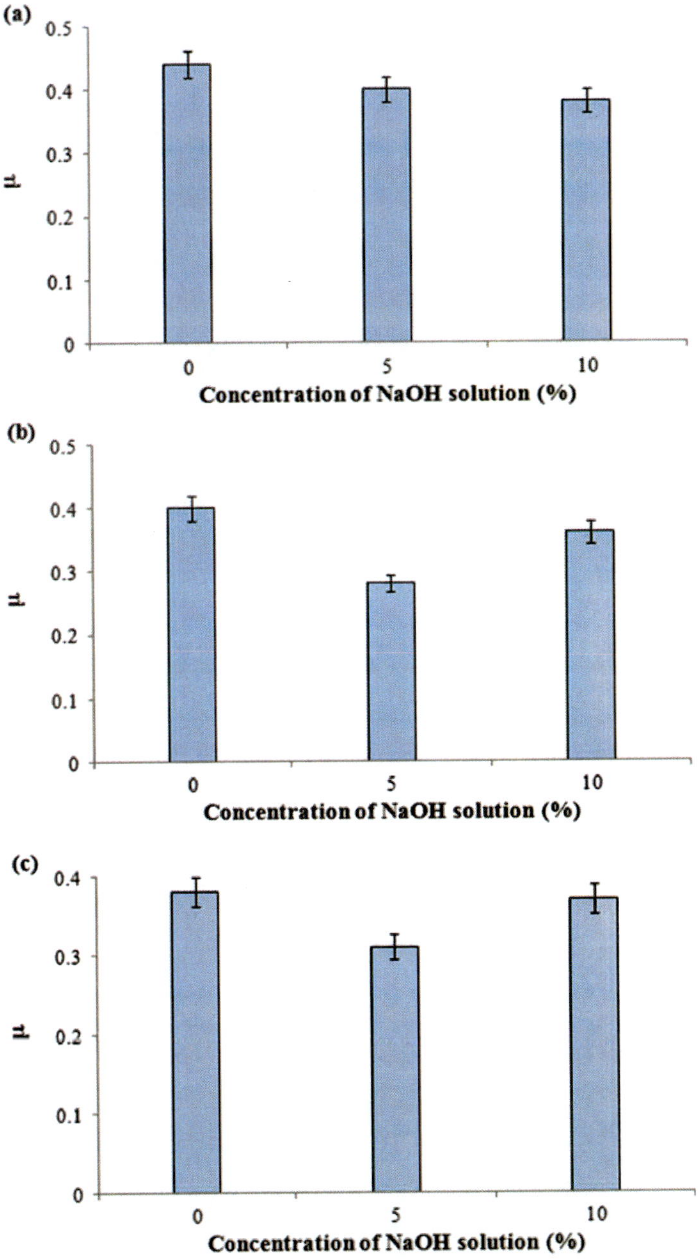


**Fig. 5** Effect of NaOH treatment on $\mu$ at **a** 10 N, **b** 20 N and **c** 30 N

Fig. 6a–c for 10, 20 and 30 N, respectively. It is worth pointing that the 5 wt.% of nanoclay loaded composites had lower coefficient of friction when compared with 1,

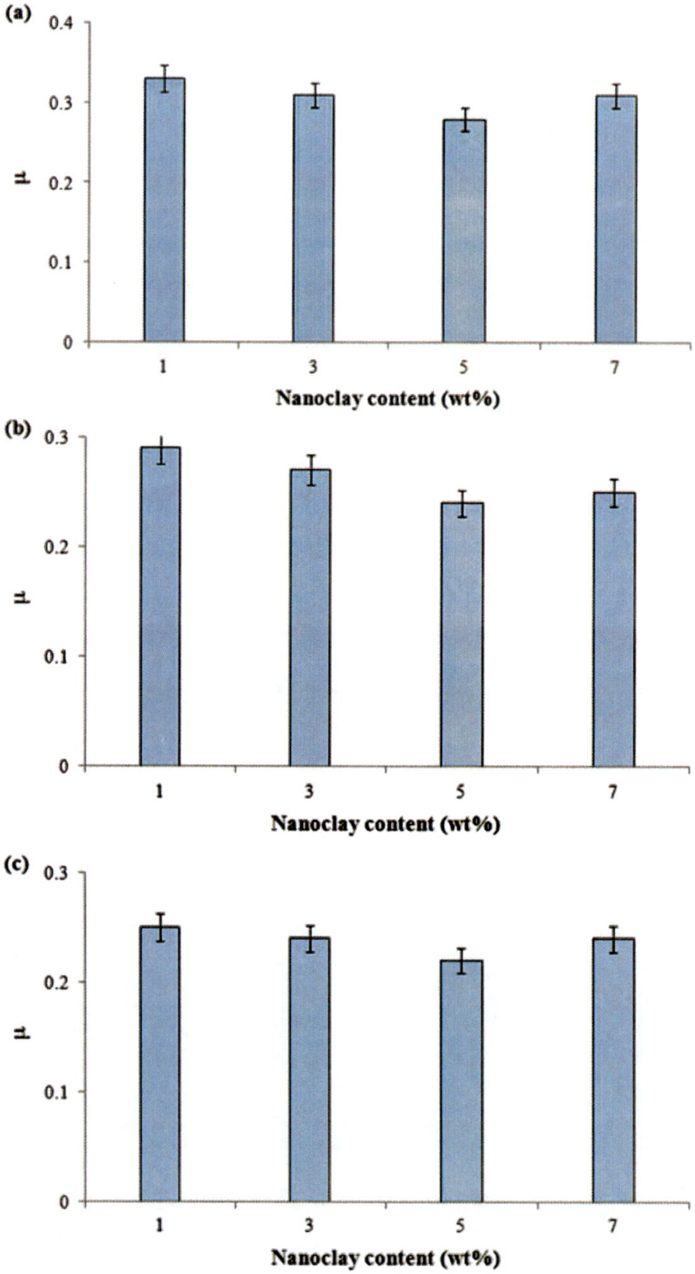


**Fig. 6** Effect of nanoclay addition on $\mu$ at **a** 10 N, **b** 20 N and **c** 30 N

3 and 7 wt.% nanoclay loaded samples. The improvement in tribological behaviour (minimum $\mu$) is attributed to the boosted interfacial adhesion bond existing between the mercerized fibers and matrix in the presence of nanoclay. However, the reinforcing effect of nanoclay is moderately reduced due to the agglomeration effect and $\mu$ value moderately increases at 7 wt.% of nanoclay.

## 4 Conclusion

The jute fiber-reinforced epoxy composites (JFRECs) were successfully fabricated by using the compression moulding method and the tribological properties (SWR and $\mu$) of these composites were probed. The effects of fiber length, fiber content, fiber surface modification and nanoclay accumulation on the aforementioned properties were evaluated and the following results were acquired.

- The wear behaviour of the JFRECs is greatly influenced by both weight fraction and length of fibers.
- Better tribological properties (lower SWR and $\mu$ values) were noted in JFRECs reinforced with 20 wt.% of 5% treated jute fibers having 20 mm length and 5 wt.% of nanoclay dispersion.
- The investigational results point out that surface modification of fiber and dispersion of nanoclay improves the interfacial adhesion existing between jute fiber and epoxy matrix and thereby, enhanced the tribological properties of the polymer composites.

The results indicate that the tribological behaviours of the JFRECs can be positively bettered with incorporation of jute fiber and nanoclay (Cloisite 20A).

## References

1. Nirmal U, Hashim J, Ahmad MM (2015) A review on tribological performance of natural fibre polymeric composites. Tribol Int 83:77–104
2. Rajeshkumar G (2020) A new study on tribological performance of phoenix Sp. fiber-reinforced epoxy composites. J Natural Fibers. https://doi.org/10.1080/15440478.2020.1724235
3. Rajeshkumar G, Hariharan V (2012) Free vibration analysis of hybrid-composite beams. In: IEEE-International conference on advances in engineering, science and management (ICAESM-2012), 30, 31 Mar 2012, Nagapattinam. IEEE, pp 165–170
4. Rajeshkumar G (2020) An experimental study on the interdependence of mercerization, moisture absorption and mechanical properties of sustainable Phoenix sp. fibre-reinforced epoxy composites. J Ind Text 49(9):1233–1251
5. Barari B, Omrani E, Moghadam AD, Menezes PM, Pillai KM, Rohatgi PK (2016) Mechanical, physical and tribological characterization of nanocellulose fibers reinforced bio-epoxy composites: an attempt to fabricate and scale the 'green' composite. Carbohyd Polym. https://doi.org/10.1016/j.carbpol.2016.03.097
6. Friedrich K, Zhang Z, Schlarb AK (2005) Effects of various fillers on the sliding wear of polymer composites. Compos Sci Technol 65(15–16):2329–2343

7. Chand N, Dwivedi U (2006) Effect of coupling agent on abrasive wear behaviour of chopped jute fibre-reinforced polypropylene composites. Wear 261(10):1057–1063
8. El-Tayeb N (2008) A study on the potential of sugarcane fibers/polyester composite for tribological applications. Wear 265(1–2):223–235
9. Yousif B, Lau ST, McWilliam S (2010) Polyester composite based on betelnut fibre for tribological applications. Tribol Int 43(1–2):503–511
10. Mylsamy K, Rajendran I (2011) Influence of fibre length on the wear behaviour of chopped agave americana fibre reinforced epoxy composites. Tribol Lett 44(1):75–80
11. Shalwan A, Yousif B (2013) In state of art: mechanical and tribological behaviour of polymeric composites based on natural fibres. Mater Des 48:14–24
12. Prabhakar PK, Debnath S, Ganesan R, Palanikumar K (2018) A review of mechanical and tribological behaviour of polymer composite materials. IOP Conf Ser Mater Sci Eng. https://doi.org/10.1088/1757-899X/344/1/01201
13. Mahesha CR, Shivarudraiah MN, Rajesh M (2017) Role of nanofillers on mechanical and dry sliding wear behavior of basalt-epoxy nanocomposites. Mater Today Proc 4:8192–8199
14. Devnani GL, Sinha S (2019) Effect of nanofillers on the properties of natural fiber reinforced polymer composites. Mater Today Proc 18:647–654
15. Mohanty A, Srivastava VK (2015) Tribological behavior of particles and fibers reinforced hybrid nanocomposites. Tribol Trans. https://doi.org/10.1080/10402004.2015.1039681
16. Ramakrishnan S, Krishnamurthy K, Rajasekar R, Rajeshkumar G (2018) An experimental study on the effect of nano-clay addition on mechanical and water absorption behaviour of jute fibre reinforced epoxy composites. J Ind Text. https://doi.org/10.1177/1528083718792915
17. Rajeshkumar G (2020) Effect of sodium hydroxide treatment on dry sliding wear behavior of Phoenix sp. fiber reinforced polymer composites. J Ind Text. https://doi.org/10.1177/1528083720918948
18. Bajpai PK, Singh I, Madaan J (2013) Tribological behavior of natural fiber reinforced PLA composites. Wear 297(1–2):829–840

# A Robust Motion Control Scheme of an Underwater Robot with Tiltable Thrusters

**Jagadeesh Kadiyam, P. S. V. S. Sai Kumar, Santhakumar Mohan, and D. Deshmukh**

**Abstract** A modern control approach for a submerged robot with tiltable thrusters is explored in this paper. The rotatable action capability of the four thrusters makes the system framework as over-actuated with only six degrees of freedom (DOF) to be carried out. The dynamic model of the submerged robot on account of rotatable angles results in highly nonlinear thrust vector mapping, where linearization of the model leads to the rejection of possibly advantageous nonlinearities. This issue can be abstained by utilizing model-based nonlinear control schemes such as backstepping in the presence of continuous rotatable action resulting in a better control action. The stability of the system is ensured by the Lyapunov stability criterion by tuning the control gain constants with the RMS values of tracking error. Finally, the numerical simulations demonstrate the cascaded nature of the derived tracking control, where the tracking errors asymptotically reduce to close vicinity of zero.

**Keywords** Rotatable/Tiltable thrusters · Robust control · Backstepping · Lyapunov stability · Gain tuning · Trajectory tracking

J. Kadiyam · D. Deshmukh
Department of Mechanical Engineering, Indian Institute of Technology Indore, Indore, Madhya Pradesh, India
e-mail: kadiyamjagadeesh@gmail.com

D. Deshmukh
e-mail: dldeshmukh@iiti.ac.in

P. S. V. S. Sai Kumar
Department of Mechanical Engineering, National Institute of Technology Tiruchirappalli, Tiruchirappalli, Tamil Nadu, India
e-mail: psvssaikumar1998@gmail.com

S. Mohan (✉)
Department of Mechanical Engineering, Indian Institute of Technology Palakkad, Palakkad, Kerala, India
e-mail: santhakumar@iitpkd.ac.in

© The Editor(s) (if applicable) and The Author(s), under exclusive license to Springer Nature Singapore Pte Ltd. 2021

S. Mohan et al. (eds.), *Materials, Design, and Manufacturing for Sustainable Environment*, Lecture Notes in Mechanical Engineering,
https://doi.org/10.1007/978-981-15-9809-8_2

# 1 Introduction

The space technology advanced to a level where rovers could land on far-off locations such as Mars and leave the footprints of man on the moon. On the other hand, oceans cover 70% of the total surface of planet earth, yet with considerable uncertainty, just five percent of seas and one percent of the ocean surface could be explored. This is because of numerous reasons, for example, unstructured, hostile marine conditions make it hard to comprehend, and, mainly, investigating these regions is costly and tedious. Human occupied underwater vehicles (HOVs) and remotely operable underwater vehicles (ROVs) are built and launched to achieve different tasks by handling these deficiencies. Few examples include seafloor studies and profiling of seas, monitoring, and maintenance of semi-submersible platforms and intercontinental submarine links. The present scenario of the advanced underwater technology demands reliable and specialized vehicles for mission-specific applications and completely autonomous operation. The demand for improved autonomy combined with the coupled dynamic behavior of the systems, which are nonlinear and time-varying expands the intricacy in the vehicle guidance, navigation, and control.

In this scenario, autonomous underwater vehicles (AUVs) are becoming popular since they are capable of undertaking different kinds of missions and explore extreme ocean depths while at the same time being a low-cost alternative to ROVs. Many observation class vehicles use control surfaces, such as rudders and diving planes, where the force/moment is dependent on the velocity of the vehicle. However, the control surfaces are ineffective during hovering, thus making it unsuitable for applications where the subsystems interfere with vehicle dynamics [1]. While a large segment of the present-day accessible AUVs from the literature are used for non-contact type missions, such as observations, seafloor mapping, or probing the ocean segments, research on work-class AUV often called intervention vehicles is still in progress. Due to the random disturbances from submerged currents and unavoidable reactions from the motions of subsystems along with physical system complexities causes the nonlinearity in system dynamics, thus resulting in a complex controller configuration [2]. Although several advanced controllers were designed and expertly executed before, the highly nonlinear dynamics and the coupled DOF in the present problem demand an alternative approach toward the design structure of a robust controller. Likewise, it is not possible to design a pure time-invariant, asymptotically stable state feedback controller to globally satisfy the required performance, due to the very behavior of the closed-loop system [3]. Thus, earlier studies have been attempting to viably deal with these complexities with the improvement of stable controllers with slight variations in already available linear and nonlinear control techniques and implemented on submerged robots [4, 5], depending on the application.

A nonlinear controller is implemented for an underwater vehicle with rotatable thrusters' in this paper, similar to the TTURT [6, 7], and Odyssey IV [8]. The vehicle comprises four thrusters mounted diagonally on a rectangular-shaped hull. These thrusters are tilted individually to attain 6-DOF motion, as shown in Fig. 1, where a

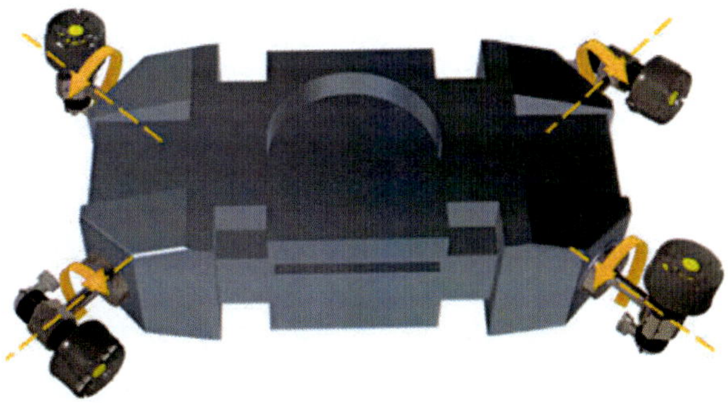

**Fig. 1** Model vehicle of the rectangular hull with rotatable thrusters

nonlinear control is needed as the rotatable angles of these thrusters [2] and results in a highly nonlinear thrust force mapping similar to that of quadcopters [9]. Thus, the backstepping is implemented for the control design due to the nonlinearity; as many wheel mobile robots [10] have successfully implemented the backstepping algorithm [11, 12].

## 2 Research Rationale

The more the number of thrusters, the better is the control and accuracy for maneuvering, hovering, and station keeping operations. However, the capacity of the power supply is a significant limitation that results in additional weight to the vehicle, thus reducing its efficiency [6]. The model for rotatable thrusters' vehicle is nonlinear due to the rotatable angles.

The rotatable thrusters lead to complex control design in comparison to the control design of a fixed thrusters version, as modeled in [13].

In this manner, taking motivation from a portion of the previously mentioned works, a nonlinear control for a submerged robotic vehicle, where the thrusters are tilted individually, is proposed in the present paper. A backstepping scheme is utilized for tracking control to follow the desired trajectory in a 3D domain. The tuning of the control gain constants is carried out using the tracking error RMS values. Unlike the vectored thrust model proposed in [6], where the forward and rear thrusters are synchronized separately, the principal attributes of the current vehicle platform are that the four thrusters are independently rotatable about the pivots passing through the diagonal axis of the vehicle hull. Another important aspect is that the previous study implemented a switching-based controller applicable only for two arrangements. This means that the thruster is either vertical or horizontal, with no control over intermittent angles, whereas the present vehicle is capable of control for the continuous thruster angle change.

## 3 The Organizational Flow of the Paper

The remainder of the paper is composed as follows. A detailed dialog on the mathematical model of underwater vehicle is introduced in Sect. 4, including hydrodynamics of the vehicle and dynamics of the thruster. Section 5 details the controller configuration using backstepping, to take care of the nonlinear issue along with the determination of system stability criterion. Gain constants tuning is portrayed in Sect. 6. Simulations of the tracking problem are performed, and the results are shown in Sect. 7. In the end, concluding remarks are given in Sect. 8.

## 4 Mathematical Model

### *4.1 Robot Design*

As the present design has eight inputs [14] to accomplish the essential response in the form of six independent outputs (pose), the present system is termed as over-actuated. The vehicle consists of four thruster arms hosting the thrusters themselves to provide inputs as thrust forces, along with four other inputs delivered by the internal servos to rotate every thruster, different from the study [6], where two servos control the tilting action of the all the thrusters. The thrusters, as shown in Fig. 1, are positioned at an angle of 45° to the vertical plane, and various movements conceivable are sketched in Fig. 2 for a six DOF vehicle represented in the vehicle frame. The submerged robot studied in the paper is a combination of five independent rigid bodies with a relative motion among each other whose physical and geometrical parameters are given in Table 1.

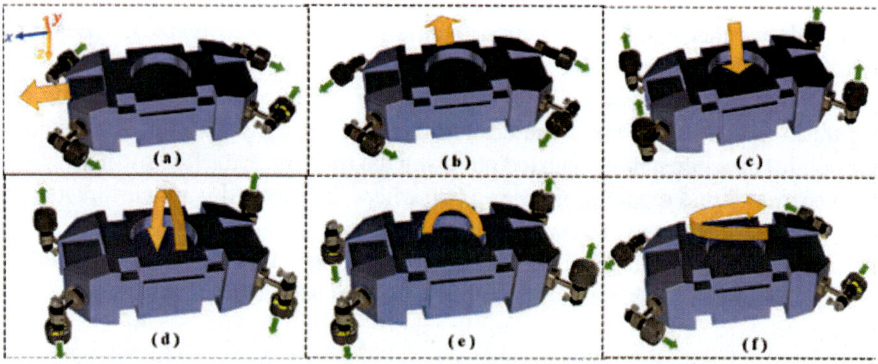

**Fig. 2** **a** Surge; **b** sway; **c** heave; **d** roll; **e** pitch; **f** yaw

**Table 1** Physical and geometrical properties [6]

| Property | Narrative |
|---|---|
| Geometric dimensions | Width = 326.00 mm<br>Length = 755.00 mm<br>Height = 280.00 mm |
| Vehicle mass | 57.10 kg |
| Vehicle bulk density | 1021.00 kg/m$^3$ |
| Center of gravity (C.G) in the vehicle frame | (0, 0, 0) |
| Center of buoyancy (C.B) in the vehicle frame | (0, 0, 0) |
| Moment of inertia of the system | $I_{xx} = 1.1$ kg-m$^2$; $I_{yy} = 3.6$ kg-m$^2$<br>$I_{zz} = 4.099$ kg-m$^2$; $I_{xy} = -0.018$ kg-m$^2$<br>$I_{yz} = -9.662 \times 10^{-5}$ kg-m$^2$; $I_{zx} = -$ 0.016 kg-m$^2$ |

## 4.2 *Assumptions*

Few sensible suppositions are contemplated to ease the mathematical modeling for generalized analysis:

- The rectangular (parallelepiped) shape has three symmetric planes. Thus, it is safe to ignore the off-diagonal elements of the inertia matrix.
- The angular momentum due to the rotatable action of the thruster is ignored.
- The restoring forces and moments are ignored as the robot's C.G coincides with the C.B. and has neutral buoyancy.
- Due to the low velocity of the vehicle, lift forces can be ignored.
- The disturbances due to density changes over a range of depth, oceanic currents, acting on the robot are ignored.
- The effects of damping due to the wave and tidal frequency are insignificant.
- Acceleration due to gravity does not change both in temporal and spatial scale (9.81 m/s$^2$).
- The same gain values are selected for position and orientation vectors, $\eta_1$, and $\eta_2$.
- The interactions between thrusters and hull-thruster are ignored.

## 4.3 *Preliminary Description of the States*

For the sake of analyzing the motion of the underwater robot in space, it is necessary to specify two reference frames: the moving reference frame or the vehicle-fixed frame conveniently located at the center of gravity of the robot. The movement of the vehicle-fixed frame is expressed about the inertial frame, which is generally attached to the earth's surface [15].

The vector $\eta \in R^6$ represents the generalized coordinates stated in the inertial frame. The linear and angular velocities vector denoted by $\nu \in R^6$ is stated in the robot frame.

$$\eta = \left[\eta_1^T\ \eta_2^T\right]^T;\ \text{where}\ \boldsymbol{\eta}_1 = (\boldsymbol{x}\ \boldsymbol{y}\ \boldsymbol{z})^T\ \text{and}\ \boldsymbol{\eta}_2 = (\boldsymbol{\varphi}\ \boldsymbol{\theta}\ \boldsymbol{\psi})^T$$
$$\boldsymbol{\nu} = \left[\boldsymbol{\nu}_1^T\ \boldsymbol{\nu}_2^T\right]^T;\ \text{where}\ \boldsymbol{\nu}_1 = \left(\boldsymbol{u}\ \boldsymbol{v}\ \boldsymbol{w}\right)^T\ \text{and}\ \boldsymbol{\nu}_2 = (\boldsymbol{p}\ \boldsymbol{q}\ \boldsymbol{r})^T$$

The vector $\eta_1$ expressed in the inertial reference frame represents the position of the robot along the surge, sway, and heave directions. Similarly, the vector $\eta_2$ represents the Euler angles in the inertial reference frame commonly identified as roll, pitch, and yaw [16]. The time derivative of $\eta$ is represented by $\dot{\eta}$ also expressed in the inertial frame, while $\nu$ indicates the linear and angular velocity vector of the vehicle with reference to inertial frame expressed in the vehicle frame of reference.

## 4.4 Motion Equations of the Submerged Robot

The mathematical model of the submerged vehicle is established based on the Euler–Lagrangian formulation derived through the principle of virtual work and the D'Alembert's principle. In general, the motion equations of the submerged robot are presented in the vehicle-fixed frame. The governing equations of the whole robot model are shown in Eqs. (1) and (2) [15]:

Dynamics equation of the submerged robot is given by

$$\boldsymbol{\tau} = \boldsymbol{M}\dot{\boldsymbol{\nu}} + \boldsymbol{C}(\boldsymbol{\nu})\boldsymbol{\nu} + \boldsymbol{D}(\boldsymbol{\nu})\boldsymbol{\nu} + \boldsymbol{g}(\boldsymbol{\eta}) \tag{1}$$

where the inertia tensor, $\boldsymbol{M}$, is the $6 \times 6$ matrix comprising rigid body and added mass terms of the robot model. $\boldsymbol{C}(\boldsymbol{\nu})$ represents a $6 \times 6$ matrix involving the centripetal and Coriolis effects on the robot. $\boldsymbol{D}(\boldsymbol{\nu})$, Damping matrix is a diagonal matrix and positive definite containing the hydrodynamic damping terms. The $6 \times 1$ vector $\boldsymbol{g}(\boldsymbol{\eta})$ represents the restoring forces and moments due to buoyancy forces and gravity. $\boldsymbol{\tau}$ is $6 \times 1$ vector of the forces/moment on the submerged robot designated in the vehicle frame.

The kinematics of the robot is presented by

$$\dot{\boldsymbol{\eta}} = \boldsymbol{J}(\boldsymbol{\eta})\boldsymbol{\nu} \tag{2}$$

A matrix $\boldsymbol{J}(\boldsymbol{\eta})$ is specified to map the velocities represented in the vehicle frame to the inertial frame by using the Euler coordinate angles. The number of rows of the $\boldsymbol{J}(\boldsymbol{\eta})$ is equal to the degrees of freedom of the system, and the number of columns represents the generalized coordinates of the vehicle pose [17]. The transformation matrix for the submerged robot in space [15], as shown in Eq. (3). Here, cos(*A*), sin(*A*), and tan(*A*) are abbreviated as *cA*, *sA*, and *tA* for brevity.

$$J(\eta) = \begin{bmatrix} c\theta c\psi & c\psi s\theta s\varphi - s\psi c\varphi & c\psi c\theta s\varphi + s\psi s\varphi & 0 & 0 & 0 \\ c\theta s\psi & s\psi s\theta s\varphi + c\psi c\varphi & s\psi s\theta c\varphi - c\psi s\varphi & 0 & 0 & 0 \\ -s\theta & s\varphi c\theta & c\varphi c\theta & 0 & 0 & 0 \\ 0 & 0 & 0 & 1 & s\varphi t\theta & c\varphi t\theta \\ 0 & 0 & 0 & 0 & c\varphi & -s\varphi \\ 0 & 0 & 0 & 0 & s\varphi/c\theta & c\varphi/c\theta \end{bmatrix} \quad (3)$$

### 4.4.1 Inertia Tensor

The mass distribution of the vehicle is represented by the inertia tensor or the mass matrix, where the vehicle is considered as a rigid body. Due to this consideration, the mass matrix in Eq. (4) consists of rigid body terms as well as added mass terms. The rigid body terms contribution is written in matrix form as $\boldsymbol{M_{RB}}$, which comprises the mass terms, mass moment of inertia, and mass products of inertia terms. The contribution of inertia due to the added mass is written in matrix form as $\boldsymbol{M_A}$ in Eq. (5) during the motion. By taking advantage of the inertia matrix properties, for example, symmetricity and positive definiteness, the inertia tensor attains a diagonal form. This is possible when the vehicle is symmetric in all the planes, and the vehicle frame is located at the C.G of the vehicle.

$$\boldsymbol{M} = \boldsymbol{M_{RB}} + \boldsymbol{M_A} \quad (4)$$

$\boldsymbol{M_A}$ is due to the inertia contribution of the added mass. The fluid encompassing the outer hull is accelerated with the vehicle. This accelerated fluid exerts a reaction force that is equal in magnitude and opposite in the direction of the robot. Thus, an extra force is necessary to reach this acceleration by the fluid [18], which is directly dependent on the form of the underwater robot. Thus, the added mass matrix terms are dependent on the added mass and potential damping, which are further reliant on the frequency and forward speed of the robot. Similar to the rigid body matrix, the added mass matrix can also be simplified into a diagonal matrix considering the symmetry, low speed, and coincidence of origin of the vehicle frame with the center of gravity of the robot [16].

$$\boldsymbol{M_A} = -\begin{bmatrix} Z_{11} & 0 & 0 & 0 & 0 & 0 \\ 0 & Z_{22} & 0 & 0 & 0 & 0 \\ 0 & 0 & Z_{33} & 0 & 0 & 0 \\ 0 & 0 & 0 & Z_{44} & 0 & 0 \\ 0 & 0 & 0 & 0 & Z_{55} & 0 \\ 0 & 0 & 0 & 0 & 0 & Z_{66} \end{bmatrix} \quad (5)$$

By using the strip theory, the inertia and added mass terms are derived [15, 19], under certain assumptions, for example, the fluid is inviscid, no circulation in the

**Table 2** Estimation of the added mass parameters

| Parameter | Approximation |
|---|---|
| $Z_{11}$ | $0.1 * m$ |
| $Z_{22}$ | $1.25\ \rho l b^2$ |
| $Z_{33}$ | $1.175\ \rho l a^2$ |
| $Z_{44}$ | $\rho l^2\ (0.086\ a^3 + 0.0875\ b^3)$ |
| $Z_{55}$ | $\rho a^2 (0.098\ l^3 + 0.115\ b^3)$ |
| $Z_{66}$ | $\rho b^2\ (0.12\ a^3 + 0.104\ b^3)$ |

Here $m$—the mass of the robot, $l$—length of the vehicle, $a$—width of the vehicle, $b$—width of the vehicle, and $\rho$—density of the robot.

fluid, negligible free surface effects due to complete submergence of the body in an unrestrained fluid domain. Using strip theory, the added mass and inertia components are derived analytically by [15, 19–21]. The added mass coefficients are shown in Eq. (5) and their estimated values are given in Table 2. These coefficients are derived by supposing the robot form as a parallelepiped.

Based on the assumptions and properties of the inertia matrix, $\boldsymbol{M}$ can be expressed as a speed independent, positive definite, and constant matrix, where

$$\boldsymbol{M} = \boldsymbol{M}^{\mathrm{T}} > 0; \quad \dot{\boldsymbol{M}} = 0$$

Thus, the estimated mass matrix [6] can be expressed as (6):

$$\boldsymbol{M} = \begin{bmatrix} 96.3 & 0 & 0 & 0 & 0 & 0 \\ 0 & 133.04 & 0 & 0 & 0 & 0 \\ 0 & 0 & 168.57 & 0 & 0 & 0 \\ 0 & 0 & 0 & 4.47 & 0 & 0 \\ 0 & 0 & 0 & 0 & 9.26 & 0 \\ 0 & 0 & 0 & 0 & 0 & 8.1 \end{bmatrix} \tag{6}$$

### 4.4.2 Coriolis and Centripetal Matrix

Like the inertia tensor, the Coriolis and centripetal matrix $C(\nu)$ is also made from two terms, a rigid body term denoted by $\boldsymbol{C_{RB}}(\boldsymbol{\nu})$ and an added mass term designated by $\boldsymbol{C_A}(\boldsymbol{\nu})$ as shown in Eq. (7). If the Coriolis and centripetal matrix is symbolized in terms of general matrix array, $C_{ij}$, then for all, $i = j$ the matrix components are characterized as Coriolis terms. For all, $i \neq j$, the matrix components are characterized as centripetal forces. The parameterization of $\boldsymbol{C}(\boldsymbol{\nu})$ as a skew-symmetric matrix is always possible for a rigid body moving through an ideal fluid.

$$\boldsymbol{C}(\boldsymbol{v}) = \boldsymbol{C}_{RB}(\boldsymbol{v}) + \boldsymbol{C}_{A}(\boldsymbol{v}) \tag{7}$$

As with the case of mass matrix, the strip theory is used to arrive at the components of the Coriolis and centripetal tensor, considering the explicit expressions of $\boldsymbol{C}_{RB}(\boldsymbol{v})$ and $\boldsymbol{C}_{A}(\boldsymbol{v})$ found in [6, 15]. Thus, the estimated Coriolis and centripetal tensor is given by Eq. (8):

$$\boldsymbol{C}(\boldsymbol{v}) = \begin{bmatrix} 0 & 0 & 0 & 0 & 168.57w & -133.04v \\ 0 & 0 & 0 & -168.57w & 0 & 96.30u \\ 0 & 0 & 0 & 133.04v & -96.30u & 0 \\ 0 & 168.57w & -133.04v & 0 & 8.10r & -9.26q \\ -168.57w & 0 & 96.30u & -8.10r & 0 & 4.47p \\ 133.04v & -96.30u & 0 & 9.26q & -4.47p & 0 \end{bmatrix} \tag{8}$$

### 4.4.3 Hydrodynamic Damping Matrix

Several factors, such as the potential damping, viscous damping, skin friction, and vortex shedding, contribute to the damping characteristics of a robot while moving underwater [15, 21]. These factors induce coupled dynamics, which is mostly due to the dissipative forces, for example, lift and drag forces arising from the viscosity of the fluid. Thus, the complexity of the model is reduced by a few approximations. The drag forces are only responsible for damping and thus neglecting any lift produced in the vehicle due to the low speed of the robot. As the coupled disturbances are assumed negligible, the hydrodynamic damping takes the form of an un-coupled diagonal matrix [15]. In the present model, only the quadratic terms are considered to derive $\boldsymbol{D}(\boldsymbol{v})$. The coefficients of drag due to linear velocity terms, i.e., the surge, sway, and heave, are arrived using the computational fluid dynamics (CFD) approach, and the coefficients of drag for rotational velocities are calculated analytically [22]. Thus, the estimated $\boldsymbol{D}(\boldsymbol{v})$, damping tensor [6, 19], is given by Eq. (9):

$$\boldsymbol{D}(\boldsymbol{v}) = \begin{bmatrix} 34.5|u| & 0 & 0 & 0 & 0 & 0 \\ 0 & 104.40|v| & 0 & 0 & 0 & 0 \\ 0 & 0 & 146.50|w| & 0 & 0 & 0 \\ 0 & 0 & 0 & 0.68|p| & 0 & 0 \\ 0 & 0 & 0 & 0 & 5.34|q| & 0 \\ 0 & 0 & 0 & 0 & 0 & 3.07|r| \end{bmatrix} \tag{9}$$

where $\boldsymbol{D}(\boldsymbol{v}) > 0, \forall\, v \in R^6$.

### 4.4.4 Restoring Forces and Moments

In a fully submerged environment, the submerged body experiences certain restoring forces and moments $g(\eta)$ due to the presence of the gravitational and buoyant forces [15]. The restoring forces and moments are ignored as the robot's C.G coincides with the C.B. and has neutral buoyancy.

## 4.5 *Thruster Dynamics*

Most under-actuated underwater vehicles are accommodated with thrusters for main propulsion and control surfaces [23] for achieving the directional change. However, the present robot uses all thrusters for controlling the vehicle in the 6 DOF domain. As already discussed, the thrusters rotate about the diagonal axes placed at an inclination 45° to the longitudinal plane of the vehicle frame. This arrangement helps to provide the thrust components in both surge and sway direction. The rotatable angles induce a nonlinear relationship between the generalized forces '$\boldsymbol{\tau}$' acting on the robot and the thrust force vector '$\boldsymbol{f}$'. The thrust force generated is a function of operational variables such as the density of the fluid, the cross-sectional area, length of the thruster, the flow rate between input–output of the thrusters, and the propeller diameter and interactions between subsystems [1]. Considering these factors is beyond the scope of the present paper.

At this juncture, we define a non-time-invariant matrix '$[\boldsymbol{B}]$', in Eq. (10), that maps the generalized forces/moment vectors expressed in the vehicle frame with a thrust force vector '$f$' expressed in the thruster frame.

$$\boldsymbol{\tau} = [\boldsymbol{B}]\boldsymbol{f} \tag{10}$$

where

$$[\mathbf{B}] = \begin{bmatrix} -1/\sqrt{2} & -1/\sqrt{2} & 1/\sqrt{2} & 1/\sqrt{2} & 0 & 0 & 0 & 0 \\ -1/\sqrt{2} & 1/\sqrt{2} & 1/\sqrt{2} & -1/\sqrt{2} & 0 & 0 & 0 & 0 \\ 0 & 0 & 0 & 0 & -1 & -1 & -1 & -1 \\ 0 & 0 & 0 & 0 & 0.5a & -0.5a & -0.5a & 0.5a \\ 0 & 0 & 0 & 0 & 0.5l & 0.5l & -0.5l & -0.5l \\ \frac{-0.5(a+l)}{\sqrt{2}} & \frac{0.5(a+l)}{\sqrt{2}} & \frac{-0.5(a+l)}{\sqrt{2}} & \frac{0.5(a+l)}{\sqrt{2}} & 0 & 0 & 0 & 0 \end{bmatrix}.$$

The thrust force components of the four rotatable thrusters *A*, *B*, *C*, *D* are defined by an $8 \times 1$ vector $\boldsymbol{f} = [f_1\, f_2\, f_3\, f_4\, f_5\, f_6\, f_7\, f_8]^{\mathrm{T}}$. Out of the $8 \times 1$ vector, $f_1$ , $f_2$ , $f_3$ and $f_4$ represents the horizontal components, and $f_5$ , $f_6$ , $f_7$ and $f_8$ denotes the vertically resolved resultant thrust forces, $F_a$, $F_b$, $F_c$, and $F_d$. The resultant thrust forces, along with their components, are shown in Fig. 3, where the tilting angle is denoted by the angle obtained by the arctangent of thrust force components. Thus,

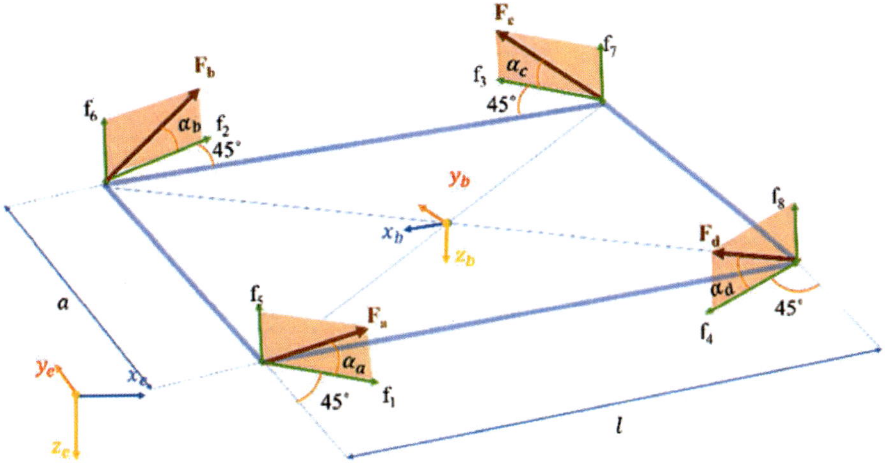


**Fig. 3** Horizontal and vertical components of resultant thruster forces

the four tilting angles subtended by the resultant thrust forces from the horizontal plane can be expressed as shown in Eq. (11)

$$\begin{aligned} \alpha_a &= \tan^{-1}\left(\tfrac{f_5}{f_1}\right) \quad \alpha_b = \tan^{-1}\left(\tfrac{f_6}{f_2}\right) \\ \alpha_c &= \tan^{-1}\left(\tfrac{f_7}{f_3}\right) \quad \alpha_d = \tan^{-1}\left(\tfrac{f_8}{f_4}\right) \end{aligned} \tag{11}$$

The presence of these tilting angles in the thruster dynamic equations results in a nonlinear model, where the tracking control with a continuous rotating angle poses a challenge in the presence of nonlinearity.

By employing the equations of motion, the thrust forces predict the resultant motion. However, to control the robot, the control inputs (thrust forces) are forced to track the trajectory based on inverse dynamics. The accelerations are solved from the control inputs using the dynamics in Eq. (1), where it is necessary for inertia matrix '$\boldsymbol{M}$' to be invertible.

Different control strategies can be employed using these equations of motion for a trajectory tracking problem, as discussed in the upcoming sections.

## 5 Nonlinear Control Strategy

The nonlinear control strategy for an underwater robot to track the desired trajectory is deliberated in this section. A controller requires to stabilize the robot within a bounded error range such that the system follows the desired trajectory within this error bound. The control system designed needs good robustness so that it can cope with uncertainties while it is suppressing noise and disturbances.

The control scheme chosen for the current problem is appropriate for the strict feedback system of equations [24]. It is a recursive procedure whose stability criteria of the state depend on the choice of a Lyapunov candidate function by designing an appropriate feedback control law called backstepping. This control law is created to handle the complex systems by breaking the design strategy into a system of reduced-order subsystems that radiate out from an irreducible subsystem, often stabilized using some other known method [25]. These lower-order systems often exhibit flexibility that can solve tracking, robust control, and complex stabilization problems [26]. In some conditions, for example, the paucity of controllability, this scheme can overwhelm singularities to a certain extent [27].

The central idea of this scheme is to exploit certain state variables as intermediate control variables, often termed as virtual controls dependent on the dynamics of the state of the system. Thus, by utilizing these intermediate controllers, the whole control scheme of the system is designed. There are two subsystems in the present system: the dynamics being outer loop and kinematics being an inner loop.

A strict feedback control-affine second-order system is written as Eq. (12):

$$\begin{aligned} \dot{z}_1 &= \boldsymbol{h}_1(z_1) + \boldsymbol{h}_2(z_1)z_2 \\ \dot{z}_2 &= \boldsymbol{h}_1(z_1, z_2) + \boldsymbol{h}_2(z_1, z_2)\boldsymbol{u} \end{aligned} \tag{12}$$

Let us consider the kinematics Eq. (2) as a first subsystem and dynamics Eq. (1) as a second subsystem. The control objective is to stabilize the system and regulate the state to the origin for any initial condition. After separating the dynamical system as a set of irreducible sub systems, each is dealt separately for the stabilization of its variable.

From Eqs. (1) to (2), $\boldsymbol{\eta}$ is the state variable which is required to be stabilized to the origin using $\boldsymbol{v}$. The controller is designed to stabilize the state of the vehicle by defining the tracking error term $\tilde{\boldsymbol{\eta}}$ in Eq. (13) and driving it to zero. To do this, $\boldsymbol{\eta}_d, \dot{\boldsymbol{\eta}}_d, \ddot{\boldsymbol{\eta}}_d$ are defined as the known, continuous, and bounded functions.

$$\tilde{\boldsymbol{\eta}} = \boldsymbol{\eta}_d - \boldsymbol{\eta} \tag{13}$$

By considering the tracking error $\tilde{\boldsymbol{\eta}}$ as a decreasing exponential function in Eq. (14), the state $\boldsymbol{\eta}$ reaches asymptotically stability with the constant decay rate of $K_1$ and a scaling factor vector '$A$'.

$$\tilde{\boldsymbol{\eta}} = \mathbf{A}\mathbf{e}^{-K_1 t} \tag{14}$$

This leaves us with an ordinary differential equation, Eq. (15)

$$\dot{\tilde{\boldsymbol{\eta}}} + K_1\tilde{\boldsymbol{\eta}} = 0 \tag{15}$$

Now, the derivative of tracking error Eq. (13) can be written as Eq. (16)

$$\dot{\tilde{\eta}} = \dot{\eta}_d - J(\eta)\nu \tag{16}$$

A virtual desired velocity controller $\nu_c$ is inserted into Eq. (16) by defining a backstepping error $\tilde{\nu}$ to be written as Eq. (17)

$$\dot{\tilde{\eta}} = \dot{\eta}_d - J(\eta)\nu_c + J(\eta)\tilde{\nu} \tag{17}$$

where $\tilde{\nu}$ is defined as

$$\tilde{\nu} = \nu_c - \nu \tag{18}$$

and from Eqs. (2) to (16) $\nu_c$ is chosen as Eq. (19) and by incorporating Eqs. (15), (19) is expanded to Eq. (20)

$$\nu_c = J_d^{-1}(\eta)\dot{\eta} \tag{19}$$

$$\nu_c = J_d^{-1}(\eta)\left[\dot{\eta}_d + K_1\tilde{\eta}\right] \tag{20}$$

and, therefore, Eq. (2) leaves us with Eq. (21)

$$\therefore \dot{\eta} = J(\eta)J_d^{-1}(\eta)\left[\dot{\eta}_d + K_1\tilde{\eta}\right] - J(\eta)\tilde{\nu} \tag{21}$$

Since $\nu$ and $\nu_c$ are initiated from different values, the control effort requires to force $\nu$ to follow $\nu_c$. Accordingly, the control $\tau_c$ could be designed to track the following output. By taking the derivative of the backstepping error from Eq. (18), we have Eq. (22)

$$\dot{\tilde{\nu}} = \dot{\nu}_c - \dot{\nu} \tag{22}$$

Also, from Eq. (20), the derivative virtual controller is written as Eq. (23)

$$\dot{\nu}_c = \dot{J}_d^{-1}(\eta)\left[\dot{\eta}_d + K_1\tilde{\eta}\right] + J_d^{-1}(\eta)\left[\ddot{\eta}_d + K_1\dot{\tilde{\eta}}\right] \tag{23}$$

The second subsystem of the dynamical system can be written as Eq. (24)

$$\dot{\boldsymbol{\nu}} = \boldsymbol{M}^{-1}\{\boldsymbol{\tau} - \boldsymbol{o}(\boldsymbol{\nu}, \boldsymbol{\eta})\} \tag{24}$$

where $\boldsymbol{o}(\boldsymbol{\nu}, \boldsymbol{\eta}) = \boldsymbol{C}(\boldsymbol{\nu})\boldsymbol{\nu} + \boldsymbol{D}(\boldsymbol{\nu})\boldsymbol{\nu} + \boldsymbol{g}(\boldsymbol{\eta})$.

Therefore, choosing the thrust force vector $\boldsymbol{f}$ from the vector of control forces/moments $\boldsymbol{\tau}_c$, from Eq. (10) is done by implementing the stability analysis.

***Lyapunov Stability Analysis***

Consider the following Lyapunov candidate function $\boldsymbol{V}$ for the chosen dynamical system as in Eq. (25), which is positive definite:

$$\boldsymbol{V}(\tilde{\boldsymbol{\eta}}, \tilde{\boldsymbol{\nu}}) = \frac{1}{2}(\tilde{\boldsymbol{\eta}}^{\mathrm{T}}\tilde{\boldsymbol{\eta}} + \tilde{\boldsymbol{\nu}}^{\mathrm{T}}\tilde{\boldsymbol{\nu}}) \tag{25}$$

where $\boldsymbol{V}(\tilde{\boldsymbol{\eta}}, \tilde{\boldsymbol{\nu}}) > 0, \forall\, \tilde{\boldsymbol{\eta}}$ and $\tilde{\boldsymbol{\nu}} \neq 0$;

Thus, to assure the system with global asymptotic stability, it is required to satisfy that time derivative of Lyapunov candidate function, $\dot{\boldsymbol{V}}$ is negative definite.

This is possible by choosing the control vector input $\boldsymbol{\tau}_c$, as in Eq. (26) such that $\dot{\boldsymbol{V}} \leq 0$.

$$\begin{aligned}\boldsymbol{\tau}_c = \boldsymbol{o}(\boldsymbol{\eta}, \boldsymbol{\nu}) + \boldsymbol{M}\Big\{\boldsymbol{J}_d^{-1}(\boldsymbol{\eta})\Big[\ddot{\boldsymbol{\eta}}_d + K_1\dot{\tilde{\boldsymbol{\eta}}}\Big] + \boldsymbol{J}_d(\boldsymbol{\eta})\big[\dot{\boldsymbol{\eta}}_d + K_1\tilde{\boldsymbol{\eta}}\big]\Big\} \\ + \boldsymbol{M}\big[K_2\tilde{\boldsymbol{\nu}} + \boldsymbol{J}^{\mathrm{T}}(\boldsymbol{\eta})\tilde{\boldsymbol{\eta}}\big]\end{aligned} \tag{26}$$

By substituting Eq. (26) in Eq. (10), the vector of thrust components is written, as shown in Eq. (27):

$$\begin{aligned}\boldsymbol{f} = [\boldsymbol{B}]^{+}\boldsymbol{o}(\boldsymbol{\eta}, \boldsymbol{\nu}) + [\boldsymbol{B}]^{+}\boldsymbol{M}\Big\{\boldsymbol{J}_d^{-1}(\boldsymbol{\eta})\Big[\ddot{\boldsymbol{\eta}}_d + K_1\dot{\tilde{\boldsymbol{\eta}}}\Big]\Big\} \\ + \boldsymbol{J}_d(\boldsymbol{\eta})\big[\dot{\boldsymbol{\eta}}_d + K_1\tilde{\boldsymbol{\eta}}\big] + [\boldsymbol{B}]^{+}\boldsymbol{M}\big[K_2\tilde{\boldsymbol{\nu}} + \boldsymbol{J}^{\mathrm{T}}(\boldsymbol{\eta})\tilde{\boldsymbol{\eta}}\big]\end{aligned} \tag{27}$$

As the non-square matrix '[$\boldsymbol{B}$]' is not invertible, the vector of thrust components is arrived by calculating the Moore–Penrose pseudo-inverse $[\boldsymbol{B}]^{+}$ [28].

$$\dot{\boldsymbol{V}}(\tilde{\boldsymbol{\eta}}, \tilde{\boldsymbol{\nu}}) = -\tilde{\boldsymbol{\eta}}^{\mathrm{T}}K_1\tilde{\boldsymbol{\eta}} - \tilde{\boldsymbol{\nu}}^{\mathrm{T}}K_2\tilde{\boldsymbol{\nu}} \leq 0$$

Hence, it is proved that $\dot{\boldsymbol{V}}(\tilde{\boldsymbol{\eta}}, \tilde{\boldsymbol{\nu}}) \leq 0; \forall t, \tilde{\boldsymbol{\eta}}, \tilde{\boldsymbol{\nu}}$; when the gain constants $K_1$ and $K_2 > 0$. Thus, the nonlinear controller designed in the present paper ensures global asymptotic stability. The next section details the tuning of gain constants by using the RMS error of the pose.

# 6 Tuning of Gain Constants

## 6.1 Trajectory Generation

The gains chosen for the controller are to be robust enough to assure the stability and performance fit for any trajectory. Thus, a reference trajectory, as shown in Fig. 4, which consists of motion along *x*-, *y*-, and *z*-directions, is chosen such that there is an influence of motion in the surge, sway, and heave directions. It is constructed by interpolation along nine Cartesian coordinates using Hermite cubic curves. Each of the segment is connected to its adjacent segments with $C^2$ continuity. The endpoint of the first curve is the start point for the second curve. By forcing the acceleration at the via points as continuous, the system automatically chooses the velocities at the via points [17].

The cubic polynomial defining each segment can be written as Eq. (28):

$$\eta(t) = a_0 + a_1 t + a_2 t^2 + a_3 t^3 \tag{28}$$

where

$$\begin{aligned}
a_0 &= \eta_0 \\
a_1 &= \dot{\eta}_0 \\
a_2 &= \frac{3}{t_f^2}(\eta_f - \eta_0) - \frac{2}{t_f}\dot{\eta}_0 - \frac{1}{t_f}\dot{\eta}_f \\
a_3 &= -\frac{2}{t_f^3}(\eta_f - \eta_0) + \frac{1}{t_f^2}(\dot{\eta}_f + \dot{\eta}_0)
\end{aligned}$$

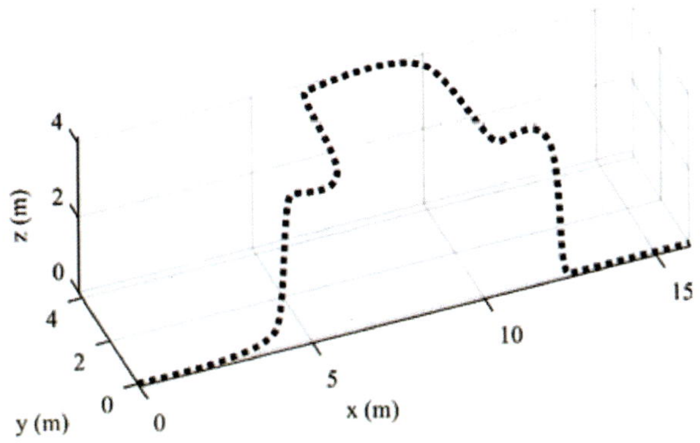


surge-heave-surge-sway-surge-sway-surge-heave-surge

**Fig. 4** Reference trajectory for gain optimization

where $\eta(t)$ is the desired pose vector dependent on time, $\eta_0$ and $\eta_f$ are desired position vector at the starting and ending point of that corresponding segment. $\dot{\eta}_0$ and $\dot{\eta}_f$ are the desired velocity vectors at starting and ending points, respectively. The insistence on continuous acceleration at the via points offers to calculate the velocities at the via points. The matrix turns out to be tridiagonal and can be easily solved [29].

Apart from the initial location, at every other intermediate point, the robot is forced to track the trajectory with a finite velocity. It follows the trajectory, as shown in Fig. 4.

### 6.2 Tuning of Gain Constants

The RMS values of pose error are computed for an extensive range of values of $K_1$ and $K_2$. The $K_1$, $K_2$, $e_{\text{rms}}$ values are plotted along $x$-, $y$-, $z$-directions of the coordinate system, respectively. The desired values for $K_1$ and $K_2$ are for the least RMS value of $\tilde{\eta}$. The estimated values for $K_1$ and $K_2$ have accuracy within 0.2. First, the set of values of $K_1$ and $K_2$ are chosen from (0 to 15) and (0 to 30), respectively, with a step size of 2. The surface and contour plots of the same are shown in Fig. 5a. A peak is observed for values beyond $K_1 = 12$ and $K_2 = 24$.

It can also be observed that there is a peak for $K_1 < 2$ and $K_2 < 1$ from Fig. 5b. The values for $K_1$ and $K_2$ are further refined to a step size of 0.2. Later, the minimum value for $e_{\text{rms}}$ is found to be 0.08848 at $K_1 = 1.6$ and $K_2 = 2.9$, as shown in Fig. 5c.

## 7 Results of Spatial Tracking Simulations

The simulation outcomes for the trajectory tracking problem where the backstepping control is applied to the underwater robot model is presented. In this section, we show and confirm the implementation and efficacy of the proposed control policy.

The backstepping control scheme is employed in the dynamic model presented earlier with the tuned gain constants. Two different trajectories are considered for numerical simulation, as shown in Fig. 6. The first profile is a circular trajectory in space with a radius of 2 m with its center at (0, 2, 2) in the inertial frame. An infinity-shaped trajectory as a second profile is a combination of two sinusoidal functions with a breadth of two meters, where the start point of one function is the terminating location of the other. The origin of the inertial frame is the deployment location of the vehicle. The results of the simulation for tracking the circular and infinity trajectories are presented in Fig. 7.

A vital concern in the submerged robot control is the associated error at the initial position and orientation at the beginning of the launch. Since most of the underwater vehicles are launched from the floating platforms, there is always an initial pose error at the deployment stage. From the above surface plots, the errors settle down quickly, but there is a caveat. The previous results were based on a few assumptions:

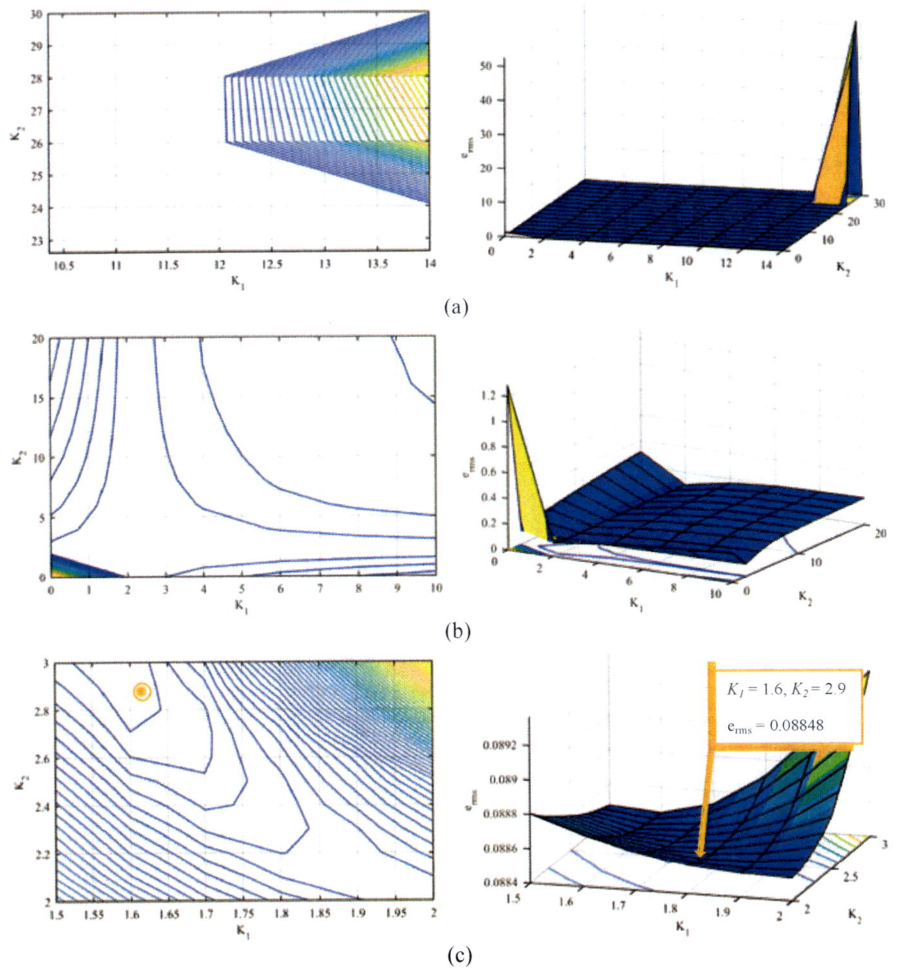


**Fig. 5** $e_{\text{rms}}$ plots for a range of values of $K_1$ and $K_2$

The response of the thrusters is instantaneous; there is no limit to the maximum thrust provided by the thruster. The thrust forces variation with time is shown in Fig. 8.

Due to the high tracking errors at the origin, high thrust values are required at the start of the mission simulation, as shown in Fig. 8. The thrust generated by each thruster during the start reaches 30 N in circular and 90 N in infinity-shaped trajectories, which is practically impossible as they exceed the maximum thrust provided by the thrusters. Thus, for accurate simulation, the maximum thrust forces obtained from thrusters are restricted, and the response time is increased using an increasing function.

Maximum thrust force provided by each thruster = 15 N

$$\text{Response rate: } r = 0.9(1 - \mathrm{e}^{-4t}) \tag{29}$$

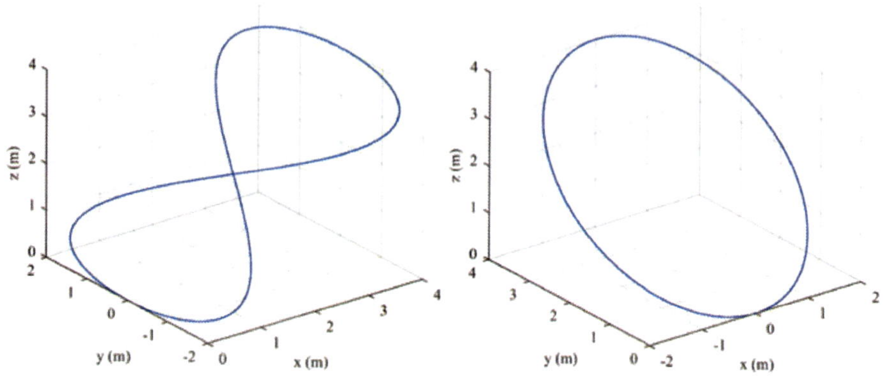


**Fig. 6** Infinity and circular trajectory generated

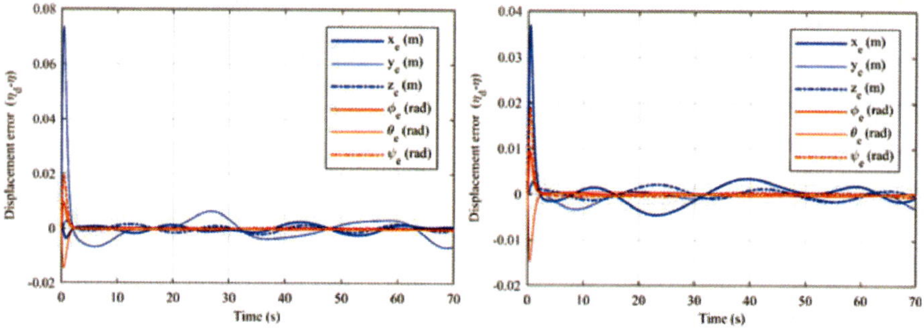


**Fig. 7** Tracking errors for infinity and circular trajectory from left to right

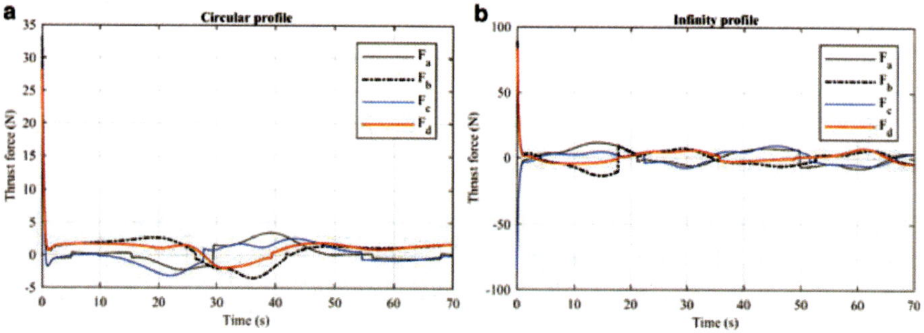


**Fig. 8** Thrust forces for **a** circular, **b** infinity trajectories

The increasing function used here is an exponential function, which is incorporated in the thrust force vector formulation to simulate the exponential increment of thrust values. The response rate in Eq. (29) implies that the thrusters accomplish 90% of the maximum threshold in 0.57 s.

After the above modifications, the thrust generated for circular and infinity profiles are replotted as in Fig. 9, where the thrust rises exponentially to a threshold of 15 N and stay there till they eliminate this error. The period for the initial tracking error to reduce to the neighborhood of zero is observed as 3 and 5 s for circular and infinity trajectories, respectively, which is directly dependent on the initial tracking error, threshold of thrust values, and thruster response to the control input. This can also be observed in the tracking error plots, as presented in Figs. 10, and 11 for circular and infinity profiles, respectively. The simulation results show that the inclusion of maximum thrust limits and response time for the thruster, the initial error is increased considerably. The numerical simulations demonstrated the controller capability by exploiting the proposed backstepping scheme in tracking the given complex trajectories.

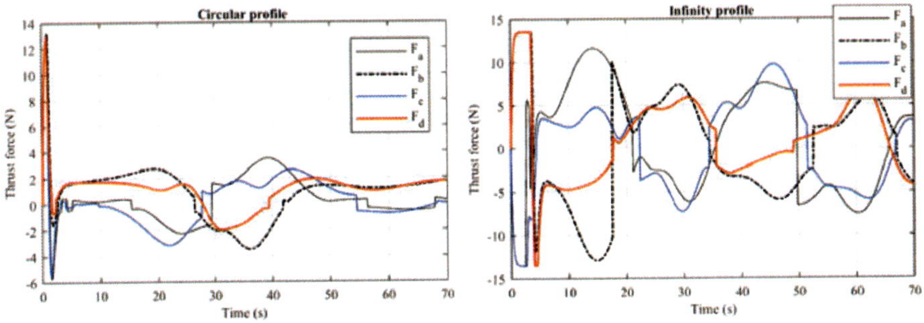


**Fig. 9** Thrust forces after restricting its maximum value for circular and infinity trajectories

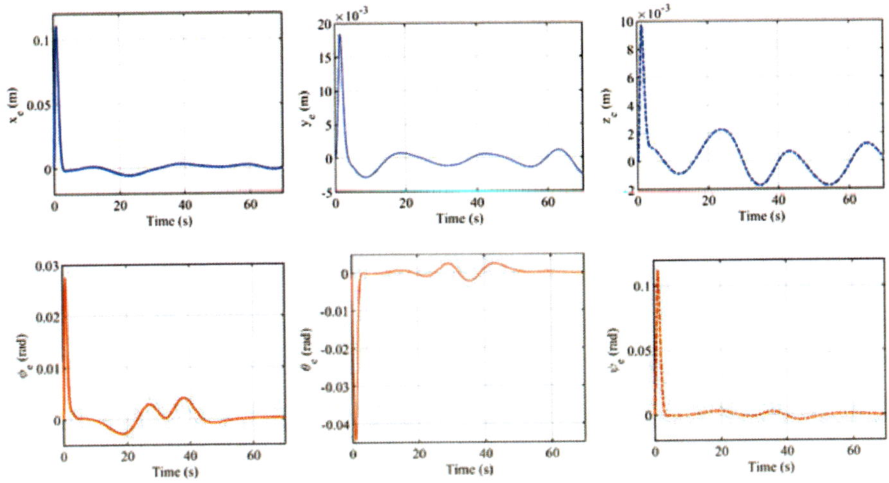


**Fig. 10** Tracking errors in circular trajectory

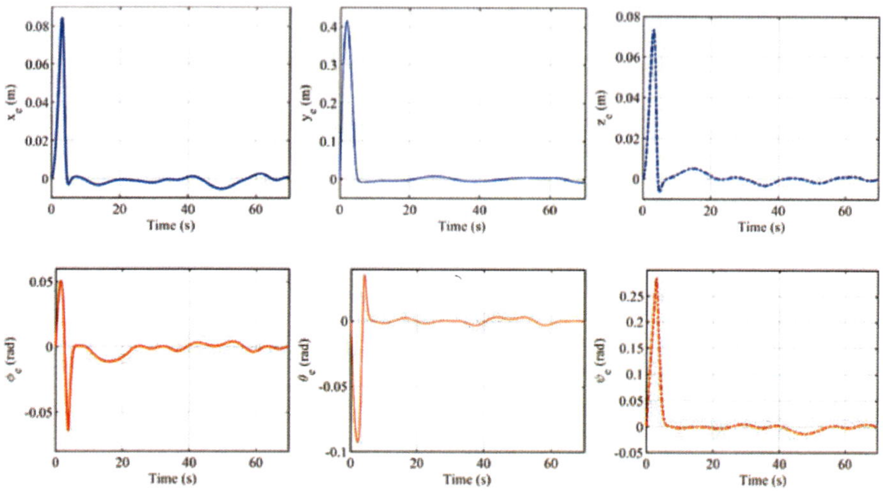


**Fig. 11** Tracking errors in infinity trajectory

# 8 Concluding Remarks

The proposed model is a nonlinear model of an eight input neutrally buoyant underwater robot with four tilting thrusters. The mathematical model is derived using the hydrodynamics of the hull and thrusters separately. The control law is derived, and robot stability is ensured based on the Lyapunov criterion of stability. The backstepping controller gain constants have been tuned using RMS of pose error. The performance of the control algorithm was investigated for two different hovering trajectories: circular trajectory and infinity trajectory. Simulation results have shown that the backstepping technique can be used for stabilizing the AUV and bringing the tracking error to the neighborhood of zero. We arrive at realistic results by considering the thrust delay and limiting the maximum thrust.

As a further study, the controller stability and robustness can be proved in the presence of unavoidable parametric uncertainties and external disturbances. Explicit study of the thruster dynamics will be undertaken in the future.

**Acknowledgements** This work was partially supported by the International Collaboration Research Program through the National Research Foundation of Korea (NRF) funded by the Ministry of Science and ICT (#NRF-2017K1A3A1A19071037) and the Department of Science and Technology (DST), India (#INT/Korea/P-43). The work is also supported by the Science and Engineering Research Board (SERB), India (#CRG/2018/000400).

## References


1. Vervoort JHAM (2009) Modeling and control of an unmanned underwater vehicle
2. Jin S, Kim J, Kim JW, Bae JH, Bak J, Kim J, Seo TW (2015) Back-stepping control design for an underwater robot with tilting thrusters. In: Proceedings of the 17th international conference on advanced robotics ICAR 2015, pp 1–8. https://doi.org/10.1109/ICAR.2015.7251425
3. Yuh J (2000) Design and control of autonomous underwater robots: a survey 8:7–24. https://doi.org/10.1023/A:1008984701078
4. Yoerger DR, Slotine J-JE (1985) Robust trajectory control of underwater vehicles. IEEE J Oceanic Eng 10:462–470. https://doi.org/10.1109/JOE.1985.1145131
5. Han J, Chung WK (2014) Active use of restoring moments for motion control of an underwater vehicle-manipulator system. IEEE J Oceanic Eng 39:100–109. https://doi.org/10.1109/JOE.2013.2241931
6. Jin S, Kim J, Kim J, Seo T (2015) Six-degree-of-freedom hovering control of an underwater robotic platform with four tilting thrusters via selective switching control. IEEE/ASME Trans Mechatron 20:2370–2378. https://doi.org/10.1109/TMECH.2014.2378286
7. Jin S, Bak J, Kim J, Seo TW, Kim HS (2018) Switching PD-based sliding mode control for hovering of a tilting-thruster underwater robot. PLoS ONE 13:1–16. https://doi.org/10.1371/journal.pone.0194427
8. Eskesen J, Owens D, Soroka M, Morash J (2009) Design and performance of Odyssey IV: a deep ocean hover-capable AUV. JGE 617:253–3438
9. Ryll M, Bülthoff HH, Giordano PR (2012) Modeling and control of a quadrotor UAV with tilting propellers. In: 2012 IEEE international conference on robotics and automation, pp 4606–4613. https://doi.org/10.1109/ICRA.2012.6225129
10. De Luca A, Oriolo G, Robuffo Giordano P (2010) Kinematic control of nonholonomic mobile manipulators in the presence of steering wheels. In: Proceedings—IEEE International conference on robotics and automation, pp 1792–1798. https://doi.org/10.1109/ROBOT.2010.5509570
11. Maruki Y, Kawano K, Suemitsu H, Matsuo T (2014) Adaptive backstepping control of wheeled inverted pendulum with velocity estimator. Int J Control Autom Syst 12:1040–1048. https://doi.org/10.1007/s12555-013-0402-4
12. Bouabdallah S, Siegwart R (2005) Backstepping and sliding-mode techniques applied to an indoor micro quadrotor. In: Proceedings—IEEE International conference on robotics and automation, pp 2247–2252 . https://doi.org/10.1109/ROBOT.2005.1570447
13. Hagan MT, Menhaj MB (1994) Brief papers. IEEE Trans Neural Networks 5:2–6. https://doi.org/10.1006/brcg.1996.0066
14. Kadiyam J, Parashar A, Mohan S, Deshmukh D (2020) Actuator fault-tolerant control study of an underwater robot with four rotatable thrusters. Ocean Eng 197. https://doi.org/10.1016/j.oceaneng.2020.106929
15. Fossen TI (1994) Guidance and control of ocean vehicles. Wiley, New York
16. Antonelli G, Fossen T, Yoerger D (2008) Underwater robotics
17. Craig JJ (2005) Introduction to robotics: mechanics and control. Pearson/Prentice Hall, Upper Saddle River
18. Mohan S, Thondiyath A (2011) A non-linear tracking control scheme for an under-actuated autonomous underwater robotic vehicle. Int J Ocean Syst Eng 1:120–135. https://doi.org/10.5574/IJOSE.2011.1.3.120
19. Newman JN (1977) Marine hydrodynamics. MIT Press, Cambridge
20. Dong RG (1978) Effective mass and damping of submerged structures. Lawrence Livermore Lab, California University, Livermore (USA)
21. Triantafyllou MS, Hover FS (2003) Maneuvering and control of marine vehicles, pp 105–110
22. Georgiades C (2005) Simulation and control of an underwater hexapod robot
23. Kadiyam J, Mohan S (2019) Conceptual design of a hybrid propulsion underwater robotic vehicle with different propulsion systems for ocean observations. Ocean Eng 182:112–125. https://doi.org/10.1016/j.oceaneng.2019.04.069

24. Raptis IA, Valavanis KP (2010) Linear and nonlinear control of small-scale unmanned helicopters. Springer Science & Business Media, Berlin
25. Krstic M, Kanellakopoulos I, Kokotovic P (1995) Nonlinear and adaptive control design. Mechatronics handbook, 576 p
26. Khalil HK (1996) Nonlinear systems, 2nd edn. Prentice-Hall, New Jersey
27. Krstic M, Kanellakopoulos I, Kokotovic P (1995) Nonlinear and adaptive control design. Wiley, Hoboken
28. Golub G, Kahan W (1964) Calculating the singular values and pseudo-inverse of a matrix. J Soc Ind Appl Math Ser B Numer Anal 2:205–224
29. Rogers DF, Adams JA (1989) Mathematical elements for computer graphics. McGraw-Hill Higher Education, New York

# Characterization of Pneumatic Air Muscle (PAM) Under Unloaded and Loaded Conditions

**S. Udhayakumar**

**Abstract** In today's world, pneumatic air muscle (PAM) has become a best alternative over fluid power actuators, since it possesses significant properties like high force-to-weight ratio, flexible structure, less compressed air consumption, various installation possibilities, no chance of fire hazard, ease in control, safe and less cost. They are commonly applied in orthosis and in robotic automation. For application of PAM in any field, its characteristics are to be understood. In this paper, the behaviors of PAM under no-load and loaded conditions are studied experimentally. The relation between input pressure, length and diameter of PAM is determined. This study of characteristics of PAM will enable them to be used effectively in industrial applications.

**Keywords** Pneumatic air muscle · Contraction ratio · Hysteresis

## 1 Introduction

Pneumatic air muscle (PAM) was developed in the 1950s for artificial limb research [1]. Since PAM possesses significant properties like high force-to-weight ratio, flexible structure, requirement of less compressed air consumption, choices for installation possibilities, safe from fire hazard, ease in control, low cost, it is considered a best alternative over fluid power actuators [2]. It has chloroprene rubber inner tube surrounded by braided shell [3] and metal fittings attached at the ends. When the inner tube is supplied with compressed air, it expands laterally since its longitudinal expansion is constrained by braided shell. The characteristics of PAM are turning it into a promising actuator choice in medical and bio-robotic applications [4]. The characteristics of PAM should be experimented at working conditions before selecting it for specific application. The importance of determining the mechanical behavior and characteristics of PAM before its application is extensively discussed

S. Udhayakumar (✉)
Department of Mechanical Engineering, PSG College of Technology, Coimbatore 641004, India
e-mail: suk.mech@psgtech.ac.in

© The Editor(s) (if applicable) and The Author(s), under exclusive license to Springer Nature Singapore Pte Ltd. 2021

S. Mohan et al. (eds.), *Materials, Design, and Manufacturing for Sustainable Environment*, Lecture Notes in Mechanical Engineering,
https://doi.org/10.1007/978-981-15-9809-8_3

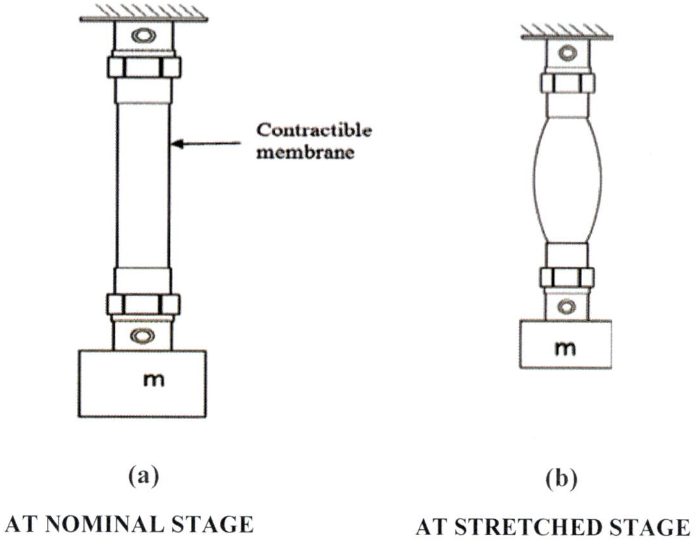


**Fig. 1** Working of PAM

in the literature [5–13]. In this work, an attempt is made to study the characteristics of PAM through experimental studies.

## 2 Working Principle of PAM

When compressed air is supplied, the contractible membrane expands in circumference. The tensile force thereby rises, as well as a contracting movement in the longitudinal direction. This movement is transferred to the workload with the aid of connecting elements which are screwed into thread. PAM is simply a pulling actuator which cannot transmit pressure forces and does not have a guide. The pulling force is at its maximum at the beginning of the contraction and drops almost linearly with the stroke to zero. The PAM at the nominal stage and stretched stage is shown in Fig. 1. The length and diameter of the PAM vary with respect to the input pressure and the load.

## 3 Experimental Procedure

The specifications of the PAM considered for the experimental studies are given in Table 1.

**Table 1** Technical specifications of PAM

| | |
|---|---|
| Manufacturer | FESTO |
| Model name | RM-CM |
| Length of PAM | 248 mm |
| Maximum force @ maximum permissible pressure | 630 N |
| Weight | 58 g |
| Maximum permissible pressure | 8 bar |

In this experiment, the specified PAM is clamped to the table and the following steps were performed:

1. The diameter of PAM at three points (Fig. 2) is measured.
2. Then, compressed air is supplied to the PAM at 2 bar.
3. The corresponding contraction in length of the PAM is measured using LVDT, and subsequent increase in diameter at three points (Fig. 2) is measured using a digital vernier.
4. Step 2 to step 3 are repeated with increment of 0.5 bar up to 6 bar.
5. The input air pressure is now reduced from 6 bar to 2 bar in decrements of 0.5 bar, and the diameter of the PAM is recorded. This is to determine the hysteresis of the PAM.
6. Steps 2 to 5 are repeated ten times to obtain reliable readings.
7. The average of the ten trials is recorded.

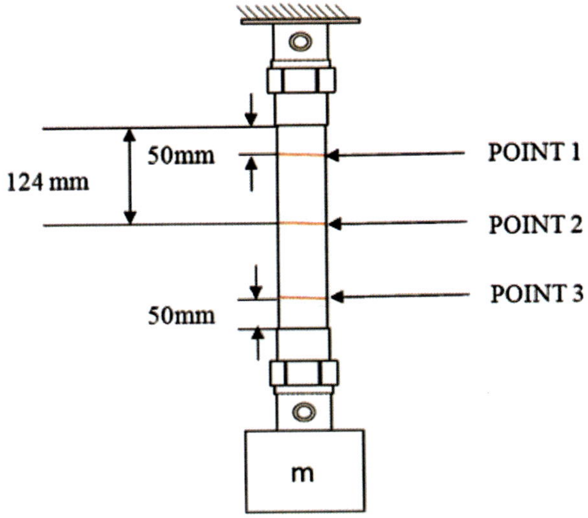


**Fig. 2** PAM with diameter measuring points

## 4 Experimental Results

The results of the experimental studies are discussed in this section.

### *4.1 Input Air Pressure Versus Length of PAM*

The relation between the input air pressure and contraction in length of PAM is plotted in Fig. 3.

From Fig. 3, it can be observed that as input air pressure increases, length of PAM decreases. At pressure of 0 bar, the length of the PAM is 248 mm, and at pressure of 6 bar, the length of PAM is 200.26 mm, almost 19.25% reduction in length. The outer braided shell of PAM constrains its longitudinal elongation, and hence it expands in radial manner which results in reduction of its length.

From the experimental results, the input air pressure and length of the PAM can be related by Eq. (1)

$$L = -0.023P^2 - 8.889P + 251.2 \tag{1}$$

where $L$ is Length of PAM in mm and $P$ is the Input air Pressure to PAM in bar.

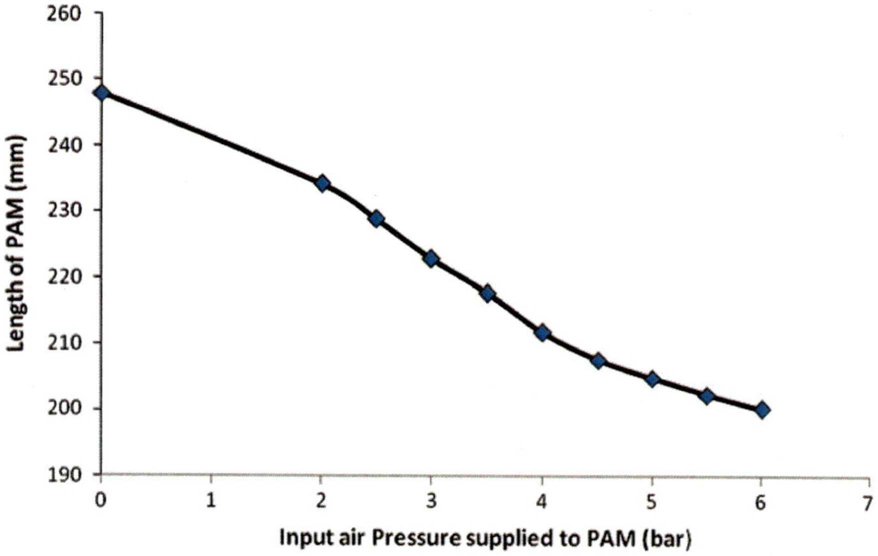


**Fig. 3** Input air pressure versus length of PAM

**Table 2** Contraction ratio% of PAM

| Input air pressure (bar) | Initial length of PAM, $L_1$ (mm) | Final length of PAM at pressurized condition, $L_2$ (mm) | Contraction % ratio $(L_1 - L_2) * 100/L_1$ |
|---|---|---|---|
| 0 | 248 | 248 | 0 |
| 2 | 248 | 243.3 | 1.88 |
| 2.5 | 248 | 228.92 | 7.93 |
| 3 | 248 | 222.92 | 10.10 |
| 3.5 | 248 | 217.80 | 12.17 |
| 4 | 248 | 211.89 | 14.55 |
| 4.5 | 248 | 207.76 | 17.43 |
| 5 | 248 | 204.95 | 17.35 |
| 5.5 | 248 | 202.42 | 18.37 |
| 6 | 248 | 200.26 | 19.25 |

### 4.2 *Input Air Pressure Versus Contraction Ratio %*

Contraction ratio % is calculated using the formula given by Eq. (2)

$$C\% = \frac{L_1 - L_2}{L_1} \times 100 \tag{2}$$

where $L_1$ is the initial length and $L_2$ is the length after pressurized air is provided to PAM.

The contraction ratio % of the PAM with respect to input air pressure is shown in Table 2.

The plot between input air pressure to PAM and contraction ratio% is shown in Fig. 4.

As observed from Fig. 4, the contraction ratio% is not linear up to input pressure of 2 bar and then becomes nonlinear. Further, the contraction ratio increases with respect to increase in inlet air pressure.

### 4.3 *Input Air Pressure Versus Diameter of PAM*

Figure 5 shows the relation between input air pressure and diameter of PAM.

The diameter of the PAM increases as the inlet pressure increases because of the expansion of the chloroprene rubber tube present inside the braided shells of the PAM. Based on the experimental results, the diameter and input air pressure to PAM

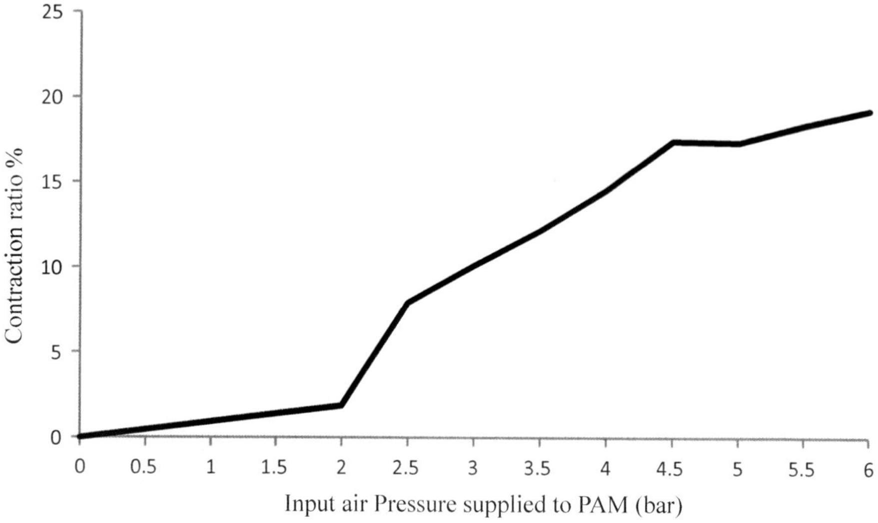


**Fig. 4** Pressure versus contraction ratio% of PAM

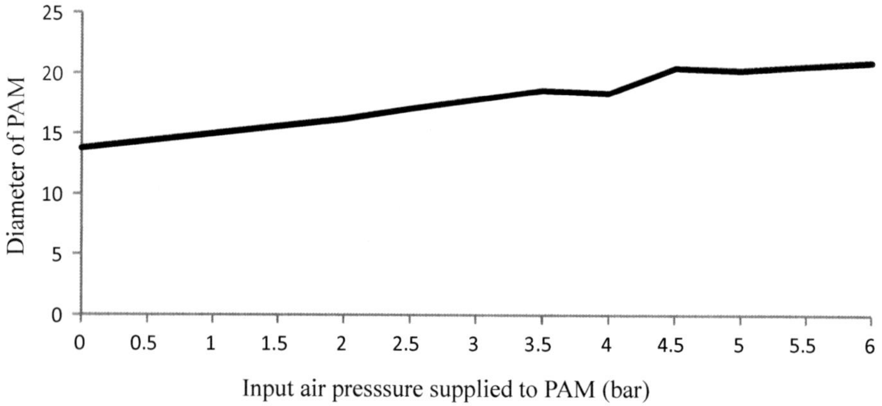


**Fig. 5** Pressure versus diameter of PAM

can be related by Eq. (3).

$$D = -0.041P^2 + 1.509P + 13.64 \tag{3}$$

where $D$ is the diameter of PAM in mm and $P$ is input air pressure to PAM in bar.

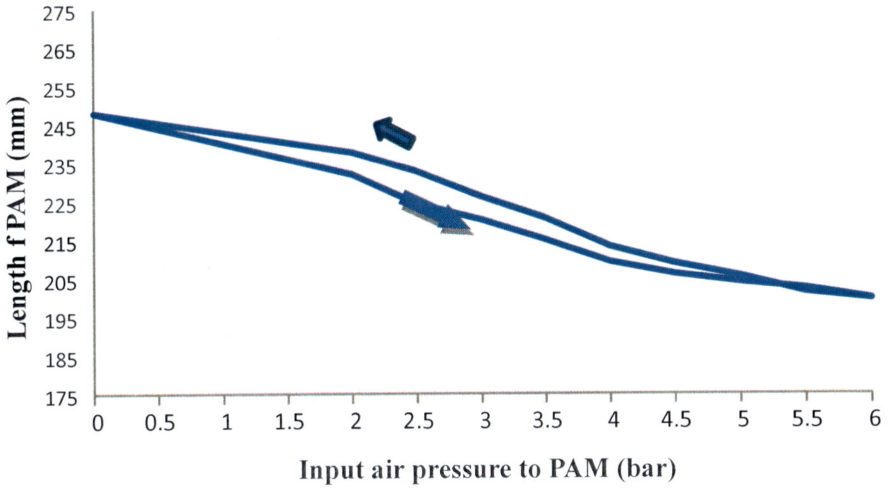


**Fig. 6** Hysteresis graph of PAM

The relationship between the pressure, length and diameter can be related as

$$P = -21.73(L/P) - 12.19(D/P) + (5577.2/P) - 174.83 \tag{4}$$

where $P$ is pressure (bar), $L$ is the length of PAM (mm) and $D$ is the diameter of the PAM (mm).

### *4.4 Hysteresis of PAM*

The hysteresis curve of PAM is presented in Fig. 6. The hysteresis is due to the friction between the fibers in the outer braided shells of the PAM [5]. The PAM hysteresis occurs due to the relative motions taking place when the muscle is inflated or deflated.

The maximum deviation occurs at 2.5 bar which is about 8.27 mm. The hysteresis area is found to be 21.9 mm$^2$ which represents the amount of energy dissipated in the form of heat. Hysteresis in PAM is disadvantageous since it limits its usage in precise positioning of systems, and hence it has to be minimized.

## 5 Characteristics Under Loaded Condition

Having studied the characteristics of PAM under no-load conditions, the next step is to determine the characteristics of PAM under loaded condition.

Figure 7 represents the stretched and unstretched lengths of PAM, where L is the original length of PAM. Unstretched length ($L_u$) of PAM is the length of PAM at pressurized condition, with no any mechanical load. Stretched length ($L_s$) of PAM is the increase in length of PAM at pressurized condition with the mechanical load. $L - (L_s + L_u)$ is the difference between the original length and the sum of stretched length and unstretched length.

To determine these characteristics under load conditions, the PAM was clamped to the table as shown in Fig. 8 and the mechanical weights were screwed to the PAM. The experimental procedure is as follows:

1. Mechanical load of 10 N is directly added to the PAM at the bolted end of PAM as shown in Fig. 8.
2. Compressed air is supplied to the PAM from 0 to 6 bar with the increment of 0.5 bar.
3. The corresponding contraction in length of the PAM is recorded in a PC using a LVDT interfaced with NI DAQ 6009 and LabVIEW software.
4. Steps 1, 2 and 3 are repeated by increasing the mechanical load in increments of 10 N up to 80 N.
5. Every experiment was repeated ten times to obtain reliable readings.
6. The average of the ten trials is recorded.

The experiment results under loaded conditions are plotted in Figs. 9 and 10.

Figure 9 shows the relation between force and stretched length ($L_s$) of PAM at input pressure of 6 bar. As the load increases, the stretched length increases due to the increase in the pulling force. Figure 10 shows the relation between the difference

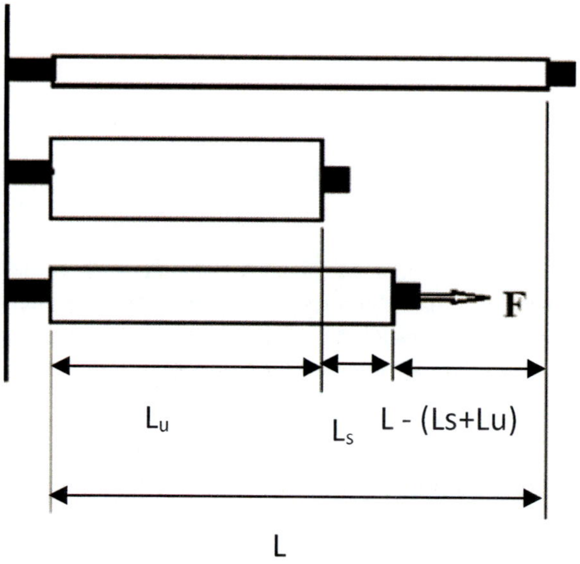


**Fig. 7** Stretched and unstretched lengths of PAM

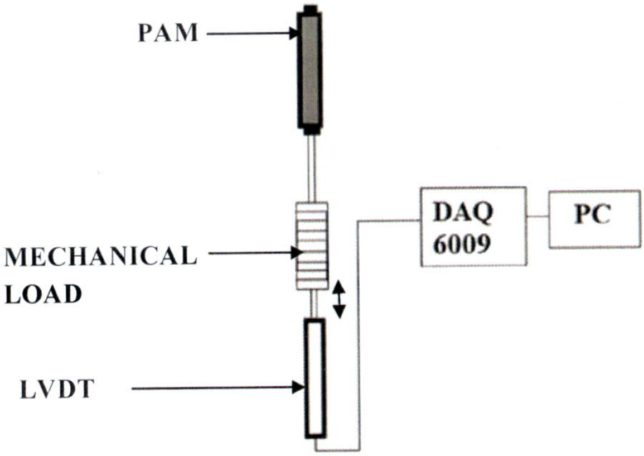


**Fig. 8** PAM under loaded condition

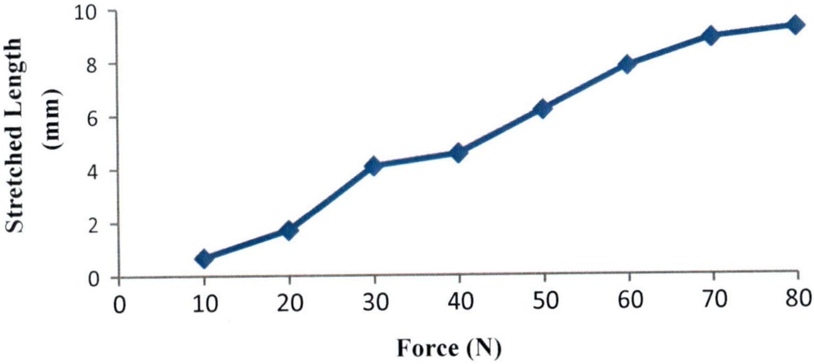


**Fig. 9** Force versus stretched length of PAM

between original length ($L$) and (unstretched ($L_u$) + stretched length ($L_s$)) versus force. As the load increases, $L - (L_s + L_u)$ reduces due to the increase in the pulling force. The study of above characteristics will enable the user to identify the characteristics of PAM and use them effectively for industrial applications.

## 6 Conclusion

In this paper, the change in length and diameter of PAM with respect to input air pressure, and change in length and diameter were studied experimentally under no-load conditions. The maximum contraction ratio% was found to be 19.25%. The relation between input air pressure, length and diameter of PAM was determined using experimental results. From hysteresis curve, it was observed that maximum deviation occurred at 2.5 bar and the hysteresis area is 21.9 mm$^2$. The characteristics

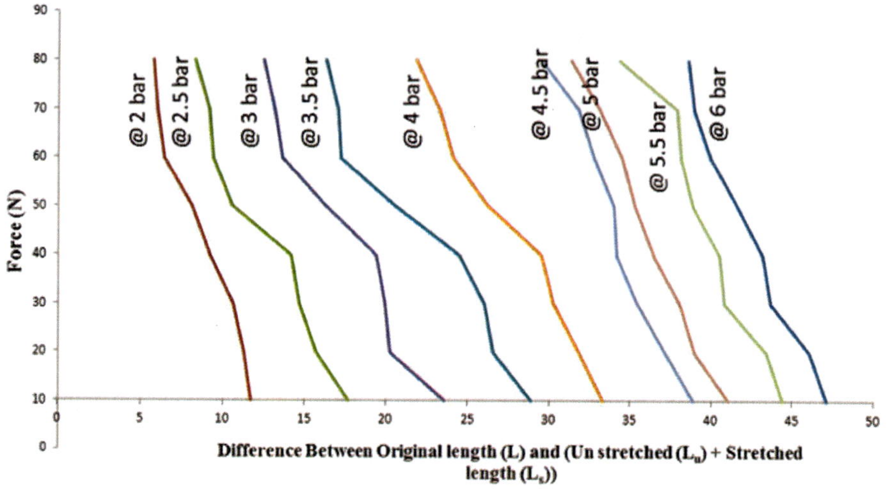


**Fig. 10** Difference between original length ($L$) and (unstretched ($L_u$) + stretched length ($L_s$)) versus force

of PAM under loaded conditions were also tested. As the load increased, the stretched length also increased. The understanding of these characteristics of PAM will help in effective usage in industrial applications such as robotics and part feeders.

## References


1. Yeh T-J, Wu M-J, Lu T-J, Wu F-K, Huang C-R (2010) Control of McKibben pneumatic muscles for a power-assist, lower-limb orthosis. Mechatronics 20(6):686–697
2. Li J, Kawashima K, Kagawa T (2015) A method to suppress temperature increase in pneumatic artificial rubber muscles. Exp Thermal Fluid Sci 61:59–65
3. Wickramatunge KC, Leephakpreeda T (2010) Study on mechanical behaviors of pneumatic artificial muscle. Int J Eng Sci 48(2):188–198
4. Andrikopoulos G, Nikolakopoulos G, Manesis S (2013) Pneumatic artificial muscles: a switching model predictive control approach. Control Eng Pract 21(12):1653–1664
5. Minh TV, Tjahjowidodo T, Ramon H, Van Brussel H (2010) Cascade position control of a single pneumatic artificial muscle–mass system with hysteresis compensation. Mechatronics 20(3):402–414
6. Bertetto M, Ruggiu M (2004) Characterization and modeling of air muscles. Mech Res Commun 31(2):185–194
7. Ganguly S, Garg A, Pasricha A, Dwivedy SK (2012) Control of pneumatic artificial muscle system through experimental modeling. Mechatronics 22(8):1135–1147
8. Wang X, Zhang Y, Fu X, Xiang G (2008) Design and kinematic analysis of a novel humanoid robot eye using pneumatic artificial muscles. J Bionic Eng 5(3):264–270
9. Wickramatunge KC, Leephakpreeda T (2013) Empirical modeling of dynamic behaviors of pneumatic artificial muscle actuators. ISA Trans 52(6):825–834
10. Gordon KE, Sawicki GS, Ferri DP (2006) Mechanical performance of artificial pneumatic muscles to power an ankle–foot orthosis. J Biomech 39(10):1832–1841
11. Jia-Fan Z, Can-Jun Y, Ying C, Yu Z, Yi-Ming D (2008) Modeling and control of a curved pneumatic muscle actuator for wearable elbow exoskeleton. Mechatronics 18(8):448–457

12. Pujana-Arrese A, Mendizabal A, Arenas J, Prestamero R, Landaluze J (2010) Modelling in Modelica and position control of a 1-DoF set-up powered by pneumatic muscles. Mechatronics 20(5):535–552
13. Lau CY, Chai A (2012) The development of a low cost pneumatic air muscle actuated anthropomorphic robotic hand. Procedia Eng 41:737–742

# Experimental Studies on Biomachining of Copper and Its Behavioural Characteristics

**B. Hemalatha, B. Rajeswari, T. Sekar, and V. Rajasekar**

**Abstract** The methods being utilized for micromachining have adverse effect on the environment. They probably cause porosity, heat-affected zone and also damages the metallurgical properties. There have been severe effects due to chemicals on human health and environment. To achieve high efficiency, researches have evolved a new field in micromachining as biomachining. The fungal species of *Aspergillus niger* has been used for metal removal. The biomachining is done using the fungal culture on the workpiece of copper. The material removal rate was calculated. The microstructural analysis and hardness test were carried out. The colour change was also analysed. In this paper, the results of the samples before and after machining were compared.

**Keywords** Biomachining · *Aspergillus niger* · Copper · Material removal rate · Microstructural analysis · Hardness

## 1 Introduction

Micromachining process like chemical machining and electric discharge machining (EDM) is used in the construction of microfeatures on various metals. These processes damage the metallurgical properties and the use of chemical acids is

B. Hemalatha (✉) · B. Rajeswari · T. Sekar
Department of Mechanical Engineering, Government College of Technology, Coimbatore, India
e-mail: hemidurai@gmail.com

B. Rajeswari
e-mail: rajesaravind@gmail.com

T. Sekar
e-mail: drtsekar76@gmail.com

V. Rajasekar
Department of Industrial Biotechnology, Government College of Technology, Coimbatore, India
e-mail: rajashekardv@gmail.com

© The Editor(s) (if applicable) and The Author(s), under exclusive license to Springer Nature Singapore Pte Ltd. 2021
S. Mohan et al. (eds.), *Materials, Design, and Manufacturing for Sustainable Environment*, Lecture Notes in Mechanical Engineering,
https://doi.org/10.1007/978-981-15-9809-8_4

unavoidable [1]. Biomicromachining is an unconventional manufacturing process used for culturing for future experiments. It is a controlled microbiological process on a workpiece by metal removal or dissolution by the use of microorganisms [2]. Istiyanto et al. [3] investigated the biomachining processes of copper using *Acidithiobacillus ferrooxidans* and compared the profiles or cross-sectional forms of copper for different parameters.

The process of biomachining has wide applications in creating intricate parts, integrated circuits and printed circuit boards [4]. Singh et al. [5] made a comparison study between biomicromachining and photochemical machining. Gan et al. [3] showed that the Cu(II) adsorption capacity increased over a certain pH range. Diaz-Tena et al. [6] studied the copper machining in oxygen-free copper by extremophile bacteria. From the above literature survey, the idea of utilizing the fungal species of *Aspergillus niger* for biomachining of copper is carried out in this work.

## 2 Experimental Details

### 2.1 Creating a Suitable Biomachining Environment

Microbial species are prone to contamination. Hence, a suitable biological environment maintaining the optimal pH, temperature, light, etc., is needed to be created. This ensures the proper growth of bacterial culture.

### 2.2 Culture Preparation

*Aspergillus niger* spores was being acquired from TNAU, Coimbatore. The sample was revived in both slant and plate media (potato dextrose agar). To further isolate spores, 0.5 to 1 ml of sterile water was added to the plate. 5 ml of the sterile water containing the spores was transferred to a 250 ml conical flask containing 100 ml of potato dextrose broth (24 g/L). The fungal species was put into the solid media like potato dextrose agar, and it was incubated till the growth became apparent on the surface media. The culture was incubated at 37 °C for 10 days. The flasks were incubated at 37 °C and were shaken at 180 rpm for 5–12 days. The samples were removed and used for culturing for future experiments.

### 2.3 Sample Preparation

For experiment, pure copper was taken. The copper was cut into dimension of 30 × 30 × 0.15 mm using normal cutting methods. Once polished, every sample was

**Fig. 1** Fungal spore

**Fig. 2** Copper specimen

contained in a sealed plastic bag to reduce oxidation before use in the experiment. The samples were preserved in a sealed plastic bag to avoid oxidation. The samples were successively rinsed with ethanol and dried in the oven prior to biomachining. The initial weight of the specimen is calculated. Figures 1 and 2 show the fungal spore and copper specimen, respectively.

## 2.4 *Biomachining of Copper*

Biomachining involves a set of process steps which involves preparation of microorganism culture, preparation of samples and introduction of copper specimen into the culture. Once the culture and the specimen were prepared, 5 ml of the prepared culture was transferred to the broth. The copper specimen is introduced into the prepared culture, and it has been kept in the shaker. The specimen is kept in the culture for about ten days for machining to happen. The workpiece was removed after the biomachining process, rinsed with ethanol, dried and weighed. Figures 3, 4 and 5 show the biomachining of copper.

**Fig. 3** Copper specimen in culture

**Fig. 4** Shaker

**Fig. 5** Machined copper specimen

# 3 Results and Discussion

## 3.1 Metal Removal Rate

The weight of the copper sample is noted before introduction of the specimen into the fungal culture. After biomachining, the samples were rinsed with sterile distilled water and ethanol.

Weight of the copper specimen before machining = 0.932 g = 932 mg.
Weight of the copper specimen after machining = 0.868 g = 868 mg.

The metal removal rate of the copper samples by *A. niger* was calculated from the following equation

$$\begin{aligned} \text{MRR} &= \frac{\text{Amount of metal removed (mg)}}{\text{Time (h)}} \\ &= \frac{932 - 868}{15 \times 24} = \frac{64}{360} \\ \text{MRR} &= 0.177\ \text{mg/h} \end{aligned} \tag{1}$$

## 3.2 Colour Change Indication

During the biomachining process, the culture media change its colour and this indicates that the machining is being carried out and it is evident from samples the initial colour of media was light yellow or off white and as the machining progress the colour became dark wine red and it changes to dark greyish shade which indicates the biomachining of copper by the *A. niger*. The colour change from pale yellow to dark greyish is due to action of *A. niger* on the copper specimen and subsequent oxidation and metal removal. Figure 6 represents the colour change during the biomachining process.

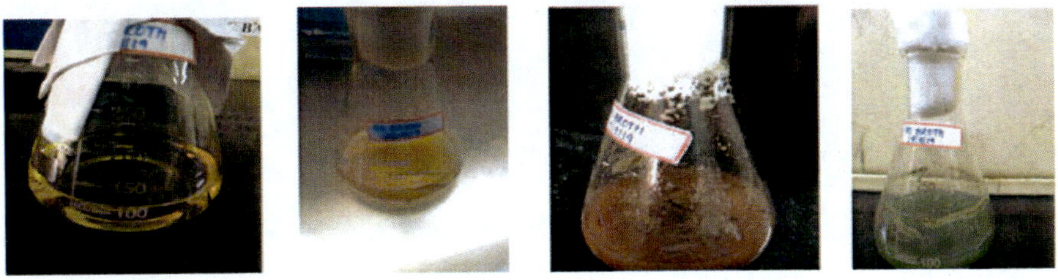

(a) DAY 1 (b) DAY 1 (c) DAY 7 (d) DAY 15

**Fig. 6** Colour change indication

**Table 1** Vickers microhardness test results

| S. No. | Sample | Observed values, HV 0.5 | | | Average, HV 0.5 |
|---|---|---|---|---|---|
| 1 | Before machining | 103.90 | 102.30 | 104.40 | 103.53 |
| 2 | After machining | 94.60 | 97.28 | 100.60 | 97.50 |

## 3.3 *Microhardness Test*

Vickers microhardness testing is done to compare the hardness of the copper specimen before biomachining and after biomachining.

From the results in Table 1, it has been shown that there is a slight decrease in hardness of the copper specimen after machining. The decrease in hardness is due to the grain boundary interaction and the surface energy required in the deformation of grain boundaries.

## 3.4 *SEM Analysis*

The results obtained from SEM at different magnifications before and after machining are shown in the figures below.

From the SEM images shown in Figs. 7b and 8b at magnification of 2000× before machining and after machining, it is clearly evident that the *A. niger* interacts with the copper and carries out metal removal. It presents the enlarged version where the individual copper grains can be observed. It shows the regions which are oxidized by *A. niger*. It shows the relation between the surface topography, size of the grains and their crystallography orientations. It is also evident that the material removal by *A. niger* is progressed through grain boundaries inwards.

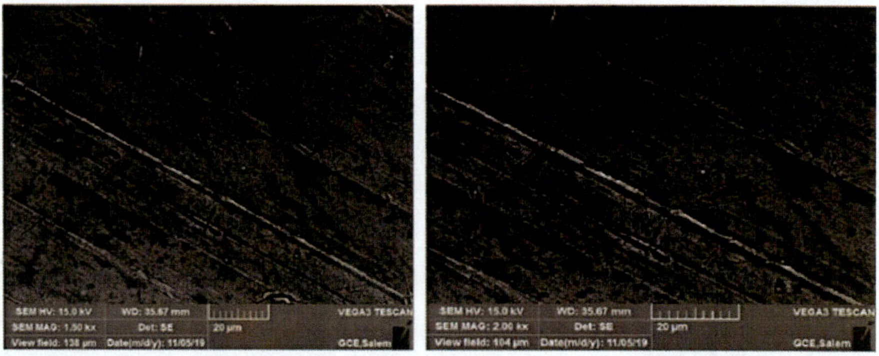


(a) (b)

**Fig. 7** SEM images of copper specimen before machining

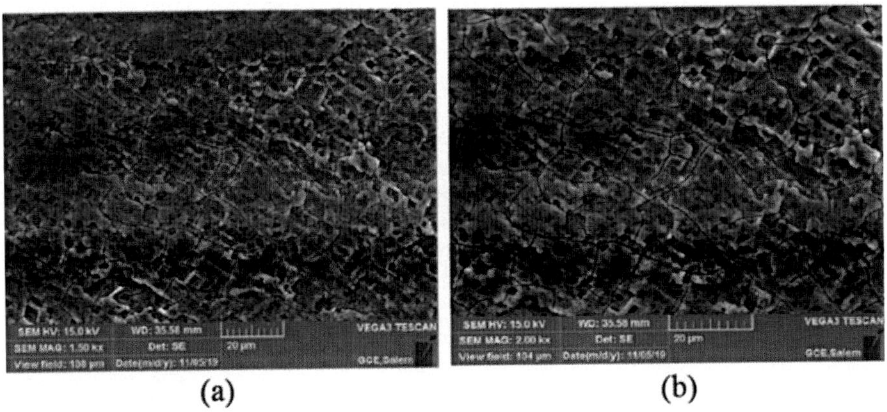


**Fig. 8** SEM images of copper specimen after machining

## 4 Conclusion

The use of *A. niger* to remove copper and hence the biomachining characteristics of copper were investigated. The performance of the biomachining process has been assessed in terms of the metal removal rate, colour change, microhardness and SEM analysis. The material removal rate obtained was 0.177 mg/h which indicates that the metal was progressively removed from the copper sample by the action of the *A. niger*. The colour change from pale yellow to dark greyish indicates that the metal was found dissolved.

As expected the microhardness of the copper sample before machining is 103.53 HV and after machining is 97.50 HV. This decrease in hardness is due to the grain boundary formation. The SEM analysis showed that the workpiece was exposed for 15 days to the *A. niger* for metal removal which progresses through grain boundaries.

## References

1. Hocheng H, Chang J, Jadhav UU (2011) Micromachining of various metals by using *Acidithiobacillus ferrooxidans* 13820 culture supernatant experiments. J Cleaner Prod
2. Diaz-Tena E, Gallastegui G, Hipperdinger M, Donati ER, Ramirez M, Rodriguez A, de Lacalle LNL, Elias A (2016) New advances in copper biomachining by iron-oxidizing bacteria. J Corros Sci
3. Gan M, Song Z, Jie S, Zhu J, Zhu Y, Liu X (2015) Biosynthesis of bifunctional iron oxyhydrosulfate by *Acidithiobacillus ferrooxidans* and their application to coagulation adsorption. Mater Sci Eng
4. Istiyanto J, Saragih A-S, Ko TJ (2012) Metal based micro-feature fabrication using biomachining process. J Microelectron Eng
5. Singh A, Arul Manikandan N, Ravi Sankar M, Pakshirajan K, Roy L (2017) Experimental investigations and surface morphology of bio micromachining on copper. ICMPC Mater Today 4225–4234

6. Diaz-Tena E, Rodriguez-Ezquerro A, de Lacalle Marcaide LNL, Gurtubay Bustinduy L, Elias Saenz A (2014) A sustainable process for material removal on pure copper by use of extremophile bacteria. J Cleaner Prod 1–9
7. Istiyanto J, Kim MY, Ko YJ (2011) Profile characteristics of biomachined copper. J Microelectron Eng
8. Diaz-Tena E, Barona A, Gallastegul G, Rodriguez-Ezquerro A, de Lacalle Marcaide LNL, Elias Saenz A (2016) Biomachining: metal etching via microorganisms. Crit Rev Biotechnol
9. Diaz-Tena E, Rodriguez-Ezquerro A, de Lacalle Marcaide LNL, Gurtubay Bustinduy L, Elias Saenz A (2013) Use of extremophiles microorganisms for metal removal. In: Conference paper on the manufacturing engineering society international conference, MESIC

# Experimental Investigation of Tensile Strength and Hardness in GMAW/GTAW Butt Welded Joints with Various Shielding Gas Compositions

**K. V. Satheesh Kumar, K. A. Srikishore, A. Shek Mohammed Asiq, and R. Sudharsan**

**Abstract** Welding plays a big part in the global environment in the manufacturing sector. The GTAW/GMAW technique dominates other welding methods for using shielding gas is the big challenge the industry faces in welding technology. The usage of shielding gases has raised interest in better efficiency, strong weld consistency to protect the welding pool from emissions. Replacing the gas cylinders that shield this atmosphere is a time-causing task. In this research, the choice of better shielding gas for the GTAW/GMAW method is important because of the varying current for different base metals and filler wires. The impact of shielding gas compositions in 316L austenitic stainless steel was studied in this research. For selected shielding gas blends, the values of tensile strength and hardness were compared in both GTAW and GMAW processes.

**Keywords** GMAW · GTAW · Shielding gas · Tensile strength · Hardness

## 1 Introduction

The welding methodology comprises connecting two metal pieces such that at the initial boundary surfaces the attachment of substantial atomic penetration occurs. The welded pieces unite into one entity. Especially regarding material use and manufacturing costs, welding is also the most efficient way to join parts. Traditional methods

K. V. Satheesh Kumar (✉) · K. A. Srikishore · A. Shek Mohammed Asiq · R. Sudharsan
Department of Mechanical Engineering, Kongu Engineering College, Erode, Tamil Nadu, India
e-mail: kvs@kongu.ac.in

K. A. Srikishore
e-mail: srikishore2300@gmail.com

A. Shek Mohammed Asiq
e-mail: shekasiq1234@gmail.com

R. Sudharsan
e-mail: sudharsan20r@gmail.com

© The Editor(s) (if applicable) and The Author(s), under exclusive license to Springer Nature Singapore Pte Ltd. 2021
S. Mohan et al. (eds.), *Materials, Design, and Manufacturing for Sustainable Environment*, Lecture Notes in Mechanical Engineering,
https://doi.org/10.1007/978-981-15-9809-8_5

of mechanical integration require alternating more complex forms and insertion of fasteners. This mechanical assembly is usually heavier than the other element. Welding is also not limited to businesses nearby. The following literature focuses on shielding gases in GMAW/GTAW process, their properties, and weld metal strength. Improvements were then made to the welding process for improved welding and significantly lower welding time to increase efficiency.

### *1.1 Studies on Shielding Gas for GMAW*

Pires listed some of the key features of the seven gas blenders used (Ar + 2% $CO_2$, Ar + 8% $CO_2$, Ar + 55% $CO_2$ + 18% $CO_2$), Ar + 5% $O_2$, Ar + 8% $O_2$, Ar + 3% $CO_2$ + 1% $O_2$, and Ar + 5% $CO_2$ + 3% $O_2$ and its effects on the processes of metal transmission [14]. In GMAW metal transfer modes for each shielding gas mixture, they experimented with the effect of operational parameters (arc and voltage).

Improved thermal conductivity and active mixture components reduce the number of parameters for spray transfer. The duration of the arc reduces with the oxidant potential because of the absence of other influences. Gulenç studied the effect of argon and hydrogen as a shielding gas on GMAW welding [3].

The analysis focused on the mechanical and microstructural properties of the welded pieces. Specific current loads were checked as (140, 180 and 240 A) in different shielding medium such as (Ar + 1.5% H, pure Ar, and Ar + 5% H). The material was welded under 1.5% H-Ar shielding mixture and the 240A welding current was found to give the highest tensile strength. The strength of the welding increases here, with the hydrogen volume in Ar and the increased welding speed. The tests of the hardness show that the base metal was longer-lasting than weld metal and heat-affected zone on certain welding criteria.

The impact of gas shielding on efficiency and gases emissions were studied by Pires, to balance six blends (Ar + 3% $CO_2$ + 1% $O_2$, Ar + 2% $CO_2$, Ar + 8% $O_2$, Ar + 8% $CO_2$, Ar + 18% $O_2$, Ar + 18% $CO_2$) on the weld bead curve, related directly to profits, utilizing fillet joint [13].

Ultimately, when the mixture amount of $O_2$ reduces, the defects in welding often increase. The weld formed by the Ar + $CO_2$ mixture induces lateral penetration that is minimized by a decrease in the $CO_2$ content in the mixture. The weld made of a ternary mixture showed good efficiency, especially with short-circuit transfer.

Wang has researched hybrid plasma in hybrid fiber laser-MIG welding [20] as just a significant physical phenomena. The reliability of process, welding efficiency, and power coupling performance are greatly affected. The high-speed video clips simply and explicitly illustrate these features in this section.

## 1.2 Studies on Mechanical Properties of Weld Under Different Shielding Gas Blends

Liao studied the effect of the mixture on the microstructure and mechanical properties of stainless steel AISI 304 [9]. For the five blends (Ar + 2% $CO_2$, Ar + 2% $O_2$, Ar + 10% $CO_2$, Ar + 20% of $CO_2$, and Ar + 5% $CO_2$ + 2% $O_2$) here it indicates that the scatter rate rises with an increase in $CO_2$ per unit in the mixture from 2 to 20%. Each vermicular and lathy ferrite is identified and the concentration of ferrite decreases when $CO_2$ rates increase from 2 to 20% in the Ar + $CO_2$ mixture.

Srivastava was analyzed in order to prove that mild steel is the most commonly used material for commercial and industrial purposes [16]. In order to achieve better mechanical properties of welded plates, the work focuses on the impact of process welding parameters. A variance analysis was then performed to assess the relevant coefficient of mechanical properties of each input factor and to estimate the prediction of mechanical properties using regression analysis. In this experiment, GMAW welded with pure carbon blends, for the protection of the carbon, Ar + 20% $CO_2$, Ar + 10% $O_2$ in IS2062 Mild Steel 15 mm. The preventive gas mix Ar + 20 percent $CO_2$ provides optimum tensile strength value as well as optimum toughness. In relation to other shielding gas blends, Pure (100%) $CO_2$ achieves the highest hardness. The calculated values indicate the remarkable effect on the mechanical properties of the welding current and the shielding gas.

## 1.3 Studies on Melting Rates in GMAW with Different Gas Blends

Suban researched the impact of the melting rate in MAG/MIG welding using shielding gas blends [17]. They investigated the influence of the shielding medium on the quality of the filler material. There was a contrast between the solid film welding level and the cored metal. The effect of the wire extension length on the melting rate was also analyzed to use a mathematical model for estimating the melting rate.

Kah talked about the importance of protecting gases from atmospheric pollution for the safety of molten metal during welding [5]. Therefore, the effect of protecting gases on different metals is significant and extensive.

In this research, Ebrahimnia examined the effect on the weld properties of ST 37-2 steel [2] of a deviation in four different gas shielding compositions. After several mechanical and metallographic tests, it has been found that the energy absorbed by the Charpy effect test has decreased and tends to increase with the carbon dioxide content of the shielding gas composition.

To research the impact of welding efficiency on the fatigue strength of the lap filleted joint [4], Hwang used the GMAW technique. The base material of the experiment was a galvanized steel sheet with a standard of 590 MPa, 2.3–2.6 mm wide.

## 1.4 Studies on Shielding Gas for GTAW

Lu studied the behavior of weld form variability in electrode oxidation in SUS304 stainless steel under the mixture (Ar + $CO_2$ and pure Ar) using GTAW [10]. For this test, the welding torch was slightly modified and the procedure was performed. The experimental findings demonstrate that argon gas is effectively prevented by the active external layer air from reacting and oxidizing the tungsten electrode during the welding process.

Kim and Son [6] also investigated the variation of gas blends in GTAW with the finite element method. The GTAW technique performed experimental as well as numerical research to find the link between the alternative supply of shielding gas and the welding efficiency to evaluate welding properties according to the variability of the alternative mixture source.

Ming researched the impact of gas shielding on welding penetration of $CO_2$ laser-GTA hybrid weld [11]. The experiment was conducted using 316L stainless steel with a specific shielding medium. The weld penetration of the hybrid weld was determined by the plasma shape of the parameters of the gas shield, especially the plasma height interfering with the laser incident.

Kah has researched on gases that shield the molten metal from ambient emissions during welding [5]. In various welding aspects, such gases play an important role, including the arc characteristics and microstructure welding.

Rizvi has explored the role of mechanical properties such as tensile, yield and impact strength, etc. and the shielding gases strongly influencing the various welding steel microstructures [15]. The configuration of the welding bead and penetration has been found to be affected by the usage of shielding gas. This short analysis demonstrates the effect on the welding performance of safe gases.

## 1.5 Comparison of Shielding Gas Blends Studies in Both GMAW/GTAW

Through utilizing the GTAW, Mittal proposed comparing the mechanical properties of the welding methods with the cost of welding the different gas blinding methods for commercially pure titanium grades 2 (as per AWS D10.6M) [12]. This study utilizes four forms of shielding gases, i.e., 100% pure argon and 100% pure helium. This alternate gas shielding method improves arc welding performance in all aspects. Complex motion in the welding pool occurs because of the different ionizing characteristics due to the supply of alternate gas. Alternative gas shielding can increase the strength of the weld metal, its weld speed, ductility, and minimize fracture, porosity, and welding processes.

Kutelu described welding as one of the main manufacturing routes for most of the manufacturing industries [8]. The choice of welding method for specific applications involves many factors, such as the composition range of the material to be welded, the thickness of the base materials, and the current form. Most metals oxidize easily in their molded state and therefore secure the welding area from ambient pollution, accomplished by shielding the GTAW tungsten arc welding process gas (argon, helium, nitrogen).

## 1.6 Dynamic Behavior Studies on GMAW

Choi reports that the action of a GMAW system is simulated by means of a short-circuit model with power supply, welding wire, and arc feature [1]. The influence of current surface tension and electronic force is expected to adjust the short-circuit parameters in a short-circuit design. This research was performed by Tipi on an improved design of characteristic equations for the process of GMAW [18]. In this paper, a complex electrical model is applied as a power rectifier source rather than as typical version.

In the area of high-speed welding, higher welding quality, and superior strength [7], Kumar conducted an experiment on both GMAW and GTAW experiment. The impact of the shielding gas was observed in four different compositions in this study. This was done on 3 and 6 mm thick surfaces in both the processes. The optimization of shielding gas blends, current intensities, flow rates, and weld speed is necessary for improving the overall efficiency of the welding process. For GTAW and GMAW specimens mechanical properties are examined. The findings reveal that GTAW welding specimens are better than GMAW welding specimens. The tensile strength and hardness values often yield superior results by reducing the $CO_2$ percentage.

In order to test an entire dynamic process of short-circuit transmission (SCT), Wang was suggested to be focused on the relationship between arcing and circuit phases [21], and improved 'mass-spring' model. This paper analyzes the basic features of the displacement of a droplet mass center and the oscillation rate of droplet variance. The results of the simulations are well in line with the results of the experiments.

## 1.7 Experimental and Computational Simulation Research in GMAW

Wahab states that metal joining is used in a cycle of gas metal arc welding (a two-component diffusion joining method); in various industrial applications [19]. In this study, both numerically and experimentally, the restraining force applied during the welding cycle is measured. The Finite Element Program (ANSYS) was used for numerical analysis.

Zhu has researched the welding of gas metal arc as one of the most significant rapid forming technology [23]. It is a rather complex procedure with many parameters and has a significant effect on training efficiency. This paper established two statistical models to analyze those parameters. One is a 3D mathematical model for a GMAW metal transfer process that is not adequate to describe the entire GMAW method, however, the transient solution for that model may be used for the description of droplet behavior.

Winczek explored the effect of electrode inclination on temperature distribution by overlaying welds [22]. In the Analytical Temperature Definition, a volumetric surface-direction heat source configuration with an inclination axis was adopted. Approximate entropy (ApEn) has been developed to measure the effects of welding parameter on the cycle stability of NGHW. The lower the arc current, the stronger the weld. Finally, the stability process addressed the impact of space restrictions on laser keyhole and arc droplet transition behavior.

## 2 Materials and Methods

Stainless Steel of 316L standard is considered for the study. Grade 316 is most widely used by different grades of austenitic stainless steel as it is the basic molybdenum-bearing class. Hence, heavy welded sections are widely used. Table 1 displays 316L stainless steel chemical composition.

**Table 1** 316L stainless steel chemical composition

| Grade | Range | C | Mn | Si | Cr | Mo | Ni |
|---|---|---|---|---|---|---|---|
| 316L | Min% | – | – | – | 16.0 | 2.0 | 10 |
| | Max% | 0.03 | 2.0 | 0.75 | 18.0 | 3.0 | 14.0 |

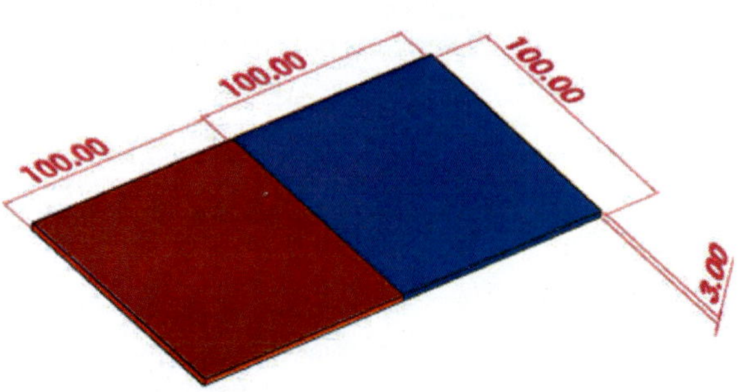


**Fig. 1** 3D model of welded specimen

**Table 2** 316L stainless steel mechanical properties

| Grade | Tensile stress (MPa) min | Yield stress 0.2% proof (MPa) min | Elongation (% in 50 mm) min | Hardness | |
|---|---|---|---|---|---|
| | | | | Rockwell (HRB) max | Brinell (HB) max |
| 316L | 720 | 346 | 40 | 95 | 217 |

**Table 3** Welding process parameters

| Process parameter | Values |
|---|---|
| Type of electrode | ER 316L |
| Diameter of electrode (mm) | 1.2 |
| Gas flow (lpm) | 15 |
| Base material thickness (mm) | 3.6 |
| Weld speed (mm/min) | 200 |
| Wire speed (mm/min) | 1.5 |
| Current used (A) | 155 |
| Voltage (V) | 23 |
| Welding speed (mm/s) | 2.2 |
| Arcing gap (mm) | 2 |
| Diameter of welding rod (mm) | 1.2 |
| Angle of 'V' butt joint | 30° (each) |

Figure 1 depicts the 3D model of welded specimen by using solid works 2018 SP5. The design denotes 3 and 6 mm plate joining made of stainless steel. Tables 1 and 2 indicate 316L stainless steel chemical composition and mechanical properties.

The higher silicone amount improves welding capabilities including wetting. The alloy is widely used in industrial, food processing, shipbuilding, and architectural structures. Table 3 represents the welding process parameter used in GMAW and GTAW for the Butt joint specimen analysis.

Four different shielding gases used for experimentation are

1. Pure $CO_2$
2. Pure Ar
3. Mixture 1—(92% Ar + 8% $CO_2$)
4. Mixture 2—(88% Ar + 12% $CO_2$).

## 3 Results and Discussions

Tensile properties including tensile strength and thickness percentage were tested for welded plates when applied to the base metal. The specimen of stainless steel

316L was checked, and the findings of the specimens are described as the result of this analysis.

## 3.1 *Tensile Strength of GMAW Welded Specimens*

The below-tabulated values are observed experimental tensile strength values for GMAW process.

### 3.1.1 Pure $CO_2$ in GMAW

For 3 and 6 mm plates, Table 4 indicates a comparison of the tensile strength of the sample being tested with the base metal. Pure $CO_2$ on a 3 mm thick plate provides 79.77% of joining.

Figure 2 showing tensile strength graph values as shown in Table 4 for Pure $CO_2$ welded form. This indicates a 3 mm plate tensile strength is greater than 6 mm due to its thickness. The specimen should be designed as higher thickness double 'V' joints. When connecting single 'V' joints, weld metal dilution will be lower.

**Table 4** Tensile strength of GMAW welded specimen with pure $CO_2$

| Sl. No. | Specimen thickness (mm) | UT load in kN | Tensile strength in MPa | | % of joining | Fracture position |
|---|---|---|---|---|---|---|
| | | | Base metal | Tested sample | | |
| 1 | 3 | 40.45 | 720 | 574.41 | 79.77 | Weld area |
| 2 | 6 | 38.65 | 700 | 530.6 | 75.8 | Weld area |

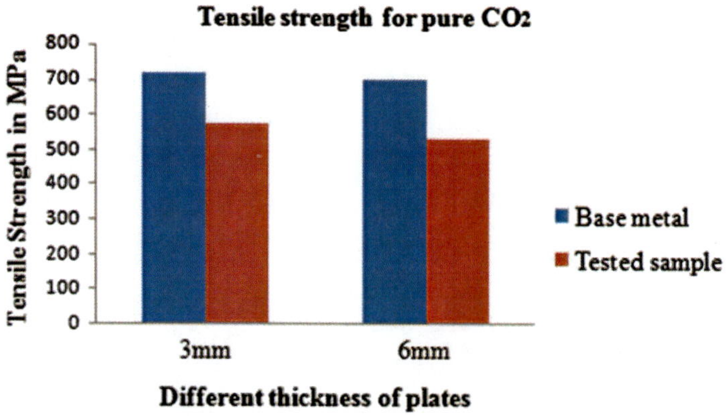


**Fig. 2** Tensile strength for pure $CO_2$ in GMAW

In Fig. 2, the blue color represents the base metal and the orange color represents the sample tested. For 3 mm thick plates, the tensile strength of the tested base metal sample is 720 MPa, while the tensile strength of the tested sample is 574.41 MPa. As a consequence, the base metal's tensile strength is considerably greater than the sample tested in the 3 mm plate. Meanwhile, in the 6 mm plate, base metal's tensile strength is 700 MPa, while the tensile strength of the tested sample is 530.6 MPa. Thus the results show that base metal's tensile strength is stronger than the tested sample.

### 3.1.2 Pure Argon in GMAW

Table 5 displays the ultimate tensile loads for 3 and 6 mm plates. The table below shows base metal comparison with tensile strength for the sample tested. For 3 mm plates, a maximum joining ratio is 92.9% in pure Ar.

Figure 3 representing the graph values of tensile strength as per Table 5 Which is the only Ar-welded case. This reflects that the tensile strength of 3 mm is greater than 6 mm owing to its size. The specimen should be fitted as double ' V' joints for higher thickness.

In Fig. 3, blue color represents the base metal and orange color represents the sample tested. In 3 mm the base metal sample tensile strength is 720 MPa while the

**Table 5** Tensile strength of GMAW welded specimen with pure Ar

| Sl. No. | Specimen thickness (mm) | UT load in kN | Tensile strength in MPa | | % of joining | Fracture position |
|---|---|---|---|---|---|---|
| | | | Base metal | Tested sample | | |
| 1 | 3 | 52.36 | 720 | 668.92 | 92.9 | Weld area |
| 2 | 6 | 42.94 | 700 | 628.56 | 89.79 | Weld area |

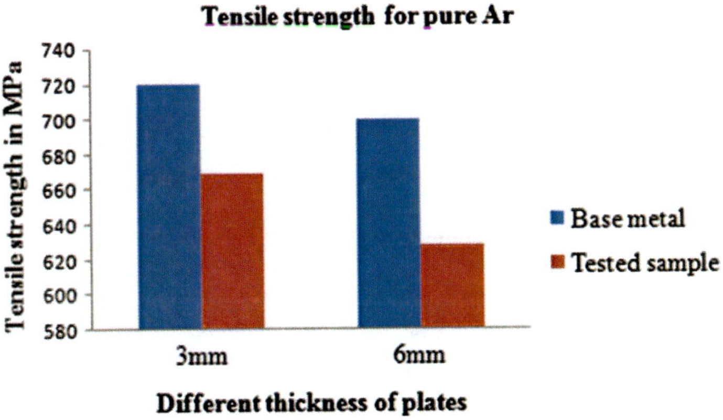


**Fig. 3** Tensile strength for pure Ar in GMAW

tensile strength of the tested sample is 668.92 MPa. As a result, the base metal's tensile strength is considerably greater than the sample tested in the 3 mm plate. Meanwhile, in the 6 mm plate, base metal's tensile strength is 700 MPa, while the tensile strength of the tested sample is 628.56 MPa. Thus test revealed that base metal's tensile strength is greater than the tested sample.

### 3.1.3 Mixture 1—(92% Ar + 8% $CO_2$) in GMAW

Plates 3 and 6 mm, the maximum tensile loads were applied. It demonstrates base metal comparison and tensile strength test samples. It gives 90.73% joining in small thickness plates as seen in Table 6.

Figure 4 representing the graph values of tensile strength as per Table 6 which is the welded specimen with mixture 1. It represents that the tensile strength of 3 mm is better than 6 mm because of its thickness.

In Fig. 4, blue color represents the base metal and orange color represents the sample tested. In 3 mm, the base metal sample tensile strength is 720 MPa while the tensile strength of the tested sample is 653.28 MPa. As a response, the base metal's tensile strength is considerably greater than the sample tested in the 3 mm plate. Meanwhile, in the 6 mm plate, base metal's tensile strength is 700 MPa, whereas the

**Table 6** Tensile strength of GMAW welded specimen with mixture 1

| Sl. No. | Specimen thickness (mm) | UT load in kN | Tensile strength in MPa | | % of joining | Fracture position |
|---|---|---|---|---|---|---|
| | | | Base metal | Tested sample | | |
| 1 | 3 | 62.36 | 720 | 653.28 | 90.73 | Weld area |
| 2 | 6 | 53.86 | 700 | 618.97 | 88.42 | Weld area |

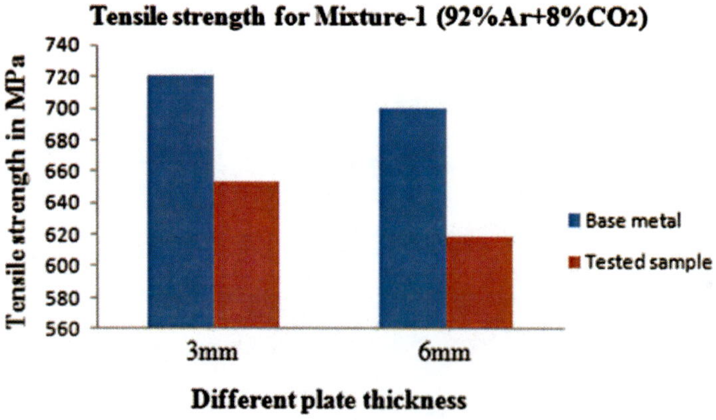


**Fig. 4** Tensile strength for Mixture-1 (92% Ar + 8% $CO_2$) in GMAW

tensile strength of the tested sample is 618.97 MPa. Thus results suggest that base metal's tensile strength is greater than the tested sample.

#### 3.1.4 Mixture 2—(88% Ar + 12% $CO_2$) in GMAW

Two thickness plates of 3 and 6 mm ultimate loads are mentioned in Table 7. The table compares base metal tensile strength and tested samples. In 3 mm plates, 89.68% of joining were obtained in mixture 2.

Figure 5 represents the tensile strength values as seen in Table 7 which is welded under mixture 2. This shows that 3 mm tensile strength is greater than 6 mm plate due to its size.

In Fig. 5, blue color represents the base metal and orange color represents the sample tested. In 3 mm the base metal sample tensile strength is 720 MPa while the tensile strength of the measured material is 645.74 MPa. As a result, the base metal's tensile strength is considerably greater than the sample tested in the 3 mm plate. While, in the 6 mm plate, base metal's tensile strength is 700 MPa, while the tensile strength of the measured sample is 606.05 MPa. Thus the tests indicate that base metal's tensile strength is greater than the tested sample.

**Table 7** Tensile strength of GMAW welded specimen with mixture 2

| Sl. No. | Specimen thickness (mm) | UT load in kN | Tensile strength in MPa | | % of joining | Fracture position |
|---|---|---|---|---|---|---|
| | | | Base metal | Tested sample | | |
| 1 | 3 | 48.42 | 720 | 645.74 | 89.68 | Weld area |
| 2 | 6 | 40.86 | 700 | 606.05 | 86.57 | Weld area |

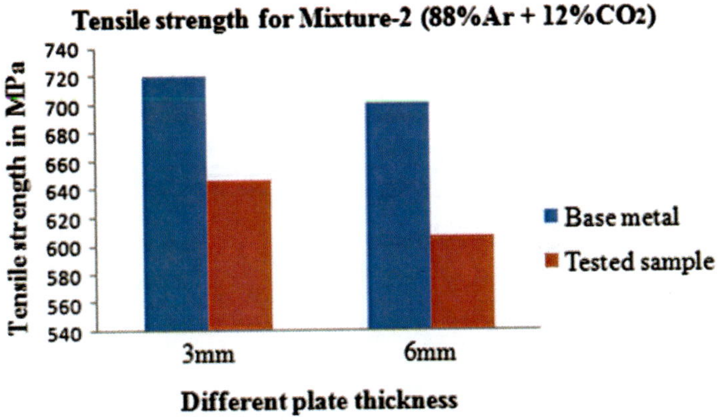


**Fig. 5** Tensile strength for mixture-2 (88% Ar + 12% $CO_2$) in GMAW

## 3.2 *Tensile Strength of GTAW Welded Specimens*

### 3.2.1 Pure $CO_2$ in GTAW

Two plates with varying thicknesses like 3 and 6 mm and the ultimate tensile loads are mentioned in Table 8. It connects the tensile strength of base metal and tested samples. It represents the 3 mm plate's tensile strength is greater than 6 mm plate.

Figure 6 Showing tensile strength graph values as per Table 8. which is welded under pure $CO_2$ welded. This indicates a 3 mm tensile strength greater than 6 mm due to its size. Of higher thickness, the specimen will be designed as double 'V' joints. When single 'V' joints are fitted, there is less weld metal dilution.

In Fig. 6, blue color represents the base metal and orange color represents the sample tested. In 3 mm the base metal sample tensile strength is 720 MPa while the tensile strength of the tested material is 580 MPa. As a response, the base metal's tensile strength is considerably greater than the sample tested in the 3 mm plate. Meanwhile, in the 6 mm plate, base metal's tensile strength is 700 MPa, whereas the tensile strength of the tested sample is 515 MPa. Thus the findings suggest that base metal's tensile strength is stronger than the checked sample.

**Table 8** Tensile strength of GTAW welded specimen with pure $CO_2$

| Sl. No. | Specimen thickness (mm) | UT load in kN | Tensile strength in MPa | | % of joining | Fracture position |
|---|---|---|---|---|---|---|
| | | | Base metal | Tested sample | | |
| 1 | 3 | 38.75 | 720 | 580 | 80.55 | Weld area |
| 2 | 6 | 36.91 | 700 | 515 | 73.57 | Weld area |

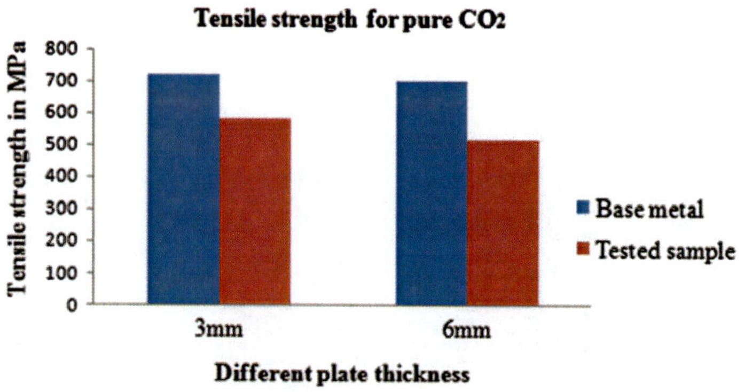


**Fig. 6** Tensile strength for pure $CO_2$ in GTAW

### 3.2.2 Pure Ar in GTAW

The thickness of 3 and 6 mm plates, the ultimate tensile loads are mentioned in Table 9. It relates the base metal and tested sample of tensile strength. For heavier ultimate tensile load percentage of a tensile loads percentage of joining is less than the lighter ultimate loads.

Figure 7 Representing tensile strength graph values as per Table 9. which is welded under pure Ar. This represents a 3 mm tensile strength greater than 6 mm due to its thickness. The specimen will be designed as higher thickness double 'V' joints. When preparing single 'V' joints, weld metal dilution should be smaller. Here when welding the thick metal indicates less strength than the GTAW welding process

In Fig. 7, blue color represents the base metal and orange color represents the sample tested. In 3 mm the base metal sample tensile strength is 720 MPa, while the tensile strength of the tested material is 684.67 MPa. As a consequence, the base metal's tensile strength is considerably greater than the sample tested in the 3 mm plate. Meanwhile, in the 6 mm plate, base metal's tensile strength is 700 MPa, while the tensile strength of the tested sample is 551.95 MPa. Thus the results show that base metal's tensile strength is stronger than the tested sample.

**Table 9** Tensile strength of GTAW welded specimen with Pure Ar

| Sl. No. | Specimen thickness | UT load in kN | Tensile strength in MPa | | % of joining | Fracture position |
|---|---|---|---|---|---|---|
| | | | Base metal | Tested sample | | |
| 1 | 3 mm | 44.75 | 720 | 684.67 | 95.09 | Weld area |
| 2 | 6 mm | 54.75 | 700 | 551.95 | 78.85 | Weld area |

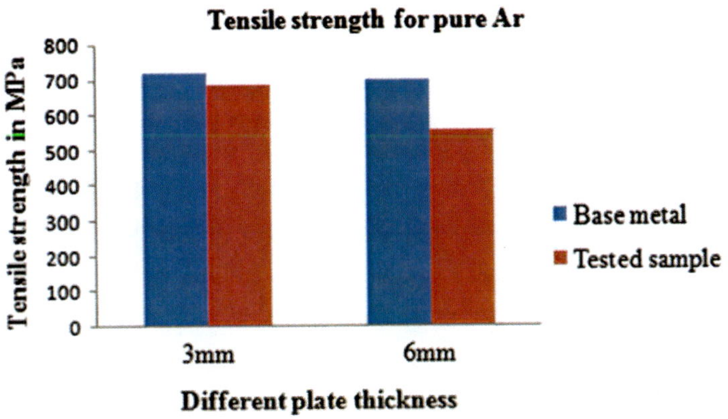


**Fig. 7** Tensile strength for pure Ar in GTAW

### 3.2.3 Mixture 1—(92% Ar + 8% $CO_2$) in GTAW

For plates of 3 and 6 mm thickness, ultimate loads are listed in Table 10. The table compares base metal tensile strength and tested samples. In 3 mm plates, 93 percent of joining were obtained using mixture 1.

Figure 8 representing the graph values of tensile strength as per Table 10. which is the Mixture welded test 1. This reflects that the tensile strength of 3 mm is greater than 6 mm owing to its size. The specimen should be designed as double 'V' joints with higher thickness. When single 'V' joints are prepared, the weld metal dilution should be smaller.

In Fig. 8, blue color represents the base metal and orange color represents the sample examined. In 3 mm the base metal sample tensile strength is 720 MPa while the tensile strength of the measured material is 669.64 MPa. As a consequence, the base metal's tensile strength is comparatively higher than the sample tested in the 3 mm plate. Meanwhile, in the 6 mm plate, base metal's tensile strength is 700 MPa, while the tensile strength of the tested sample is 524.68 MPa. Thus results suggest that base metal's tensile strength is greater than the tested sample.

**Table 10** Tensile strength of GTAWwelded specimen with mixture 1

| Sl. No. | Specimen thickness (mm) | UT load in kN | Tensile strength in MPa | | % of joining | Fracture position |
|---|---|---|---|---|---|---|
| | | | Base metal | Tested sample | | |
| 1 | 3 | 40.75 | 720 | 669.64 | 93 | Weld area |
| 2 | 6 | 53.86 | 700 | 524.68 | 74.95 | Weld area |

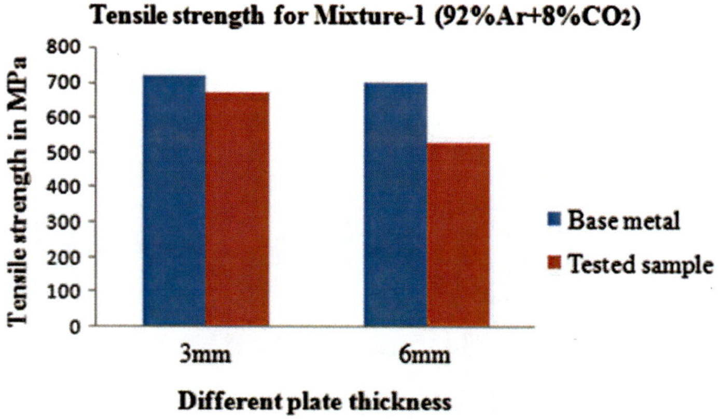


**Fig. 8** Tensile strength for Mixture-1 (92% Ar + 8% $CO_2$) in GTAW

### 3.2.4 Mixture 2—(88% Ar + 12% $CO_2$) in GTAW

The thickness of 3 and 6 mm plates, the ultimate tensile loads are to be acted and their values are in Table 11. It relates to the tensile strength of parent metal and tested samples. For mixture 2, if the ultimate tensile loads are heavier the percentage of joining should be less than the lighter loads.

Here Fig. 9 represents the tensile strength of gas tungsten arc welding with specific shielding gas blends then the comparison shows pure Ar demonstrates greater tensile strength relative to other blends. Next to that the lesser amount of concentration pressure on $CO_2$ is safer from the blends. Another aspect is that the result obtained from welding the gas metal arc is the lower intensity for 3 mm thickness compared only with welding the gas tungsten arc.

In Fig. 9, blue color represents the base metal and orange color represents the sample tested. In 3 mm, the base metal sample tensile strength is 720 MPa while the tensile strength of the tested material is 650.26 MPa. As a result, the base metal's tensile strength is significantly greater than the sample tested in the 3 mm plate. Meanwhile, in the 6 mm plate, base metal's tensile strength is 700 MPa, while the tensile strength of the tested sample is 516.28 MPa. Thus tests indicate that base metal tensile strength is greater than the tested sample.

**Table 11** Tensile strength of GTAW welded specimen with mixture 2

| Sl. No. | Specimen thickness (mm) | UT load in kN | Tensile strength in MPa | | % of joining | Fracture position |
|---|---|---|---|---|---|---|
| | | | Base metal | Tested sample | | |
| 1 | 3 | 46.23 | 720 | 650.26 | 90.03 | Weld area |
| 2 | 6 | 50.68 | 700 | 516.28 | 73.75 | Weld area |

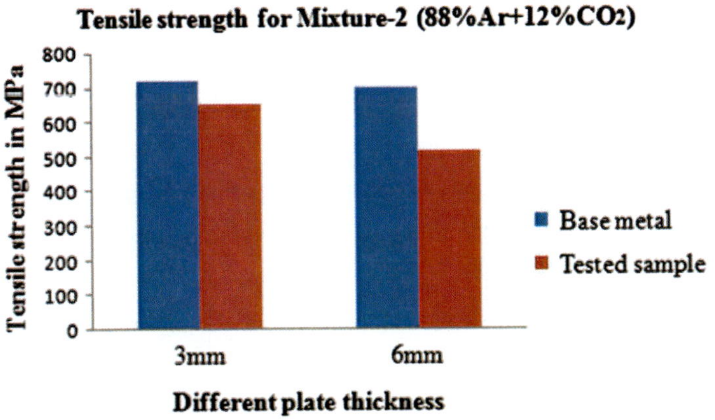


**Fig. 9** Tensile strength for mixture-2 (88% Ar + 12% $CO_2$) in GTAW

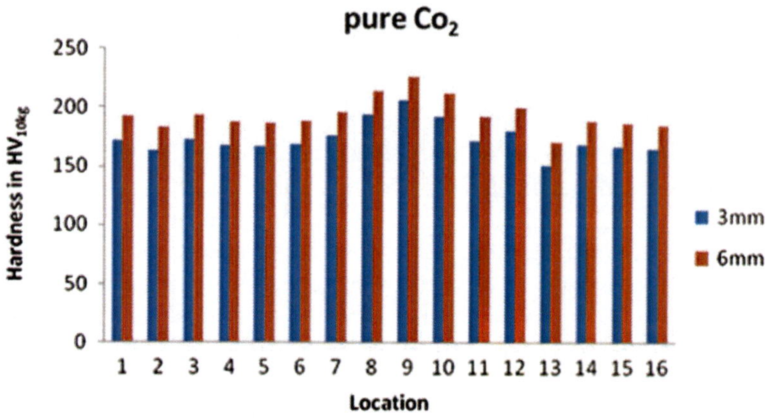


**Fig. 10** Hardness value for pure $CO_2$ in GMAW

## 3.3 *Hardness of GMAW Welded Specimens*

Hardness value was measured for welded plates and compared to the parent metal. For stainless steel 316L specimen, it was tested and the result of this analysis is reported. The graph provides the Hardness values in HV 10 kg.

### 3.3.1 GMAW Hardness Results

These are hardness values obtained from the GMAW process for different shielding medium.

### 3.3.2 Pure $CO_2$ in GMAW

Here Fig. 10 shows Hardness, welded with pure $CO_2$ for both 3 and 6 mm. For each position, 16 locations were identified as having a specified hardness value.

Here, the welded region position shows the higher toughness relative to the other area, the heat-influenced zone implies the lower toughness value. During the welding process, the heat distributed in this area, the grains were changed and the hardness relative to the parent metal region decreased.

### 3.3.3 Pure Ar in GMAW

Here Fig. 11 shows the Hardness of both 3 and 6 mm welded with pure Ar. For each position, 16 locations were identified as having a specified hardness value.

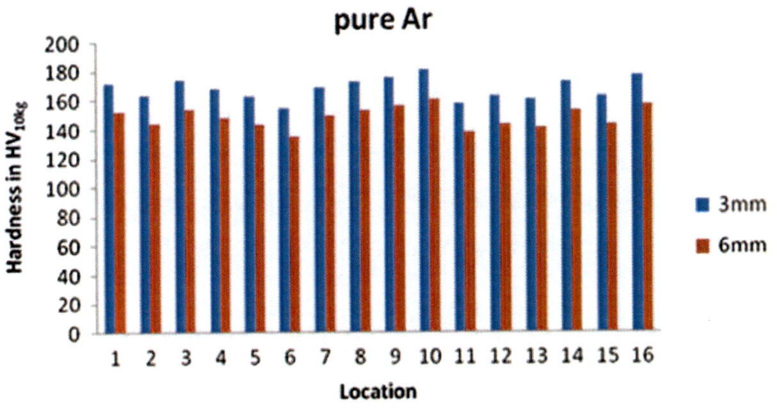


**Fig. 11** Hardness value for pure Ar in GMAW

Here, the welded region position shows the higher toughness relative to the other region, the heat-influenced zone reveals the lower toughness value. During the welding process, the heat dispersed in this field, the grains were changed and the hardness relative to the parent metal region decreased.

### 3.3.4 Mixture 1 in GMAW

Here Fig. 12 shows Hardness performance welded with Mixture 1 for both 3 and 6 mm Sixteen locations were defined as the hardness value for each area.

Here also the welded region position reveals the best toughness compared to the other area in which the heat-influenced zone indicates the lower hardness value.

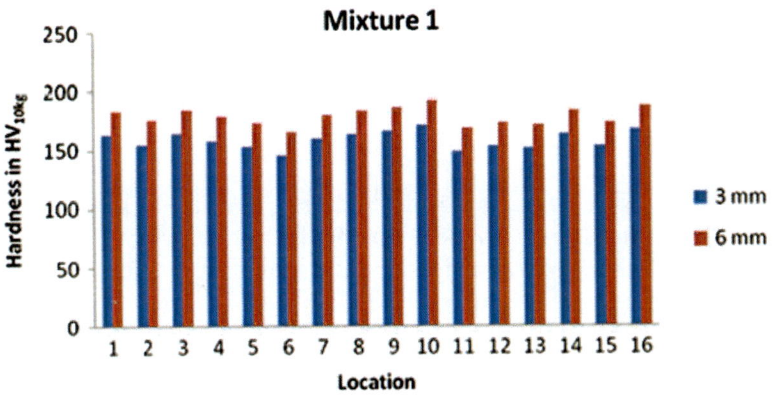


**Fig. 12** Hardness value for Mixture-1 (92% Ar + 8% $CO_2$) in GMAW

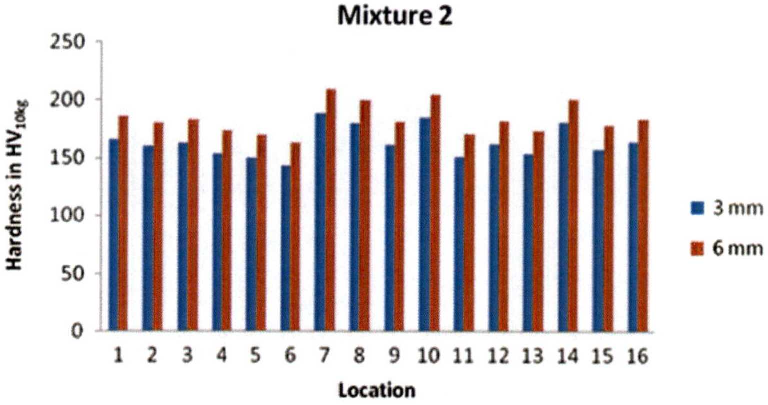


**Fig. 13** Hardness value for Mixture-2 (88% Ar + 12% $CO_2$) in GMAW

#### 3.3.5 Mixture 2 in GMAW

Here Fig. 13 shows the Hardness rating for Mixture 2 welded in both 3 and 6 mm. For each position, 16 locations were identified as having a specified hardness value.

The welded region position indicates the higher toughness relative to the other field where the heat affected area reveals the poorer hardness value here during the welding process, the heat was dispersed in this area, the grains were modified and the strength was decreased.

### *3.4 Hardness of GTAW Welded Specimens*

Below shown graphs are displaying the hardness value according to different shielding mediums in GTAW process.

#### 3.4.1 Pure $CO_2$ in GTAW Process

Here Fig. 14 shows the Hardness of both 3 and 6 mm welded with Pure $CO_2$. Sixteen locations were defined as the hardness value for each area.

Here, the welded region location shows the higher toughness relative to the other area, the heat-influenced zone suggests the lower toughness value.

#### 3.4.2 Pure Ar in GTAW Process

Here Fig. 15 shows the hardness of both 3 and 6 mm welded with pure Ar Sixteen locations were defined as the hardness value for each area.

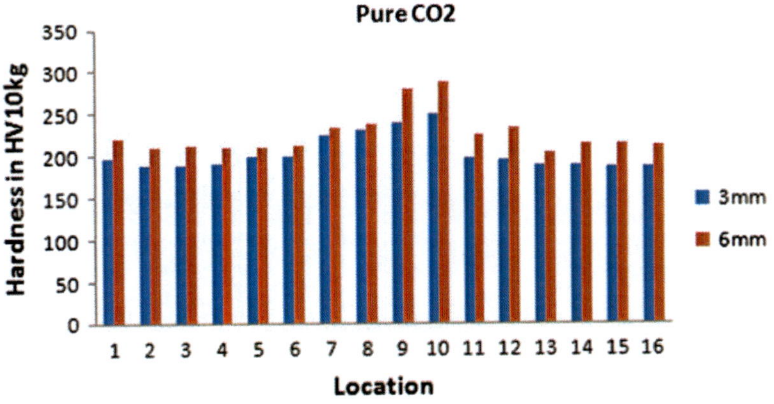


**Fig. 14** Hardness value for pure $CO_2$ in GTAW

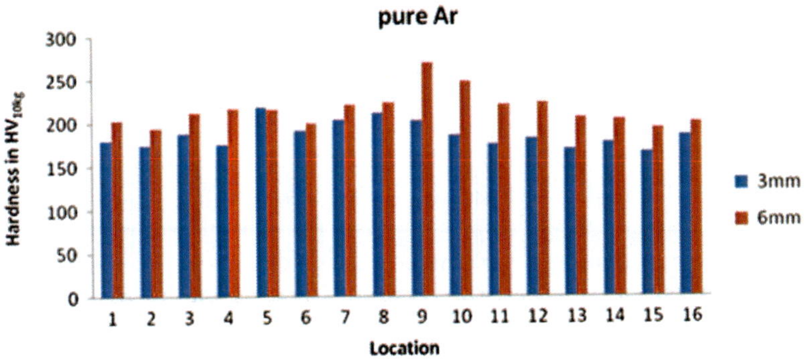


**Fig. 15** Hardness value for pure Ar in GTAW

Here the welded region position reveals the higher toughness relative to the nearest area the heat-influenced zone indicates the poor value of hardness. During the welding process the heat was dispersed in this area the grains were modified and the strength compared to the parent metal area was decreased.

### 3.4.3 Mixture 1 in GTAW

Here Fig. 16 shows Hardness performance welded with Mixture 1 for both 3 and 6 mm For each position, 16 locations were identified as having a specified hardness value.

Here the welded region position reveals the higher toughness relative to the other area the HAZ indicates the poorer hardness value here during the welding process the heat was dispersed in this area the grains were modified and the strength compared to the parent metal area was decreased.

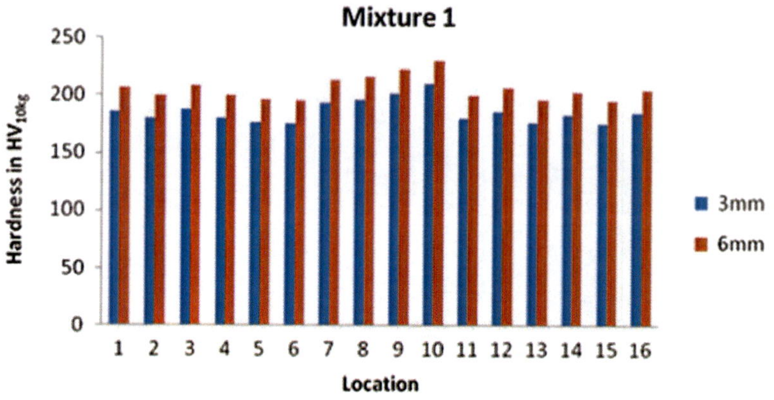


**Fig. 16** Hardness value for Mixture-1 (92% Ar + 8% $CO_2$) in GTAW

### 3.4.4 Mixture 2 in GTAW

Here Fig. 17 shows the Hardness rating for Mixture 2 welded in both 3 and 6 mm. For each position, 16 locations were identified as having a specified hardness value.

Here the welded region position shows the higher toughness compared to the other area, the heat-influenced zone indicates the lower toughness value here during the welding process, the heat was dispersed in this area, the grains were modified and the strength compared to the parent metal area decreased.

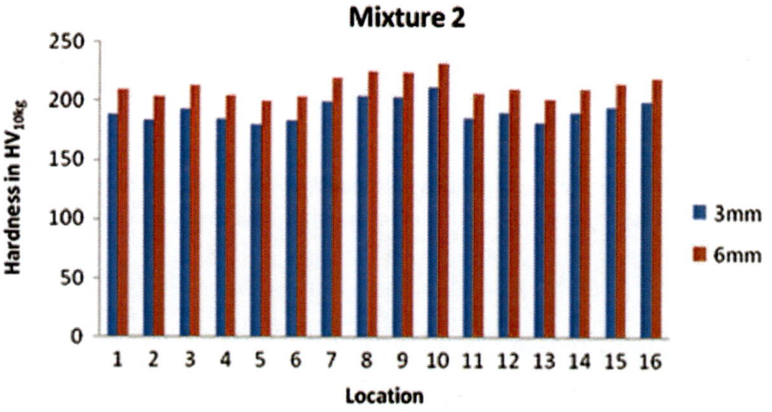


**Fig. 17** Hardness value for Mixture-2 (88% Ar + 12% $CO_2$) in GTAW

## 4 Conclusion

To analyze the mechanical properties for different gas shielding blends, a butt welded model is developed. The sequential coupling method was used to evaluate a suitable gas mixture for GMAW and GTAW. In GMAW and GTAW, for 3 and 6 mm thick plates, the tensile strength of pure argon is superior to other shielding gas compositions with the same current intensity. The best hardness performance was obtained from pure $CO_2$ in both GMAW and GTAW methods. In this work, samples of the mixture 1 with (92% Ar + 8% $CO_2$) and mixture 2 (88% Ar + 12% $CO_2$) were used. Mixture 1 offers the maximum tensile strength and toughness in contrast with mixture 2 for both GMAW and GTAW. This is also evident that the elevated amounts of Ar improve the tensile strength and hardness of stainless steel plates between 3 and 6 mm.

## References


1. Choi JH, Lee JY, Yoo CD (2001) Simulation of dynamic behavior in a GMAW system. Weld J New York 80(10):239S-246S
2. Ebrahimnia M, Goodarzi M, Nouri M, Sheikhi M (2009) Study of the effect of shielding gas composition on the mechanical weld properties of steel ST 37–2 in gas metal arc welding. Mater Des 30(9):3891–3895
3. Gülenç B, Develi K, Kahraman N, Durgutlu A (2005) Experimental study of the effect of hydrogen in argon as a shielding gas in MIG welding of austenitic stainless steel. Int J Hydrogen Energy 30(13–14):1475–1481
4. Hwang I, Kim DY, Jeong G, Kang M, Kim D, Kim YM (2017) Effect of weld bead shape on the fatigue behavior of GMAW lap fillet joint in GA 590 MPa steel sheets. Metals 7(10):399
5. Kah P, Martikainen J (2013) Influence of shielding gases in the welding of metals. Int J Adv Manuf Technol 64(9–12):1411–1421
6. Kim IS, Son JS, Kim HJ, Chin BA (2006) Development of a mathematical model to study the variation of shielding gas in GTA welding. J Achievements Mater Manuf Eng 19(2):73–80
7. Kumar KS, Gejendhiran S, Prasath M (2014) Comparative investigation of mechanical properties in GMAW/GTAW for various shielding gas compositions. Mater Manuf Processes 29(8):996–1003
8. Kutelu BJ, Seidu SO, Eghabor GI, Ibitoye AI (2018) Review of GTAW welding parameters. J Miner Mater Charact Eng 6(5):541–554
9. Liao MT, Chen WJ (1998) The effect of shielding gas compositions on the microstructure and mechanical properties of stainless steel weldments. Mater Chem Phys 55(2):145–151
10. Lu S, Fujii H, Nogi K (2010) Weld shape variation and electrode oxidation behavior under Ar-(Ar-CO2) double shielded GTA welding. J Mater Sci Technol 26(2):170–176
11. Ming G, Xiaoyan Z, Qianwu H (2007) Effects of gas shielding parameters on weld penetration of CO2 laser-TIG hybrid welding. J Mater Process Technol 184(1–3):177–183
12. Mittal A, Kumar A (2016) Effect of shielding gas on titanium CP (Gr-2) by using gas tungsten arc welding. Int J Sci Eng Technol 5(6):339–345
13. Pires I, Quintino L, Miranda RM (2007) Analysis of the influence of shielding gas blends on the gas metal arc welding metal transfer modes and fume formation rate. Mater Des 28(5):1623–1631
14. Pires I, Rosado T, Costa A, Quintino L (2007) Influence of GMAW shielding gas in productivity and gaseous emissions. In: Proceedings of the 10th international aachen welding conference, Aachen, Germany, pp 22–25

15. Rizvi SA, Tewari SP, Ali W (2015) Effect of shielding gases on weld quality in GTA & GMA welding-A
16. Srivastava S, Garg RK (2017) Process parameter optimization of gas metal arc welding on IS: 2062 mild steel using response surface methodology. J Manuf Process 25:296–305
17. Suban M, Tušek J (2001) Dependence of melting rate in MIG/MAG welding on the type of shielding gas used. J Mater Process Technol 119(1–3):185–192
18. Tipi AD, Pariz N (2015) Improving the dynamic metal transfer model of gas metal arc welding (GMAW) process. Int J Adv Manuf Technol 76(1–4):657–668
19. Wahab MA, Alam MS, Painter MJ, Stafford PE (2006) Experimental and numerical simulation of restraining forces in gas metal arc welded joints. Weld J New York 85(2):35
20. Wang J, Wang C, Meng X, Hu X, Yu Y, Yu S (2011) Interaction between laser-induced plasma/vapor and arc plasma during fiber laser-MIG hybrid welding. J Mech Sci Technol 25(6):1529
21. Wang Y, Wang L, Lv X (2016) Simulation of dynamic behavior and prediction of optimal welding current for short-circuiting transfer mode in GMAW. J Manuf Sci Eng 138(6)
22. Winczek J, Gucwa M, Mičian M, Makles K (2019) Numerical analysis of the influence of electrode inclination on temperature distribution during GMAW overlaying. Math Probl Eng
23. Zhu S, Liang Y, Xia D, Wang Q (2012) Research on numerical simulation for welding process in gas metal arc welding rapid forming. J Comput Theor Nanosci 9(9):1218–1221

# Double-Loop Robust Motion Control of a Ground-Based Vehicle-Manipulator System

**Swati Mishra and Santhakumar Mohan**

**Abstract** This research work emphasizes the equations of motion of a ground-based vehicle-manipulator system that constitutes of a spatial six degree of freedom serial manipulator and is modelled considering the dynamic coupling and other effects. Subsequently, a robust double-loop motion control scheme is applied for the end effector trajectory tracking of the vehicle-manipulator system. Simulation experiments based on real-time system parameters are accomplished to elucidate the essence of the recommended control scheme. The outcomes of this work emphasize that the proffered motion control scheme is suited to the similar kinematically redundant robotic systems and is effective even in the existence of unknown external disturbances and system uncertainties.

**Keywords** Double-loop control · Task-space control · Vehicle-manipulator system · Robust control · Kinematically redundant system

## 1 Introduction

Ground-based vehicle-manipulators are majorly found in warehouses, service applications, building constructions, industrial purposes such as painting, welding and many more applications. Vehicle-manipulator consists of a robotic arm mounted on a vehicle base. It has to follow the user-defined path safely and perform the strenuous task of picking and placing or other intervention tasks. In past years, many research have been performed for tracking the given trajectories and controlling the

S. Mishra
Department of Metallurgy Engineering and Material Science, Indian Institute of Technology Indore, Indore, Madhya Pradesh, India
e-mail: phd1501181004@iiti.ac.in

S. Mohan (✉)
Department of Mechanical Engineering, Indian Institute of Technology Palakkad, Palakkad, Kerala, India
e-mail: santhakumar@iitpkd.ac.in

© The Editor(s) (if applicable) and The Author(s), under exclusive license to Springer Nature Singapore Pte Ltd. 2021

S. Mohan et al. (eds.), *Materials, Design, and Manufacturing for Sustainable Environment*, Lecture Notes in Mechanical Engineering,
https://doi.org/10.1007/978-981-15-9809-8_6

motion of the vehicle-manipulator system [1, 2]. Various motion control schemes have been developed which illustrates the behaviour of the system. Reference [3] presents the cascaded control scheme which comprises of kinematic controller and a disturbance compensator for tracking the trajectory of the vehicle-manipulator along with obstacle avoidance. Reference [4] discusses a new scheme of redundancy control which considers the task priority in association to the inverse kinematic problem of redundant vehicle-manipulators. For distributed real-time systems, reference [5] focuses on redundancy resolution schemes and gives the solution for inverse kinematics problem for redundant vehicle-manipulator. Reference [6] illustrates that the internal loop aims for the robustness, and the outer loop helps in obtaining the desired trajectory tracking performance. In [7], the task–space tracking errors caused due to mechanical errors can be reduced by the proffered dual-integral sliding mode control scheme. Reference [8] describes that the feedback can be taken from the motors as well as the end effector. The actual feedback is necessary for tracking the performance of the system. Dual-loop control scheme is providing much better results than other controller schemes in terms of reduced tracking errors. Reference [9] discusses the PID controller in inner loop with PID sliding mode in outer loop. Reference [10] focuses on recent developments and overview of the robust control scheme of uncertain systems. But still there is a scope of improvement in the real-time applications and motion control of kinematically redundant systems.

The main target is to introduce a robust double-loop motion control scheme for tracking the end effector of the vehicle-manipulator system under dynamic variations with the help of redundant feedback. The outer loop consists of the task-space kinematics, and the inner loop consists of the configuration-space system dynamics. This paper also discusses the tracking performance when uncertainties and disturbances arise during the motion and contact of the vehicle-manipulator with the environment.

The structure of the paper as follows: System description and mathematical model are described in Sect. 2. Section 3 focuses on motion control design. Simulation results and outcomes are discussed in Sect. 4. At the end, Sect. 5 closure of the paper.

## 2 System Description and Mathematical Model

### *2.1 Description of the Ground-Based Vehicle-Manipulator System*

The vehicle-manipulator studied for the motion analysis integrates a six-link serial manipulator attached on a four-mecanum wheeled mobile platform. The mobile platform of the vehicle-manipulator is driven by the four independent motored wheels. In Fig. 1, the photography image and the kinematic frame arrangement of the vehicle-manipulator are shown, where the inertial frame (earth fixed) is considered as $O$ (0, 0, 0); $B$ ($x_B$, $y_B$, $z_B$) is the vehicle base frame, and end effector frame is denoted

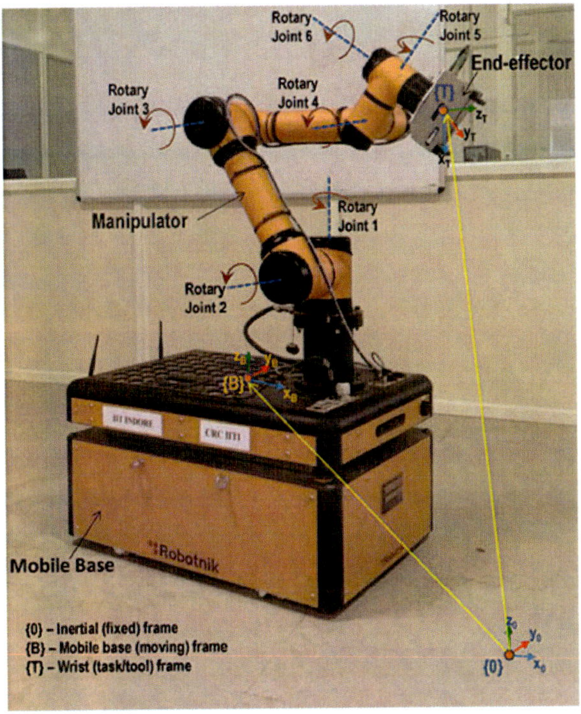


**Fig. 1** Photography image of the JR2 vehicle-manipulator

as the $T$ $(x_t, y_t, z_t)$. $\boldsymbol{\eta} \in \Re^{9\times1}$ is the configuration (joint) space vector of position variables; $\boldsymbol{\eta} = \begin{bmatrix} \boldsymbol{\eta}_v & \boldsymbol{\eta}_m \end{bmatrix}^T$. $\boldsymbol{\eta}_v \in \Re^{3\times1}$ is the vehicle base positions and orientation vector and defined as: $\boldsymbol{\eta}_v = \begin{bmatrix} x_v & y_v & \theta_v \end{bmatrix}^T$. $\boldsymbol{\eta}_m \in \Re^{6\times1}$ is the manipulator rotary joint angles vector and stated as: $\boldsymbol{\eta}_m = \begin{bmatrix} \theta_1 & \theta_2 & \theta_3 & \theta_4 & \theta_5 & \theta_6 \end{bmatrix}^T$ where $\theta_1, \theta_2, \theta_3, \theta_4, \theta_5$ and $\theta_6$ are the joint angles of the manipulator interconnected with serial links. However, the actuator inputs are with respect to the vehicle frame; therefore, the vector of configuration-space velocity can be depicted as:

$$\dot{\boldsymbol{\eta}} = \mathbf{J}_1(\boldsymbol{\eta})\boldsymbol{\xi} \tag{1}$$

$\boldsymbol{\xi} = \begin{bmatrix} \boldsymbol{\xi}_b & \boldsymbol{\xi}_m \end{bmatrix}^T \in \Re^{9\times1}$ is the control inputs vector in body-fixed coordinate frame or command velocities, where $\boldsymbol{\xi}_b \in \Re^{3\times1}$ is the velocity inputs vector of the vehicle, and $\boldsymbol{\xi}_m \in \Re^{6\times1}$ is the joint velocities vector of the serial manipulator. $\mathbf{J}_1(\boldsymbol{\eta}) \in \Re^{9\times9}$ is the Jacobean matrix of the vehicle-manipulator which maps the configuration-space velocities from the body-fixed frame to the inertial frame.

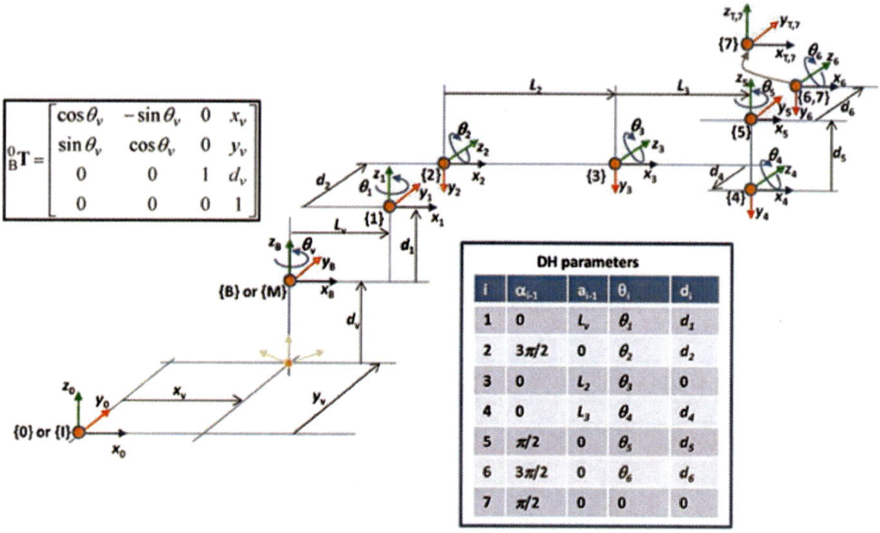


**Fig. 2** Kinematic frame arrangement of the JR2 vehicle-manipulator

## 2.2 *Kinematic Model*

In this research work, proposed vehicle-manipulator aggregates three degree of freedom (dof) of vehicle base and a six-dof manipulator arm. Figure 2 presents the kinematic frame arrangement and the system configuration of the JR2 vehicle-manipulator. The cylindrical-shaped manipulator links are considered with serial arrangement. The vehicle-manipulator's kinematic model is formulated, and the forward kinematic solution using the Denavit-Hartenberg formulation and the forward kinematic solution of the vehicle- manipulator are expressed below:

$$\boldsymbol{\mu} = \begin{bmatrix} L_v - (d_2 - d_4)\sin\theta_1 + (L_2\cos\theta_2 + L_3\cos(\theta_2 + \theta_3))\cos\theta_1 \\ (d_2 + d_4)\cos\theta_1 + (L_2\cos\theta_2 + L_3\cos(\theta_2 + \theta_3))\sin\theta_1 \\ d_1 - L_2\sin\theta_2 - L_3\sin(\theta_2 + \theta_3) \end{bmatrix} \tag{2}$$

## 2.3 *Dynamic Model*

The Newton-Euler recursive method is used for deriving the equation of motion of the vehicle-manipulator, and it can be presented in the matrix and vector form as follows:

$$\mathbf{M}(\boldsymbol{\eta})\dot{\boldsymbol{\xi}} + \mathbf{n}(\boldsymbol{\eta}, \dot{\boldsymbol{\xi}}) + \mathbf{g}(\boldsymbol{\eta}) = \boldsymbol{\sigma} \tag{3}$$

$$\ddot{\boldsymbol{\eta}} = \mathbf{J}_1(\boldsymbol{\eta})\dot{\boldsymbol{\xi}} + \dot{\mathbf{J}}_1(\boldsymbol{\eta})\dot{\boldsymbol{\eta}} \tag{4}$$

$\dot{\boldsymbol{\eta}} \in \Re^{9\times1}$, $\ddot{\boldsymbol{\eta}} \in \Re^{9\times1}$ are the configuration (joint) space velocities, and accelerations vector $\mathbf{M}(\boldsymbol{\eta})\dot{\boldsymbol{\xi}}$ is the inertial forces and moments vector of the vehicle-manipulator; $\mathbf{n}(\boldsymbol{\eta}, \dot{\boldsymbol{\xi}})$ is the dissipative and non-conservative forces vector which includes frictional, Coriolis and centripetal effects of the vehicle-manipulator; the vector of gravity effects of the vehicle-manipulator is written as $\mathbf{g}(\boldsymbol{\eta})$. $\boldsymbol{\sigma} = \left[\boldsymbol{\sigma}_v\ \boldsymbol{\sigma}_m\right]^T \in \Re^{9\times1}$ is the control inputs vector, where $\boldsymbol{\sigma}_v \in \Re^{3\times1}$ is the inputs of the mobile base vector, and $\boldsymbol{\sigma}_m \in \Re^{6\times1}$ is the input torques vector of the serial manipulator attached on a mobile base. The vector of inputs can be further considered as two variables by considering control inputs and disturbances.

$$\boldsymbol{\sigma} = \boldsymbol{\sigma}_{\text{con}} + \boldsymbol{\sigma}_{\text{dis}} \tag{5}$$

$$\boldsymbol{\sigma}_{\text{dis}} = \boldsymbol{\sigma}_{\text{edis}} + \boldsymbol{\sigma}_{\text{idis}} \tag{6}$$

$$\boldsymbol{\sigma}_{\text{idis}} = \left(\hat{\mathbf{M}}(\boldsymbol{\eta}) - \mathbf{M}(\boldsymbol{\eta})\right)\dot{\boldsymbol{\xi}} + \left(\hat{\mathbf{n}}(\boldsymbol{\eta}, \boldsymbol{\xi}) - \mathbf{n}(\boldsymbol{\eta}, \boldsymbol{\xi})\right) + \left(\hat{\mathbf{g}}(\boldsymbol{\eta}) - \mathbf{g}(\boldsymbol{\eta})\right) + \mathbf{F}(\boldsymbol{\eta}, \boldsymbol{\xi}) + \boldsymbol{\delta} \tag{7}$$

where $\hat{\mathbf{M}}(\boldsymbol{\eta})$, $\hat{\mathbf{n}}(\boldsymbol{\eta}, \boldsymbol{\xi})$ and $\hat{\mathbf{g}}(\boldsymbol{\eta})$ are the noted (inaccurate) model equations of the vehicle-manipulator. $\mathbf{F}(\boldsymbol{\eta}, \boldsymbol{\xi})$ is the frictional effects vector which comprises of static, coulomb and viscous frictional effects. $\boldsymbol{\delta}$ is the internal disturbances vector acquainted with the system due to measurement and process noises. $\boldsymbol{\sigma}_{\text{edis}}$ is the external disturbances vector acting on the vehicle-manipulator $\boldsymbol{\sigma}_{\text{idis}}$ is the internal disturbances vector due to parametric and frictional effects; disturbances and system uncertainties arise due to measurement and process noises.

## 2.4 Actuator Inputs and Its Allocations

While correlating the single force actuator inputs with the generalized input vector of the suggested kinematically redundant vehicle-manipulator, the input (control) vector can be rephrased below:

$$\boldsymbol{\sigma}_{\text{con}} = \mathbf{B}\boldsymbol{\kappa} \tag{8}$$

where $\mathbf{B} \in \Re^{6\times10}$ is the actuator configuration matrix, and $\boldsymbol{\kappa} \in \Re^{10\times1}$ is the vector of actuator input. The recommended vehicle-manipulator has four actuator (individual wheel motors) inputs in its mobile platform and three rotary actuators as actuator inputs at the manipulator.

Putting value of (6) into (2) and reorganizing the terms, we have,

$$\dot{\boldsymbol{\xi}} = \mathbf{M}(\boldsymbol{\eta})^{-1}\left(\mathbf{B}\boldsymbol{\kappa} + \mathbf{n}\left(\boldsymbol{\eta}, \dot{\boldsymbol{\xi}}\right) + \mathbf{g}(\boldsymbol{\eta}) + \boldsymbol{\sigma}_{\text{dis}}\right) \tag{9}$$

The desired manipulator behaviour is well chosen and presented in the Cartesian (operational) space. The operational-space position, velocity and acceleration vectors can be insinuated below:

$$\begin{aligned}\boldsymbol{\mu} &= \text{fun}(\boldsymbol{\eta})\\ \dot{\boldsymbol{\mu}} &= \mathbf{J}_2(\boldsymbol{\eta})\dot{\boldsymbol{\eta}}\\ \ddot{\boldsymbol{\mu}} &= \mathbf{J}_2(\boldsymbol{\eta})\ddot{\boldsymbol{\eta}} + \dot{\mathbf{J}}_2(\boldsymbol{\eta})\dot{\boldsymbol{\eta}}\end{aligned} \tag{10}$$

where $\boldsymbol{\mu} \in \Re^{6\times 1}$ is the operational-space position vector, and $\boldsymbol{\mu} = \left[x\ y\ z\ \alpha\ \beta\ \gamma\right]^T$. $\mathbf{J}_2(\boldsymbol{\eta}) \in \Re^{6\times 9}$ is the Jacobian matrix. However, the vehicle-manipulator has two coordinate frames, and the configuration (inertial fixed) space velocities can be mapped with body-fixed (moving) frame velocities as:

$$\dot{\boldsymbol{\eta}} = \mathbf{J}_1(\boldsymbol{\eta})\boldsymbol{\xi} \tag{11}$$

where $\boldsymbol{\xi} \in \Re^{6\times 1}$ is the body-fixed frame velocities. Therefore, the operational-space velocities can be rewritten with body-fixed velocities as follows:

$$\dot{\mu} = \mathbf{J}_2(\boldsymbol{\eta})\mathbf{J}_1(\boldsymbol{\eta})\boldsymbol{\xi} = \mathbf{J}(\boldsymbol{\eta})\boldsymbol{\xi} \tag{12}$$

## 3 Motion Control Design

An improved backstepping control design scheme is proposed for the vehicle-manipulator so that it can follow accurately the desired task-space position trajectory in the presence of system uncertainties and external disturbances. This motion control scheme has an advantage of rejecting all the unstable nonlinearities and preserves the nonlinearities which help in stabilizing the system. The main aim of the proffered controller is that the tracking errors must converge to zero and controller should overcome and adapt itself from all the issues associated with the system parameter variations, frictional effects, external and internal disturbances, unmodelled dynamics, etc. The recommended observer estimates the disturbance vector which compensates the next step response based on current state measurements. The Lyapunov's method is used for designing the proposed control strategy which also helps to prove the closed-loop system stability. To display the asymptotic convergence property of the proffered controller, following presumptions are considered in the design of the controller.

**Assumption 1** The controller and observer gain matrices are assumed to be symmetric positive definite matrices, i.e.

$$K_1 = K_1^T > 0; K_2 = K_2^T > 0; K_3 = K_3^T > 0; \tag{13}$$

For simplicity, in this numerical investigation, these gains are assumed as positive diagonal matrices, as follows:

$$K_1 = k_1 I_{3\times3}; K_2 = k_2 I_{6\times6}; K_3 = k_3 I_{6\times6}; k_1 > 0, k_2 > 0, k_3 > 0. \tag{14}$$

**Assumption 2** The total lumped disturbance vector value is arbitrarily large, bounded and slowly varying with time, i.e. $\dot{\sigma}_{\text{dis}} \approx 0$. The system is bounded to follow the given task- space position trajectory, and the desired task-space trajectory is considered as $\sigma_d$.

## *3.1 Stability Analysis*

Consider the system of which the governing equations are given by (2). The system dynamic model can be rewritten as two single-order sub-systems in a control-affine form which is expressed as:

$$\begin{aligned} \dot{\mathbf{x}}_1 &= \dot{\boldsymbol{\mu}} = \mathbf{J}_1(\mathbf{x}_1)\mathbf{x}_2 \\ \dot{\mathbf{x}}_2 &= \dot{\boldsymbol{\xi}} = \mathbf{M}(\mathbf{x}_1)^{-1}(\mathbf{B}\boldsymbol{\kappa} - \sigma(\mathbf{x}_1, \mathbf{x}_2) + \sigma_{\text{dis}}) \end{aligned} \tag{15}$$

Here, $\mathbf{x}_1 = \boldsymbol{\mu}$ and $\mathbf{x}_2 = \boldsymbol{\xi}$ are the state variables, and they will be available as state feedback signals to the motion controller. $\mathbf{x}_1 = \begin{bmatrix} x & y & z \end{bmatrix}^T$ and $\mathbf{x}_2 = \begin{bmatrix} u & v & r & \dot{\theta}_1 & \dot{\theta}_2 & \dot{\theta}_3 \end{bmatrix}^T$ For the proper choice of $\mathbf{x}_2$ can stabilize the first sub-system and allow the sub-system $\boldsymbol{\mu}$ to track the given desired position trajectory, $\boldsymbol{\mu}_d$. However, $\mathbf{x}_2$ is the state vector and available as feedback to the controller and controller cannot choose any values. Therefore, the controller chooses a virtual control vector called $\mathbf{x}_2^{\text{vc}}$, and the state $\mathbf{x}_2$ should follows the given, $\mathbf{x}_2^{\text{vc}}$. This action can be controlled by the second sub-system with a proper input vector. A positive definite Lyapunov's candidate function is defined as:

$$V(e_1, e_2, e_3) = \frac{1}{2}\left(e_1^T e_1 + e_2^T e_2 + e_3^T K_3^{-1} e_3\right) \tag{16}$$

where $K_3$ is a design matrix and assumed as a symmetric positive definite matrix. On differentiating the Lyapunov function when relating to time along with state trajectories, it presents as:

$$\dot{V}(e_1, e_2, e_3) = e_1^T \dot{e}_1 + e_2^T \dot{e}_2 + e_3^T K_3^{-1} \dot{e}_3 \tag{17}$$

where $\dot{e}_1$, $\dot{e}_2$ and $\dot{e}_3$ are the error time derivatives. By choosing proper stabilizing function to the virtual control input and substituting virtual control in error derivatives. $\mathbf{J}^{+}(\mathbf{x}_1)$ is the pseudo inverse of the Jacobian matrix. Choose a control vector as presented:

$$\sigma_{\text{con}} = \hat{\mathbf{M}}(\mathbf{x}_1)\big(\mathbf{J}^{+}(\mathbf{x}_1)(\ddot{\mathbf{x}}_{1d} + K_1\dot{e}_1) + K_2 e_2 + \mathbf{J}^{T}(\mathbf{x}_1)e_1\big) + \hat{\sigma}(\mathbf{x}_1, \mathbf{x}_2) - \hat{\sigma}_{\text{dis}} \quad (18)$$

$K_2$ is the controller gain matrix and presumed as a symmetric positive definite matrix, i.e. $K_2 = K_2^T > 0$ choose an adaptive law based on velocity feedback as follows:

$$\begin{aligned} \hat{\sigma}_{\text{dis}} &= K_3\hat{\mathbf{M}}(\mathbf{x}_1)\mathbf{x}_2 + \mathbf{x}_3 \\ \dot{\mathbf{x}}_3 &= -K_3\big(\sigma_{\text{con}} - \hat{\sigma}(\mathbf{x}_1, \mathbf{x}_2) + \hat{\sigma}_{\text{dis}} + e_2\big) - K_3\dot{\hat{\mathbf{M}}}(\mathbf{x}_1)\mathbf{x}_2 \end{aligned} \quad (19)$$

$$\dot{e}_3 = \dot{\sigma}_{\text{dis}} - K_3(e_3 - e_2) \quad (20)$$

Since the vehicle-manipulator moves slowly and its disturbance vector is also slowly varying, i.e. $\dot{\sigma}_{\text{dis}} \approx 0$. This assumption reduces (17) as follows:

$$\dot{e}_3 = -K_3(e_3 - e_2) \quad (21)$$

Substituting all error derivatives it gives,

$$\dot{V}(e_1, e_2, e_3) = -\big(e_1^T K_1 e_1 + e_2^T K_2 e_2 + e_3^T e_3\big) \quad (22)$$

The Lyapunov candidate function's time derivative is negative definite which means that chosen control design is globally asymptotically reliable and asymptotically the error converges to zero. If the disturbance vector $\dot{\sigma}_{\text{dis}}$ is not slowly varying and it is bounded, then the choice of $K_3$ can guarantees the stability of the system.

$$\dot{V}(e_1, e_2, e_3) = -\big(e_1^T K_1 e_1 + e_2^T K_2 e_2 + e_3^T e_3\big) + e_3^T K_3^{-1}\dot{\sigma}_{\text{dis}} \quad (23)$$

## 4 Simulation Experiments and Outcomes

### *4.1 Description of the Simulation Experiments*

In this paper to verify the recommended double-loop motion control design scheme, performance investigations for tracking the trajectory of vehicle-manipulator have been done using MATLAB/SIMULINK package. Figure 3 describes with the help of block diagram, double-loop motion control scheme. Double-loop controller consists

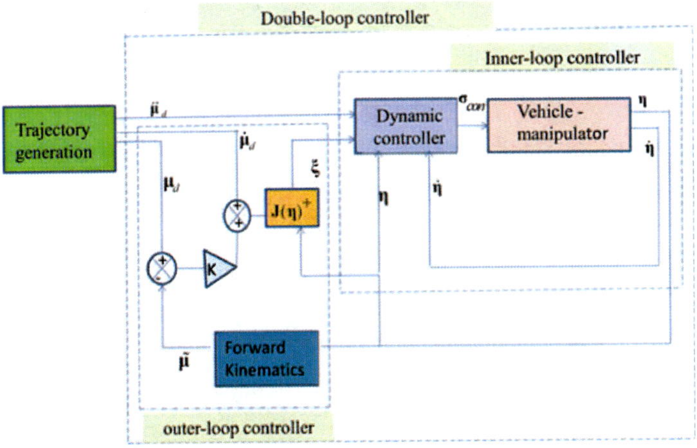


**Fig. 3** Block diagram of the double-loop controller

of inner-loop controller and outer-loop controller. The outer-loop controller consists of the operational-space kinematics whereas inner-loop controller consists of the configurational-space system dynamics. Real-time vehicle-manipulator, namely JR2 is considered for performing the dynamic simulations. However, the derived dynamic model is verified in the virtual robot model in gazebo package environment, and the JR2 vehicle manipulator in the virtual background is presented in Fig. 4. The derived model is almost matching with the virtual system motion in both forward and inverse dynamic cases. For the performance evaluation, manipulator has to begin

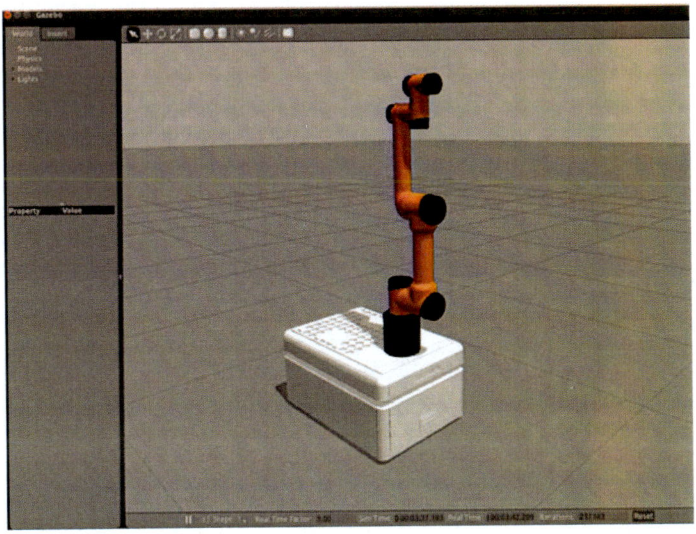

**Fig. 4** JR2 vehicle-manipulator in gazebo environment

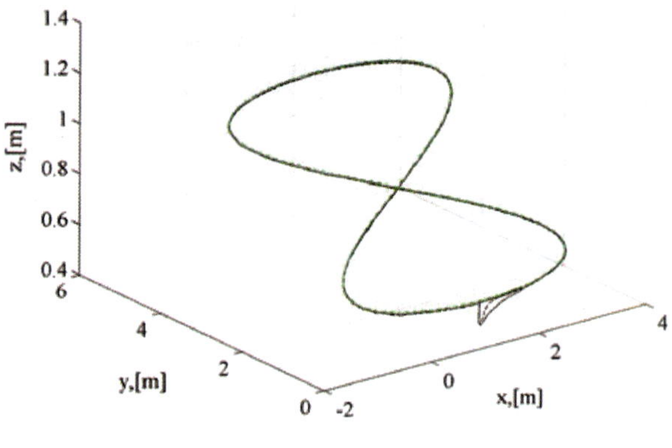


**Fig. 5** Desired eight-shaped complex trajectory

from a given initial position and end to the original position. While travelling, the manipulator will track an eight-shaped spatial position trajectory. Under the existence of the internal and external disturbances, the effectiveness and feasibility are verified by the recommended motion control design. Figure 4 shows the photography image of JR2 vehicle-manipulator in gazebo environment.

### 4.2 *Simulation Outcomes and Discussion*

The computer-based dynamic simulation results are obtained based on the proposed double-loop controller as shown from Figs. 5, 6, 7, 8 and 9. To testify the controller, the dynamic simulations are performed at uncertain conditions. Uncertain conditions comprises of process noises, sensor noises such as white Gaussian noises and external effect like unknown payload can also cause dynamic variations in the system. Figure 5 shows the desired eight-shaped complex trajectory.

## 5 Conclusions

An inverse dynamics of the ground-based vehicle-manipulator system are proposed by virtue of a double-loop motion control scheme in addition with a nonlinear disturbance observer. The proposed motion control scheme achieves asymptotic stability for the slowly varying unknown external disturbances, and system dynamic changes by means of the feedback of the positions, velocities of the mobile robot and the manipulator joints. The control rule is capitulated agreeable tracking assets regardless of the system/parameter uncertainties, disturbing and actuator dynamic effects.

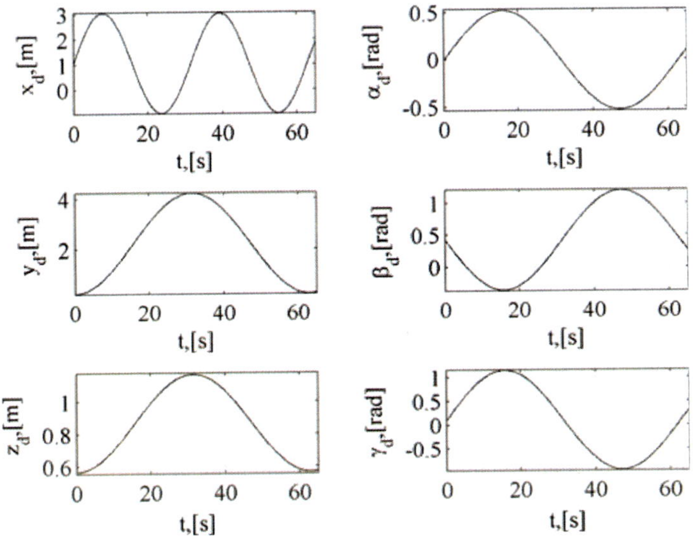


**Fig. 6** Time trend of the proposed spatial operational-space trajectory

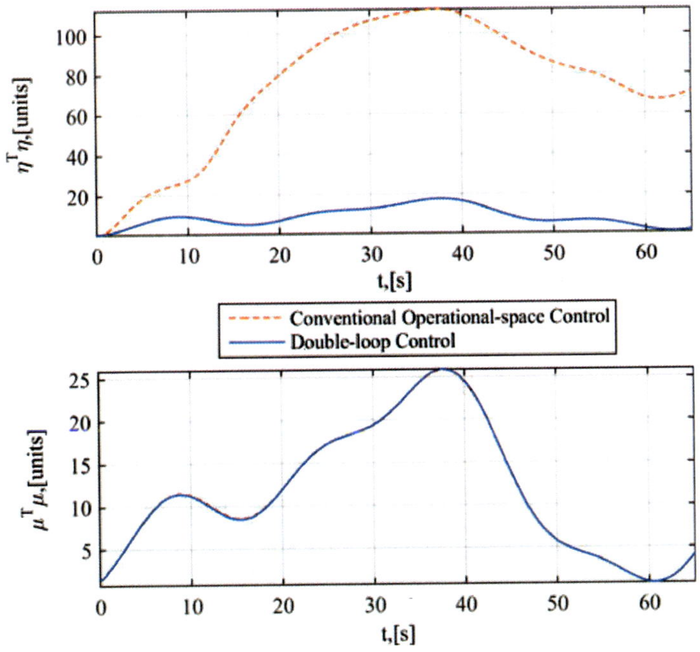


**Fig. 7** Time trend of configurational-space and operational-space trajectories

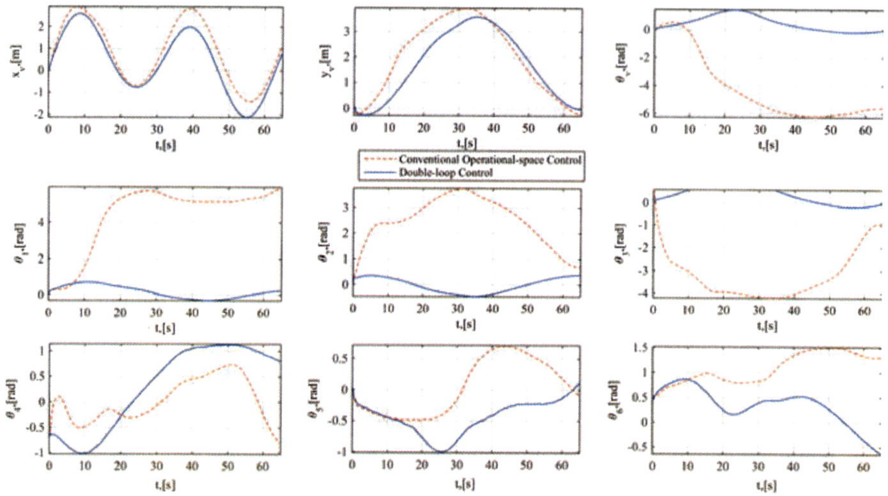


**Fig. 8** Time trend of joint angles in the operational-space trajectory

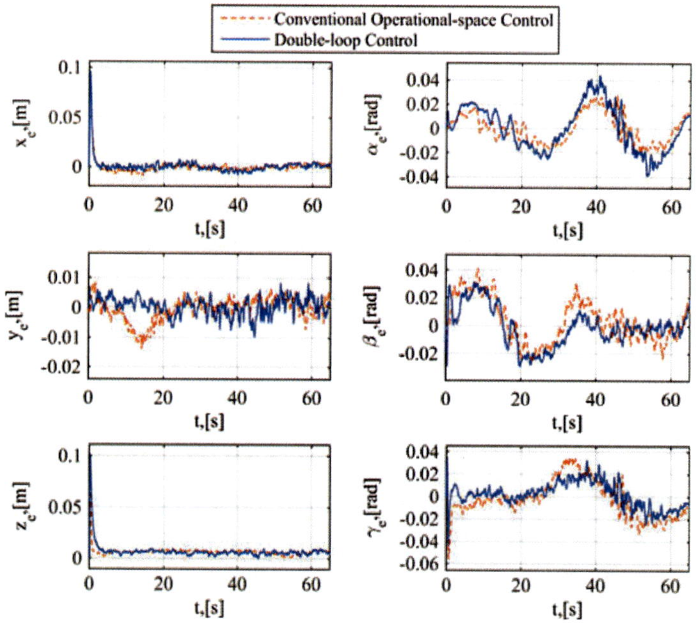


**Fig. 9** Error norm of the operational-space trajectory

The end effector tracking pose errors increase, specifically in $z$-axis position (heave) and y-axis rotation (pitch) when the system dynamic changes in terms of modelling errors, and the unknown external effects are introduced. Nevertheless, these error values are procured to be in acceptable design limits as the requirement for low energy operation of the overall system is concerned. Further, these errors can be decreased if higher values of gains are used at the expenditure of a higher sampling rate of the controller and lower response time of the actuator. Acknowledgement This work is partially supported by the Department of Higher Education, Ministry of Higher Resource Development (MHRD),India, Research Grant#1-36/2016-PN-II to Santhakumar Mohan.

**Acknowledgement** This work is partially supported by the Department of Higher Education, Ministry of Higher Resource Development (MHRD), India, Research Grant#1-36/2016-PN-II to Santhakumar Mohan.

## References

1. Sabanovic A, Ohnishi K (2011) Motion control systems. Wiley
2. Tal J (1999) Two feedback loops are better than one. Machine Design 71
3. Andaluz V, Roberti F, Carelli R (2010) Robust control with redundancy resolution and dynamic compensation for mobile manipulators. In: IEEE International Conference on Industrial Technology (ICIT), Vina del Mar, Chile, 14–17 March, 2010
4. Nakamura Y, Hanafusa H, Yoshikawa T (1987) Task-priority based redundancy control of robot manipulators. Int J Robot Res 6(2):3–15
5. Chiaverini S, Oriolo G, Walker ID (2008) Kinematically redundant manipulators. Springer Handbook of Robotics. Springer Verlag, pp 245–268
6. Tsai MC, Yang FY, Chen CL (2014) A double-loop control structure for tracking control and disturbance attenuation. In: Proceedings of the 19th world congress. The International Federation of Automatic Control Cape Town. South Africa. August 24–29
7. Mohan S, Mohanta JK (2018) Dual integral sliding mode control loop for mechanical error correction in trajectory-tracking of a planar 3-PRP parallel manipulator. J Intell Robot Syst 89(3–4):371–385
8. Agarwal A, Nasa C, Bandyopadhyay S (2011) Dual-loop control for backlash correction in trajectory-tracking of a planar 3-RRR manipulator In: Proceedings of 15th national conference on machines and mechanisms NacoMM2011 189, Chennai 2011, pp 1–8
9. DeSantis RM (2000) PID dual loop control for industrial processes. IFAC Proc 33(4):397–402
10. Petersen IR, Tempo R (2014) Robust control of uncertain systems: classical results and recent developments. Automatica 50(5):1315–1335

# Trend Plot Analysis of Dry Sliding Wear in Al/SiC Co-continuous Ceramic Composites

**R. Ramesh, A. S. Prasanth, and P. Gopalakrishnan**

**Abstract** Composites consisting of an interpenetrated structure of ductile metal and hard ceramic phase are termed as co-continuous ceramic composites (C4). Since C4 can contain a higher fraction of ceramic phase when compared with conventional particulate composites, they are suitable for applications ranging from ballistic protection armour to brake discs. In the present study, wear studies were performed on C4 comprising of three-dimensionally continuous silicon carbide (SiC) foam as ceramic phase and three distinct aluminium alloys. The effect of the applied load on the dry sliding friction coefficient ($\mu$) and specific wear rate ($W_s$) of C4 was evaluated for the three soft Al penetrant-based C4 utilizing their trend plots. It was inferred that, for all the C4 samples, the $\mu$ decreases and $W_s$ increases as the applied load increases. Additionally, the trend plots revealed that, in all loading conditions, the C4 with the highest hardness possessed the lowest $\mu$ and $W_s$. The outcome of this study is of significance to wear performance of C4 in friction and braking applications.

**Keywords** Co-continuous ceramic composite · C4 · Interpenetrated composite · Friction · Braking · Wear performance · Friction coefficient · Specific wear rate

R. Ramesh
Department of Mechanical Engineering, PSG Institute of Technology and Applied Research, Coimbatore, Tamil Nadu, India
e-mail: ramesh@psgitech.ac.in

A. S. Prasanth (✉)
Department of Mechanical Engineering, PSG College of Technology, Coimbatore, Tamil Nadu, India
e-mail: asp.mech@psgtech.ac.in

P. Gopalakrishnan
Department of Metallurgical Engineering, PSG College of Technology, Coimbatore, Tamil Nadu, India
e-mail: pgk.metal@psgtech.ac.in

© The Editor(s) (if applicable) and The Author(s), under exclusive license to Springer Nature Singapore Pte Ltd. 2021

S. Mohan et al. (eds.), *Materials, Design, and Manufacturing for Sustainable Environment*, Lecture Notes in Mechanical Engineering,
https://doi.org/10.1007/978-981-15-9809-8_7

# 1 Introduction

Composites consist of a combination of multitude of materials. This results in a material with enhanced properties than its constituents. Unlike in an alloy, the integrant materials in a composite are distinct and distinguishable from each other [1]. Generally, composites are composed of a discontinuous reinforcement phase, interspersed within a continuous matrix phase in order to strengthen the composite. Metal, polymer, ceramic and carbon are the four primary types of matrix phases utilized. Among these, metal matrix composites (MMCs), wherein a light metal is loaded with hard ceramic phase, have been studied extensively to establish lightweight and cost-effective materials for aerospace, automotive and electronic industries [2]. MMCs are further classified as continuously reinforced composites (CRCs) and discontinuously reinforced composites (DRCs). Continuous reinforcements comprise boron, $Al_2O_3$, Gr and SiC in the form of filaments or fibres, whereas discontinuous reinforcements consist of whiskers, chopped fibres or particulate forms of $Al_2O_3$, SiC and $TiB_2$ [3, 4]. Although a wide range of alternatives exist for the choice of reinforcement particulates or fibres, the homogeneity of distribution within the metal matrix is a crucial factor affecting the mechanical properties of the synthesized DRC. Studies on microstructure of DRCs have revealed non-uniform distribution [5] and agglomeration of reinforcements resulting in weak zones which act as sites for fracture initiation leading to a deterioration of mechanical properties [6]. This is a limiting factor for using DRC-MMCs in numerous applications. This limitation can be overcome by utilizing a continuous network of ceramic instead of discontinuous whiskers or particles [7]. The clustering effect of DRCs can be avoided by infiltrating a continuous network of ceramic in the shape of a preform with a suitable metallic phase, thereby creating an interpenetrated, continuous network of both the metal and ceramic phases. Such three-dimensionally interconnected composites are referred to as C4 [8].

C4 composed of Al as metal and SiC as the ceramic phase are widely utilized to develop new class of materials with high strength-to-weight ratio. Such composites possess beneficial properties for applications entailing better wear resistance, high temperature and strength [9, 10]. Specific advantages of C4 include high stiffness, improved thermal and electrical conductivity, elevated hardness and wear resistance over discontinuous phase composites [11]. A multitude of fabrication techniques for C4 have been outlined in the literature. Among these, infiltration of porous SiC preforms with Al alloy in the absence of pressure referred to as gravity infiltration, is the foremost technique to fabricate low-cost intricate shape C4 [7, 12].

In the present study, C4 were manufactured by gravity infiltration of three aluminium alloys into porous SiC foams. Thereafter, the correlation between the hardness and wear behaviour of the three C4 was evaluated using Taguchi-based methods and trend plots. Such systematic studies on wear performance of C4 are vital for extensive use of C4 in real-world applications.

## 2 Materials and Methods

The method of fabrication, hardness and dry sliding wear tests on C4 specimens is delineated in this section.

### *2.1 Fabrication of C4*

The C4 samples were manufactured using a three-dimensionally porous SiC ceramic foam with 75% purity. The foam was composed of 20% of alumina ($Al_2O_3$), 5% of silica and 75% of SiC with a porosity of 83.66% and thickness of 22 mm. Three commercial grade aluminium alloys (AA2024, AA6063 and AA7068) were utilized to infiltrate the pores in the SiC foam placed inside a resistance-type furnace at 800 °C (~ 150 °C of superheat), using gravity infiltration technique [13] as shown in Fig. 1.

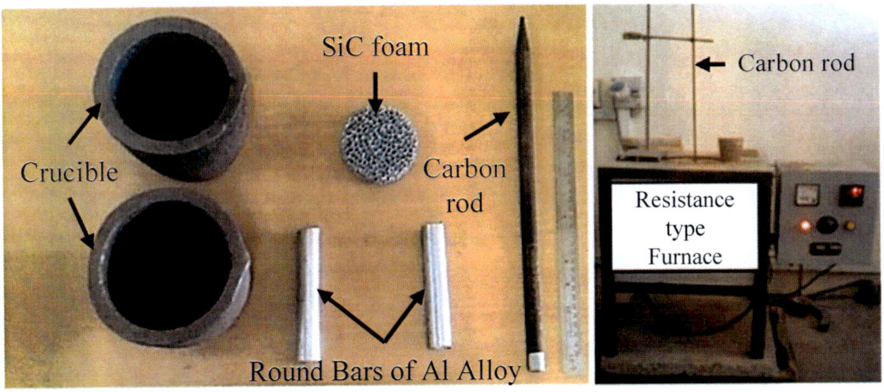


**Fig. 1** Materials and equipment used to manufacture C4 samples

**Table 1** Elemental composition of Al alloys

| *Constituents of AA2024* | | | | | | |
|---|---|---|---|---|---|---|
| Element | Al | Cu | Mg | Mn | Zn | Cr |
| Composition % | 92.54 | 5.274 | 1.232 | 0.747 | 0.076 | 0.029 |
| *Constituents of AA6063* [14] | | | | | | |
| Element | Al | Mg | Si | Fe | Cr | Cu |
| Composition % | 98.46 | 0.642 | 0.387 | 0.210 | 0.034 | 0.026 |
| *Constituents of AA7068* [15] | | | | | | |
| Element | Al | Zn | Mg | Cu | Fe | Si |
| Composition % | 87.1 | 7.75 | 2.74 | 2.05 | 0.14 | 0.11 |

The Al alloy composition as shown in Table 1 was estimated by optical emission spectroscopy.

During the progress of the infiltration, atmospheric interaction inside the furnace chamber was avoided by utilizing argon gas to purge at 0.5 lpm. The infiltrated composite was then solidified by furnace cooling.

## 2.2 Hardness Tests

The hardness of the gravity-infiltrated C4 samples was evaluated using a Brinell hardness tester. The specimens were prepared according to ASTM E10-15 standards at a Brinell hardness scale of HBW 5/250. The hardness number was determined using the established formula shown in Eq. (1).

$$\mathrm{HB} = \frac{2F}{\pi D\left(D - \sqrt{D^2 - d^2}\right)} \tag{1}$$

where $F$ is the applied load in kgf, and $D$ and $d$ represent the diameter of the ball and the mean diameter of the indentation in mm, respectively. The Brinell hardness tester was used to create the indentations on the surface of the monolithic alloys and the C4. Ten indentations were performed on the Al phase as well as on the interface of the C4. The average of the values is reported as the hardness of the composite.

## 2.3 Dry Sliding Wear Tests

Wear studies were conducted on specimens extracted by water jet machining of the three composites using a nozzle of diameter 1.1 mm, a feed rate of 80 mm/min and abrasive silica sand of mesh size 80. Tribological studies were then performed on the three fabricated C4 using a Ducom TR20M36 pin-on-disc tribometer following ASTM G99-17 standards [14]. In order to perform systematic wear studies on the composites, three control factors, each at three levels, were selected as shown in Table 2.

**Table 2** Control factors and levels for dry sliding wear studies

| Control factors | Applied load (L), N | Sliding speed (S), m/s | Sliding distance (D), m |
|---|---|---|---|
| Level 1 | 20 | 1 | 1000 |
| Level 2 | 40 | 2 | 2000 |
| Level 3 | 60 | 3 | 3000 |

The evaluation of the wear behaviour of the C4 was performed from experimental observations of two responses, the friction coefficient ($\mu$, no unit) and specific wear rate ($W_s$, $mm^3$/Nm).

$$\text{Friction Coefficient } (\mu) = \frac{F_f}{F_n} \tag{2}$$

where $F_f$ and $F_n$ represent the force due to kinetic friction and normal force, respectively.

$$\text{Specific Wear Rate } (W_s) = \frac{\Delta W}{\rho L D} \tag{3}$$

where $\Delta W$ and $\rho$ represent the weight loss during wear of the pin (g) and density of the C4 specimen (g/$mm^3$), respectively. The wear mechanism of the C4 was observed using micrographs from a Carl Zeiss EVO 18 Scanning Electron Microscope (SEM).

## 3 Results and Discussion

As delineated in Sect. 2.1, the three C4 specimens were manufactured by gravity infiltration of Al alloy into porous SiC foams. Brinell hardness tests were performed on the monolithic alloy and the gravity-infiltrated C4 samples. Figure 2 displays the interpenetrated structure of the Al/SiC C4 and the hardness indentations. The hardness of the monolithic alloy and the C4 samples is given in Table 3. It can be inferred from the table that, the interpenetrating composite of AA7068/SiC has the highest hardness among the three C4 samples considered in this study.

In an effort to comprehend the wear behaviour of the C4 specimens, the wear performance of the fabricated C4 samples was evaluated using a Taguchi array with nine experimental runs as exhibited in Table 4.

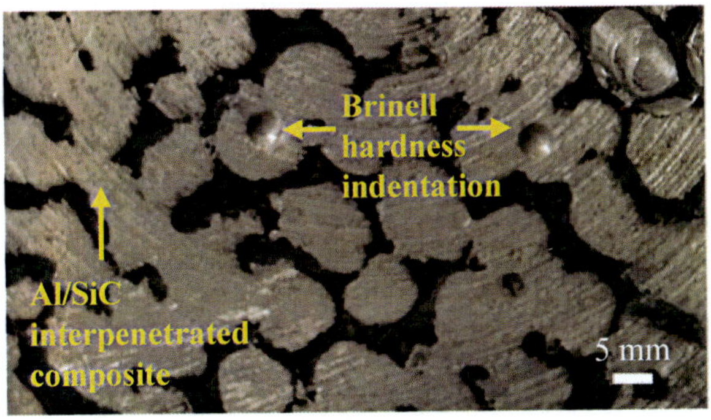


**Fig. 2** C4 interpenetrated structure and Brinell indentation

**Table 3** Hardness of Al alloys

| S. No. | Alloy system | Hardness (BHN) | |
|---|---|---|---|
| | | Monolithic alloy | C4 |
| 1 | AA2024 | 48 | 103 |
| 2 | AA6063 | 26 | 74 |
| 3 | AA7068 | 71 | 122 |

The typical worn surface morphology of the C4, studied using SEM micrograph, is presented in Fig. 3. During pin-on-disc tests, the surface exhibits excessive adhesion of the worn Al particles to the wear tracks. This results in undesired roughness of the wear tracks [16] leading to higher levels of wear rate for the three C4 at elevated loads. The wear mechanism of these composites suggests that, after initial wear of the soft Al metal phase at low loads and low speeds, the brittle ceramic struts stand exposed. When subjected to further wear, the SiC struts fragment into particles and abrade the surface of the composite, thereby forming uniform wear tracks. Therefore, the SEM micrograph reveals that both adhesive and abrasive wear mechanisms contribute to dry sliding wear in C4.

The wear studies on C4 aided in comprehending the mechanism of progression of wear in C4. An understanding of the variation of $\mu$ and $W_s$ with the applied load is paramount for beneficial use of such composites in practical applications. Figures 4 and 5 present the variation of $\mu$ and $W_s$ with the applied load for the three composites. It can be inferred from Fig. 4 that, for all the three considered C4 samples, an increase in the applied load leads to a decrease in $\mu$. Further, at all loading conditions, AA7068/SiC C4 exhibits the lowest $\mu$ followed by AA2024/SiC and AA6063/SiC, respectively. This trend follows the hardness of the three composites reported in Table 3. The AA7068/SiC C4 with the highest hardness exhibits lowest $\mu$ for all applied loads.

The plot of $W_s$ in Fig. 5 reveals that at elevated loads, the wear rate increases for all the three composites. However, among the three C4 samples, the wear rate for the hardest C4, namely AA7068/SiC, is the lowest at all loading conditions. The highest intensity of wear rate is exhibited by the relatively soft AA6063/SiC C4. This substantiates earlier reports in the literature that a hard material exhibits diminished wear rate and thereby offers improved wear resistance [17, 18].

## 4 Conclusions

This study explored the wear behaviour of Al/SiC co-continuous composites. Initially, the composites were manufactured by gravity infiltration and their Brinell hardness was established. SEM micrographs revealed that the wear mechanism in C4 follows both adhesive and abrasive wear. Analysis of the trend plots revealed that highest intensity of wear rate is manifested by the relatively soft AA6063/SiC

**Table 4** Taguchi array and responses for C4

| Run | Control factors | | | AA2024/SiC C4 | | AA6063/SiC C4 [14] | | AA7068/SiC C4 | |
|---|---|---|---|---|---|---|---|---|---|
| | L (N) | S (m/s) | D (m) | $\mu$ | $W_s \times 10^{-3}$ ($mm^3$/Nm) | μ | $W_s \times 10^{-3}$ ($mm^3$/Nm) | $\mu$ | $W_s \times 10^{-3}$ ($mm^3$/Nm) |
| 1 | 20 | 1 | 1000 | 0.511 | 2.24 | 0.531 | 2.31 | 0.486 | 2.08 |
| 2 | 20 | 2 | 2000 | 0.525 | 2.52 | 0.559 | 2.79 | 0.51 | 2.48 |
| 3 | 20 | 3 | 3000 | 0.542 | 2.33 | 0.57 | 2.53 | 0.538 | 2.33 |
| 4 | 40 | 1 | 2000 | 0.426 | 2.99 | 0.457 | 3.22 | 0.402 | 2.65 |
| 5 | 40 | 2 | 3000 | 0.443 | 3.18 | 0.463 | 3.52 | 0.416 | 2.94 |
| 6 | 40 | 3 | 1000 | 0.467 | 3.05 | 0.471 | 3.38 | 0.431 | 2.82 |
| 7 | 60 | 1 | 3000 | 0.383 | 3.36 | 0.411 | 3.79 | 0.363 | 3.21 |
| 8 | 60 | 2 | 1000 | 0.403 | 3.8 | 0.434 | 3.63 | 0.395 | 3.09 |
| 9 | 60 | 3 | 2000 | 0.436 | 3.42 | 0.452 | 4.11 | 0.427 | 3.31 |

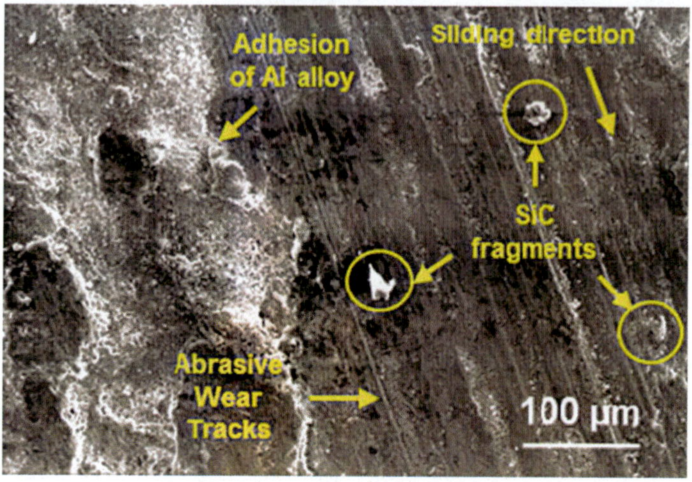


**Fig. 3** Typical worn surface morphology of C4

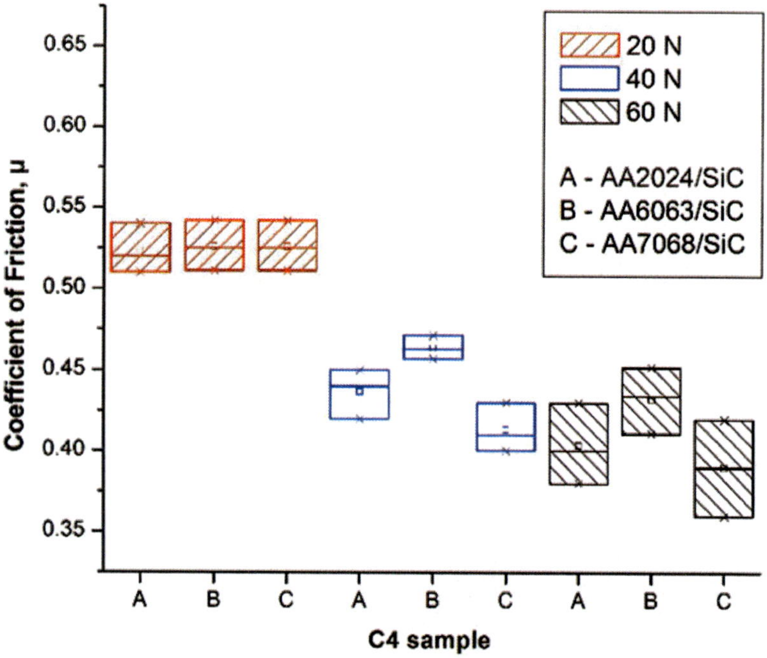


**Fig. 4** Change of $\mu$ with applied load

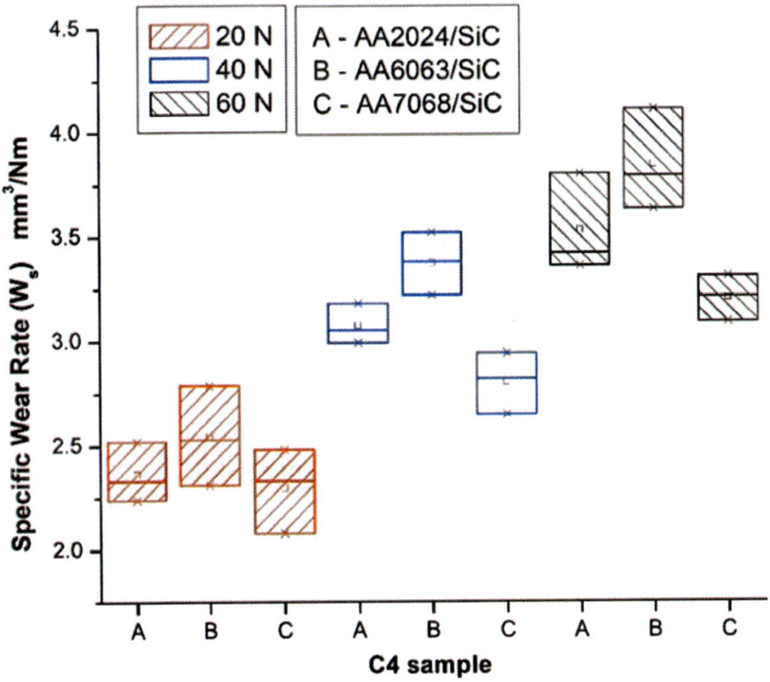


**Fig. 5** Change of $W_s$ with applied load

C4 and the lowest wear rate is displayed by the hardest AA7068/SiC C4. This trend followed the hardness of the monolithic alloys used to create the interpenetrating composites.

## References


1. Park SJ, Seo MK (2011) Types of composites. In: Interface Sci. Technol. Elsevier, New York, pp 501–629
2. Haghshenas M (2016) Metal–matrix composites. In: Ref Modul Mater Sci Mater Eng Elsevier, Waterloo, pp 1–28
3. Foltz JV, Blackmon CM (1990) Metal-matrix composites. In: ASM Handbook, vol. 2 Prop Sel Alloy Spec Mater ASM International, Ohio, pp 903–912
4. Mistry JM, Gohil PP (2018) Research review of diversified reinforcement on aluminum metal matrix composites: fabrication processes and mechanical characterization. Sci Eng Compos Mater 25:633–647
5. Meena KL, Manna AD, Banwait D, Jaswanti D (2013) An analysis of mechanical properties of the developed Al/SiC-MMC's. Am J Mech Eng 1:14–19
6. Raj R, Thompson LR (1994) Design of the microstructural scale for optimum toughening in metallic composites. Acta Metall Mater 42:4135–4142
7. Rosso M (2006) Ceramic and metal matrix composites: routes and properties. J Mater Process Technol 175:364–375

8. Yang D, Zhou Y, Yan X, Wang H, Zhou X (2020) Highly conductive wear resistant Cu/Ti3SiC2(TiC/SiC) co-continuous composites via vacuum infiltration process. J Adv Ceram 9:83–93
9. Lu Y, Yang J, Lu W, Liu R, Qiao G, Bao C (2010) The mechanical properties of co-continuous Si3N4/Al composites manufactured by squeeze casting. Mater Sci Eng, A 527:6289–6299
10. Lei C, Zhai H, Huang Z, Hu W, Cai L, Chen S, Yu W, Zhou Y (2019) Fabrication, microstructure and mechanical properties of co-continuous TiCx/Cu-Cu4Ti composites prepared by pressureless-infiltration method. Ceram Int 45:2932–2939
11. Manfredi D, Pavese M, Biamino S, Antonini A, Fino P, Badini C (2010) Microstructure and mechanical properties of co-continuous metal/ceramic composites obtained from reactive metal penetration of commercial aluminium alloys into cordierite. Compos Part A Appl Sci Manuf 41:639–645
12. Breslin MC, Ringnalda J, Xu L, Fuller M, Seeger J, Daehn GS, Otani T, Fraser HL (1995) Processing, microstructure, and properties of co-continuous alumina- aluminum composites. Mater Sci Eng, A 195:113–119
13. Prasanth AS, Ramesh R (2017) Investigation of surface roughness and tool wear in end milling of Al7075-SiC Co-continuous composite. Adv Struct Mater 315–328
14. Prasanth AS, Ramesh R, Gopalakrishnan P (2018) Taguchi grey relational analysis for multi-response optimization of wear in co-continuous composite. Mater (Basel) 11:1743
15. Prasanth AS, Ramesh R, Gopalakrishnan P, Ramu M (2019) Multi-response optimization of end milling parameters for Al-Zn-Mg/SiC co-continuous composite using response surface methodology. Medziagotyra 25:471–477
16. Parthasarathi NL, Borah U, Albert SK (2013) Correlation between coefficient of friction and surface roughness in dry sliding wear of AISI 316L (N) stainless steel at elevated temperatures. Comput Model New Technol 17:51–63
17. Ravindran P, Manisekar K, Rathika P, Narayanasamy P (2013) Tribological properties of powder metallurgy—processed aluminium self lubricating hybrid composites with SiC additions. Mater Des 45:561–570
18. Kumar TS, Subramanian R, Shalini S, Anburaj J, Angelo PC (2016) Synthesis, microstructure and mechanical properties of Al-Si-Mg alloy hybrid (zircon + alumina) composite. Indian J Eng Mater Sci 23:20–26

# Study on Fibre Behaviour for Chemical Treatment and Fabrication of ABS-Based Fibre Composite

**T. Ramesh Kumar, T. Guruprakash, P. Nandha Kumar, R. Gokul, and A. Ramakrishnan**

**Abstract** There has been a growing demand for new material discovery especially in composites. Natural reinforced fibre composites are in great demand due to its eco-friendly characteristic and reliable usage. Polymer-based composites play a major participation in the numerous synthetic composites. Utilization of natural fibres in composites has recently emerged with the advancement in material science. With the immense success of the use of composites, the use of natural fibres has arisen. Among the various natural fibres, the addition of banana and sisal fibres is shown positive improvements in material property. The present paper gives us glimpse about how the fibres react to various chemical treatment and its influential properties and fabrication technique. The fibre is treated with different chemical solutions like NaOH, $Ba(OH)_2$, $H_2O_2$, KOH, $NH_4OH$ and $CaCO_3$ for various timings, chemical concentration and tested for its mechanical strength. The best fibre combinations are chosen for the fabrication of the hybrid composite using additive manufacturing process (3D printing).

**Keywords** Sisal fibre · Banana fibre · Chemical treatment · Additive manufacturing

T. Ramesh Kumar (✉) · T. Guruprakash · P. Nandha Kumar · R. Gokul · A. Ramakrishnan
Department of Mechanical Engineering, Bannari Amman Institute of Technology, Sathyamangalam, Tamil Nadu, India
e-mail: rameshkumart@bitsathy.ac.in

T. Guruprakash
e-mail: guruprakasht.me16@bitsathy.ac.in

P. Nandha Kumar
e-mail: nandhakumarp.me16@bitsathy.ac.in

R. Gokul
e-mail: gokulr.me16@bitsathy.ac.in

A. Ramakrishnan
e-mail: ramakrishnana@bitsathy.ac.in

© The Editor(s) (if applicable) and The Author(s), under exclusive license to Springer Nature Singapore Pte Ltd. 2021
S. Mohan et al. (eds.), *Materials, Design, and Manufacturing for Sustainable Environment*, Lecture Notes in Mechanical Engineering,
https://doi.org/10.1007/978-981-15-9809-8_8

# 1 Introduction

In recent decades, natural fibres have played a vital role in preparation of composites providing the good substitute for synthetic fibres. Natural fibre-reinforced composites are prepared by reinforcing the naturally available fibres with suitable composition. Commonly used natural fibres for making composites are sisal, coir, jute, bamboo, date and flax. Each fibre has the different specialized properties of its own. The sisal and banana are most commonly available fibres which have advantages such as good weight to strength ratio, low cost, easy in extraction, abundantly present, high tensile strength, etc. Synthetic fibres like polyester-based fibre are also being used in preparing the hybrid- or single-reinforced composites. But natural fibre is different from polyester fibre with its biodegradability and its weight. In this natural fibre sisal and banana, fibre is easily available on natural, especially in India. Although these have their own strength when chemically treated it, strength is improved in a huge level and further processing is made better.

Patel et al. [1] developed the natural fibre-reinforced epoxy composite with the help of epoxy resin using the sisal and banana fibres including the chemical treatment using NaOH at 2% concentration for 24 h. By using hand-lay-up technique, the mixture of resin and fibres is proportionally laid on one another uniformly up to required thickness on the prepared mould. Rollers are used to eliminate the air gaps and concluded that there is a high stability in bending and flexural test. Oladele et al. [2] have analysed the polyester-based sisal fibre composite for its mechanical behaviour under different chemical treatment. The obtained sisal fibre was treated with different combination of chemical solution to underpin the enhancement of fibre composite properties. The fibres were subjected to tensile stress and hardness test at different loaded conditions. It is to be noted that KOH gave the best result in overall mechanical behaviour of fibres.

Rajan et al. [3] manufactured biocomposites using sisal and banana fibres. The fibre is treated with 2 wt% of NaOH for 2 h to eliminate hemicelluloses. The treated banana and sisal fibre-reinforced composite with PLA has relatively higher impact strength, flexural strength and tensile strength and concluded that the chemical treated improved fibre matrix composite interaction by removing of layer called lignin. Kumaresan et al. [4] described the mechanical properties of sisal fibres and banana fibres with other natural fibres. He also classified the types of fibres and also the types of polymer matrix and given rule for mixing the two composites and the assumptions made for polymer composites. He also proposed the different methods of manufacturing the polymer composites, and also mentioned pretreatment will improve the interfacial adhesion between the matrix and the fibre, thereby increasing the mechanical behaviour of resultant composite.

Mohammed et al. [5] discussed the natural fibre composite and its chemical composition of common natural fibres. They examined the effect of composite performance based on the orientation, physical properties and adhesion property. They also explained chemical treating of fibres with alkali solution and gave the comparison between treated and untreated fibres. Passos et al. [6] have investigated the fibre characterization for pseudostem of banana in par with data obtained from jute and sisal fibre. The fibre was subjected to tensile strength followed by study of its morphological structure. Pseudostem banana fibre showed better mechanical properties when compared to raw and other compared fibres. The morphological study unveiled the structural characters of the enhanced fibre.

Dharmalingam et al. [7] had shown the comparison between the chemically treated and untreated fibre of sisal and kenaf fibre composite. The composite is prepared by hand-lay-up process using epoxy resin and performed test like tensile, flexural impact and compression test. The outcome reveals that treated fibre has improved properties. Hariprasad et al. [8] have explained chemical treatment of banana fibre under 2% NaOH for 2 h and coir fibre in 10% NaOH for 3 h. The treated fibres have an improvement in the properties like Tensile, Impact, and Strength but the Flexural remains less and also showed that the reduction in the diameter of the fibre after treatment.

Madhuri et al. [9] showed that sisal fibre is treated with 5% NaOH solution and dried in oven at 70 °C for 4 h, and then composite is prepared with epoxy sisal glass fibres by hand-lay-up method and given the comparison between the treated and untreated fibres for various mechanical properties and stated that treated fibre has greater mechanical properties. Santhosh et al. [10] have studied the treatment of banana with 5% NaOH for one hour, and the specimen is produced by moulding process for untreated and treated fibre which is implemented as reinforcing material for both vinyl ester resin matrix and epoxy. The test results between treated and untreated NaOH concluded that alkali treatment has provided better mechanical properties.

Bachtiar et al. [11] have studied the treatment of fibres with alkaline solution on the flexural properties of sugar palm fibre-reinforced epoxy composites. The alkali solution used is NaOH with 0.25 and 0.5 M concentration treating for 1, 4 and 8 h and enhanced the interfacial bonding between matrix and fibre surfaces and viewed under SEM visualization. Srisuwan et al. [12] have explained the effects of salinized and alkalized woven sisal fibre on mechanical behaviour. The composite is prepared using epoxy resin by hand-lay-up process. The alkaline treatment is done on 2 wt% of NaOH for 2 h and 7 wt% salinized sisal fibre showed enhancement of 230% impact strength of the composite.

Jordan et al. [13] have described the treatment of dicumyl peroxide and potassium permanganate to improve the bonding between interfacial surfaces of banana fibres and LDTE. Peroxide had the improvement of effect of increasing the tensile properties of banana fibres, and pseudostem banana fibres give reasonable increase to LDTE properties and its tensile stiffness. Li et al. [14] have studied the different chemical modification on natural fibres for use in natural fibre-reinforced composite. Chemicals include alkali, permanganates, silane, benzoylation, isocyanates, maleated coupling agents, arylation and acetylation. These treatments with chemicals have secured various levels of improvement in fibre fitness, strength of fibre and its adhesion property in natural fibre-reinforced composite. Mazzanti et al. [15] had used mould less technique with high-dimensional quality using fused deposition modelling 3D printing. Using this technique, fibres are printed using the different structural and printing parameters and obtained considerable even on small parameter change include nozzle diameter, extrusion temperatures, matrix used, etc. and also analysed different mechanical properties.

The objective of this paper is to find out the best chemical treated banana and sisal fibre through B-force testing using UniStretch 250 Multi-strength Tester Machine. The selected fibre will be further integrated with ABS material to produce polymer hybrid composite using 3D printing technique.

## 2 Experimental Procedure

This chapter explains the process underwent during the fibre extraction, chemical treatment, testing and fabrication of specimen.

### 2.1 *Fibre Extraction Process*

The processing of extracting sisal fibres includes soaking accompanied with scraping and Decorticators. After the leaves are cut down from the plant, soaking the leaves for a week with the water helps to separate fibres from the stalk. The fleshy pulp scraped away using the blunt knife and the remaining component is dispatched between rollers called decorticators. Then the fibre is cleaned in distilled water, and it is dried in sunlight as long as moisture content is present. Banana stem is collected from the local cultivators; the barks are separated from the inner layer manually and cleaned with water. Then the barks are dried in room temperature. Using decorticators, fibres are extracted. Finally, fibres are cleaned and dried in room temperature (Figs. 1, 2, 3 and 4).

**Fig. 1** Fibre extraction using decorticators

**Fig. 2** Fibre drying process

**Fig. 3** Extracted sisal fibre

**Fig. 4** Extracted banana fibre

## 2.2 *Chemical Treatment of Fibres*

The fibres are treated with chemical solution to remove the lignin and wax present in the surface of the fibres as well as increase its interfacial boding between them.

**Fig. 5** Sisal fibre with KOH

**Fig. 6** Banana fibre with KOH

Fibres are treated with different solutions like KOH, NaOH, $NH_4OH$, $H_2O_2$ and $CaCO_3$. The solution is prepared for three different concentrations as 2%, 4% and 6%. Then fibres are immersed in solution and kept for three different timings 30 min, 60 min and 90 min, respectively. Then the fibres are washed in distilled water and dried in sunlight for 48 h. Above procedure is followed for both the banana and sisal fibres. Then these fibres are subjected to tensile test to understand its elongation and mechanical strength (Figs. 5, 6, 7, 8, 9, 10, 11, 12, 13, 14, 15 and 16).

## 2.3 Testing of Fibres

The tensile test is done in UniStretch (250) Multi-strength Tester Machine. It works on the principle of constant rate extension (CRE). The test is done for single yarn fibre strength. As per the standard procedure, fibre length is taken and fixed between two jaws of the machine. Once the load is applied, lower jaw moves downwards and

**Fig. 7** Sisal fibre with NaOH

**Fig. 8** Banana fibre with NaOH

**Fig. 9** Sisal fibre with $NH_4OH$

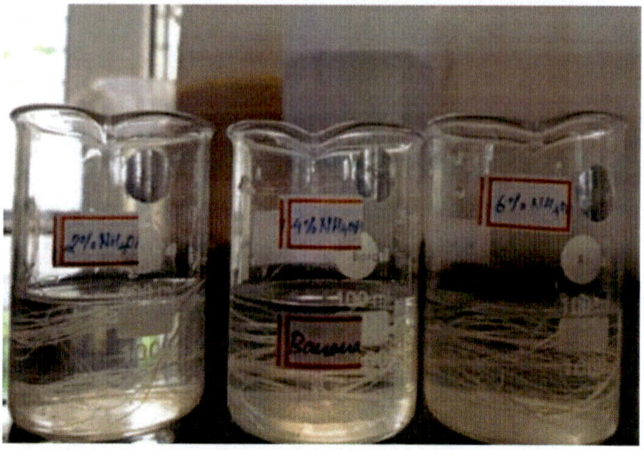

**Fig. 10** Banana fibre with $NH_4OH$

**Fig. 11** Sisal fibre with $CaCO_3$

**Fig. 12** Banana fibre with $CaCO_3$

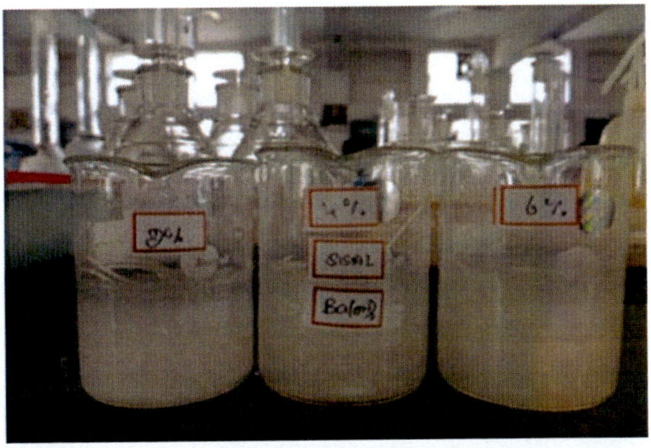


**Fig. 13** Sisal fibre with $Ba(OH)_2$

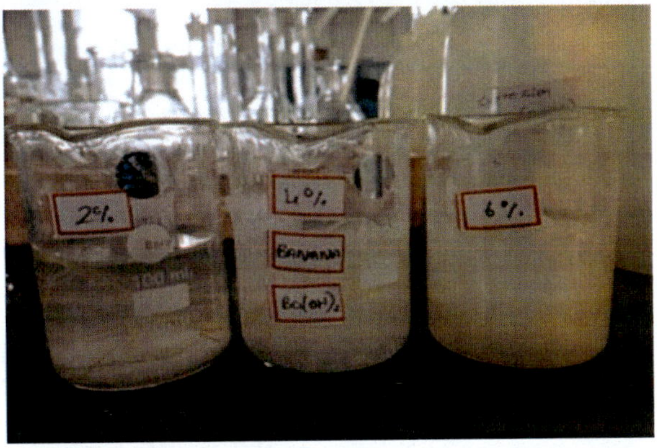


**Fig. 14** Banana fibre with $Ba(OH)_2$

**Fig. 15** Sisal fibre with $H_2O_2$

**Fig. 16** Banana fibre with $H_2O_2$

fibre gets elongated along the direction of load. It elongates till its breaking point, and its breaking force is noted down from the monitor. The results are noted down for further evaluation (Fig. 17).

**Fig. 17** UniStretch 250 multi-strength tester machine

## 2.4 *Fabrication of Sample Specimen*

The best combination of the fibre is chosen for fused deposition modelling (FDM) [15]. The model is drawn using CREO 5.0 software as per the ASTM Standards (D638-Tensile Test, D7264-Flexural Test). The created model is converted into STL file format for compatibility with 3D printer. The specimen is fabricated by layers as of the requirements and fibres are laid in between them. The fabricated specimen is further taken for post-processing to get a fine finished product.

# 3 Results and Discussion

This chapter deals with the outcome of the research from the chemical treatment of fibre.

## 3.1 *Effect of Chemical Treatment on Sisal and Banana Fibre*

### 3.1.1 Chemical Treatment Using KOH Solution

Results from the above graph in Figs. 18 and 19 clearly show that for 2% concentration of KOH and treatment of fibre for 30 min gave the best breaking force for both banana and sisal fibres. This clearly tells us that for longer period and higher concentration, chemical treatment with fibre reduces the strength of fibre.

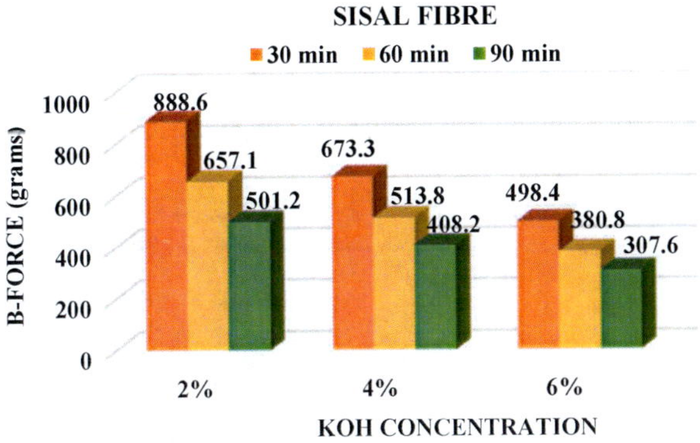


**Fig. 18** KOH treated sisal fibre

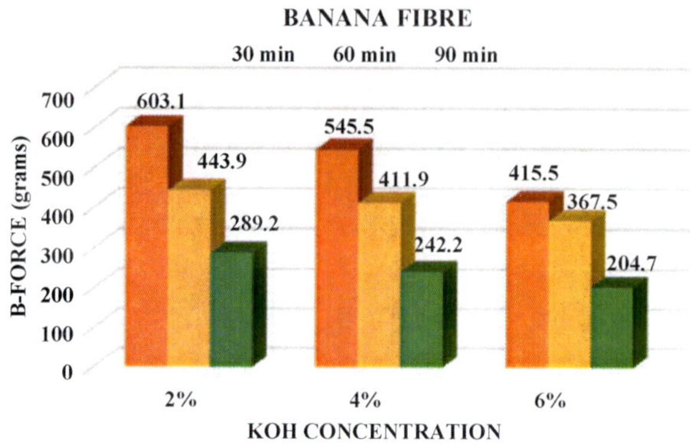


**Fig. 19** KOH treated banana fibre

### 3.1.2 Chemical Treatment Using NaOH Solution

Results from the above graph in Figs. 20 and 21 clearly show us that for 2% concentration of $N_aOH$ and treatment of fibre for 30 min gave the best breaking force for

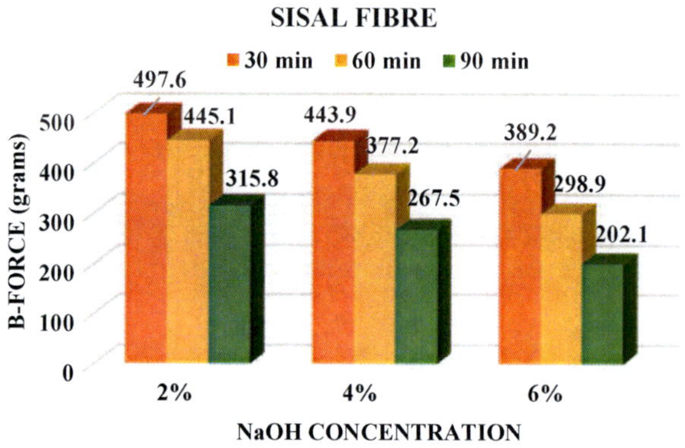


**Fig. 20** NaOH treated sisal fibre

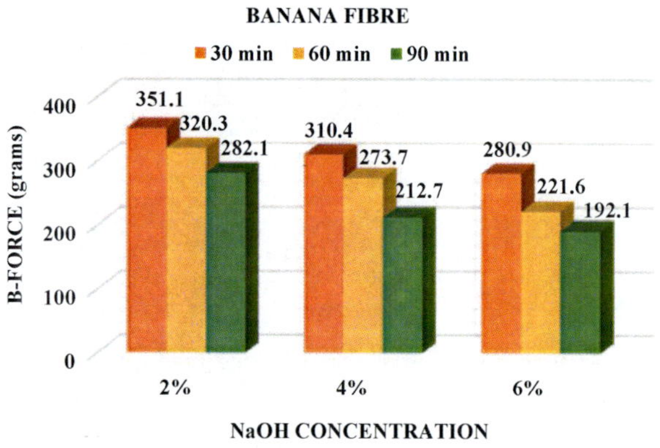


**Fig. 21** NaOH treated banana fibre

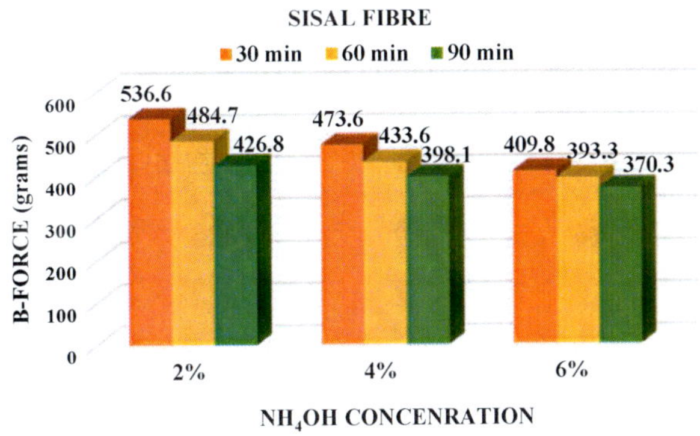


**Fig. 22** $NH_4OH$ treated sisal fibre

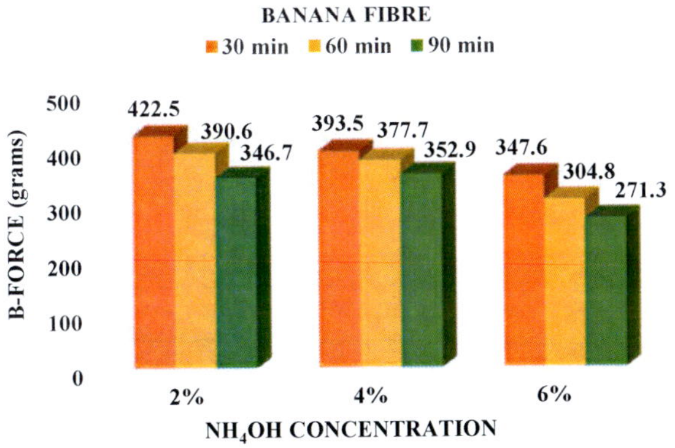


**Fig. 23** $NH_4OH$ treated banana fibre

both banana and sisal fibres. This clearly tells us that for longer period and higher concentration, chemical treatment with fibre reduces the strength of fibre.

### 3.1.3 Chemical Treatment Using NH4OH Solution

Results from the above graph in Figs. 22 and 23 clearly show us that for 2% concentration of $NH_4OH$ and treatment of fibre for 30 min gave the best breaking force for both banana and sisal fibres. This clearly tells us that for longer period and higher concentration, chemical treatment with fibre reduces the strength of fibre.

### 3.1.4 Chemical Treatment Using $CaCO_3$ Solution

Results from the above graph in Figs. 24 and 25 clearly show us that for 2% concentration of $CaCO_3$ and treatment of fibre for 30 min gave the best breaking force for both banana and sisal fibres. This clearly tells us that for longer period and higher concentration, chemical treatment with fibre reduces the strength of fibre.

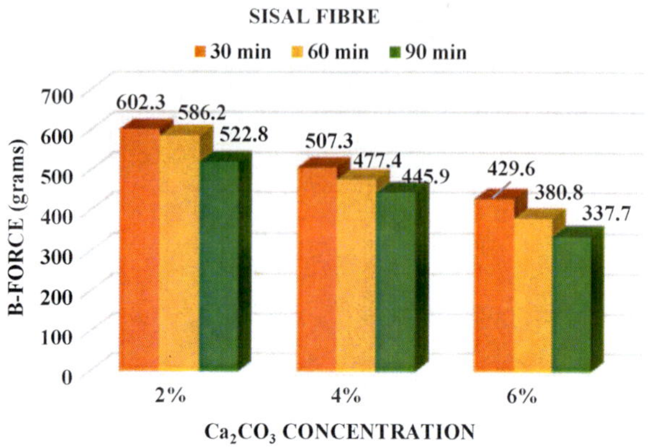


**Fig. 24** $CaCO_3$treated sisal fibre

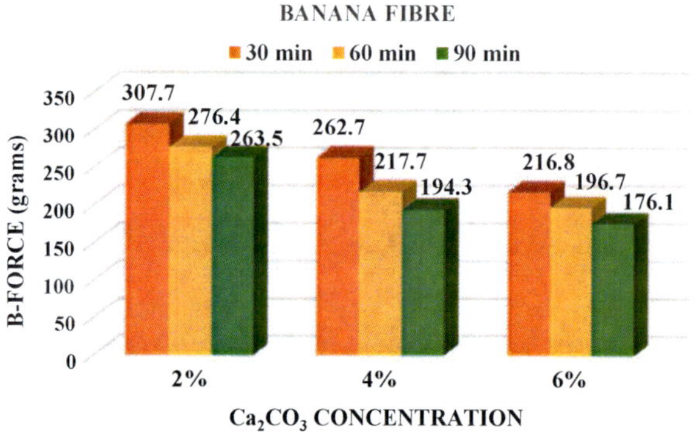


**Fig. 25** $CaCO_3$ treated banana fibre

### 3.1.5 Chemical Treatment Using $Ba(OH)_2$ Solution

Results from the above graph in Figs. 26 and 27 clearly show us that for 2% concentration of $B_a(OH)_2$ and treatment of fibre for 30 min gave the best breaking force for

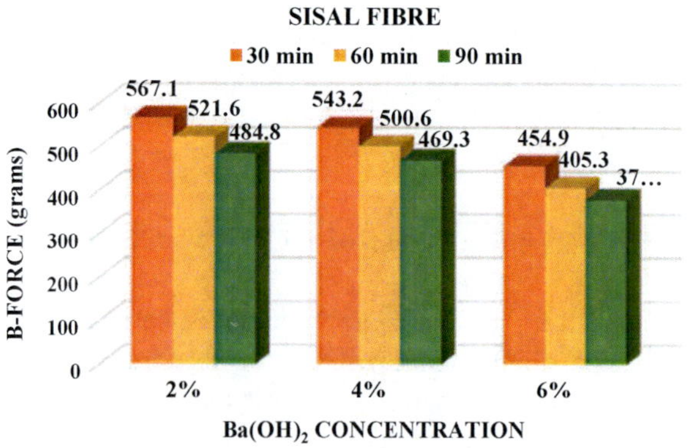


**Fig. 26** $Ba(OH)_2$ treated sisal fibre

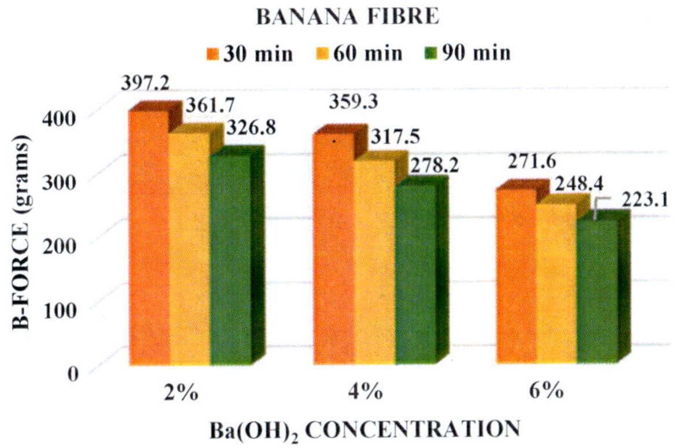


**Fig. 27** $Ba(OH)_2$ treated banana fibre

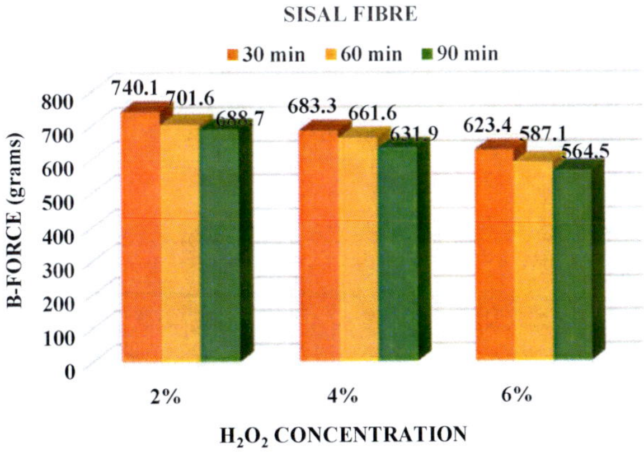


**Fig. 28** $H_2O_2$ treated sisal fibre

both banana and sisal fibres. This clearly tells us that for longer period and higher concentration, chemical treatment with fibre reduces the strength of fibre.

#### 3.1.6 Chemical Treatment Using $H_2O_2$ Solution

Results from the below graph in Figs. 28 and 29 clearly show us that for 2% concentration of KOH and treatment of fibre for 30 min gave the best breaking force for both banana and sisal fibres. This clearly tells us that for longer period and higher concentration, chemical treatment with fibre reduces the strength of fibre.

### *3.2 Specimen Preparation Using FDM*

The best combination of the fibre is taken, and it is integrated into 3D printing. ABS material is chosen for fabrication of the specimen. The fill percentage of filament,

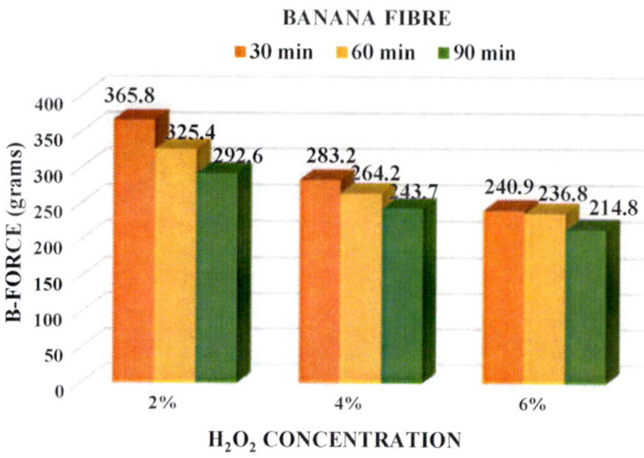


**Fig. 29** $H_2O_2$ treated banana fibre

machine orientation of FDM and layers of fibre to be laid are chosen according to the application of usage. This composite will help break new barrier in material science discoveries as well as find wide application usage in aerospace, robotics and machine buildings (Fig. 30).

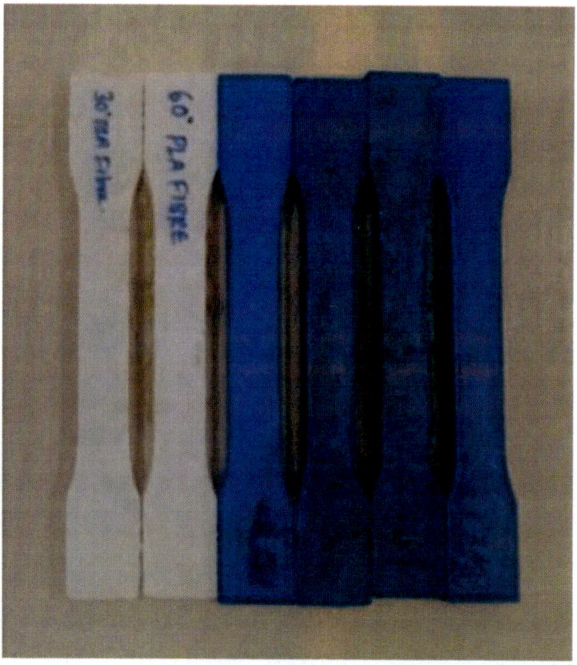

**Fig. 30** Fabricated specimens (ASTM-D638)

## 4 Conclusions

From the above research conducted, the various results have been concluded to choose the best treated fibre. From the above results, it is clear that minimum timing (30 min) and minimum concentration (2%) give the best result for KOH treatment compared to other chemical treatments. This treatment strengthens the fibre as well as removing the lignin and wax coating from the fibre [2]. For the further fabrication of hybrid composite, KOH treated fibre of sisal and banana will be chosen in base of 2% concentration in 30 min duration. Fabrication of hybrid composite is done using additive manufacturing process (3D printing) using ABS material. This will help us discovering new material composite applications.

## References

1. Patel H, Parkhe A, Shrama PK (2016) Mechanical behaviors of banana and sisal hybrid composites reinforced using of various parameter. Int J Eng Technol Manage Res 3(1):1–4
2. Oladele IO, Daramola OO, Fasooto S (2014) Effect of chemical treatment on the mechanical properties of sisal fibre reinforced polyester composites. Leonardo Electron J Pract Technol 1(24):1–12
3. Ranjan R, Bajpai PK, Tyagi RK (2013) Mechanical characterization of banana/sisal fibre reinforced PLA hybrid composites for structural application. Asian Bus Consortium Eng Int 1(1):39–49
4. Kumaresan M, Ramesh N, Ramesh S, Vijay Benjamin Lazarus S (2017) Review on mechanical behavior of sisal & banana fibre reinforced polymer composites. Int J Adv Manage Technol Eng Sci 2(1):350–355
5. Mohammed L, Ansari MNM, Pua G, Jawaid M, Saiful Islam M (2015) A review on natural fiber reinforced polymer composite and its applications. Int J Polym Sci 1:1–15
6. Cecci RRR, Passos AA, de AguiarNeto TC, Silva LA (2019) Banana pseudostem fibers characterization and comparison with reported data on jute and sisal fibers. Int J Current Eng Technol 2(20):121–126
7. Dharmalingam G, Kumar V, Sengolerayan A (2018) Mechanical behaviour of treated and untreated sisal-kenaf hybrid composite materials. Int J Mech Prod Eng Res Dev 8(3):39–50
8. Hariprasad T, Dharmalingam G, Praveen Raj P (2013) A study of mechanical properties of banana-coir hybrid composite using experimental and fem techniques. J Mech Eng Sci 4:518–531
9. Madhuri KS, Rao DHR (2014) An investigation of mechanical and thermal properties of reinforced sisal-glass fibers epoxy hybrid composites. Int J Eng Res 3(1):112–115
10. Santhosh J, Balanarasimman N, Chandrasekar R, Raja R (2014) Study of properties of banana fiber reinforced composites. Int J Res Eng Technol 3(11):144–149
11. Bachtiar D, Sapuan SM, Hamdan MM (2010) Flexural properties of alkaline treated sugar palm fibre reinforced epoxy composites. Int J Automot Mech Eng 1:79–90
12. Srisuwan S, Prasoetsopha N, Suppakarn N, Chumsamrong P (2014) The effects of alkalized and silanized woven sisal fibers on mechanical properties of natural rubber modified epoxy resin. 11th Eco-Energy Mater Sci Eng 56:19–25

13. Jordan W, Chester P (2017) Improving the properties of banana fiber reinforced polymeric composites by treating the fibers. Int Conf Nat Fibers 7:283–289
14. Li X, Tabil LG, Panigrahi S (2007) Chemical treatments of natural fiber for use in natural fiber-reinforced composites: a review. J Polym Environ 15(1):25–33
15. Mazzanti V, Malagutti L, Mollica F (2019) FDM 3D printing of polymers containing natural fillers. Polymers 11(7):3–22

# A Study on Influence of Frictional Coefficient on Stresses in AISI-1045 Forging Using DEFORM-3D

**Silpy Suresh Kumar, Nirmal Jose, Sooryanath KU, Jobin Varghese, and Sam Joshy**

**Abstract** The traditional process of forging is often determined based on experiences which require repeated testing and repairing of dies. It also involves long production cycle and high prototype cost. In recent times, the finite element approach has evolved as an effective technique for virtual process testing and simulation-based design. The aim of this study is to model and conduct FEA on metal die using SolidWorks and DEFORM-3D software packages. This would result in the enhancement of the overall process competence and efficiency at a lower production cost. In this research paper, an effort has been made to gain insight into the process variables that affects closed-die hot forging and its interaction. As sample case, metal (AISI 1045) component is considered. Commercial software (DEFORM-3D) that uses a simulation-driven approach based on the finite element method (FEM) has been implemented. The result will be useful for tool and die developers as well as work developers for optimizing the forging process and to analyse deformation in the components especially in automotive parts.

**Keywords** AISI 1045 · AISI H-13 · Coefficient of friction · DEFORM-3D · Finite element analysis (FEA)

## 1 Introduction

Forging is characterized as a metal working method in which the required form of the work piece is obtained in a solid state by means of compressive forces applied by utilizing dies, tools and equipment. Forging is usually classified on the basis of its working temperature: 'cold', 'hot', or 'warm' forging. Forged pieces will vary in weight from less than one kilogram to 170 metric tons. Forged parts typically need more processing in order to produce a finished product [1]. The distinct advantage

S. Suresh Kumar · N. Jose (✉) · S. KU · J. Varghese · S. Joshy
Department of Mechanical Engineering, SCMS School of Engineering and Technology, Vidya Nagar, Palissery, Karukutty, Kerala 683582, India
e-mail: nirmaljoe14@gmail.com

© The Editor(s) (if applicable) and The Author(s), under exclusive license to Springer Nature Singapore Pte Ltd. 2021
S. Mohan et al. (eds.), *Materials, Design, and Manufacturing for Sustainable Environment*, Lecture Notes in Mechanical Engineering,
https://doi.org/10.1007/978-981-15-9809-8_9

of hot die forging is when the metal is deformed, and the work hardening effects are counteracted by the recrystallization process. Typically, the shapes of the products developed by forging process are sophisticated; and many deficiencies and imperfections are triggered during the forging process, such as incomplete forging penetration, surface cracking, cold shutting, under filling, flakes and improper grain flow [2]. Simulation and development of a connecting rod made from hyper-fine-grained material and isothermal forging [3] help in the elimination of crankshaft defects [4]. The authors examined the results of differing coefficients of friction, variations in geometry and temperature field in the forging process [5] and conducted a die stress assessment. The researchers have presented both the distribution of stress and strains in the various regions of the component. The authors have been able to investigate the material flow of the forging component using DEFORM-3D [6]. Simulation of stresses, strains and temperature in different regions has been conducted for the study of defects [7]. The objective of this study is to provide an introduction to the finite element analysis of the forging process of a sample metal part employing DEFORM-3D software [8]. This study will be valuable for a beginner designer and would provide an insight into the FEA [9]. This paper has been divided into five sections. The first section discusses the development of forging and a description on the scope and purpose of this research paper. Various literature reviews are covered in the second section. The methodology and approach applied in the paper is addressed in the third section. Various findings and observations from computational simulations are presented in the fourth section. The fifth section summarizes the findings of the study.

## 2 Literature Survey

The broad objective of the literature review is to include the essential context details on the proposed research. The areas of the survey reported include:

### *2.1 Closed-Die Forging (CDF) Process and the Process Parameters*

The forging mechanism is used to turn the basic structure of the component into the ideal final form by manipulating the plastic deformation. In closed-die forging, the die exerts stress on the material via the interface resulting in the creation of the cavity-shaped part. The hot forging parameters, together with the closed-die state, can achieve a higher degree of deformation with a decent geometric accuracy, making it the desired method for mass production of complex-moulded components. The closed-die hot forging operation is ideally adapted for the production of sophisticated formed work pieces with acceptable profile accuracy. The method is very complicated from the point of view of analysis, since there are many variables that may affect the mechanism and the rate of rejection [10]. Closed-die hot forging process involves multiple input parameters that can be listed in the following groups [11].

- Product variables
- Material variables
- Tooling parameters
- Machine parameters
- Deformation zone characterization
- Tool and work interface behaviour.

These parameters may be more broadly grouped into two categories: design and process parameters. Our analysis focuses primarily on the process parameter of variable friction coefficients and its diverse impact on the AISI-1045 work piece. Design parameters such as preform shape design, die design and process variables like input billet geometry, flash thickness, draft angle, flash width, corner and fillet radius, etc., are beyond the scope of this paper.

## 3 Design Methodology

See Fig. 1.

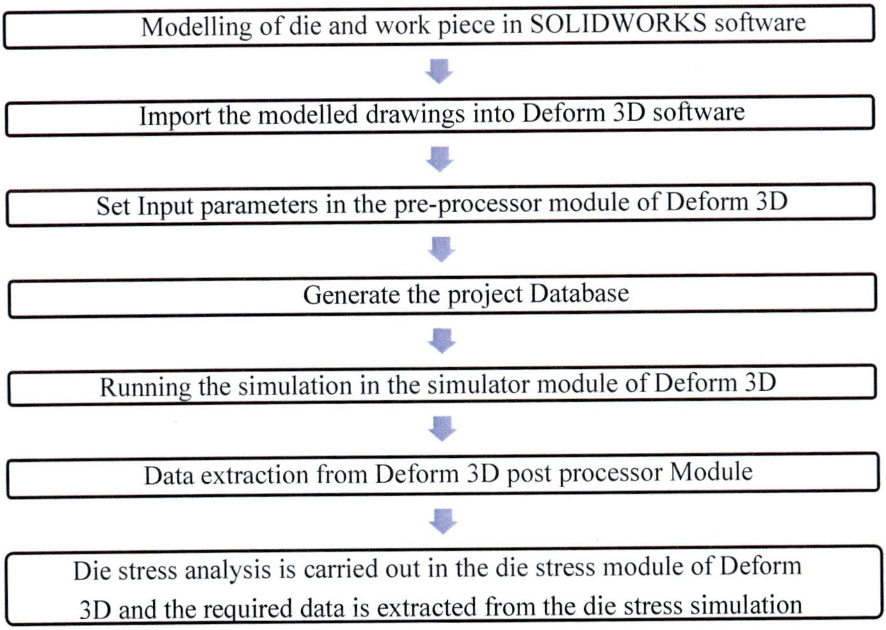


**Fig. 1** Overview of design methodology

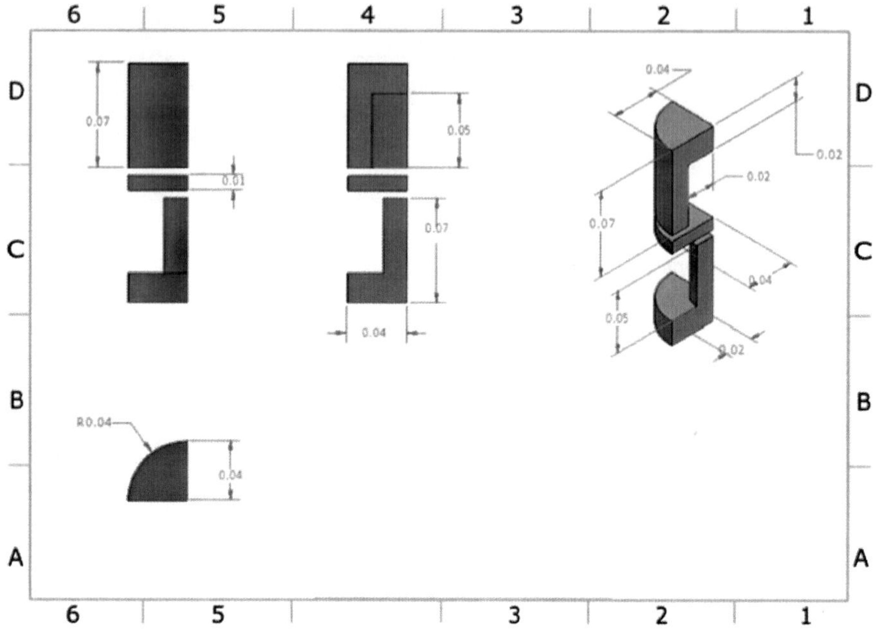


**Fig. 2** Mechanical drawing of component and dies

## 3.1 *Drawing of Forged Component and Die in SolidWorks*

See Fig. 2.

## 3.2 *Steps*

### 3.2.1 Selection of Component

First select the component that is to be analysed.

### 3.2.2 Preparation of 3D Modelling of Component

The preparation of CAD or 3D modelling of selected component and its dies is done by using SolidWorks software.

**Table 1** Process parameters

| S. No. | Parameter | Value |
|---|---|---|
| 1 | Work piece material | AISI 1045 |
| 2 | Die material | AISI H-13 |
| 3 | Work piece temperature | 1200 °C |
| 4 | Die temperature | 300 °C |
| 5 | Work piece volume | 18,832 $mm^3$ |
| 6 | Coefficient of friction | 0.3, 0.4, 0.5, 0.6, 0.7 |
| 7 | Heat transfer coefficient | 11 $W/m^2$ K |

### 3.2.3 Selection of Equipment

Depending upon the complexity of the component and availability, suitable equipment is selected.

### 3.2.4 Specification of Parameters

Specification of parameters is done in DEFORM-3D. Different parameters are like material of billet, proper die filling, forging temperature, friction factor, die velocity and lubricant (Table 1).

### 3.2.5 Die Design

The construction of a mechanical component is basically the method of providing form to a part needed to execute certain useful functions. Alternate technologies and techniques should be used to satisfy the specifications of the construction of metals. It is the mission of the designer to find the appropriate combination of material and method that optimizes the configuration, properties and costs factors. Forging offers the best approach to an expanding list of design applications. Forging die development is driven by the characteristics of the metal component being manufactured and the capabilities and the available technologies. The cost of dies is between 15 and 20% of the entire cost of production. Hence, it should therefore be economical and safe for both the user as well as the manufacturer. Materials of high forge ability and design limits are often quite narrow. Knowledge of the material behaviour is crucial for the designer to prepare the design of the fabricated components.

# 4 Results and Discussions

## 4.1 *Effective Stress Analysis*

The stress distribution of the forged component is shown in Fig. 3. The stress distribution across various regions of the forged component is evaluated using point tracking (P1–P10).

As shown in Fig. 3, the points are taken both radially and axially, and the corresponding stress values are shown in Fig. 4. From Fig. 4, we can see that P8 is the maximum stress point with a value of 79.9 MPa, followed by P5, P4, P3, P1 and P7 which indicate that the region of contact between the upper and lower die is a region of high stress in the work piece. From Fig. 4, we can also conclude that P6 (42.3 MPa) is the point of minimum stress, and this is due to the fact that this area actually rests on the bottom die surface.

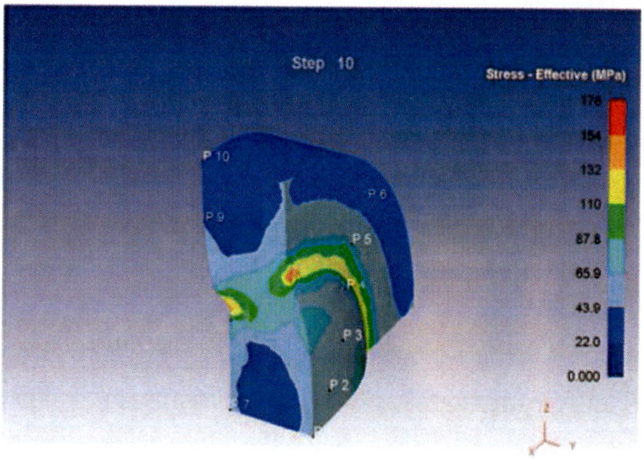


**Fig. 3** Effective stress in work piece

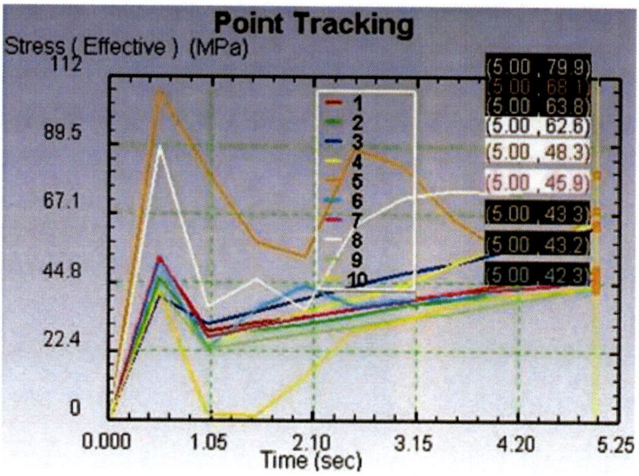


**Fig. 4** Graph of effective stress in work piece

## 4.2 *Effective Strain Analysis*

The strain distribution of the forged component is as shown in Fig. 5. The strain variation across various regions of the component is studied by utilizing the graph obtained from point tracking (P1–P10) as shown in Fig. 6.

From Figs. 5 and 6, we can infer that P3 has maximum strain with a value of 0.183 mm, followed by P2, P4 and P1. And we can also see that P10 is the point of minimum strain whose values is 0.00296 mm. Hence, we can say that the region which comes in contact with the top die surface is the region of maximum strain, whereas the region of work piece which rests on the bottom die is the region of minimum strain.

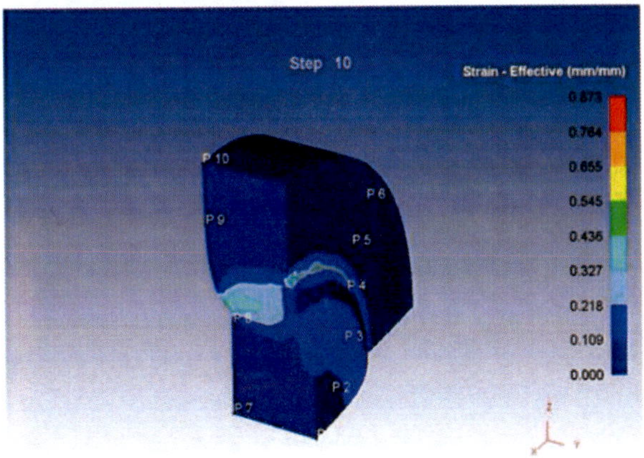


**Fig. 5** Effective strain in work piece

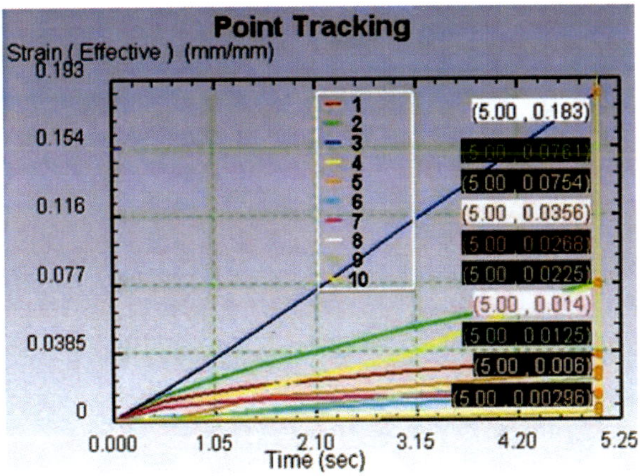


**Fig. 6** Graph of effective strain in work piece

## 4.3 Damage Analysis

The maximum damage to the work piece is at P3, P2, P4 and P1 from Figs. 7 and 8, and as we can infer from the effective stress and strain analysis, these points are points of high stress and strain values. Hence, we can concur that the maximum damage to the work piece occurs at points which lie in between the region of contact of top and bottom die. With varying friction coefficients, the stress and strain magnitudes vary. When the friction coefficient value increases, the stress values at high stress points of the work piece often increase.

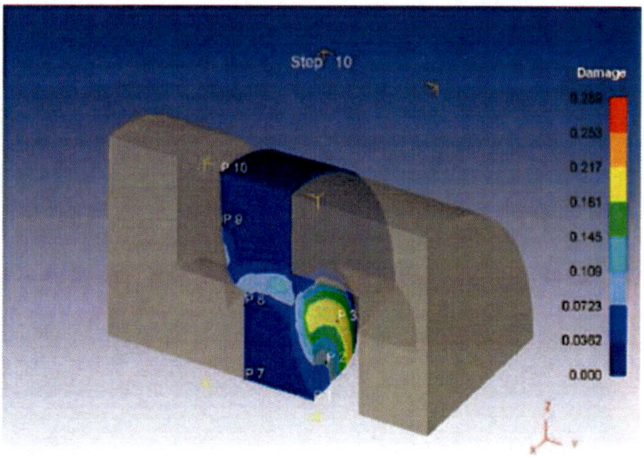


**Fig. 7** Damage distribution in work piece

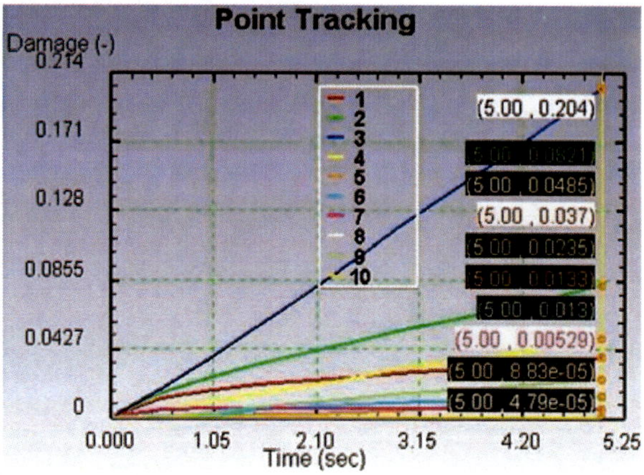


**Fig. 8** Damage analysis graph

# 5 Conclusion

Thus, with the aid of DEFORM-3D, the simulation was done at multiple friction coefficient values. Via this analysis, we infer that stress on the work piece can vary with varying coefficient of friction, and we can find out all the resulting stress, strain and damage at various points on work piece at any time using the graphs obtained. Some of the old analytical techniques show that some of the idealized calculation assumptions hinder their ability to handle intricate metal forming cases. Moreover, neither of the techniques considers the influence of the temperature gradient on the working material and its consequences during the process. Potential research could also be based on process optimization of the forging operation, the effect of various other forging parameters, such as blank and die temperature, could also be researched, and FEA study of automobile parts, such as crankshaft, crown wheel, gears and pitman neck, could also be carried out.

# References


1. Rathi MG, Jakhade NA (2014) An effect of forging process parameters on filling the job weight:—an industrial case study Int J Sci Res Publ 4(6) ISSN 2250-3153
2. Hawryluk M, Jakubik J (2015) Analysis of forging defects for selected industrial die forging processes. Eng Fail Anal (Elsevier)
3. Fuertes JP, Luis CJ, Luri R, Salcedo D, Leon J, Puertas I (2016) Design simulation and manufacturing of a connecting rod from ultra-fine grained material and isothermal forging. J Manuf Proc 21:56–68 (Elsevier)
4. Chandna P, Chandra A (2009) Quality tools to reduce crankshaft forging defects: an industrial case study. J Ind Syst Eng 3(1):27–37 (Spring)
5. Kima H, Altanb T (2014) Effects of surface finish and die temperature on friction and lubrication in forging. Proced Eng 81:1848–1853 (Elsevier)
6. Thakkar L, Chauhan H, Sheladiya M (2017) A review on unfilling defect found in hot forged 42CrMo4 connecting rod. Trends in Mech Eng Technol 7(1) (STM Journals)
7. Patel BV, Thakkar HR, Mehta SB (2014) Review of analysis on forging defects for quality improvement in forging industries. J Emerg Technol Innovative Res 1(1407054)
8. Zhang ZJ, Dai GZ, Wu SN, Dong LX, Liu LL (2009) Simulation of 42CrMo steel billet upsetting and its defects analyses during forming based on the software DEFORM- 3D. Mater Sci Eng A 499:49–52 (2008 Elsevier)
9. Altan T, Vazquez V (1997) Status of process simulation using 2D and 3D finite element method 'What is practical today? What can we expect in the future? J Mater Proc Technol 71:49–63 (Elsevier)
10. Thottungal AP, Sijo MT (2013) Controlling measures to reduce rejection rate due to forging defects. Int J Sci Res Publ 3(3), ISSN 2250-3153
11. Joshy S, Jayadevan KR, Ramesh A, Mahipal D (2019) Influence of in-service thermal softening on wear and plastic deformation in remanufactured hot forging dies. Sam Joshy et al 2019 Eng. Res. Express 1(025024)

# Influence of *Phoenix* sp. Fiber Content on the Viscoelastic Properties of Polymer Composites

**G. Rajeshkumar, Arvindh Seshadri, K. R. Sumesh, and K. C. Nagaraja**

**Abstract** In this work, the influence of fiber content on viscoelastic properties of *Phoenix* sp. fiber-incorporated composites was investigated. The composites were fabricated using compression molding process. The results revealed that the composite samples incorporated with 40% volume of *Phoenix* sp. fiber have better viscoelastic properties when compared to other composites investigated. A positive shift in glass-transition temperature was also noted with respect to the increase in fiber volume fraction. Moreover, the phase behavior of the composites was analyzed using Cole–Cole plot.

**Keywords** Natural fibers · Cellulose · Viscoelastic properties · *Phoenix* sp. fiber · Polymer composites

G. Rajeshkumar (✉) · A. Seshadri
Department of Mechanical Engineering, PSG Institute of Technology and Applied Research, Coimbatore, India
e-mail: rajesh@psgitech.ac.in

A. Seshadri
e-mail: 17me009@psgitech.ac.in

K. R. Sumesh
Department of Mechanical Engineering, KPR Institute of Engineering and Technology, Coimbatore, India
e-mail: sumesh2311@gmail.com

K. C. Nagaraja
Department of Mechanical Engineering, Acharya Institute of Technology, Bangalore, Karnataka, India
e-mail: nagarajkc15@gmail.com

© The Editor(s) (if applicable) and The Author(s), under exclusive license to Springer Nature Singapore Pte Ltd. 2021

S. Mohan et al. (eds.), *Materials, Design, and Manufacturing for Sustainable Environment*, Lecture Notes in Mechanical Engineering,
https://doi.org/10.1007/978-981-15-9809-8_10

# 1 Introduction

Over the past two decades, natural fibers have garnered popularity as reinforcement in thermoplastic and thermosetting matrices [1]. The prime reason for the rise in the interest in cellulose fibers is due to the concept of sustainable development, ecological awareness and environmental regulations. These fibers have a plethora of merits such as lightweight, cheap, biodegradable, abundance, low $CO_2$ emissions, non-toxic and high specific properties. However, their usage is limited due to hydrophilicity, low fire resistance, low thermal stability and poor fiber/matrix bonding [2]. Polymeric composites that use these natural cellulosic fibers as reinforcement are nowadays employed in wide range of industries like aerospace, automobile, textile, machinery, consumer goods, etc., and its usage rate is continuously increasing [3–6].

Viscoelastic test or dynamic mechanical analysis is an effective technique to characterize a composite's viscous response, elastic response and damping behavior as a function of time, temperature, stress, frequency or a combination of these [7]. The elastic response or the storage modulus tells us about how stiff the composite is and it slumps with the rise in temperature. It is connected to the modulus of elasticity. The viscous response or the loss modulus deals with the amount of heat energy that is dissipated from the material and is associated with internal friction. The damping factor (Tan δ) is expressed as the ratio of loss modulus to the storage modulus. A composite that has high value of damping factor refers to high, non-elastic strain, and a low damping factor refers to high elasticity. The value of Tan δ is linked to dislocations, grain and phase boundaries, molecular movements, etc. [8, 9].

Several works have been carried out by researchers to estimate the viscoelastic properties of natural fiber-reinforced polymer matrix composites. A comparative analysis of evaluating the damping properties was carried out using thermoplastic and thermosetting resins reinforced with flax, glass and carbon fibers. It was found that using flax fibers provided enhanced damping characteristics as compared to synthetic ones [10]. The viscoelastic behavior of jute/epoxy composites was studied, and it was reported that jute fibers enhance the viscoelastic properties of epoxy resin [11]. The viscoelastic test of poly ethylene-co-vinyl-acetate reinforced with cellulose microfibers was done, and it was concluded that fiber loading resulted in the increase of storage modulus and a reduction is damping factor [12]. Moreover, there has been a substantial amount of research work carried out in determining the viscoelastic properties of natural fiber composites [13–17].

Even though there are numerous works that study the viscoelastic behavior of natural-fiber-based composites, there is always a need to explore the dynamic mechanical characteristics of novel materials so that we can employ them efficiently in various applications to yield maximum efficacy. Keeping this mind, our present work has been planned to study the viscoelastic properties of *Phoenix* sp. fiber-based composites. In this paper, the influence of fiber volume fraction on storage modulus, loss modulus and damping factor was investigated and reported at different temperatures. Furthermore, a Cole–Cole plot for different volume fractions of the fiber is also presented.

## 2 Materials and Methods

### 2.1 *Materials*

In the present work, a mixture of epoxy resin LY556 and its hardener HY951 was used as the matrix system. They are mixed in a ratio of 9:1 (by weight). Both the resin and its hardener were purchased from a wholesale trader called Covai Seenu and Company situated in Coimbatore, India.

Manual extraction of *Phoenix* sp. fiber was carried out. These fibers were obtained from the leaf stems of the *Phoenix* sp. plant, and water retting process was used to peel off the fibers from the stems. The extraction procedure is explained in detail in our previous works [18, 19].

### 2.2 *Fabrication of Composites*

In the current study, the composite panels were produced with the help of hot pressing technique in a mild steel die with dimensions 300 × 300 × 3 $mm^3$. The epoxy matrix was reinforced with 20 mm long Phoenix sp. fibers at different volume fractions ranging from 10 to 50%. As aforementioned, the epoxy resin and its hardener were mixed in the ratio of 9:1 as per the manufacturer's recommendation. A releasing agent was applied on the mold, and the fibers were randomly placed on it. The matrix was then poured onto the mold, and the closed mold was subjected to pressure of 5 bar using a hydraulic press for 6 h. Then, post-curing was done at 60 °C for 4 h to instigate the homogeneity. A diamond saw was used to cut the composite panels into samples as per ASTM standards.

### 2.3 *Viscoelastic Properties*

The viscoelastic test of the composite (of size 50 × 10 × 3 $mm^3$) was done in a Seiko dynamic mechanical analyzer model DMS SII EXSTAR 6100, in presence of nitrogen gas. The test was carried out under the following conditions: temperature range, 30–300 °C; heating rate, 2 °C/min; frequency, 5 Hz.

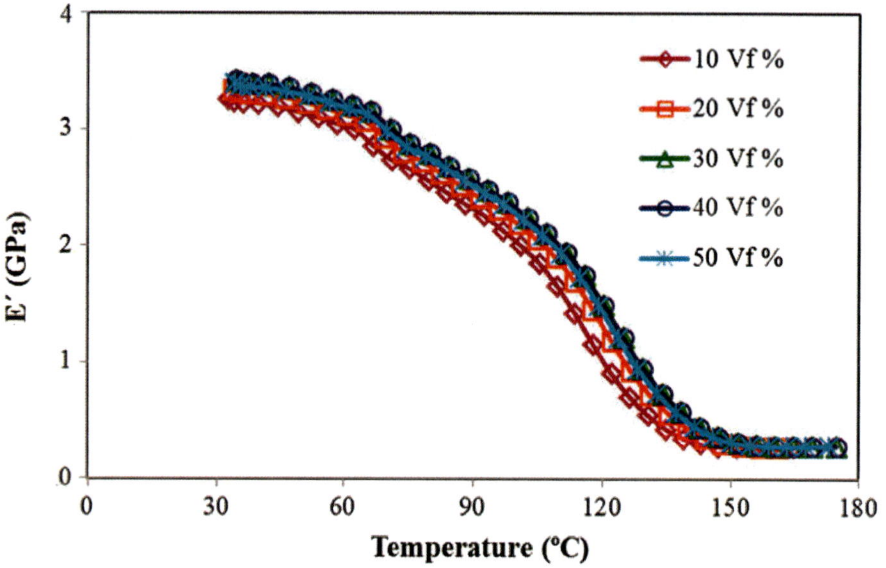


**Fig. 1** Effect of fiber volume fraction on E′

# 3 Results and Discussion

## 3.1 *Storage Modulus (E′)*

Figure 1 depicts the E′ curves of PFREC with varying volume fraction as a function of temperature. It is observed that the E′ increases with the increase in the fiber content up to 40% and then decreases with the further increase in the fiber content. E′ is higher for the composites at glassy region, irrespective of fiber volume fraction, confirming the influence of fiber stiffness at lower temperatures. At the temperature range from 80 to 130 °C, the fiber volume fraction slightly influences the E′ values, and for other temperature values, the contribution of fiber volume fraction toward E′ is very minimum or negligible. Moreover, a light shift in Tg toward higher temperature is observed with the increase in fiber volume fraction due to the constraint imposed by the *Phoenix* sp. fibers toward matrix molecular motions. In all cases, E′ decreases with the increase in temperature, which is attributed to lowering of fiber stiffness at higher temperatures.

## 3.2 *Loss Modulus (E″)*

The E″ curve of PFREC with varying volume fraction is presented in Fig. 2. It is noticed that at all volume fraction, the E″ increases with the increase in temperature

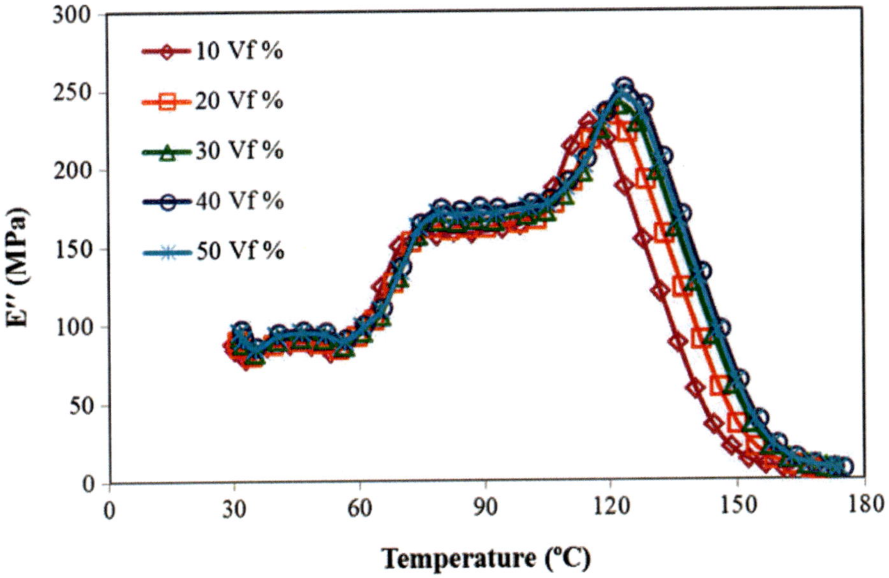


**Fig. 2** Effect of fiber volume fraction on E″

up to Tg and then decreases with further increase in temperature. Results reveal that the volume fraction of fibers has marginal effect on E″ values. The shift in Tg values of various composites from composites incorporated with 40% of fiber and E″ values for different volume fractions is shown in Table 1. A positive shift in Tg values is perceived with respect to the increase in the volume fraction up to 40%. It is indicative of better bonding between the matrix and the fiber at higher temperatures [20]. It is inferred from this that the composites incorporated with 40% of fibers have good interfacial bonding at transition region compared to the other composites. It is further noted that the addition of *Phoenix* sp. fibers introduces more free volume and chain segments, which may contribute to the flattening of E″ curves at a temperature range from 80 to 100 °C.

**Table 1** E″ and tan δ peak of composites

| Fiber volume fraction (%) | E″ (MPa) | Shift in Tg (°C) | Tan δ | Shift in Tg (°C) |
|---|---|---|---|---|
| 10 | 228 | – | 0.309 | – |
| 20 | 233 | 4 | 0.296 | 4 |
| 30 | 240 | 7 | 0.292 | 7 |
| 40 | 251 | 9 | 0.284 | 12 |
| 50 | 247 | 8 | 0.305 | 10 |

### 3.3 Damping Factor (Tan δ)

The influence of fiber volume fraction on tan δ of PFREC as a function of temperature is represented in Fig. 3. It is noted that the tan δ peak decreases with the increase in the fiber content up to 40%. This is because the increase in the fiber volume fraction decreases the volume fraction of matrix, thereby increasing the effective stress transfer between the composite constituents. Moreover, the lowest tan δ peak is observed for the composites incorporated with 40% volume of fibers, which confirms the attainment of better interfacial bonding between the constituents at this fiber loading. In addition, at higher fiber volume fraction, the crack propagation is prevented by neighboring fibers, which is impossible in the case of composites with lower fiber content. When the fiber content is lower, the packing of fibers is inefficient resulting in matrix-rich area. This makes the failure of bonding easy at the interfacial region. The fiber volume fraction has significant effect on Tg values. The shift in Tg and tan $\delta$ values at peak for the fabricated composites is shown in Table 1. A positive shift in Tg value is observed, with respect to the increase in fiber volume fraction, which shows the effectiveness of *Phoenix* sp. fiber as a reinforcing agent. The shifting of Tg to higher temperature range indicates the control of polymer chain mobility due to the incorporation of fibers [21]. In this aspect, the PFREC with 40% volume of fibers shows maximum shift in Tg of 12 °C resulting in lower tan δ of 0.284.

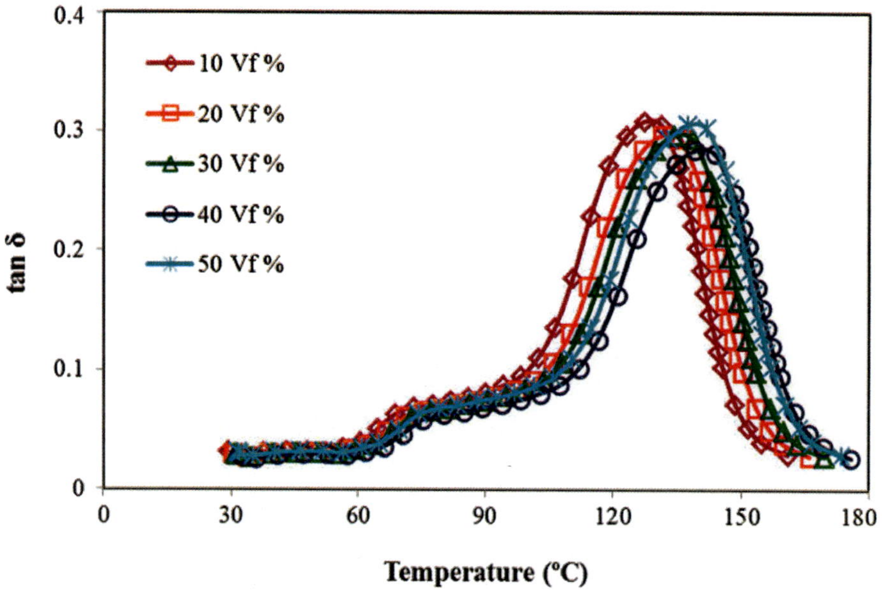


**Fig. 3** Effect of fiber volume fraction on tan δ

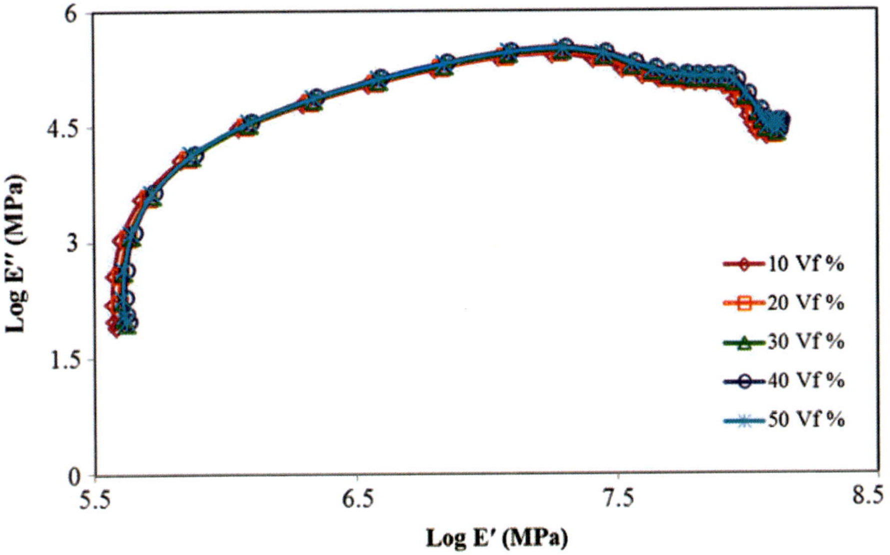


**Fig. 4** Cole–Cole plot

### *3.4 Cole–Cole Plot*

Figure 4 shows the Cole–Cole plot for composites reinforced with *Phoenix* sp. fibers of varying volume fractions. It is observed that for any volume fraction, the curve at the peak is improper semicircular, which indicates the heterogeneity of the composite system.

## 4 Conclusions

The polymeric composites were fabricated using compression molding technique and investigated its viscoelastic properties such as storage modulus, loss modulus and damping factor. The results disclosed that the viscoelastic properties increased with the increase in fiber up to 40% and decreased with further increase in fiber content due to wide agglomeration of fibers. The shift in Tg values toward the higher temperature denotes the increased stiffness of composites. Finally, it can conclude that good composites can be fabricated using 40% volume of *Phoenix* sp. fibers for diverse applications.

## References


1. Pothan LA, Oommen Z, Thomas S (2003) Dynamic mechanical analysis of banana fiber reinforced polyester composites. Compos Sci Technol 63(2):283–293
2. Nagarjun J, Kanchana J, Rajesh Kumar G (2020) Improvement of mechanical properties of coir/epoxy composites through hybridization with sisal and palmyra palm fibers. J Nat Fib, 1–10. https://doi.org/10.1080/15440478.2020.1745126
3. Mochane MJ, Mokhena TC, Mokhothu TH, Mtibe A, Sadiku ER, Ray SS, Ibrahim ID, Daramola OO (2019) Recent progress on natural fiber hybrid composites for advanced applications: a review. Express Polym Lett 13(2): 159–198
4. Mohan TP, Kanny K (2019) Tribological properties of nanoclay-infused banana fiber reinforced epoxy composites. J Tribol, 141(5)
5. Nagaraja KC, Rajanna S, Prakash GS, Rajeshkumar G (2020) Mechanical properties of polymer matrix composites: effect of hybridization. Mater Today: Proc https://doi.org/10.1016/j.matpr.2020.03.108
6. Nagaraja KC, Rajanna S, Prakash GS, Rajeshkumar G (2020) The role of stacking order on mechanical properties of glass/carbon reinforced epoxy hybrid composites prepared by resin infusion technique. Mater Today: Proc 22:2446–2451
7. Ravikumar P, Suresh AR, Rajeshkumar G (2020) An Investigation into the tribological properties of bidirectional jute/carbon fiber reinforced polyester hybrid composites. J Nat Fib 1–11. https://doi.org/10.1080/15440478.2020.1764444
8. Romanzini D, Lavoratti A, Ornaghi HL Jr, Amico SC, Zattera AJ (2013) Influence of fiber content on the mechanical and dynamic mechanical properties of glass/ramie polymer composites. Mater Des 47:9–15
9. Naveen J, Jawaid M, Zainudin ES, Sultan MT, Yahaya R, Majid MA (2019) Thermal degradation and viscoelastic properties of Kevlar/Cocos nucifera sheath reinforced epoxy hybrid composites. Compos Struct 219:194–202
10. Zhang Z, Wang P, Wu J (2012) Dynamic mechanical properties of EVA polymer-modified cement paste at early age. Phys Proced 25:305–310
11. Duc F, Bourban PE, Plummer CJG, Månson JA (2014) Damping of thermoset and thermoplastic flax fibre composites. Compos Part A 64:115–123
12. Gassan J, Bledzki AK (1999) Possibilities for improving the mechanical properties of jute/epoxy composites by alkali treatment of fibres. Compos Sci Technol 59(9):1303–1309
13. Sonia A, Dasan KP, Alex R (2013) Celluloses microfibres (CMF) reinforced poly (ethylene-co-vinyl acetate) (EVA) composites: dynamic mechanical, gamma and thermal ageing studies. Chem Eng J 228:1214–1222
14. Shinoj S, Visvanathan R, Panigrahi S, Varadharaju N (2011) Dynamic mechanical properties of oil palm fibre (OPF)-linear low density polyethylene (LLDPE) biocomposites and study of fibre–matrix interactions. Biosyst Eng 109(2):99–107
15. Jabbar A, Militk J, Wiener J, Karahan M (2016) Static and dynamic mechanical properties of novel treated jute/green epoxy composites. Text Res J 86(9):960–974
16. Selvakumar K, Meenakshisundaram O (2019) Mechanical and dynamic mechanical analysis of jute and human hair-reinforced polymer composites. Polym Compos 40(3):1132–1141
17. Gheith MH, Aziz MA, Ghori W, Saba N, Asim M, Jawaid M, Alothman OY (2019) Flexural, thermal and dynamic mechanical properties of date palm fibres reinforced epoxy composites. J Mater Res Technol 8(1):853–860
18. Rajeshkumar G (2020) An experimental study on the interdependence of mercerization, moisture absorption and mechanical properties of sustainable *Phoenix* sp. fibre-reinforced epoxy composites. J Ind Text 49(9):1233–1251
19. Rajeshkumar G (2020) Effect of sodium hydroxide treatment on dry sliding wear behavior of *Phoenix* sp. fibre-reinforced polymer composites. J Ind Text, https://doi.org/10.1177/1528083720918948

20. Pothan LA, Thomas S, Groeninckx G (2006) The role of fibre/matrix interactions on the dynamic mechanical properties of chemically modified banana fibre/polyester composites. Compos Part A 37(9):1260–1269
21. Idicula M, Malhotra SK, Joseph K, Thomas S (2005) Dynamic mechanical analysis of randomly oriented intimately mixed short banana/sisal hybrid fibre reinforced polyester composites. Compos Sci Technol 65(7):1077–1087

# Development of Visionless Flexible Part Feeder for Handling Shock Absorbers

**S. Udhayakumar, A. Mohan, J. Gowthamachandran, R. Prakash, and P. Shanmugam**

**Abstract** Present trend in automating the batch production industries which manufactures a family of parts uses the flexible part feeder with the vision system to feed the part into the system. The vision system makes the system complex and costly which limits the use of flexible part feeders. To increase the use of the flexible part feeder low-cost flexible part feeders have to be used. To reduce the cost, replacement of the vision system and the robotic arm has to be done. Thus designing a flexible part feeder with an alternative to the vision system and robotic arm for a family of parts could serve as an initiative to reduce the cost of the currently used flexible part feeder. This article focuses on the development of a flexible part feeder without vision system and robotic arm. The approach involves finding the replacement of the vision system, designing the flexible part feeder for a family of parts in three steps, development of identification unit with the help of the replacement found and integrating the identification unit with the controller and controlling the orienting tools with the help of pneumatic actuator followed by testing of the setup for the functional requirement. The work throws light on the replacement of vision system with the capacitive proximity sensors and to replace the robotic arm with the help of orienting tools and thereby reducing the cost and complexity.

S. Udhayakumar (✉) · A. Mohan · J. Gowthamachandran · R. Prakash · P. Shanmugam
Department of Mechanical Engineering, PSG College of Technology, Peelamedu, Coimbatore, Tamil Nadu, India
e-mail: suk.mech@psgtech.ac.in

A. Mohan
e-mail: mohan3mmm@gmail.com

J. Gowthamachandran
e-mail: gowtham.psgmech@gmail.com

R. Prakash
e-mail: pkh.psg@gmail.com

P. Shanmugam
e-mail: shanpsg146@gmail.com

© The Editor(s) (if applicable) and The Author(s), under exclusive license to Springer Nature Singapore Pte Ltd. 2021

S. Mohan et al. (eds.), *Materials, Design, and Manufacturing for Sustainable Environment*, Lecture Notes in Mechanical Engineering,
https://doi.org/10.1007/978-981-15-9809-8_11

**Keywords** Flexible feeders · Shock absorber · Singularising unit · Orienting tool · Capacitive sensors

# 1 Introduction

Every industry is being automated to increase productivity and in some cases to avoid accidents. In an automated assembly to load the parts into the machine with the proper orientation, part feeders are used. These part feeders orient the parts and send them to a machine or an assembling unit. The part feeder that is to be used in the batch production industry must have very less changeover time or zero changeover time if possible. For the same reason, flexible part feeders have been developed [1]. The commonly available flexible part feeders use a costlier vision system to identify the part and its orientation [2, 3].

In case of vision-based systems, the complexity begins with the camera, which includes the field of vision, camera mounting error, and lighting of the parts. When the part is in motion it is also difficult to move the robot arm to the part since it has been identified by the camera before the processing has been done. An adaptive part feeding system for handling sector-shaped parts without the use of vision system was discussed by Udhayakumar et al. [4]. This makes the system very complex and the camera along with the robotic arm makes the system costlier. The aim of this work is to design, fabricate and test a flexible part feeder for a family of parts without the help of vision system and arm and thereby making the system cheaper and simpler [5]. The design part of the work includes designing the orienting tools for each part of the family and then making it flexible. The part considered for experimental studies is a part of shock absorber assembly (Fig. 1) and its family, used in Indian two-wheelers. The family of this part contains three variants with different dimensions and the geometry is asymmetric [6]. The flexible part feeder unit without vision system has been designed for the family of the parts, fabricated, and tested.

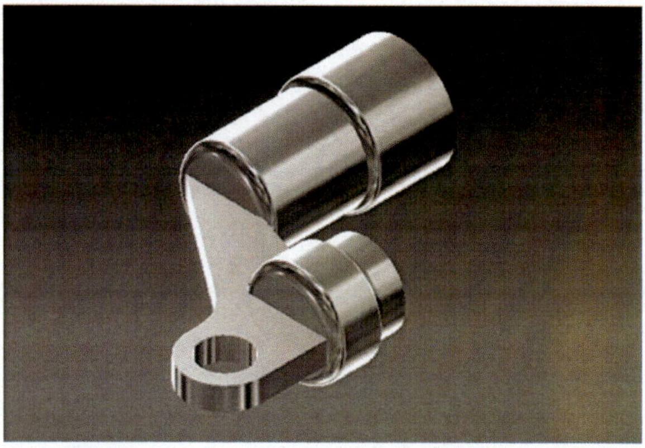

**Fig. 1** Part of shock absorber

## 2 Methodology

The flexible part feeder is to be designed for the family of parts having the geometry shown in Fig. 1. The capacitive proximity sensors when mounted in the proper positions can be used to identify the part and its orientation. This serves the function of replacing the vision system at a very low cost. This sensor gives out the pulse signals when an object comes within the range of the sensor. These pulse signals are to be used for the part identification. The orienting tools to be used could serve the function of the robotic arm at a very low cost. But to make the system flexible the orienting tools have to be altered or moved which necessitates the use of the actuators in the system. This is the outline of the solution to the problem that has been dealt.

The replacement of the vision system with the help of a capacitive proximity sensor is done by interfacing the sensor with the LabVIEW software which constantly monitors the signal from the sensors and identifies the part depending on the program that has been written. The simple concept that could be used to keep the velocity of the conveyor constant and measuring the time of the pulse signals from each of the sensors. But this will limit the use of the part feeder to the constant velocity feeding. The other workaround to this issue can be by closely looking into the variation in the geometry of the part in different orientations. The sensors can be mounted in different positions and the signals at an instant when the part travels across the sensor may be enough to find the orientation of the part.

Once the system identifies the orientation of the part, when a robot arm is used it simple reorients the part by rotating the arm with the help of actuators. But depending on the input orientation of the part the orienting tools have to be adjusted, either brought into the path of the conveyor or to be moved out of the conveyor. This is to be done with the help of interfacing the pneumatic actuator with the LabVIEW software. The orienting part of the part has been decided based on the family of parts chosen for which the flexible part feeder is to be designed. The orientation has been divided into three different phases and the three phases are shown in Fig. 2.

The three phases have been done in three different units having the following names in order, Singularizing unit, Orienting unit, and Toppling unit. The idea is to first singularize the number of possible orientation in two possible resting aspects of the part into one of the four orientations with the help of the orienting tools. From these four the required orientation (orientation is shown last) is achieved in two further steps in the orienting unit and toppling unit. In these three units, singularizing unit does not have an identification unit the other two have since these two units work depending on the input orientation of the part coming into the unit.

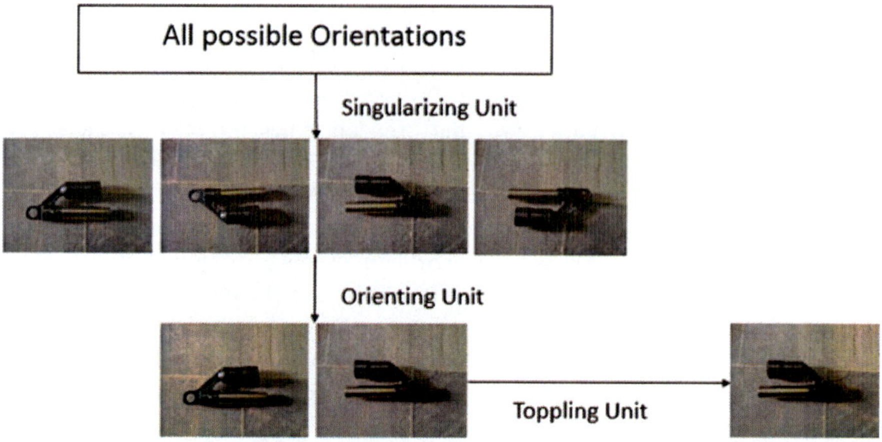


**Fig. 2** Three phases of orientation

# 3 Singularizing Unit

As the name implies the functional requirement of the singularizing unit is to converge the number of the possible orientation into very few with the help of the orienting tools [7]. These tools cannot be designed with the help of any calculations as normally done with the help of some simulation software. The only way for designing orienting tools is to experiment. In the flexible part feeder that is to be designed for the family of parts chosen, the number of possible resting orientations of the part is very large. The part can lie on two surfaces and can have variations of 360° in each of the resting surfaces. This will make the identification unit very complex and thus these orientations have to be brought down to four possible orientations at the end of the singularizing unit.

The design process involves (1) Selection and the type of orienting tools that are to be used and altering the parameters of the orienting tool to meet the functional requirement and (2) Finalizing the design and testing for the functional requirement.

## *3.1 Selection of Type of Orienting Tools to Be Used*

As discussed in the previous section there are many orienting tools that can be used for singularizing the parts. Some of them are fences, deflector blades, edge risers, and slotted tracks [8]. Among these the most commonly used orienting tool used for singularizing unit is deflector blades [7]. These deflector blades have the following parameters that affect the orientation of the parts.

1. Number of blades
2. Length of the blades

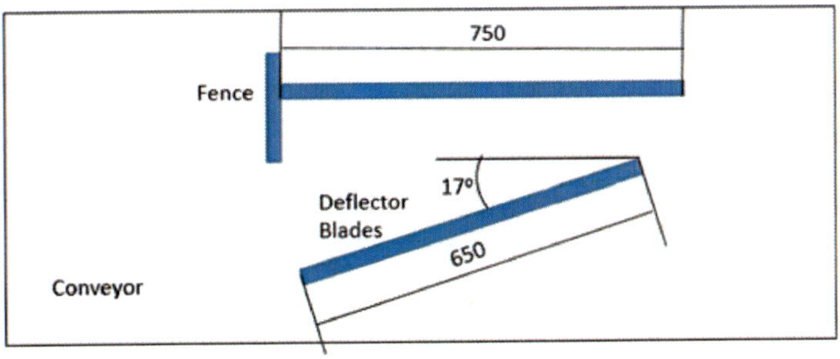


**Fig. 3** Layout of the singularizing unit

3. Angle of blades
4. Height of the blades
5. Distance between the blades.

These parameters are the major concerns that are to be considered and altered when experimenting with them for the required parts. These deflector blades have been selected as the first choice for the singularizing unit. With the help of trial and error method, the layout of the singularizing unit is arrived and shown in Fig. 3.

## 3.2 *Testing of the Singularizing Unit*

The unit after fabrication (Fig. 4) has been tested for the functional requirement and out of all the orientations that are being fed into the conveyor within the velocity the singularizing unit succeeded in converting them into the four orientations with their axis parallel to the motion of the conveyor. This singularizing unit forms the beginning stage of the flexible part feeder that has been designed.

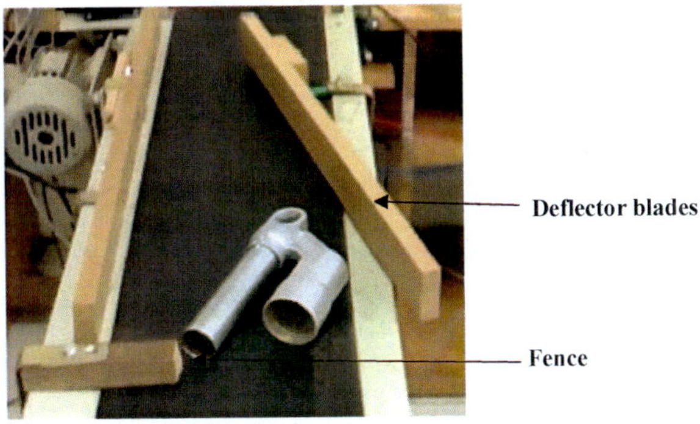


**Fig. 4** Fabricated singularizing unit

# 4 Orienting Unit

The function of the orienting unit is to transform the four orientation into one of the two orientations that form the output set of this unit. Among the four orientations, only two orientations are to be altered and the remaining two are sent to the output without any orientations because they are already in the required two orientations [9]. Thus in simple words, it transforms the four input orientations into two output orientations. Figure 5 shows the functional requirement of the orienting unit (i.e. the input orientations and output orientations).

## 4.1 Design of the Orienting Unit

Unlike the singularizing unit, this unit has to orient only two orientations of the four inputs to the system. Thus this unit has to be flexible to perform the required function. This flexibility is achieved by introducing the identification unit [10] at the beginning of the unit to distinguish the pair of orientation unit to be orientated and the pair of orientations that is to leave undisturbed throughout the time the part traverses through the unit. The design of this unit involves the following steps

1. To integrate the identification unit into the system.
2. Designing the orienting tools to orient the pair of orientations that requires orientation.
3. Interfacing actuators with the help of control circuit using LabVIEW to make the unit flexible.

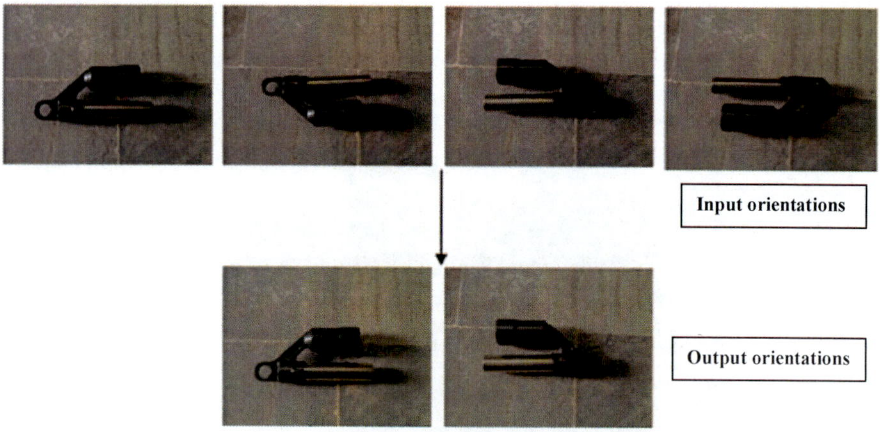


**Fig. 5** Functional requirement of the orienting unit

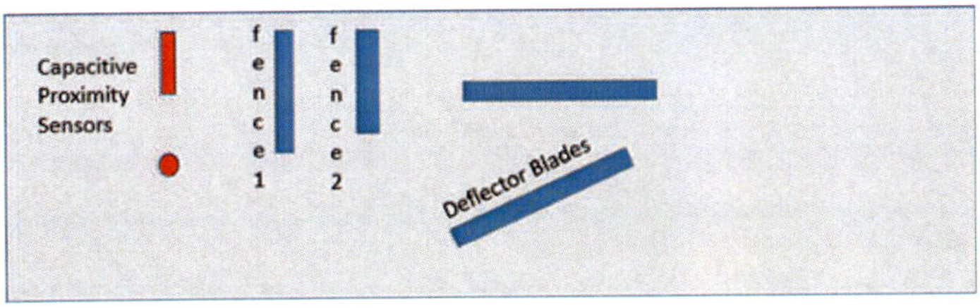


**Fig. 6** Setup for the orienting unit

## *4.2 Designing the Orienting Tools to Orient the Pair to Be Altered*

The pair that has to be altered has to be rotated 180° to achieve one of the orientations that are desired at the output of the orienting unit. This is done with the help of having fences that can rotate the part. But the problem is that this fence has to be removed from the path of the part motion when the pair to be un-altered enters the unit. This is where the help of the identification unit along with the help of the pneumatic operated actuators is required. Initially, a single fence has been used which showed that the part is not rotated completely and then two fences have been used and at the end, deflector blades have been used to perfectly straighten the parts. Figure 6 shows the setup layout of the orienting unit.

## *4.3 Control System*

The logic from the sensor signals has been shown in Fig. 7 for the parts that need to be oriented (i.e., the piston fixed with the fence has to be retracted). Figure 8 shows

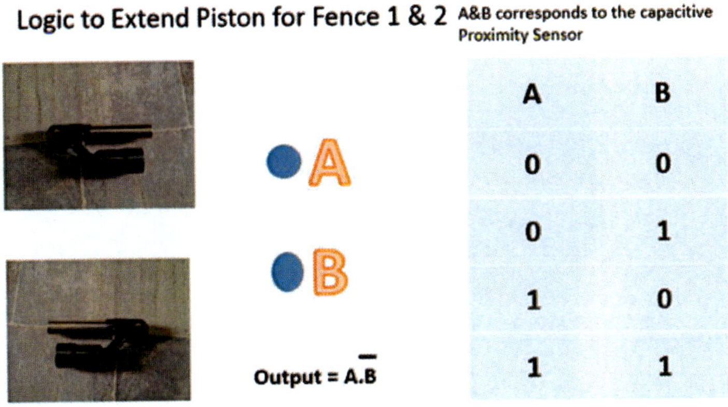


| A | B | EXTENSION |
|---|---|---|
| 0 | 0 | 0 |
| 0 | 1 | 0 |
| 1 | 0 | 1 |
| 1 | 1 | 0 |

**Fig. 7** Logic for piston extension

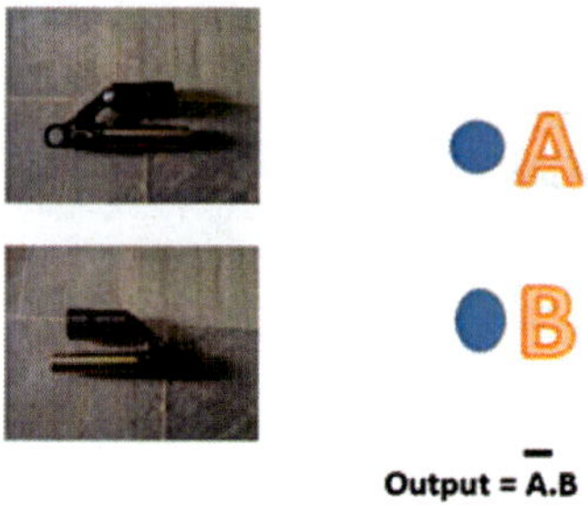


| A | B | RETRACTION |
|---|---|---|
| 0 | 0 | 0 |
| 0 | 1 | 1 |
| 1 | 0 | 0 |
| 1 | 1 | 0 |

**Fig. 8** Logic for piston retraction

the logic for the part that does not require orienting (i.e., the piston is to be retracted position). The deflector blades do the function of straightening the parts.

The logic reprogrammed in LabVIEW for controlling the circuit. The fences that are attached to the pneumatic actuator comes in the track when the logic for the parts to be oriented is sensed by the LabVIEW through the capacitive proximity sensors. This logic helps in identifying the orientation even when the parts are fed with varying conveying speeds.

## 4.4 Fabrication and Testing

The control circuit has to be connected with the external devices including the transistor circuit, DCV, and capacitive proximity sensors. These components are connected and the setup is shown in Figs. 9, 10, and 11.

One of the most important parts in this setup is the sensor mounting, the reason being the fact in senses the part flowing underneath it and errors in mounting the sensors make the whole system to fail in functional requirement. It is the basic part of the control circuit. But the range of the capacitive proximity sensor that has been

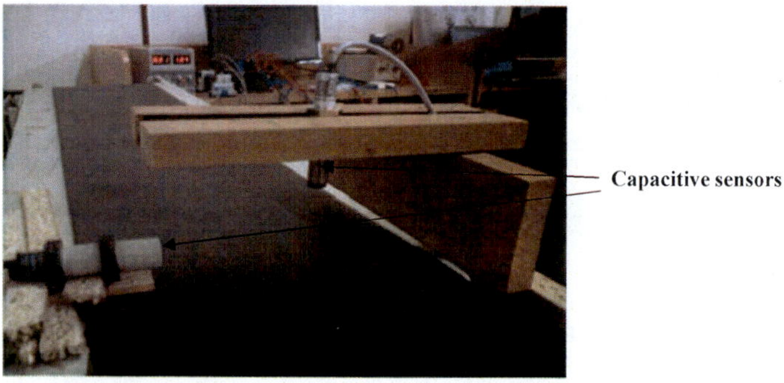


**Fig. 9** Sensor mounting

**Fig. 10** DCV and transistor connection

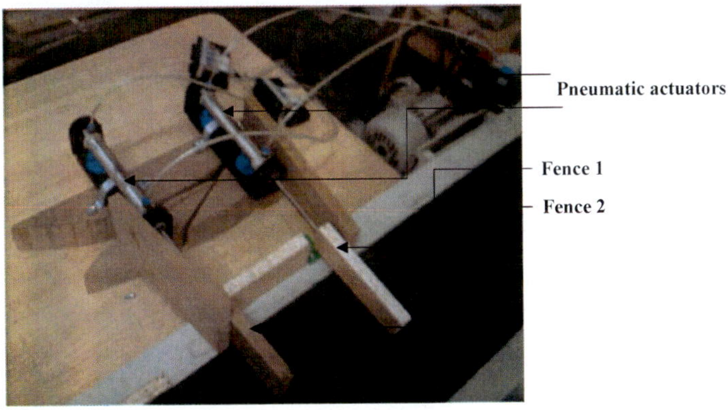


**Fig. 11** Fences connected with the pneumatic actuators

used is very less 8 and 15 mm which increased the error in this part of the system. When tested for about 120 times randomly in each orientation the number failures in the sensor detection is given in Table 1.

Looking closely into the test results and the way the sensors sensed during the experiment. The cause for the failure is mainly due to the sensor range which is very lesser than what is required. Though few times the sensors failed in the current setup when the manual error in mounting the sensor is eliminated and the range increase

**Table 1** Test results of the orienting unit

| Number of failures in sensing the part | Part 1 | Part 2 | Part 3 |
|---|---|---|---|
| Orientation 1 | 3 | 0 | 0 |
| Orientation 2 | 2 | 1 | 2 |
| Orientation 3 | 1 | 2 | 0 |
| Orientation 4 | 1 | 1 | 2 |

will nullify the failure numbers in the results. Out of 120 trials, failure happened in 15 trials which means the setup has been able to pass 87.5% of the parts.

## 5 Toppling Unit

The functional requirement of the toppling unit is to transform the two orientation into the final required orientation. This is achieved with the help of orienting tools. This unit is the last and important unit of the part feeder because it gives out the parts in the final required orientation. The functional requirement for the toppling unit is shown in Fig. 12.

### *5.1 Design of Orienting Tools in the Toppling Unit*

There are several possible designs for toppling the part by 180° to achieve the final required orientation. Among the two orientations which are given to the system. One of the orientation is already the required orientation. Thus the part that is coming in the required orientation has to be left alone without any orienting tools in its path. Whereas the other orientation has to be rotated 180° along the axis perpendicular to the motion of the conveyor and also lies on the plane of the conveyor. This would have done in a single step with the help of the rotary actuator which can clamp the part and rotate it. But this makes the system complex. To make it much simpler it has been done in two steps.

Step 1: Toppling it the axis along with the motion of the conveyor with the help of an orienting tool similar to the edge riser.
Step 2: Then rotating it by a fence as in the case of the orienting unit.

Figure 13 shows the edge riser like tool that is used to topple the part along the axis parallel to the motion of the conveyor.

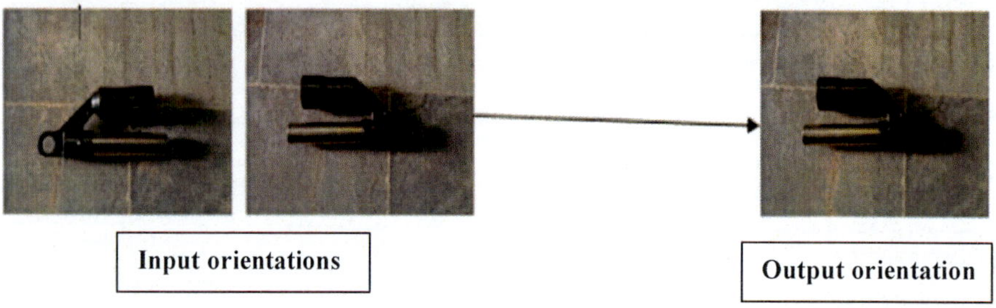


**Fig. 12** Functional requirement of the toppling unit

**Fig. 13** Edge riser to topple the part

The problem is that this tool lies in the path of the conveyor and cannot be moved in the pneumatic actuator as the clearance between the conveyor and the edge riser is very less and making it a fixed part. Thus the part that is already in the required orientation has to be moved away from the path in which edge riser is to be placed. For this the identification unit is has been used and the logic for distinguishing the part in two orientations is discussed in the next section.

## 5.2 *Development of Identification Unit for Toppling*

The identification unit uses the capacitive proximity sensor as in the case of the orienting unit. The capacitive proximity sensor uses the hole in the part which will

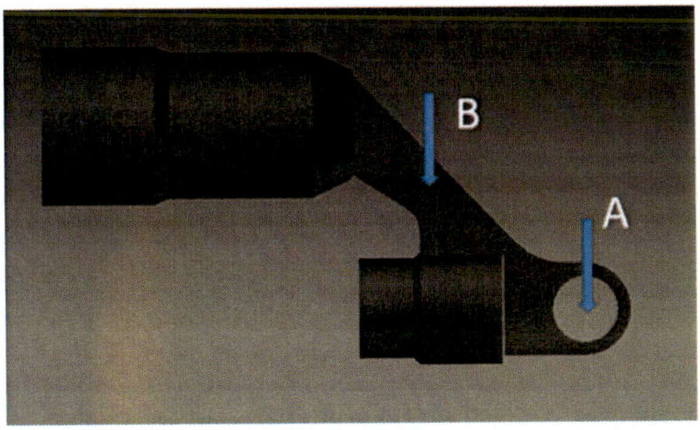


**Fig. 14** Sensor mounting point

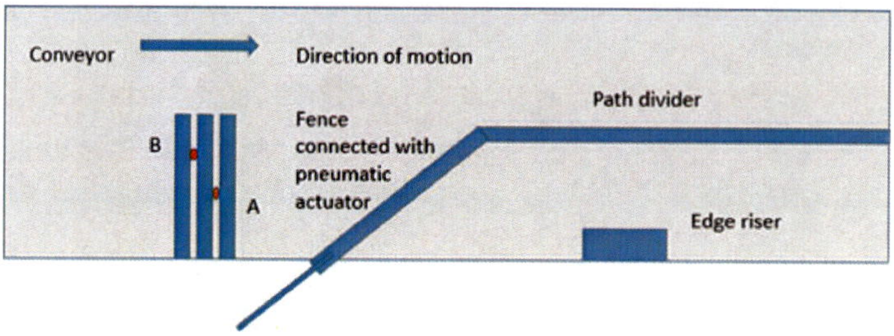


**Fig. 15** Path divider

**Table 2** Logic for the piston extension and retraction

| A | B | Retraction | Extension |
|---|---|---|---|
| 0 | 0 | 0 | 0 |
| 0 | 1 | 0 | 1 |
| 1 | 0 | 0 | 0 |
| 1 | 1 | 1 | 0 |

not be sensed as well as web joining the two cylinders as shown in Fig. 14. Figure 14 shows the sensor mounting points.

Figure 15 shows how the part in the required orientation is to be sent into another path.

The fence that has been connected with the pneumatic actuator guides the part in the required orientation to the path with no tools and it comes to the output in the required orientation. When the sensor signal recognized by the LabVIEW corresponds to that of the part that has to be oriented then the actuator retracts making the part to pass through the edge riser followed by the fence to rotate the part.

Logic for sensing the orientation is given in Table 2.

## 5.3 Fabrication and Testing of the Toppling Unit

The fabricated toppling setup is shown in Fig. 16. This setup when tested shows that the identification unit had the same errors as in the case of the previous unit. But the edge riser worked perfectly toppling the part and the final orientation is achieved.

The success rate was around 85%. The failure in the case of the identification unit in this system is about 15% of the total values and is attributed to an error in sensor mountings and at higher velocities, the system response is too slow to capture the signals thereby failing to actuate the pneumatic actuator properly.

**Fig. 16** Toppling setup

## 6 Conclusions

The objective of the work is to design and fabricate a flexible part feeder without vision system. The flexible part feeder for the canister family in the shock absorber assembly has been designed by trial and error method for each of the parts in the family. The design has three stages which are singularizing unit, orienting unit, and toppling unit. The singularizing unit is the part where the random orientations are singularized which contains no automated parts. The second unit is flexible and adapts itself with the fence as an orienting tool depending upon the input orientation of the part coming into the unit. The toppling unit divides the feeder into two paths and the parts are sent into the appropriate path depending on the input orientation.

The units have been fabricated and LabVIEW has been used for controlling the entire setup. From testing and results of the functional requirement of the flexible part, feeder proposed the following conclusions are drawn.

1. The identification unit has been developed with the help of two capacitive proximity sensors which replaces the vision system of the part feeder which serves the function of the vision system in the flexible part feeder that has been designed.
2. The part feeding unit has been developed for the family of parts chosen and the units are found to be passing 87.5% of the parts to the system and the source of error is found to be due to the range of the capacitive proximity sensor and the manual error in the sensor mounting.
3. The test for the units developed has been done which gave the result of passing 105 parts of 120 parts when fed to the system.
4. The toppling unit that has been developed also passed the 85% of parts that is fed into the system and the cause of the error is due to the limitation in the response time to detect sensor signal and control the circuit in the higher feeding velocities and the manual error in the sensor mounts.

## References

1. Causey GC, Quinn RD, Barendt NA, Sargentand DM, Newman WS (1997) Design of a flexible parts feeding system. Proc IEEE Int Conf Robot Autom 2:1235–1240
2. Quinn Roger D, Causey Greg C (1996) Design of an agile manufacturing work cell for light mechanical applications. Robot Autom IEEE Int Conf 1:858–863
3. Han L, Li CB, Hu GP (2011) A Study on the vision-based flexible vibratory feeding system. Adv Mater Res 279:434–439
4. Udhayakumar S, Mohanram PV, Yeshwanth S, Manas Ranjan B, Sabareeswaran A (2014) Development of an adaptive part feeder for handling sector shaped parts. Assem Autom 34(3):227–236
5. Udhayakumar S, Mohanram PV, Anand PK, Srinivasan R (2014) Trap based part feeding system for stacking sector shaped parts. J Brazilian Soc Mech Sci Eng 36(2):421–431
6. Sadasivam U (2015) Development of vibratory part feeder for material handling in manufacturing automation: a survey. J Auto Mob Robot Intell Syst 9
7. Chua Patrick SK (2007) Novel design and development of an active feeder. Assemb Autom 27(1):31–37
8. Janeja A, Lee N (1998) A modular, parametric vibratory feeder: a case study for flexible assembly tools for mass customization. IIE Trans 30(10):923–931
9. Berretty RP, Goldberg Kenneth Y, Overmars MH, van der Stappen AF (2001) Trap design for vibratory bowl feeders. Int J Robot Res 20(11):891–908
10. Suresh M, Jagadeesh KA, Varthanan PA (2013) Determining the natural resting orientation of a part using drop test and theoretical methods. J Manuf Syst 32(1):220–227

# Influence of Micro $B_4C$ Particles Reinforced Al 4043 Composite Filler Wires on Structural Properties of Al 6061 Weldment

S. Ramani, K. Leo Dev Wins, and R. Robinson Gananadurai

**Abstract** Components of metal matrix composites influence its strength and stability of structures fabricated through them. Aluminium boron carbide (Al 4043/$B_4C$) composite filler rods with varying (0, 2, 4, 6, 8 and 10) wt% of $B_4C$ microparticles and 2 wt% of $B_4C$ nanoparticles were fabricated successfully by stir casting. The required size of filler rods was produced using electrical discharge machine. TIG welding of Al 6061 plates were carried out using $B_4C$ reinforced composite filler rods under optimized conditions. The mechanical and microstructural properties of fabricated weldments were analysed. The optical microscope, FESEM, and TEM images revealed the uniform dispersion of particles and grain refinement of reinforced weldment. The presence of $B_4C$ particles in the weldment was also confirmed using X-ray diffraction technique. The obtained results proved that the inclusion of $B_4C$ particles into the weldment improved the tensile strength and microhardness of the reinforced weldment by 117% and 21%, respectively.

**Keywords** Metal matrix composites · Stir casting · Reinforced composite filler rod · Weldment · Tensile strength · Microhardness

S. Ramani
School of Engineering, Avinashilingam Institute for Home Science and Higher Education for Women, Coimbatore, India
e-mail: ramaniks@gmail.com

K. Leo Dev Wins (✉)
Karunya Institute of Technology and Sciences, Coimbatore, India
e-mail: leo@karunya.edu

R. Robinson Gananadurai
Institute of Technology, University of Gondar, Gondar, Ethiopia
e-mail: rrobinsongnanadurai@gmail.com

© The Editor(s) (if applicable) and The Author(s), under exclusive license to Springer Nature Singapore Pte Ltd. 2021

S. Mohan et al. (eds.), *Materials, Design, and Manufacturing for Sustainable Environment*, Lecture Notes in Mechanical Engineering,
https://doi.org/10.1007/978-981-15-9809-8_12

# 1 Introduction

Welding plays a major role in aircraft manufacturing, ship building, satellite construction and service repair of structural parts and machinery. A lot of difficulties are encountered during aluminium welding processes due to high coefficient of thermal expansion and high thermal conductivity [1]. Metal strengthening mechanisms like solid solution alloying, grain size reduction, strain hardening, annealing, etc., are used to enhance the strength of the alloy materials. However, the enhanced strength of alloy materials is reduced at the fusion zone/weldment during welding process [2, 3]. The tensile strength of materials which undergo traditional welding processes like metal inert gas welding (MIG) and friction stir (FS) welding reduces by 35% and 30%, respectively, compared to the base metal [4]. Also, yield strength of materials which undergo FS welding and gas tungsten arc welding decreases by 20% and 50%, respectively, compared to the base metal [5–7]. This research work is focused on the enhancement of weldment strength by reinforcing particles at the heat-affected zone. Accordingly, boron carbide ($B_4C$) particles were introduced into the weldment in order to improve the hardness and strength of the substrate during fusion welding process and found that fine dispersion of carbide particles in the alloy matrix delayed the free movement of dislocations and enhanced the hardness of weldment [8–10].

Aluminium matrix composites (AMCs) possess superior mechanical and microstructural properties due to their low density, enhanced strength, and good wear resistance due to the existence of microsized and nanosized reinforcement particles [11–13]. Al 4043 is commonly used as a filler metal for welding Al 6061 base metal [14]. In this research, aluminium boron carbide (Al 4043/$B_4C$) composite filler rods with varying (0, 2, 4, 6, 8 and 10) wt% of $B_4C$ microparticles and 2 wt% of $B_4C$ nanoparticles have been fabricated, and TIG welding was performed on Al 6061 substrate. The mechanical and metallurgical characterisation were performed on microsized and nanosized $B_4C$ reinforced weldment.

# 2 Materials and Methods

## *2.1 Composite Fabrication*

Al 4043 alloy, the preferred filler material for welding Al 6061 base metal, was used as matrix material for the fabrication of Al/$B_4C$ composites. Table 1 gives

**Table 1** Chemical composition (wt%) of base metal and filler metal

| Type of material | Cu | Fe | Mg | Mn | Si | Ti | Zn | Al |
|---|---|---|---|---|---|---|---|---|
| Base metal (Al 6061) | 0.2 | 0.5 | 1 | 0.1 | 0.7 | 0.1 | 0.2 | Bal. |
| Filler metal (Al 4043) | 0.08 | 0.26 | 0.15 | 0.07 | 5.8 | 0.06 | 0.07 | Bal. |

**Fig. 1** **a** SEM micrographs of $B_4C$ particles and **b** filler rods with varying wt% of $B_4C$

the chemical composition of base metal and filler material. Stir casting technique was used to prepare Al 4043/$B_4C$ composites reinforced with varying (0, 2, 4, 6, 8 and 10) wt% of $B_4C$ microparticles and 2 wt% of $B_4C$ nanoparticles. $B_4C$ particles with average size of 10 μm was used for the fabrication of microcomposite and is confirmed by SEM micrograph (Fig. 1a). It is one of the hardest particles (third-hardest material) used as reinforcement which has exclusive properties like high strength, good wear resistance, and extremely high hardness [15]. To synthesize nanoparticles, $B_4C$ microparticles were ball milled with zirconium vial about 6 h to reduce its size upto 150 nm. The $B_4C$ particles were mixed with hexafluoro titanate ($K_2TiF_6$) to acquire good wettability with metal matrix.

The stir casting set-up consisted of an electrical resistance heating furnace with inert gas protection systems. A clay graphite crucible with a capacity of 1 kg was used for melting the metal matrix in which a batch of 200 g Al 4043 alloy was melted at about 730 to 750 °C for processing molten aluminium. The specific (0, 2, 4, 6, 8 and 10) wt% of microparticles and 2 wt% of nanoparticles were added for formulating individual batch of melt. The molten metal was stirred with the help of a graphite-coated mechanical stirrer. High-power ultrasonic probe (Hangzhou Success, China) made of Ti–6Al–4V horn was employed to treat nanoparticle reinforced melt for attaining better particle dispersion. The well-dispersed melt was poured into a rectangular mild steel mould of size 200 mm × 10 mm × 10 mm.

## 2.2 *Filler Rod Fabrication*

Al 4043/$B_4C$ composites with different wt% of particle content were machined to the desired dimensions of filler rod using wire EDM (Make: M/s Electronica) [16]. Figure 1b shows the fabricated filler rods with varying wt% of $B_4C$ particles. The filler rods have been fabricated with a cross section of 4 mm × 4 mm and a length of 200 mm.

## 2.3 TIG Welding

Wrought Al 6061–T6 sheets of 6 mm thickness were machined to the desired size of 150 mm × 100 mm. Table 1 shows the chemical composition of the Al 6061 alloy. Before welding, the plates were cleaned using stainless steel brush and wiped with acetone solution [17]. Initial positioning of the base plates was done by tack welding, and process has been directed to the normal rolling direction with a single pass. TIG welding was carried out on Al 6061 substrates using Al 4043/$B_4C$ composite filler rods with optimized welding parameters (220 A of current and manual welding speed ~4 mm/s). Different weldments were prepared using Al 4043/$B_4C$ composite filler rods of different wt%.

## 2.4 Sample Preparation

The tensile and microhardness test specimens with different wt% $B_4C$ were prepared according to the ASTM E8M04 standard using wire EDM process. The tensile test was carried out using computerized universal testing machine (TMC Chennai—50 kN). Figure 2 shows the tensile specimens of TIG welded joint; (a) before and (b) after the tensile test. The hardness test was carried out using Vickers hardness tester (Daksh systems-HV 1000DT, 200 g load and 10 s dwell time) [18]. The polished specimens were etched by colour etchant (Ethyl alcohol (122 ml), HCl (122 ml), $FeC_{l2}$ (8.5 g) and $CuC_{l2}$ (2.4 g) $HNO_3$ (6 ml)). The prepared and etched specimens were observed with the help of XJL-17QS metallurgical microscope with a magnification of 100×.

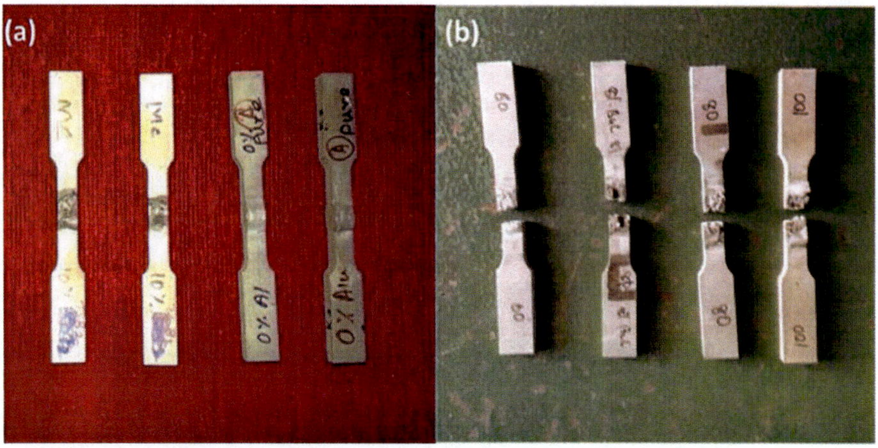


**Fig. 2** Tensile specimens of TIG welded joint; **a** before **b** after the tensile test

# 3 Results and Discussion

## 3.1 Hardness

Figure 3a shows the microhardness values recorded on the specimens with different wt% of $B_4C$ in the weldment. It was observed that the microhardness values of weldment was gradually increasing with increase in weight percentages of $B_4C$ particles in the weldment. Also, the increase in wt% of $B_4C$ in the weldment reduced the grain size of the particles in the weldment [19]. The existence of a reinforced particle in the weldment offered more resistance to grain refinement and plastic deformation of weldment and also led to enhanced microhardness of the welded joints [20]. The reinforcement of $B_4C$ ceramic particles in elastic and ductile matrix reduced the ductile properties of the composite due to the rationalizing of ductile metal content by which caused significant improvement in hardness [21]. In the present investigation, increase in weight percentage of reinforcement increased the microhardness from 72 to 87 VHN in the weldment. It was also observed that increase in weight percentage of $B_4C$ particles beyond 8 wt%, leads to agglomeration in the grain surfaces and reduces the microhardness to 80 VHN for 10 wt% $B_4C$ particles. Further, the scope of this research has been taken forward as an attempt to find the effect of nanoparticle reinforcement in increasing weld strength. Accordingly, 2 wt% of nanosized $B_4C$ particles have been reinforced into the weldment through the filler rods, and the results were compared with weldment produced with 2 wt% of microsized reinforcement particles (Table 2). Al 6061/2 wt% $B_4C$ nanocomposite reinforced filler rod produced weldment with microhardness 153 VHN against the microhardness (87 VHN) achieved through 2 wt% $B_4C$ microcomposite filler. Thus, an increase in microhardness of 76% could be achieved through Al 6061/2 wt% $B_4C$ nanocomposite reinforced filler rod.

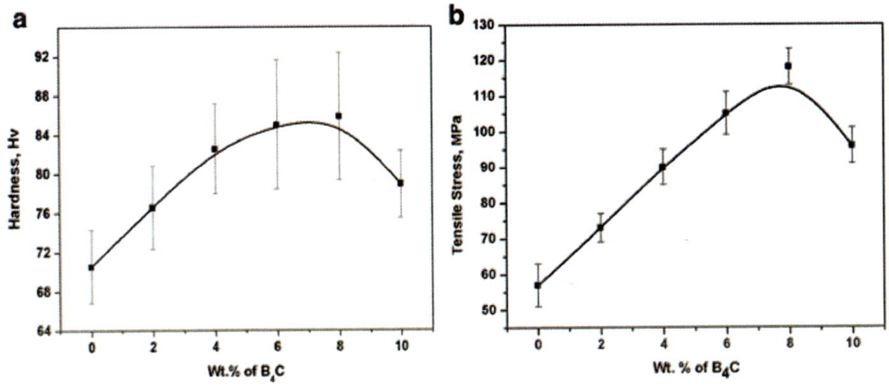


**Fig. 3** **a** Microhardness of Al/$B_4C$ weld with different wt% $B_4C$ and **b** tensile stresses of Al/$B_4C$ weld with different wt% $B_4C$

**Table 2** Tensile strength of Al 6061 weld with Al 4043 microcomposite (MC) and nanocomposite (NC) filler

| Samples | Al 6061 base plate | Al 6061/4043 weld without $B_4C$ | Al 6061/4043 weld with 2 wt% $B_4C$ MC | Al 6061/4043 weld with 2 wt% $B_4C$ NC |
|---|---|---|---|---|
| UTS (MPa) | 310 | 119 | 76 | 130 |
| Elongation (%) | 11 | 12.8 | 13.6 | 16.9 |
| Hardness (VHN) | 70 | 78 | 87 | 153 |

## 3.2 *Tensile Test*

The tensile strength of Al 6061 weld joints got improved due to the incorporation of $B_4C$ microparticles. Figure 3b shows the relation between the tensile strength and the wt% of $B_4C$ microparticles in the weldment. The tensile strength of weldment increased from 57 to 118 MPa due to the addition of different wt% of $B_4C$ particles. The enhancement in tensile strength has been achieved through the addition of microparticles into the weldment. It was found that tensile strength and deformation to fracture of tensile specimen produced by 8 wt% $B_4C$ particles were higher than other tensile specimens. While examining the fractured surface of the tensile specimen, it was found that the fracture of tensile specimen occurred at the interface between weld bead and HAZ (Fig. 2b). The results confirmed that the strength of the weldment was largely enhanced by the addition of 8 wt% $B_4C$ reinforcement. Table 2 shows the strength comparison of microsized and nanosized $B_4C$ reinforced weld joint. Moreover, Al 6061/4043 2 wt% nanocomposite weld has reported 130 MPa tensile strength (given in Table 2). The reduction of the particle size of the reinforcement increased the strength of the weldment for the same wt% of reinforced composite filler weldment.

## 3.3 *Microstructural Examination*

The microstructure of different (0, 4, 8 and 10) wt% of $Al/B_4C$ welded joints were examined by optical microscopy (QS Metrology–XJL 17). The optical micrographic images of HAZ with different weight percentages of $B_4C$ are shown in Fig. 4a–d. However, the type of reinforced particles influences the refinement of grains at the welded region, and the grain size of particles plays a significant role in the properties of weldment. A well-deformed coarse grains were observed in the HAZ of the sample without $B_4C$ particle content in Fig. 4a. After the reinforcement of $B_4C$ particles, considerable grain refinement of the matrix was observed in the weld bead, and the presence of reinforcement particles caused elongated grains in the HAZ as shown in Fig. 4b–d as a whole. The selected diffraction patterns of 4, 8 and 10 wt% of Al

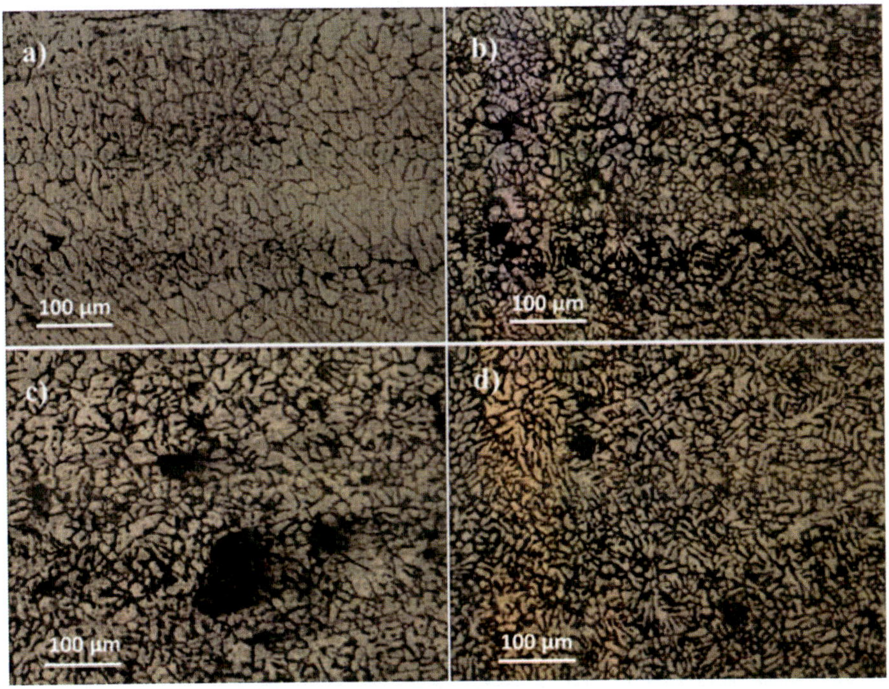


**Fig. 4** Optical microscopic images of **a** Al 6061/0% $B_4C$, **b** Al 6061/4% $B_4C$, **c** Al 6061/8% $B_4C$ and **d** Al 6061/10% $B_4C$ reinforced microcomposite weldment

6061/$B_4C$ weld confirm the existence of microparticles in the reinforced weldment (Fig. 5).

The fusion zone of TIG welded joints contained wider dendritic structure due to rapid heating of base metal and slow cooling of molten metal during the TIG welding process. Strength and ductile nature reveals direct relationship with the dimple size of the joint microstructure, i.e., if the dimple size of the weld is finer, the strength and ductile property of the particular joint would be superior and vice versa [22].

The FESEM micrographs of Al 6061/2 wt% $B_4C$ microcomposite and nanocomposite filler fused weldment are shown in Fig. 6. It reveals the particle size and dispersion of $B_4C$ particles in the weldment even after the fusion welding of composite fillers. The effect of stirring action and optimized process parameters attribute the uniform dispersion of $B_4C$ particles in the aluminium matrix. However, conventional stir casting method can produce ceramic particle reinforcement with poor wettability. So, the flux ($K_2TiF_6$) was used to improve the wettability of Al 6061/$B_4C$ composite matrix by which heat is generated around $B_4C$ particles through exothermic reaction. This localized increase in temperature improves the reinforcement of particles into the aluminium melt and provides better bonding with the matrix [23].

Figure 6a reveals the dispersion and agglomeration of $B_4C$ particles in the inner dendritic region. The ultrasonic treatment improves dispersion of particles and refines the grain size of the metal matrix. It can be attributed that the ultrasonic treatment

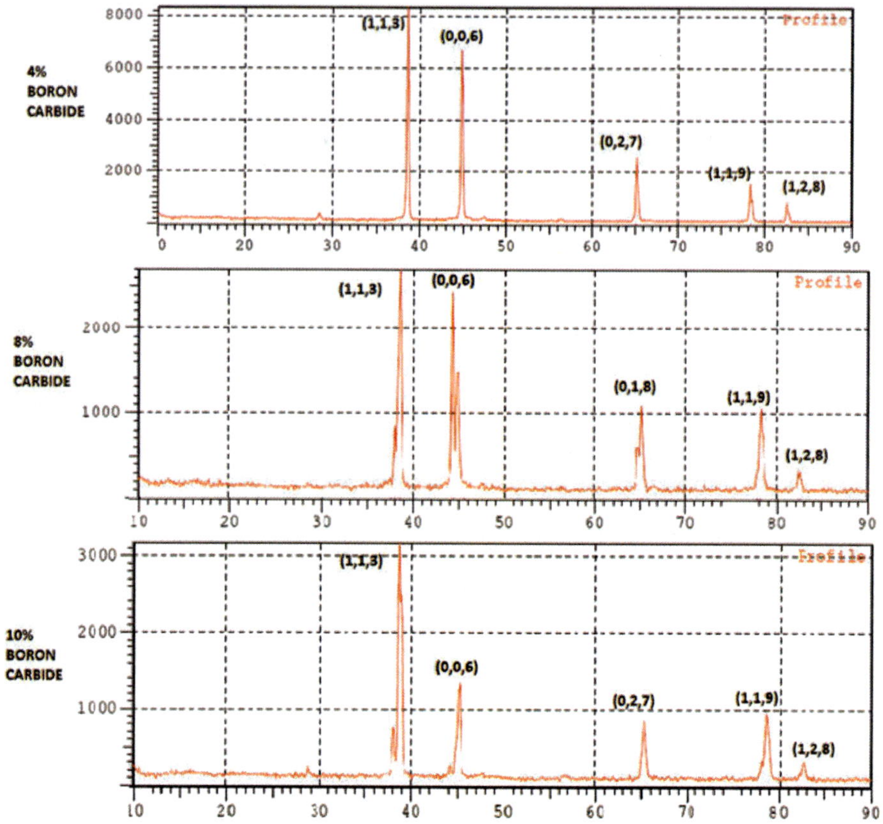


Fig. 5 XRD pattern of 4, 8 and 10 wt% of Al 6061/$B_4C$ weld

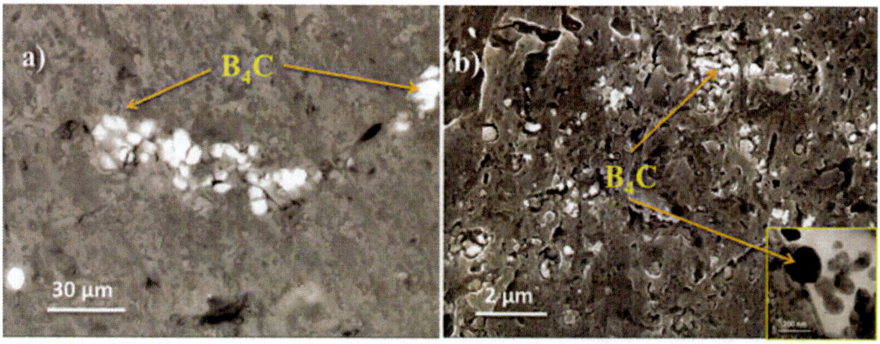


**Fig. 6** FESEM images of Al 6061/2 wt% $B_4C$ **a** microcomposite and **b** nanocomposite-reinforced weldment

improves wettability with molten aluminium and act as nucleant for grain refinement [24]. The magnified image in Fig. 6b validates uniform dispersion of $B_4C$ particles and proves the presence of the nanosized $B_4C$ particles in the metal matrix weldment (inset of Fig. 6b).

## 4 Conclusion

In this research, the effect of inclusion of microsized and nanosized $B_4C$ particles into Al 6061 TIG welded joints through Al/$B_4C$ composite filler rods on metallurgical and mechanical properties of weldment is given. The microcomposite-filler-fused weldment revealed higher tensile strength of 118 MPa and hardness of 85 VHN for Al 6061/8 wt% micro $B_4C$. The nanocomposite filler of Al 6061/2 wt% nano $B_4C$ produced 71% higher tensile strength and 75.86% higher microhardness as compared with same wt% of microparticle-reinforced weldment. The optical micrographic, FESEM, TEM studies and XRD analysis revealed grain refinement, particle distribution, size and presence of $B_4C$ reinforcement particles in the welded joints.

## References


1. Embury JD, Lloyd DJ, Ramachandran TR (1989) Strengthening mechanisms in aluminum alloys. Treatise Mater Sci Technol 31:579–601
2. Ahmad R, Abdulmalik SS (2014) The effect of microstructure and mechanical properties of aluminium AA6061 before and after heat treatment using TIG. Appl Mech Mater 465–466:881–885
3. Dewan MW, Wahab MA, Okeil AM (2016) Effect of weld defects on tensile properties of lightweight materials and correlations with phased array ultrasonic nondestructive evaluation. In: International manufacturing science and engineering conference, Detroit, Michigan, USA, 9–13 June 2014
4. Moreira PMGP, de Figueiredo MAV, de Castro PMST (2007) Fatigue behaviour of FSW and MIG weldments for two aluminium alloys. Theoret Appl Fract Mech 48:169–177
5. Muñoz AC, Rückert G, Huneau B, Sauvage X, Marya S (2008) Comparison of TIG welded and friction stir welded Al-4.5Mg-0.26Sc alloy. J Mater Process Technol 197:337–343
6. Verma RP, Pandey KN (2017) Fracture behavior of GMA welded joints of dissimilar and similar aluminum alloys of 6061–T6/5083-O. J Fail Anal Prev 17:248–254
7. Lakshminarayanan AK, Balasubramanian V, Elangovan K (2009) Effect of welding processes on tensile properties of AA6061 aluminium alloy joints. Int J Adv Manuf Technol 40:286–296
8. Sahin Y (2014) Preparation and some properties of SiC particle reinforced aluminium alloy composites 24:671–679
9. Selvakumar M, Rajamani GP, Kalaiselvan K (2014) Synthesis and characteristic of AA6061/SiC sand cast composite. Appl Mech Mater 591:43–46
10. Sudhakar I, Reddy GM, Rao KS (2015) Ballistic behavior of boron carbide reinforced AA7075 aluminium alloy using friction stir processing—an experimental study and analytical approach. Defence Technol
11. Ceschini L, Boromei I, Minak G, Morri A, Tarterini F (2007) Effect of friction stir welding on microstructure, tensile and fatigue properties of the AA7005/10 vol.%$Al_2O_{3p}$ composite. Compos Sci Technol 67:605–615

12. Nagaral M, Auradi V, Kori SA (2015) Microstructure and mechanical properties of Al6061-graphite composites fabricated by stir-casting process. Appl Mech Mater 766–767:308–314
13. Abenojar J, Martínez MA, Velasco F (2009) Effect of boron carbide filler on the curing and mechanical properties of an epoxy resin. J Adhes 85:216–238
14. Wire AW, Technical information sheet 4043 aluminum welding wire & rod. https://www.harrisproductsgroup.com/
15. Mohanty RM, Balasubramanian K, Seshadri SK (2008) Boron carbide-reinforced alumnium 1100 matrix composites: fabrication and properties 498:42–52
16. Jia Y, Kim BS, Hu DJ, Ni J (2010) Parametric study on near-dry wire electrodischarge machining of polycrystalline diamond-coated tungsten carbide material. Proc Inst Mech Eng Part B J Eng Manuf 224:185–193
17. Meng C, Cui H, Lu F, Tang X (2013) Evolution behavior of $TiB_2$ particles during laser welding on aluminum metal matrix composites reinforced with particles. Trans Nonferrous Met Soc China 23:1543–1548
18. Yigezu BS, Venkateswarlu D, Mahapatra MM, Jha PK, Mandal NR (2014) On friction stir butt welding of Al + 12Si/10 wt%TiC in situ composite. Mater Des 54:1019–1027
19. Mohanavel V, Rajan K, Senthil Kumar KR (2015) Study on mechanical properties of AA6351 alloy reinforced with titanium di-boride ($TiB_2$) composite by in situ casting method. Appl Mech Mater 787:583–587
20. Salih OS, Ou H, Wei X, Sun W (2019) Microstructure and mechanical properties of friction stir welded AA6092/SiC metal matrix composite. Mater Sci Eng A 742:78–88
21. Ramesh CS, Keshavamurthy R, Channabasappa BH, Ahmed A (2009) Microstructure and mechanical properties of Ni-P coated $Si_3N_4$ reinforced Al6061 composites. Mater Sci Eng A 502:99–106
22. Lin DC, Wang GX, Srivatsan TS (2003) A mechanism for the formation of equiaxed grains in welds of aluminum-lithium alloy 2090. Mater Sci Eng A 351:304–309
23. Kalaiselvan K, Murugan N, Parameswaran S (2011) Production and characterization of AA6061–$B_4C$ stir cast composite. Mater Des 32:4004–4009
24. Meti VKV, Shirur S, Nampoothiri J, Ravi KR, Siddhalingeshwar IG (2017) Synthesis, characterization and mechanical properties of AA7075 based MMCs reinforced with $TiB_2$ particles processed through ultrasound assisted in-situ casting technique. Trans Indian Inst Met

# Ballistic Performance Simulation of Graphene–Dyneema Multi-layered Armor

S. Vignesh, R. Surendran, T. Sekar, and B. Rajeswari

**Abstract** Dyneema (unidirectional bullet-resistant fiber) is a crucial ballistic ingredient for the finest life-protecting ballistic applications, protecting law enforcement officials and soldiers. Since dyneema requires material quantity much less than others to achieve a given performance and due to its high durability and long service life, it has been a prime option for ballistic material. Also, dyneema leaves the lowest carbon impression per unit tenacity. Dyneema-made vests have the ability to protect against a large variety of ballistic threats including a direct barrage from AK47 but fail when bullets from armor-piercing Type-V rifles like 0.30–0.06 Springfield or 0.50 caliber had a direct impact on the vests. On the other hand, graphene is one of the strongest materials ever found having a single-atom-thick layer of 2D carbon atoms arranged in a hexagonal lattice. It exhibits properties like high tensile strength, thermal conductivity, toughness, stiffness, transparency and surface area. This high combination of a variety of properties is due to the covalent trigonal bonding of the carbon atoms arranged in $sp^2$ hybridization. This study is a performance evaluation of graphene-sandwiched dyneema laminates against a 0.50 caliber using ANSYS 18.1. The resulting deformation of the armor laminates and stresses induced in the vest and bullet are studied and analyzed for improving the efficiency of the armors.

**Keywords** Ballistic performance · Deformation · Dyneema · Equivalent stress · Graphene

S. Vignesh (✉) · R. Surendran · T. Sekar · B. Rajeswari
Department of Mechanical Engineering, Government College of Technology, Coimbatore, Tamil Nadu, India
e-mail: vickytechie03@gmail.com

R. Surendran
e-mail: surendran_r@gct.ac.in

T. Sekar
e-mail: drtsekar76@gmail.com

B. Rajeswari
e-mail: rajes.aravind@gmail.com

© The Editor(s) (if applicable) and The Author(s), under exclusive license to Springer Nature Singapore Pte Ltd. 2021

S. Mohan et al. (eds.), *Materials, Design, and Manufacturing for Sustainable Environment*, Lecture Notes in Mechanical Engineering,
https://doi.org/10.1007/978-981-15-9809-8_13

**Table 1** Material properties of common armor ceramic materials

| Material | Silicon carbide | Boron carbide | Alumina | Titanium di-boride |
|---|---|---|---|---|
| Hardness (HV) | 1800–2800 | 2800–3400 | 1500–1900 | 2100–2600 |
| Young's modulus (GPa) | 380–430 | 420–460 | 350–390 | 520–550 |
| Poisson's ratio | 0.14–0.18 | 0.14–0.19 | 0.22–0.26 | 0.05–0.15 |
| Fracture toughness (MPa $m^{1/2}$) | 3–5 | 2–3 | 3–5 | 5–7 |
| Bulk density ($kg/m^3$) | 3090–3220 | 2500–2520 | 3810–3920 | 4450–4520 |

# 1 Introduction

## *1.1 Body Armors*

Body armors are any protective material which protects the human body from trauma due to firearms, shrapnel and other threats during combats [1, 2]. With the development of new weapon systems, it is necessary to renovate the ballistic technology to improve the protection. Based on the mechanism and materials used, body armors are classified as passive armors and reactive armors [3]. Passive armors are regular non-plated armor for moderate to substantial protection, and reactive armors are one with high-strength material plates inserted in between composites for maximum protection. Passive armors are typically constructed by a series of dense ballistic fibers woven together with an adhesive which uses constituent material properties to absorb the kinetic energy of the projectiles [4]. They are usually used for personnel protection against firearms from handguns due to proximity. While hard armors are constructed with ceramics, compressed laminates, metallic plates or other composites that incorporate more than one material uses the kinetic force to counter the ballistic threat. Usually, high-strength steels, titanium alloys and ceramic plates are used due to their high hardness and stiffness [5]. The material properties of common ceramic materials used in armors are given in Table 1.

## *1.2 Materials Used*

The development of the ultra-high-molecular-weight polyethylene (UHMWPE) was one of the most significant advancements in the field of dynamic impact materials [6]. The vital choices of materials potentially available for ballistic applications now are aramid fibers and UHMWPE [7]. Although aramid fibers like Kevlar are a seemingly good material choice, their inability to cope with compression, degradation to environmental conditions and comparatively low strength make UHMWPE the first choice [8]. Honeywell's spectra and DSM's dyneema are the commercially available ballistic UHMWPE materials. Dyneema was chosen for the performance evaluation analysis due to its higher energy absorption capacity than spectra [9].

### *1.3 Graphene*

Since it was discovered in 2003, researches into this wonder material had extended rapidly [10]. The material is well known for its ultra-high strength and ability to conduct heat and electricity with great efficiency [11]. This unique combination of excellent properties offers a captivating material platform for the development of newer technologies in many areas such as superfast electronics, multi-functional composites and coatings, ultra-high sensitive sensors, medicine, biotechnology energy and storage devices [12–15]. However, very little research had been done into the material's ability to absorb dynamic impacts. Despite its low weight, the material is remarkably strong and durable, even more so than steel and diamond. A pristine mono-layer of graphene offers a strength that would require a force exerted by a mass of 2000 kg to puncture it with a sharp object [16]. With 100 times the strength of steel, graphene can scatter kinetic energy 10 times than that of steel for the same weight. This kinetic energy absorption of the projectile is done by stretching the graphene fibers into a fine funnel shape at the impact point, then by cracking the bullet outward radially [17].

## 2 Analysis

### *2.1 Design Specifications*

The current objective is to design a hybrid armor comprising of both ballistic fabric material and collapsible energy absorbers that were discussed. The armor assembly was proposed as a dyneema fabric–graphene laminate sandwich. The main function of dyneema fiber is to capture the projectile and to disrupt and diffuse its kinetic energy over a sufficiently large area to avoid regional failures at the point of impact. The function of graphene laminates was to act as an energy diffuser to dissipate energy into the ballistic system after partial deformation to prevent total failure. The final layer of protection was a thin layer of dyneema behind the graphene sheets to form a backing spall liner. The schematic diagram of the cross-sectional construction of the dyneema plate and dyneema–graphene plate was shown in Fig. 1.

The impact analysis was carried out on an armor composed of dyneema fibers of thickness 12 mm at the front to reduce the projectile's kinetic energy, and a backing layer of 3 mm was at the end of the body armor. A graphene laminate of 5 mm was sandwiched between this dyneema fabric setup. The dimensions of the vest are given in Fig. 2, and the 3D model of vest and bullet is shown in Fig. 3.

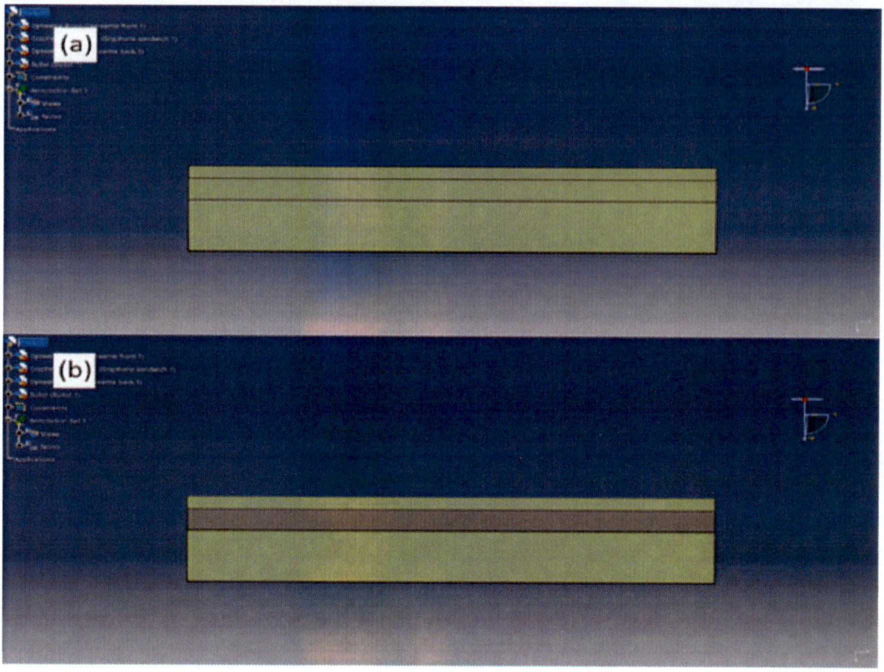


**Fig. 1** Schematic diagram of cross-sectional construction of **a** dyneema plate and **b** dyneema–graphene composite plate

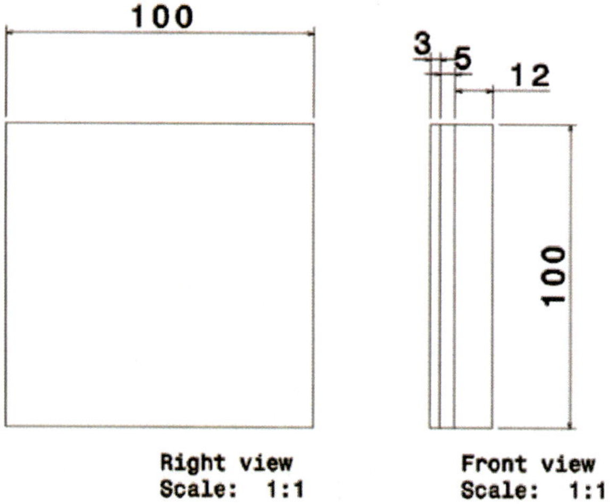


**Fig. 2** Dimensions of bulletproof vest

**Fig. 3** 3D model of 0.5 caliber and bulletproof vest

To compare the ballistic performance, the same performance evaluation was carried out in a plain dyneema 12 and 3 mm fabrics. The bullet chosen for the performance evaluation is 0.50 caliber which is Type-IV bullet as specified by NIJ-0101.06. The 0.50 caliber or 0.50 Browning machine gun (BMG) is an iron-based bullet that travels with a velocity of about 878 m/s. It has a high ballistic co-efficient producing 15,000–20,000 J muzzle energies depending upon the type of bullet.

### 2.2 *Boundary Conditions*

The body armor laminates and bullet were modeled using CATIA V5R20, which is given in Fig. 2. The models were imported into ANSYS 18.1 as stp files. The models were then given initial conditions such as the boundaries of the armor are fixed and the bullet was given a velocity of 880 m/s. Then, the models were meshed and analyzed for ballistic performance (Fig. 4).

## 3 Results and Discussion

### 3.1 *Analysis of Plain Dyneema Fibers*

The dynamic impact analysis was performed on plain dyneema fabric. The total deformation and stresses developed in the armor material were calculated.

**Fig. 4** Zoomed view of meshed vest and bullet

Figure 5 shows the result of the dynamic impact analysis of 0.50 caliber on a plain dyneema armor. The total deformation induced in the plain dyneema armor during the impact is about 0.07624 m. The maximum stress developed is about 1816.8 MPa, and maximum principal stress is 1466.9 MPa which is induced in the ruptured bullet parts after the ballistic impact.

## 3.2 Analysis of Dyneema–Graphene Fibers

The dynamic impact analysis was performed on graphene–dyneema hybrid armor. The total deformation and the stresses developed in the armor material were calculated.

Figure 6 shows an analysis of the impact of 0.50 caliber on a graphene–dyneema armor. The total deformation induced in the plain dyneema armor during the impact is about 0.075462 m which is less than that of the deformation produced in plain dyneema fibers. The maximum stress developed is about 1816.8 MPa, and maximum principal stress is 1466.9 MPa which is induced in the ruptured bullet parts after the impact. The maximum stresses are developed in the bullet causing the bullet to rupture immediately after high impact with the armor.

Table 2 compares the after analysis ballistic properties of the material that is the total deformation, equivalent stress and maximum principal stress induced in the material.

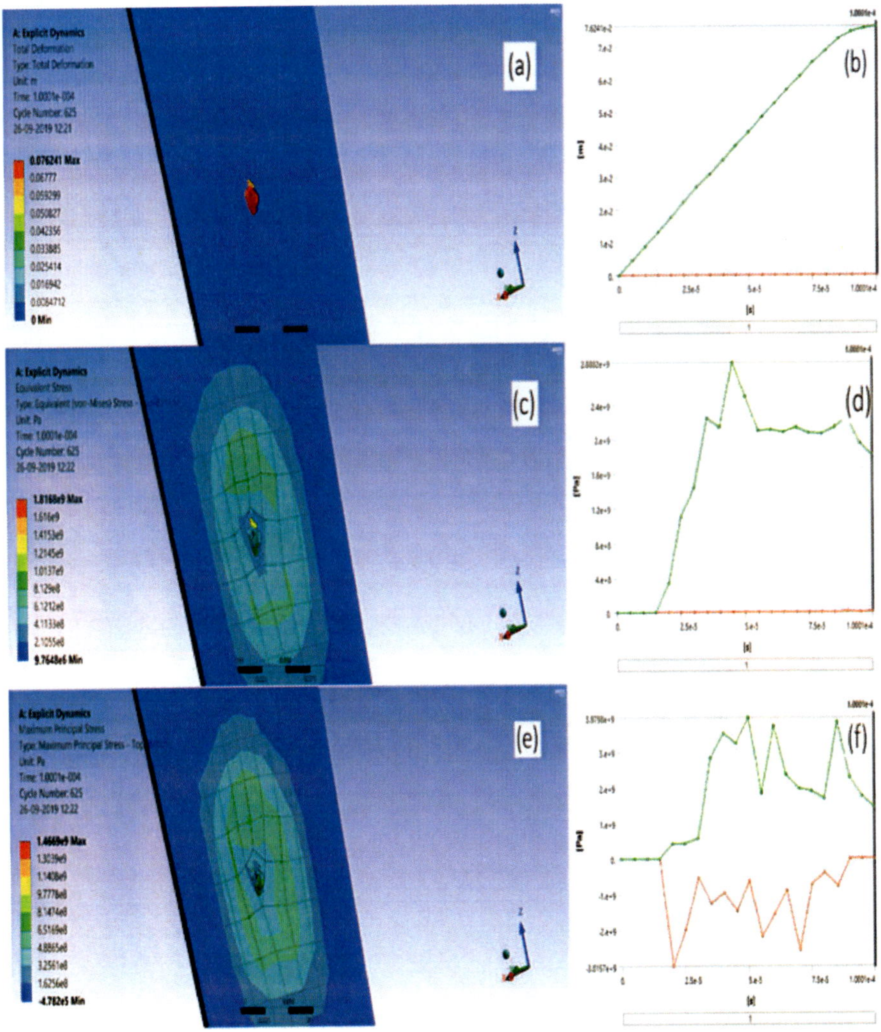


Fig. 5 Dynamic impact analysis of dyneema armor

## 4 Conclusion

The standards for bulletproofing were increasing every day due to the advancement of high-level ballistic technology. The need for high-strength low-weight armor is required to increase the soldier's mobility during combats. The conclusion of the above analysis can be summarized as

- Performance evaluation of plain and hybrid graphene–dyneema armor was done according to the IV level standard NIJ-0101.06. A finite element analysis of bullet

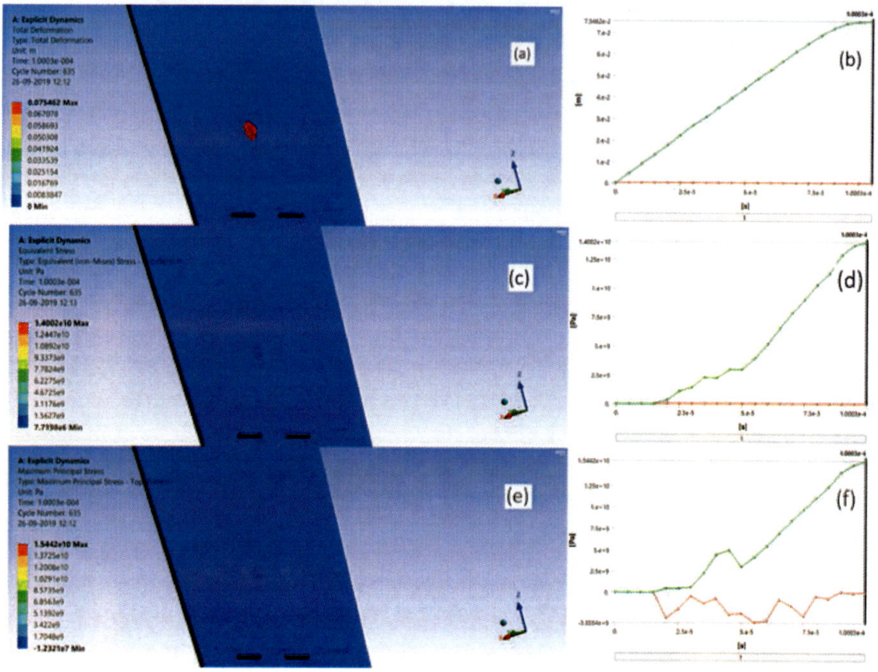


**Fig. 6** Dynamic impact analysis results of graphene inserted hybrid dyneema armor

**Table 2** Comparison of ballistic analysis properties of dyneema and graphene–dyneema armor

| Parameters | Dyneema armor | Graphene–dyneema armor |
|---|---|---|
| Equivalent stress | 1816 MPa | 4264 MPa |
| Total deformation | 0.076 m | 1.091 m |
| Maximum principal stress | 1466 MPa | |

impact analysis of multi-layered graphene–dyneema armor was done, and the results were plotted.

- The deformation induced in the graphene–dyneema laminate armor is less than that of plain dyneema armor, which means inserting a graphene laminate of 5 mm in between the dyneema layers absorbs the kinetic energy of the bullet more efficiently than that of the plain dyneema fiber.
- Also, the stresses induced in the bullet by hybrid armor were much greater than plain dyneema convincing that the hybrid armor material was stronger and stiffer than the plain dyneema fiber.
- Thus, graphene laminates can be sandwiched in between the fabrics of ballistic armor to increase the efficiency of the armor.

## References


1. Wu K-K, Chen Y-L, Yeh J-N, Chen W-L, Lin C-S (2020) Ballistic impact performance of SiC ceramic-dyneema fiber composite materials. Adv Mater Sci Eng 2020:1–9
2. Cline J (2019) The effect of in-plane properties on the ballistic response of polyethylene composites. In: Proceedings of the 2018 annual conference on experimental and applied mechanics, pp 251–253
3. Singh P, Malik V, Lather P, Analysis of composite materials used in bullet proof vests using fem technique. Int J Sci Eng Res 4(5):1789–1796 (ballistic impact)
4. Pundhir N, Goyal D, Singh P, Pathak H, Zafar S (2019) Numerical simulation of composite armor subjected to ballistic impact. In: Materials today proceedings, pp 34–41
5. Rolc S, Buchar J, Krestan JAN, Ridky R, Berankova I (2019) Numerical simulation of ballistic impact on different types of helmets
6. Fernado E, Niles S, Morrison A, Pranavan P, Godakanda I, Mubarak M, Design of a bullet-proof vest using shear thickening fluid
7. Grunwald C, Lässig T, Werff H, Heisserer U, Nguyen L, Riedel W (2017) Numerical sensitivity study of ballistic impact on UHMWPE composites
8. Werff H, Heisserer U (2016) High performance ballistic fibres: ultra-high molecular weight polyethylene (UHMWPE), pp 71–108
9. May M, Lässig T (2015) Hypervelocity impact simulation on hard ballistic composites
10. Werff H, Vlasblom M, Balzano L, Engels T, Heisserer U, Oosterlinck F, Coussens B (2014) New developments of the dyneema® ultra high molecular weight polyethylene fiber
11. Shánĕl V, Španiel M (2014) Ballistic impact experiments and modelling of sandwich armor for numerical simulations. Procedia Eng 79:230–237
12. Wang L, Kanesalingam S, Nayak R, Padhye R (2014) Recent trends in ballistic protection text. Light Ind Sci Technol 3:37
13. Heisserer U, Werff H, Hendrix J (2013) Ballistic depth of penetration studies in dyneema® composites, vol 2
14. Greenhalgh E, Bloodworth V, Iannucci L, Pope D (2013) Fractographic observations on dyneema (R) composites under ballistic impact. Compos Part A Appl Sci Manuf 44:51–62
15. Naik N, Kumar S, Dharmane R, Joshi M, Akella K (2013) An energy-based model for ballistic impact analysis of ceramic-composite armors. Int J Damage Mech 22:145–187
16. Kasano H (2011) Ballistic impact performance of composite plate with and without bonding
17. Jena P, Mishra B, Kumar K, Bhat T (2010) An experimental study on the ballistic impact behavior of some metallic armour materials against 7.62 mm deformable projectile. Mater Des 31:3308–3316

# Performance Analysis of Ball Bearing with Solid Contaminants Using Vibration Analysis

**K. A. Ibrahim Sheriff, V. Hariharan, and B. Varunesh**

**Abstract** Bearing is a machine element that constrains relative motion and reduces friction between moving parts to only the desired motion. These are very commonly used component in all rotating machineries. There are several reasons for the failure of bearing. The contamination of solid particles in bearing parts is one of the main causes for the bearing failure in machinery parts which lead to the catastrophic failure, less productivity and more downtime. Vibration measurement technique is used for the diagnosis of bearing defects. This work deals with the analysis on the effect of lubricant contamination with solid particles like Greensand in the 6205 type ball bearings using vibration signals. The contaminant particle sizes are considered as 75, 106 and 150 μm. The experimental work was done at different speeds such as 900, 1500 and 2100 rpm. The applied load on the bearing is 1, 3, 5 kg and no load. The experiment is carried out in a defect-free bearing. The test is carried out with solid contaminant at different concentration levels (5, 15 and 25%) under various load and speed conditions. Dewesoft acquisition software was used for acquiring vibration signals. RMS was the signal parameter considered for the analysis, and signals were compared for different contaminant concentrations.

**Keywords** Ball bearing · Solid particle · Condition monitoring signal

---

K. A. Ibrahim Sheriff (✉) · V. Hariharan · B. Varunesh
Department of Mechanical Engineering, Kongu Engineering College, Perundurai, Erode, Tamil Nadu, India
e-mail: ibusheriffmech786@gmail.com

V. Hariharan
e-mail: hariharanvag@gmail.com

B. Varunesh
e-mail: varunmech77@gmail.com

© The Editor(s) (if applicable) and The Author(s), under exclusive license to Springer Nature Singapore Pte Ltd. 2021
S. Mohan et al. (eds.), *Materials, Design, and Manufacturing for Sustainable Environment*, Lecture Notes in Mechanical Engineering,
https://doi.org/10.1007/978-981-15-9809-8_14

## 1 Introduction

Bearings are widely used in all rotating equipment. The bearings are let down under dynamic loading circumstance. Solid contaminants (SCs) in the lubricant majorly affect the performance of ball bearing (BB). The vibration level (VL) due to the bearing wear depends on the SC; the VL due to the bearing wear depends on the impurity piece. The VL due to the presence of elements was related to the vibration of the worn bearing as element attention increases [1]. When the contamination level (CL) increases in lubrication by small amount, vibration signatures and acceleration values also go on increasing. As the element size is augmented, the equivalent acceleration is also growing up to certain limit, and later, it starts getting decreased [2]. The amplitude level is depending on the CL in the BB, and the bearings are failed to rotate at 30 percent of CL [3]. Vibration investigation in time domain signal (TDS) can designate, and there is an atypical operation of a bearing and shows the trend of the growth in amplitude. While ductile impurities were used, the VL was greater, and as an element size rises, damage size also increased [4]. The VL increased with CL, tending to stabilize in a limit. As the element size is greater than before, the VL first improved and then reduced. Element relaxing effect was the feasible factor for VL decrease [5]. The vibration, stator current, acoustic emission and shock pulse method measurements performed on lubricant polluted bearings are substantially improved as CL and contaminant dimension growths [6]. A growth in the width of the outer race defect resulted in a growth in ratio between the extreme amplitude of the AE transient burst for the defect to the critical operational noise levels [7]. The severe growth in the value of counts for the smaller defect size indicates that this may serve as a good constraint for emerging error discovery in bearings, and for larger defect size, AE counts are incapable to deliver any data about the growth of the fault [8]. Acoustic emission TDS is the best suitable method for detection of hardness of contaminant particles. Small size particles generate a high AE pulse count level than larger size elements [9]. The vibration RMS and extreme amplitude values vary with fault state, and the rate of such changes highlights the greater sensitivity of the AE technique to early defect detection. The AE transitory ruptures might be allied to the defect source while the frequency domain signal (FDS) of vibration readings failed to identify the defect frequency or source [10]. The effect of 10% water by figure in a water-in-oil emulsion on the fixed performance of the hydrostatic bearing growths both the film thickness and friction torque. The film thickness of the bearing, increases with the speed and falls with the load, while the friction torque and the temperature increase with both speed and load [11]. The vibration signal is acquired for healthy and various simulated faulty bearings under different load conditions. The distributed faults are found through RMS and peak-peak in vibration signals. Also kurtosis is a sensitive parameter for bearing condition monitoring (CM) [12]. Vibration retort of the bearings to the defects on outer race, inner race and the rolling elements is examined. Kurtosis, one of the statistic indicators, is also estimated for different state of bearing [13]. When compared to conventional machine learning (ML) algorithms, deep learning algorithms can easily learn the most important features from data [14].

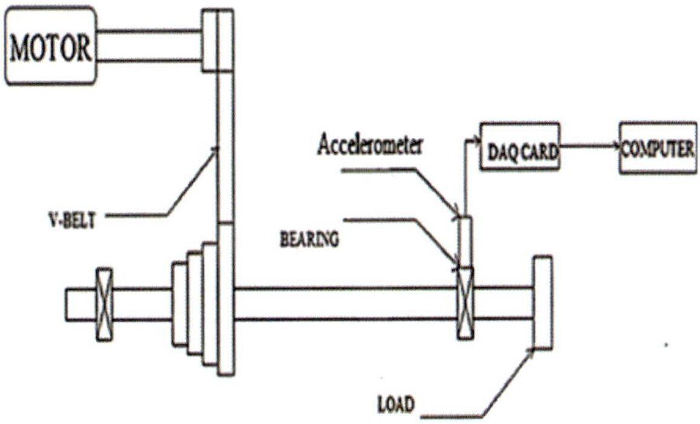


**Fig. 1** Experimental setup

With an increase of temperature in the contamination medium, the effect of solid particles on surface wear and the abrasive wear becomes very severe. The existence of solid elements in the lubricant has a dual effect. It increases the friction between the surfaces with a relative sliding and leads to severe abrasive wear [15].

The above literature analysis indicated the area of work on the vibration analysis was based on the wear due to contaminants and type of material (ductile and brittle). Some of the authors were concentrating on the type of method used for the analysis like TDS and FDS approach. The research work on the single particle at different concentration and particle size using TDS approach was not common in earlier cases.

## 2 Experimental Setup

The new setup is contrived as shown in Fig. 1 to test the bearing. Good bearings (GB) and contamination bearing (CB) are placed at either side of the shaft. The speed of the shaft was controlled by a motorized and pulley. The ball bearing type 6205 is used for the study, and system is run at different speeds such as 900, 1500 and 2100 rpm which are termed as $N_1$, $N_2$ and $N_3$. The loading measures are categorized into four which series from 0 to 5 kg. The AE signal is acquired using mike and data acquisition card (DAC) in Dewesoft software.

## 3 Results and Discussion

TDS of the GB with 5 kg load at 900, 1500 and 2100 rpm is shown in Fig. 2a–c, respectively. There are no fault frequencies in the signals. The peaks are present in the signals obtained from the good bearing. This is due to the increase of speed and load in the system.

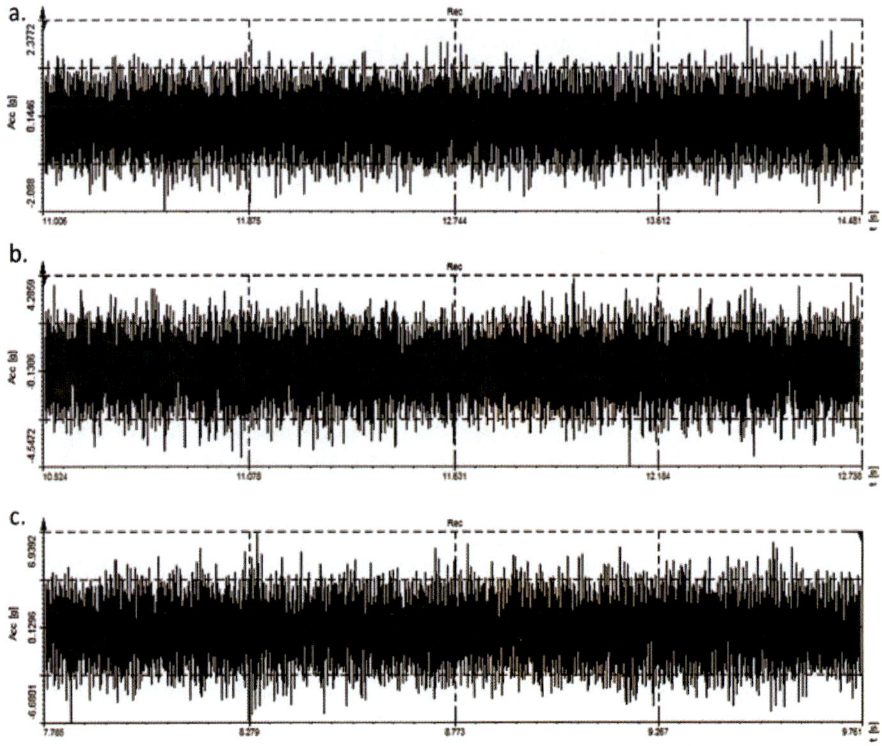

**Fig. 2** **a** TDS for GB for 5 kg load at 900 rpm **b** TDS for GB for 5 kg load at 1500 rpm **c** TDS for GB for 5 kg load at 2100 rpm

### *3.1 Comparison of GB with 5% Contaminated GB*

The comparison of GB signal with solid contaminated GB signal is done, at low (N1) to high speed ($N_3$). The RMS value increases as load and speed increases. When the CL of grease is increased, the signal parameters are slightly increased. The reason for slight variation in the bearing does not have any defect, the increase in value is due to contamination. The particle sizes 75, 106 and 150 μm are used in 5% CL.

The RMS is the root-mean-square value of the signal. When the GB is compared with contamination, at low ($N_1$) and high speed ($N_3$), the RMS value increases in smaller value, and at intermediate speed ($N_2$), the RMS value increases in higher value which is shown in Fig. 3. This indicates that grease contaminants are sensitive to the intermediate speed. When concentration increases from 0 to 5%, the RMS value increases at lower speed itself which is shown in Fig. 3. At 5% CL, the value increases with the particle size. At the higher speed and load, the rate of increase is high. This is due to presence of higher particle size.

Figure 3 shows the comparison of RMS values of GB without and with SC at CL of 5% under particle of sizes 75, 106 and 150 μm. At low concentration of 5%,

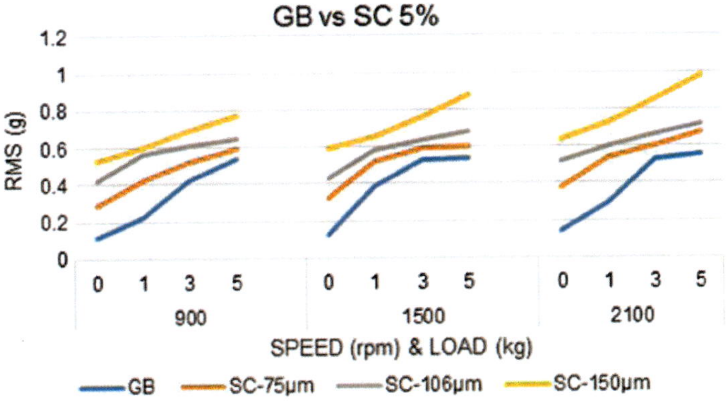


**Fig. 3** Vibration RMS of greensand at 5% concentration level

the RMS value variation was less and similar. At 5% CL, at low speed (900 rpm), the RMS range for good bearing at no-load condition and bearing contaminated with 150 μm running with 5 kg load is 0.1127–0.7733 g. At the intermediate speed (1500 rpm), the RMS range for GB at no-load condition and bearing contaminated with 150 μm running with 5 kg load is 0.1256–0.8833 g. Similarly, for bearings running at high speed (2100 rpm), the RMS range for GB at no-load condition and bearing contaminated with 150 μm running with 5 kg load is 0.1453–0.9836 g.

## 3.2 *Comparison of GB with 15% Contaminated GB*

The CL of the contaminants was increased from 5 to 15%, in order to study the bearing's behaviour at 15%. As we have seen in the 5% bearings, the RMS value increases as the speed and the load were increased gradually. The rate of increase of the RMS value increases with the speed and load. This is clearly visible in the graph as shown in Fig. 4.

Figure 4 shows the comparison of RMS values of GB without and with solid contamination at CL of 15% under particle of sizes 75, 106 and 150 μm. When compared to 5% CL, 15% CL shows a significant change in the parameter ranges. At 15% CL, at low speed (900 rpm), the RMS range for GB at no-load condition and bearing contaminated with 150 μm running with 5 kg load is 0.1127–0.7935 g. At the intermediate speed (1500 rpm), the RMS range for GB at no-load condition and bearing contaminated with 150 μm running with 5 kg load is 0.1256 g – 0.8937 g. Similarly, for bearings running at high speed (2100 rpm), the RMS range for GB at no-load condition and bearing contaminated with 150 μm running with 5 kg load is 0.1453–1.004 g.

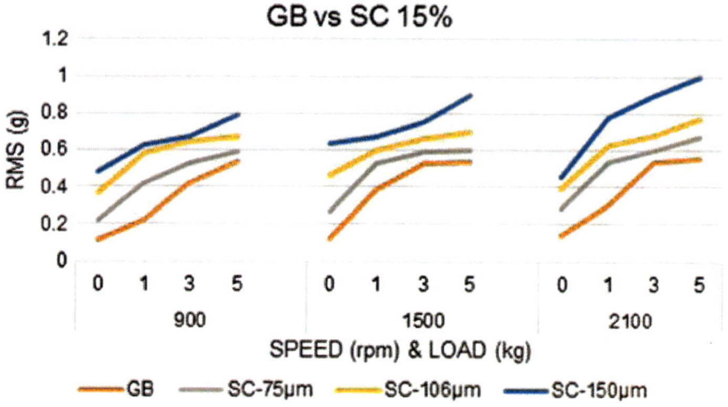


**Fig. 4** Vibration RMS of greensand at 15% concentration level

## 3.3 Comparison of GB with 25% Contaminated GB

The CL of the contaminants was increased from 15 to 25%, in order to study the bearing's behaviour at 25%. As we have seen in the 15% bearings, the RMS value increases as the speed and the load were increased gradually. The rate of increase of the RMS value increases with the speed and load; this is clearly visible in the graph as shown in Fig. 5.

Figure 5 shows the comparison of RMS values of GB without and with SC at CL of 25% under particle of sizes 75, 106 and 150 μm. Among these concentration levels, 25% shows the maximum range between the RMS values. At 25% CL, at low speed (900 rpm), the RMS range for GB at no-load condition and bearing contaminated with 150 μm running with 5 kg load is 0.1127–0.8736 g. At the intermediate speed (1500 rpm), the RMS range for GB at no-load condition and bearing contaminated with 150 μm running with 5 kg load is 0.1256–0.9557 g. Similarly, for bearings

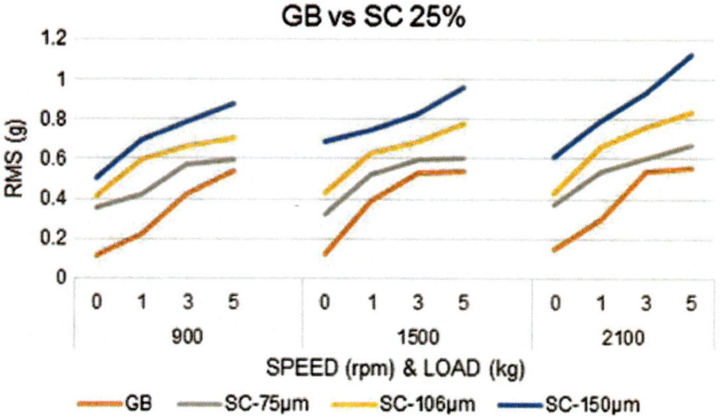


**Fig. 5** Vibration RMS of greensand at 25% concentration level

running at high speed (2100 rpm), the RMS range for GB at no-load condition and bearing contaminated with 150 μm running with 5 kg load is 0.1453–1.1235 g.

All the above discussion said about the increase in the vibration of bearing with the increase in the particle size, speed and load. The reason for the above increase was because of the wear on the surface of the bearing caused by the contaminant. The literature work also shows the same reason for the increase in RMS values [1, 7].

## 4 Conclusions

The defective bearing signals are compared with GB for various speed and load conditions for 5, 15 and 25% levels of different particle sizes. The results found from the experimental work are as follows. The RMS value is increased from 17.1% to 78.2% when compared to the GB, with the increase of particle size, speed and load at 5% CL. Similarly, for 15% and 25%, they were from 19.5% to 81.3% and from 22.1% to 103.6%, respectively. The above percentage comparison also shows an increase in the percentage value as the CL is increased. From the above analysis, the energy level is high due to wear flaws on the superficial of bearing parts by contaminants. The superior element size and CL of impurity affect the bearing concert heavier than the other two element sizes.

## References


1. Maru MM, Castillo RS, Padovese LR (2007) Study of solid contamination in ball bearings through vibration and wear analyses. Tribol Int 40:433–440
2. Mahajan OL, Utpat AA (2012) Study of effect of solid contaminants in the lubricant on ball bearings vibration. Int J Instrum Cont Autom (IJICA) 1(3):4
3. Srinivasan PSS, Hariharan V (2010) Condition monitoring studies on ball bearings considering solid contaminants in the lubricant. Proc Inst Mech Eng Part C: J Mech Eng Sci 224:1727–1748
4. Koulocheris D, Stathis A, Costopoulos T, Tsantiotis D (2014) Experimental study of the impact of grease particle contaminants on wear and fatigue life of ball bearings. Eng Fail Anal 39:164–180
5. Singotia G, Jain A (2013) Effect of solid contamination in ball bearings-a review. Int J Current Res Rev 5(12):119–124
6. Tandon N, Yadava G, Ramakrishna K (2007) A comparison of some condition monitoring techniques for the detection of defect in induction motor ball bearings. Mech Syst Signal Process 21:244–256
7. Al-Dossary S, Hamzah R, Mba D (2009) Observations of changes in acoustic emission waveform for varying seeded defect sizes in a rolling element bearing. Appl Acoust 70:58–81
8. Choudhury A, Tandon N (2000) Application of acoustic emission technique for the detection of defects in rolling element bearings. Tribol Int 33:39–45
9. Miettinen J, Andersson P (2000) Acoustic emission of rolling bearings lubricated with contaminated grease. Tribol Int 33:777–787
10. Al-Ghamd AM, Mba D (2006) A comparative experimental study on the use of acoustic emission and vibration analysis for bearing defect identification and estimation of defect size. Mech Syst Signal Process 20:1537–1571

11. Harika E, Bouyer J, Fillon M, Hélène M (2013) Measurements of lubrication characteristics of a tilting pad thrust bearing disturbed by a water-contaminated lubricant. Proc Inst Mech Eng, Part J: J Eng Tribol 227:16–25
12. Kulkarni S, Bewoor A (2016) Vibration based condition assessment of ball bearing with distributed defects. J Meas Eng 4(2):87–94
13. Dube Abhinay V, Dhamande L, Kulkarni P (2013) Vibration based condition assessment of rolling element bearings with localized defects. Int J Sci Technol Res 2(4):149–155
14. Hoang D-T, Kang HJ (2019) A survey on deep learning based bearing fault diagnosis. Neurocomputing 335:327–335
15. Sari MR, Haiahem A, Flamand L (2007) Effect of lubricant contamination on gear wear. Tribol Lett 27:119–126

# Water Hyacinth (*Eichhornia Crassipes*) Natural Fiber Composite Properties—A Review

**A. Ajithram, J. T. Winowlin Jappes, and I. Siva**

**Abstract** This review paper briefly discussed the water hyacinth plant fiber production and compare to some chemical and other types of treatments of other fibers and chemical treatment values, improvement of mechanical, thermal properties at the same time engineering and commercial applications and their surface topography. Nowadays, all of the researchers were moved and focused upon their doing and upcoming works in cost-effective manner and considered the environmental aspects also. So only they were choosing the natural fibers instead of using synthetic fibers. Especially now most of the works involved the water hyacinth plant fibers, and this paper involves detail analyze with some researcher's works belongs to the water hyacinth plant in chemical treatments and alkali treatments and before and after immersion some chemical treatments like LDPE and compare to other natural fibers also. In all of their works, the final results may be improved in certain percentage of the treatments, and simultaneously, the properties improved because these chemical treatments maintained and become to better their property in effective manner at the same time attained the good results. A lot of testing techniques and test machines were used in these fields like SEM, TEM, XRD, FTIR and some other tests also.

**Keywords** Water hyacinth natural fiber · Mechanical strength

A. Ajithram · J. T. Winowlin Jappes (✉) · I. Siva
Department of Mechanical Engineering, Kalasalingam Academy of Research and Education, Virudhunagar, Tamil Nadu, India
e-mail: winowlin@klu.ac.in

A. Ajithram
e-mail: ajithram867@gmail.com

I. Siva
e-mail: isiva@klu.ac.in

© The Editor(s) (if applicable) and The Author(s), under exclusive license to Springer Nature Singapore Pte Ltd. 2021
S. Mohan et al. (eds.), *Materials, Design, and Manufacturing for Sustainable Environment*, Lecture Notes in Mechanical Engineering,
https://doi.org/10.1007/978-981-15-9809-8_15

# 1 Introduction

Water hyacinth was in past ages from South America, and it is a universally present resource for various uses. Now, the water hyacinth plants were available from all over the world especially in lake, pond and areas [1, 2]. The air is fully filled in stem and leaves of these plants, so they were bear the water surfaces. These plants contain several advantageous characteristics at the same time disadvantages also. From the advantage's point of view, it has been used to local people in variety of ways and used to industries' several treatments also [3]. At the disadvantage manner, they were highly grown-up type of plants. So, it covers all over the lake, pond water surface in one or two days. After covering, the sun race and the brig light will not go beyond the water [4]. But, on the water surface, a lot of the fishes and so many domestic organisms were lived. They were directly affected because of water hyacinth plant. So mainly, the local business peoples, water departments and fisheries departments were heavily affected by these types and these families of various plants. So, all over the world, the local government were spent more number of money to remove and to cultivate the water hyacinth plant. In the winter times, these plants occupy more number of water densities and areas, and this is the main reason that the government have to take immediate action to cultivate these water hyacinth plants [5]. In the application manner, these plants are more useful to local people in various applications like ornamental, local business, etc. in some cosmetic, decorative items and house-holding devices and particleboard analyzers, and in industry sector, wastewater treatment and biomass production, these water hyacinth plants were used. And also these pants were corporate with cement materials used as a building agent. In the upcoming future, the engineering agenda is the sustainable development on the human safety and cost. The key point of the main idea is renewable source which was combined with the local materials. The novel approach is an essential one to the natural fiber composites. Consider the above concept, and the water hyacinth plant fiber is one of the important sides of a waste management topic [6, 7]. At the same time, these plants were marked by the world's top 100 dangerous plants lost. These plants were mainly established in tropical and subtropical areas. From that, so many literature reviews and other sources briefly explained these plant extraction processes about per day in the different areas. And then the food and agricultural departments spent so many dollars for these plants [8]. In India, mainly Kerala state spent more number of money to remove these types of the invasive plants. This hyacinth plant creates so many nuisance problem to the environment and near by peoples. In the technical consideration, the water hyacinth plant fiber has attained very low density [9]. These low density properties were directly proportional to the structure placed inside the plant. This low density properties plays a vital role in many natural fibers. Compared to the other types of the natural fiber, the water hyacinth fiber has gained very low density property. These low density properties were important to the thermal properties. Low density properties of water hyacinth plant is directly related to thermal properties. These water hyacinth plants were used as the superabsorbent material. The superabsorbent characteristic means they acquaint in water and survive in a gel state, and also they engage the large amount of water [10, 11]. These water hyacinth (*Eichhornia Crassipes*) plants were treated in so many ways like water treatment manufacturing of ethanol, production of biogases, local business-related paper making, man of craft items, etc. Nowadays, water hyacinth fiber is a modern approach to get the affluence out of the waste. The water hyacinth plant was the good water-absorbent properties compared to others like cotton, etc. These fibers contain very good tensile properties, and at the same time, they have more C.V% value. So these are the main reasons for the water hyacinth fiber which was not suitable for tensile-related products [12–14]. Now, they are used mainly in disposal products and decorative works. These plants were defined and considered the waste plant, but they have their own very useful properties and manufacturing so many different products [15].

## 2 Processing of Water Hyacinth Plant Fiber [15]

Initially, these water hyacinth plants were found majorly on ponds and canals and the local water bodies. Once these plants were identified, then immediately the plant stem and petioles were removed from these plants. After that, these plant stems and petioles were dried into sunlight 1 week or 10 days. After that, these dried petioles and stems were chopped into the small amount of water hyacinth plant fibers as per the guides of Fig. 1.

### *2.1 Natural Fiber Properties [16, 17, 19]*

See Table 1.

## 3 Mechanical Testing

### *3.1 Tensile Strength Properties*

One of the most ancient and conventional and then rottenly used methods of measured properties in composite materials is the tensile strength. That means these natural fiber composite materials were restricted to the tensile loads. These strengths were measured the computerized universal testing machine, and also the standard ASTM D 638 10 was strictly followed. The average crosshead speed of universal testing machine is 5 mm/min [16].

Figure 2 shows the various percentages of fibers with respect to several matrix materials. And then, it will reach some different values on different types of the

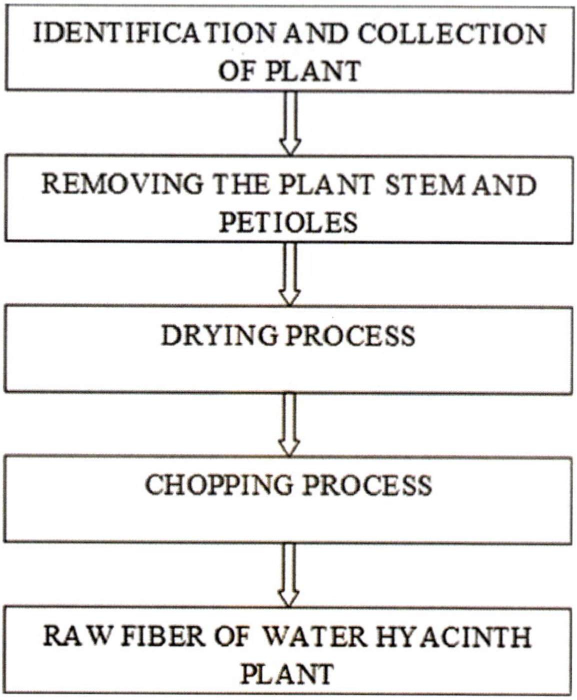


**Fig. 1** Process of water hyacinth fiber

**Table 1** Natural fiber intermolecular properties

| Fiber | Cellulose | Hemi cellulose | Lignin | Pectin | Wax |
|---|---|---|---|---|---|
| Sisal | 50–75 | 10–14 | 8–11 | 1 | 3 |
| Jute | 55 | 12 | 10 | 0–4 | 0–5 |
| Hemp | 68 | 16–18 | 3–5 | 0–8 | 0–7 |
| Flax | 64–84 | 16–18 | 3–5 | 0–8 | 0–7 |
| Bagasse | 50 | 25 | 25 | – | 0–8 |
| Water hyacinth | 60 | 8 | 17 | – | – |

matrix materials. The above figure shows the researchers' different types of works. In these works, the sisal, coir and water hyacinth plant fibers were taken. They were treated into some of the matrix materials like epoxy, polyester and polypropylene. And all the trails were attained some important properties [17, 18]. In these, the above diagram shows that the 25 wt% fibers with epoxy resin matrix and 35 wt% fibers were added to the polyester resin matrix material and 30 wt% fibers with respect to the polypropylene resin matrix materials of sisal, coir and water hyacinth fibers, and simultaneously, they attained some tensile strength and modulus values. Epoxy, polyester, polypropylene matrix material on this diagram values of the fiber volume fractions only improving their strength of the composites. After that adding more percentage of the fibers, these tensile strengths and modulus values decreased

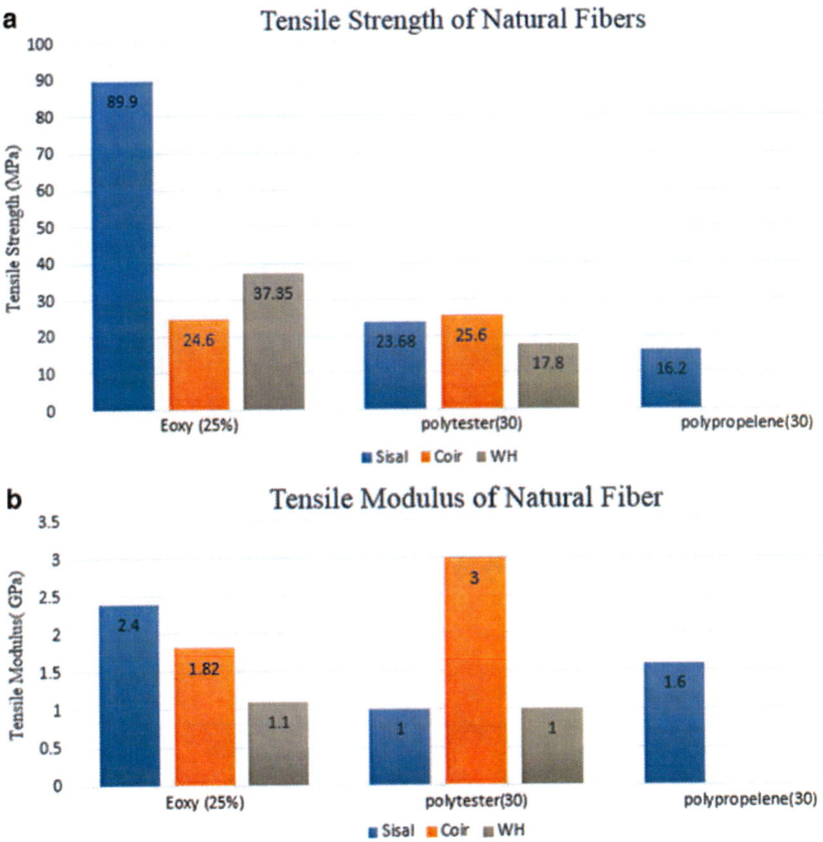


**Fig. 2** **a**, **b** Tensile properties of natural fibers

rapidly. So these values of sisal, coir and water hyacinth fiber were the optimum value. Comparing the three different matrix materials, the sisal fiber optimum tensile value 89 MPa attained the epoxy matrix with 25 wt% of fiber volume fractions and the coir fiber attained the optimum tensile value 2368 MPa and the water hyacinth fibers attained the good results in epoxy resin matrix and gained the 16.2 MPa value. But the tensile modulus was completely different compare to tensile strength values. The sisal and coir optimum tensile modulus was attained in polyester composite with respect to the 2.4 and 3 GPa [18, 19].

## 3.2 *Flexural Strength Properties*

Figure 3 clearly explained the various percentages of fibers added to some different types of matrix materials. And then, it will reach some different values on different types of the matrix materials. The above figure shows the researchers' different

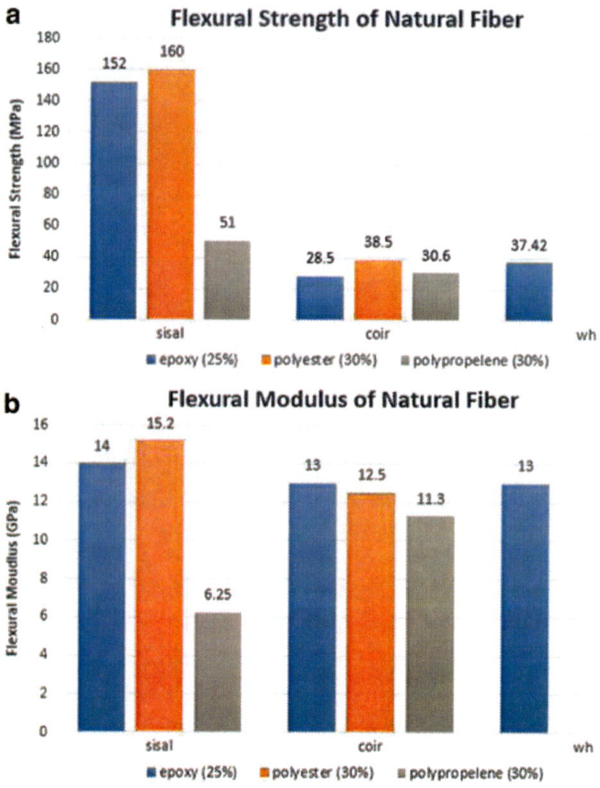


**Fig. 3** **a, b** Flexural properties of natural fibers

types of works. In these works, the sisal, coir and water hyacinth plant fibers were taken. They were treated into some of the matrix materials like epoxy, polyester and polypropylene. And all the trails were attained some important amount of properties. In these, the above diagram shows that the 25 wt% fibers with epoxy resin matrix and 35 wt% fibers were added to the polyester resin matrix material and also 30 wt% fibers were added to the polypropylene resin matrix materials of sisal, coir and water hyacinth fibers, and simultaneously, they attained the flexural strength and flexural modulus values [19, 20]. Three matrix materials of this diagram values of the fiber volume fractions only improving their strength of the composites. After that adding more percentage of the fibers, these flexural strengths and modulus values decreased rapidly. So these values of sisal, coir and water hyacinth fibers value were the optimum value. Comparing the three different matrix materials, the sisal fiber optimum flexural value 152 MPa attained the epoxy matrix with 25 wt% of fiber volume fractions and the coir fiber attained the optimum flexural value 160 MPa and the water hyacinth fibers attained the good results in epoxy resin matrix and gained the 37.42 MPa value. The flexural modulus values always high and gain more values in epoxy matrix only compare to other resin matrix. The sisal and coir optimum flexural modulus was attained in polyester composite with respect to the 14 and 15.2 GPa [20–23].

From the pervious works the flexural test is conducted the three point bending fixture on INSTON 3382 universal testing machine with the standard format of ASTM D 790 and then flexural test were taken with the 5 mm/min cross head speed. The dimensions of these test specimens are 127 mm × 12.7 mm × 4 mm. In these tests, the five specimens are taken to test, and then the average value was calculated and reported the flexural strength of the specimen [24–26].

## 3.3 Thermal Testing

In the natural fibers, commonly the mechanical and thermal testing was taken. Most thermal testing, namely the differential scanning calorimetry (DSC) and thermo-gravimetric analysis (TGA), were taken. The DSC is one of the thermos analytical technique to measure the function of temperature of the samples. The TGA is to measure the mass of the sample with respect to the time on the observation of temperature change. These thermal tests were usually taken by finding the thermal stability and thermal degradation of the fiber composite [27–29].

Earlier, researchers are conducted one test in a NCO—POLYOL is a filler content with the LDPE and water hyacinth fiber. In this experiment, DSC and TGA tests were taken. These tests have certain two types of fiber, one is the modified LDPE/WH fiber and another one is the unmodified LDPE/WH fiber. The fillers were adding both the modified and the UN modified fibers in the TGA tests. These two types of residual mass were increased while increasing the filler content. At the final stage of the experiment, the modified fiber gains the ore temperature and more final decomposition temperature (FDT) compared to the UN modified powder. In DSC tests, percentage of crystallinity was observed. On this test, the modified fiber filler (25%) content also has the higher value compared to unmodified fiber filler content. Finally, the modified fiber TGA residual mass of 25% filler content was 2.39%. The crystallinity percentage of fillers was 26.20% [20].

## 3.4 Surface Testing

These review papers focused on most of the research works based on the water hyacinth plant fiber. In these fibers, all the experiments finally end up with checking the surface topography. There are several techniques used to find the surface morphology studies. Most of the works checking the SEM and TEM high-quality images are doing these fibers with the particular interior nanometer size [30]. As all the work results with the different treatments of the fibers were improved, there surface properties were compared. At the same time, these water hyacinth fibers were machined, and also undergoing some different chemical or other treatments, these surface smoothness and other properties were improved in effective manner.

The porosity of the surface was decreased and gets lower value compared to other types of the fibers [31–33].

## 4 Commercial and Engineering Application

Mainly, these water hyacinth plants were in an aquatic weed and hydrophilic nature fastest growing characterized, and they were used in wastewater treatments, and also it is used in industrial pollution removal function [34, 35]. These plant fibers and plants were produced in renewable and sustainable energy production. In these fiber plants, the water treatment compelled with the biogas systems. In industries' wastewater treatments, they were combined with the rhizofiltration systems, and the contaminated water was removed. And these plants are mainly used for some agricultural purposes [36]. In the aspects of commercial, they were a lot of uses in surrounding peoples of local ponds and lakes. In these water hyacinth plants, a lot of aesthetic items were made, and also some commercial household products were produced. Even though this water hyacinth plant is used to produced particle board production aspects [37, 38]. The main focus of many researchers in that water hyacinth field means they were cost-effective because most of the state governments considered these plants as wastes only. So nowadays, many of the researchers attained the good research results and commercial and engineering applications used in water hyacinth plants [39–41]. These plant fibers were mainly used in textile industries and used in clothes and handicrafts and fiberboards, and some construction industries were used these water hyacinth plant fibers [41].

## 5 Discussion and Conclusion

Nowadays, the modern world found a lot of issues to developing new technologies treated in solid wastes. In these waste management aspects, the water hyacinth plant was the main issue in all over the world. Based upon the hydrophilic nature, they occupy more places and densities in the water bodies. The researchers were focused on the way to convert these plants and improve these plants in a scientific manner. There are so many commercial product is produced to this water hyacinth aquatic plants. So, nowadays, scientific aspects of the properties of water hyacinth plants were studied and improved by many researchers. The water hyacinth plant getting opportunity is plenty of available in nature. This review paper is clearly analysed the mechanical, thermal properties of water hyacinth natural fiber. Past to nowadays, many researchers conduct varying tests from the water hyacinth fiber. These fibers were analyzed, treated, untreated, modified, unmodified, before and after and also some various treatment aspects. Through this review all the results are clearly explained the raw material of water hyacinth fiber is provided good strength to the both mechanical and thermal properties. If any researcher includes the fiber in some treatments like adding fillers, etc., then definitely the fibers' property range will become growing at a certain level, then the treatment was higher, and then their properties were decreased. So, all the treatments are done in water hyacinth fiber, the properties were strongly proved in a certain way and a certain range.

## References


1. Bledzki AK, Reihmane S, Gassan J (1998) Thermoplastics reinforced with wood fillers: a literature review. Polym Plast Technol Eng 37(4):451–468. https://doi.org/10.1080/03602559808001373
2. Singha AS, Thakur VK (2014) Physical, chemical and mechanical properties of *Hibiscus sabdariffa* fiber/polymer composite. Int J Polym Mater Polym Biomater 37–41. https://doi.org/10.1080/00914030802639999
3. Methacanon P, Weerawatsophon U, Sumransin N, Prahsarn C, Bergado DT (2010) Properties and potential application of the selected natural fibers as limited life geotextiles. Carbohydr Polym 82(4):1090–1096. https://doi.org/10.1016/j.carbpol.2010.06.036
4. Taylor P, Abral H, Putra H, Sapuan SM, Ishak MR (2014) Effect of alkalization on mechanical properties of water hyacinth fibers-unsaturated polyester composites effect of alkalization on mechanical properties of water hyacinth fibers-unsaturated polyester composites. Polym Plast Technol Eng 37–41. https://doi.org/10.1080/03602559.2012.748804
5. Natural fibers, plastics and composites (2004). https://doi.org/10.1007/978-1-4419-9050-1
6. Rokbi M, Osmani H, Imad A, Benseddiq N (2011) Effect of chemical treatment on flexure properties of natural fiber-reinforced polyester composite. Procedia Eng 10:2092–2097. https://doi.org/10.1016/j.proeng.2011.04.346
7. Nekkaa S, Guessoum M, Grillet AC, Haddaoui N (2012) Mechanical properties of biodegradable composites reinforced with short *Spartium junceum* fibers before and after treatments. Int J Polym Mater Polym Biomater 61(13):1021–1034. https://doi.org/10.1080/00914037.2011.617332
8. Mohanty AK, Misra M (1995) Studies on jute composites—a literature review. Polym Plast Technol Eng 34(5):729–792. https://doi.org/10.1080/03602559508009599
9. Gañán P, Mondragon I (2004) Influence of compatibilization treatments on the mechanical properties of fique fiber reinforced polypropylene composites. Int J Polym Mater Polym Biomater 53(11):997–1013. https://doi.org/10.1080/00914030490516648
10. Goud G, Rao RN (2011) The effect of alkali treatment on dielectric properties of *Roystonea regia*/ epoxy composites. Int J Polym Anal Charact 16(4):239–250. https://doi.org/10.1080/1023666X.2011.570039
11. Abral H, Gafar MF, Andriyanto H et al (2012) Alkali treatment of screw pine (*Pandanus odoratissimus*) fibers and its effect on unsaturated polyester composites. Polym Plast Technol Eng 51(1):12–18. https://doi.org/10.1080/03602559.2011.593090
12. Vilay V, Mariatti M, Mat Taib R, Todo M (2008) Effect of fiber surface treatment and fiber loading on the properties of bagasse fiber-reinforced unsaturated polyester composites. Compos Sci Technol 68(3–4):631–638. https://doi.org/10.1016/j.compscitech.2007.10.005
13. Asdrubali F, D'Alessandro F, Schiavoni S (2015) A review of unconventional sustainable building insulation materials. Sustain Mater Technol 4:1–17. https://doi.org/10.1016/j.susmat.2015.05.002
14. Salas-Ruiz A, del Mar Barbero-Barrera M (2019) Performance assessment of *Water hyacinth*–cement composite. Constr Build Mater 211:395–407. https://doi.org/10.1016/j.conbuildmat.2019.03.217
15. Zhou W, Zhu D, Langdon A, Li L, Liao S, Tan L (2009) The structure characterization of cellulose xanthogenate derived from the straw of *Eichhornia crassipes*. Bioresour Technol 100(21):5366–5369. https://doi.org/10.1016/j.biortech.2009.05.066

16. Kumar A, Srivastava A (2017) Preparation and mechanical properties of jute fiber reinforced epoxy composites. Ind Eng Manag 06(04). https://doi.org/10.4172/2169-0316.1000234
17. Obele C (2017) Mechanical properties of coir fiber reinforced epoxy resin composites for helmet shell. Ind Eng Lett 5(7)
18. Joseph PV, Joseph K, Thomas S (2002) Short sisal fiber reinforced polypropylene composites: the role of interface modification on ultimate properties. Compos Interfaces 9. https://doi.org/10.1163/156855402760116094
19. Kursun A, Ferreira LFP, Bayraktar E, Miskioglu I (2018) Mechanics of composite and multi-functional materials, vol 7, pp 225–231. https://doi.org/10.1007/978-3-319-63408-1
20. Gupta MK, Srivastava RK (2014) Tensile and flexural properties of sisal fibre reinforced epoxy composite: a comparison between unidirectional and mat form of fibres. Procedia Mater Sci 5:2434–2439. https://doi.org/10.1016/j.mspro.2014.07.489
21. Tan L, Zhu D, Zhou W, Mi W, Ma L, He W (2008) Preferring cellulose of *Eichhornia crassipes* to prepare xanthogenate to other plant materials and its adsorption properties on copper. Bioresour Technol 99(10):4460–4466. https://doi.org/10.1016/j.biortech.2007.08.022
22. Sundari MT, Ramesh A (2012) Isolation and characterization of cellulose nanofibers from the aquatic weed water hyacinth—*Eichhornia crassipes*. Carbohydr Polym 87(2):1701–1705. https://doi.org/10.1016/j.carbpol.2011.09.076
23. Supri AG, Ismail H (2010) The effect of NCO-polyol on the properties of low-density polyethylene/water hyacinth fiber (*Eichhornia crassiper*) composites. Polym Plast Technol Eng 49(8):766–771. https://doi.org/10.1080/03602551003749569
24. Athijayamani A, Thiruchitrambalam M, Natarajan U, Pazhanivel B (2009) Effect of moisture absorption on the mechanical properties of randomly oriented natural fibers/polyester hybrid composite. Mater Sci Eng A 517(1–2):344–353. https://doi.org/10.1016/j.msea.2009.04.027
25. Sapuan SM, Pua FL, El-Shekeil YA, Al-Oqla FM (2013) Mechanical properties of soil buried kenaf fibre reinforced thermoplastic polyurethane composites. Mater Des 50:467–470. https://doi.org/10.1016/j.matdes.2013.03.013
26. Padmanabhan RG, Arun N, Kolli Bala Sivarama Reddy S (2016) Investigation of mechanical behavior of water hyacinth fiber/polyester with aluminium powder composites. Int J Appl Innov Eng Manag 5(2):56–62
27. Abral H, Kadriadi D, Rodianus A et al (2014) Mechanical properties of water hyacinth fibers—polyester composites before and after immersion in water. Mater Des 58:125–129. https://doi.org/10.1016/j.matdes.2014.01.043
28. Ismail H, Edyham MR, Wirjosentono B (2002) Bamboo fibre filled natural rubber composites: the effects of filler loading and bonding agent. Polym Test 21(2):139–144. https://doi.org/10.1016/S0142-9418(01)00060-5
29. Adhikary KB, Pang S, Staiger MP (2008) Dimensional stability and mechanical behaviour of wood–plastic composites based on recycled and virgin high-density polyethylene (HDPE) 39:807–815. https://doi.org/10.1016/j.compositesb.2007.10.005
30. Aji IS, Zainudin ES, Abdan K, Sapuan SM, Khairul MD (2013) Mechanical properties and water absorption behavior of hybridized kenaf/pineapple leaf fibre-reinforced high-density polyethylene composite. J Compos Mater 47(8):979–990. https://doi.org/10.1177/0021998312444147
31. Bhanawat PD, Patil VM (2017) A review on advance material with jute fiber-polyester reinforced composites. Int J Text Eng Process 3:24–30
32. Morreale M, Scaffaro R, Maio A, La Mantia FP (2008) Effect of adding wood flour to the physical properties of a biodegradable polymer. Compos Part A Appl Sci Manuf 39(3):503–513. https://doi.org/10.1016/j.compositesa.2007.12.002
33. Gozdecki C, Wilczyński A, Kociszewski M, Zajchowski S (2014) Properties of wood–plastic composites made of milled particleboard and polypropylene. Eur J Wood Wood Prod 73(1):87–95. https://doi.org/10.1007/s00107-014-0852-2
34. Supri A, Lim B (2009) Effect of treated and untreated filler loading on the mechanical, morphological, and water absorbent properties of water hyacinth fibers-low density polyethylene composites. J Phys Sci 20(2):85–96

35. Xie Y, Hill CAS, Xiao Z, Militz H, Mai C (2010) Silane coupling agents used for natural fiber/polymer composites: a review. Compos Part A Appl Sci Manuf 41(7):806–819. https://doi.org/10.1016/j.compositesa.2010.03.005
36. Zadorecki P, Flodin P (1986) Surface modification of cellulose fibers III. Durability of cellulose–polyester composites under environmental aging. J Appl Polym Sci 31(6):1699–1707. https://doi.org/10.1002/app.1986.070310616
37. Gunnarsson CC, Petersen CM (2007) Water hyacinths as a resource in agriculture and energy production: a literature review. Waste Manag 27(1):117–129. https://doi.org/10.1016/j.wasman.2005.12.011
38. Rahmawati W, Haryanto A, Suharyatun S (2018) Development of biodegradable board using water hyacinth (*Eichornia crassipes*). Int J Environ Agric Biotechnol 3(1):170–174. https://doi.org/10.22161/ijeab/3.1.21
39. Wang Z, Calderon MM (2012) Environmental and economic analysis of application of *Water hyacinth* for eutrophic water treatment coupled with biogas production. J Environ Manage 110:246–253. https://doi.org/10.1016/j.jenvman.2012.06.031
40. Gajalakshmi S, Abbasi SA (2002) Effect of the application of water hyacinth compost/vermicompost on the growth and flowering of *Crossandra undulaefolia*, and on several vegetables. Bioresour Technol 85(2):197–199. https://doi.org/10.1016/S0960-8524(02)00096-2
41. Rezania S, Ponraj M, Fadhil M (2015) The diverse applications of water hyacinth with main focus on sustainable energy and production for new era: an overview. Renew Sustain Energy Rev 41:943–954. https://doi.org/10.1016/j.rser.2014.09.006

# Automatic Body Posture Corrector for Spinal Cord Patients

**V. G. Pratheep, E. B. Priyanka, S. Thangavel, K. Heenalisha, M. Ariya Manickam, and A. P. Logaram**

**Abstract** In the human comfort, the spine dislocation paves most important unhealthy symptoms which leads to greatest worst impact on the bone spur. Because the change in the cervical spine is the starting point till reaches the lumbar spine where it could produce degenerative breakout in the position changes due to the improper body posture views. The present work mainly focuses on the patients who are all facing critically this body posture problem. Our idea is to create a belt containing all the controls and indicators that are needed to correct body posture. By using Arduino gyroscope 6050 to get an exact change in body posture, using this output it is planned to make a code to run the vibrator to alert the people who wear the belt through voice commands. The application of vibrator and voice command changes with respect to people's comfortability. The main work of this belt is to correct body posture, so we need to find the exact position of the belt (fix with a person), to correct their posture so that we need MPU6050 gyroscope, used to find the exact position. If any changes happen, it gives digital input to the Arduino controller. Then, the Arduino controller reacts based on the program to give output to the vibrator or

V. G. Pratheep · E. B. Priyanka (✉) · S. Thangavel · K. Heenalisha · M. Ariya Manickam · A. P. Logaram
Department of Mechatronics Engineering, Kongu Engineering College, Perundurai, Erode 638060, India
e-mail: priyankabhaskaran1993@gmail.com

V. G. Pratheep
e-mail: pratheep.vg@gmail.com

S. Thangavel
e-mail: thangavel.mts@kongu.ac.in

K. Heenalisha
e-mail: heenalisha@gmail.com

M. Ariya Manickam
e-mail: ariyamanickam@gmail.com

A. P. Logaram
e-mail: aplogaram@gmail.com

© The Editor(s) (if applicable) and The Author(s), under exclusive license to Springer Nature Singapore Pte Ltd. 2021

S. Mohan et al. (eds.), *Materials, Design, and Manufacturing for Sustainable Environment*, Lecture Notes in Mechanical Engineering,
https://doi.org/10.1007/978-981-15-9809-8_16

voice command (SD card module). Based on the users' need, the actuation changes to vibrator or voice command. In the case of voice command, a headset can be connected with the patients.

**Keywords** Body posture · Belt · Vibrator · Gyroscope · Arduino

## 1 Introduction

A straight, upstanding stance is constantly connected with great wellbeing and with the picture of a unique individual. Then again, a bowed stance is related to unforeseen weakness [1–4]. Poor stance means that poor muscle tone. Admonishments to "stand straight" or "pull your shoulders back" do not improve body pose. A few issues may result from mistaken body act. For instance, upper back agony is frequently connected with round back (kyphosis); thoracic kyphosis can deliver a reduction of chest volume [5]. The essential spotlight is on the patients who are on the whole confronting fundamentally this body act issue. Poor body pose has numerous unsafe impacts to one's wellbeing, for instance, ill-advised sitting stance prompts short understanding separation and henceforth causes or declines nearsightedness; off base sitting stance is additionally a significant underlying driver for neck torment, neck shoulder muscle strain, just as lower back torment [6]. Other than the damage to wellbeing, poor body act is additionally not attractive for stylish reasons; for instance, constant hunchback influences one's quirks and certainty level. There is, in this manner, a requirement for gadgets to address poor body act. Most stance revision gadgets right and additionally forestall (1) ongoing hunchback and additionally, (2) round shoulders conditions, yet not many rectification gadgets right as well as forestall (3) shoulder unevenness, (4) scoliosis, (5) constant forward head position, and (6) routine brought down head position [7–9].

Figure 1 shows the different body postures representing the variations in the movements of dislocations associated with various parts of the spinal cords. It is often the case that people have bad postures because they pay little attention to their sitting postures or stances. It is especially hard for people to perceive their own bad postures because they have been accustomed to such postures. Consequently, such bad postures tend to affect the body's health by causing a hunchback after an extended period of time. Currently, a variety of products for preventing hunchbacks and keeping proper postures can be found in the market, such as posture-correcting bands, posture-correcting waistbands and hunchback-preventing desks and chairs [10]. In this research work, it is mainly focused on the spinal cord patients. Spinal cord patients mainly have the problems are

- Muscle atrophy. Muscle decay is the squandering or loss of muscle tissue
- Contractures. Contractures occur due to loss of motion in a joint
- Osteopenia/Osteoporosis
- Fractures
- Heterotrophic ossification (HO)

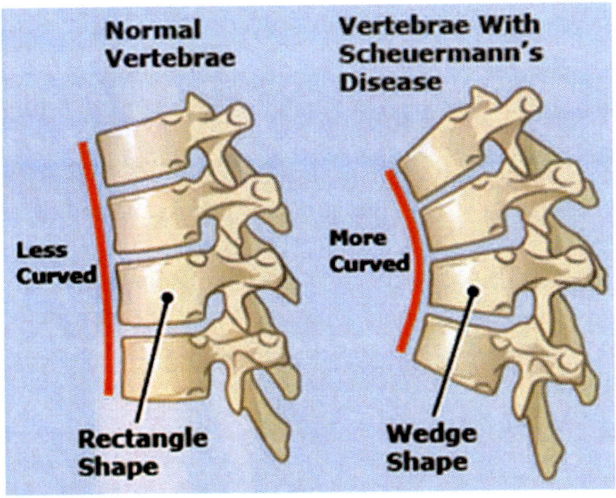


**Fig. 1** Body postures with defect highlighting in the spinal cord

- Overuse syndromes
- Changes in posture.

One of the important problems for the spinal cord patients is the changes in body posture. They are struggling to correct their own body posture. Some patients are do not have a sense in the lower body. They do not know which position they are tilting. They need some other person to change their own body posture [11–14].

## 2 Literature Review

### *2.1 Posture Warning Device*

A belt has a stance cautioning gadget looked after in that. The belt incorporates a lash with a clasp toward one side thereof. The tie incorporates pockets that get a battery and an electric ringer or signal. Electrical contact is kept up on the lash and associated with the battery, while the ringer is electrically associated with the clasp [15]. Development of the clasp against the electrical contact finishes the circuit between the battery and signal to emanate a sound demonstrating to a wearer that he has permitted his muscular strength to unwind to a point confirming awful stance. The clasp is one-sided away from the electrical contact by methods for springs interconnected between the tie and clasp with the end goal that, when the wearer confirms great stance and stomach muscles control, the bell is dormant [16].

## 2.2 Posture Sensing Alert Apparatus

A posture sensing alert apparatus is provided. The posture sensing alert apparatus comprises an attachment element, a detecting element, a processing element and an alert element. The attachment element is adapted to attach to a human body. The detecting element is disposed on the attachment element and is adapted to sense a posture change from the human body. The processing element is disposed on the attachment element and connects to the detecting element [17]. The processing element is adapted to output a signal to the alert element in response to the posture change for a predetermined period so that the alert element is adapted to output an alert accordingly.

## 2.3 Existing System

Existing stance revision gadgets ordinarily have structures containing two flexible shoulder ties and a different abdomen tie. The midriff lash is secured immovably around the midsection of the wearer [18, 19]. There are commonly two gatherings of structures, one in which non-customizable flexible shoulder lashes associated with a back tie are circled around the shoulders and the back tie is then associated with the abdomen tie, and another where length-movable versatile shoulder ties are extended around the shoulders from a back tie and joined straightforwardly back to the midsection tie. Be that as it may, both structure bunches have certain weaknesses. Then again, interfacing movable versatile shoulder lashes legitimately to the midriff tie can bring about extreme pulling powers causing the uprooting of the midsection tie, which accordingly brings about loss of in general body pose revising impact and furthermore makes distress the client [20, 21]. A need, along these lines, exists to give a stance vest that addresses in any event one of the previously mentioned issues.

In this research work, the proposed system is to correct the existing belt by adding new additional features. The additional feature is to create an electrical circuit design with several control and indicators. In this work, the correction of the body posture of the spinal cord patients is done using voice command or vibrator with interfacing with a mobile phone [19]. The mobile has the application to control the voice command.

The advantages of this system are:

- Presence of the indication system
- Good user flexibility compared to the existing belt
- Improved technology is implemented.

# 3 Methodology of the Proposed Work

The system consists of the following sections

- Finding of posture changes
- Actuating vibrator
- Playing voice command
- Control using a mobile phone.

## 3.1 *Finding of Posture Changes*

One main component is needed to find the body posture changes. On the backside of the belt, a gyroscope needed to be fixed at the center. It gives six qualities as yield: three qualities from the accelerometer and three from the gyrator. The MPU 6050 is a sensor dependent on MEMS (smaller-scale electro-mechanical frameworks) innovation. It executes based on the gyroscope value to find the exact position of the patient. When the belt is switched on by the patient, the gyroscope has a power supply. Once it gets the power supply, it gets finalized [22]. The gyroscope gives the value of *x*-, *y*- and *z*-axis value changes. When the patient moves left direction, it gives value in the *y*-axis in a negative direction. When the patient moves the right direction, it gives positive *y* value and *x* value greater than 10,000. When the patient moves backside, the *z*-axis value gets negative and finally, when the patient moves front, it gives *x* values greater than 16,000 and *y* value greater than 5000. So, by these values changes, it is possible to find the exact position of the patient [23].

## 3.2 *Actuating Vibrator*

For indicating the patient about their changes in posture, the belt must need an indicator for indicating the patient first actuating mechanism is to create a vibrating mechanism. A vibrating motor is used to vibrate the system to indicate the spinal cord patient. In Arduino controller, the program for the vibrator is already fed. In this work, the Arduino controller gets the output from the gyroscope 6050 to turn on the vibrator to indicate the spinal cord patients [24].

## 3.3 *Playing Voice Command*

Some of the spinal cord patients do not have a sense in the lower body. They do not even know which position they are tilting. That kind of person needs some other person to rectify their body posture. The other person needs to tell the patient like

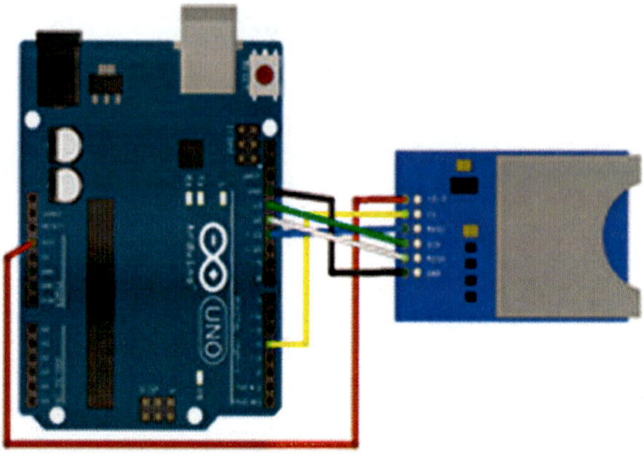

**Fig. 2** Connection calibrated by wiring the SD card module on the organized input channel available on the Arduino Uno

move right or move left or move front or move back. To rectify this problem, this corrector belt gives voice command to the patient like moving front or backward or through the adjacent-side mobility actions. To do this, belt has the gyroscope 6050 interconnected with the SD card module interface toward on the input port of the Arduino controller [25]. Based on the Arduino programming, the changes in the value from the gyroscope sensor are used to play the voice command through the headset associated with the SD card module. Figure 2 shows the connection calibrated by wiring the SD card module on the organized input channel available on the Arduino Uno.

### *3.4 Control Using Mobile Phone*

The indication of both voice command and vibrator is present in the belt, but some of the patients do not need voice command and some of the spinal cord patients do not need a vibrator; to solve that kind of problem, this belt is connected with mobile phone through Bluetooth module (HC-05). To mainly perform voice command performances, the mobile control applications were installed which were activated by appropriate pointing the voice control mode [22]. When we switch off the voice command mode using a mobile phone, by default the vibrator only indicates the changes in the body posture. For interfacing the Bluetooth module with a mobile phone, this belt has to use Arduino nano to rectify the communication problem.

## 4 Operational and Technical Feasibility

The operation of the entire model can be achieved by using gyroscope, SD card module, Arduino Uno controller circuit. Gyroscope gives values of *x*-, *y*- and *z*-axis changes of the patient posture measured data communicated through the Arduino Uno. Along with this action, the other triggering action by SD card module gives voice command to the patient through headset based on the controller program. Bluetooth module interfacing with Arduino Uno is used to control the voice command through mobile phone. Till now no other model comes with this technology, so operation-wise it is feasible enough. The operation is also easily understandable by the normal persons and easily operated by the patient himself [16, 17, 21].

The components required and the technologies adapted for the project are available in easy sufficient technology and is available to carry out this project. Globally available software packages fulfill the requirements adapted for developing the project into the product. The design which has to be implemented can be simulated through the available software, and the prototype can be developed according to the concept [14]. The sample prototype represents the concept or the process that is going to be implemented. After attaining the results of the prototype, the concept which has designed can be fabricated into a product. The proposed system is technically feasible and makes the existing system incompetent. The additional features like mobile control for voice control and vibrator control and belt control on/off are enhances the belt and makes it technically feasible to the patient as well as normal persons [22].

## 5 Results and Discussions

In this section, all the required components are assembled in simulation software and verified. Software called PROTEUS is being used for the simulation. The design consists of proteus simulation, an electric circuit for SD card connection and an electrical circuit for Bluetooth module connection. In Fig. 3, in the proteus model only the necessary components are connected. No other additional features are added. The SD card module is added at the port numbers of (digital) 10, 11, 12, 13. The gyroscope is connected through the TX and RX ports of the Arduino Uno. The program is dumped using the HEX file created by using Keil software. After all the connections and necessary steps are done, the simulation runs successfully.

Figure 4 shows the fabricated body posture corrector belt entire view holding Arduino Uno incorporated with SD card module. This is the core concept of this project and makes the belt to check body posture correction. The power supply for all the components is provided with the respective battery. The SD card module has a total of 16 ports. In this project, only six ports have been used: two ports for power supply, another two ports are for headset connection, and the last two ports are used to communicate with the Arduino Uno controller. The gyroscope 6050 connects with

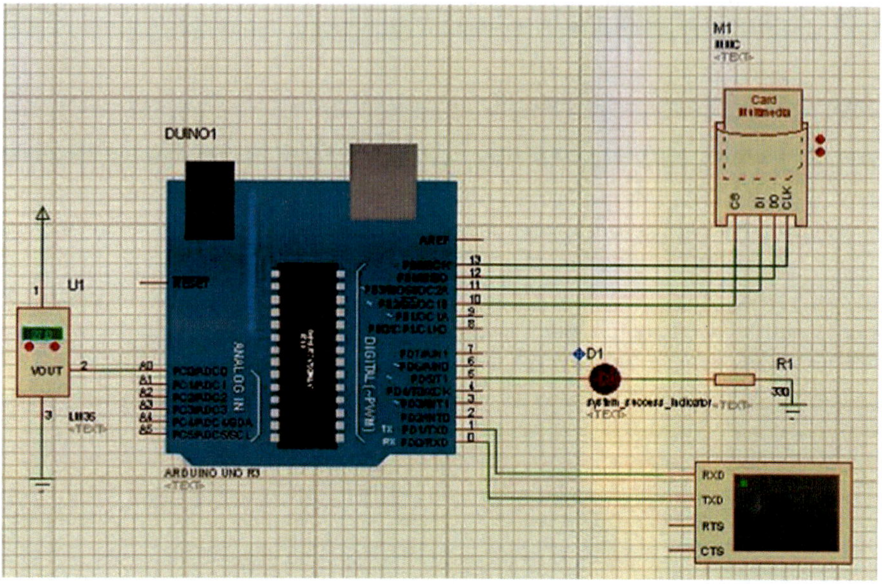


Fig. 3 Proteus design of the proposed model layout

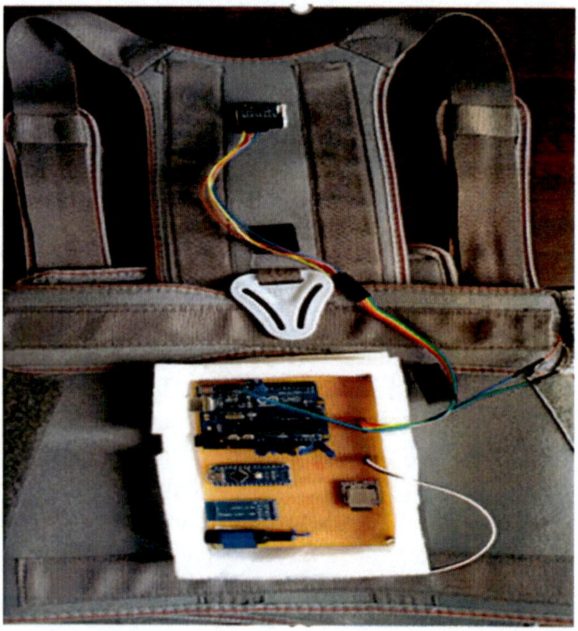

Fig. 4 Body posture corrector belt

the Arduino controller with A0 and A1 ports, respectively. It is all combinable formed to monitor the body posture and give feedback to the patient.

In this Bluetooth module connection, Arduino nano is used to make a connection with the Bluetooth module to communicate with Arduino Uno as well as a mobile phone. The power supply for all the components present in the system is given using a battery. The transmitter and receiver pins of Arduino nano are connected to the receiver and transmitter of the Bluetooth module. Based on the program dumped on the Arduino nano, it gives supply to the D13 pin. This pin gives the power supply to the SD card module. Using mobile phone application, the control for voice command is connected. If the patient does not need voice command, they can switch off the voice command using the mobile phone.

## 6 Working Methodology

When the belt is switched on by the patient, the gyroscope has a power supply. Once it gets the power supply, it gets finalized. The gyroscope gives the values of $x$-, $y$- and $z$-axis value changes. For indicating the patient about their changes in posture, the belt must need an indicator. For indicating the patient, first actuating mechanism is to create a vibrating mechanism. A vibrating motor is used to vibrate the system to indicate the spinal cord patient. In Arduino controller, the program for the vibrator is already fed. In this proposed model, the Arduino controller receives the output from the gyroscope 6050 to turn on the vibrator to indicate the spinal cord patients.

Some of the spinal cord patients do not have the sense in the lower body. They do not even know which position they are tilting. That kind of person needs some other person to rectify their body posture [26]. Based on the Arduino programming, the changes in the value from the gyroscope sensor are used to play the voice command through the headset using the SD card module. The indication of both voice command and vibrator is present in the belt, but some of the patients do not need voice command, and some of the spinal cord patients do not need a vibrator [27]. When we switch off the voice command mode using a mobile phone, by default the vibrator only indicates the changes in the body posture. For interfacing the Bluetooth module with the mobile phone, this belt has to use Arduino nano to rectify the communication problem.

## 7 Conclusion

A compact instrument, initially intended to gauge elbow firmness, can be utilized to quantify lower leg solidness. High-speed solidness estimations all around related to identical firmness estimations got by a torque engine, and the spinal posture habit can likewise separate between low (probably speaking to detached firmness) and high (apparently reflecting dynamic solidness) movement speed. At the point when utilized with the foot, the convenient gadget has high intra-regulators and also possess better adaptive control quality and is adaptable enough to separate neurological members from controls. The gadget does not be that as it may, account in its current structure for the inertial part of impedance. This belt is not only for spinal cord patients, if the voice command is switched off it will become a normal belt with a vibrator. Hence, normal people who need to change their posture also can be useful with this belt. Single Arduino mega is a good replacement for both Arduino Uno and Arduino nano using printed circuit board (PCB) on circuit connection make the design compatible and feasible.

## References


1. Chopra S, Kumar M, Sood S (2016) Wearable posture detection and alert system. In: 2016 international conference system modeling & advancement in research trends (SMART), Moradabad, pp 130–134. https://doi.org/10.1109/SYSMART.2016.7894504
2. Luo X, Du W, Zhang J (2015) Safety benefits of belt predesigning in conjunction with precast braking in a frontal crash. In: 2015 IEEE intelligent vehicles symposium (IV), Seoul, pp 871–876. https://doi.org/10.1109/IVS.2015.7225794
3. Nam H, Kim J-H, Kim J-I (2016) Smart belt: a wearable device for managing abdominal obesity. In: 2016 international conference on big data and smart computing (BigComp), Hong Kong, pp 430–434. https://doi.org/10.1109/BIGCOMP.2016.7425964
4. Kalaiselvi P, Pratheep VG (2015) Analysis of interval system using model order reduction. In: 2015 international conference on innovations in information, embedded and communication systems (ICIIECS), Coimbatore, pp 1–6. https://doi.org/10.1109/ICIIECS.2015.7193142.6
5. Pragadeesh M, Pratheep VG, Saravana Kumar M, Santhosh M (2016) Automatic pot making machine. Int J Eng Res Technol (IJERT) 4(26):52–54
6. Ramesh K, Pratheep VG, Venkatachalam K (2014) Design and analysis of interval system using order reduction technique. Int J Control Theor Appl 7(2):85–98
7. Pratheep VG, Priyadharshini K (2013) Analysis of cross layer based multicasting in manet. Comput Sci Telecommun 37(1)
8. Pratheep VG, Venkatachalam K, Ramesh K (2014) Model order reduction of interval systems by pole clustering technique using GA. J Theor Appl Inf Technol 66(1)
9. Priyanka EB, Maheswari C, Thangavel S (2020) A smart-integrated IoT module for intelligent transportation in oil industry. Int J Numer Model Electron Netw Devices Fields e2731. https://doi.org/10.1002/jnm.2731
10. Priyanka E, Maheswari C, Thangavel S (2019) Remote monitoring and control of LQR-PI controller parameters for an oil pipeline transport system. Proc Inst Mech Eng Part I J Syst Control Eng 233(6):597–608. https://doi.org/10.1177/0959651818803183
11. Priyanka EB, Maheswari C, Thangavel S (2019) Proactive decision making based IoT framework for an oil pipeline transportation system. In: International conference on computer networks, big data and IoT. Springer, Cham, pp 108–119
12. Priyanka EB, Maheswari C, Thangavel S (2018) IoT based field parameters monitoring and control in press shop assembly. Internet Things 3:1–11
13. Priyanka EB, Maheswari C (2016) Parameter monitoring and control during petrol transportation using PLC based PID controller. J Appl Res Technol 14(5):125–131
14. Priyanka E, Maheswari C, Ponnibala M, Thangavel S (2019) SCADA based remote monitoring and control of pressure & flow in fluid transport system using IMC-PID controller. Adv Syst Sci Appl 19(3):140–162. https://doi.org/10.25728/assa.2019.19.3.676

15. Priyanka EB, Krishnamurthy K, Maheswari C (2016) Remote monitoring and control of pressure and flow in oil pipelines transport system using PLC based controller. In: 2016 online international conference on green engineering and technologies (IC-GET). IEEE, pp 1–6
16. Pratheep VG, Priyanka EB, Raja R (September, 2019) Design and fabrication of 3-axis welding robot. Int J Innov Technol Explor Eng (IJITEE) 8(11):1588–1592
17. Pratheep VG, Priyanka EB, Prasad PH (2019) Characterization and analysis of natural fibre-rice husk with wood plastic composites. In: IOP conference series: materials science and engineering, vol 561, no 1. IOP Publishing, UK, p 012066
18. Priyanka EB, Maheswari C, Thangavel S, Bala MP (2020) Integrating IoT with LQR-PID controller for online surveillance and control of flow and pressure in fluid transportation system. J Ind Inf Integr 17:100127
19. Subramaniam T, Bhaskaran P (2019) Local intelligence for remote surveillance and control of flow in fluid transportation system. Adv Model Anal C 74(1):15–21. https://doi.org/10.18280/ama_c.740102
20. Priyanka EB, Maheswari C, Thangavel S (2018) Remote monitoring and control of an oil pipeline transportation system using a fuzzy-PID controller. Flow Meas Instrum 62:144–151
21. Priyanka EB, Thangavel S, Parameswari P (September, 2019) Collision waring system using RFID in automotives. Int J Innov Technol Explor Eng (IJITEE) 8(11S):153–158
22. Laval JCC (1967) U.S. Patent No. 3,359,976. U.S. Patent and Trademark Office, Washington, DC
23. Latimer J, Lee M, Goodsell M, Maher C, Wilkinson B, Adams R (1996) Instrumented measurement of spinal stiffness. Manual Ther 1(4):204–209
24. Antonya C, Butnariu S, Pozna C (2016) Real-time representation of the human spine with absolute orientation sensors. In: 2016 14th international conference on control, automation, robotics and vision (ICARCV). IEEE, pp 1–6
25. Wolf SG, Lossing WW (1986) U.S. Patent No. 4,602,619. U.S. Patent and Trademark Office, Washington, DC
26. Priyanka EB, Thangavel S, Pratheep VG (2020) Enhanced digital synthesized phase locked loop with high frequency compensation and clock generation. Sens Imaging 21(1):1–12. https://doi.org/10.1007/s11220-020-00308-0
27. Maheswari C, Priyanka EB, Thangavel S, Parameswari P (November, 2018) Development of unmanned guided vehicle for material handling automation for industry 4.0. Int J Recent Technol Eng 7(4):428–432

# Design and Development of Augmented Reality Application for Manufacturing Industry

**D. J. Hiran Gabriel and A. Ramesh Babu**

**Abstract** In the past decade, the field of education has transformed tremendously with the advent of technology. Due to the advance of Smartphones, a new paradigm is known as M-learning has been developed. M-learning is an interactive learning technique by which students use a smartphone to study. In the presented work M-learning using the Augmented reality process has been discussed. The software development kit used for augmented reality development in Unity and Vuforia. The work focuses on the development of the interactive book and vital factors like dimensions of the page, the distance of the book page from VST display, orientation, and effect of curvature caused by binding. Experiments are conducted and results were obtained on the mentioned parameters. From the obtained results the work suggests the desired which works more effectively than a traditional layout followed for preparation of the interactive book for mobile augmented reality.

**Keywords** M-learning · Interactive books · Augmented reality · Unity · Vuforia · VST displays

## 1 Introduction

Traditionally educations was a knowledge-sharing procedure that was carried out face to face between Instructor and the student [1]. The knowledge transfer took place through static materials like books and papers [2]. Though the traditional methods of knowledge transfer are effective they fail to provide any additional information to the student and do not increase the interest of the students in the learning process. In recent times due to the advent of technologies, the methods of teaching and learning have

D. J. Hiran Gabriel (✉) · A. Ramesh Babu
Department of Mechanical Engineering, PSG College of Technology, Coimbatore, Tamil Nadu 641004, India
e-mail: victorrajan1996@gmail.com

A. Ramesh Babu
e-mail: dr.rameshbabua@gmail.com

© The Editor(s) (if applicable) and The Author(s), under exclusive license to Springer Nature Singapore Pte Ltd. 2021
S. Mohan et al. (eds.), *Materials, Design, and Manufacturing for Sustainable Environment*, Lecture Notes in Mechanical Engineering,
https://doi.org/10.1007/978-981-15-9809-8_17

been revolutionized [3]. This transformation in methods of education has designed the environments of education such that they have become more realistic and engaging [4]. The addition of dynamic features like animations and videos has increased the understanding of concepts by the students and helped them in achieving better in academics [3, 5].

In context to the educational tools, Marshall et al. in his works mention the importance of a physical book and its advantages. Physical books are simple, flexible, transportable, and provide the user with the feeling of satisfaction due to their presence [6]. Recent advantages in Augmented Reality, Mixed Reality, and Virtual reality has an impact on the current education system. Thus, researchers aimed to improve the experience of reading a physical book using such immersive technologies. A physical book generally consists of images and text which are static, while an interactive book or a smart book consists of similar images and text but while viewed through a viewing device, they can display dynamic information that consists of 3D models, videos, and images [7].

The motivation of the work is to design and develop an interactive book for mobile-based augmented reality purposes. Augmented Reality is a technology in which a computer is used to generate a mixed environment of real and virtual objects which can be viewed by the user through technological devices like smart glasses, PDA's, smartphones, etc., [8]. The distinctive characteristics of Augmented Reality from other immersive technologies are the interaction of the user with the awareness of the real world without immersing him in a virtual world, Usage of multiple formats of virtual objects which include text, image, videos, and 3D models, and cost-effectiveness [9]. A new paradigm in the field of education called Mobile learning or M-learning has developed in recent days since no education happens completely in classrooms due to the advent of the Internet and other similar technologies [10]. Smartphone's small form factor and their increased usage these days have made them a primary device of recommendation for learning purposes [9].

## 2 Related Works

Several books for education has been prepared for education based on Augmented reality, LIRA is an Augmented Reality based book which is aimed at increasing the response to stimuli for special children. LIPRA is a book which is developed mainly for educating chess, the book consists of different topics that are discussed through animations. SOLRA is a book that is developed to introduce concepts of the solar system to students using features like image, text, audio, video, and 3D models. GeoAR is a book used to introduce geometrical shapes to children using a web camera of a computer [11]. Thus, it is evident that several interactive books have been developed based on Augmented Reality.

In all the above-mentioned works the developers have mainly focused on the development of technology and its implementation. In the presented work focus on designing a book page for Mobile Augmented Reality, the purpose will be discussed.

## 3 Considerations Design Space for an Interactive Book

An interactive book or a smart book is the same as the ordinary book but is visually enhanced using immersive technologies when viewed through a display. They can be used for normal reading as well as to display virtual pop up's over them. Thus, while designing an Interactive book both the reader's perspective as well as video see-through (VST) display perspective. The following are the criteria are considered for designing an interactive book:

1. The size of the page.
2. The orientation of the page.
3. The flatness of the page.

The experiments were conducted based on the above-mentioned criteria. They are discussed in detail below.

## 4 Software's Development Kits Used in the Experiments

For the experimental work, the following software tools are being used.

**Unity**—Unity a game development engine is preferred for the application development process for the project since it has support for more than 25 software platforms which includes android.

**Vuforia**—Vuforia is an Augmented reality software development kit that uses computer vision technology to detect 2D images and 3D objects. Vuforia supports Augmented reality in mobile devices.

## 5 Effect of the Size of the Page

The size of the page determines the dimension of the book, it's handling, and the transportability of the book by the user. But while designing a book page for the augmented reality it should be completely visible in the VST display thus the user will be able to completely see the virtual object augmented on the page. Conventionally books are printed in either size of 210 mm $\times$ 297 mm, 254 mm $\times$ 304 mm, and 210 mm $\times$ 148 mm which gives a good in-hand feel to the user. But pages of 210 mm $\times$ 297 mm, 254 mm $\times$ 304 mm these sizes are not completely visible in a VST display thus distance has to be increased to accommodate them completely into a VST display. Increasing the distance decreases the image recognition capabilities of the AR SDK. Thus, to facilitate handling by the user without affecting the image recognition capability of the AR SDK pages are recommended to be designed. A 210 mm $\times$ 148 mm sized paper is preferred for Augmentation purposes since it lies

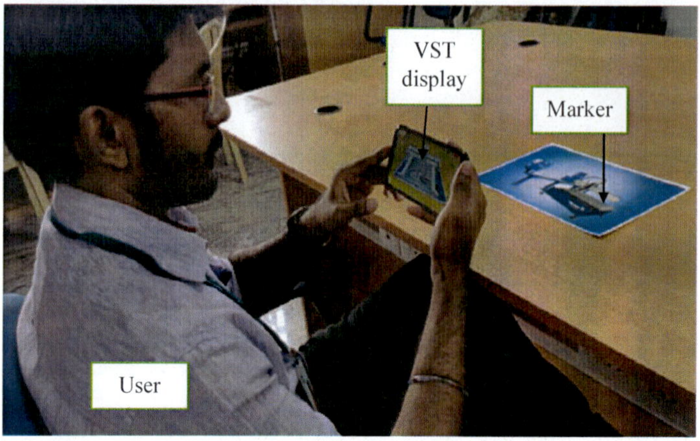


**Fig. 1** AR view through a VST display

in the field of view of the user and lesser the distance comfortable to use. Figure 1 shows the experiment conducted in which the user is made to sit on a standard chair and the target is placed on a standard study table. The figure shows how 210 mm × 297 mm, 254 mm × 304 mm, and 210 mm × 148 mm size pages are seen through VST displays and their corresponding distances from VST display. Figure 2 shows the distance of different sizes of targets from the VST display found experimentally.

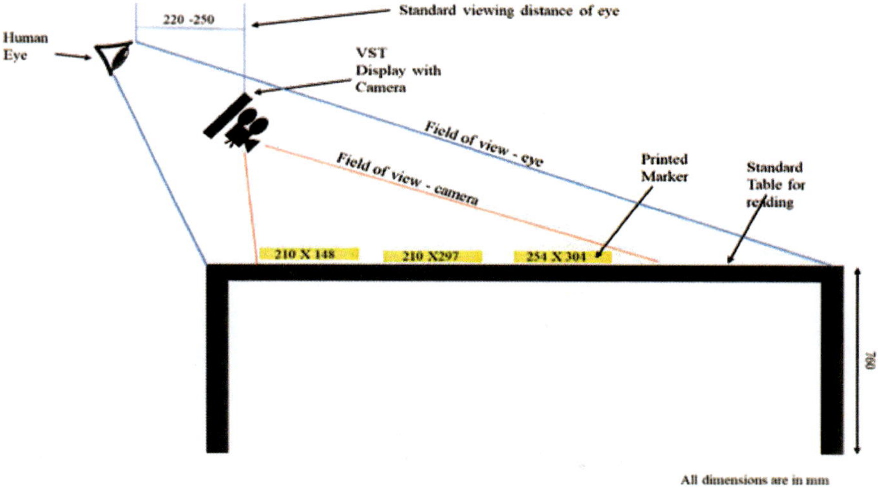


**Fig. 2** Image showing the field of view of the human eye and VST display

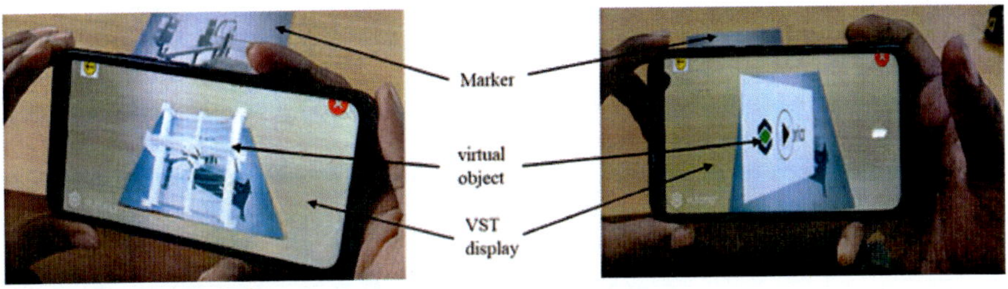


**Fig. 3** Unstable display of virtual objects in portrait orientation of marker

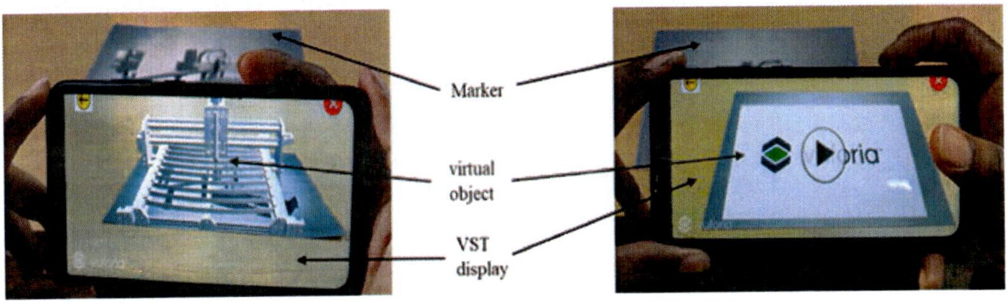


**Fig. 4** Stable display of virtual objects in landscape orientation of the marker

## 6 Effect of Orientation of the Page

The orientation of the book though doesn't affect the reading capability of the reader, but when it comes to displaying virtual objects like videos and 3D models on the VST display must be in landscape position. Landscape orientation is preferred because videos do have a greater width than height e.g.: 1024 × 768 pixels, 1920 × 1080 pixels, etc. It is essential to place both target and VST display to be in a landscape position to display the complete frame and details of the video. The experiment was conducted using the same marker printed in both portrait and landscape orientation it was found game objects had good stability in landscape mode. Especially videos and 3D models as virtual objects were found to be more stable in landscape target than the portrait one. Figures 3 and 4 show the stability of virtual objects in portrait and landscape orientation, respectively.

## 7 Effect of Curvature of the Page

A virtual object augmented should have perfect stability for visualization i.e. the augmented virtual objects should not fluctuate while it is displayed on the VST display. The stability of the virtual object aids in improved interpretation of the provided information to the user. For maintaining the stability of the virtual object placed over the marker it is essential to ensure its flatness. Flat markers are preferred

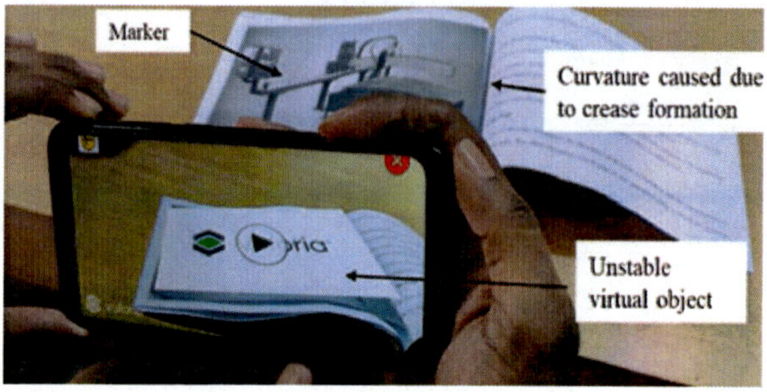


**Fig. 5** Lack of stability the virtual object due to curvature

**Fig. 6** More stable virtual object in coil binding

by Augmented reality SDK for easier image recognition. An experiment was conducted with different types of binding available, from the investigations it is evident that in spiral binding and wire coil binding pages remained flat, while in other bindings there's a curvature at the binding edges which reduces the flatness of the page.

From Figs. 5 and 6, it is clear that curvatures at creases of the book cause the virtual object to fluctuate and the virtual objects become unstable in VST display. Hence it is essential to maintain a coil or a spiral binding in order to avoid such fluctuations.

## 8 Results and Conclusion

From the above experiments, it is clear that factors like the size of the page, the orientation of the page, and the flatness of the page have an impact on the stability and trackability of virtual objects that are placed over the marker or the target image. The study suggests that it is necessary to maintain the size of an interactive book as 210 mm × 148 mm which aids in the handling and trackability of the tracker while other sizes can be used in other applications like the poster. While the orientation can be maintained in the landscape according to the experimental results. Since from the experiment, it is found that flatness also affects the stability of the virtual object Coil binding is preferred over other bindings for an interactive book.

## References

1. De Freitas S, Rebolledo-mendez G, Liarokapis F, Magoulas G, Poulovassilis A (2010) Learning as immersive experiences: using the four-dimensional framework for designing and evaluating immersive learning experiences in a virtual world. Br J Educ Technol 41(1):69–85
2. Chao P, Chen G (2009) Interacting with computers augmenting paper-based learning with mobile phones. Interact Comput 21(3):173–185
3. Nincarean D, Alia MB, Halim NDA, Rahman MHA (2013) Mobile augmented reality: the potential for education. Procedia Soc Behav Sci 103:657–664
4. Kirkley BSE, Kirkley JR (2002) Creating next generation blended learning environments using mixed reality. Video Games Simul 49(3):42–54
5. Kühl T, Scheiter K, Gerjets P, Gemballa S (2011) Computers & education can differences in learning strategies explain the benefits of learning from static and dynamic visualizations? Comput Educ 56(1):176–187
6. Marshall CC, Reading and Interactivity in the Digital Library: Creating an experience that transcends paper
7. Grasset R, Dünser A, Billinghurst M (2008) The design of a mixed-reality book: is it still a real book? In: Proceedings of 7th IEEE international symposium on mixed and augmented reality 2008, ISMAR 2008, pp 99–102
8. Azuma R, Baillot Y, Behringer R, Feiner S, Julier S, MacIntyre B (2001) Recent advances in augmented reality. IEEE Comput Graph Appl 21(6):34–47
9. Corrêa AGD (2015) Interactive books in augmented reality for mobile devices. Human Comput Interact 1:1084–1101
10. Zhang W, Su L (2011) Mobile-learning (M-learning) apply to physical education in colleges. In: Proceedings PACCS 2011, 2011 3rd Pacific-Asia conference on circuits, communications and system, pp 1–3
11. Altinpulluk H, Kesim M (2016) The classification of augmented reality books: a literature review. In: INTED2016 proceedings, vol 1, pp 4110–4118

# Evaluation of Drilled Hole Quality in Aluminum $MoS_2$ Metal Matrix Composite

K. Renuga Devi and K. Somasundara Vinoth

**Abstract** Metal matrix with self-lubrication is studied to replace conventional composites in tribo problems eliminating the need for lubricants. In this work, the machinability of self-lubricating metal matrix (MMC) materials were fabricated using stir casting process which is studied. The drilling examination was conducted using CNC vertical machining center under dry drilling condition. Response surface methodology (RSM) was employed to conduct experiments, based on it suitable mathematical equation that was evolved. To determine outcomes of various drilling factors on the surface roughness, thrust force and hole size of AlMMC analysis of variance (ANOVA) were utilized. The result by changing the production factors such as spindle speed (2000, 2250, 2500 rpm) and feed rate (0.2, 0.25, 0.3 mm/rev) were investigated, and their effects were presented. The test result reveals that the spindle speed and feed rate were the major parameter influencing the surface roughness, thrust force, and feed rate which is the major parameter influencing the hole size of the AlMMC.

**Keywords** Drilling · Aluminum metal matrix composites · Thrust force · Surface roughness · Molybdenum disulfide

## 1 Introduction

Recently, there has been much development in the area of composite materials. Metal matrix composite is one of the emerging materials in the industrial application as it is having better specific stiffness, specific strength, improved tribo properties, and corrosion resistance, etc. However, the major difficulty with metal matrix composite

K. Renuga Devi (✉) · K. Somasundara Vinoth
Department of Production Engineering, PSG College of Technology, Peelamedu, Coimbatore, Tamil Nadu, India
e-mail: renugadevi.kp97@gmail.com

K. Somasundara Vinoth
e-mail: vinothks@yahoo.com

© The Editor(s) (if applicable) and The Author(s), under exclusive license to Springer Nature Singapore Pte Ltd. 2021

S. Mohan et al. (eds.), *Materials, Design, and Manufacturing for Sustainable Environment*, Lecture Notes in Mechanical Engineering,
https://doi.org/10.1007/978-981-15-9809-8_18

is machining, because of the properties of reinforcement. It is essential for establishing efficient machining parameters for processing advanced composite materials. Among various machining processes, drilling is one of the metal removal process and as well as a major process in the assembly section. Due to an increase in the application of these materials, the need for establishing machinability parameters plays a vital role in manufacturing field. Among the several drilling factors, feed rate, spindle speed, drilling tool, tool material, and its geometry have significant effects on the machinability of the material.

The manufacturing technique of aluminum metal matrix composite plays a vital role in increasing the mechanical properties. There are many methods available for manufacturing of particulate reinforced aluminum MMCs, such as solid-state processing such as powder metallurgy and liquid-state processing such as stir casting. Based on literature survey, stir casting process was identified as the easy and economical process for manufacturing AlMMCs. Experiment shows that stir casting specimens have better mechanical properties than other specimen [1].

The process starts when the Al6061 ingots were melted in the furnace, and the liquefied metal is transferred into the die. As soon as the metal starts solidifying, the die will be released after solidification.

Demir carried out drilling tests using HSS, carbide, tin-coated carbide under the dry sliding condition and studied material microstructure and mechanical properties, tool wear, thrust force, and torque in wet grinder stone dust-reinforced Al6063 [2]. Cinar studied the impact of changed cutting conditions on quality of the drill and chip formation in Al7075 alloy, and conditions were compared [3].

Singh has evaluated the impact of various drilling parameters on Al6063 metal matrix composite and evaluated the cutting force [4].

Eapen et al. studied microstructure, hardness, surface roughness, diameter error, and roundness error of drilled surface [5]. Road et al. evaluated thrust force, surface roughness, and ovality during drilling of hybrid metal matrix composite [6].

Radhika et al. investigated microstructure, thrust force, surface roughness, tool wear, and damage in the workpiece in metal matrix composite [7]. Ranjan et al. tested thrust force and torque using ANOVA and regression model under various drilling parameters [8]. Rath et al. conducted a drilling experiment by varying speed and feed and analyzed surface roughness and thrust force using Minitab statistical analysis software [9].

The influence of the drilling cycle was studied by Renevier et al. by means of the G3 peck drill as well as G1 drilling cycles by taking into account of variation in spindle speed and also the feed rate [10]. The impact of different drilling factors on the quality of the hole was studied by Said et al. in aluminum hybrid matrix material, and the various outcomes of it in terms of surface roughness and circularity error [11] were reported.

Benny et al. conducted experiments to identify surface roughness, chip morphology, and tool reaction during machining of iron-based alloy [12].

Though enormous research that we performed to test thrust force, surface roughness, and tool wear for drilling study of various metal matrix composite, there was no contribution to AlMMC/$MoS_2$ metal matrix composite. Therefore, in this study,

**Table 1** Chemical composition of Al6061 (in wt%)

| Al | Mg | Si | Fe | Cu | Cr | Zn | Ti | Mn |
|---|---|---|---|---|---|---|---|---|
| 95.8–98.5 | 0.8–1.2 | 0.4–0.8 | 0–0.7 | 0.15–0.4 | 0.04–0.35 | 0–0.25 | 0–0.25 | 0–0.15 |

research work based on the effect of various drilling parameters on testing drilled hole quality was selected for drilling study on aluminum metal matrix composite.

The significance of this work is to use molybdenum disulfide as reinforcement and test surface roughness, thrust force, and drilled hole size. Aluminum reinforced with molybdenum disulfide is prepared using the stir casting technique, and a drilling study was carried out using an HSS drill bit. The main objectives of the work are to understand the application of drilling process in MMC, to understand the effect of drilling parameters in the quality of drilled holes of the MMC, and to maintain the quality of drilled holes.

# 2 Materials and Methods

## 2.1 *Materials*

In this study, material selected for investigation is 1 wt% molybdenum disulfide-reinforced aluminum metal matrix composite. Aluminum 6061 is most commonly used in automotive components. Table 1 shows the chemical composition of Al6061. Molybdenum disulfide is having a self-lubricating property in metal matrix composite. The test specimen was fabricated using stir casting process using the cylindrical die with a square section. Initially, aluminum 6061 is melted in a furnace at 800 °C, and reinforcements are preheated and added with selected 1 wt% $MoS_2$. Then, care is taken to maintain the temperature and stirred continuously. Melted metal was then poured into the die and after the solidification specimen was taken out.

The mold specimen taken out is then machined for the required dimension of 40 × 40 × 130 square rod using a milling machine.

## 2.2 *Experimental Setup*

Drilling experiments was executed using CNC vertical machining center as shown in Fig. 1 using a 10 mm HSS drill bit under dry condition. The selection of process parameters was done using the design of experiments under the central composite design, and a mathematical model was developed. Experiments were carried out by varying spindle speed and feed rate [13].

**Fig. 1** Drilling setup

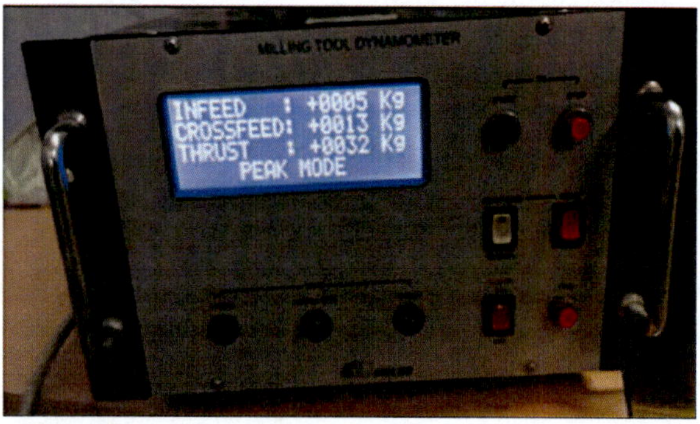


**Fig. 2** Drilling dynamometer

The Syscon dynamometer as shown in Fig. 2 was utilized to measure thrust force while conduction of experiments. Surface roughness of the drilled holes in terms of arithmetic average of the roughness profile (Ra) was estimated using a Mitutoyo SJ-201 surface roughness tester and the diameter of drilled holes was measured using a Mitutoyo Coordinate Measuring Machine (CMM).

## 2.3 Selection of Factors and Levels

The factors selected to conduct experiments were spindle speed (2000, 2250, 2500 rpm) and feed rate (0.2, 0.25, 0.3 mm/rev) as given in Table 2. RSM is utilized to fix the test runs. Design of experiments was created with two factors (Spindle speed and feed rate) with three levels [14]. A central composite design with 13 runs was developed, and experiments were carried out based on the combination of drill parameters as given in Table 3, which was developed using response surface methodology in Minitab 19 software.

**Table 2** Machining factors and their levels

| Drilling parameters | Notation | Levels | | |
|---|---|---|---|---|
| | | −1 | 0 | 1 |
| Spindle speed (rpm) | *N* | 2000 | 2250 | 2500 |
| Feed rate (mm/rev) | *F* | 0.20 | 0.25 | 0.30 |

**Table 3** Experimental matrix and results

| Ex. No. | Spindle speed (rpm) | *F* (mm/rev) | Ra (μm) | Thrust force (N) | Hole dia (mm) |
|---|---|---|---|---|---|
| 1 | 2000 | 0.25 | 6.97 | 890 | 9.940 |
| 2 | 2250 | 0.25 | 9.10 | 970 | 10.031 |
| 3 | 2500 | 0.25 | 5.97 | 970 | 9.934 |
| 4 | 2500 | 0.20 | 8.74 | 780 | 9.728 |
| 5 | 2250 | 0.25 | 9.10 | 970 | 10.031 |
| 6 | 2250 | 0.30 | 9.44 | 1120 | 9.832 |
| 7 | 2000 | 0.30 | 7.55 | 1070 | 9.703 |
| 8 | 2500 | 0.30 | 5.89 | 1060 | 9.664 |
| 9 | 2250 | 0.25 | 9.10 | 970 | 10.031 |
| 10 | 2250 | 0.20 | 11.70 | 800 | 9.873 |
| 11 | 2250 | 0.25 | 9.10 | 970 | 10.031 |
| 12 | 2250 | 0.25 | 9.10 | 970 | 10.031 |
| 13 | 2000 | 0.20 | 9.46 | 760 | 9.708 |

# 3 Results and Discussion

The experiments for this work were planned using RSM, and the mathematical equations were evolved for thrust force, Ra, and hole size of significant values, which is affected by different drilling parameters. The regression equations were calculated using Minitab 19 software, and mathematical models were developed.

## 3.1 Study on Surface Roughness

The surface finish of the composite materials has been measured in terms of Ra. Mitutoyo SJ-201 surface roughness tester as shown in Fig. 3 was used to test the coarseness of the drilled holes in aluminum composite. Ra is measured using a diamond-tipped probe moving horizontally inside the drilled holes.

**Fig. 3** Surface roughness tester

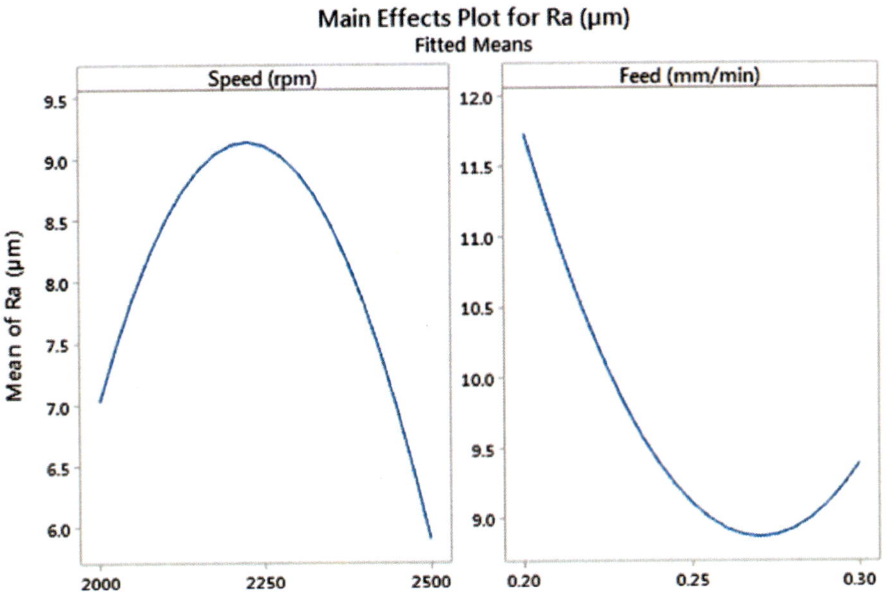


**Fig. 4** Drilling parameters affecting surface roughness. **a** Spindle speed, **b** feed rate

### 3.1.1 Impact of Spindle Speed on Surface Roughness

The impact of *N* on Ra is illustrated in Fig. 4a. Minimum and maximum Ra was noted at *N* of 2500 and 2250 rpm, respectively. Initially, Ra increased by 30% from 2000 to 2250 rpm. As the *N* increases from 2250 to 2500 rpm, Ra gradually decreased by 34% due to an increase in burnishing and honing impact created by reinforcement particles in composite material. With increase in *N*, coefficient of friction decreases between tool and specimen, improves the surface quality. A similar behavior was observed by Yasir et al., during the machining of AISI 316L SS.

### 3.1.2 Impact of Feed Rate on Surface Roughness

F has an impact on Ra as shown in Fig. 4b. For the low *F* of 0.2 mm/rev, Ra is very high. During increase in *F* from 0.2 to 0.25 mm/rev, Ra gradually decreased by 25%. Minimum Ra was noted when the *F* was 0.25 mm/rev, and furthermore an increase of 4% was noted from the *F* of 0.25–0.3 mm/rev. But a different behavior was observed by Jailani et al. wherein the *F* increase resulted in increase of material removal rate affects the Ra.

Yasir et al. observed easy material removal from the specimen when the *F* increases, which tends to decrease in Ra [15]. Koura and Sayed observed decrease in Ra as the *F* increases while machining metals like aluminum. The regression equation for Ra in terms of uncoded units is given below.

$$\text{Surface Roughness, Ra }(\mu\text{m}) = -168.13 + 0.19264\ \text{Speed} - 272.18\ \text{Feed} - 0.000042\ \text{Speed} * \text{Speed} + 581.9\ \text{Feed} * \text{Feed} - 0.01874\ \text{Speed} * \text{Feed}$$

## 3.2 *Study on Thrust Force*

Thrust force leads to high tool wear, which affects the dimensional accuracy. During drilling, the operation was performed by varying drilling parameters, and thrust was measured using drilling dynamometer. The Syscon dynamometer was utilized to quantify thrust force during drilling of AlMMC.

### 3.2.1 Impact of Spindle Speed on Thrust Force

The impact of various parameters on thrust force is illustrated in Fig. 5a. The thrust force increases by 7% from *N* of 2000–2250 rpm and decreases by 3% from *N* of 2250–2500 rpm. At lower *N*, axial pressure was less which reduces the thrust force. A similar effect was reported by Kumar et al. that self-lubricating reinforcements reduces the thrust force when compared with other reinforcements.

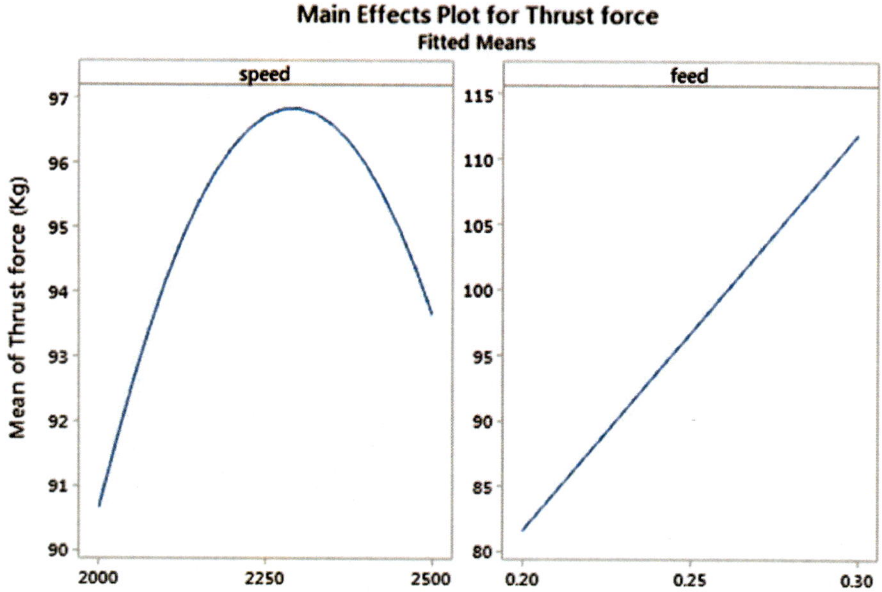


**Fig. 5** Drilling parameters affecting thrust force. **a** Spindle speed, **b** feed rate

#### 3.2.2 Impact of Feed Rate on Thrust Force

The impact of $F$ on thrust force is illustrated in Fig. 5b. Thrust force gradually increased from 0.2 to 0.3 mm/rev by 37%. Minimum thrust force was noted at the $F$ of 0.2 mm/rev, and maximum thrust force was noted when the $F$ is 0.3 mm/rev. As the $F$ is having direct relation to the area of cut, change in $F$ amplifies the thrust force [15]. A similar observation was noted by Anarghya et al., that minimum thrust force was obtained at a lower $F$. The regression equation for thrust force is given below.

$$\text{Thrust force(N)} = -361.0 + 0.3334\ \text{Speed} + 303.3\ \text{Feed} \\ -0.000073\ \text{Speed} * \text{Speed}$$

### 3.3 *Study on Hole Size*

#### 3.3.1 Impact of Spindle Speed on Hole Size

The impact of $N$ on hole size is shown in Fig. 6a. Minimum and maximum hole size were observed at a $N$ of 2000 and 2500 rpm, respectively. A decrease by 0.5% hole size was observed from the $N$ of 2000–2500 rpm.

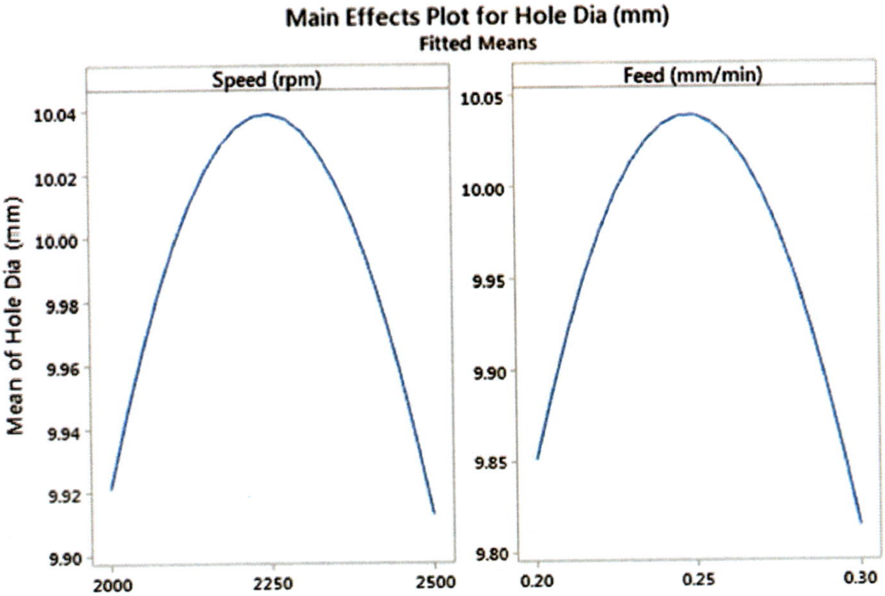


**Fig. 6** Drilling parameters affecting hole size. **a** Spindle speed, **b** feed rate

### 3.3.2 Impact of Feed Rate on Hole Size

The influence of $F$ on hole size was shown in Fig. 6b. Minimum and maximum hole size was noticed at $F$ 0.2 and 0.25 mm/rev. A gradual increment by 1.5% was observed from $F$ of 0.2–0.25 mm/rev and gradual decrease by 1% from $F$ 0.25 to 0.3 mm/rev. The regression equation for hole size is given below.

$$\begin{aligned}\text{Hole Dia(mm)} = & -4.871 + 0.008768\ \text{Speed} + 40.88\ \text{Feed} \\ & -0.000002\ \text{Speed} * \text{Speed} \\ & -82.49\ \text{Feed} * \text{Feed}\end{aligned}$$

## 4 Conclusion

The effects of various drilling parameters are studied in Al-$MoS_2$ MMC. The following theories were concluded from this work:

1. Design and experimentation were carried out and analyzed effectively using RSM for the drilling process.
2. Surface quality of the drilled holes obtained by changing the drilling factors namely spindle speed and feed rate was observed.
3. Better surface roughness and hole size was obtained at maximum spindle speed and feed rate (2500 rpm, 0.3 mm/rev), and lesser thrust force was obtained at minimum spindle speed and feed rate (2000 rpm, 0.2 mm/rev).

## References

1. Prakash KS, Kanagaraj A, Gopal PM (2015) Dry sliding wear characterization of Al 6061/rock dust composite. Trans Nonferrous Met Soc China 25(12):3893–3903
2. Demir Z (2018) Investigation of the fluctuation size in thrust force and chip morphology in drilling. Celal Bayar Univ J Sci 14(4):385–397
3. Cinar S, Venkatesh B, Baris B (2020) Investigation of surface roughness and chip morphology of aluminum alloy in dry and minimum quantity lubrication machining. Mater Today Proc 1–5
4. Singh S (2016) Effect of modified drill point geometry on drilling quality characteristics of metal matrix composite (MMCs). J Mech Sci Technol 30(6):2691–2698
5. Eapen J, Murugappan S, Arul S (2017) A study on chip morphology of aluminum alloy 6063 during turning under pre cooled cryogenic and dry environments. Mater Today Proc 4(8):7686–7693
6. Road L, District G (2014) Chip morphology and surface roughness in high-speed milling of nickel-based superalloy Inconel 718. Int J Mach Mach Mater 15
7. Radhika N (2013) Machining parameter optimisation of an aluminium hybrid metal matrix composite by statistical modelling. Ind Lubr Tribol 6:425–435

8. Ranjan S, Panda A, Dhupal D (2018) Hard turning of AISI 4340 steel using coated carbide insert: surface roughness, tool wear, chip morphology and cost estimation. Mater Today Proc 5(2):6560–6569
9. Rath D, Panda S, Pal K (2018) Prediction of surface quality using chip morphology with nodal temperature signatures in hard turning of AISI D3 steel. Mater Today Proc 5(5):12368–12375
10. Renevier NM, Hamphire J, Fox VC, Witts J, Allen T, Teer DG (2001) Advantages of using self-lubricating, hard, wear-resistant MoS2-based coatings. Surf Coat Technol 142–144:67–77
11. Said MS (2014) Surface roughness and chip formation of AlSi/AIN metal matrix composite by end milling machining using the Taguchi method. Sci Technol 68(4):13–17
12. Benny SMN, Venkatesan K, Manivannan K, Devendiran S, Mathew AT (2019) Study of forces, surface finish and chip morphology on machining of Inconel. Procedia Manuf 30:611–618
13. Krishnaraj V, Prabukarthi A, Ramanathan A, Elanghovan N, Kumar SM, Davim JP, Zitoune R (2012) Optimization of machining parameters at high speed drilling of carbon fiber reinforced plastic (CFRP) laminates. Compos Part B 43(4):1791–1799
14. Babu KV, Uthayakumar M, Jappes JTW, Rajan TPD (2015) Optimization of drilling process on Al-SiC composite using grey relation analysis. Int J Manuf Mater Mech Eng 5(4):17–31
15. Jailani HS, Rajadurai A, Mohan B, Sornakumar T (2015) Evaluation of drilled hole quality of Al-Si alloy/fly ash composites produced by powder metallurgical technique. Indian J Eng Mater Sci 22(4):414–420

# A Comprehensive Review on Mechanical Properties of Natural Cellulosic Fiber Reinforced PLA Composites

**G. Rajeshkumar, K. Naveen Kumar, M. Aravind, S. Santhosh, T. K. Gowtham Keerthi, and S. Arvindh Seshadri**

**Abstract** Due to the alarming rise in environmental regulations and ecological threats, research on natural fiber-based biodegradable polymer composites has burgeoned. One such polymer which has received a plethora of attention in recent years is Poly-Lactic Acid (PLA). Natural fibers possess distinct physical, chemical, and mechanical properties. In addition, it is recyclable, less expensive, and easily available. However, natural fibers have some drawbacks like moisture absorption, subsequent swelling, poor chemical resistance, and poor interfacial interactions with polymer matrices. These shortcomings can be overcome through surface modification techniques which include the usage of plasma technology and chemical agents such as sodium chloride, silane, stearic acid, etc. This paper presents a review of the mechanical properties of PLA based natural fiber composites. This review also demonstrates the influence of fiber surface treatment and manufacturing methods on the properties of PLA composites.

G. Rajeshkumar · K. Naveen Kumar · M. Aravind · S. Santhosh · T. K. Gowtham Keerthi (✉) · S. Arvindh Seshadri
Department of Mechanical Engineering, PSG Institute of Technology and Applied Research, Coimbatore, Tamil Nadu, India
e-mail: gowthampsg2018@gmail.com

G. Rajeshkumar
e-mail: grajeshkumar.me@gmail.com

K. Naveen Kumar
e-mail: knk321naveenkumar@gmail.com

M. Aravind
e-mail: aravindmohan486@gmail.com

S. Santhosh
e-mail: santhoshkumaar41@gmail.com

S. Arvindh Seshadri
e-mail: arvindhseshadri@gmail.com

© The Editor(s) (if applicable) and The Author(s), under exclusive license to Springer Nature Singapore Pte Ltd. 2021

S. Mohan et al. (eds.), *Materials, Design, and Manufacturing for Sustainable Environment*, Lecture Notes in Mechanical Engineering,
https://doi.org/10.1007/978-981-15-9809-8_19

**Keywords** Poly-lactic acid · Natural fibers · Surface modification · Mechanical properties · Manufacturing methods

## 1 Introduction

Poly-lactic acid, most commonly known as (PLA) is a thermoplastic polymer obtained from natural resources like maize or sugarcane which can be an efficient substitute to polymers produced from petroleum reserves. PLA is one of the front-runners in the bio-polymer market due to its connatural properties and it also has a production capacity of 140,000 tonnes per annum. This is also the second-largest production volume of any bioplastic [1]. Furthermore, it can be manufactured using existing equipment that is designed and currently used to produce petroleum-based plastics. Thus, it is reasonably cost-efficient to produce PLA [2].

PLA is exceptionally adaptable and can be extruded into sheet or film, injection molded into plastic parts, thermoformed into packaging items, or spun into fibers for non-woven and textiles [3]. The peculiar properties for end applications can be offered by blending PLA with other polymers. These blends are investigated for applications like sutures, implants, drug delivery systems, and tissue engineering in the medical field. In addition, PLA is commonly used in plastic films, bottles, and biomedical prototypes [4]. Also, PLA constricts while heating thereby making it suitable to employ it as a shrink wrap material and in some 3-D printing applications (lost PLA casting) [5].

The end-of-life options for PLA include reusing, renewable energy recovery (incineration), biodegradation, and feed-stock recovery. The production of PLA utilizes 65% less energy than the production of conventional plastics. Therefore, consumers and manufacturers can use resources more efficiently by reducing waste. Polymers that can degrade naturally have led researchers with a panacea to waste disposal problems by obtaining the same level of utility. It produces non-toxic compounds when burned, unlike many plastics, and also it doesn't break down quickly on land or in the ocean [6–8]. On the other hand, the major short-comings in its commercial usage is its high price, low impact resistance, hydrophobicity, and inability to withstand increased temperature. These issues are solved by using natural fibers as reinforcements, which enables us to obtain specific desirable properties of the composites for specific end-user applications. By considering the increasing environmental concerns, the interest of the public for biopolymers and natural fiber-reinforced polymers has proliferated. The sisal, banana, flax, coir, jute, etc. are some of the naturally occurring fibers that can be reinforced with PLA [9]. The fiber content as well as the processing parameters will have a significant impact on the characteristics of natural fiber-reinforced composites. Therefore, reasonable processing methods like direct injection, extrusion injection, extrusion compression moulding, etc. and the corresponding parameters should be conscientiously chosen to obtain an ideal end product. This paper will act as an effective source to academicians, industry personnel, scientists, and researchers about the mechanical properties of untreated and treated natural fibers based PLA composites [10].

## 2 Natural Fibers Reinforced in PLA

Natural fibers are generally extracted from plant sources as well as animal sources and some of these include banana, sisal, flax, jute, coir, silk, wool, etc., [11, 12]. In particular, the plant-based natural fibers that are reinforced in PLA to improve the properties of the resulting composite are discussed below.

### *2.1 Banana Fiber*

Banana fiber (*botanical name: Musa ulugurensiswarb*) is popularly called Musa fiber. This fiber is lightweight, has good strength with adequate elongation, and is biodegradable. The fiber is extracted from the banana stem, these stems are thereafter dried in sun for 7–8 h and then in a hot oven for 1 day at 75–80 °C, thereby removing the water absorbed in the banana fiber [13]. Banana fiber reinforced PLA composites show higher flexural and tensile properties on comparing to neat PLA. These composites can be employed in areas where lightweight is of prime importance and strength is not an issue [2].

### *2.2 Sisal Fiber*

Sisal fiber (*botanical name: Agave sisalana*) is the most prominently used natural fiber and it is cultivated widely across the globe. Some of its useful properties include good toughness, low cost, high durability, minimum wear, and recyclability. They were extracted from the sisal plant from the outer skin of the sisal leaf and kept under the sun and dried. This resulted in obtaining good moisture resistance and sufficient load absorbing properties. Table 1 enlists the different mechanical and chemical properties of sisal [14]. Sisal fiber-reinforced in PLA provides moderate tensile, superior impact, and flexural properties [15].

### *2.3 Flax Fiber*

Flax (*botanical name: Linumusitatissimum*) is a cellulose fiber, crystalline in the structure which absorbs and releases water quickly and is stiffer in handling. Flax is extracted by soaking it for 7–10 days and drying it in the atmosphere. In doing so, the

**Table 1** Mechanical and chemical properties of natural fibers

| Fibers | Banana [13] | Sisal [14] | Flax [16] | Jute [18] | Coir [20] |
|---|---|---|---|---|---|
| *Mechanical properties* | | | | | |
| Density (g/cm$^3$) | 750–950 | 1.45 | 1.5 | 1.4 | 1.4 |
| Tensile strength (MPa) | 392–677 | 350–360 | 600–1100 | 400–800 | 17–20 |
| Young's modulus (GPa) | 27–32 | 3.8 | 60–80 | 30 | 3.79 |
| Poisson ratio | 0.28 | 0.4 | 0.24 | 0.27 | 0.3 |
| *Chemical properties* | | | | | |
| Cellulose (%) | 60–65 | 65–70 | 64.1 | 50–57 | 43–44 |
| Hemi cellulose (%) | 6–19 | 14 | 16.7 | 20–24 | 25 |
| Lignin (%) | 5–10 | 9.9 | 2.0 | 8–10 | 45–84 |
| Moisture content (%) | – | 10 | 10 | 65 | 5–20 |

fiber is broken down, brushed, and finally it is crimped. The compressive strength is about 80% of tensile strength [16]. Flax fiber with PLA matrices produces promising mechanical properties. However, PLA/flax composites are not widely used as the bonding between PLA and flax is poor [17].

## 2.4 *Jute Fiber*

Jute fiber (*botanical name: Corchorusolitorius*) is widely used cellulose and a bast fiber for various industrial applications. Initially, it is dried in the sunlight, and then it is soaked in water, simultaneously breaking and crimping is done and jute fiber is extracted [18]. Jute reinforced with PLA has shown better fiber/matrix bonding, higher production efficacy, and considerably good mechanical properties. In addition to these, jute fibers also enhance the Izod strength and tensile modulus of the polymer [19].

## 2.5 *Coir Fiber*

Coir fiber (*botanical name: Cocos nucifera*) are available in India in large amounts, especially in the southern states of India. Coir is extracted from the coconut husk with the help of a disintegrator machine. The husk is then beaten and combined using a revolver screener, and it is kept in the sun and dried [20]. Coir fibers show greater elongation at break (15–40%) in contrast to sisal, jute, etc. which has low elongation. Therefore ductility of PLA can be improved when coir is reinforced in PLA [21].

## 3 Surface Modification of Natural Fibers

Due to environmental issues, several studies are taken up by the researchers to find a substitute for synthetic fibers. Natural fibers are biodegradable, less expensive, abundant, and have low density. But they have some downsides such as moisture absorption, swelling property, and poor compatibility [22, 23]. To overcome these drawbacks, physical (plasma treatment) and chemical (alkali, silane) treatments are made on the surface of the natural fibers. Physical techniques like plasma treatment will introduce different functional groups on natural fiber surfaces and will form strong covalent bonds with the matrix. This leads to strong adhesion and enhancement in properties. Chemical techniques can be done by introducing chemical agents like peroxides, alkali, water-repelling agents, etc. [24]. It has been demonstrated that chemical treatments can remarkably enhance the properties of natural fibers by changing their crystalline structure and eliminating weak constituents like hemicelluloses and lignin from the fiber structure [25].

### *3.1 Surface Modified Banana Fiber Reinforced PLA*

Banana fibers were taken and cut to the required length and soaked in detergent to remove the impurities. The fibers were then dried in sunlight for minimum 2 days. Mercerization is then carried out by dipping the fiber into sodium hydroxide solution (NaOH) 1 N for one hour at ambient temperature. Finally, the fiber was taken out and dried at ambient temperature for 24 h and in Oven at 80 °C for 12 h. The composite material was produced by the twin-screw extrusion process (acceleration 50 rpm) and then it was subjected to the injection process. Finally, a composite was obtained. Then Mechanical, Thermal analysis, Heat deflection test was conducted using the sample. The result obtained is shown in Table 2. From the above result, it was clear that surface-modified banana fiber shows good bonding with PLA and the composite obtained shows a considerable rise in the tensile modulus, tensile strength, and impact strength [26].

### *3.2 Surface Modified Sisal Fiber Reinforced PLA*

Sisal fibers were taken and cut into the required amount and soaked into di-oxane solution of MPS-g-PLA at 1 wt% for 48 h. Acetic acid was used to control the pH. After this, the fibers were taken out and dried at ambient temperature for 48 h. Then the fibers were allowed to react with MPS-g-PLA at 120 °C for 2 h. The composite material was prepared by twin-screw extrusion process at 75 rpm and 190 °C for 10 min. After that samples were prepared by injection moulding. Then tests like heat deflection, Mechanical, Thermal analysis, were conducted using the sample. The

**Table 2** Properties of natural fibers before and after surface modifications

| Fibers | Properties | Tensile strength (MPa) | Flexural strength (MPa) | Impact strength ($kJ/m^2$) |
|---|---|---|---|---|
| Banana [26] | Untreated | 14.61 | – | 19.1 |
| | Treated | 16.01 | – | 19.69 |
| | Percent increase (%) | 9.58 | – | 3.09 |
| Sisal [27] | Untreated | 56.68 | 97 | 3.25 |
| | Treated | 59.71 | 104 | 3.1 |
| | Percent increase (%) | 5.35 | 7.22 | −4.84 |
| Jute [28] | Untreated | 55 | 67 | 12.98 |
| | Treated | 62 | 78 | 14.25 |
| | Percent increase (%) | 12.73 | 16.42 | 9.78 |
| Coir [21] | Untreated | 4 | 25 | – |
| | Treated | 7 | 35 | – |
| | Percent increase (%) | 75 | 40 | – |
| Flax [29] | Untreated | 55 | – | – |
| | Treated | 63.4 | – | – |
| | Percent increase (%) | 11.23 | – | – |

result obtained is shown in Table 2, from the above result, it was clear that surface-modified sisal fiber showed good bonding with PLA and the composite obtained shows a significant improvement in properties like impact strength, tensile strength, and tensile modulus [27].

### 3.3 *Surface Modified Jute Fiber Reinforced PLA*

Jute fiber is obtained from the bark of the jute plant. It has 3 major chemical compounds cellulose (58–63%), Hemicellulose (20–24%), lignin (12–15%). First, the jute fibers were extracted by the retting process. After the retting process, the fiber was soaked in water for one hour. Then it is taken out and dried at 50 °C. After this, the Mercerization process was done by immersing the fiber in the required percentage of NaOH for 6 h at 70 °C with stirring and shaking. Acetic acid was used as the neutralizing agent to absorb excess alkali. Then the alkali-treated jute fiber was passed for the bleaching process. Here the fiber was added with hydrogen peroxide with occasional stirring for 45 min. Finally, the fiber was taken out and dried in the Oven at 50 °C until it reaches constant weight. The composite material was prepared by a vertical injection moulding process with the required mixture of treated jute fiber. Then Mechanical, Thermal analysis, heat deflection tests were conducted using the samples obtained. The result obtained is shown in Table 2. We can conclude from the result that surface-modified Jute fiber shows good bonding with PLA and the composite obtained shows a considerable increase in mechanical properties [28].

## 3.4 Surface Modified Coir Fiber Reinforced PLA

Coir fibers are extracted from the husk of coconut It has three main chemical compounds cellulose (36–43%), Hemicellulose (0.2%), lignin (41–45%). Coir fibers were taken and cut to the required length and it is washed well with distilled water. The prepared fiber was immersed in the NaOH solution for 1 h at 70 °C; it was then washed again to remove the excess NaOH. In order to remove the absorbed alkali, Acetic acid was used as the neutralizing agent. After this, the fiber was dried in the oven at 60 °C for one day. The composite sample was prepared by compression moulding process. Then Mechanical, Thermal analysis, Heat Deflection tests were conducted using the sample. The result obtained is shown in Table 2. It is understood from the obtained results that surface-modified coir fiber shows good bonding with PLA and the composite obtained shows a considerable increase in mechanical properties [21].

## 3.5 Surface Modified Flax Fiber Reinforced PLA

A flax fiber comprised of a middle lamella region, which is composed of pectin and a small quantity of lignin. Flax fibers were initially soaked in NaOH (5%) solution for 20 min at 23 °C. The fibers were removed, filtered, and thoroughly washed in distilled water. It was then rinsed with a diluted solution of HCl (neutralizing agent) to remove excessive NaOH and it was washed again with distilled water. Then it was dried in a vacuum at 65 °C for 3 h. The hot pressing method was used to manufacture the composite. It was then subjected to a tensile test in a universal testing machine. The results obtained were Tensile strength (MPa) untreated Flax fiber = 55, treated = 63.4. From the above result, it is clear that surface-modified flax fiber removed certain weakly adhering amorphous polysaccharides, most importantly pectins and hemicelluloses from the middle lamellae. Other noticeable treatments show that NaOH reaches the secondary layer and attacks the surrounding polymers and decreases the strength of the fiber thereby affecting the mechanical properties of a composite. By adding the proper amount of NaOH (%) and proper drying we can avoid the intrusion of NaOH into the inner layers [29].

# 4 Manufacturing Methods

Natural fiber-based PLA composites can be produced by a wide range of processes. Some of the most common manufacturing methods are discussed below. Figure 1 depicts the classification of the common processing techniques used to fabricate PLA-based composites.

## *4.1 Direct Injection Moulding (DIM)*

In direct injection moulding, the PLA and natural fibers were thoroughly mixed and were fed into the direct injection moulding machine. Various parts in DIM are hopper, nozzle, heaters, driving unit, and the mould assembly. The temperature of the barrel can be set according to our requirements like 160 °C, 170 °C, etc. The temperature from hopper to the nozzle will remain almost the same. The pressure can also be set according to our requirements. The screw is provided in the injection moulding machine which will be rotated using the driving unit. This screw is useful for mixing PLA and natural fiber. PLA and natural fiber which is fed in the hopper reaches the barrel and both these get mixed and heated. During heating, PLA gets converted into liquid. This mixed reinforced composite enters into the nozzle and from the nozzle, it enters into the mould-assembly. These composites are held and cooled for a particular time. Then it is removed from the mould assembly and can be cut into a number of pieces to prepare samples [30].

## *4.2 Extrusion Injection Moulding (EIM)*

In this type of moulding, PLA and fiber were fed into the hopper. The parts present in this moulding are hoppers, heaters, nozzle, cooling tank, guide rollers, and pelletizer. From the hopper, PLA and fiber enter into the barrel and it is compounded at a required temperature. Then these reinforced composites enter into the cooling tank where it

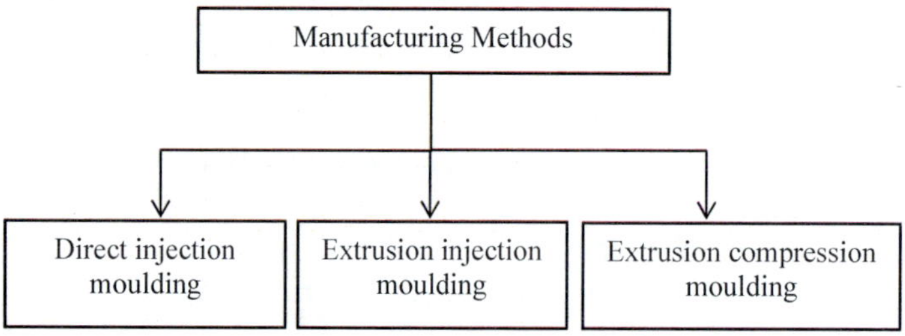


**Fig. 1** Classification of manufacturing methods for PLA based composites

is cooled and it further enters into pelletizer. Pelletizer is used to cut the composite pellets into a number of pieces. Now the composite pellets coming out of pelletizer were fed into the injection moulding machine and it is fabricated [30].

### *4.3 Extrusion Compression Moulding (ECM)*

The compounding in this process is similar to extrusion injection moulding. The various parts are also similar to injection moulding. The melted PLA and natural fiber were collected in the preheated mould. From the preheated mould, it is transferred to the compression moulding machine where it is pressed under the required temperature and pressure. Then it is taken from the mould and was cut into various pieces [2].

## 5 Conclusion

There has been a tremendous increase in the demand for newer materials that are environmentally safe and also reduces the burden on petroleum-based products. One such material that has gained a lot of attention is natural fiber-reinforced bio-polymer composites. Researchers are working on identifying different natural fibers that can act as a replacement for synthetic fibers like carbon, glass and Kevlar. The bio-polymer composites fabricated by using the natural fibers have properties comparable to the synthetic fiber-based composites. Furthermore, recent research works are concentrating on the surface treatment of natural fibers to enhance the interfacial bonding between the matrix and the fiber thereby good quality composites can be obtained. The cost associated with using bio-polymers as matrix material is considerably high and this makes it difficult for the commercialization of green composites. Therefore, further research should be explored to develop low-cost manufacturing techniques for fabricating green composites. This will ensure the reduction in the usage of synthetic fiber-based non-biodegradable composites that contribute to a sustainable environment.

## References

1. Bax B, Müssig J (2008) Impact and tensile properties of PLA/Cordenka and PLA/flax composites. Compos Sci Technol 68(7–8):1601–1607
2. Komal UK, Lila MK, Singh I (2020) PLA/banana fiber-based sustainable biocomposites: a manufacturing perspective. Compos Part B Eng 180:107535
3. Gurunathan T, Mohanty S, Nayak SK (2015) A review of the recent developments in biocomposites based on natural fibers and their application perspectives. Compos Part A Appl Sci Manuf 77:1–25

4. Pervaiz M, Sain MM (2003) Carbon storage potential in natural fiber composites. Resour Conserv Recyl 39(4):325–340
5. Murariu M, Dubois P (2016) PLA composites: from production to properties. Adv Drug Deliv Rev 107:17–46
6. Oksman K, Skrifvars M, Selin JF (2003) Natural fibers as reinforcement in polylactic acid (PLA) composites. Compos Sci Technol 63(9):1317–1324
7. Mohanty AK, Misra M, Drzal LT (eds) (2005) Natural fibers, biopolymers, and biocomposites. CRC Press, Boca Raton
8. Henton DE, Gruber P, Lunt J, Randall J (2005) Polylactic acid technology. In: Natural fibers, biopolymers, and biocomposites, vol 16, pp 527–577
9. Ku H, Wang H, Pattarachaiyakoop N, Trada M (2011) A review of the tensile properties of natural fiber reinforced polymer composites. Compos Part B Eng 42(4):856–873
10. Bajpai PK, Singh I, Madaan J (2014) Development and characterization of PLA-based green composites: a review. J Thermoplast Compos Mater 27(1):52–581
11. Rajeshkumar G (2020) Characterization of surface modified *Phoenix* sp. fibers for composite reinforcement. J Nat Fibers 1–12. https://doi.org/10.1080/15440478.2019.1711284
12. Rajeshkumar G (2020) An experimental study on the interdependence of mercerization, moisture absorption and mechanical properties of sustainable *Phoenix* sp. fibre-reinforced epoxy composites. J Ind Text 49(9):1233–1251
13. Kamble DP, Bhoomkar M, Vispute P (2020) Design and analysis of strength of banana fiber composite material. In: Pawar PM, Ronge BP, Balasubramaniam R, Vibhute AS, Apte SS (eds) Techno-societal 2018. Springer, Cham, pp 457–469
14. Senthilkumar K, Saba N, Rajini N, Chandrasekar M, Jawaid M, Siengchin S, Alotman OY (2018) Mechanical properties evaluation of sisal fiber reinforced polymer composites: a review. Constr Build Mater 174:713–729
15. Hagstrand PO, Oksman K (2001) Mechanical properties and morphology of flax fiber reinforced melamine-formaldehyde composites. Polym Compos 22(4):568–578
16. Batra SK (1998) Other long vegetable fibers. In: Lewin M, Pearce EM (eds) Handbook of fiber science and technology: fiber chemistry, vol 4. Marcel Dekker, New York, pp 505–575
17. Yan L, Chouw N, Jayaraman K (2014) Flax fiber and its composites—a review. Compos Part B Eng 56:296–317
18. Ramesh M, Palanikumar K, Reddy KH (2013) Mechanical property evaluation of sisal–jute–glass fiber reinforced polyester composites. Compos Part B Eng 48:1–9
19. Yang Y, Murakami M, Hamada H (2012) Molding method, thermal and mechanical properties of jute/PLA injection molding. J Polym Environ 20(4):1124–1133
20. Satyanarayana KG, Kulkarni AG, Rohatgi PK (1981) Structure and properties of coir fibers. Proc Indian Acad Sci Sect C Eng Sci 4(4):419–436
21. Dong Y,Ghataura A, Takagi H, Haroosh HJ, Nakagaito AN, Lau KT (2014) Polylactic acid (PLA) biocomposites reinforced with coir fibers: evaluation of mechanical performance and multifunctional properties. Compos Part A Appl Sci Manuf 63:76–84
22. Kumar GR, Hariharan V, Saravanakumar SS (2019) Enhancing the free vibration characteristics of epoxy polymers using sustainable *Phoenix* sp. fibers and nano-clay for machine tool applications. J Nat Fibers 1–8. https://doi.org/10.1080/15440478.2019.1636740
23. Rajeshkumar G (2020) A new study on tribological performance of *Phoenix* sp. fiber-reinforced epoxy composites. J Nat Fibers 1–12. https://doi.org/10.1080/15440478.2020.1724235
24. Rajeshkumar G (2020) Effect of sodium hydroxide treatment on dry sliding wear behaviour of *Phoenix* sp. fibre reinforced polymer composites. J Ind Text. https://doi.org/10.1177/1528083720918948
25. Rajeshkumar G, Ramakrishnan S, Pugalenthi T, Ravikumar P (2020) Performance of surface modified pineapple leaf fiber and its applications. In: Jawaid M, Asim M, Md. Tahir P, Nasir M, Pineapple leaf fibers. Springer, Singapore, pp 309–321
26. Jandas PJ, Mohanty S, Nayak SK, Srivastava H (2011) Effect of surface treatments of banana fiber on mechanical, thermal, and biodegradability properties of PLA/banana fiberbiocomposites. Polym Compos 32(11):1689–1700

27. Li Z, Zhou X, Pei C (2011) Effect of sisal fiber surface treatment on properties of sisal fiber reinforced polylactide composites. Int J Polym Sci 2011
28. Rajesh G, Prasad AVR (2014) Tensile properties of successive alkali-treated short jute fiber reinforced PLA composites. Procedia Mater Sci 5:2188–2196
29. Raj G, Balnois E, Baley C, Grohens Y (2011) Role of polysaccharides on mechanical and adhesion properties of flax fibers in flax/PLA biocomposite. Int J Polym Sci 2011
30. Chaitanya S, Singh I (2017) Processing of PLA/sisal fiber biocomposites using direct-and extrusion-injection molding. Mater Manuf Processes 32(5):468–474

# Electrical Properties of Cement and Geopolymer Composite Under Cyclic Compressive Loading

**B. Nivetha and D. Suji**

**Abstract** In this study, the electrical property of cement and geopolymer-based composites was evaluated under cyclic compressive loading. The electrical resistivity ($\rho$) of conventional cement-based and novel geopolymer composites was compared to assess its ability to sense the change of stress under external load. The sensing ability is measured in terms of Ohm-cm ($\Omega$ cm), which can be determined from the measured voltage (V) for the applied current (1 A). The change of resistance is observed for both the composites at the age of 7 and 28 days. The experimental results reveal that, in general, the electrical resistivity at the age of 28 days will be more than that in 7 days, and also, the resistivity of geopolymer composite was less when compared to cement-based composite. Form the test results, it can be inferred that geopolymer composites will be very effective in responding to the external stresses than cement-based composite and can be used for further research.

**Keywords** Electrical resistivity · Cement composite · Geopolymer composite · Cyclic loading

## 1 Introduction

Many civil engineering structures, namely bridges, pavements, structural components such as foundation, beam, and column demand periodic maintenance. Improper maintenance leads to accumulation of distress in the structure, thus resulting in an investment of huge money for repair. In some cases, lack of maintenance can even cause loss of life due to sudden failure of structures. Such periodic maintenance helps us to predict the causes inducing strains and displacements, reflected in form

B. Nivetha (✉) · D. Suji
Department of Civil Engineering, PSG College of Technology, Coimbatore, Tamil Nadu, India
e-mail: nivethabalasubramanian221@gmail.com

D. Suji
e-mail: sujimohan2002@gmail.com

© The Editor(s) (if applicable) and The Author(s), under exclusive license to Springer Nature Singapore Pte Ltd. 2021

S. Mohan et al. (eds.), *Materials, Design, and Manufacturing for Sustainable Environment*, Lecture Notes in Mechanical Engineering,
https://doi.org/10.1007/978-981-15-9809-8_20

of cracks which occur internally as well as externally. But it will be difficult to investigate them regularly due to many practical constraints. Thus, this situation demands a health monitoring system that can sense the distresses due to external loads.

Structural health monitoring (SHM) refers to the periodic monitoring of state of stress at a point in a structural element by a structural health monitoring system [1–6]. This input data after appropriate modification can be used to evaluate the extent of the damage. Though the buildings are built with the utmost care, many inherent problems such as improper compaction, aggressive environment, prolonged usage, overloading may lead to the formation of nano cracks, which later gets developed into micro and macro-cracks, failing the structural component. The so-mentioned health monitoring system should be compatible with the components to which it is to be installed [7]. Thus, cement-based and geopolymer-based composites can be used as self-sensing elements which can show variations in its electrical resistance in response to the distress developed. This property of showing variations in electrical resistance under external load is known as piezo-resistivity [8].

Geopolymer-based composites will be an eco-friendly material when compared to cement-based composites. Cement-based products will emit more carbon dioxide which will increase the percentage of greenhouse gases in the atmosphere, thus affecting the environment. Also, their manufacturing consumes more energy and raw materials, thus affecting their sustainability [9–11]. Whereas geopolymers are composites which have aluminosilicate-rich industrial by-products such as fly ash, rice husk ash, slag, mud as their main component [12, 13]. Thus, this type composite not only uses the waste materials but also emits less amount of greenhouse gas, maintaining equilibrium with the environment. Previous researches have focused in producing cement-based sensing materials according to which, cementitious composites can be used to sense the distresses, which shows variation in their electrical resistance [10, 14, 15]. Here, the electrical resistivity of geopolymer-based composites was evaluated and comparing with conventional cement composite for its conductivity performance.

## 2 Components of SHM

A structure health monitoring system comprises of sensor, data acquisition system, communication system, data processing, and storage systems. Among these components, the sensor can be of fiber optic, piezo-electric, and piezo-resistive type [5, 11, 16–19]. Among these types of sensors, piezo-resistive type will be more compatible and durable with the structural system [18]. The various components of SHM are represented in Fig. 1.

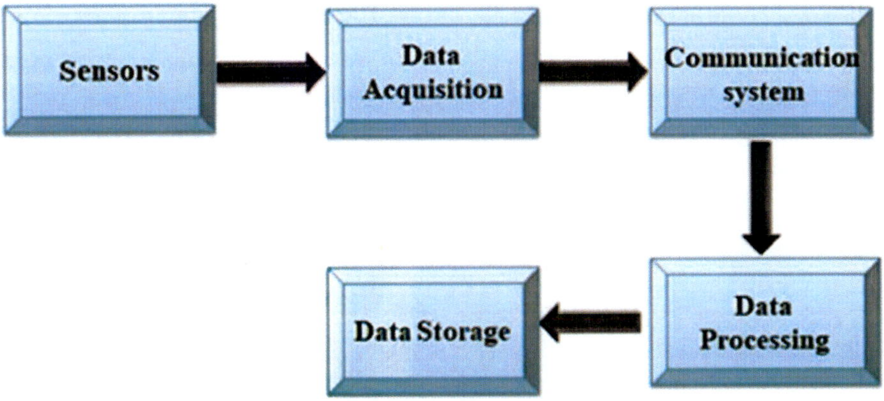


**Fig. 1** Components of SHM

## 3 Need for Current Study

Sensor is the main component which measures the strains, temperature variations, moisture variations, etc. [20]. A conventional non-destructive method assesses the quality of the element only at the surface level. Also, they will not be compatible with the structural components to which they are installed, whereas the cementitious composite-based SHM system will be compatible and helps us to assess the damages at the interior of the elements. The service life of existing sensors such as strain gauges and fiber optic sensors will be less due to their incompatibility [21]. Here comes the need for a composite which will be compatible with the concrete structures. In this study, cement-based and geopolymer-based composites were evaluated for its conductive ability.

## 4 Piezo-Resistivity

It is the property of a specimen to show the changes in the voltage, i.e., electrical resistivity for the change in deformation (strain). This property helps to determine the strain variations in the specimen. A composite with this property will be more suitable to be used as a self-sensing composite [20, 21].

## 5 Characteristics of Composite

The composite must be compatible with the structural element. It must have good conductivity. To evaluate the conductive property of the composite, its resistivity values were determined for the applied current (1 A) and configuration of the copper plates (electrode).

## 6 Geopolymer Composite

Geopolymer is a composite, which comprises of aluminosilicate-rich cementitious binder which is activated by alkaline hydroxides [22]. The activator can be of sodium or potassium-based. The aluminosilicate material can be fly ash, rice husk ash, ground granulated blast furnace slag (GGBS), silica fume, mica, metakaolin, clay or kaolinite [23–26]. The poly-condensation reaction between aluminosilicate binder and alkaline hydroxide yields the reaction products [27–29].

## 7 Materials

For cement-based composite, cement of grade OPC 53 was used as binder. Portable water was used for mixing. For geopolymer composite, fly ash and ground granulated blast furnace slag (GGBS) were used as binder. Sodium-based alkaline solutions were used as alkaline activators. For preparation of alkaline solution, 98% pure sodium hydroxide pellets and sodium silicate solutions were used. Polycarboxylate ether was used as superplasticizer.

### *7.1 Classification of Fly Ash*

According to IS 3812 (Part 1), fly ash is categorized into class F and class C depending upon the variation in the percentages of various constituents. Class F and class C fly ash are differentiated depending upon the percentage of calcium oxide. As per IS 3812 (Part 1), class F fly ash is one in which the percentage of calcium oxide will be less than 10 and is categorized as siliceous type (low calcium) and class C fly ash is one in which the percentage of calcium oxide will be more than 10 and is categorized as calcareous type (high calcium) [30].

The fly ash to be used in the geopolymer composite was characterized by XRD analysis and was found to be siliceous type (low calcium) fly ash [30]. The percentages of various compositions of the binder arrived based on XRD analysis (Fig. 2) and listed in Table 1.

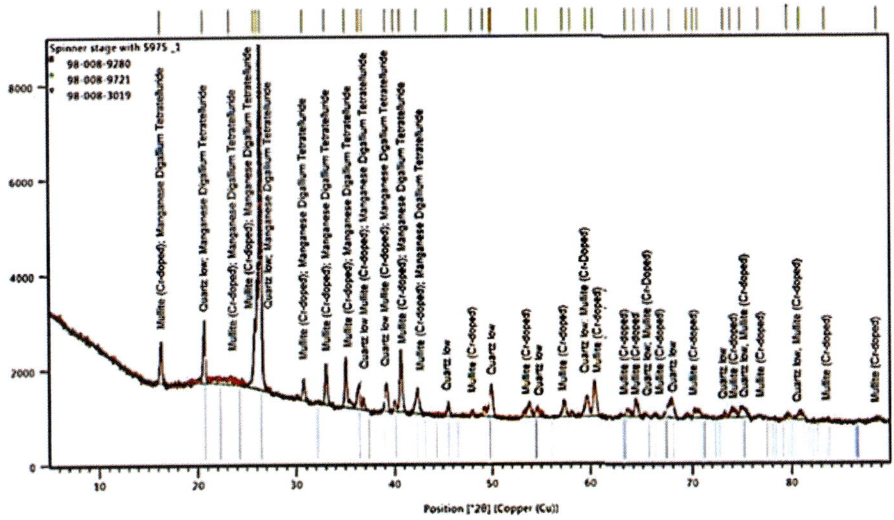

Fig. 2 XRD graph for fly ash

**Table 1** Percentage of oxide components

| S. No. | Compound | Composition (%) |
|---|---|---|
| 1 | $SiO_2$ | 66.94 |
| 2 | $Fe_2O_3$ | 13.59 |
| 3 | CaO | 4.19 |
| 4 | $Al_2O_3$ | 16.99 |
| 5 | MgO | 2.95 |

## 7.2 *Mix Proportion*

### 7.2.1 Mix Proportion for Cement Composite

Cement paste was prepared using cement of grade OPC 53 with water to cement ratio as 0.6. To increase the workability of the paste, superplasticizer (0.8%) was added.

### 7.2.2 Mix Proportion for Geopolymer Composite

Geopolymer paste was prepared incorporating low calcium fly ash and ground granulated blast furnace slag (GGBS) as binders. Sodium hydroxide (10 M) and sodium silicate solutions were used as alkali activators. Based on various trials, the percentage of fly ash and GGBS was fixed as 85% and 15% with liquid to binder ratio as 0.6 and sodium silicate to sodium hydroxide ratio as 1. To increase the workability of

**Table 2** Mix proportions for geopolymer composite

| S. No | Material | Quantity (g) |
|---|---|---|
| 1 | Binder | 170 |
| 2 | Fly ash (85% of binder) | 144.5 |
| 3 | GGBS (15% of binder) | 25.5 |
| 4 | Sodium hydroxide | 51 |
| 5 | Sodium silicate | 51 |
| 6 | Superplasticizer (0.8% of binder) | 1.36 |

the paste, superplasticizer (0.8%) was added. The mix proportions of geopolymer composite are listed in Table 2.

## 8 Methods of Measurement

There are various methods to measure the electrical resistance of the specimen. Among such methods, two-probe and four-probe methods are most commonly used. Two-probe method gives the bulk resistance of the specimen where the electrodes will be attached to the edges of the specimen, whereas the four-probe method gives the surface resistance. In four-probe method, four electrodes will be embedded into the specimen. The electrodes used in either of the method can be of solid plates or meshes or wires. Depending upon the electrode type, their spacing and mesh size (in the case of electrode of mesh type) the resistance measured will vary [4, 31]. In general, four-probe method is preferred, as it eliminates the contact resistance [2].

### *8.1 Specimen Preparation*

Acrylic mold of size 50 mm × 50 mm × 50 mm was used to cast the specimens. Cement specimens were cured in water, and geopolymer specimens were cured at ambient temperature. To measure the electrical resistivity, the electrical resistances of the specimens were determined at the age of 7 and 28 day with and without loading conditions. For electrical resistance measurements, four-probe method was used. Thin copper plates of size 20 mm × 70 mm × 3 mm were used as electrodes. These copper plates were placed with a spacing of 10 mm at the time of casting [32]. Figure 3 shows the specimen with copper plates.

Four-probe method is one in which current will be applied at the outer electrodes and the voltage drop between inner plates will be measured. Figure 4 represents four-probe method of electrical resistance measurement, and Figs. 5 and 6 illustrate the test setup for electrical resistivity measurement without loading.

**Fig. 3** Sample specimen (geopolymer) with copper plates as electrode

## 8.2 *Electrical Resistivity*

The electrical resistivity ($\rho$) of the composites was determined by measuring the voltage value from the specimen for the applied current. For the voltage and current values, electrical resistance value was determined.

$$R = \frac{V}{I} \tag{1}$$

where $R$ is the resistance in Ohm, $V$ is the voltage in Volt, and $I$ is the applied current in Ampere.

From the electrical resistance value, the resistivity in Ohm-cm is computed as follows

$$\rho = \frac{RA}{L}$$

where $\rho$ is the resistivity in Ohm-cm, $R$ is the resistance in Ohm, $A$ is the area of electrode in contact with the specimen (20 mm × 50 mm), and $L$ is the length (10 mm) between the electrodes across which the voltage is measured.

The electrical resistivity of cement-based and geopolymer-based composites was determined at 7th day and 28th day. At the age of 28 day, cyclic compressive loading

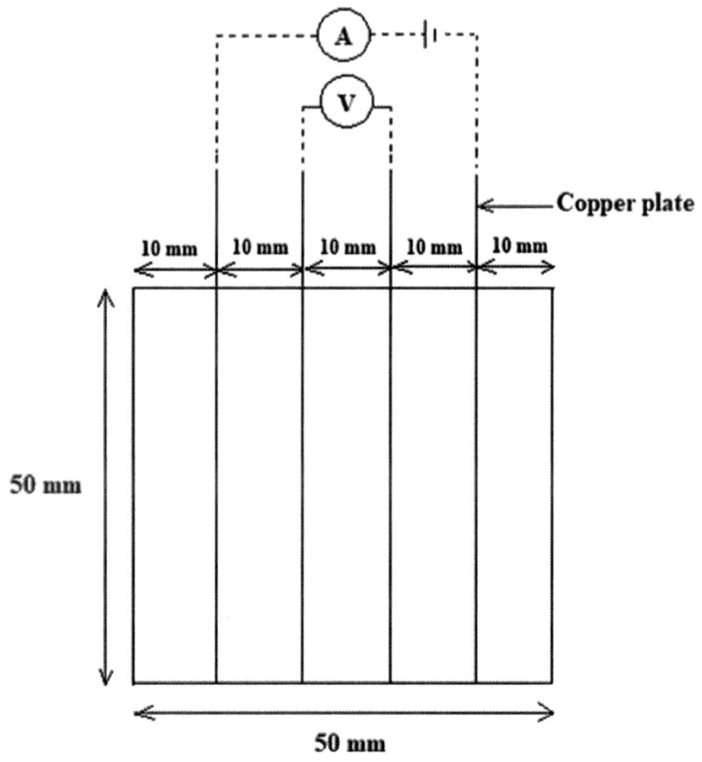


**Fig. 4** Four-probe method of electrical resistance measurement

was applied, and the respective voltage and strain reading were noted. For the different loading cycles, a plot between, load and time was made. At 28th day, the resistivity values were computed using the measured voltage values for the applied current. The cyclic compressive loading was applied using universal testing machine (UTM) of capacity 1000 kN. A load versus time is a graph which was plotted and is shown in Fig. 7.

### 8.2.1 Electrical Resistivity of Cement-Based Composite

Resistivity of cement-based composite varied from $3.98 \times 10^2$ to $5.10 \times 10^2$ $\Omega$ cm at the age of 7 days and 28 days without loading. At the age of 28 days, under cyclic compressive loading the average electrical resistivity was found to be $8.23 \times 10^2$ $\Omega$ cm. This increase in resistivity value is due to the formation of hydration products and usage of pore water for product formation.

 

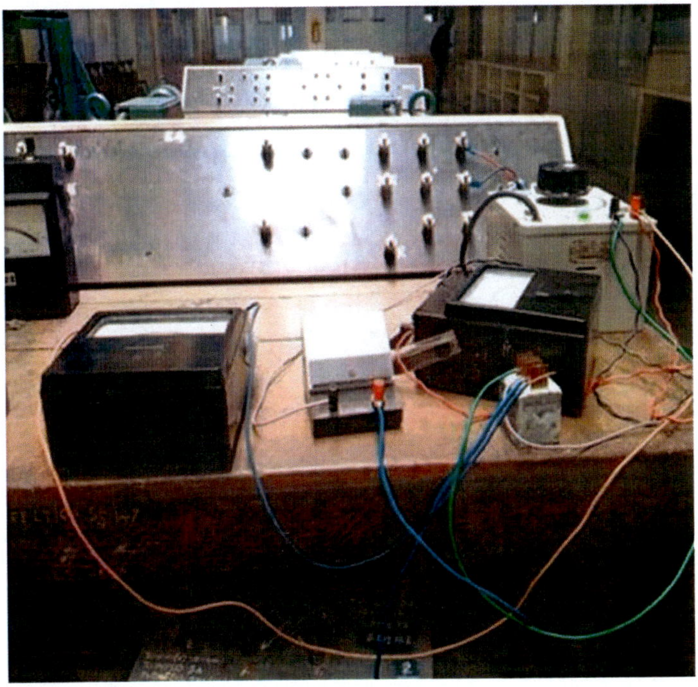

**Fig. 5** Four-probe setup for electrical resistivity measurement

### 8.2.2 Electrical Resistivity of Geopolymer-Based Composite

Resistivity of geopolymer-based composite varied from $1.108 \times 10^2$ to $3.60 \times 10^2$ $\Omega$ cm at the age of 7 days and 28 days without loading. At the age of 28 days, under cyclic compressive loading the average electrical resistivity was found to be $5.23 \times 10^2$ $\Omega$ cm. This increase in resistivity value is due to the formation of polymerization products and usage of alkaline solution for product formation.

## 9 Conclusion

The results reveal that geopolymer-based composites show lesser resistivity on comparison with cement-based composites. Electrical conductivity is inversely proportional to electrical resistivity. Thus, geopolymer-based composite will be more conductive than cement-based composite. This is due to the presence of conductive metal salt (sodium hydroxide) in geopolymer-based composite. The alkaline solution fills the pores, which acts as electrolyte and helps in conductivity [33].Geopolymer-based composite is more effective in responding to the change in distress than the cement-based composite and can be successfully used as base composite in sensor.

Fig. 6 Wires connected to copper electrodes

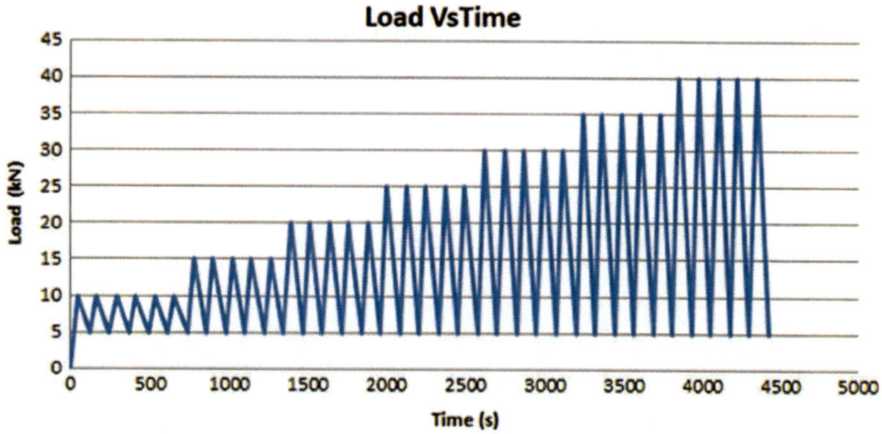


Fig. 7 Load versus time plot

Thus, this base composite after incorporation with electrically conductive materials at optimum percentage can be used as a sensor for SHM applications.

## References


1. Aggelis DG, Alver N, Chai HK (2014) Health monitoring of civil infrastructure and materials. Hindwai Publishing Corporation, UK
2. Han B, Ding S, Yu X (2015) Intrinsic self-sensing concrete and structures: a review. Measurement 59:110–128
3. Chang PC, Flatau A, Liu SC (2003) Review paper: health monitoring of civil infrastructure. Struct Health Monit 2:257–267
4. Han BG, Guan XC, Ou JP (2007) Electrode design, measuring method and data acquisition system of carbon fiber cement paste piezoresistive sensors. Sens Actuators A 135(2):360–369
5. Lee SJ, You I, Zi G, Yoo DY (2017) Experimental investigation of the piezoresistive properties of cement composites with hybrid carbon fibers and nanotubes. Sensors 17(11):2516
6. Rana S, Subramani P, Fangueiro R, Correia AG (2016) A review on smart self-sensing composite materials for civil engineering applications. AIMS Mater Sci 3(2):357–379
7. Han B, Ding S, Yu X (2015) Intrinsic self-sensing concrete structures: a review. Measurement 59(2015):110–112
8. Zhu SR, Chung DDL (2007) Numerical assessment of the methods of measurement of the electrical resistance in carbon fiber reinforced cement. Smart Mater Struct 16:1164–1170
9. Kupaei RH (2013) Mix design for flyash based oil palm shell geopolymer light weight concrete. Constr Build Mater 43:490–496
10. Smart sensor phenomena, technology, networks, and systems. In: Proceedings of SPIE, vol 6933, San Diego, Calif, USA, p 10
11. Rees CA, Provis JL, Lukey GC, Van Deventer JSJ (2007) Attenuated total reflectance Fourier transform infrared analysis of fly ash geopolymer gel ageing. Langmuir 23(15):8170–8179
12. Duxson P, Fernandez-Jimenez A, Provis JL, Lukey GC, Palomo A, Van Deventer JSJ (2007) Geopolymer technology: the current state of the art. J Mater Sci 42(9):2917–2933
13. Saafi M, Tang L, Fung J, Rahman M, Sillars F, Liggat J, Zhou X (2014) Graphene/fly ash geopolymeric composites as self-sensing structural materials. Smart Mater Struct 23
14. Dong W, Li W, Tao Z, Wang K (2019) Piezoresistive properties of cement-based sensors: review and perspective. Constr Build Mater 203:146–163
15. Monteiro AO, Loredo A, Costa P, Oeser M, Cachim PB (2017) A pressure-sensitive carbon black cement composite for traffic monitoring. Constr Build Mater 154:1079–1086
16. Lim YY, Bhalla S, Soh CK (2006) Structural identification and damage diagnosis using selfsensing piezo-impedance transducers. Smart Mater Struct 15(987–995):9
17. Ou J, Zhou Z (2008) Applications of optical fiber sensors of SHM in infrastructures. Smart Sens Phenom Technol Networks Sys:693311
18. Sun M, Staszewski WJ, Swamy RN (2010) Smart sensing technologies for structural health monitoring of civil engineering structures. Adv Civil Eng 2010:1–13
19. Wallah SE, Rangan BV (2006) Low-calcium fly ash based geopolymer concrete: long term properties. Technical Report No. GC 2, Curtin University of Technology, Perth, Australia
20. Azhari F (2008) Cement-based sensors for structural health monitoring. Post graduate thesis, The University of British, Columbia
21. Ackermann KC (2018) Self-sensing concrete for structural health monitoring of smart infrastructures. Open access master's thesis. Paper 1285
22. Xu H, Van Deventer JSJ (2000) The geopolymerisation of alumino-silicate minerals. Int J Miner Process 59(3):247–266
23. Rizzo P, Di Scalea FL (2006) Wavelet-based feature extraction for automatic defect classification in strands by ultrasonic structural monitoring. Smart Struct Syst 2:253–274
24. Sun MQ, Staszewski WJ, Swamy RN et al (2008) Application of low-profile piezo-ceramic transducers for health monitoring of concrete structures. NDT E Int 41:589–595
25. Wan KT, Leung CKY (2007) Applications of a distributed fiber optic crack sensor for concrete structures. Sens Actuators A 135(458–464):11

26. Payakaniti P, Pinitsoonthorn S, Thongbai P, Amornkitbamrung V, Chindaprasirt P (2018) Effects of carbon fiber on mechanical and electrical properties of fly ash geopolymer composite. Mater Today Proc 5:14017–14025
27. Davidovits J (1994) Properties of geopolymer cements. In: Proceedings of first international conference on alkaline cements and concretes. KIEV, Ukraine
28. Mackenzie KJD (2003) What are these things called geo-polymers? A physio-chemical perspective. Ceram Trans 153:175–186
29. Payakaniti P, Pinitsoonthorn S, Thongbai P, Amornkitbamrung V, Chindaprasirt P (2017) Electrical conductivity and compressive strength of carbon fiber reinforced fly ash geopolymeric composites. Constr Build Mater 135:164–176
30. IS 3812(Part 1):2013 Pulverised fuel ash—specification
31. Ye XW, Su YH, Han JP (2014) Structural health monitoring of civil infrastructure using optical fiber sensing technology: a comprehensive review. Sci World J 2014
32. Vaidya S, Allouche EN (2011) Strain sensing of carbon fiber reinforced geopolymer concrete. Mater Struct 44:1467–1475
33. Hanjitsuwan S, Chindaprasirt P, Pimraksa K (2011) Electrical conductivity and dielectric property of fly ash geopolymer pastes. Int J Min Metall Mater 18:94–99

# Techniques on Corrosion Prevention and Rust Removal on Different Steels

**K. S. Gowri Shankar and K. R. Ponnsahana**

**Abstract** The paper aims to fetch data on corrosion prevention and rust removal in steel structures for engineers. Corrosion, a natural process that occurs in the moist atmosphere in which the chemically active metals get corrode, whereas rusting is a process in which iron corrodes due to atmospheric exposure. This study fulfils the knowledge gap through reviewing the correlation between corrosion prevention and rust removal process in various steel structures. The steels under concern are mild steel, carbon steel, weathering steel, base metal and stainless steel. The comparisons help in representing the data gathered from many research reports and journal papers in graphs and tables under different environmental conditions. By taking those environmental conditions and cleaning efficiency into account, the techniques (use of inhibitors, physical and chemical methods) are chosen serving as a foundation for the upcoming research.

**Keywords** Corrosion prevention · Rust removal · Techniques · Inhibitors

## 1 Introduction

The stability of a steel structure is affected mainly due to corrosion, which then further progressively weakens the structure. Cracks in external walls are the effects of extensive corrosion of a steel works. The heavy corrosion of a structure may lead to sudden collapse. Deficiency in design, presence of salts in the brickwork and inadequate painting of the steel leads to steel corrosion [1]. Thus, rust prevention is mainly essential to maintain the stability. The protocols for the prevention of corrosion must be based on the steel type and rust condition. Different methods and strategies were

K. S. Gowri Shankar (✉) · K. R. Ponnsahana
Department of Civil Engineering, Kongu Engineering College, Perundurai, Erode, Tamil Nadu, India
e-mail: gowrishankarks@gmail.com

K. R. Ponnsahana
e-mail: ponnsahana42@gmail.com

© The Editor(s) (if applicable) and The Author(s), under exclusive license to Springer Nature Singapore Pte Ltd. 2021
S. Mohan et al. (eds.), *Materials, Design, and Manufacturing for Sustainable Environment*, Lecture Notes in Mechanical Engineering,
https://doi.org/10.1007/978-981-15-9809-8_21

proposed for the metals protection against corrosion. Among the widespread techniques, the most affordable method of application is usage of inhibitors. As a first step, the corrosion inhibitors were developed. The corrosion inhibitors are categorized into three types based on nature, namely organic, inorganic and mixed inhibitor. Earliest inhibitor developed was inorganic inhibitor which was followed by organic inhibitors. Though organic compounds are highly efficient, they produced some hazardous substances causing harm to environmental as well as human health. Later on, mixed inhibitor was developed to overcome the drawbacks in both inorganic and organic inhibitors showing excellent performance. In addition, they were of less cost and eco-friendly [2].

The study intends to review the rust removal and preventive methods in steel structures made of different steel material and also exposed to different environmental conditions. The techniques used will be taken into consideration based on the efficiency obtained as a result of comparing inhibitory steel and non-inhibitory steel with the help of scanning electron microscopy (SEM), weight loss, electrochemical analysis, X-ray photoelectron spectroscopy (XPS), computational calculations and energy dispersive spectroscopy (EDS) examination. The study aims to serve as a foundation for the upcoming research.

# 2 Different Steels and Their Techniques

## *2.1 Mild Steel*

Mild steel corrosion is a crucial affair in the industry. Mild steel finds it wide application in construction industry as building materials to install the big structures like bridges, high-rise buildings and dams. Extensive use of mild steel in most industrial sector is because of its availability and less cost [3]. The composition of chemicals in mild steel is as shown in Table 1.

Ageing of the embedded steel in harsh environment prospects the corrosion problems. Long-term atmospheric corrosion was low in countryside and metropolitan cities but high in industrial and marine areas [4]. The problem of getting corroded when exposed to acidic environmental conditions is overcome with the help of mixed type corrosion inhibitors [2–6]. Nature of the inhibitor was demonstrated with the help of potentiodynamic [2–12] and Tafel polarization curves [13]. The *Sesamum indicum* natural oil [2], 4 Amino-3-butyl-5-mercapto-1,2,4-triazole (ABMT) [6] and 4-Aminotriazole derivative (4-MAT) [5] were some of the good performing mixed

**Table 1** Composition of chemicals in mild steel [2]

| Element | C | Mn | Si | P | S | Al | Ni | Fe |
|---|---|---|---|---|---|---|---|---|
| % composition | 0.15 | 0.45 | 0.18 | 0.01 | 0.031 | 0.005 | 0.008 | Balance |

inhibitor seen adsorbed on the metal surface by following Langmuir [2, 5], El-Awady [4], Frumkin's [13] and Temkins adsorption isotherm [6].

The inhibitors are chosen according to the environmental nature. In hydrochloric acid-rich (HCl) environment, indicators like *Sesamum indicum* natural oil, poly (vinyl alcohol-cysteine), 4-Aminotriazole derivative (4-MAT) shows an excellent inhibition efficiencies of about 85.74%, 95.39%, 98% at concentrations 20 v/v, 0.6% (in g), $4 \times 10^{-4}$ M [2, 4, 5, 14]. Sesamum indicum natural oil [2], 4 Amino-3-butyl-5-mercapto-1,2,4-triazole (ABMT) [6] had 98.04%, 89% efficiency on exposure to sulphuric acid rich environment with 50 v/v and 1000 ppm concentrations. Few inhibitors formed a film on the steel by degradation of its molecules [10] and also by using Langmuir adsorption model [2, 6]. The upcoming inhibitors, namely chitosan and gum Arabic of concentration 0.125 g/L to 1 g/L, showed 77%, 83.85% inhibition efficiency (IE), respectively, were of filmy nature [7, 13]. One such includes, polyethylene Glycol (PEG) of 2.5 g/L concentration with cerium nitrate formed a $CeO_2$ coating increased the corrosion inhibition capacity and stability of the obtained films was non-toxic and respectful [8].

Among all the inhibitors, Schiffs base of 0.219 mM added to K1 of 0.602 mM concentrations yielded the atmost efficiency of 99.03% on exposure to severe sulphuric acid environment [5]. Whereas in marine conditions, Neodymium–Benzimidazole on addition to salt water yielded the maximum corrosion inhibition efficiency of 83.6% and 9 times greater response than the non-inhibitory [3, 11, 15, 16]. Similarly, thermally sprayed aluminium coatings served as a barricade to electrolyte penetration in marine conditions showed better results preventing the intrusion of salt inside the steel [9, 11, 17, 18]. Likewise, in case of manufacturing a waterproof and corrosion resistance steel, simple spraying and thermal curing techniques played important role [3, 15, 17]. One such includes the incorporation of the amino modified silica nanoparticles into the cardanol-based polybenzoxazine (PC-a/$SiO_2$) [16, 19–21]. Elseways, bacterial action in steel can be prevented by suggesting a coating of modified FCC catalyst, i.e. adhesive and FCC catalyst mixed in 1:1 proportion inhibited the bacterial growth by forming an inhibition zone with 94% efficiency after removal from buried soil for 30 days [4, 10, 22].

Scanning electron microscopy (SEM) [2–4, 10, 11, 15, 17, 19–21], EDS analysis [2, 3, 11, 20, 21], XPS [3, 20–24], energy dispersive X-ray analysis (EDX) [4, 22], contact angle measurements [3], X-ray diffraction (XRD) [9, 10, 15], weight loss [4, 6, 7], electrochemical analysis [6, 23], atomic force microscope (AFM) [11], X-ray photoelectronscopy (XPS) [3, 11, 24], electrochemical impedance spectroscopy (EIS) methods [4, 7–11, 15, 16, 21, 24, 25], single sine electrochemical impedance spectroscopy (SSEIS) [13], surface analysing techniques [22] and computational calculations [4–6] were used for analysing the IE of the steel. Whereas the inhibitor effect was jointly allied to the $E_{HOMO}$ [26], $E_{LUMO}$ [26], hardness, polarizability, softness, electrophilicity index, dipole moment and charges [27]. It was observed in 4 Sulphanoamides [Sulfacetamide (SAM), Sulfapyridine (SPY), Sulfamerazine (SMR), Sulfathiazole (STI)] inhibitor which was effective and harmless. The %IE was in the order SMR > SPY > STI > SAM close to the experimental corrosion

**Table 2** Composition of chemicals in carbon steel [29]

| Composition of chemicals | | | | |
|---|---|---|---|---|
| C | Mn | P | Si | Fe |
| 0.2 | 0.35 | 0.024 | 0.003 | Rest |

efficiency of the inhibitor [26]. The examination revealed that the %IE and corrosion resistance increases with increase in concentration and time of exposure [2, 6].

## 2.2 *Carbon Steel*

The galvanostatic polarization helped to determine the nature of the inhibitor [28]. The composition of chemicals in carbon steel is shown in Table 2.

Poly(*N*-methylpyrrole) prevented corrosion by forming a passive film inhibiting the anodic dissolution at 3.38 $mA/cm^2$ current density producing 70% IE at 1.4 and 2.4 pH [30, 31]. 4-phenylazo-3-methyl-2-pyrazolon-5-one, a mixed type inhibitor inhibited corrosion in severe hydrochloric acid-rich condition with 77.5% IE at a concentration of about $11 \times 10^{-6}$ at 30 °C [32]. Similar to weathering steel, nanoparticles like magnetite nanoparticles with 5–20 nm thickness showed excellent inhibition in severe HCl medium [14, 33]. Currently for highly corrosive surroundings, an average of 70–80% IE was achieved by the Al-Zn base sacrificial anode method for both horizontal and vertical elements [34–37]. The horizontal element proved to resist more. The inhibitors used were commonly of adsorption nature and are adsorbed by Frumkin's adsorption isotherm. DC polarization [30], EIS [14, 27–30, 38–41], XRD [42], weight loss [31, 38, 43, 44], galvanostatic polarization techniques [14, 28, 29, 32, 38], SEM [31] and XPS analysis [38, 43] were used for finding the inhibition efficiency. Inhibitor efficiency increased with inhibitor concentration and decreased with temperature [29, 40, 45–47].

## 2.3 *Weathering Steel*

Weathering steels or low alloyed steels or low carbon steels were formed by increasing the proportions of Cu, Cr, Ni, P, Si, Mn, etc., found normally in mild steel [48, 49]. Their usage is increasing in applications like transmission towers, bridges, roofings. Low alloyed steel does not need painting and can also be painted for other normal conditions [50]. X-ray diffraction, infrared spectroscopy [51], Raman spectroscopy [51], electron probe micro-analysis, optical microscopy, scanning electron microscopy, and transmission electron microscopies discussed that the products formed after the modification of the composition of weathering steel produced packed

**Table 3** Composition of chemicals in weathering steel [52]

| Specimen | Composition of chemicals (mass %) | | | | | | |
|---|---|---|---|---|---|---|---|
| | C | Si | Mn | P | Cr | Cu | Ni |
| Advanced weathering steel | 0.05 | 0.04 | 1.02 | 0.008 | – | 0.40 | 3.03 |
| Conventional weathering steel | 0.10 | 0.42 | 1.54 | 0.004 | 0.52 | 0.30 | 0.32 |

nano αFeOOH inhibiting further corrosion [38, 52]. The composition of chemicals in advanced and conventional weathering steels is shown in Table 3.

This paved the concept of "the evolution of the Fe $(O,OH)_6$ nano-network" explained that the $CrO_2^-$ rusts formed as a result of corrosion with a thickness of 5–20 nm after long-term atmospheric exposure acted as a high corrosion resistor. Experimental results proved that the weathering steel mentioned with the above composition of chemicals showed corrosion loss lesser than 1/20 of the conventional weathering steel after 9 years of exposure. Synchroton radiation helped in having a view on the formation of rust [20, 21, 52]. The concept paved the way for nano-structured sol–gel hybrid coatings with Zr/Si of 1:9 and 1:4 atomic ratios producing $ZrO_2$ nanoparticle of diameter 40 and 200 nm with perfect blocking effect [20, 53]. The bonding performance of the base-rust and rust density plays significant part in rust protection. Ultra low carbon steel lacks the corrosion resistance nature [48]. With those bases, corrosion formed under severe marine condition was removed with the help of UV illumination by following the photovoltaic effect. In such case the thickness, conductive nature, properties and composition of the corrosion products were considered [50, 54, 55]. In the above cases, the inner layer of the rust remained dense and crack-free, whereas the outer layer was loose and lacked continuity. This problem was overcome by removing the outer layer of rust showing reduction in corrosion compared with those with rust having water holding nature [51, 54, 56]. The analysing yielded that the increase in corrosion resistance is achieved by adding 3 wt% of Ni [48]. At present, the weathering steels are protected from corrosion by using inhibitor coatings. One such includes, $CeO_2/Mg(OH)_2$ mixed coating was used for corrosion inhibition in severe marine conditions [57].

## 2.4 *Base Metals*

Copper, lead, nickel, tin, aluminium and zinc generally come under base metals having wide application in construction and manufacturing. Similar to mild steel the potentiodynamic curves played important role in determining the nature of inhibitor and in addition analyses were done by using SEM–EDX, XRD and Mossbauer analyses [36, 58]. Epoxy sealing applied by arc thermal metal spraying method with 0.60 V potential difference showed the ultimate anticorrosion property [58]. Ionic

**Table 4** Composition of chemicals in stainless steel [60, 61]

| Chemical composition | | | | | | | | | |
|---|---|---|---|---|---|---|---|---|---|
| Ni | Cr | Fe | Mo | C | Mn | Si | S | P | Others |
| 25.09 | 20.77 | 43.97 | 4.39 | 0.0013 | 1.84 | 1.45 | 0.007 | 0.0029 | 2.36 CU |

liquid, a mixed type inhibitors, was used as a coating by using Langmuir adsorption isotherm due to its lower volatility, fireproof, non-toxic, rich thermal and chemical stability [59].

## 2.5 *Stainless Steel*

Tafel polarization, potentiodynamic polarization, open-circuit potential measurement, EIS, SEM and optical microscopy were used for determining the nature of the inhibitor [35, 60, 61]. The inhibitor was adsorped by obeying the Langmuir adsorption model [61]. The composition of chemicals in stainless steel is shown in Table 4.

Inhibitor efficiency increased with increment in inhibitor concentration [60–64]. Graphene oxide with $NaNO_3$ (oxidizing agent) provided inhibition in hydrochloric acid-rich environment by electrophoresis [65]. $NaNO_2$ acted as a corrosion inhibitor for the ease of $CaCO_3$ scaling [66]. Artemisia herba-alba essential oil AHAO, a plant extract and mixed-type inhibitor yielded 88% inhibition efficiency at 1 g/L with temperature ranging between 298 to 353 K in acidic media rich in phosphor [61]. Nano-composite coating of mixture $ZrO_2$ and $SiO_2$ enhanced passive oxide film acting as a barrier [62, 67]. Analysis was done by energy dispersion spectroscopy, X-ray diffraction, Raman spectra and infrared spectra [61–64, 67].

# 3 Conclusions

In short, this over review attempts to show the state of knowledge about the techniques to inhibit corrosion. Steels are indispensable components of construction and corrosion being one of the dire ultimatum to the durability performance. Aiming at this controversy, the rust removal techniques have been applied to enhance its corrosion resistance. This study examined the prevention of corrosion development and rust removal in five types of non-identical steels under different climatic conditions. The cardinal inferences of this study are shown below:

1. In many instances, the assay of corrosion inhibition in steel is carried out in hydrochloric acid, the most hostile acidic medium. Numerous techniques have been isolated to prevent the acid from direct contact with the steel.

2. Coating is the main entente to amplify the anticorrosion effect of the steel. The coatings generally preferred are paints acting as inhibitor. The study reviews some other inhibitors preventing corrosion development and rust clearance.
3. Carbon steels are normally coated to inflate corrosion and regular recoating is needed, whereas weathering steel can be boldly exposed to the surrounding without coatings. But toppling of chloride ions, deteriorate the protective layer.
4. In all the cases mentioned above, a similarity has been observed irrespective of their nature of steel and environmental condition; i.e. increase in concentration increases the inhibition efficiency and is vice versa in case of temperature with efficiency. Thus, the concentration of inhibitor with their efficiencies and environment used has been critically reviewed.
5. Works in sulphuric condition are limited since such studies are still under investigation and results were comparatively lower than in hydrochloric medium.

There are more keen works to understand. With huge confidence that this paper helps in allowing more pieces of the enigma to be filled in.

## References


1. Refait P, Jeannin M, Sabot R, Antony H, Pineau S (2015) Corrosion and cathodic protection of carbon steel in the tidal zone: products, mechanisms and kinetics. Corros Sci 90:375–382
2. Abdulwahab M, Fayomi O (2012) Corrosion inhibition of mild steel in *Sesamum indicum*-2M $HCl/H_2SO_4$ interface. Int J Electrochem Sci 7:5805–5816
3. Ralkhal S, Shahrabi T, Ramezanzadeh B (2019) Synthesis and construction of a highly potent hybrid organic/inorganic anti-corrosive pigment for effective corrosion control of mild steel in simulated seawater. Constr Build Mater 222:400–413
4. Rahiman AFSA, Sethumanickam S (2017) Corrosion inhibition, adsorption and thermodynamic properties of poly (vinyl alcohol-cysteine) in molar HCl. Arab J Chem 10:S3358–S3366
5. Bentiss F, Jama C, Mernari B, El Attari H, El Kadi L, Lebrini M et al (2009) Corrosion control of mild steel using 3, 5-bis (4-methoxyphenyl)-4-amino-1, 2, 4-triazole in normal hydrochloric acid medium. Corros Sci 51(8):1628–1635
6. Quraishi M, Sharma HK (2003) 4-amino-3-butyl-5-mercapto-1, 2, 4-triazole: a new corrosion inhibitor for mild steel in sulphuric acid. Mater Chem Phys 78(1):18–21
7. Azzaoui K, Mejdoubi E, Jodeh S, Lamhamdi A, Rodriguez-Castellón E, Algarra M et al (2017) Eco friendly green inhibitor gum arabic (GA) for the corrosion control of mild steel in hydrochloric acid medium. Corros Sci 129:70–81
8. Boudellioua H, Hamlaoui Y, Tifouti L, Pedraza F (2019) Effects of polyethylene glycol (PEG) on the corrosion inhibition of mild steel by cerium nitrate in chloride solution. Appl Surf Sci 473:449–460
9. Esfahani EA, Salimijazi H, Golozar MA, Mostaghimi J, Pershin L (2012) Study of corrosion behavior of arc sprayed aluminum coating on mild steel. J Therm Spray Technol 21(6):1195–1202
10. Devikala S, Kamaraj P, Arthanareeswari M, Pavithra S (2019) Green corrosion inhibition of mild steel by asafoetida extract in 3.5% NaCl. Mater Today Proc 14:590–601
11. Majd MT, Shahrabi T, Ramezanzadeh B (2019) Production of an eco-friendly anti-corrosion ceramic base nanostructured hybrid-film based on Nd (III)-C7H6N2 on the mild steel surface; electrochemical and surface studies. Constr Build Mater 221:456–468

12. Muralidharan S, Quraishi M, Iyer S (1995) The effect of molecular structure on hydrogen permeation and the corrosion inhibition of mild steel in acidic solutions. Corros Sci 37(11):1739–1750
13. Srivastava M, Srivastava S, Ji G, Prakash R (2019) Chitosan based new nanocomposites for corrosion protection of mild steel in aggressive chloride media. Int J Biol Macromol 140:177–187
14. Khamis EA, Hamdy A, Morsi RE (2018) Magnetite nanoparticles/polyvinyl pyrrolidone stabilized system for corrosion inhibition of carbon steel. Egypt J Petrol 27(4):919–926
15. Laourayed M, El Moudane M, Khachani M, Boudalia M, Guenbour A, Bellaouchou A, Zarrouk A (2019) Thermal, structural and corrosion inhibition performances of a new phosphate glasses on mild steel in HCl medium. Chem Data Collections 100305
16. Huang J, Lou C, Xu D, Lu X, Xin Z, Zhou C (2019) Cardanol-based polybenzoxazine superhydrophobic coating with improved corrosion resistance on mild steel. Prog Org Coat 136:105191
17. Singh AAMM, Franco PA, Binoj J (2019) Enhancement of corrosion resistance on plasma spray coated mild steel substrate exposed to marine environment. Mater Today Proc 15:84–89
18. Chaves IA, Jeffrey R, Melchers RE (2015) Rust removal from steel coupons after short-term marine immersion. Corrosion 71(7):811–818
19. Xu D, Lou C, Huang J, Lu X, Xin Z, Zhou C (2019) Effect of inhibitor-loaded halloysite nanotubes on active corrosion protection of polybenzoxazine coatings on mild steel. Prog Org Coat 134:126–133
20. Zheludkevich M, Serra R, Montemor M, Yasakau K, Salvado IM, Ferreira M (2005) Nanostructured sol–gel coatings doped with cerium nitrate as pre-treatments for AA2024-T3: corrosion protection performance. Electrochim Acta 51(2):208–217
21. Zheludkevich M, Serra R, Montemor M, Salvado IM, Ferreira M (2006) Corrosion protective properties of nanostructured sol–gel hybrid coatings to AA2024-T3. Surf Coat Technol 200(9):3084–3094
22. Thukkaram M, Cools P, Nikiforov A, Rigole P, Coenye T, Van Der Voort P et al (2019) Antibacterial activity of a porous silver doped $TiO_2$ coating on titanium substrates synthesized by plasma electrolytic oxidation. Appl Surf Sci 144235
23. Hashim NZN, Kassim K, Zaki HM, Alharthi AI, Embong Z (2019) XPS and DFT investigations of corrosion inhibition of substituted benzylidene Schiff bases on mild steel in hydrochloric acid. Appl Surf Sci 476:861–877
24. Li P, Lin J, Tan K, Lee J (1997) Electrochemical impedance and X-ray photoelectron spectroscopic studies of the inhibition of mild steel corrosion in acids by cyclohexylamine. Electrochim Acta 42(4):605–615
25. Collazo A, Nóvoa X, Pérez C, Puga B (2008) EIS study of the rust converter effectiveness under different conditions. Electrochim Acta 53(25):7565–7574
26. Ebenso EE, Arslan T, Kandemirlı F, Love I, Öğretır Cl, Saracoğlu M, Umoren SA (2010) Theoretical studies of some sulphonamides as corrosion inhibitors for mild steel in acidic medium. Int J Quantum Chem 110(14):2614–2636
27. Cook DC (2005) Spectroscopic identification of protective and non-protective corrosion coatings on steel structures in marine environments. Corros Sci 47(10):2550–2570
28. González-García Y, González S, Souto R (2007) Electrochemical and structural properties of a polyurethane coating on steel substrates for corrosion protection. Corros Sci 49(9):3514–3526
29. El-Katori EE, Al Angari YM, Abousalem AS (2019) Corrosion mitigation of carbon steel by spin coating with Ag–$TiO_2$ gel films in acidic solution: fabrication, characterization, electrochemical and quantum chemical approaches. Surf Coat Technol 374:852–867
30. Iroh JO, Su W (2000) Corrosion performance of polypyrrole coating applied to low carbon steel by an electrochemical process. Electrochim Acta 46(1):15–24
31. Ayukayeva VN, Boiko GI, Lyubchenko NP, Sarmurzina RG, Mukhamedova RF, Karabalin US, Dergunov SA (2019) Polyoxyethylene sorbitan trioleate surfactant as an effective corrosion inhibitor for carbon steel protection. Colloids Surf A 579:123636

32. Fouda A, Al-Sarawy A, El-Katori E (2006) Pyrazolone derivatives as corrosion inhibitors for C-steel in hydrochloric acid solution. Desalination 201(1–3):1–13
33. Cen H, Chen Z, Guo X (2019) N, S co-doped carbon dots as effective corrosion inhibitor for carbon steel in $CO_2$-saturated 3.5% NaCl solution. J Taiwan Inst Chem Eng 99:224–238
34. Charrier C, Jacquot P, Denisse E, Millet J, Mazille H (1997) Aluminium and Ti/Al multilayer PVD coatings for enhanced corrosion resistance. Surf Coat Technol 90(1–2):29–34
35. Fushimi K, Naganuma A, Azumi K, Kawahara Y (2008) Current distribution during galvanic corrosion of carbon steel welded with type-309 stainless steel in NaCl solution. Corros Sci 50(3):903–911
36. Vu A, Vuillemin B, Oltra R, Allély C (2011) Cut-edge corrosion of a Zn–55Al-coated steel: a comparison between sulphate and chloride solutions. Corros Sci 53(9):3016–3025
37. Marder A (2000) The metallurgy of zinc-coated steel. Prog Mater Sci 45(3):191–271
38. Guo J, Yang S, Shang C, Wang Y, He X (2009) Influence of carbon content and microstructure on corrosion behaviour of low alloy steels in a Cl− containing environment. Corros Sci 51(2):242–251
39. Almeida E, Santos D, Uruchurtu J (1999) Corrosion performance of waterborne coatings for structural steel. Prog Org Coat 37(3–4):131–140
40. Abdou M, Fadl A (2019) Assessment of nano-$FeTiO_3$/non crystalline silica cold galvanizing composite coating as a duplex corrosion guard system for steel electricity transmission towers in severe aggressive media. Constr Build Mater 223:705–723
41. Akkouche R, Rémazeilles C, Jeannin M, Barbalat M, Sabot R, Refait P (2016) Influence of soil moisture on the corrosion processes of carbon steel in artificial soil: active area and differential aeration cells. Electrochim Acta 213:698–708
42. Barchiche C, Sabot R, Jeannin M, Refait P (2010) Corrosion of carbon steel in sodium methanoate solutions. Electrochim Acta 55(6):1940–1947
43. Ghareba S, Omanovic S (2010) Interaction of 12-aminododecanoic acid with a carbon steel surface: towards the development of 'green' corrosion inhibitors. Corros Sci 52(6):2104–2113
44. Dall'Agnol LT, Cordas CM, Moura JJ (2014) Influence of respiratory substrate in carbon steel corrosion by a sulphate reducing prokaryote model organism. Bioelectrochemistry 97:43–51
45. Hansen DC, Zimlich KR, Bennett BN (2019) Inhibition of flash rusting of HY80 by a mussel adhesive protein: characterizing the interaction of MeFP-5 with a high strength low alloy steel. Electrochim Acta 301:411–420
46. Humble H (1949) Cathodic protection of steel piling in sea water. Corrosion 5(9):292–302
47. Hurley MF, Scully JR (2006) Threshold chloride concentrations of selected corrosion-resistant rebar materials compared to carbon steel. Corrosion 62(10):892–904
48. Díaz I, Cano H, Lopesino P, de la Fuente D, Chico B, Jiménez JA et al (2018) Five-year atmospheric corrosion of Cu, Cr and Ni weathering steels in a wide range of environments. Corros Sci 141:146–157
49. Liu C, Revilla RI, Zhang D, Liu Z, Lutz A, Zhang F et al (2018) Role of Al2O3 inclusions on the localized corrosion of Q460NH weathering steel in marine environment. Corros Sci 138:96–104
50. Albrecht P, Hall TT Jr (2003) Atmospheric corrosion resistance of structural steels. J Mater Civ Eng 15(1):2–24
51. Zhang X, Yang S, Zhang W, Guo H, He X (2014) Influence of outer rust layers on corrosion of carbon steel and weathering steel during wet–dry cycles. Corros Sci 82:165–172
52. Kimura M, Kihira H, Ohta N, Hashimoto M, Senuma T (2005) Control of Fe (O, OH) 6 nano-network structures of rust for high atmospheric-corrosion resistance. Corros Sci 47(10):2499–2509
53. Zohdy K, El-Shamy A, Kalmouch A, Gad EA (2019) The corrosion inhibition of (2Z, 2′ Z)-4, 4′-(1, 2-phenylene bis (azanediyl)) bis (4-oxobut-2-enoic acid) for carbon steel in acidic media using DFT. Egypt J Petrol 28(4):355–359
54. Song L, Ma X, Chen Z, Hou B (2014) The role of UV illumination on the initial atmospheric corrosion of 09CuPCrNi weathering steel in the presence of NaCl particles. Corros Sci 87:427–437

55. López-Ortega A, Bayón R, Arana J, Arredondo A, Igartua A (2018) Influence of temperature on the corrosion and tribocorrosion behaviour of high-strength low-alloy steels used in offshore applications. Tribol Int 121:341–352
56. Kubzova M, Krivy V, Kreislova K, Hong HPT (2019) Amount of chlorides in corrosion products of weathering steel. Transp Res Procedia 40:751–758
57. Aggoun K, Chaal L, Creus J, Sabot R, Saidani B, Jeannin M (2019) Marine corrosion resistance of $CeO_2$/Mg (OH) 2 mixed coating on a low alloyed steel. Surf Coat Technol 372:410–421
58. Choe H-B, Lee H-S, Shin J-H (2014) Experimental study on the electrochemical anti-corrosion properties of steel structures applying the arc thermal metal spraying method. Materials 7(12):7722–7736
59. Verma C, Ebenso EE, Quraishi M (2017) Ionic liquids as green and sustainable corrosion inhibitors for metals and alloys: an overview. J Mol Liq 233:403–414
60. de Lima HMLF, Tavares SSM, Martins M, Araújo WS (2019) The effect of copper addition on the corrosion resistance of cast duplex stainless steel. J Mater Res Technol 8(2):2107–2119
61. Boudalia M, Fernández-Domene R, Tabyaoui M, Bellaouchou A, Guenbour A, García-Antón J (2019) Green approach to corrosion inhibition of stainless steel in phosphoric acid of *Artemesia herba* albamedium using plant extract. J Mater Res Technol 8(6):5763–5773
62. Farhadian M, Raeissi K, Golozar M, Labbaf S, Hajilou T, Barnoush A (2019) Electrophoretic deposition and corrosion performance of zirconia-silica composite coating applied on surface treated 316L stainless steel: toward improvement of interface structure. Surf Coat Technol 125015
63. Gedge G (2008) Structural uses of stainless steel—buildings and civil engineering. J Constr Steel Res 64(11):1194–1198
64. Loto RT (2019) Corrosion resistance and morphological deterioration of 316Ti austenitic, GX4CrNiMo16-5-1 martensitic and 444 ferritic stainless steels in aqueous corrosive environments. Results Phys 102423
65. Ho C-Y, Huang S-M, Lee S-T, Chang Y-J (2017) Evaluation of synthesized graphene oxide as corrosion protection film coating on steel substrate by electrophoretic deposition. Appl Surf Sci 226–231
66. Zhu T, Wang L, Sun W, Wang M, Tian J, Yang Z et al (2018) The role of corrosion inhibition in the mitigation of $CaCO_3$ scaling on steel surface. Corros Sci 140:182–195
67. Zhang Z, Bell T (1985) Structure and corrosion resistance of plasma nitrided stainless steel. Surf Eng 1(2):131–136

# Pull-Out and Bond Degradation of Rebars in Reinforced Concrete Structures

K. S. Navaneethan, B. Kiruthika Nandhini, S. Anandakumar, and K. P. Jayakrishna

**Abstract** This review paper outlines the recent research works on bond properties of reinforced concrete specimens. Bond strength depends mainly on corrosion of the main bar and stirrups, type of concrete, concrete cover and corrosion rate. The environmental effects also play a dynamic role in the degradation of bond stability within reinforced rods and concrete. To integrate the bond behaviour amidst reinforced rods and concrete, several testing methods have been used and the pull-out experiment is used worldwide among them due to its simplicity. The study is gone through bond toughness amid reinforcement and concrete in different environments in different types of concretes such as high strength concretes, fibre-reinforced concretes, reinforcement corroded concretes, lightweight concrete and recycled aggregate concrete. Bond-slip behaviour is also studied for the lightweight concrete, and the graph visualizes the fact that the curve of bond-slip in lightweight concrete is comparable with the standard concrete. From the literature study, it is remarked that very few investigations have organized on the behaviour of bond in corroded reinforced steel rods. The span of cracks on the surface of concrete that occurs due to corrosion performs a prominent part in estimating the bond stability. Also, the bond toughness varies in real-time corroded structures compared with laboratory corroded techniques. Still, there is a shortage of study in those areas, and advanced tests and investigations are required.

K. S. Navaneethan (✉) · B. Kiruthika Nandhini · S. Anandakumar
Department of Civil Engineering, Kongu Engineering College, Perundurai, Erode, Tamil Nadu, India
e-mail: navaneethan.kec@gmail.com

B. Kiruthika Nandhini
e-mail: kiruthikanandhini7@gmail.com

S. Anandakumar
e-mail: anandkeccivil15@gmail.com

K. P. Jayakrishna
Department of Civil Engineering, Manipal Institute of Technology, Eshwar Nagar, Manipal, Karnataka, India
e-mail: kpjayakrishna24@gmail.com

© The Editor(s) (if applicable) and The Author(s), under exclusive license to Springer Nature Singapore Pte Ltd. 2021
S. Mohan et al. (eds.), *Materials, Design, and Manufacturing for Sustainable Environment*, Lecture Notes in Mechanical Engineering,
https://doi.org/10.1007/978-981-15-9809-8_22

**Keywords** Bond degradation · Bond-slip curve · Pull-out experiment · Bond stability

## 1 Introduction

Bond stress is a stress that acts on the outer layer of the reinforced bar and the concrete that surrounds the reinforcement. This stress resists the force which attempts to pull out the reinforced material from the concrete. The communication between the concrete and the reinforcing member is over bonding that usually permits force to transfer from the reinforcing member to the concrete. It also deals with the composite action of embedded concrete members. Three components that embrace the bond stress are chemical cohesion, friction and mechanical interlace [1]. Chemical cohesion is the bond due to chemicals that exists at the initial loading of the structure and the bond relationship fails when the load is increased. Thus, the chemical adherence, which is a resisting property, is not steady. The frictional bond of the rebars built up over the peculiarity of rebar's surface. The interfacial zone of the rebar and concrete has a roughness where the frictional forces get rolled out. This force has an eloquent role in the tie between concrete and rods. Mechanical interlacing is the elementary component of the relationship amidst the embedded ribs and the concrete keys. When the ultimate bond is applied, the occurrence of shear cracks in the middle of concrete and ribs takes place. This cracking ends in a slip where the significant bearing stresses are caused due to interlocking forces [2]. Figure 1 shows the influence of bond components between reinforced rods and concrete, where 1-chemical cohesion, 2-mechanical interlacing and 3-friction.

The exposure of concrete to moisture, temperature cycles and marine or other dynamic environmental condition that may decrease the integrity and also limits the durability of the concrete structures [3]. Technological parameters due to ageing and environmental effects such as steel corrosion, rust, bond improvement and bond decay due to low and high temperatures, respectively, perform an indicative role over the behaviour of the bond [4]. Bond deterioration of concrete and reinforcement reduces the serviceability as well as the extreme load-carrying ability of the structure. In addition, it changes the condition of failure due to ductility to the failure due to

Fig. 1 Schematic representation of bond components

catastrophic brittleness [5–9]. The corroded reinforcement has an esteemed reaction on the bending tests of concrete beams those fail in bond [10]. The corrosion of steel rods reinforced in concrete specimens for the laboratory tests is done artificially, which differs a lot from the reinforcement corrosion in concrete structures that occurs under natural condition [11].

Pull-out tests are executed for measuring the bond strength amid reinforced rods and concrete. Bond-slip curve has obtained at intervals of the application of load and the slippage of the steel rod that occurred in the test specimen. The capacity of the structure to carry new loads is unpredictable due to an unaccustomed increase or decrease in load on the structures [12]. Additive or alternate materials are practising in concrete to boost the strength of concrete and to beat the demand for raw materials that are used in concrete, respectively. Fibre-reinforced concrete (FRC) emerges as a superior alternative as it prevents the earlier bonding failure of critical regions in a structure [13], where the dangerous areas in a structure include the base of a column, a middle span in a beam and the beam-column joint. The other such alternate is concrete using recycled aggregate. Studies were done in the reuse of waste concrete expose that the concrete comprised of coarse aggregate obtained by recycling possess identical mechanical properties corresponding to nominal concrete, and the base idea for this eco-friendly practice is in the process of discovering high strength concrete [14].

## 2 Bond Degradation

Bond strength is the resisting force that exerts during the separation of concrete from the reinforcing bars. The bond strength lay on the primary reinforcement bar corrosion, lateral confinement, type of concrete and corrosion rate. When the bonding in between concrete and reinforced bar gets fail, there occurs degradation of the bond. Figure 2 gives the flow of bond components in bond degradation.

### *2.1 Influence of Main Reinforcement Corrosion*

The major concern for the bond deterioration within reinforcing steel and concrete from past decades is by cause of the main reinforcement corrosion. To overcome the problem of reinforcement corrosion, many techniques were conducted in the laboratory by accelerating artificial corrosion. The corrosion in the main reinforcing bar is first explored in 1990 [15]. It is found that the bond strength has a confident impact found in the earlier phase of corrosion. As the level of corrosion increased, there is a gradual decrease in the durability of the bond in the middle of the concrete and the reinforcing substance. It is also ended up that the visibility of cover cracking due to corrosion exhibits no negative strike over the stability of the bond [16]. There

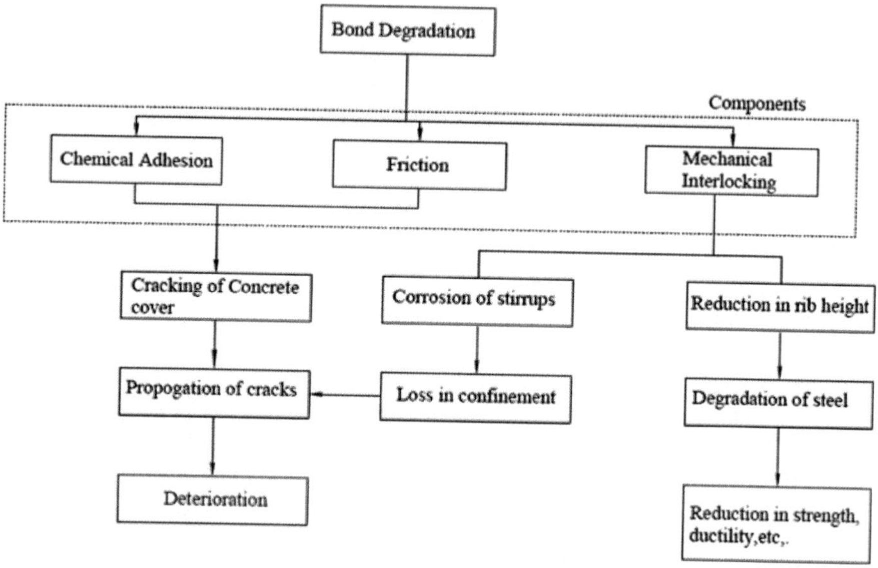


**Fig. 2** Influence of bond components in bond degradation

is an origin of radial pressure created over the interface of the concrete and reinforcement rod due to products of corrosion at the initial stage, which means that the occurrence of corrosion begins. The roughness of the surface in the reinforcing bar is increased even at the stage of minor corrosion. There is a reduction in a mechanical bond component when the corrosion gets initiated, as the corrosion-induced radial pressure fights with the bursting stresses created by the action of bond interlocking [17]. In the higher level of corrosion, there is a development of longitudinal cracks as the capacity of concrete on tension is fully used up. After the concrete cover gets cracked, the radial pressure, which is induced by the corrosion, has got released up at the space amidst the concrete and the reinforced material. The area of the rib of reinforcing material got reduced due to severe corrosion, and the material gets weakened. The corrosion products are mounted over the surface of the interface, where those two effects cause a reasonable loss in strength of the bond.

## 2.2 *Effects Concerning Stirrups*

### 2.2.1 Stirrups Without Corrosion

The observation from literature about bond degradation is that the enormous bond tests have been conducted on specimens that are cast without stirrups. But the real-time concrete structural elements are built up with the stirrups. The comparative results of the bond test on specimens with stirrups and without stirrups show different values of bond deterioration. It is compiled that when steel gets corroded it would not create any impact on the strength of bonds when there is a presence of stirrups [18, 19]. The strength of the bond is higher in the specimens cast with stirrups than the specimens without stirrups [20, 21]. It is detected that the strength of the bond increased gradually as increases in corrosion of the specimens cast with stirrups. The occurrence of cover cracking leads to the production of passive pressure on the anchorage by the stirrups. Width of cracks can be reduced with the help of stirrups [22]. After cracking of cover, the stirrups carry the internal stress that is generated by the enlargement of products of corrosion. Thus, the stirrups help in improving the bond protection of the corroded main bar.

### 2.2.2 Stirrups with Corrosion

The experimental investigations reveal that the stirrups in the specimen will corrode initially, and also the rate of corrosion is higher in the stirrups when compared with the main reinforcement bars [17, 23, 24]. This difference happens because of the smaller diameter of the stirrups than the main bars and also due to the location of stirrups, which is adjacent to the exposed outer face of the specimen. The composition of chemicals in the main rod is low alloy steel, and stirrups are mild steel, which permits the creation of macro-cell corrosion in which main bars act as a cathode and the stirrups acts as an anode. In real-time structures, the stirrups are placed adjacent to the shallow of bending cracks in flexural members, which leads to the rapid corrosion of the stirrups. The relentless corrosion of the stirrups forces it in the loss of factional area of stirrups. Pull-out tests have performed on the samples that are cast with corrosion in main bars and stirrups in which the test results prove that the bond resistance is superior in samples cast by stirrups than the samples without stirrups [25]. Beam-end specimens were tested experimentally, and the numerical modelling was done in which the results confirm that the loss of bond takes place when severe corrosion starts [25, 26]. Likewise, several experimental results suggested that stirrup corrosion at the initial stage increases bond strength and bond deterioration occurs at the extreme corrosion [27, 28]. The results of the pull-out test by eccentric loading have exposed that the bond endurance is diminished at the stage of severe corrosion of the stirrups [29].

## 2.3 *Effect Concerning Concrete Cover*

Failure in the cover of concrete where the steel rod is without corrosion influences the bond stability amid the reinforcement and concrete. The technique of deterioration either occurs by splitting cover or crushed concrete keys. In steel bars with corrosion, there exists a confident outcome over the large cover of concrete after the initiation of crack. The experimental tests result that when the cover of concrete in a specimen is larger, then the specimen possesses high bond strength [15, 30], which happens even at the severe corrosion stage. This varying bond strength occurs by virtue of two reasons, which is when the cover of concrete is larger, and then, the time taken to initiate the corrosion of concrete is larger. This action permits the products of corrosion to get into microscopic openings of concrete and postpone the degradation of the bond. The other noticeable thing is that the confinement exists even after the cracking of cover; this is by cause of tensile capacity of concrete not loosening suddenly after the cracking, where the shear stress and tensile stress are retained through the fine cracks.

## 2.4 *Effect Concerning Type of Concrete*

The action of the bond in the middle of the reinforcing rod and concrete gets varied as the type of concrete varies. The type of concrete influence here is the addition or replacement of other materials in concrete. The action of bond in concrete strengthened with fibre and concrete fabricated with recycled aggregate are taken in the study.

### 2.4.1 Fibre-Reinforced Concrete

Fibres are added as the additional material in the concrete, which intensifies the resilience of the concrete. The use of fibre in the concrete will keep up the relationship within the concrete and reinforced member, and it further helps in reducing or preventing the cracking that occurs in the area of provided concrete cover. The fibres, namely polypropylene fibre, steel fibre, nylon fibre and polyvinyl alcohol fibre were studied. Those study results prove that apart from the type of fibre used, the fibre-reinforced concrete delays the cracking of concrete cover, and provides higher strength in the bond when correlated with the typical concrete [31–34]. When the fibre is added in concrete, the experimental assessment outcomes prove that the stability of the bond is no more altered even when the corrosion stratum of steel reached from 10 to 15% [35].

### 2.4.2 Recycled Aggregate Concrete

Due to the demand for raw materials, the replacement of material is done where the aggregates are supplanted by artificial recycled aggregate. The experimental test outcomes illustrate that there is no great discrepancy in deterioration of bond due to corrosion in both concrete with usual aggregates and with recycled aggregates. In unconfined experimental samples, the toughness of bond initially gets increased and then shortened, whereas the bond disintegration is restricted in the test specimens with stirrups. The test on beams cast using coarse aggregates after recycling ended with the reduction of the bond as the rate of supplantation of recycled coarse aggregate gets intensified [36]. The other experimental results revealed that the tenacity of the bond gets raised as there is an upsurge in the percentage of replacement of recycled aggregate and also showed off the delayed cracking of concrete cover [37]. Since controversial results were obtained, advanced studies are mandatory for the clarification.

## 2.5 *Effect Concerning the Rate of Corrosion*

Accelerated corrosion techniques were done because of the shorter duration of the period of corrosion, and the level of corrosion could be controllable. There is a major discrepancy in the deterioration of bond amidst the stimulated corrosion and normal corrosion. On comparing the current densities of corrosion adopted in field condition and laboratory condition, it is seen that the density of current for corrosion in field condition fluctuate through 0.1 and 1 $\mu A/cm^2$ [22]. In contrast, the current density of corrosion in laboratory conditions is a thousand times higher. When the rate of corrosion is larger, then there occurs different oxidation when comparing natural corrosion and accelerated corrosion, and the expansion volume of corrosion products resulted from corrosion will vary. When the corrosion takes place in a faster manner, there will be an only fewer moment for the compounds of corrosion to get into the finite orifice of the concrete or the cracks. Thus, the bond deterioration varies concerning the time taken for the corrosion.

Ayop and Cairns [38] have conducted an accelerated test with two different densities of current such as 80 and 400 $\mu A/cm^2$ that are noted in the act of 'slow' and 'fast' densities of current. The after effect from the experiment exposed that the crack width is more comprehensive, and the bond strength is lower during the fast current when compared with slow current, and the resistance of bond decreases as the corrosion rate gets increased. The capacity to carry the load is lower, and the reduction in the bond is higher in the accelerated corrosion while comparing with typical corrosion [39]. It is found that there is severe bond deterioration in the reinforced concrete beams that are corroded using the galvanostatic method in comparison with the ordinary corrosive ambience [40]. Pull-out experiments were handled on the original concrete samples that are obtained from the demolition of a bridge in Norway, and its service life is calculated as 29 years and the outcomes display that the ability of the bond is not distressed until the corrosion level reaches up to 10% [41]. It is crucial to broaden the awareness over the degradation of bond in steel–concrete structures that are under the natural corrosive experiment, so it is vital to conduct bond tests of the structures under the natural corrosive environment.

## 3 Pull-Out Test

There are various methods for measuring the stability of the bond within concrete and reinforcing material. Among them, the pull-out experiment is a technique which is used among worldwide [2, 42–49].The pull-out test can also be done by mounting reinforcing rods near the outer surface [50–55]. In the pull-out mechanism, a steel rod or any other reinforcing material is embedded in the block of concrete, which may be cube[56, 57] or rectangular [18] or cylindrical [58] specimen. While the load is imposed, the rod pulls out from the sample with the controlled load. LVDT is used to measure the displacement in the rod. From the experiment, the load at which the rod slips out from the specimen is taken as tenacity of the bond amidst reinforced material and concrete. This test results also help in obtaining bond-slip relativity. The strength of binding is predicted by using Eq. 1 when the applied load and embedded length are known [56].

$$T = P/(L * d) \tag{1}$$

where

$T$ Bond stability (MPa).
$P$ Maximum enforced load (N).
$d$ Circumference of the embedded bar (mm).
$L$ Embedded segment's length (mm).

The samples for the pull-out experiment can be cast under the recommendations of RILEM [52, 59–62] or ASTM [63, 64]. Different testing methods are used for testing of which universal testing machine is commonly used [65, 66]. The loading may be eccentric loading or focus point loading or two-point loading. The occurrence of slippage of the rod from the specimen is deliberated by linear variable differential transducer (LVDT) [50, 67–69], and it can also be monitored and by acoustic emission [42, 70–72]. The simplest schematic portrayal of the pull-out experiment for a cubical specimen using INSTRON which is a machine used to perform tensile tests is presented in Fig. 3.

The conditions of failure that occur during the pull-out experiment of the samples are of three types [56]. (i) Pull-out failure—this mode of failure occurs when the bar reaches its peak load value, and the rod gets split out from the specimen. (ii) Splitting failure—when the bar reaches its maximum, the bar pulls out with cracks parallel to the appliance of force in the front face of the specimen. (iii) Yielding of the rod—this occurs when the load carried by the samples stretches higher than the load needed for the yielding. Yielding of the rod is literally not technical failure. In many cases, when the yielding gets visible, the test would be stopped to prevent the fracture of the embedded bar.

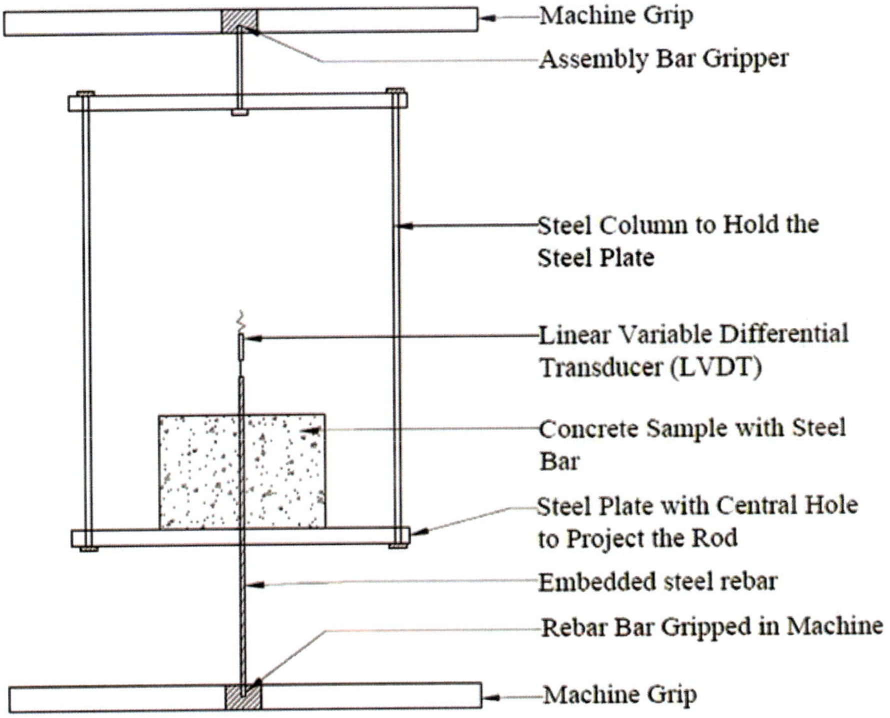


**Fig. 3** Schematic representation of pull-out test

## 4 Bond-Slip Relationship

Bond-slip curves are mainly comprised of two parts; they are ascending and descending curves [10, 73, 74]. At the initial stage, the binding property within reinforced rod and concrete is due to chemical adhesiveness, and when load gets increased, adhesion undergoes failure; hence, mechanical interacting plays an integral role in bonding mechanism. There is a formation of minute cracks in diagonal direction that occurs at the apex of the rib of rods due to the elongation stresses closer to the apex of rib induced by the bearing stresses at the front face of the rib surface [10, 20, 73, 75–78]. These cracks are called bond cracks, which permit the slip of bar along the direction of application of load, which results in a bond-slip curve in a softer and nonlinear way. Splitting of the specimen occurs over the whole concrete cover when the installation of the rod in concrete is with low confinement [67]. At the maximum load, there will be a plateau over the bond-slip curve, and a linear line is endured in a decreased manner corresponding to maximum frictional strength of bond at a slip. The value of slip is almost nearer to clear distance that occurs between the legs of the reinforced bars [79].

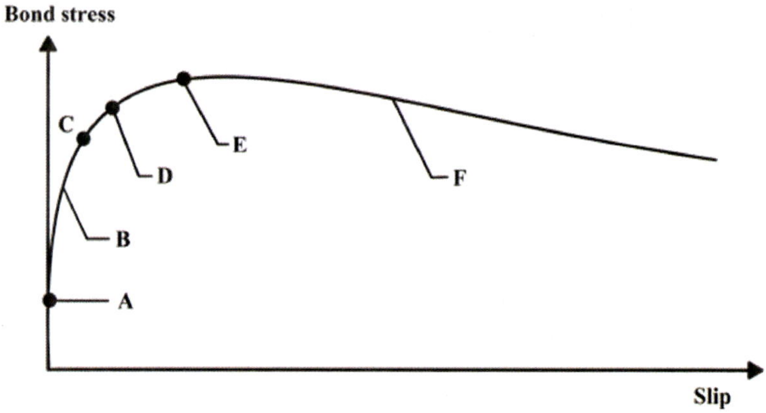


**Fig. 4** Relationship between bond stress-slip in well-confined concrete

Thus,

- Unconfined concrete fails by splitting of the concrete specimen,
- Confined concrete fails by pulling out of the steel rod in which failure occurs due to splitting happens in the concrete sample,
- Well-confined concrete fails purely by pulling out of the rod, which is embedded in the specimen [80].

Figure 4 depicts the affiliation amidst bond stress and slip in well-confined concrete, where

A Initial bond due to chemical adhesion.
B Failure of chemical adhesion.
C Visibility of splitting crack.
D Initiation of slip.
E Occurrence of transverse cracks.
F Shear action of concrete between ribs completed.

For example, Figs. 5 and 6 have plotted for the diameter of the bar to the highest load and the slip at that maximal load for lightweight aggregate concrete for two varying mixes, respectively, where the value of slip observed for all mixes lies between 0.8 and 2.15 mm [67].

## 5 Conclusion

This paper gave an overall view about the reasons for degradation of bond and the pull-out analysis through which the stability of the bond is determined and also covered the basic concept of bond-slip relationship with an example of lightweight aggregate concrete. It is noticed that the corrosion of the main reinforcement rod

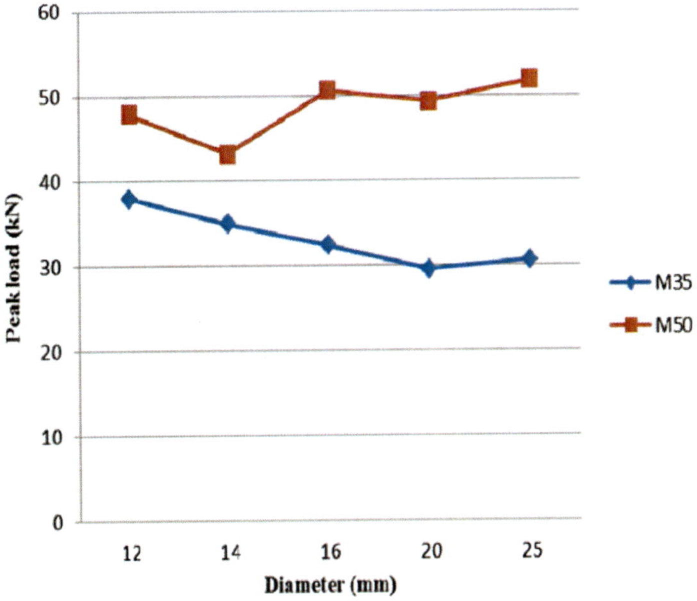


**Fig. 5** Graph for varying diameter of the bar to the peak load

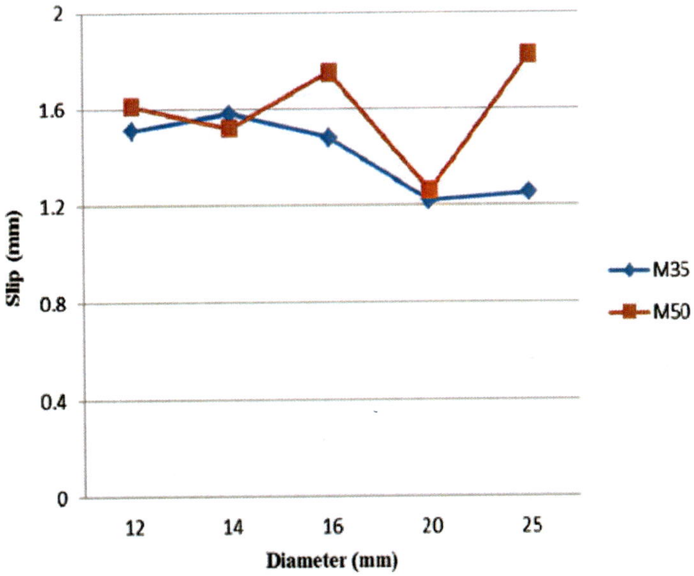


**Fig. 6** Graph for varying diameter of the rod to slip

pointed to boost up in toughness of bond at the initial stage, and later as the degree of corrosion enlarges, the toughness of bond gets decreased. The grade of corrosion of reinforcement rod varied within usual corrosive situation and stimulated corrosion technique, where the rate of corrosion is slow under a natural corrosive environment. The study revealed the law of raise in the diameter of bars had reduced bond strength and the shallow temperature would have a antipathetic strike on the bond relationship amid the concrete and reinforced rod. The exposure environment of the concrete specimen or concrete structure has more considerable sway on the stability of the bond.

It is also observed that the tenacity of the bond amidst the reinforced rod and concrete is related to the transverse pressure over the specimen, and also the cover density of concrete had a proficient impact over the bond strength. The stability in binding differs in the samples, which were cast with stirrups and without stirrups. It also proved that the width of the crack is lesser in samples cast with stirrups when compared with the samples without stirrups. It is finalized that the stirrups with corrosion at the initiative phase upsurges the tenacity of bond. The stirrups get corroded easily, as it is placed near the surface of bending cracks in flexural members and the diameter of the stirrups is lower than the main reinforcement bar, which is also the reason for quick corrosion of stirrups.

The type of concrete used also hold a considerable outcome on the bond stability of the specimen. As discussed, the reinforced concrete with fibre and recycled aggregate delays in the cracking of the concrete. Studies show that the simplest and common method used for regulating bond capacity is the pull-out experiment, where the usual equipment used for measuring the slip of bar is a linear variable differential transducer (LVDT). For real-time corroded samples, there is deficient in investigation over the strength of the bond. As the rate of corrosion varies in real-time structures and laboratory specimens, it will be helpful if more investigations have done in real-time corroded structures.

## References

1. Tepfers R (1979) Cracking of concrete cover along anchored deformed reinforcing bars. Mag Concr Res 31(106):3–12
2. Hollý I, Bilčík J, Keseli O, Gažovičová N (2016) Bond of GFRP reinforcement with concrete. Key Eng Mater 691:356–365
3. Benzarti K, Chataigner S, Quiertant M, Marty C, Aubagnac C (2011) Accelerated ageing behaviour of the adhesive bond between concrete specimens and CFRP overlays. Constr Build Mater 25(2):523–538
4. Bilcik J, Holly I (2013) Effect of reinforcement corrosion on bond behaviour. Procedia Eng 65:248–253
5. Ahmad S (2003) Reinforcement corrosion in concrete structures, its monitoring and service life prediction—a review. Cement Concr Compos 25(4–5):459–471
6. Lachemi M, Al-Bayati N, Sahmaran M, Anil O (2014) The effect of corrosion on shear behavior of reinforced self-consolidating concrete beams. Eng Struct 79:1–12
7. Kallias AN, Rafiq MI (2010) Finite element investigation of the structural response of corroded RC beams. Eng Struct 32(9):2984–2994

8. Khan I, François R, Castel A (2012) Structural performance of a 26-year-old corroded reinforced concrete beam. Eur J Environ Civ Eng 16(3–4):440–449
9. Recupero A, Spinella N, Tondolo F (2018) Failure analysis of corroded RC beams subjected to shear-flexural actions. Eng Fail Anal 93:26–37
10. Bhargava K, Ghosh A, Mori Y, Ramanujam S (2006) A proposed model for corrosion-induced bond degradation in reinforced concrete. In: Structures congress 2006: 17th analysis and computation specialty conference, pp 1–15
11. Tahershamsi M, Fernandez I, Lundgren K, Zandi K (2017) Investigating correlations between crack width, corrosion level and anchorage capacity. Struct Infrastruct Eng 13(10):1294–1307
12. Toutanji HA, Gomez W (1997) Durability characteristics of concrete beams externally bonded with FRP composite sheets. Cement Concr Compos 19(4):351–358
13. Mansour FR, Bakar SA, Ibrahim IS, Marsono AK, Marabi B (2015) Flexural performance of a precast concrete slab with steel fiber concrete topping. Constr Build Mater 75:112–120
14. Ajdukiewicz A, Kliszczewicz A (2002) Influence of recycled aggregates on mechanical properties of HS/HPC. Cement Concr Compos 24(2):269–279
15. Al-Sulaimani G, Kaleemullah M, Basunbul I (1990) Influence of corrosion and cracking on bond behavior and strength of reinforced concrete members. Struct J 87(2):220–231
16. Altowaiji WA, Darwin D, Donahey RC (1986) Bond of reinforcement to revibrated concrete 83(6)
17. Tastani S, Pantazopoulou SJ (2007) Behavior of corroded bar anchorages. ACI Struct J 104(6):756
18. Fang C, Lundgren K, Chen L, Zhu C (2004) Corrosion influence on bond in reinforced concrete. Cem Concr Res 34(11):2159–2167
19. Fischer C (2010) Experimental investigations on the effect of corrosion on bond of deformed bars. In: Proceedings of the 8th fib Ph.D. symposium, Kgs, Lyngby, Denmark
20. Zandi Hanjari K, Coronelli D, Lundgren K (2011) Bond capacity of severely corroded bars with corroded stirrups. Mag Concr Res 63(12):953–968
21. Rodriguez J, Ortega L, Casal J (1994) Corrosion of reinforcing bars and service life of reinforced concrete structures: corrosion and bond deterioration. In: International conference on concrete across borders, vol 2, Odense, Denmark, pp 315–326
22. Lin H, Zhao Y (2016) Effects of confinements on the bond strength between concrete and corroded steel bars. Constr Build Mater 118:127–138
23. Fu C, Jin N, Ye H, Jin X, Dai W (2017) Corrosion characteristics of a 4-year naturally corroded reinforced concrete beam with load-induced transverse cracks. Corros Sci 117:11–23
24. Otsuki N, Miyazato S-i, Diola NB, Suzuki H (2000) Influences of bending crack and water-cement ratio on chloride-induced corrosion of main reinforcing bars and stirrups. Mater J 97(4):454–464
25. Tondolo F (2015) Bond behaviour with reinforcement corrosion. Constr Build Mater 93:926–932
26. Coronelli D, Hanjari KZ, Lundgren K (2012) Severely corroded RC with cover cracking. J Struct Eng 139(2):221–232
27. Zhou H, Liang X, Zhang X, Lu J, Xing F, Mei L (2017) Variation and degradation of steel and concrete bond performance with corroded stirrups. Constr Build Mater 138:56–68
28. Zhou H, Lu J, Xv X, Dong B, Xing F (2015) Effects of stirrup corrosion on bond–slip performance of reinforcing steel in concrete: an experimental study. Constr Build Mater 93:257–266
29. Lin H, Zhao Y, Yang J-Q, Feng P, Ozbolt J, Ye H (2019) Effects of the corrosion of main bar and stirrups on the bond behavior of reinforcing steel bar. Constr Build Mater 225:13–28
30. Amleh L, Mirza M, Ahwazi B (2000) Bond deterioration of reinforcing steel in concrete due to corrosion, vol 3
31. Farhan NA, Sheikh MN, Hadi MN (2018) Experimental investigation on the effect of corrosion on the bond between reinforcing steel bars and fibre reinforced geopolymer concrete. Structures 14:251–261

32. Haddad RH, Ashteyate AM (2001) Role of synthetic fibers in delaying steel corrosion cracks and improving bond with concrete. Can J Civ Eng 28(5):787–793
33. Xinhua C, Shilang X, Shiping Y, Zhen H (2012) Experimental research on bond behaviors of corroded rebar and ultra high toughness cementitious composites (UHTCC). J Chin Soc Corros Prot 32(3):228–234
34. Berrocal CG, Fernandez I, Lundgren K, Löfgren I (2017) Corrosion-induced cracking and bond behaviour of corroded reinforcement bars in SFRC. Compos B Eng 113:123–137
35. Hou L, Liu H, Xu S, Zhuang N, Chen D (2017) Effect of corrosion on bond behaviors of rebar embedded in ultra-high toughness cementitious composite. Constr Build Mater 138:141–150
36. Yang H, Lan W, Qin Y, Wang J (2016) Evaluation of bond performance between deformed bars and recycled aggregate concrete after high temperatures exposure. Constr Build Mater 112:885–891
37. Fernandez I, Etxeberria M, Marí AR (2016) Ultimate bond strength assessment of uncorroded and corroded reinforced recycled aggregate concretes. Constr Build Mater 111:543–555
38. Ayop SS, Cairns JJ (2015) The influence of impressed current on residual bond strength of corroded structure. Appl Mech Mater 773–774:984–989
39. Sæther I, Antonsen A, Vennesland Ø (2007) Effect of impressed anodic current density applied to accelerated corrosion laboratory results. In: International RILEM workshop on integral service life modeling of concrete structures, pp 307–314
40. Yuan Y, Ji Y, Shah SP (2007) Comparison of two accelerated corrosion techniques for concrete structures. ACI Struct J 104(3):344
41. Horrigmoe G, Saether I, Antonsen R, Arntsen B (2007) Laboratory investigations of steel bar corrosion in concrete: sustainable bridges background document SB3, p 10
42. Abouhussien AA, Hassan AA (2017) Acoustic emission monitoring for bond integrity evaluation of reinforced concrete under pull-out tests. Adv Struct Eng 20(9):1390–1405
43. Barbosa MTG, Sánchez Filho S (2013) Investigation of bond stress in pull out specimens with high strength concrete. J Res Eng Civ Struct Eng 13(3):55–64
44. Ihekwaba N, Hope B, Hansson C (1996) Pull-out and bond degradation of steel rebars in ECE concrete. Cem Concr Res 26(2):267–282
45. Maranan GB, Manalo AC, Karunasena W, Benmokrane B (2015) Pullout behaviour of GFRP bars with anchor head in geopolymer concrete. Compos Struct 132:1113–1121
46. Chang X, Yue G, Lin H, Tang C (2010) Modeling the pullout behavior of fiber reinforced polymer bars from concrete. Constr Build Mater 24(4):431–437
47. Chourasia A, Singhal S, Chourasia A (2019) Pull-out behaviour of headed bars embedded in concrete. 8th International Engineering Symposium - IES 2019 , Kumamoto, Japan
48. Chu S, Kwan A (2018) A new method for pull out test of reinforcing bars in plain and fibre reinforced concrete. Eng Struct 164:82–91
49. Cunha VM, Barros JA, Sena-Cruz JM (2009) Pullout behavior of steel fibers in self-compacting concrete. J Mater Civ Eng 22(1):1–9
50. De Lorenzis L, Rizzo A, La Tegola A (2002) A modified pull-out test for bond of near-surface mounted FRP rods in concrete. Compos B Eng 33(8):589–603
51. Kalupahana W, Ibell T, Darby A (2013) Bond characteristics of near surface mounted CFRP bars. Constr Build Mater 43:58–68
52. Kotynia R (2012) Bond between FRP and concrete in reinforced concrete beams strengthened with near surface mounted and externally bonded reinforcement. Constr Build Mater 32:41–54
53. Soliman SM, El-Salakawy E, Benmokrane B (2010) Bond performance of near-surface-mounted FRP bars. J Compos Constr 15(1):103–111
54. Yu B, Kodur V (2014) Effect of high temperature on bond strength of near-surface mounted FRP reinforcement. Compos Struct 110:88–97
55. Zhang H, He L, Li G (2015) Bond failure performances between near-surface mounted FRP bars and concrete for flexural strengthening concrete structures. Eng Fail Anal 56:39–50
56. Mansoor YA, Zhang ZQ (2013) The reinforcement bond strength behavior under different corrosion condition. Res J Appl Sci Eng Technol 5:2346–2353

57. Ali A, Iqbal S, Holschemacher K, Bier TA (2016) Effect of fibers on bond performance of lightweight reinforced concrete. Periodica Polytech Civ Eng 60(1):97
58. Németh OI, Lublóy É, Farkas G (2014) Bond of reinforcement in polymer concrete. Periodica Polytech Civ Eng 58(2):137–141
59. Achillides Z, Pilakoutas K (2004) Bond behavior of fiber reinforced polymer bars under direct pullout conditions. J Compos Constr 8(2):173–181
60. Silva Filho L, Vale Silva B, Dal Bosco V, Gomes L, Barbosa M, Lorrain M (2012) Analysis of the influence of rebar geometry variations on bonding strength in the pull-out test. Bond Concr 978-88-907078-1-0:63–68
61. Yalciner H, Marar K (2017) Experimental study on the bond strength of different geometries of corroded and uncorroded reinforcement bars. J Mater Civ Eng 29(7):05017002
62. Weiße D, Holschemacher K (2003) Some aspects about the bond of reinforcement in ultra high strength concrete. Leipzig annual civil engineering report (LACER), no 8, pp 251–263
63. Hassan M, Benmokrane B, ElSafty A, Fam A (2016) Bond durability of basalt-fiber-reinforced-polymer (BFRP) bars embedded in concrete in aggressive environments. Compos B Eng 106:262–272
64. Ali A, Iqbal S, Holschemacher K, Bier TA (2017) Bond of reinforcement with normal-weight fiber reinforced concrete. Periodica Polytech Civ Eng 61(1):128–134
65. Krstulovic-Opara N, Watson KA, LaFave JM (1994) Effect of increased tensile strength and toughness on reinforcing-bar bond behavior. Cement Concr Compos 16(2):129–141
66. Sydenstricker TH, Mochnaz S, Amico SC (2003) Pull-out and other evaluations in sisal-reinforced polyester biocomposites. Polym Testing 22(4):375–380
67. Al-Shannag MJ, Charif A (2017) Bond behavior of steel bars embedded in concretes made with natural lightweight aggregates. J King Saud Univ Eng Sci 29(4):365–372
68. Kotynia R, Szczech D, Kaszubska M (2017) Bond behavior of GRFP bars to concrete in beam test. Procedia Eng 193:401–408
69. Wu Y-Z, Lv H-L, Zhou S-C, Fang Z-N (2016) Degradation model of bond performance between deteriorated concrete and corroded deformed steel bars. Constr Build Mater 119:89–95
70. Di B, Wang J, Li H, Zheng J, Zheng Y, Song G (2019) Investigation of bonding behavior of FRP and steel bars in self-compacting concrete structures using acoustic emission method. Sensors 19(1):159
71. Gallego A, Benavent-Climent A, Suarez E (2015) Concrete-galvanized steel pull-out bond assessed by acoustic emission. J Mater Civ Eng 28(2):04015109
72. Wang L, Yi J, Xia H, Fan L (2016) Experimental study of a pull-out test of corroded steel and concrete using the acoustic emission monitoring method. Constr Build Mater 122:163–170
73. Firmo JP, Correia JR, França P (2012) Fire behaviour of reinforced concrete beams strengthened with CFRP laminates: protection systems with insulation of the anchorage zones. Compos B Eng 43(3):1545–1556
74. Gudonis E, Kacianauskas R, Gribniak V, Weber A, Jakubovskis R, Kaklauskas G (2014) Mechanical properties of the bond between GFRP reinforcing bars and concrete. Mech Compos Mater 50(4):457–466
75. Cairns J, Du Y, Law D (2006) Residual bond strength of corroded plain round bars. Mag Concr Res 58(4):221–231
76. Harajli M (2010) Bond behavior in steel fiber-reinforced concrete zones under static and cyclic loading: experimental evaluations and analytical modeling. J Mater Civ Eng 22(7):674–686
77. Haskett M, Oehlers DJ, Ali MM (2008) Local and global bond characteristics of steel reinforcing bars. Eng Struct 30(2):376–383
78. Hassan A, Hossain K, Lachemi M (2010) Bond strength of deformed bars in large reinforced concrete members cast with industrial self-consolidating concrete mixture. Constr Build Mater 24(4):520–530
79. Lundgren K (1999) Three-dimensional modelling of bond in reinforced concrete theoretical model, experiments and applications. Chalmers Univer Technol 91-7197-853-4:1-50
80. Magnusson J (2000) Bond and anchorage of ribbed bars in high-strength concrete. Chalmers Univer Technol 9171978798

# Structural Behavior of Glulam Beams with Proper Reinforcement Bars—A Review

**S. Suchithra and S. Jayashree**

**Abstract** Glulam materials were commonly used and are ease in construction materials. Glulam timber provides strength and stiffness, to increase the properties of it while using in construction, along with reinforcement bars. The reinforcement material used by most of the researchers was Fiber-Reinforced Polymers (FRPs) Since this provides fatigue resistance, high tensile strength, stiffness and, durability properties. This paper reviews the structural behavior of glulam beams with different reinforcement materials. The parameters discussed were the strength of the connections, compressive strength, flexural strength, and tensile strength of beams. Result exhibits that FRP material is the best reinforcement material for glulam beams. Notch connections result in fail earlier and fire test also exhibits early failure than expected.

**Keywords** Glulam · Fiber-reinforced polymer (FRP) · Connections · Adhesives · Timber

## 1 Introduction

At an early age, constructions were carried out using timber materials. Even nowadays, timbers were used for more construction works. It is easily available material. Timber is renewable, economical, gives a good aesthetic appearance and ease in construction. It also has good strength properties like low weight, a more tensile strength in bending, and environmentally friendly due to the absorption of carbon [1, 2]. The strips of wood are jointly glued to form a laminated timber. The laminated timber is of two types, Glued Laminated Timber (GLT) and Cross Laminated Timber (CLT). The CLT is obtained from the lamella of the softwood [3]. The strips of wood

S. Suchithra · S. Jayashree (✉)
Department of Civil Engineering, Kongu Engineering College, Perundurai, Erode, Tamil Nadu, India
e-mail: jayashreesiva22@gmail.com

S. Suchithra
e-mail: suchithra.civil@kongu.edu

© The Editor(s) (if applicable) and The Author(s), under exclusive license to Springer Nature Singapore Pte Ltd. 2021
S. Mohan et al. (eds.), *Materials, Design, and Manufacturing for Sustainable Environment*, Lecture Notes in Mechanical Engineering,
https://doi.org/10.1007/978-981-15-9809-8_23

are arranged perpendicularly and parallel alternatively and glued to form a Cross Laminated Timber. Glued laminated (GLULAM) is an engineered wood, consists of laminations with moisture-resistant and durable adhesives. The laminations and adhesives used in the glulam provide durability and bond strength. The common adhesive used in the timber is polyurethane which gives colorless joints and cured by exposing it to moisture. The glulam can be used as beams and purlins in construction projects. Glulam beams, provided with reinforcement bars increases the strength of the structure. The strength of the glulam beams varies with the climatic conditions at the time of construction. For example, the load-carrying capacity of glulam beams purely depends on the season of the year, i.e. moisture gradients of the glulam members [4].

The connections in glulam members are important to transfer the forces to the members. The connections used in the glulam members were shear connections, bolt and dowel connectors, nail type fasteners, notched shear key, finger joints, inclined screws, carbon epoxy patch, small grooves, steel joist, self-tapping screws, bolted connection, FRP laminates, nailed connections, anchors, notches as a connection.

Research has been conducted on glulam beams with different Fiber-Reinforced Polymers (FRPs) are Carbon Fiber-Reinforced Polymer (CFRP), Hybrid Fiber-Reinforced Polymer (HFRP), Glass Fiber-Reinforced Polymer (GFRP), and Basalt Fiber-Reinforced Polymer (BFRP). The FRP consists of advanced composite materials, which provide more advantages in recent years in the construction fields. The FRP materials having a high strength to weight ratio and high stiffness hence, it is suitable rebar for timber. Additionally, it provides durability and the best fatigue resistance [1]. The FRP that uses carbon fiber is known as CFRP. The CFRP is used as the perfect material for the flexural reinforcement of wood, which reduces the deflection due to continuous loading [5]. The glass fiber is a low-cost composite. The FRP in which glass fiber is used called GFRP. The strength of the timber increases by increasing the stiffness, when this GFRP is used as a rebar [6]. The HFRP comprises two or more fiber polymers. The BFRP is formed by using very fine basalt fibers, which consists of olivine, plagioclase and pyroxene chemicals.

This paper reviews the structural behavior of glulam beams with different reinforcement bars. The parameters discussed are compressive strength, flexural strength, and tensile strength of glulam members with different rebars. Additionally, the strength of different connections used in glulam beams and the tests performed on the members are also reviewed. The codebook used for the design of timber structures is Eurocode 5.

# 2 Materials

## *2.1 Glulams*

The pine/spruce timber is commonly used timber [3]. The tropical wood named cosipo also known as redwood which is red-brown. This wood gives good resistance to insects of dry wood and fungus [7]. The connections used in glulam members are douglas fir, screws, metallic fasteners, mesh connectors, grooves, bolt and dowel type connections. To determine the strength of the connections, various tests were also carried out [8–21].

## *2.2 Adhesives*

To resist the moisture and provide durability adhesives were used. Epoxy adhesives are commonly used as structural adhesives. This is because of their ability to stick fast to the materials and gives more strength. The smooth stress-displacement plot was produced by these epoxy plates [22]. The other adhesives were polymer, phenol resorcinol formaldehyde (PRF), and melamine formaldehyde [1, 3].

## *2.3 Fiber-Reinforced Polymers (FRPs)*

The FRP reinforcement consists of unidirectional fibers embedded in the polymeric resins which provide stiffness, ultimate moment resisting capacity. The resins protect the fibers due to the transfer of loads. The FRP helps to overcome the defects of timber and also increases the performance of glulam beams [1, 23].

### 2.3.1 Carbon Fiber-Reinforced Polymer (CFRP)

In CFRP rods the grains of laminates are arranged either orthogonal or parallel when loaded, shows good performance in shear strength [24]. The CFRP rods reduce ductility in repaired corroded RC beam [17]. The specimen failure occurs immediately after the failure of CFRP rods which occurs after the failure of wood [25].

### 2.3.2 Glass Fiber-Reinforced Polymer (GFRP)

The woven fabric prepreg and unidirectional tape are used to manufacture Graphite fiber-reinforced polymers (GFRP) [26]. The use of GFRP affects the flexural strength

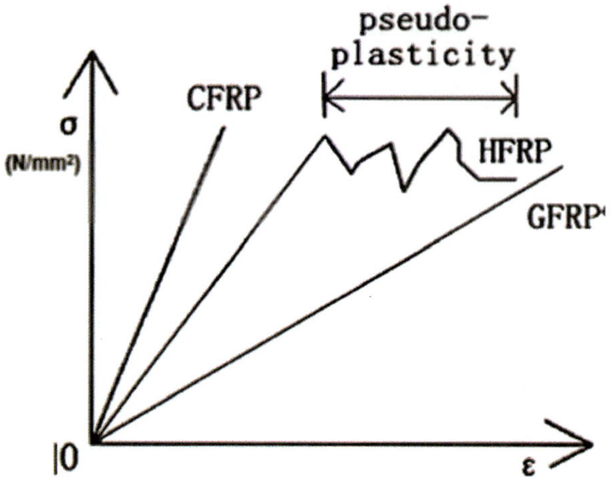


**Fig. 1** Stress–strain curve for different FRPs (Source [29], Reproduced from Elsevier, License Number- 4,813,711,295,851)

of the beam which leads to tension in piles [27]. To prevent the beam from slippage and stress the GFRP rods are provided with polymer resins and steel tubes. Even though it exhibits failure mode in tension [28].

### 2.3.3 Hybrid Fiber-Reinforced Polymer (HFRP)

Two or more than two different reinforcing materials are combined as a single matrix to form a Hybrid fiber-reinforced polymer. This improves the mechanical properties (i.e.) stiffness and modulus of elasticity [29]. The stress–strain behavior of CF, GF, and HF are denoted in Fig. 1, which shows CF provides low elongation compared to GF. Whereas HFRP provides more elongation and exhibits bilinear pseudoplastic behavior by the combination of high strength glass fiber and carbon fiber (Fig. 1) [29].

### 2.3.4 Basalt Fiber-Reinforced Polymer (BFRP)

On comparing with GFRP, the BFRP gives more benefits and also cost-effective when compared with CFRP. BFRP is a potential resource for timber reinforcement [30].

**Table 1** Tests performed in glulam beams

| S. No. | Test | Parameters investigated | Size of the specimen | Load applied | References |
|---|---|---|---|---|---|
| 1 | Fire test | Capacity of the specimen under fire | 116*116*272 mm | 300 °C | [8] |
| 2 | Acoustic test | Sound insulations | 3.5*3.4 m | 60 dB | [53] |
| 3 | V-notch test, notch tensile test | Stress at the time of fracture | | Impact loading | [54] |

## 3 Mechanical Properties

To determine the behavior of the beam the mechanical tests like the shear test, compressive test, push-out test, pull-out test, and bending tests were conducted. The compressive test was conducted on CTM by considering various sizes of reinforced glulam beam to determine the compressive strength [27, 31–33]. The shear test was conducted at different climatic conditions to determine its response at various seasons [4]. The push-out test was conducted on UTM to determine the relative slip concerning time, relative slip concerning load, and strength concerning time by plotting a graph [7, 10, 16, 34]. To determine the performance of FRP rods with glulam and bond strength, pull-out tests were conducted [24, 35–37]. The bending tests in the form of two-point, three-point, and four-point loading on beams were conducted and the load–deflection curve was analyzed [6, 14, 16, 25, 28, 38–52].

Beyond the mechanical strength tests, other tests conducted on the reinforced glulam beams were given in Table 1.

## 4 Behavior of Connections

For the larger span of glulam members the failure occurs mainly at the connections. Hence the type of connection should be selected properly and proper detailing should be provided [55]. Notches in the glulam members reduce the load-carrying capacity. If notches in the members cannot be avoided, then reinforcement bars should be provided at the ends to prevent brittle failure. For the composite action of timber beams, the most commonly used connection is dowel type connections [56]. Various connections were used for glulam timbers, which were formulated in Table 2.

**Table 2** Connections used for glulam timbers

| S. No. | Connection type | Description | References |
|---|---|---|---|
| 1 | Shear | Helps to achieve an efficient structure | [7, 13, 57–59] |
| 2 | Dowel connectors | Used to prevent splitting shear failure | [8, 19, 57, 60–64] |
| 3 | Nail type fasteners | To provide increased lateral load-carrying capacity | [7, 32, 65, 66] |
| 4 | Notched shear key | Minimize the relative displacement and Interlayer force transfer between the timber and concrete | [2, 56] |
| 5 | Finger joints | It provides dimensional stability and straightness | [67] |
| 6 | Inclined screws | The nonlinear behavior helps to increase the strength of the glulam member | [61] |
| 7 | Carbon epoxy patch | To repair the wood effectively | [39] |
| 8 | Small grooves | To strengthen the existing glulam beams | [16, 24, 68, 69] |
| 9 | Hole in a joist | It helps to change the failure mechanism and the hole acts as a stress concentrator | [12, 38] |
| 10 | Self-tapping screws | The screw arrangement achieves high performance and suitable to predict the capacity of CLT connections | [14, 16, 46, 70] |
| 11 | FRP laminates | Provides durability | [24] |
| 12 | Nailed connections | To connect the timber pieces | [62, 71, 72] |
| 13 | Anchors | Durable joint, corrosion-resistant | [13, 56] |
| 14 | Notches as connection | Provides high stiffness and helps to achieve the ductile compression failure of timber | [60, 61, 73, 74] |
| 15 | Bolt | To repair the wooden beams | [8, 16, 18, 20, 36, 72, 75, 76] |
| 16 | Lag screws | To resist the tendency of split at the square-cornered notches | [53] |
| 17 | Glued in rod | Stiff and strong joints, which can be used for frame corners and column foundations | [57, 62] |
| 18 | Threaded steel rods | To transfer the shear forces in composite beams | [77] |
| 19 | Steel hook | To produce high slip modulus | [78] |
| 20 | Round dovetail | Strongest connection without glue | [46] |

## 5 Discussions

On comparing with unreinforced beams, the flexural capacity and stiffness of CFRP or GFRP strengthened timber beams is increased up to 44–63% and 10% respectively [79]. Likewise, comparing with un-strengthened control beams, the CFRP and GFRP strengthened beams that provide an increased load-carrying capacity of 17.7–77.3%. The elastic modulus of GF is 30% that of CF. The elongation of the fracture of carbon fiber (CF) is 1.2–1.8% and of the Glass Fiber (GF) is 4.3–5.4%. The HFRP comprising of CF and Short Glass Fiber (SGF) provides more ductility and more elongation respectively and it bears a heavy load when compared with the CFRP [29]. During the entire test, the GFRP strengthened timber provides good bond strength between FRP, adhesive, and timber [63]. From the shear test of glulam members, the strength of the glulam member increases 60% by reducing the RH 80–40% [4].

Heating the timber reduces the capacity to resist the shear, which was produced by dowel or bolt connectors. Hence the connection fails earlier than expected [8]. During the bending test, the nonlinear response is given by the wood in terms of the sudden drop of the load due to CFRP rupture. The rupture in CFRP is followed by the failure in wood at the tension zone, which leads to failure in specimen [25, 33]. The mode of failure is changed from brittle on the tension side to ductile on the compression side. This change in the mode of failure helps to improve the safety of the structure [79]. The load-carrying capacity of the glulam beams is more when compared with the control beam. The bending test result shows that the prestressed beams have more load-carrying capacity and stiffness compared to the beams without prestressing [42]. From a two-point loading test, the behavior of the reinforced and unreinforced beams vary. The deflection is high for the unreinforced beam for small loading and the deflection for the reinforced beam is small for large loading. This can be confirmed by the graph in Fig. 2 using the compression test [33].

The pull-out test shows that the ratio of loaded end configuration corresponding to the bond length reduces the slip percentage [30]. It concludes that the splitting of the bond failure is very critical for the bars parallel to the grains. Even the cover is less or thicker, the bond strength is more if the bars are perpendicular to the grains of the beams [37]. The acoustic test shows that the timber floors gives poor performance in both airborne and impact sound installation [53].

## 6 Conclusions

- Comparing unreinforced and reinforced glulam beams, the reinforced beams provide more strength compared to the unreinforced beams. Because failure initially occurs in the rebar then it transfers to glulam.
- Since HFRPs are comprised of two or more fibers it provides greater stiffness when compared with other FRPs. The displacement of HFRP beams comprising of CF and SGF is 106% greater than the CFRP strengthened beams.

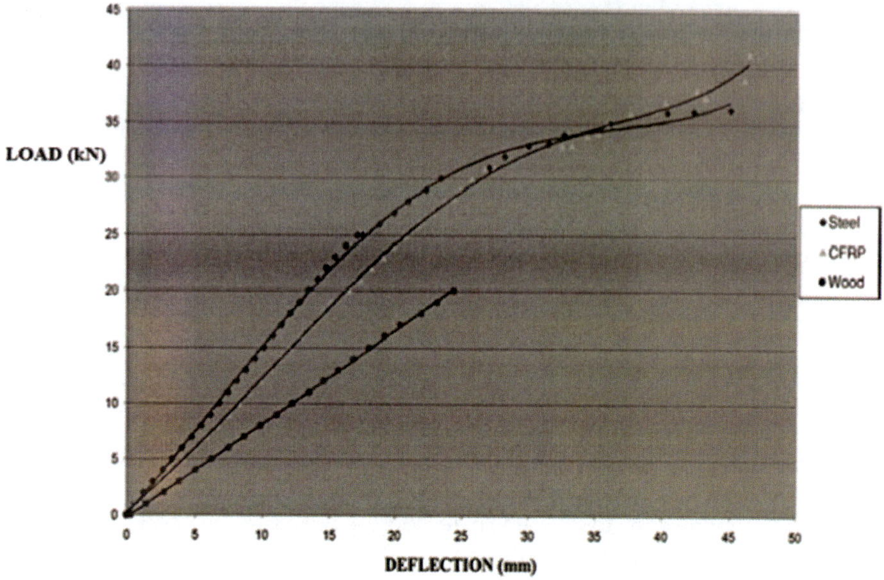


**Fig. 2** Load–deflection curve for three different beams (Source [33], Reproduced from Elsevier, License Number 4814820803428)

- Although CFRP exhibits high tensile strength, it provides low elongation. HFRP beams exhibit good elongation and tensile strength by the action of two or more fibers.
- The connectors used in glulam construction must be fire-resistant, for better performance. Compared to all the connections, the dowel connection provides a good bond.
- Steel bars can be used in prestressed glulam beams to improve flexural properties.
- When compared with the control beam, glulam beams provide more strength.
- Even though the strength of the beam satisfies the design requirements, it also exhibits certain disadvantages. Like,
- The strength of the glulam members depends on the moisture gradient. A decrease in relative humidity of glulam increases the tension capacity.

## References

1. Raftery GM, Harte AM (2013) Nonlinear numerical modelling of FRP reinforced glued laminated timber. Compos Part B Eng 52:40–50
2. Gutkowski RM, Brown K, Shigidi A, Natterer J (2004) Investigation of notched composite wood–concrete connections. J Structural Eng 130(10):1553–1561
3. Lineham SA, Thomson D, Bartlett AI, Bisby LA, Hadden RM (2016) Structural response of fire-exposed cross-laminated timber beams under sustained loads. Fire Safety J 85:23–34

4. Jönsson J, Thelandersson S (2003) The effect of moisture gradients on tensile strength perpendicular to grain in glulam. Holz Als Roh-Und Werkstoff 61(5):342–348
5. Taheri F, Nagaraj M, Khosravi P (2009) Buckling response of glue-laminated columns reinforced with fiber-reinforced plastic sheets. Compos Struct 88(3):481–490
6. Alhayek H, Svecova D (2012) Flexural stiffness and strength of GFRP-reinforced timber beams. J Compos Constr 16(3):245–252
7. Djoubissie DD, Messan A, Fournely E, Bouchaïr A (2018) Experimental study of the mechanical behavior of timber-concrete shear connections with threaded reinforcing bars. Eng Structures 172:997–1010
8. Barber D (2017) Determination of fire resistance ratings for glulam connectors within US high rise timber buildings. Fire Safety J 91:579–585
9. D'Ambrisi A, Focacci F, Luciano R (2014) Experimental investigation on flexural behavior of timber beams repaired with CFRP plates. Compos Structures 108:720–728
10. Jiang Y, Hong W, Hu X, Crocetti R, Wang L, Sun W (2017) Early-age performance of lag screw shear connections for glulam-lightweight concrete composite beams. Constr Build Mater 151:36–42
11. Pošta J, Hataj M, Jára R, Ptáček P, Kuklík P (2019) Comparison of the use of angle brackets in timber joints with eurocode 5. Constr Build Mater 205:611–621
12. Sebastian W (2019) Switch from connection ductility to reinforcement ductility with curvature reversal in timber-concrete composites. Constr Build Mater 229:116886
13. Yeoh D, Fragiacomo M, Carradine D (2013) Fatigue behaviour of timber–concrete composite connections and floor beams. Eng Structures 56:2240–2248
14. Jockwer R, Steiger R, Frangi A, Serrano E (2016) Load-carrying capacity and failure modes of glulam beams with reinforced notches. Eur J Wood Wood Prod 74(3):481–482
15. Dias A (2011) Analysis of the nonlinear behavior of timber-concrete connections. J Structural Eng 138(9):1128–1137
16. Khorsandnia N, Valipour H, Schänzlin J, Crews K (2016) Experimental investigations of deconstructable timber–concrete composite beams. J Structural Eng 142(12):04016130
17. Kreit A, Al-Mahmoud F, Castel A, François R (2011) Repairing corroded RC beam with near-surface mounted CFRP rods. Mater Struct 44(7):1205–1217
18. Wang M, Song X, Gu X, Tang J (2019) Bolted glulam beam-column connections under different combinations of shear and bending. Eng Structures 181:281–292
19. Rosowsky DV, Bulleit WM (2002) Load duration effects in wood members and connections: order statistics and critical loads. Struct Saf 24(2–4):347–362
20. Smith I, Vasic S (2003) Fracture behaviour of softwood. Mech Mater 35(8):803–815
21. Raftery G M, Whelan C (2014) Low-grade glued laminated timber beams reinforced using improved arrangements of bonded-in GFRP rods. Constr Build Mater 52:209–220
22. Lavoie JA, Kellas S (1996) Dynamic crush tests of energy-absorbing laminated composite plates. Compos Appl Sci Manuf 27(6):467–475
23. Schober K-U, Harte A M, Kliger R, Jockwer R, Xu Q, Chen J-F (2015) FRP reinforcement of timber structures. Constr Build Mater 97:106–118
24. Fava G, Carvelli V, Poggi C (2013) Pull-out strength of glued-in FRP plates bonded in glulam. Constr Build Mater 43:362–371
25. Plevris N, Triantafillou TC (1992) FRP-reinforced wood as structural material. J Mater Civ Eng 4(3):300–317
26. Jeong H (1997) Effects of voids on the mechanical strength and ultrasonic attenuation of laminated composites. J Compos Mater 31(3):276–292
27. Gentry TR (2011) Performance of glued-laminated timbers with FRP shear and flexural reinforcement. J Compos Constr 15(5):861–870
28. Yang H, Liu W, Lu W, Zhu S, Geng Q (2016) Flexural behavior of FRP and steel reinforced glulam beams: experimental and theoretical evaluation. Constr Build Mater 106:550–563
29. Yang Y-L, Liu J-W, Xiong G-J (2013) Flexural behavior of wood beams strengthened with HFRP. Constr Build Mater 43:118–124

30. Yeboah D, Taylor S, McPolin D, Gilfillan R (2013) Pull-out behaviour of axially loaded Basalt Fibre Reinforced Polymer (BFRP) rods bonded perpendicular to the grain of glulam elements. Constr Build Mater 38:962–969
31. Gilfillan J, Gilbert S, Patrick G (2003) The use of FRP composites in enhancing the structural behavior of timber beams. J Reinf Plast Compos 22(15):1373–1388
32. Taazount M, Amziane S, Molard D (2013) Tangential behavior of nailed composite timber–concrete floor structures. Constr Build Mater 40:506–513
33. Issa CA, Kmeid Z (2005) Advanced wood engineering: glulam beams. Constr Build Mater 19(2):99–106
34. Rad AR, Weinand Y, Burton H (2019) Experimental push-out investigation on the in-plane force-deformation behavior of integrally-attached timber Through-Tenon joints. Constr Build Mater 215:925–940
35. Yeboah D, Taylor S, McPolin D, Gilfillan R, Gilbert S (2011) Behaviour of joints with bonded-in steel bars loaded parallel to the grain of timber elements. Constr Build Mater 25(5):2312–2317
36. Micelli F, Scialpi V, La Tegola A (2005) Flexural reinforcement of glulam timber beams and joints with carbon fiber-reinforced polymer rods. J Compos Constr 9(4):337–347
37. De Lorenzis L, Scialpi V, La Tegola A (2005) Analytical and experimental study on bonded-in CFRP bars in glulam timber. Compos B Eng 36(4):279–289
38. Ardalany M, Fragiacomo M, Carradine D, Moss P (2013) Experimental behavior of Laminated Veneer Lumber (LVL) joists with holes and different methods of reinforcement. Eng Structures 56:2154–2164
39. Dourado N, Pereira F, De Moura M, Morais J (2012) Repairing wood beams under bending using carbon–epoxy composites. Eng Structures 34:342–350
40. Chen Q, Lang J, Wang J (2018) Flexural properties of Bamboo-log composite beam. J Eng Sci Technol Rev 11(3)
41. She J, Inoue T, Ueno K (2000) Multilayer $Al_2O_3$/SiC ceramics with improved mechanical behavior. J Eur Ceram Soc 20(11):1771–1775
42. Anshari B, Guan Z, Kitamori A, Jung K, Komatsu K (2012) Structural behaviour of glued laminated timber beams pre-stressed by compressed wood. Constr Build Mater 29:24–32
43. De Luca V, Marano C (2012) Prestressed glulam timbers reinforced with steel bars. Constr Build Mater 30:206–217
44. Yang H, Ju D, Liu W, Lu W (2016) Prestressed glulam beams reinforced with CFRP bars. Constr Build Mater 109:73–83
45. Ferreira J G, Cruz H, Silva R (2017) Failure behaviour and repair of delaminated glulam beams. Constr Build Mater 154:384–398
46. Li H, Lam F, Qiu H (2019) Comparison of glulam beam-to-beam connections with round dovetail and half-lap joints reinforced with self-tapping screws. Constr Build Mater 227:116437
47. Glišović I, Stevanović B, Petrović M (2015) Bending behaviour of glulam beams reinforced with carbon FRP plates. J Civil Eng Manage 21(7):923–932
48. Raftery G M, Kelly F (2015) Basalt FRP rods for reinforcement and repair of timber. Compos Part B Eng 70:9–19
49. Skuturna T, Valivonis J (2016) Experimental study on the effect of anchorage systems on RC beams strengthened using FRP. Compos Part B Eng 91:283–290
50. Borri A, Corradi M, Grazini A (2005) A method for flexural reinforcement of old wood beams with CFRP materials. Compos B Eng 36(2):143–153
51. Borri A, Corradi M, Speranzini E (2013) Reinforcement of wood with natural fibers. Compos Part B Eng 53:1–8
52. Yusof A, Saleh A (2010) Flexural strengthening of timber beams using glass fibre reinforced polymer. Electronic J Structural Eng 10:45–56
53. Martins C, Santos P, Almeida P, Godinho L, Dias A (2015) Acoustic performance of timber and timber-concrete floors. Constr Build Mater 101:684–691
54. Baker J, Tipper C (1956) The value of the notch tensile test. Proc Inst Mech Eng 170(1):65–93
55. Fawwaz M, Hanna A (2012) Structural behavior of notched glulam beams reinforced by means of plywood and FRP

56. Gutkowski R, Brown K, Shigidi A, Natterer J (2008) Laboratory tests of composite wood–concrete beams. Constr Build Mater 22(6):1059–1066
57. Lyu Z, Málaga-Chuquitaype C, Ruiz-Teran A (2019) Design of timber-concrete composite (TCC) bridges with under-deck stay cables. Eng Structures 189:589–604
58. Ali Y A Z (2018) Flexural behavior of FRP strengthened concrete-wood composite beams. Ain Shams Eng J
59. Harris B, Bunsell A (1975) Impact properties of glass fibre/carbon fibre hybrid composites. Composites 6(5):197–201
60. Gutkowski RM, Miller NJ, Fragiacomo M, Balogh J (2010) Composite wood-concrete beams using utility poles: time-dependent behavior. J Structural Eng 137(6):625–634
61. Dias AM, Jorge LF (2011) The effect of ductile connectors on the behaviour of timber–concrete composite beams. Eng Struct 33(11):3033–3042
62. Buchanan A, Fairweather R (1993) Seismic design of glulam structures. Bull N Z Natl Soc Earthq Eng 26(4):415–436
63. Svecova D, Eden R (2004) Flexural and shear strengthening of timber beams using glass fibre reinforced polymer bars—an experimental investigation. Can J Civ Eng 31(1):45–55
64. Dias AM, Kuhlmann U, Kudla K, Mönch S, Dias AMA (2018) Performance of dowel-type fasteners and notches for hybrid timber structures. Eng Structures 171:40–46
65. Hong W, Jiang Y, Li B, Qin Z, Hu X (2018) Nonlinear parameter identification of timber-concrete composite beams using long-gauge fiber optic sensors. Constr Build Mater 164:217–227
66. De Baere I, Jacques S, Van Paepegem W, Degrieck J (2012) Study of the Mode I and Mode II interlaminar behaviour of a carbon fabric reinforced thermoplastic. Polym Testing 31(2):322–332
67. Lacroix DN, Doudak G (2018) Determining the dynamic increase factor for glued-laminated timber beams. J Structural Eng 144(9):04018160
68. Sebaey T, Lopes C, Blanco N, Costa J (2011) Ant colony optimization for dispersed laminated composite panels under biaxial loading. Compos Struct 94(1):31–36
69. Fossetti M, Minafò G, Papia M (2015) Flexural behaviour of glulam timber beams reinforced with FRP cords. Constr Build Mater 95:54–64
70. Fragiacomo M, Lukaszewska E (2013) Time-dependent behaviour of timber–concrete composite floors with prefabricated concrete slabs. Eng Structures 52:687–696
71. Sartori T, Tomasi R (2013) Experimental investigation on sheathing-to-framing connections in wood shear walls. Eng Structures 56:2197–2205
72. Kim YJ, Harries KA (2010) Modeling of timber beams strengthened with various CFRP composites. Eng Struct 32(10):3225–3234
73. Boccadoro L, Steiger R, Zweidler S, Frangi A (2017) Analysis of shear transfer and gap opening in timber–concrete composite members with notched connections. Mater Struct 50(5).231
74. Branco J M, Descamps T, Tsakanika E (2018) Repair and strengthening of traditional timber roof and floor structures. In: Strengthening and retrofitting of existing structures, pp 113–138
75. Pindera J (1989) Local effects and defect criticality in homogeneous and laminated structures
76. Shu Z, Li Z, Yu X, Zhang J, He M (2019) Rotational performance of glulam bolted joints: Experimental investigation and analytical approach. Constr Build Mater 213:675–695
77. Kuhlmann U, Aldi P (2008) Fatigue of timber-concrete-composite beams: characterisation of the connection behaviour through push-out tests. In: Proceedings of the 10th world conference on timber engineering, Miyazaki, Japan, p 580
78. Miotto J L, Dias A A (2015) Structural efficiency of full-scale timber–concrete composite beams strengthened with fiberglass reinforced polymer. Compos Structures 128:145–154
79. Johnsson H, Blanksvärd T, Carolin A (2007) Glulam members strengthened by carbon fibre reinforcement. Mater Struct 40(1):47–56

# Review of Cold-Formed Steel Columns

**K. Sivasathya and S. Vijayanand**

**Abstract** Cold-formed steel (CFS) involves forming of thin steel sections at room temperature. These sections play a significant role in construction due to its efficiencies like lightweight, durability, stability, the economy in transport and handling, corrosion and fire resistance, recyclability, etc., The CFS can be used as built-up (connecting multiple separate members viz., channels and angles) or single sections. The built-up compression members can be formed either giving space at uniform intervals or closely placed one. Various researchers have worked on the performance of CFS sections. Generally, the compression members under axial loading have a chance to buckle about their major axis. The buckling behavior is more in the slender column which is the major drawback due to its thinner section. Therefore, by overcoming such behavior, this paper attempts to review the different CFS sections.

**Keywords** Cold-formed steel · Built-up sections · Single sections · Buckling behavior

## Abbreviations

| | |
|---|---|
| NAS | North American Specification |
| FEA | Finite Element Analysis |
| FEM | Finite Element Method |
| AISI | American Iron and Steel Institute |
| DSM | Direct Strength Method |
| EWM | Effective Width Method |

K. Sivasathya · S. Vijayanand (✉)
Department of Civil Engineering, Kongu Engineering College, Perundurai, Erode, Tamil Nadu, India
e-mail: atmvijay.anand@gmail.com

K. Sivasathya
e-mail: sivasathyacivil97@gmail.com

© The Editor(s) (if applicable) and The Author(s), under exclusive license to Springer Nature Singapore Pte Ltd. 2021
S. Mohan et al. (eds.), *Materials, Design, and Manufacturing for Sustainable Environment*, Lecture Notes in Mechanical Engineering,
https://doi.org/10.1007/978-981-15-9809-8_24

AS/NZS Australian/New Zealand Standard

## 1 Introduction

Hot rolled steel is produced by rolling steel at higher temperatures and by further processing in mills where cold reduction takes place and CFS is produced. Shaping of CFS is brought about by employing either press brake or cold roll-forming manufacturing processes. The usage of CFS started at the end of the nineteenth century. The CFS has wide applications in the construction of decks, buildings, sewers, culverts, compartments in railways, and so on [1]. It was proved in an assessment that CFS construction has many advantages like low capital cost, good seismic answer, and environmental performance [2]. However, there is a noticeable risk of buckling in open sections. In the place of columns made of single sections, built-up sections are employed, which supports in carrying heavy loads [3]. But there were many research works on CFS single sections than built-up sections. Thus, researches can be extended towards CFS built-up sections.

## 2 Selection of CFS Sections

In designing a CFS section, suitable sections were chosen in line with NAS-2016. The single section's geometric properties are mentioned in NAS. Therefore, for built-up sections too, the same is applicable. Figure 1 shows the geometric limitations of a single section in NAS [4]. Further, the suitability of the selected sections for a structure was recognized employing experimental, numerical, and theoretical approaches. Lastly, the appropriateness of these approaches was checked.

## 3 Experimental Investigations

This section discusses the outcomes of the tests conducted in the laboratory. The experimental studies that are mainly engaged in investigating CFS columns are tensile coupon tests and axial compression tests. Properties of CFS were found by tensile coupon tests [5, 6]. The column's resistance to axial loading was found by the compression test. Besides these tests, to anticipate the member capacity at a high temperature undergoing simultaneous bending and twisting an elevated temperature test was performed [7].

| Section | Geometric limitations |
|---|---|
| 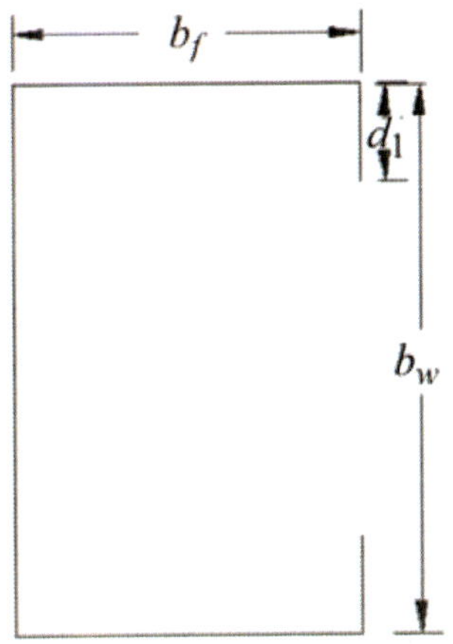 | $b_w/t<472$<br>$b_f/t<159$<br>$4<d_1/t<33$<br>$0.7<b_w/b_f<5.0$<br>$0.05<d_1/b_f<0.41$<br>$\Theta = 90°$<br>$E/f_y>340$ ($f_y < 593$MPa) |



**Fig. 1** Lipped channel—geometric limitations

## 4 Numerical Investigations

A numerical investigation is done by FEA where there are several packages available. FEA is the discretization of a structure into a large number of very small elements [8]. Satisfactory results can be produced for the complex structural elements by means of FEM tools. FEA also saves time and resources required for the experimental investigation [9]. FEA is often accomplished through a program called ABAQUS. FEA through ABAQUS is of two types. First, buckling modes and loads are determined which is called eigenvalue analysis. The second one, load-displacement non-linear analysis, follows the prediction of eigenvalues [10]. It determines the ultimate loads of CFS columns [11]. FEM using ABAQUS closely determined the experimental results [12, 13]. Chen et al. [14] used ABAQUS-UMAT in which the simulations matched the test results of CFS of a thick wall under cyclic loading. Though ABAQUS is very efficient in its function, it has modeling challenges in certain circumstances [15]. Other programs were also rarely used for FEA as follows. Sivakumaran [16] employed a program called ADINA, a non-linear FEM for analyzing local buckling behavior of CFS having web opening in their sections. The crippling effect in CFS members having openings in the web was investigated by another program called ANSYS, a non-linear elastoplastic FEM [17]. Its advanced capabilities include nonlinear material characteristics, geometrical nonlinearities, contact elements, etc., needed for analysis [8]. ANSYS was found effective in predicting ultimate strength for CFS members [18], for lipped channel section with or without perforations [19], and cold-formed I and box (built-up) sections [20] against test results. Nguyen et al. [21] used Marc software for analyzing the dimpled CFS columns. In addition to ABAQUS, MSC/PATRAN was adopted to anticipate the modes of failure in cold-formed hollow

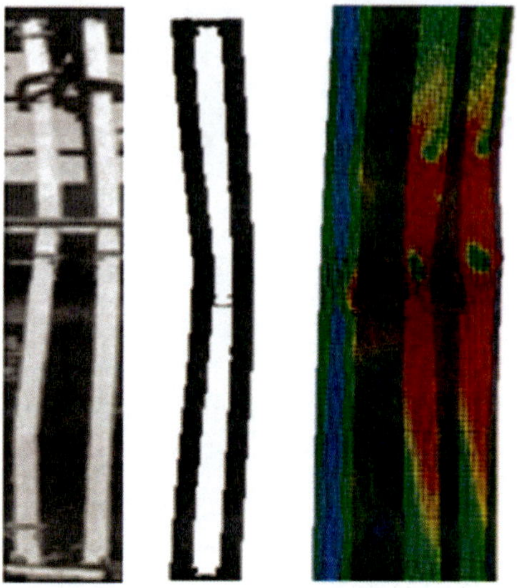

**Fig. 2** Failure modes represented experimentally and numerically (ABAQUS)

flange channels [22]. Yet another FEM is the Finite Strip Method, which uses strips in place of finite elements, particularly worthy for prismatic structures [8].

FEM had shown a good agreement in calculating distortional buckling critical stress of various web stiffened lipped CFS channels [23]. Figure 2 shows the simulations of lipped back-to-back CFS channels with spacers [24]. In the design and research of CFS, modeling computationally will substantially contribute more in the future [25].

## 5 Theoretical Investigations

Various international codes and standards are involved in investigating the behavior of CFS sections theoretically. It covers American standards, Australian standards, and Eurocode primarily. AISI specification constitutes DSM and EWM. During 1940, Prof. G. Winter presented Winter's effective width equations from which EWM got developed. Later, DSM was adopted by NAS which is an efficacious alternative to EWM [26]. One of the greatest benefits of the DSM is the prediction of the structural behavior of complex folded sections efficiently [27]. On doing FEA [28] and experimental study [29] in built-up CFS closed columns, modified DSM and DSM were conservative in predicting the ultimate strengths. Xingyou et al. [30] investigated CFS lipped channels by EWM intending to improve the codal practice for distortional buckling strength. Australian specification (AS/NZS 4600) was preferred as it includes high strength steels. Further, Eurocode (EN 1993) was also utilized for the design purpose. Recently, for cold-formed members, European procedures were evaluated [31].

# 6 CFS Single Sections

CFS single sections include channels, angles, hat sections, sigma sections, zed sections, etc., Besides these sections, CFS was also formed in new creative shapes using CFS grammar which used scripting languages for forming new shapes and improving its strength capabilities [32]. The following were the works done in creative sections. Designing hollow compact oval-shaped sections was conservative on using NAS, AS/NZS, and European specifications [33]. Elchalakani et al. [34] attempted to give precise slenderness limits for cold-formed hollow circular sections. Sani et al. [35] studied a CFS section (channel-curved) for a truss used in the roof. For getting better bending capacity results of CFS hollow stainless steel sections, the continuous strength method was suggested [36]. The authors modified the current DSM equation to determine the cold-formed hollow flange channel stub column capacity [22]. Then, by using modified NAS and DSM, revised guidelines for designing were put forward for tubular members made of ferritic cold-formed stainless steel [37]. Later, a CFS square column was investigated for its material properties, residual stresses, etc., [38]. Followingly, hollow cold-formed square sections were studied for measuring residual stresses in it [39]. For the cold-formed YSt-310 hollow steel section, a design equation was proposed recently which is subjected to torsion [40]. There was also an attempt made in providing new recommendations for stainless steel hollow cold-formed circular sections [41] and high strength hollow cold-formed rectangular sections [42]. On semi-oval cold-formed hollow sections, experimental and numerical [43] and cross-sectional behavior were studied [44]. Later, Fang et al. [45] numerically investigated the CFS tubular columns on its structural performance. Yet another creative finding was the dimpled geometry in CFS columns that reduced the lateral impact [46]. Singh et al. [47] on investigating the perforated CFS tubular stub columns experimentally found that sections with a circular perforation at the center provided scattered design predictions.

## 6.1 *Ultimate Capacity of the Column*

Moldovan [48] conducted columns and stub columns test on the channel and channel with lips and compared them with the theoretical ultimate results obtained from AISI, European Convention for Conventional Steelwork, and Eurocode 3-Annex A specifications. The outcomes showed greater conservativeness. Shanmugam et al. [49] then developed a design equation based on parametric studies using FEM for CFS channel stub columns with perforations to predict the maximum load capacity. At times, new design rules were proposed as the present rules for the design were unconservative. One of the researchers proposed new design rules to find the column strengths of CFS intermediate and short column lengths of unequal angle members on the comparison between parametric study and NAS [12]. Strengths evaluated using NAS for CFS non-symmetric angle sections with lips were conservative on comparing the test strengths and then reliability analysis was carried out further [50]. Australian and European codes provided only elastic design for CFS. However, CFS structures are well suited for plastic design too. Thus, new slenderness limits were suggested for the Australian standard to get the CFS section capacity having a channel profile accurately in bending [51].

## 6.2 *Buckling Effects*

Sivakumaran et al. [52] developed an FEA model for CFS members under axial compression to predict the post-local buckling behavior which in comparison with experimental results of channel (lipped) sections with and without perforations were reliable. Young et al. [53] done an experimental demonstration of local buckling effects on the singly symmetric lipped and plain channel fixed and pin-ended cold-formed columns.

Feng et al. [54] suggested that by modifying the current code design methods (BS 5950 Part 5, Eurocode 3, and AISI), advanced modes of behavior like distortional buckling, effects of elevated temperature, and service holes could be found. On computing the distortional buckling for innovative CFS members, a suggestion was put forward to AS/NZS 4600, for improving its preciseness [55]. Muftah et al. [56] predicted that the CFS stub columns failed by distortional buckling at post elevated temperature and it's tested predictions were in good agreement with DSM.

Various researchers investigated the buckling effects at elevated temperatures as follows. Ranawaka et al. [57] studied the behavior of C-shaped lipped CFS sections without and with additional lips at elevated and ambient temperatures. At ambient temperature, the results were found to be accurate on comparing with AS/NZS 4600 and DSM. On slightly modifying the codal provisions with reduced mechanical properties, ultimate loads at elevated temperatures were found. Similarly, on doing modifications in the current distortional design curves in DSM, distortional buckling with non-uniform elevated temperature in CFS columns were evaluated [58]. The same suggestion as in [57] was given to the CFS pined ended and fixed ended slender columns suffering flexural–torsional buckling [7] and for compression members with and without lips undergoing local buckling [59] along with some additional recommendations.

## 6.3 *Effects of Geometrical Imperfections*

Lipped C-section CFS cut stub columns were tested for the effects of geometric imperfections under axial compression. It indicated that there was a reduction in the column's maximum strength because of geometric imperfections caused by cutting [60]. Thus, initial imperfections had a great influence on the column strength [61, 62].

Similar to geometric imperfections, slotted holes too have an impact on CFS sections. It was seen that the holes influenced ductility and peak response of the CFS structural studs greatly but had only a minimal effect on ultimate compressive strength [63]. The CFS circular and polygonal cross-sections with openings affected the resistance to loading [64]. Channel sections with web openings and edge stiffeners increased the compression resistance to about 22%, whereas for unstiffened openings there was a decrement of about 20% on comparing with plain channels [65].

Geometric interactions and the flange and lip sizes are some very important factors that influence the buckling characteristics. Bedair [66] on considering the geometric interactions and influence of flange and lip sizes of lipped CFS channels under compression, proposed a new numerical and theoretical approach. It was found to be economical as it saves material. The effects of column lengths and end conditions too were considered to find a distortional buckling formula [67].

### *6.4 Optimization*

Buckling loads produce nonlinear behavior of CFS members, which makes it difficult to optimize their cross-sectional shapes [68, 69]. Micro genetic algorithms presented extremely good results for less weight design [70]. Li et al. [71] later suggested a new algorithm to develop CFS lipped channels in less weight.

## 7 CFS Built-Up Sections

CFS built-up sections may be spaced at uniform intervals or closely spaced.

### *7.1 Closely Placed CFS Built-Up Sections*

Young et al. [72] conducted a reliability analysis on the suitability of DSM in finding the buckling stresses of a single section, single-restrained section, and closed CFS built-up column with intermediate web stiffener. DSM was good for a single section only. Further, Reyes et al. [73] applied a modified slenderness ratio approach on 48 samples of cold-formed box sections seam welded (spacing 100–900 mm) composed of two C-sections with rigid and flexible boundary conditions. It was seen that the capacity was not reduced except for the weld spacing 900 mm with flexible supports. Conservative results were found on comparing the experimental results of 4 distinct cross-sections of CFS columns (built-up) with axial capacities from DSM predictions [74]. The test results showed uncertain buckling capacities calculated by AISI for CFS columns (built-up) - one with closed and the other with open orientation. This was because the maximum load capacity was influenced by the orientation [75] whereas, the results were conservative for CFS built-up open sections on using modified DSM [76]. Kesawan et al. [77] performed a compression test on CFS hollow flange built-up sections. Subsequently, Liao et al. [78] on testing and analyzing the CFS built-up multi limb stub columns arrived at the failure modes. Following the multi limb sections, cold-formed built-up T-section triple-lambs open columns were investigated for axial compression capacity by using FEM, Chinese design code, and AISI [79].

As discussed in single sections, some factors influence the load-bearing capacity of built-up columns too. One of the factors was the fastener spacing ratio [80]. 10–25% of ultimate strength was reduced in CFS built-up I sections due to geometrical imperfections [81, 82]. The consequence of thickness [83] and screw spacing [84] was examined on the built-up CFS column. Fratamico et al. [85] put forward that a column's capacity was increased to 33% when it was installed with Effective Fastener Grouping. The more the number of rows of fastener more is the flexural rigidity [86].

### *7.2 CFS Built-Up Sections with Gap*

Anbarasu et al. [3] investigated built-up battened CFS columns (web stiffened) on the behavior and strength characteristics. An equation was then proposed for the battened built-up column design based on parametric studies. CFS built-up columns with battens were investigated both experimentally [87] and numerically [88] then. The results in comparison with NAS, AS/NZS, and Eurocode were unconservative when failing by local buckling and safe when failing by elastic flexural buckling. Eurocode 3 and NAS were inappropriate in finding the load-bearing capacity of built-up columns (laced) but numerical validation produced good results with test results [89] whereas, EWM predicted better results for CFS built-up lipped columns [24].

On investigating the compression behavior of the CFS unstiffened angle laced section, a column's behavior was found to get affected by the width of the endplate, configuration, and slenderness of lacing [90]. Battens enhanced the strength of double angle members by giving the compression eccentrically [91].

## 8 Conclusions

Cold-formed steel has been gained worldwide popularity and developments have been taken place. The intention of this paper is to provide an overview of the vastly different CFS columns and their structural design adequacies. It also summarizes the investigations involved in predicting the axial capacities. In that, newly emerged were the software that will surely do a great role in the future because of its simplicity, preciseness, and the economics involved in it. Many modifications were proposed to the codes and guidelines for the improvement in performance for the CFS single and built-up section columns. Hence, this paper provides valuable guidance on designing the structures using CFS sections. It was also noted that there were not much research works on CFS built-up section columns when compared to CFS single sections. Thus, the researchers may have a vast area of exploring their findings on different CFS built-up section columns.

## References


1. Online (Accessed: 18 11 2019) Cold-formed steel. Available at: https://en.wikipedia.org/w/index.php?title=Cold-formed_steel&oldid=922072387
2. Vitale P, Spagnuolo A, Lubritto C, Arena U (2018) Environmental performances of residential buildings with a structure in cold formed steel or reinforced concrete. J Cleaner Prod 189:839–852
3. Anbarasu M, Kanagarasu K, Sukumar S (2015) Investigation on the behaviour and strength of cold-formed steel web stiffened built-up battened columns. Mater Struct 48(12):4029–4038
4. Vijayanand S, Anbarasu M (2019) Strength and behavior of cold-formed steel built-up battened columns: tests and numerical validation. J Structural Eng 46(2):154–164
5. Singh TG, Singh KD (2019) Mechanical properties of YSt-310 cold-formed steel hollow sections at elevated temperatures. J Constr Steel Res 158:53–70
6. Chen M-T, Young B (2019) Material properties and structural behavior of cold-formed steel elliptical hollow section stub columns. Thin-Walled Structures 134:111–126
7. Gunalan S, Heva YB, Mahendran M (2014) Flexural–torsional buckling behaviour and design of cold-formed steel compression members at elevated temperatures. Eng Struct 79:149–168
8. Macdonald M, Heiyantuduwa M, Rhodes J (2008) Recent developments in the design of cold-formed steel members and structures. Thin-Walled Structures 46(7–9):1047–1053
9. Kulatunga M, Macdonald M, Rhodes J, Harrison D (2014) Load capacity of cold-formed column members of lipped channel cross-section with perforations subjected to compression loading–Part I: FE simulation and test results. Thin-Walled Structures 80:1–12
10. Chen J, Young B (2007) Cold-formed steel lipped channel columns at elevated temperatures. Eng Struct 29(10):2445–2456
11. Gunalan S, Mahendran M (2013) Improved design rules for fixed ended cold-formed steel columns subject to flexural–torsional buckling. Thin-Walled Structures 73:1–17
12. Young B, Ellobody E (2007) Design of cold-formed steel unequal angle compression members. Thin-Walled Structures 45(3):330–338
13. Anbarasu M (2019) Numerical investigation on behaviour and design of cold-formed steel built-up column composed of lipped sigma channels. Adv Structural Eng 22(8):1817–1829
14. Chen Z, Yang S, Liu H (2019) Research on the constitutive model of cold-formed thick-walled steel under cyclic loading. J Constr Steel Res 160:457–475
15. Kyvelou P, Kyprianou C, Gardner L, Nethercot D (2019) Challenges and solutions associated with the simulation and design of cold-formed steel structural systems. Thin-Walled Structures 141:526–539
16. Sivakumaran K (1987) Analysis for local buckling capacity of cold-formed steel sections with web opening. Comput Struct 26(1–2):275–282
17. Uzzaman A, Lim JB, Nash D, Rhodes J, Young B (2012) Cold-formed steel sections with web openings subjected to web crippling under two-flange loading conditions—part I: tests and finite element analysis. Thin-Walled Structures 56:38–48
18. Ren W-X, Fang S-E, Young B (2006) Analysis and design of cold-formed steel channels subjected to combined bending and web crippling. Thin-Walled Structures 44(3):314–320
19. Kulatunga MP, Macdonald M (2013) Investigation of cold-formed steel structural members with perforations of different arrangements subjected to compression loading. Thin-Walled Structures 67:78–87

20. Li Y, Li Y, Wang S, Shen Z (2014) Ultimate load-carrying capacity of cold-formed thin-walled columns with built-up box and I section under axial compression. Thin-Walled Structures 79:202–217
21. Nguyen VB, Mynors D, Wang C, Castellucci M, English M (2016) Analysis and design of cold-formed dimpled steel columns using finite element techniques. Finite Elem Anal Des 108:22–31
22. Kesawan S, Mahendran M (2017) Section compression capacity of high strength cold-formed hollow flange channels. J Constr Steel Res 133:202–213
23. Huang X-h, Yang J, Liu Q-f, Zhu J, Bai L, Wang F-l, Wang J-h (2018) A simplified flange–lip model for distortional buckling of cold-formed steel channel-sections with stiffened web. Int J Mech Sci 136:451–459
24. Vijayanand S, Anbarasu M (2017) Effect of spacers on ultimate strength and behavior of cold-formed steel built-up columns. Proc Eng 173:1423–1430
25. Schafer BW, Li Z, Moen CD (2010) Computational modeling of cold-formed steel. Thin-Walled Structures 48(10–11):752–762
26. Li Y-Q, Shen Z-Y, Wang L, Wang Y-M, Xu H-W (2007) Analysis and design reliability of axially compressed members with high-strength cold-formed thin-walled steel. Thin-Walled Structures 45(4):473–492
27. Schafer BW (2019) Advances in the direct strength method of cold-formed steel design. Thin-Walled Structures 140:533–541
28. Zhang J-H, Young B (2018a) Finite element analysis and design of cold-formed steel built-up closed section columns with web stiffeners. Thin-Walled Structures 131:223–237
29. Zhang J-H, Young B (2018b) Experimental investigation of cold-formed steel built-up closed section columns with web stiffeners. J Constr Steel Res 147:380–392
30. Xingyou Y, Yanli G, Yuanqi L (2016) Effective width method for distortional buckling design of cold-formed lipped channel sections. Thin-Walled Structures 109:344–351
31. Baldassino N, Bernuzzi C, Simoncelli M (2019) Evaluation of European approaches applied to design of TWCF steel members. Thin-Walled Structures 143:106186
32. Franco JMS, Duarte JP, de Miranda BE, Landesmann A (2014) Shape Grammar of steel cold-formed sections based on manufacturing rules. Thin-Walled Structures 79:218–232
33. Zhu J-H, Young B (2012) Design of cold-formed steel oval hollow section columns. J Constr Steel Res 71:26–37
34. Elchalakani M, Zhao XL, Grzebieta R (2002) Bending tests to determine slenderness limits for cold-formed circular hollow sections. J Constr Steel Res 58(11):1407–1430
35. Sani MSHM, Muftah F, Tan CS, Tahir MM (2015) Structural behaviour of cold-formed cut-curved channel steel section under compression. Proc Eng 125:1008–1014
36. Zheng B, Shu G, Xin L, Yang R, Jiang Q (2016) Study on the bending capacity of cold-formed stainless steel hollow sections. In: Structures, pp 63–74
37. Li H-T, Young B (2017) Cold-formed ferritic stainless steel tubular structural members subjected to concentrated bearing loads. Eng Struct 145:392–405
38. Liu D, Liu H, Chen Z, Liao X (2017) Structural behavior of extreme thick-walled cold-formed square steel columns. J Constr Steel Res 128:371–379
39. Somodi B, Kövesdi B (2017) Residual stress measurements on cold-formed HSS hollow section columns. J Constr Steel Res 128:706–720
40. Devi SV, Singh TG, Singh KD (2019) Cold-formed steel square hollow members with circular perforations subjected to torsion. J Constr Steel Res 162:105730
41. Buchanan C, Real E, Gardner L (2018) Testing, simulation and design of cold-formed stainless steel CHS columns. Thin-Walled Structures 130:297–312
42. Li H-T, Young B (2019) Behaviour of cold-formed high strength steel RHS under localised bearing forces. Eng Struct 183:1049–1058
43. Chen M-T, Young B (2018a) xperimental and numerical investigation on cold-formed steel semi-oval hollow section compression members. J Constr Steel Res 151:174–184
44. Chen M-T, Young B (2018b) Cross-sectional behavior of cold-formed steel semi-oval hollow sections. Eng Struct 177:318–330

45. Fang H, Chan T-M, Young B (2018) Structural performance of cold-formed high strength steel tubular columns. Eng Struct 177:473–488
46. Liang C, Wang CJ, English M, Mynors D (2018) Behaviour of cold-formed dimpled columns under lateral impact. Eng Struct 163:167–176
47. Singh TG, Singh KD (2018) Experimental investigation on performance of perforated cold–formed steel tubular stub columns. Thin-Walled Structures 131:107–121
48. Moldovan A (1994) Compression tests on cold-formed steel columns with monosymmetrical section. Thin-Walled Structures 20(1–4):241–252
49. Shanmugam N, Dhanalakshmi M (2001) Design for openings in cold-formed steel channel stub columns. Thin-Walled Structures 39(12):961–981
50. Young B, Chen J (2008a) Column tests of cold-formed steel non-symmetric lipped angle sections. J Constr Steel Res 64(7–8):808–815
51. Maduliat S, Bambach MR, Zhao XL (2012) Slenderness limits for cold-formed channel sections in bending by experimental methods. J Constr Steel Res 76:75–82
52. Sivakumaran K, Abdel-Rahman N (1998) A finite element analysis model for the behaviour of cold-formed steel members. Thin-Walled Structures 31(4):305–324
53. Young B, Rasmussen KJ (1999) Behaviour of cold-formed singly symmetric columns. Thin-Walled Structures 33(2):83–102
54. Feng M, Wang Y, Davies J (2003) Structural behaviour of cold-formed thin-walled short steel channel columns at elevated temperatures. Part 2: design calculations and numerical analysis. Thin-Walled Structures 41(6):571–594
55. Narayanan S, Mahendran M (2003) Ultimate capacity of innovative cold-formed steel columns. J Constr Steel Res 59(4):489–508
56. Muftah F, Sani MSHM, Osman AR, Mohammad S (2015) Mechanical behaviour of the cold-formed steel channel stub column under post elevated temperature. Proc Eng 125:1015–1022
57. Ranawaka T, Mahendran M (2009) Distortional buckling tests of cold-formed steel compression members at elevated temperatures. J Constr Steel Res 65(2):249–259
58. Shahbazian A, Wang YC (2012) Direct strength method for calculating distortional buckling capacity of cold-formed thin-walled steel columns with uniform and non-uniform elevated temperatures. Thin-Walled Structures 53:188–199
59. Gunalan S, Heva YB, Mahendran M (2015) Local buckling studies of cold-formed steel compression members at elevated temperatures. J Constr Steel Res 108:31–45
60. Lam S, Chung K, Wang X (2006) Load-carrying capacities of cold-formed steel cut stub columns with lipped C-section. Thin-Walled Structures 44(10):1077–1083
61. Dinis PB, Camotim D, Silvestre N (2007) FEM-based analysis of the local-plate/distortional mode interaction in cold-formed steel lipped channel columns. Comput Struct 85(19–20):1461–1474
62. Al Ali M, Tomko M, Demjan I, Kvočák V (2012) Thin-walled cold-formed compressed steel members and the problem of initial imperfections. Proc Eng 40:8–13
63. Moen CD, Schafer B (2008) Experiments on cold-formed steel columns with holes. Thin-Walled Structures 46(10):1164–1182
64. Tran AT, Veljkovic M, Rebelo C, da Silva LS (2016) Resistance of cold-formed high strength steel circular and polygonal sections—Part 1: experimental investigations. J Constr Steel Res 120:245–257
65. Chen B, Roy K, Uzzaman A, Raftery GM, Nash D, Clifton GC, Pouladi P, Lim JB (2019) Effects of edge-stiffened web openings on the behaviour of cold-formed steel channel sections under compression. Thin-Walled Structures 144:106307
66. Bedair O (2009) A cost-effective design procedure for cold-formed lipped channels under uniform compression. Thin-Walled Structures 47(11):1281–1294
67. Zhou T, Lu Y, Li W, Wu H (2017) End condition effect on distortional buckling of cold-formed steel columns with arbitrary length. Thin-Walled Structures 117:282–293
68. Liu H, Igusa T, Schafer B (2004) Knowledge-based global optimization of cold-formed steel columns. Thin-Walled Structures 42(6):785–801

69. Ma W, Becque J, Hajirasouliha I, Ye J (2015) Cross-sectional optimization of cold-formed steel channels to Eurocode 3. Eng Struct 101:641–651
70. Lee J, Kim S-M, Park HS (2006) Optimum design of cold-formed steel columns by using micro genetic algorithms. Thin-Walled Structures 44(9):952–960
71. Li Z, Leng J, Guest JK, Schafer BW (2016) Two-level optimization for a new family of cold-formed steel lipped channel sections against local and distortional buckling. Thin-Walled Structures 108:64–74
72. Young B, Chen J (2008b) Design of cold-formed steel built-up closed sections with intermediate stiffeners. J Structural Eng 134(5):727–737
73. Reyes W, Guzmán A (2011) Evaluation of the slenderness ratio in built-up cold-formed box sections. J Constr Steel Res 67(6):929–935
74. Georgieva I, Schueremans L, Vandewalle L, Pyl L (2012) Design of built-up cold-formed steel columns according to the direct strength method. Proc Eng 40:119–124
75. Kang TH-K, Biggs KA, Ramseyer C (2013) Buckling modes of cold-formed steel columns. Int J Eng Technol 5(4):447
76. Zhang J-H, Young B (2015) Numerical investigation and design of cold-formed steel built-up open section columns with longitudinal stiffeners. Thin-Walled Structures 89:178–191
77. Kesawan S, Mahendran M, Dias Y, Zhao W-B (2017) Compression tests of built-up cold-formed steel hollow flange sections. Thin-Walled Structures 116:180–193
78. Liao F, Wu H, Wang R, Zhou T (2017) Compression test and analysis of multi-limbs built-up cold-formed steel stub columns. J Constr Steel Res 128:405–415
79. Liu X, Zhou T (2017) Research on axial compression behavior of cold-formed triple-lambs built-up open T-section columns. J Constr Steel Res 134:102–113
80. Abbasi M, Khezri M, Rasmussen K, Schafer B (2018) Elastic buckling analysis of cold-formed steel built-up sections with discrete fasteners using the compound strip method. Thin-Walled Structures 124:58–71
81. Abu-Hamd M, Abdel-Ghaffar MM, El-Samman BM (2018) Buckling strength of axially loaded cold formed built-up I-sections with and without stiffened web. Ain Shams Eng J 9(4):3151–3167
82. Dobrić J, Marković Z, Buđevac D, Spremić M, Fric N (2018) Resistance of cold-formed built-up stainless steel columns–part I: experiment. J Constr Steel Res 145:552–572
83. Roy K, Ting TCH, Lau HH, Lim JB (2018) Effect of thickness on the behaviour of axially loaded back-to-back cold-formed steel built-up channel sections-experimental and numerical investigation. In: Structures, pp 327–346
84. Ting TCH, Roy K, Lau HH, Lim JB (2018) Effect of screw spacing on behavior of axially loaded back-to-back cold-formed steel built-up channel sections. Adv Structural Eng 21(3):474–487
85. Fratamico DC, Torabian S, Zhao X, Rasmussen KJ, Schafer BW (2018) Experiments on the global buckling and collapse of built-up cold-formed steel columns. J Constr Steel Res 144:65–80
86. Phan DK, Rasmussen KJ (2019) Flexural rigidity of cold-formed steel built-up members. Thin-Walled Structures 140:438–449
87. Dabaon M, Ellobody E, Ramzy K (2015a) Experimental investigation of built-up cold-formed steel section battened columns. Thin-Walled Structures 92:137–145
88. Dabaon M, Ellobody E, Ramzy K (2015b) Nonlinear behaviour of built-up cold-formed steel section battened columns. J Constr Steel Res 110:16–28
89. Dar MA, Sahoo DR, Pulikkal S, Jain AK (2018) Behaviour of laced built-up cold-formed steel columns: experimental investigation and numerical validation. Thin-Walled Structures 132:398–409
90. Dar MA, Sahoo DR, Jain AK (2019) Axial compression behavior of laced cold-formed steel built-up columns with unstiffened angle sections. J Constr Steel Res 162:105727
91. Maia W, Vieira L Jr, Schafer B, Malite M (2016) Experimental and numerical investigation of cold-formed steel double angle members under compression. J Constr Steel Res 121:398–412

# Analysis of Tool Wear in Micro-EDM Drilling Using Response Surface Methodology

**M. Parthiban and M. Harinath**

**Abstract** This paper presents the effect of micro-drilling parameters such as pulse on time, pulse off time and servo voltage to analyze tool wear in electrical discharge machine (EDM). In this work, hard material beryllium copper is selected as a workpiece to study the optimal parameter in micro-drilling processes in EDM to find the optimal tool wear for the copper electrode. Micro-drilling experiments were performed by changing the parameter levels. This paper deals with the response surface methodology approach for optimizing the tool wear. The experiments were designed based on response surface design where the Box–Behnken design method was used, and the results were analyzed using a microscope. The outcome of the parameters on tool wear was determined by performing an analysis of variance (ANOVA) and regression analysis is performed with significant interactions. The results indicate that the predicted values are closer to the experimental values with a maximum deviation of 12.03%. It shows clearly that the experimental models are in good agreement with the mathematical model and optimal values.

**Keywords** EDM · Micro-drilling · Tool wear · Response surface methodology · ANOVA

## 1 Introduction

In this current scenario, demand for hard metal machining and complex profile, products are increasing day by day. In aerospace applications and die manufacturing applications where the hard metals like vanadium and beryllium with more complex profiles and microstructures are used. Some of the potential applications of beryllium

M. Parthiban (✉) · M. Harinath
Department of Mechanical Engineering, PSG College of Engineering, Coimbatore, Tamil Nadu 641004, India
e-mail: mpn.mech@psgtech.ac.in

M. Harinath
e-mail: harinath1039@gmail.com

© The Editor(s) (if applicable) and The Author(s), under exclusive license to Springer Nature Singapore Pte Ltd. 2021

S. Mohan et al. (eds.), *Materials, Design, and Manufacturing for Sustainable Environment*, Lecture Notes in Mechanical Engineering,
https://doi.org/10.1007/978-981-15-9809-8_25

copper in the present world are consumer electronics and telecommunications, industrial components and commercial aerospace, medical industries, cryogenic equipment, armor and weapons, automobile engine valves, percussion instruments. The electrical discharge machining (EDM) is one of the unconventional machining technologies that effectively use the spark between the electrode and workpiece. The rapid repetitive spark that produces between the tool and the workpiece machines the materials regardless of its physical property (hardness) and besides that, the major advantage of the machining process is that electrically conductive material of any kind can be machined. During the EDM machining process, the tool and workpiece have no direct contact, thus eliminating high cutting forces and stress produced during the machining process, and it also eliminates vibration problems on a workpiece. The dielectric medium is a fluid that is used in the EDM process in which the electrode and workpiece are immersed. The spark is produced in the dielectric medium when the potential difference is increased, which in turn machining process takes place and a small spark gap is maintained between the tool and the material to flow away the debris formed during the machining process by the flushing system and EDM is used in various applications for machining hard and brittle materials and in manufacturing some complex profile dies. The only criterion required for the machining process is that the workpiece material must be electrically conductive. Zhang et al. [1] experimented with the WEDG method with the tangential feed to enhance the accuracy of microelectrodes, and he proposed to enhance the processing efficiency of array micro-holes. Besides the system, the position and dimensional accuracies during machining are analyzed carefully and the perfect position is selected. The charged coupled device is used to couple the online measuring system with the existing wire electric discharge grinding process. Fleischer et al. [2] suggested a new application of the micro-EDM method, for the manufacturing of micro-cutting milling tools. Tungsten carbide is machined with the help of the WEDG process to produce micro-milling cutters. Micro-EDM gives merits by producing apparatus with low cost, parts with a high aspect ratio, and possesses the capability of producing complicated 3D shapes. Therefore, the process can be greatly applicable for manufacturing small devices that require micro-parts and micro-holes. Parthiban et al. [3] in an already existing wire EDG setup analyzed the machining capability of tungsten material. Microelectrodes on tungsten material with a diameter of 2 mm are reduced to 0.4 mm diameter. The input parameters wire speed, input current, pulse off time, spindle speed are selected, and optimization is carried out to know which input parameter plays a dominant role in obtaining diameter accuracy. He concluded that the diameter accuracy is by affected spindle speed and pulse off time. Rahman et al. [4] in their work brass is selected as workpiece material and micro-drilling is performed to analyze the effect of an optimal parameter for the drilling tool. Material removal rate and surface roughness are mostly affected by spindle speed and feed rate. Burr size, tool wear are influenced by incrementing the feed rate and spindle speed. Singh et al. [5] studied the Inconel 601 material that can be machined in the EDM process. He investigated the effect on Ra and MRR of Inconel 601 with different input parameters during the EDM process. He correlated the ANOVA and the RSM with responses to the selected parameters and showed that input current influences the MRR during the machining process. Gupta et al. [6] investigated the implementation of the logical fuzzy and Taguchi method for multiple output optimizations. They concluded that the optimal results are obtained during the medium level of the nose radius of the tool and cutting speed. The level of feed and depth of cut is lower. The most favorable parameter selected all among was found to be the cryogenic environment. Fnide et al. [7] investigated the implementation of RSM for obtaining cutting force model for the hard turning of tool steel. Under dry machining conditions, the ceramic tool insert is used for machining. The result yielded during the machining process is that the depth of cut is the most influencing parameter which affects cutting force. The rated feed and speed mostly influences in the tangential and radial cutting forces. Ashok Kumar et al. [8] in his work investigated the optimum process parameter to get a low tool wear rate in Die-Sinker EDM. Deionized water and kerosene are utilized to improve surface quality without changing the drilling time. Rath [9] used gray-RSM to optimize the parameter of EDM on EN19. Based on the experiment, the model expressed the pulse on time as the highest influential parameter followed by the open-circuit current and then pulse duty factor. Ohdar [10] implemented the RSM method to get the optimal EDM output parameters. The copper electrode is used as the electrode, and optimization is carried based on the different process parameters. Ashok Kumar [11] studied the tool wear in micro-hole machining performance in EDM. Optimization was carried using the Taguchi approach by varying the process parameters. Based on the experiment, the result yields that the current plays the most influencing parameter in electrode wear followed by other parameters. Liu et al. [12] developed the surface layer model to study its effect in the micro-EDM process. The developed result showed that when the surface free energy is higher, the higher the MRR and machined hole taper is minimum. Discharge current influences the surface layer of the workpiece when it is lower. Li et al. [13] in his work studied the process capability of microelectrodes with good accuracy control in micro-EDM machine for micro-hole array drilling process. The experimental result showed that electrode wear of radial and axial are compensated to improve the dimensional accuracy of micro-holes. Li [14] in his investigation analyzed the performance of deionized water during micro-drilling in the EDM process. Used the mist jet principle to electrolyze suppression, the dispersion was analyzed. Pellegrini [15] studied the improvisation for the production system by enhancing the sustainability of micro-drilling in the EDM process. The aim of the system is used to make the manufacturing process more feasible and the tools utilized for the processing with various conditions to be easily computable.

From the literature survey, it is found that the major problem associated with the drilling of micro-holes is higher TWR. The objective of this paper is to analyze the micromachining ability of manufactured microelectrodes, on hard metal beryllium copper in EDM, and to optimize the process parameter to analyze the electrode wear during micro-drilling. In this experiment, the optimization is carried out to find the optimum parameter for getting low tool wear using response surface methodology. ANOVA is performed to know which parameter has a significant effect over tool wear, and prediction is also carried out using RSM and it shows that the mathematical model is in good relationship with experimental results.

## 2 Experimental Process

The experimental setup is carried in EDM to analyze the micromachining ability of microelectrodes. The micro-drilling is performed on the EDM machine, and electrode wear during the time of machining is observed. The hard metal beryllium copper is selected as the workpiece material, which has a great potential application in the field of health care and aerospace industries. The electrodes for micro-drilling are manufactured by the WEDG process, and their diameters are measured with the help of a polarized light optical microscope.

The copper rods are made as microelectrodes which are to be used in the EDM process for micro-drilling. The experimental data for micro-drilling on beryllium copper using copper electrodes are given in Table 1.

### 2.1 *Selection of Levels and Parameters*

The micro-hole drilling parameters for machining in the EDM machine are servo voltage, pulse on time, pulse off time are selected as process variables for micro-drilling on beryllium copper with electrode material selected as a copper. The level of parameters is chosen for a different set of machining operations and based on the overall specifications of the machine (Table 2).

Based on the parameters and their levels, the design of experiments is performed for the experiment. In that Box-Behnken design, one of the response surface methodologies is used for the design of experimentation purpose, where 15 experiments are performed for the optimization. The levels and their parameter combinations of the

**Table 1** Experimental data

| S. No | Parameter | Value |
|---|---|---|
| 1 | Diameter of the hole (mm) | Ø400 μm |
| 2 | Depth of hole (mm) | 4 mm |
| 3 | Spark gap (mm) | 0.05 mm |
| 4 | Input current (Ip) | 1.5 A |
| 5 | Dielectric | EDM oil |
| 6 | Machine used | EDM EA 8 |

**Table 2** Level of parameters

| Parameters | Low (−1) | Medium (0) | High (+1) |
|---|---|---|---|
| Servo voltage (V) | −2 | 0 | 2 |
| Pulse on time (μs) | 3 | 6 | 9 |
| Pulse off time (μs) | 9 | 10.5 | 12 |

**Table 3** Design matrix using RSM

| Std. Order | Run Order | Pt. Type | Blocks | Servo voltage (V) | Pulse on time (μs) | Pulse off time (μs) |
|---|---|---|---|---|---|---|
| 12 | 1 | 2 | 1 | 2 | 6 | 12.0 |
| 10 | 2 | 2 | 1 | −2 | 6 | 12.0 |
| 5 | 3 | 2 | 1 | −2 | 3 | 10.5 |
| 11 | 4 | 2 | 1 | 2 | 6 | 9.0 |
| 1 | 5 | 2 | 1 | 0 | 3 | 9.0 |
| 8 | 6 | 2 | 1 | 2 | 9 | 10.5 |
| 6 | 7 | 2 | 1 | −2 | 9 | 10.5 |
| 9 | 8 | 2 | 1 | −2 | 6 | 9.0 |
| 15 | 9 | 0 | 1 | 0 | 6 | 10.5 |
| 4 | 10 | 2 | 1 | 0 | 9 | 12.0 |
| 7 | 11 | 2 | 1 | 2 | 3 | 10.5 |
| 14 | 12 | 0 | 1 | 0 | 6 | 10.5 |
| 2 | 13 | 2 | 1 | 0 | 9 | 9.0 |
| 3 | 14 | 2 | 1 | 0 | 3 | 12.0 |
| 13 | 15 | 0 | 1 | 0 | 6 | 10.5 |

design matrix are formed using response surface methodology and are shown in Table 3. In Table 3, the point type represents the point location within the box of Box-Behnken design, number 2 represents the corner points, and number 0 represents the center points of the box. Blocks represent that all the experiments are conducted as a single block without any change in the atmospheric conditions, and the block number 1 is used throughout experimentation purposes.

## 2.2 *Micro-Drilling*

Manufactured micro-electrode is set as a tool in EDM, and beryllium copper is set as a workpiece for machining, where it is clamped with the help of fixtures. The workpiece is set as a positive terminal, and electrode is set as the negative terminal. The generation of spark is obtained due to a huge potential difference in voltage and causes the material to be removed in the workpiece. Machining is taken place according to the order and their levels. Micro-drilling on beryllium copper is shown in Fig. 1.

Figure 2 shows the microelectrodes that are utilized for the different parameters for the micro-drilling process. The machined microelectrodes and workpiece (beryllium copper) are shown in Figs. 2 and 3, respectively.

**Fig. 1** Micro-drilling in EDM

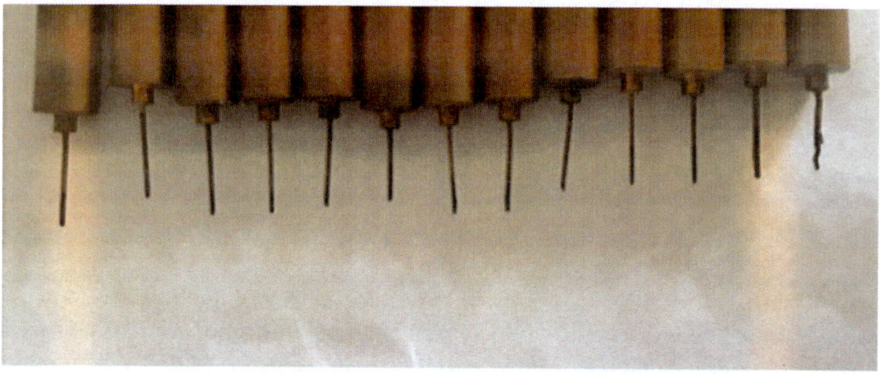

**Fig. 2** Micro-electrodes after drilling

**Fig. 3** Beryllium copper after drilling

Figure 3 clearly shows the micro-drilling of 400 μm diameter performed with the help of Ø400 μm copper electrodes. The microscope image of tool and workpiece after micro-drilling and is shown in Figs. 4 and 5, respectively.

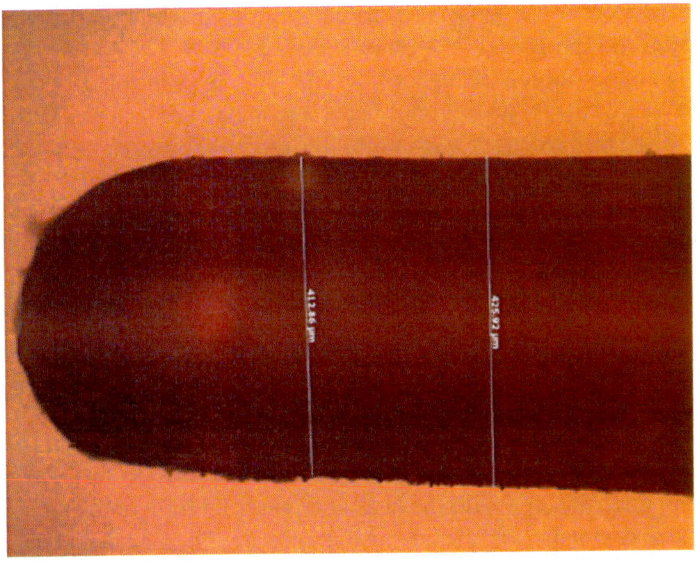

**Fig. 4** Microscope image of tool

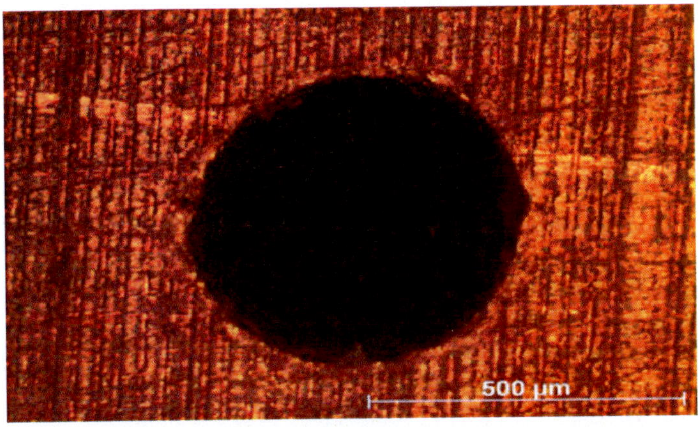


**Fig. 5** Microscope image of micro-holes in a workpiece

### 2.3 Tool Wear

After the micro-drilling is done, the tool used for micro-drilling is taken for analysis purpose, and the tool wear is chosen for analysis using statistical software. Tool wear is the most important response need to be considered while machining in EDM. In EDM, the heat generation is around 8000–12,000 °C due to the spark erosion principle, which is moreover higher than the melting point of copper electrodes. Hence, analyzing the tool wear is important for manufactured microelectrodes. There are different types of electrode wear ratios that can be expressed. In this experiment, end wear is selected for our analysis.

The shape of the electrode replicates on the workpiece material in the EDM process. Hence, the material selected as electrode should be capable of withstanding high thermal conductivity and a high melting point. In EDM, during machining the various factors that account for the responsible for electrode wear such as current, voltage, polarity and electrode material. Improper selection of these parameters will lead to the high tool wear ratios and in micromachining, even more, pronounced effects will take place.

The tool wear of microelectrodes is measured with the help of diameter reduction during micromachining. Diameters of the microelectrodes before drilling and after drilling are measured with the help of a polarized light optical microscope as discussed above with the help of Axio vision software. The diameter after the drilling has measured, and they were used for tool wear measurement as shown in Eq. 1. With the help of this equation, the tool wear for all the 15 numbers of microelectrodes is calculated and shown in Table 4.

$$\text{TW} = D_i - D_f \tag{1}$$

where,
TW—Tool wear.
$D_i$—Diameter of microelectrode before drilling.
$D_f$—Diameter of microelectrode after drilling.

## 3 Optimization of Process Parameters for Tool Wear

The output response—tool wear of the copper microelectrodes—are matched with the corresponding set of data with the help of statistical software. With the use of experimental results, the optimization is carried using RSM methodology where contour plots and S/N plots are generated with the use of these, the optimum parameters for getting low tool wear on microelectrodes are found out. The smaller values have been chosen for tool wear from the signal-to-noise ratio plot for microelectrode and are shown in Fig. 6.

**Table 4** Design matrix with results

| S. No | Servo voltage (V) | Pulse on time (μs) | Pulse off time (μs) | Tool wear (μm) |
|---|---|---|---|---|
| 1 | 2 | 6 | 12.0 | 11.74 |
| 2 | −2 | 6 | 12.0 | 29.72 |
| 3 | −2 | 3 | 10.5 | 20.94 |
| 4 | 2 | 6 | 9.0 | 7.63 |
| 5 | 0 | 3 | 9.0 | 19.39 |
| 6 | 2 | 9 | 10.5 | 1.63 |
| 7 | −2 | 9 | 10.5 | 11.79 |
| 8 | −2 | 6 | 9.0 | 6.60 |
| 9 | 0 | 6 | 10.5 | 6.37 |
| 10 | 0 | 9 | 12.0 | 8.88 |
| 11 | 2 | 3 | 10.5 | 16.67 |
| 12 | 0 | 6 | 10.5 | 6.37 |
| 13 | 0 | 9 | 9.0 | 6.09 |
| 14 | 0 | 3 | 12.0 | 18.82 |
| 15 | 0 | 6 | 10.5 | 6.37 |

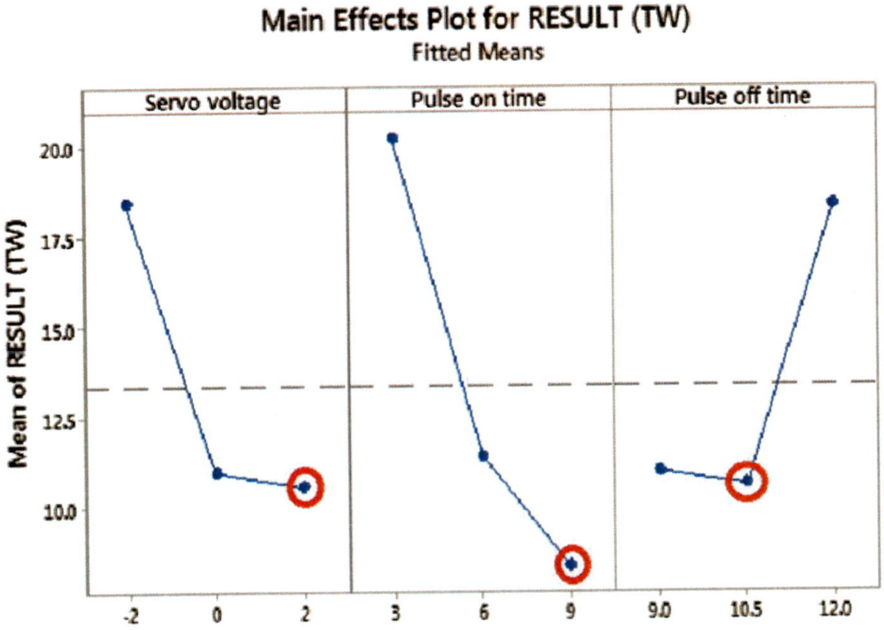


**Fig. 6** S/N ratio plot for tool wear

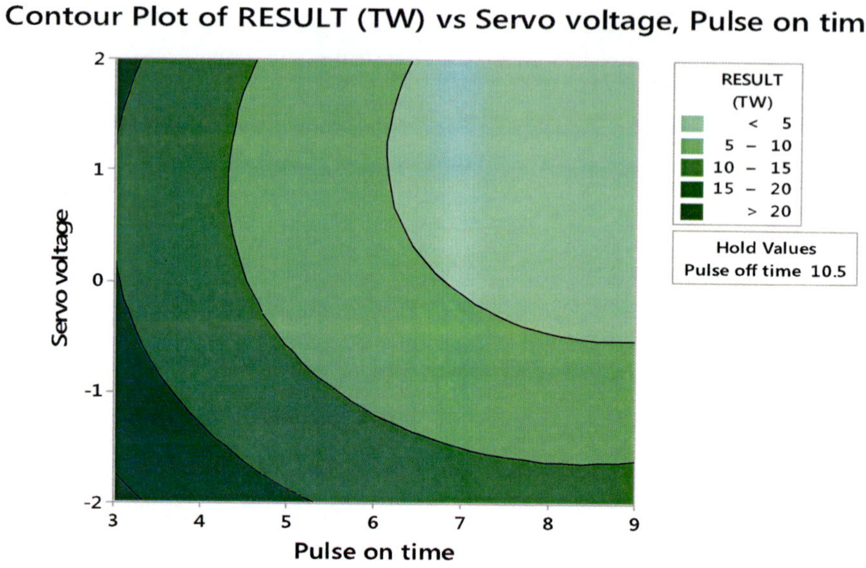


**Fig. 7** Contour plot for tool wear ($S_V$, $T_{on}$)

Figure 6 clearly shows the best optimum parameter for machining, to get low tool wear. The increased pulse on time and servo voltage give out more thermal energy which causes better removal of material. The amount of working time of the tool to reduce automatically which may lead to a reduction in tool wear. High pulse on time of 9 μs to maintain the duty cycle in a proportional way for machining and to reduce the tool wear the mid-value of pulse off time 10.5 μs is set by the machine for better material removal. The S/N ratio values are verified with the help of contour plots.

From Fig. 7, the contour plots for tool wear indicate that light green area at the top right corner, where the less tool wear is obtained at Ton around 9 μs and the voltage of above 2 V to hold the value of pulse off time 10.5 μs, and the tool wear is less than 5 μm.

From Fig. 8, the contour plots for tool wear Vs Ton and Toff shows that the light green area at the bottom right corner, where the tool wear is lower than 5 μm. During pulse on time at 9 μs and pulse off time in the mid-region 9.5–10.5 μs during the hold value of servo voltage 0 V.

From Fig. 9, the plot indicates that the light green area where the tool wear is lower with the servo voltage of around 2 V, pulse off time of above 10.5 μs and hold a value of pulse on time 6 μs in the mid-region where the tool wear is found to be below 5 μm. The optimum parameter values for obtaining low tool wear are identified with the help of both S/N curve and contour plots and shown in Table 5.

Optimum parameters for getting low tool wear shown in Table 5, and experiments are conducted with the optimum parameters for five trails, and the tool wear is measured. Numerical results show that for this optimum parameter, the tool wear is 1.63 μm; then, the experiments are conducted for five trails; the average of five trails

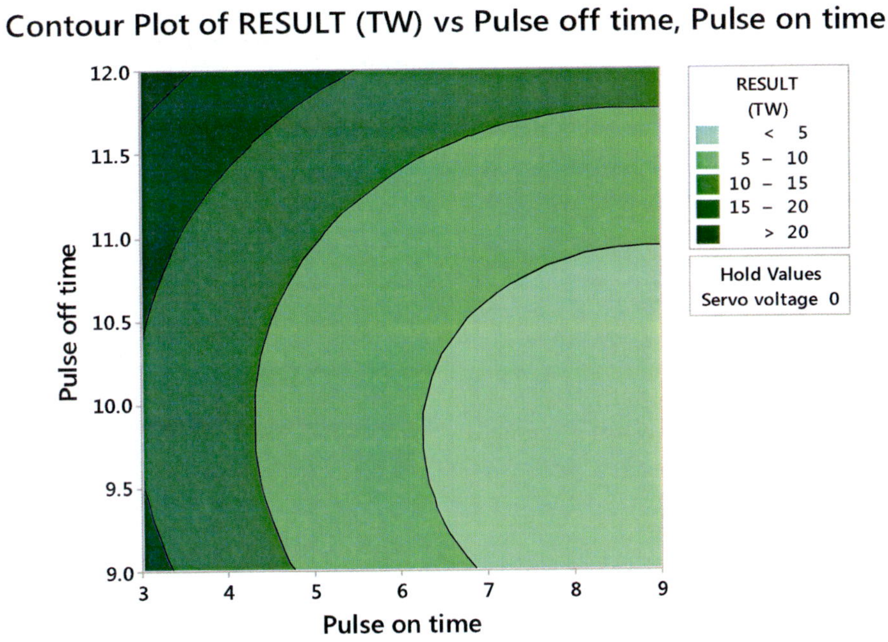


**Fig. 8** Contour plot for tool wear ($T_{on}$, $T_{off}$)

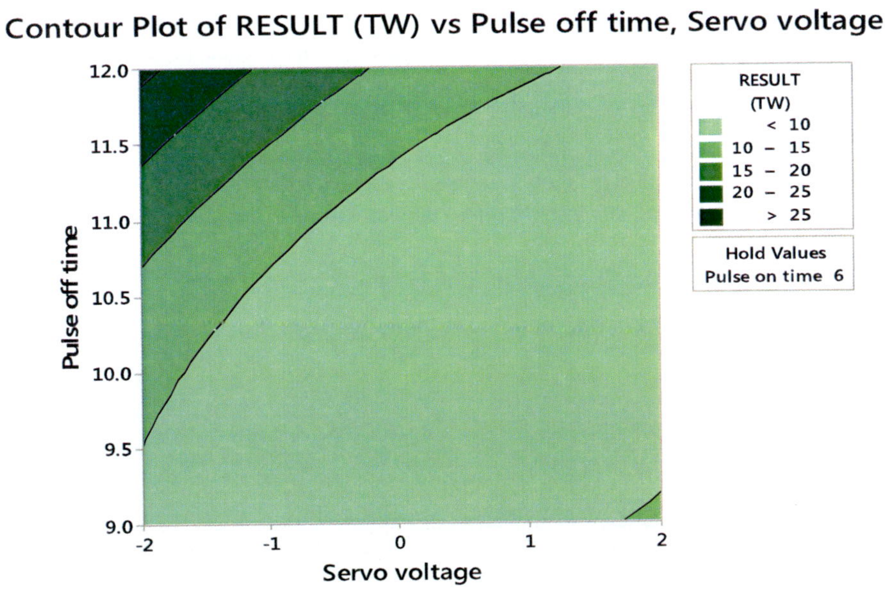


**Fig. 9** Contour plot for tool wear ($S_V$, $T_{off}$)

**Table 5** Optimal parameters

| Parameters | Pulse on time (μs) | Pulse off time (μs) | Servo voltage (V) |
|---|---|---|---|
| Optimum values | 9 | 10.5 | 2 |

**Table 6** Experimental verification results

| S. No | Optimization result (μm) | Experimental verification results (μm) | Relative error (%) |
|---|---|---|---|
| 1 | 1.63 | 1.62 | 0.61 |

is calculated and is shown in Table 6 which indicates that the experimental results are matched with numerical results formed by the statistical software with the minimum error of 0.61%.

## 4 Result

Analysis of variance (ANOVA) is a statistical method used to analyze the variation in and among the groups. Statistical analysis ANOVA was done in determining the significance level of each parameter on the wear rate. The analysis was can be performed for a different confidence level. And in this model, the level of confidence is selected as 95%. The percentage of contribution of various parameters is performed, and the result for tool wear is shown in Table 8 with the use of the statistical software.

From Table 7, it clearly shows each process parameter pulse on time with a percentage contribution of 45.63% gives the most contribution from the selected parameter to obtain the minimal tool wear, pulse off time and servo voltage comes next with percentage contributions of 23.78% and 22.32%, respectively, which has a less influence for tool wear.

**Table 7** Anova table

| Source | DF | Seq SS | Adj SS | Adj MS | *F*-Value | *P*-Value | %C |
|---|---|---|---|---|---|---|---|
| Servo voltage | 2 | 180.977 | 194.513 | 97.256 | 3.73 | 0.038 | 22.32% |
| Pulse on time | 2 | 370.711 | 400.826 | 201.913 | 7.39 | 0.008 | 45.63% |
| Pulse off time | 2 | 192.826 | 194.826 | 97.413 | 3.73 | 0.047 | 23.78% |
| Error | 8 | 67.935 | 67.935 | 33.967 | | | 8.27% |
| Lack-of-fit | 6 | 67.935 | 67.935 | 33.967 | | | 8.27% |
| Pure error | 2 | 0.000 | 0.000 | 0.000 | | | 0.00% |
| Total | 14 | 812.449 | | | | | 100.00% |

**Table 8** Predicted results

| S. No | Experimental results (TW) (μm) | Predicted result (TW) (μm) | Error % |
|---|---|---|---|
| 1 | 11.74 | 10.9287 | 6.91 |
| 2 | 29.72 | 26.2788 | 11.57 |
| 3 | 20.94 | 21.1362 | 0.928 |
| 4 | 7.63 | 8.0713 | 5.46 |
| 5 | 19.39 | 17.3825 | 10.35 |
| 6 | 1.63 | 1.4338 | 12.03 |
| 7 | 11.79 | 12.2238 | 3.54 |
| 8 | 6.60 | 7.4112 | 10.9 |
| 9 | 6.37 | 6.3700 | 0 |
| 10 | 8.88 | 9.6875 | 8.33 |
| 11 | 16.67 | 16.2363 | 2.63 |
| 12 | 6.37 | 6.3700 | 0 |
| 13 | 6.09 | 5.8450 | 4.02 |
| 14 | 18.82 | 20.0650 | 6.20 |
| 15 | 6.37 | 6.3700 | 0 |

The regression equation is obtained based on the results from the statistical software. The predicted and the experimental values are compared with the regression equation.

**Regression Equation**

**Tool Wear** = 214 + 16.15 Servo voltage − 7.78 Pulse on time − 36.4 Pulse off time + 0.877 Servo voltage × Servo voltage + 0.320 Pulse on time × Pulse on time + 1.798 Pulse off time × Pulse off time − 0.245 Servo voltage × Pulse on time − 1.584 Servo voltage × Pulse off time + 0.187 Pulse on time × Pulse off time.

Based on the experimental results, the regression equation is formed with the $R^2$ value of 98.25%, where given by the statistical software. Based on this regression equation, the prediction of final results (TW) is shown in Table 8. The percentage of error that deviated from the predicted value and experimental results is also calculated and shown in the same table.

Table 8 shows the predicted values are closer to that of the experimental values with a maximum deviation of 12.03%. It shows clearly that the calculated models are in good agreement with the mathematical model, and optimal values chosen by our experimental studies are found to be much closer to the regression equation given by the software.

## 5 Conclusion

Micro-drilling is performed on beryllium copper with the use of manufactured copper microelectrodes. The DoE is carried out to find the optimum parameter for getting low tool wear using response surface methodology. ANOVA is performed to know which parameter has a significant effect over tool wear, and prediction is also carried out using RSM, and it shows that the mathematical model is in good relationship with experimental results.

- The optimum parameters to achieve low tool wear on copper microelectrodes are pulse on time 9 $\mu$s, pulse off time 10.5 $\mu$s and servo voltage of 2 V.
- From the study, it is evident that the pulse on time is the most influencing parameter for TW of 45.63% as a contribution followed by pulse off time (23.78%) and by servo voltage (22.32%).
- Prediction of tool wear has been found out for each set of the parameter using RSM with the maximum error of 12.02% comparing with experimental results.

## References

1. Zhang L, Tong H, Li Y (2015) Precision machining of micro tool electrodes in micro EDM for drilling array micro holes. Precis Eng 39:100–106
2. Fleischer J, Masuzawa T, Schmidt J, Knoll M (2004) New applications of Micro-EDM. J Mater Process Technol 149:246–249
3. Parthiban M, Krishnaraj V, Naveen Anthuvan R (2014) Optimization of parameters for diameter accuracy in wire electric discharge grinding for micro machining of tungsten rods. Appl Mech Mater 592–594:625–629. ISSN: 1662-7482
4. Rahman AA, Mamat A (2009) Effect of machining parameters on hole quality of micro drilling for brass. Modern Appl. Sci 3(5):221–230
5. Neelesh Singh BC, Routara DD (2018) Study of machining characteristics of Inconel 601 in EDM using RSM. Mater Today Proc 5:3438–3449
6. Gupta A, Singh H, Aggarwal A (2011) Taguchi-fuzzy multi output optimization (MOO) in high speed CNC turning of AISI P-20 tool steel. Expert Syst Appl 38(6):6822–6828
7. Fnides B, Yallese MA, Mabrouki T, Rigal JF (2011) Application of response surface methodology for determining cutting force model in turning hardened AISI H11 hot work tool steel. Sadhana 36(Part 1):109–123
8. Donga S, Wanga Z, Wanga Y, Liua H (2016) An experimental investigation of enhancement surface quality of micro holes for Be-Cu alloys using micro-EDM with multi-diameter electrode and different dielectrics. In: 18th CIRP conference on electro physical and chemical machining (ISEM XVIII). Procedia CIRP, vol 42, pp 257–262
9. Rath U (2017) Parametric optimization of EDM on EN19 using grey-RSM analysis. Indian J Sci Technol 10(23)
10. Ohdar NK, Jena BK, Sethi SK (2017) Optimization of EDM process parameters using RSM method with copper electrode. Int Res J Eng Technol (IRJET) 04(04)
11. Ashok Kumara U, Laxminarayana P (2018) Optimization of electrode tool wear in micro holes machining by Die Sinker EDM using Taguchi approach. In: International conference on processing of materials, minerals and energy, materials today: proceedings, vol 5, issue 1, Part 1, pp 1824–1831

12. Liu Q, Zhang Q, Zhang M, Yang F, Rajurkar KP (2019) Effects of surface layer of AISI 304 on micro EDM performance. Precis Eng 57:195–202
13. Li Z, Bai J, YanCao YW, Zhu G (2019) Fabrication of microelectrode with large aspect ratio and precision machining of micro-hole array by micro-EDM. J Mater Process Technol 268:70–79
14. Li G, Natsu W (2020) Realization of micro EDM drilling with high machining speed and accuracy by using mist deionized water jet. Precis Eng 61:136–146
15. Pellegrini G, Ravasio C (2020) A sustainability index for the micro-EDM drilling process. J Cleaner Prod 247

# Multiple Criterion Decision-Making Technique for Optimization of Machining Parameters: A Case on Drilling of Titanium Alloy

**H. Sahul Hameed, A. Prabukarthi, P. Guhapranav, and S. Deva Surya**

**Abstract** In highly competitive market, the ultimate goal of the manufacturing industry is to manufacture low-cost product without comprising on its quality. One of the main objectives in achieving the goal is the optimization of machining parameters. Among all-titanium machining processes, drilling is a vital process carried out as a final operation in machining of mechanical components. A new trend in optimization of multiple quality characteristics is conjunction of Multiple Criterion Decision Making (MCDM) technique with Taguchi's philosophy. An attempt is made to perform Taguchi philosophy based structured experimentation combined with Grey Relational Analysis (GRA) and Technique for Order Preference by Similarity to Ideal Solution (TOPSIS) techniques were used for optimizing the machining parameters while dry drilling of Ti6Al4V by considering the correlation between the quality characteristics.

**Keywords** MCDM · Drilling · Ti-6Al-4V · GRA · TOPSIS · Taguchi's philosophy

H. S. Hameed · A. Prabukarthi · P. Guhapranav · S. Deva Surya (✉)
Department of Mechanical Engineering, PSG College of Technology, Avinashi Road, Peelamedu, Coimbatore 641004, India
e-mail: devasurya03@gmail.com

H. S. Hameed
e-mail: sahulmechanical@gmail.com

A. Prabukarthi
e-mail: prabukarthi.arumugam@gmail.com

P. Guhapranav
e-mail: guhapranav1996@gmail.com

© The Editor(s) (if applicable) and The Author(s), under exclusive license to Springer Nature Singapore Pte Ltd. 2021

S. Mohan et al. (eds.), *Materials, Design, and Manufacturing for Sustainable Environment*, Lecture Notes in Mechanical Engineering,
https://doi.org/10.1007/978-981-15-9809-8_26

# 1 Introduction

The precondition for any process optimization begins with Taguchi's optimization philosophy. It has proven itself to be effective in producing high-quality products at relatively low cost; however, it has proven to be ineffective at solving numerous quality characteristic problems. Therefore, numerous quality feature problems are converted into a single objective problem using MCDM techniques. The general structure of the evaluation of MCDM technique consists of determination of evaluation criteria, collection of data for each quality criterion, normalization, formulating the preference based on weights, analyzing the alternatives, and identify the best alternative. Several works have been carried out by researchers by combining MCDM technique with Taguchi's philosophy [1–5]. But the major concentration of work focused on techniques such as Technique for Order Preference by Similarity to Ideal Solution (TOPSIS), Gray Relational Analysis (GRA), weighted sum method, and utility theory. The quality of hole carried out by drilling operation on Ti6Al4V is considered based on the variation in hole diameter, roundness, and burr height.

The pioneers in the field of drilling of titanium alloy have investigated the various factors such as machining parameters, tool material, and coating, forces acting during machining, quality of drilled hole, and tool wear [6–9]. But the work towards optimization of machining parameters to achieve better hole quality still needs a lot of attention. In this paper, an attempt has been made to solve multi-objective problem to optimize machining parameters using a combination of GRA and TOPSIS techniques in conjunction with Taguchi's philosophy. The aforesaid method has been employed during drilling of Ti6Al4V to select the optimal machining and geometrical parameters (Cutting speed, Feed, Tool geometry) by considering the quality uniqueness such as force in axial direction, torque applied and drill hole quality characteristics (Diameter deviation, Roundness and burr height).

# 2 Drilling of Titanium Alloy

## *2.1 Experimental Details*

Experimental analysis was carried out using CNC machine, the workpiece is mounted on the strain gauge type drill tool dynamometer and it is mounted on the machine table which is presented in Fig. 1. The study was done using a coated (TiAlN) modified carbide twist drill of 5 mm diameter for drilling Ti6Al4V. Axial force and torque were constantly monitored using a two-channel digital storage oscilloscope; roundness (circularity), deviation in the hole diameter was reviewed using Coordinate Measuring Machine (CMM) (Carl Zeiss Contura G2) and exit burr height by using Digital height master (TESA microhite-600) (This is modified from the earlier version pa).

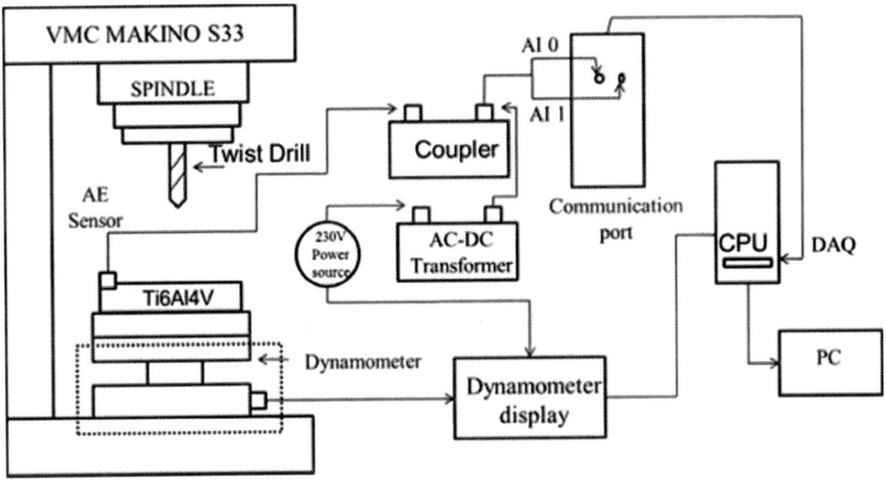


**Fig. 1** Experimental setup

**Table 1** Design of experiments (Refer Rough draft I for the update)

| Factors | 1 | 2 | 3 | 4 |
|---|---|---|---|---|
| Cutting Speed (V) (m/min) | 30 | 40 | 50 | 60 |
| Feed (F) (mm/rev) | 0.03 | 0.06 | 0.09 | 0.12 |
| TG | 1 | 2 | 3 | 4 |

The experimental effects were initiated by Taguchi's design of experiments. Thus the effect of the machining parameters could be understood by carrying out limited trials. The machining parameters taken into consideration for optimization study are cutting speed ($V$), feed ($F$), and tool geometry (TG) with each parameter subjected to four-level variations (as shown in Table 1). The experiment is carried further by L16 Orthogonal Array (OA).

## 3 Drilling Performance Appraisal Attributes

Drilling operation was performed on Ti6Al4V to evaluate performance metrics such as thrust force, torque, exit burr height, diameter deviation, and roundness. Thrust force impacts the quality of machined surfaces in an increase in thrust force lead to spindle axis vibration; untimely failure of drills is triggered by increased thrust force which would result in poor quality surface finish. Torque indicates the amount of rubbing between the drill tool and the workpiece; increased torque attributes to more heat generation at the drill tool and the work material interface.

**Table 2** Experimental measured quality characteristics

| Thrust force ($N$) | Torque (Nm) | Burr height (μm) | Diameter deviation (μm) | Roundness (μm) |
|---|---|---|---|---|
| 226.00 | 0.75 | 45.67 | 34.4 | 34.4 |
| 275.00 | 0.69 | 42.00 | 19.3 | 15.4 |
| 321.33 | 1.01 | 33.33 | 21.4 | 42 |
| 492.00 | 1.17 | 39.00 | 18.8 | 41.5 |
| 217.33 | 0.66 | 47.67 | 28.5 | 31.4 |
| 287.67 | 1.15 | 50.00 | 22 | 18 |
| 417.67 | 1.56 | 44.33 | 11.2 | 31.2 |
| 390.67 | 1.48 | 41.33 | 12.9 | 13.5 |
| 213.67 | 0.55 | 39.33 | 22.6 | 66.3 |
| 347.33 | 1.17 | 44.00 | 17 | 44.9 |
| 331.00 | 1.41 | 31.33 | 15 | 27.2 |
| 382.00 | 1.39 | 44.00 | 3.5 | 15.1 |
| 267.00 | 0.83 | 42.00 | 28 | 23.3 |
| 273.33 | 1.04 | 43.00 | 23 | 30.3 |
| 335.67 | 1.39 | 35.67 | 19.6 | 26.7 |
| 367.33 | 1.48 | 40.67 | 11.6 | 46 |

The burr that is found at the margins of the drill is called exit burr or Poisson burr. Thus deburring increases the cost estimate in aerospace applications by 30% [10]. The geometric variations in the hole is evaluated by hole quality (hole diameter, roundness). The average hole diameter must lie between tolerance bounds. The roundness deviation of the drill hole is expressed in terms of circularity. The error of circularity is well-defined as the distance amongst the least circumscribing circle diameter and the extreme inscribing circle diameter (modified) [1]. The experimentally measured quality characteristics (thrust force, torque, burr height, diameter deviation, and roundness) are presented in Table 2.

## 4 Grey Relational Analysis (GRA)

According to the Grey system theory by Deng in 1982, a system is ambiguous and the evidence regarding the system is inadequate to shape a relational analysis [11]. The steps intricate in GRA are generation of grey relation, determination of grey relational grade and estimation of overall grey relational grade.(updated).

***Step 1***: Grey relational generation

The experimental data of output responses have different units, so it is important to normalize the output responses. The normalization of output response is done based on the quality characteristic criterion.

For lower the better (L-B criterion)

$$A_i^*(k) = \frac{\max c_i(k) - c_i(k)}{\max c_i(k) - \min c_i(k)} \tag{1}$$

For higher the better (H-B criterion)

$$A_i^*(k) = \frac{c_i(k) - \min c_i(k)}{\max c_i(k) - \min c_i(k)} \tag{2}$$

where

$i = 1, \ldots, f$; $k = 1, \ldots, g$; $f$ = number of experimental data items; $g$ = number of parameters; $A_i(k)$ = original sequence.

$A_i^*(k)$ = sequence after grey relational generation; $0 \le A_i^*(k) \le 1$.

***Step 2***: Grey relational coefficient

Calculate of the grey relation coefficient

$$\xi_i(k) = \frac{\Delta_{\min} + \zeta\,\Delta_{\max}}{\Delta_{\text{oi}}(k) + \zeta\,\Delta_{\max}} \tag{3}$$

where

$\Delta_{\text{oi}}(k)$ = deviation of the comparability sequence from reference (i.e. the quality loss estimate)

$$\Delta_{\text{oi}} = \left\| A_o^*(k) - A_i^*(k) \right\| \tag{4}$$

$$\Delta_{\min} = \min_{\forall j \in i} \min_{\forall k} \left\| A_o^*(k) - A_i^*(k) \right\| \tag{5}$$

$$\Delta_{\max} = \max_{\forall j \in i} \max_{\forall k} \left\| A_o^*(k) - A_i^*(k) \right\| \tag{6}$$

$A_o^*(k)$ = the reference sequence; $A_i^*(k)$ = the comparability sequence; $\zeta$ = Identification coefficient with $\zeta \in (0, 1)$; $\zeta = 0.5$

***Step 3***: Grey relational grade

Grey relational grade represents the degree of correlation between the reference sequence and comparability sequence.

$$\gamma_i = \frac{1}{n}\sum_{k=1}^{n} w_k \xi_i(k) \tag{7}$$

wk–weight assigned to each quality characteristic.

## 5 Topsis

TOPSIS (*Technique for Order Preference by Similarity to Ideal Solution*) method is a familiar multi-attribute decision making (MADM) methodology proposed by Hwang and Yoon (1981). Principle behind this method is that the selected alternative must have the shortest distance from the positive ideal solution and the longest distance from negative ideal (anti-ideal) solution.

**Procedural steps in TOPSIS**

The calculation of TOPSIS will be lead by.

***Step 1***: Establishment of decision matrix:

$$\begin{matrix} B_1 \\ B_2 \\ . \\ B_i \\ . \\ B_r \end{matrix} \begin{pmatrix} y_{11} & y_{12} & . & y_{1j} & y_{1s} \\ y_{21} & y_{22} & . & y_{2j} & y_{2s} \\ . & . & . & . & . \\ y_{i1} & y_{i2} & . & y_{ij} & . \\ . & . & . & . & . \\ y_{r1} & y_{r2} & . & y_{rj} & y_{rs} \end{pmatrix}$$

Here $B_i$ $(i = 1, 2, \ldots, r)$ shows the viablereplacements; $y_j (j = 1, 2, \ldots, s)$ shows the attributes relating to viable replacement performance, $i = 1, 2, \ldots, s$ and $x_{ij}$ is the performance of $B_i$ with respect to $y_j$.

***Step 2***: Normalization of matrix:

$$n_{ij} = \frac{y_{ij}}{\sqrt{\sum_{i=1}^{r} y_{ij}^2}} \tag{8}$$

Here, $n_{ij}$ represents the normalized performance of $B_i$ with respect to $y_j$.

***Step 3***: Weighted decision matrix:

$$V = [v_{ij}] V = w_j r_{ij} \text{ Here, } \sum_{j=1}^{n} w_j = 1 \tag{9}$$

$$D=\begin{bmatrix} Y_{11} & Y_{12} & . & Y_{1j} & Y_{1s} \\ Y_{21} & Y_{22} & . & Y_{2j} & Y_{2s} \\ . & . & . & . & . \\ yi_{1} & y_{i2} & . & y_{ij} & . \\ . & . & . & . & . \\ y_{r1} & y_{r2} & . & y_{rj} & y_{rs} \end{bmatrix}$$

***Step 4***: Calculate the ideal (best) and negative ideal (worst) solutions:

a. The positive ideal solution (PIS):

$$A^{+}=\left\{\begin{matrix}\left(\left(\min_{i} y_{ij}\,|\,j\in J\right)\right), \\ \left(\left(\max_{i} y_{ij}\,|\,j\in J'|i=1,2,\ldots,r\right)\right)\end{matrix}\right\}=\left\{y_{1}^{+}, y_{2}^{+}, y_{3}^{+}, .., y_{s}^{+}\right\} \tag{10}$$

b. The negative ideal solution (NIS):

$$A^{-}=\left\{\begin{matrix}\left(\left(\min_{i} y_{ij}\,|\,j\in J\right)\right), \\ \left(\left(\max_{i} y_{ij}\,|\,j\in J'|i=1,2,\ldots,r\right)\right)\end{matrix}\right\}=\left\{y_{1}^{-}, y_{2}^{-}, y_{3}^{-}, \ldots, s\right\} \tag{11}$$

$J=\{j=1,2,\ldots,s|j\}$: Related with the favorable attributes $J'=\{j=1,2,\ldots,s|j\}$: Related with non-favorable attributes.

***Step 5***: Find the distance measures. The parting of each alternative on or after the perfect solution is given by $n$-dimensional using the Euclidean distance method from the following equations:

$$S_{i}^{+}=\sqrt{\sum_{j=1}^{s}\left(y_{ij}-y_{j}^{+}\right)^{2}} \quad j=1,2,\ldots,r \tag{12}$$

$$S_{i}^{-}=\sqrt{\sum_{j=1}^{s}\left(y_{ij}-y_{j}^{-}\right)^{2}} \quad j=1,2,\ldots,r \tag{13}$$

$S_{i}^{+}$ = Distance between PIS and alternative.
$S_{i}^{-}$ = Distance between NIS and alternative.
***Step 6***: Calculate the overall performance coefficient closest to the ideal solution:

$$C_{i}^{+}=\frac{S_{i}^{-}}{S_{i}^{+}+S_{i}^{-}} i=1,2,\ldots,m: 0\leq C_{i}\leq 1 \tag{14}$$

$C_{i}^{+}$ = Overall performance measure

***Step 7***: The ranking is done by conferring to the desired order. The replacement with the largest relative closeness is extremely preferable.

## 6 Methodology for GRA-TOPSIS

***Step 1***: Determination of S/N ratio for each quality characteristic based on its quality criterion.

Higher is better (H-B criterion)

$$S/N_{\text{ratio}} = -10\log_{10}\left(\frac{1}{s}\sum_{i=1}^{s}\frac{1}{x_{ij}^2}\right) \tag{15}$$

Smaller is better (S-B criterion)

$$S/N_{\text{ratio}} = -10\log_{10}\left(\frac{1}{s}\sum_{i=1}^{s}x_{ij}^2\right) \tag{16}$$

where

$i = 1, 2, \ldots, r; j = 1, 2, \ldots, s$; n- number of experimental data; $x_{ij}$—observed response; $r$—number of quality characteristic.

***Step 2***: Normalization of S/N ratio using equation(s) (1) & (2).

***Step 3***: Calculation of individual grey relational coefficient using equation(s) (3). This gives the decision matrix for TOPSIS technique. The calculated individual grey relational coefficient is mentioned in Table 3.

***Step 4***: Determination of Weighted normalized decision matrix using the Eq. (9).

The weights for the quality characteristics are calculated using Entropy method and the weighted performance measure are shown in Table 4.

***Step 5***: Determination of PIS and NIS using equation(s) (10) and (11).

The positive and negative solutions are determined and are shown in Table 5.

***Step 6***: Determination of distance measures using equation(s) (12) and (13).

***Step 7***: Calculation of overall performance measure using Eq. (14) and rank the OPI in order of preference.

The separation distance measures (S+,S−) and overall performance measure (C+) and S/N ratio of OPI corresponding to H-B criterion are shown in Table 6.

**Table 3** Individual grey relational coefficient

| Thrust force | Torque | Burr height | Diameter deviation | Roundness |
|---|---|---|---|---|
| 0.8814 | 0.6274 | 0.3828 | 0.3333 | 0.4597 |
| 0.6230 | 0.6973 | 0.4437 | 0.4009 | 0.8580 |
| 0.5054 | 0.4608 | 0.7906 | 0.3869 | 0.4121 |
| 0.3333 | 0.4090 | 0.5163 | 0.4047 | 0.4147 |
| 0.9608 | 0.7466 | 0.3577 | 0.3527 | 0.4853 |
| 0.5837 | 0.4155 | 0.3333 | 0.3833 | 0.7345 |
| 0.3835 | 0.3333 | 0.4024 | 0.4956 | 0.4872 |
| 0.4087 | 0.3449 | 0.4576 | 0.4669 | 1.0000 |
| 1.0000 | 1.0000 | 0.5068 | 0.3799 | 0.3333 |
| 0.4619 | 0.4081 | 0.4077 | 0.4196 | 0.3984 |
| 0.4879 | 0.3568 | 1.0000 | 0.4398 | 0.5318 |
| 0.4179 | 0.3604 | 0.4077 | 1.0000 | 0.8766 |
| 0.6518 | 0.5618 | 0.4437 | 0.3546 | 0.5932 |
| 0.6287 | 0.4493 | 0.4247 | 0.3777 | 0.4960 |
| 0.4800 | 0.3604 | 0.6434 | 0.3988 | 0.5385 |
| 0.4349 | 0.3454 | 0.4726 | 0.4881 | 0.3936 |

**Table 4** Weighted normalized matrix

| Thrust force | Torque | Burr height | Diameter deviation | Roundness |
|---|---|---|---|---|
| 0.0956 | 0.1124 | 0.0105 | 0.1080 | 0.1659 |
| 0.0676 | 0.1249 | 0.0122 | 0.1299 | 0.3096 |
| 0.0548 | 0.0825 | 0.0218 | 0.1253 | 0.1487 |
| 0.0362 | 0.0733 | 0.0142 | 0.1311 | 0.1497 |
| 0.1043 | 0.1337 | 0.0099 | 0.1143 | 0.1751 |
| 0.0633 | 0.0744 | 0.0092 | 0.1242 | 0.2651 |
| 0.0416 | 0.0597 | 0.0111 | 0.1605 | 0.1758 |
| 0.0443 | 0.0618 | 0.0126 | 0.1513 | 0.3609 |
| 0.1085 | 0.1791 | 0.0140 | 0.1231 | 0.1203 |
| 0.0501 | 0.0731 | 0.0112 | 0.1359 | 0.1438 |
| 0.0529 | 0.0639 | 0.0275 | 0.1425 | 0.1919 |
| 0.0453 | 0.0645 | 0.0112 | 0.3240 | 0.3163 |
| 0.0707 | 0.1006 | 0.0122 | 0.1149 | 0.2141 |
| 0.0682 | 0.0805 | 0.0117 | 0.1224 | 0.1790 |
| 0.0521 | 0.0645 | 0.0177 | 0.1292 | 0.1943 |
| 0.0472 | 0.0619 | 0.0130 | 0.1581 | 0.1420 |
| **Weights** | | | | |
| 0.11 | 0.18 | 0.03 | 0.32 | 0.36 |

**Table 5** Positive and negative ideal solution

| | Thrust force | Torque | Burr height | Diameter deviation | Roundness |
|---|---|---|---|---|---|
| A+ | 0.1085 | 0.1791 | 0.0275 | 0.3240 | 0.3609 |
| A− | 0.0362 | 0.0597 | 0.0092 | 0.1080 | 0.1203 |

**Table 6** Computation of results

| S+ | S− | C+ | S/N ratio |
|---|---|---|---|
| 0.2993 | 0.0916 | 0.2344 | −12.60 |
| 0.2125 | 0.2039 | 0.4897 | −6.20 |
| 0.3110 | 0.0463 | 0.1295 | −17.76 |
| 0.3137 | 0.0401 | 0.1133 | −18.92 |
| 0.2844 | 0.1147 | 0.2875 | −10.83 |
| 0.2499 | 0.1489 | 0.3734 | −8.56 |
| 0.2828 | 0.0767 | 0.2133 | −13.42 |
| 0.2189 | 0.2446 | 0.5277 | −5.55 |
| 0.3137 | 0.1405 | 0.3093 | −10.19 |
| 0.3121 | 0.0414 | 0.1170 | −18.64 |
| 0.2790 | 0.0834 | 0.2302 | −12.76 |
| 0.1392 | 0.2919 | 0.6771 | −3.39 |
| 0.2703 | 0.1082 | 0.2859 | −10.88 |
| 0.2921 | 0.0715 | 0.1968 | −14.12 |
| 0.2865 | 0.0793 | 0.2167 | −13.28 |
| 0.3051 | 0.0559 | 0.1549 | −16.20 |

## 7 Determination of Optimal Parameter Setting

On behalf of the various performance index estimated using GRA-TOPSIS technique, the significance for every trial run has been calculated. Optimistic parametric grouping has been determined using S/N ratio. So as to improve the quality by minimizing the several performance measures, optimal condition with the highest S/N ratio (higher-is-better criterion) has to be selected. The response table for S/N ratio is shown in Table 7.

## 8 Confirmation Test

In order to approve the factors and levels selected from the experiment confirmation test is used which causes a process to process to do in a definite manner. The selected optimal machining condition is run for a certain number of times to validate whether

**Table 7** Response table for S/N ratio

| Level | V | F | TG |
|---|---|---|---|
| 1 | −13.869 | −11.124 | −12.529 |
| 2 | −9.59 | −11.879 | −8.425 |
| 3 | −11.243 | −14.305 | −11.905 |
| 4 | −13.62 | −11.014 | −15.463 |
| Delta | 4.28 | 3.291 | 7.038 |

the results are close to the predicted value. The predicted optimal value of OPI is determined using Eq. 17.

$$\eta_{\text{opt}} = r + \sum_{j=1}^{p} \left[ \left( r_{i,j} \right)_{\max} \right] - r \tag{17}$$

where

$\left(r_{i,j}\right)_{\max}$ = response means for optimal level $i$ of factor $j$ $P$ = number of main design parameter.

The confidence interval is attained to check the closeness of predicted value and observed value is given by

$$C.I = \sqrt{F_{(1,v_e)} V_e \left( \frac{1}{n_{\text{eff}}} + \frac{1}{n} \right)}$$

where

$F_{(1,v_e)}$ = Fisher value for 95% confidence level; $v_e$ = Degree of freedom for error; $V_e$ = Mean square of pooled error.

$n_{\text{eff}} = \frac{N}{1+v}$ N = Total trail in OAV = Degree of freedom for $P$ factor; $n$ = sample size for confirmatory test.

From Table 8, it is noticed that the variances between predicted and observed values are small and lie well within the confidence interval.

**Table 8** Confirmatory test results

| Predicted Optimal process setting | $V_2 F_4 T_2$ |
|---|---|
| $\eta_{\text{opt}}$ | 0.5670 |
| Confidence interval | 0.2007–0.9333 |
| $\eta_{\text{obs}}$ | 0.3663 |

## 9 Conclusion

In this work, an effort has been made to formulate a methodology to optimize the machining parameters by conjoining GRA and TOPSIS techniques in combination with Taguchi's philosophy. The following were the outcomes of this work.

- The beneficial of combining GRA and TOPSIS is that it enables the formulation of decision matrix based on degree of association among reference and comparability sequence.
- The calculation of weights of each quality characteristic based on entropy method eliminates the problem of assuming equal weights and facilitates the determination of weights based on measured data.
- The conjunction of GRA with TOPSIS improves the precision in the calculation of overall performance measure by measuring the distance of the predicted solution from the aspired and worst solution.
- The optimal process setting for drilling of titanium alloy was found to be 40 m/min (cutting speed), 0.12 mm/rev (feed), and tool geometry 2.
- The predicted optimal setting is verified and validated by performing confirmation test and the results obtained were satisfactory at a 95% confidence level.

## References

1. Abhishek K, Datta S, Mahapatra SS (2014) Multi-Response optimization in drilling of composites: introduction of a similarity based approach in combination with Taguchi's philosophy. J Manuf Sci Prod 28; 14(3):151–70
2. Sonkar V, Abhishek K, Datta S, Mahapatra SS (2014) Multi-objective optimization in drilling of GFRP composites: a degree of similarity approach. Proc Mater Sci 6:538–543
3. Prasanna J, Karunamoorthy L, Raman MV, Prashanth S, Chordia DR (2014) Optimization of process parameters of small hole dry drilling in Ti–6Al–4V using Taguchi and grey relational analysis. Measurement 48:346–354
4. Hasani H, Tabatabaei SA, Amiri G (2012) Grey relational analysis to determine the optimum process parameters for open-end spinning yarns. J Eng Fibers Fabr 7(2):155892501200700220
5. Kuo Y, Yang T, Huang GW (2008) The use of grey relational analysis in solving multiple attribute decision-making problems. Comput Ind Eng 55(1):80–93
6. Zhang PF, Churi NJ, Pei ZJ, Treadwell C (2008) Mechanical drilling processes for titanium alloys: a literature review. Mach Sci Technol 12(4):417–444
7. Shetty PK, Shetty R, Shetty D, Rehaman NF, Jose TK (2014) Machinability study on dry drilling of titanium alloy Ti-6Al-4V using L9 orthoganal array. Proc Mater Sci 5:2605–2614
8. Sharif S, Rahim EA (2007) Performance of coated-and uncoated-carbide tools when drilling titanium alloy—Ti-6Al-4V. J Mater Process Technol 185(1–3):72–76
9. Prabukarthi A, Krishnaraj V, Santhosh M, Senthilkumar M, Zitoune R (2013) Optimisation and tool life study in drilling of titanium (Ti-6Al-4V) alloy. Int J Mach Mach Mater 13(2–3):138–157
10. Dornfeld DA, Kim JS, Dechow H, Hewson J, Chen LJ (1999) Drilling burr formation in titanium alloy, Ti-6AI-4V. CIRP Ann 48(1):73–76

11. Gwo-Hshiung T, Tzeng GH, Huang JJ (2011) Multiple attribute decision making: methods and applications. Multiple attribute decision making: methods and applications
12. Ramulu M, Branson T, Kim D (2001) A study on the drilling of composite and titanium stacks. Compos Structures 54(1):67–77

# Prediction of Surface Characteristics of Cut Surfaces Produced by Plasma Arc Cutting Process by Using Image Processing and Fuzzy Logic Technique

**A. Mohan, S. Samsudeensadham, A. Ashwin Kumar, and M. Kirubakaran**

**Abstract** Plasma arc cutting process is one of the advanced nonconventional machining processes which is frequently used in modern metal cutting industries to compete with the laser cutting process. To obtain better surface quality during the plasma arc cutting process is one of the challenging issues. The controlling parameters such as cutting speed, gas flowrate, and arc current are the most influencing parameters which affect the surface roughness of the machined material. In this research, the fuzzy logic along with image processing is used to predict the surface roughness ($R_a$) of square plates machined by the plasma arc cutting process. The process parameters considered for the experiments were cutting speed (CS), gas flow rate (GFR), and arc current ($C$). Response Surface Methodology (RSM) was used to design the experimental runs and a total of 15 sets of experiments were performed and responses were measured. Fuzzy rule-based modeling can be effectively used to predict the surface roughness. Experimental results were compared with the predicted values. Base on the study it was observed that experimental results have got in good agreement with predicted results.

**Keywords** Fuzzy logic · Cutting speed · Gas flow rate · Arc current · RSM · Surface roughness · Image processing

A. Mohan (✉) · S. Samsudeensadham · A. A. Kumar · M. Kirubakaran
Department of Mechanical Engineering, PSG College of Technology, Coimbatore, Tamil Nadu, India
e-mail: mohan3mmm@gmail.com

S. Samsudeensadham
e-mail: samsudeensadham@gmail.com

A. A. Kumar
e-mail: ashwinkumara4015@gmail.com

M. Kirubakaran
e-mail: kirubaakaranmahendran@gmail.com

© The Editor(s) (if applicable) and The Author(s), under exclusive license to Springer Nature Singapore Pte Ltd. 2021
S. Mohan et al. (eds.), *Materials, Design, and Manufacturing for Sustainable Environment*, Lecture Notes in Mechanical Engineering,
https://doi.org/10.1007/978-981-15-9809-8_27

# 1 Introduction

The unconventional thermal cutting process like a plasma arc cutting process plays an important role in cutting mild steel, stainless steel, and, alloys which are difficult to machine. The components used in the plasma arc cutting process are a power supply unit, plasma torch, arc starting circuit. This process is mainly for electrically conductive metals but nonconductive materials can also be machined. First, the material is melted by the plasma, and the pressurized air gas which flows pushes the melted material away and thus the material is removed. The input process parameters for a plasma arc cutting process are cutting speed, current, gas flow rate, stand-off distance, kerf width, etc. these parameters are included in the research according to the design of the experiments. The response parameters that could be measured are surface roughness, bevel angle, conicity, heat affected zone, dross formation, material removal rate, etc. Ramakrishnan et al. [1] used a Taguchi method for analyzing the SS321, the inference shows that lower values of stand-off distance and cutting current gives us better surface roughness and minimum heat-affected zone. Peko et al. [2] varied cutting speed and current for experimentation and used artificial neural networks for prediction of surface roughness. Hisman et al. [3] was used as a combination of cutting speed, arc voltage, current, and thickness. The Authors used S235JR material for experimentation and predicted the surface roughness and hardness using the Type-2 Fuzzy set system. Bhuvenesh et al. [4] used AISI 1017 steel and the input parameters were air pressure, current, cutting speed, and stand-off distance. The Taguchi method was used for the design of experiments and found that surface roughness is inversely proportional to the material removal rate. Salonitis et al. [5] used cutting speed, gas pressure, stand-off distance, and current are used as input parameters used Taguchi method for experimentation and found that surface roughness and conicity are mainly affected by stand-off distance and heat affected zone is influenced by cutting current. Nammi et al. [6] used a Gaussian low pass filter to preprocess the images captured using machine vision. Shanmugamani et al. [7] applied Gaussian filter to images captured from machine vision for defects classification and used classifiers like artificial neural networks, decision trees, and k-Nearest Neighbor for classifying the surface defects. This study aims to predict the surface roughness based on Fuzzy rule modeling through image processing techniques.

# 2 Plasma Arc Cutting Process for Mild Steel

## *2.1 Experimental Details*

In this research work, Messer Craft made CNC plasma arc cutting machine (as shown in Fig. 1) was used to perform the experiments. Air plasma was used for machining the mild steel plate (6 mm). The workpiece was slashed as a square plate (30 mm $\times$ 30 mm) for each machining condition.

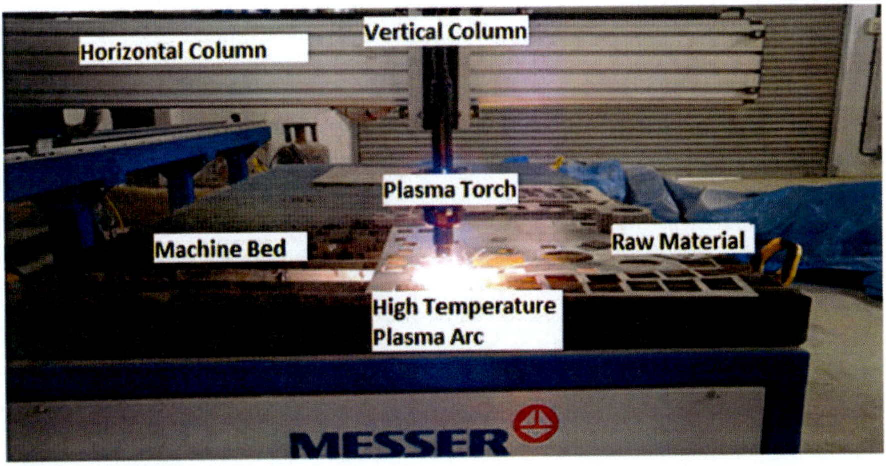


**Fig. 1** Experimental setup for plasma arc cutting

## 2.2 *Design of Experiment*

RSM is a tool to analyze problems where several variables are involved and these variables affect the response, and experimentation is done to predict the response. There are different types of design, For this experimentation Box-Behnken design was chosen. Three-factor and three levels were chosen. It has three centre points i.e. with 3 repetitive experiments. Process parameters considered for the prediction of surface roughness are cutting speed (CS), gas flow rate (GFR), and arc current ($C$). In this experimentation module, $L_{15}$Array has been used for machining a mild steel plate of 6 mm thickness.

## 2.3 *Response Measurements*

The experiments have been conducted using RSM (Box–Behnken) with combination of machining parameters as shown in Table 1. The response ($R_a$) for each side of the mild steel squared plate was measured byroughness measuring instrument (Kosaka labs SurfCoder SE1200). The square plate consists of 4 sides (as shown in Fig. 2)

**Table 1** Design of experiments

| Factor | Parameters | Low level | Medium level | High level |
|---|---|---|---|---|
| A | Cutting Speed (mm/min) | 1000 | 1300 | 1600 |
| B | Gas Flow Rate (LPM) | 68 | 72 | 76 |
| C | Current (A) | 39 | 42 | 45 |

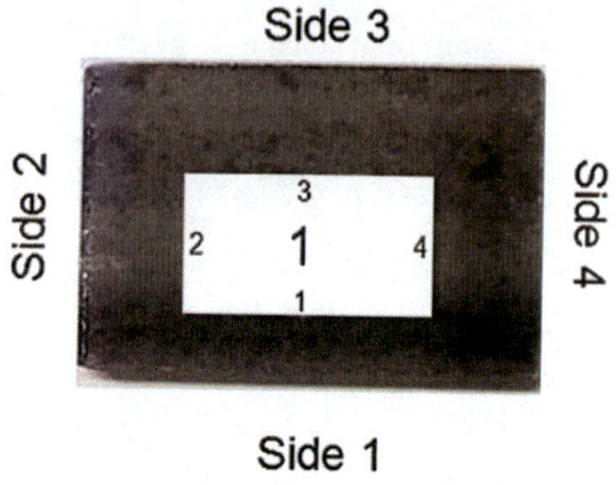


**Fig. 2** 4 sides of the workpiece

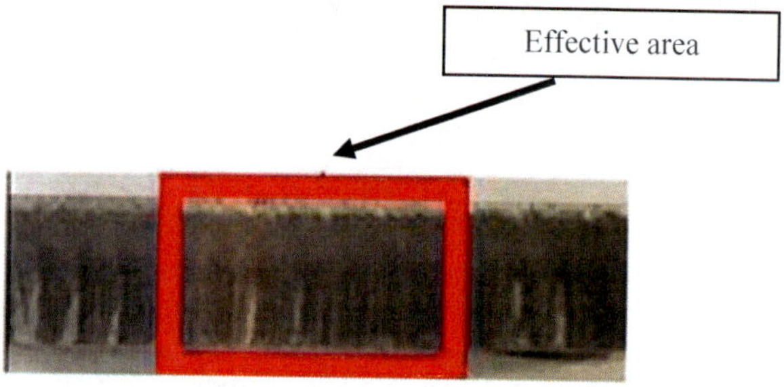


**Fig. 3** The effective areas

and surface roughness was measured only in the effective area which is shown in the Fig. 3.

The surface roughness is very high at the corners. This is because the torch movement is not uniform in the corners. This increases the roughness at the corners. Therefore the effective area as shown in Fig. 3 is considered for measurement of roughness. The experimentally measured roughness for all four sides is presented in Table 2.

## 2.4 *Image Processing*

The effective area of the specimens was captured using Mitutoyo's Tool Maker's microscope with a magnification of 6 × in 640 × 480 resolution. The captured RGB format images were converted into the greyscale format by two different stages to process the images. At the initial stage, the images were cropped for the purpose of neglecting the unwanted backgrounds and is resized to 640 × 480 pixels. In the final stage, the noises of the images were removed by the Gaussian low pass filter [6]. The Gaussian kernel is given by Eq. (1).

$$G_\sigma(X, Y; \sigma) = \frac{1}{2\pi\sigma^2} \exp^{\left(\frac{X^2+Y^2}{2\sigma^2}\right)} \tag{1}$$

**Table 2** Experimentation order and measured $R_a$

| S. No | CS (mm/min) | GFR (Lpm) | C (amps) | Surface roughness (μm) | | | |
|---|---|---|---|---|---|---|---|
| | | | | Side 1 | Side 2 | Side 3 | Side 4 |
| 1 | 1300 | 72 | 42 | 2.22 | 2.35 | 1.84 | 1.12 |
| 2 | 1300 | 76 | 45 | 2.42 | 2.20 | 2.38 | 1.20 |
| 3 | 1600 | 76 | 42 | 3.04 | 2.80 | 2.53 | 1.57 |
| 4 | 1600 | 68 | 42 | 4.26 | 2.01 | 3.16 | 1.58 |
| 5 | 1000 | 68 | 42 | 3.09 | 2.23 | 2.51 | 1.58 |
| 6 | 1600 | 72 | 45 | 3.31 | 2.21 | 1.90 | 2.83 |
| 7 | 1300 | 72 | 42 | 3.15 | 5.37 | 2.06 | 1.13 |
| 8 | 1300 | 68 | 39 | 3.92 | 1.96 | 2.30 | 2.69 |
| 9 | 1000 | 76 | 42 | 5.76 | 2.54 | 3.06 | 1.58 |
| 10 | 1300 | 68 | 45 | 4.65 | 1.89 | 4.27 | 2.18 |
| 11 | 1300 | 76 | 39 | 3.99 | 2.64 | 2.05 | 1.02 |
| 12 | 1000 | 72 | 45 | 6.46 | 1.41 | 2.61 | 1.07 |
| 13 | 1600 | 72 | 39 | 2.94 | 2.74 | 2.01 | 1.87 |
| 14 | 1000 | 72 | 39 | 3.06 | 3.92 | 2.38 | 1.91 |
| 15 | 1300 | 72 | 42 | 2.26 | 2.51 | 2.28 | 2.56 |

Each pixel of the grayscale image has a value from 0 to 255 according to the illumination and scattering of light. The grayscale image was converted into its corresponding matrix, it consists of 480 rows and 640 columns. Surface roughness ($R_a$) was measured in the middle region of the workpiece.

Therefore middle rows, i.e. row number ranging from 120 to 240 was selected. These values depend on the light illumination and scattering of light on the surface, which mainly depends on the nature of the surface and grayscale values that changes with respect to the surface, and the row matrix was calculated for row 240. The row matrix of each image is shown in Table 3.

The obtained 60 different grayscale values were shown in Table 4 which was found at the 4 sides of each specimen in their effective area. Python programming software was used for the matrix conversion. The grayscale value for each workpiece with its respective sides are shown in Table 3.

## 3 Prediction Techniques

### *3.1 Fuzzy Logic*

Fuzzy logic is one of the important techniques in artificial intelligence. The system is defined with desired inputs and outputs. It uses IF–THEN rules for predicting the

**Table 3** Gray Scale Values (GSV)of 4 sides of the workpiece

| Specimen | Side 1 GSV | Side 2 GSV | Side 3 GSV | Side 4 GSV |
|---|---|---|---|---|
| 1 | 139.33 | 124.38 | 148.64 | 138.99 |
| 2 | 127.01 | 126.47 | 147.48 | 133.85 |
| 3 | 108.93 | 102.64 | 159.82 | 132.48 |
| 4 | 110.55 | 106.05 | 135.70 | 128.46 |
| 5 | 127.61 | 124.05 | 134.86 | 132.27 |
| 6 | 133.00 | 123.31 | 128.64 | 127.16 |
| 7 | 133.64 | 127.46 | 134.79 | 128.71 |
| 8 | 139.68 | 139.60 | 128.59 | 133.21 |
| 9 | 132.54 | 131.94 | 130.72 | 138.27 |
| 10 | 130.68 | 143.84 | 141.59 | 125.77 |
| 11 | 122.42 | 144.76 | 136.12 | 134.50 |
| 12 | 132.97 | 135.15 | 132.81 | 117.29 |
| 13 | 134.97 | 135.39 | 111.88 | 128.43 |
| 14 | 130.39 | 123.57 | 127.57 | 132.13 |
| 15 | 130.37 | 122.66 | 134.92 | 133.20 |

output. In this experiment the input variables considered are Cutting Speed (CS), Gas Flow Rate (GFR), Current (CC), and the output variable is surface roughness ($R_a$). The input parameters are divided into three levels namely 'low' medium' and 'high' as we use triangular membership function [8].

Figures 4a, b, and 5a shows the low, medium, and high range of Cutting Speed, Gas Flow Rate, and Current respectively. The cutting speeds are 1000 mm/min, 1300 mm/min, and 1600 mm/min, respectively. Similarly, the three levels in Gas Flow Rate are 68 lpm, 72 lpm, and 76 lpm and Current 39 amps, 42 amps, and 45amps.

The fuzzy expressions for output parameters are given in the range from 1 μm to 7.6 μm in a 0.2 μm difference between each value is shown in Fig. 5b. For the modeling of input process variables and output process variables, the triangular membership function is chosen [8]. The concept of fuzzy reasoning for 3 inputs and one output fuzzy logic unit is described as follows:

*Rule 1: if CS is* $X_1$ *and G is* $Y_1$ *and* $C_1$ *is* $Z_1$ *then* $R_a$ *is* $D_1$

*Else*

*Rule 2: if CS is* $X_2$ *and G is* $Y_2$ *and C is* $Z_2$ *then* $R_a$ *is* $D_2$

*Else*

...

...

*Else*

*Rule n: if CS is* $X_n$ *and G is* $Y_n$ *and C is* $Z_n$ *then* $R_a$ *is* $D_n$

**Table 4** Experimental $R_a$ versus Fuzzy prediction

| Side 1 | | | Side 2 | | | Side 3 | | | Side 4 | | |
|---|---|---|---|---|---|---|---|---|---|---|---|
| Exp. Ra (μm) | Pred. Ra (μm) | Error % | Expt. Ra (μm) | Pred. Ra (μm) | Error % | Expt. Ra (μm) | Pred. Ra (μm) | Error % | Expt. Ra (μm) | Pred. Ra (μm) | Error % |
| 2.22 | 2.1 | 5.69 | 2.35 | 2.44 | 3.57 | 1.84 | 1.83 | 0.36 | 1.12 | 1.09 | 2.80 |
| 2.42 | 2.44 | 1.02 | 2.20 | 2.1 | 4.98 | 2.38 | 2.44 | 2.64 | 1.20 | 1.09 | 10.23 |
| 3.04 | 3.3 | 7.91 | 2.80 | 2.82 | 0.57 | 2.53 | 2.44 | 3.69 | 1.57 | 1.59 | 1.26 |
| 4.2 | 4.16 | 2.31 | 2.01 | 2.1 | 4.29 | 3.16 | 3.1 | 1.97 | 1.58 | 1.59 | 0.75 |
| 3.09 | 3.33 | 7.31 | 2.23 | 2.1 | 6.19 | 2.51 | 2.44 | 2.75 | 1.58 | 1.59 | 0.79 |
| 3.31 | 3.3 | 0.36 | 2.21 | 2.1 | 5.24 | 1.90 | 1.84 | 2.99 | 2.83 | 2.82 | 0.34 |
| 3.15 | 2.1 | 50.05 | 5.37 | 2.44 | 120.2 | 2.06 | 1.83 | 12.70 | 1.12 | 1.09 | 3.21 |
| 3.92 | 4.16 | 5.76 | 1.96 | 1.83 | 7.35 | 2.30 | 2.44 | 5.84 | 2.68 | 2.44 | 10.16 |
| 5.76 | 5.6 | 2.92 | 2.54 | 2.44 | 3.91 | 3.06 | 3.1 | 1.35 | 1.57 | 1.59 | 0.79 |
| 4.65 | 4 | 16.36 | 1.89 | 1.84 | 2.80 | 4.27 | 4.1 | 4.17 | 2.17 | 2.1 | 3.60 |
| 3.99 | 4.6 | 13.30 | 2.64 | 2.44 | 8.26 | 2.05 | 1.83 | 11.99 | 1.02 | 1.08 | 5.42 |
| 6.46 | 4.75 | 35.92 | 1.41 | 1.25 | 13.12 | 2.61 | 2.44 | 7.05 | 1.06 | 1.09 | 2.06 |
| 2.94 | 4 | 26.51 | 2.74 | 2.82 | 2.75 | 2.01 | 1.83 | 9.97 | 1.87 | 1.83 | 2.40 |
| 3.06 | 3 | 1.97 | 3.92 | 3.8 | 3.13 | 2.38 | 2.44 | 2.40 | 1.91 | 2.44 | 21.64 |
| 2.26 | 2.1 | 7.67 | 2.51 | 2.44 | 2.70 | 2.28 | 1.83 | 24.48 | 2.56 | 1.09 | 134.5 |

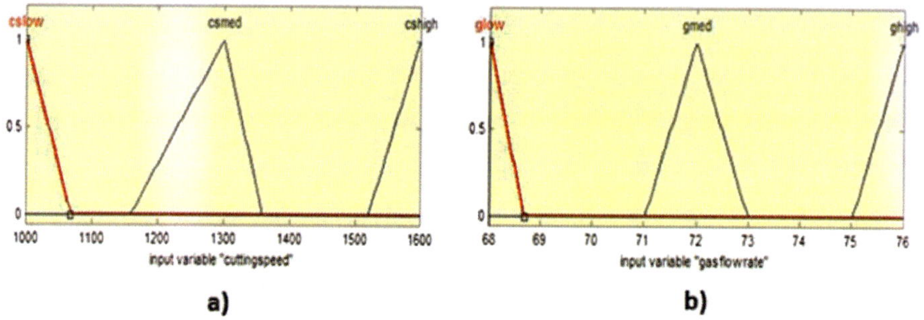


**Fig. 4** Low, medium, and high **a)** variables in cutting speed **b)** range of gas flow rate

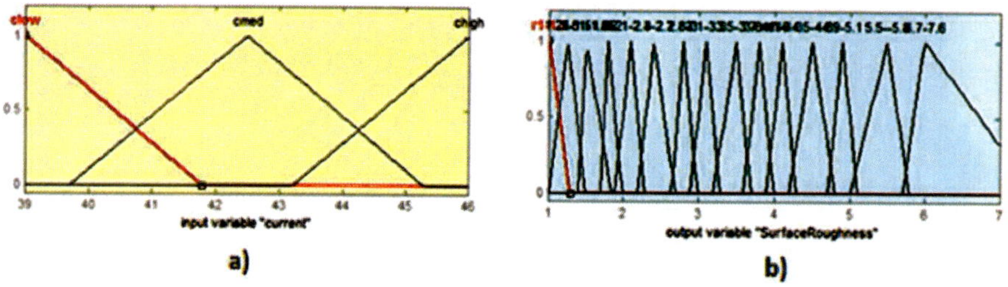


**Fig. 5** Low, medium, and high range of **a)** current **b)** $R_a$ in range 1.0–7.6 μm

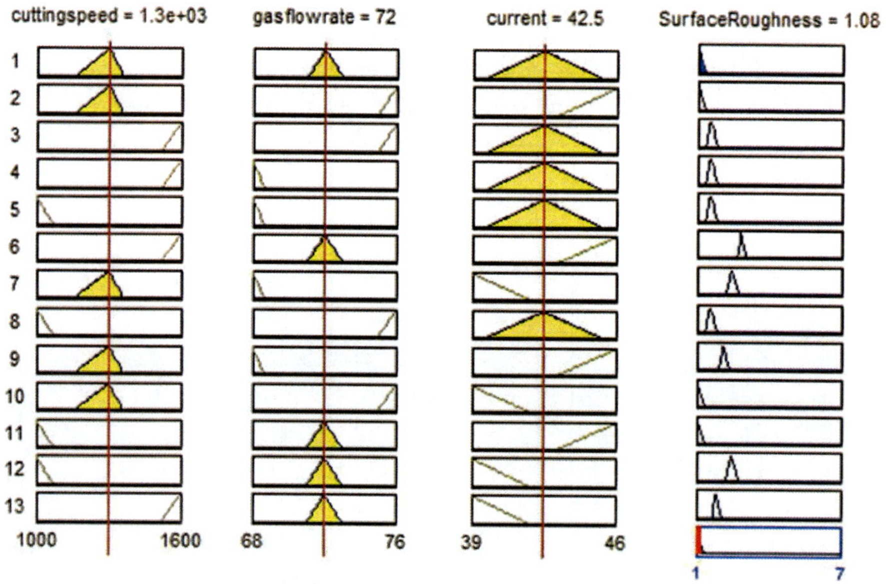


**Fig. 6** Plots of inputs and outputs

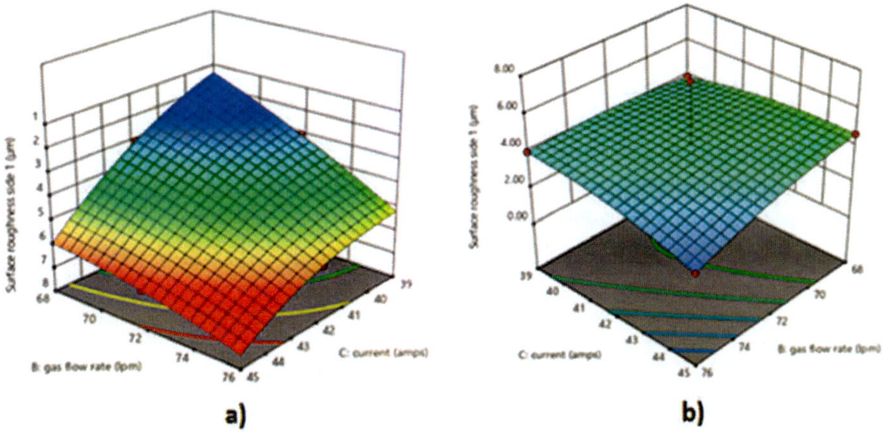


**Fig. 7** Surface plot for side 1 **a)** at 1000 mm/min **b)** at 1300 mm/min

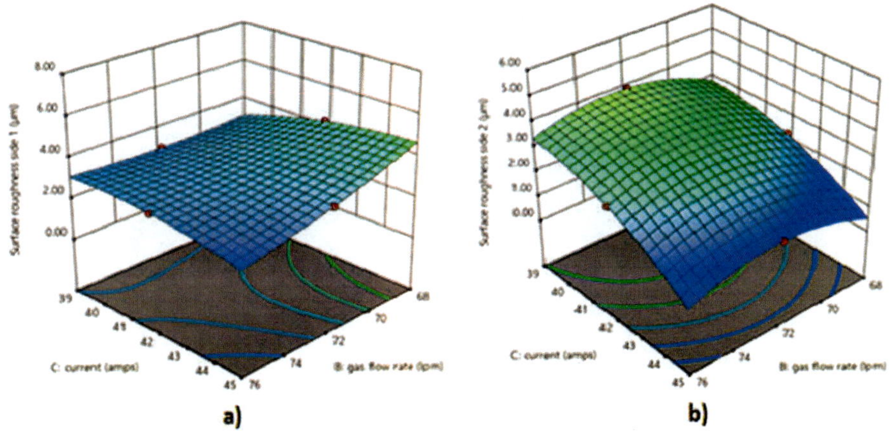


**Fig. 8** Surface plot for side 1 **a)** at 1600 mm/min **b)** at 1000 mm/min

Finally, defuzzification is carried out using the centroid defuzzification method [9]. Since there are 4 sides in square plates and the surface roughness is not uniform on all sides. Therefore individual model for each side is developed such that 4 models F1, F2, F3, and F4 for side 1, side 2, side 3, and side 4, respectively for predicting $R_a$. Fuzzy rules for each side are framed. The surface roughness value for the square plates at any given side is predicted using its respective fuzzy logic model as shown in Fig. 6.

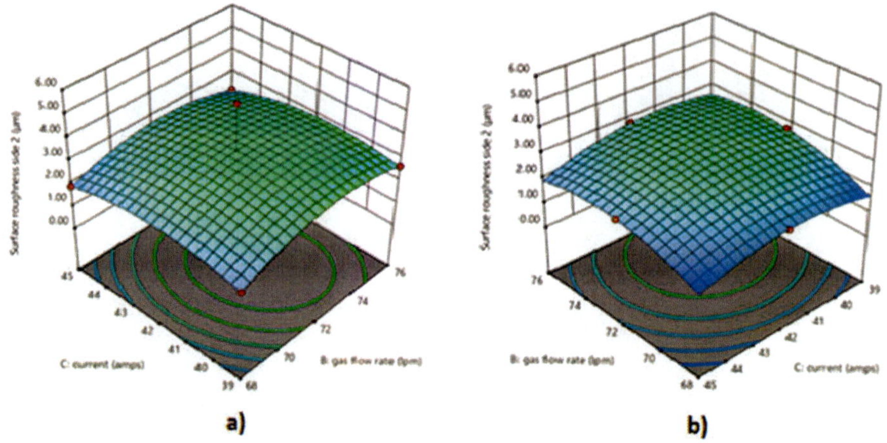


Fig. 9 Surface plot for side 2 **a)** at 1300 mm/min **b)** at 1600 mm/min

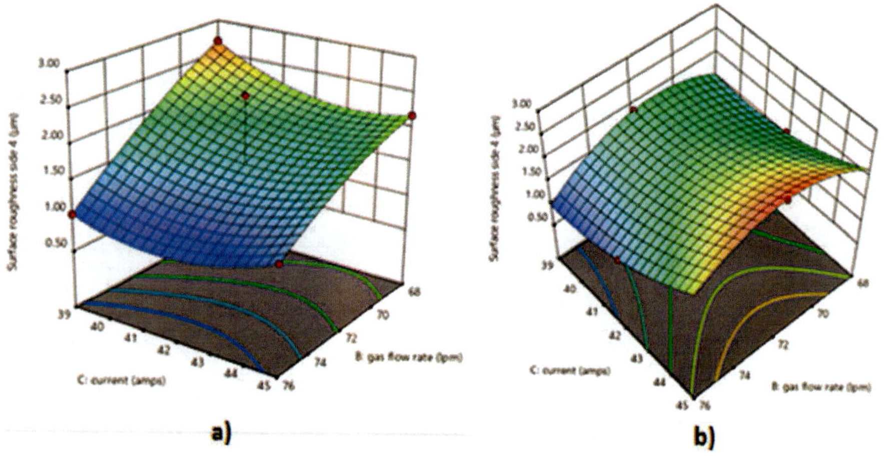


Fig. 10 Surface plot for side 3 **a)** at 1600 mm/min **b)** at 1000 mm/min

## 4 Results and Discussion

Fuzzy logic model for all the four sides are developed separately, the graph which shows the output changes with respect to the change in input parameters cutting speed, gas flow rate, and current.

The experimentally measured $R_a$ value and the value predicted by the fuzzy model for each side and the relative error is shown in Table 4.

It is found that the fuzzy model predicts the same $R_a$ value for the repetitive experiments but practically it is impossible to achieve the same roughness even though the input parameters used are same. Surface plot for the response ($R_a$) was plotted as cutting speed is the least contributing factor [10] it is made as constant at

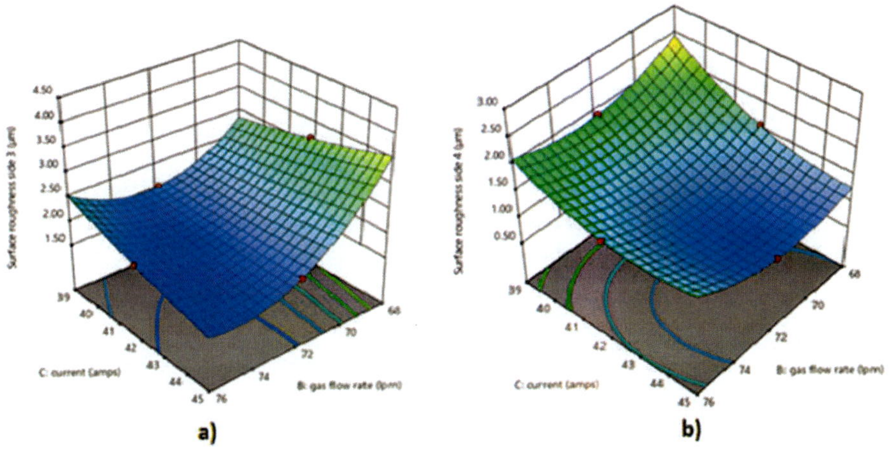


**Fig. 11** Surface plot for side 4 **a)** at 1300 mm/min **b)** at 1600 mm/min

low, medium and high speeds i.e. at 1000 mm/min, 1300 m/min and 1600 mm/min then the surface plot was computed and few of the most influenced graphs were discussed as follows.

From the 3D plot as is shown in Fig. 7a inferred that at 1000 mm/min Gas Flow Rate (GFR) is directly proportional to the current (CC) i.e. when both current and gas flow rate is low then $R_a$ around 2μm. From Fig. 7b it is observed that when there is a decrease in current and gas flow rate roughness is decreases, and it doesn't have much change in $R_a$ when CS is 1300 mm/min for side 1. To show the influenced effect of machine thermal stability with constant speed, 3D plots for high, medium, and low had been discussed for side 1.

Figure 8a shows that when there is an increase a current and decrease in gas flow rate results in higher roughness value. Figure 8b, shows that when current is lower and the gas flow rate is moderate then $R_a$ is higher. When cutting current is higher and gas flow rate is moderate then $R_a$ is coming at the mid-range values, when both CC and GFR is low (or) high then $R_a$ is low for side 2 at 1000 mm/min and from Fig. 9a and b it is inferred that when current is higher and gas flow rate is low it results in lower $R_a$for side 2 at 1300 mm/min.

Figure 10a shows that when cutting current with lower value and lower GFR gives less roughness for side 3 at 1000 mm/min and from Fig. 10b it is inferred that when CC is high GFR is low then $R_a$ is high.

Figure 11a shows that when current is increasing and gas flow rate with gradual increase produces low roughness value for side 4 at 1300 mm/min and from the Fig. 11b shows that for side 4 with increased cutting speed (In this cast: at 1600 mm/min) generates lower surface roughness.

From the surface plot, it is observed that the surface roughness of four sides of the square plates are different, even though the input parameters are the same, and from the surface roughness measurement it is observed that the $R_a$ at side 1 is high and least in side 4, this may be due to the quality of plasma produced which depends on thermal stability i.e. when cutting the first side of the square plate, the thermal stability is lower and it is sufficient to cut the plate but not good enough to produce a better quality cut when the torch reaches side 4 the system generates plasma gas which has high thermal stability when compared to the one delivered at cutting side 1.

Thus a good quality cut i.e. with minimum surface roughness is produced at side 4 when it is compared with other sides. It is also inferred that, when the cutting speed is increased, the gas flow rate should be increased to obtain minimum surface roughness. The image processing technique used for this work to predict the roughness was done offline. The future work can be made online using restoration of blurred images using distortion removal process [11] using advance machine vision techniques.

## 5 Conclusions

In this research work, models to predict the surface roughness of the square plate cut by the plasma arc cutting process was developed. It is found that the surface roughness is not uniform on all four sides of the square plate which could be inferred from the surface plots. This is because the temperature of the plasma is not stabilized initially i.e. when the plasma torch is at side 1, but the temperature of the plasma gradually increases and reaches the maximum when it reaches side 4. This affects the surface roughness and a pattern is observed, it is that side 1 is having maximum roughness and roughness decreases gradually and is minimum at side 4. Therefore 4 models F1, F2, F3, and F4 were developed with fuzzy logic for predicting surface roughness at side 1, side 2, side 3, and side 4, and the average relative error of the fuzzy models are found to be 12.34%, 13.07%, 6.29%, and 13.36%, respectively.

## References

1. Ramakrishnan H, Balasundaram R, Ganesh N, Karthikeyan N (2018) Experimental investigation of cut quality characteristics on ss321 using plasma arc cutting. J Brazilian Soc Mech Scie Eng 40(2):60
2. Peko I, Nedić B, Đorđević A, Džunić D, Janković M, Veža I (2016) Modeling of surface roughness in plasma jet cutting process of thick structural steel. Tribol Industry 38(4)
3. Çelik YH, Özek MB, Özek C (2013) Investigation of plasma arc cutting parameters with type-2 fuzzy set and system. Mater Testing 55(10):789–795
4. Bhuvenesh R, Manan A, Norizaman MH (2012) The study of surface roughness and mrr of mild steel using manual plasma arc cutting machining. In: Advanced materials research, vol 576. Trans Tech Publications, pp 3–6
5. Salonitis K, Vatousianos S (2012). Experimental investigation of the plasma arc cutting process. Proc CIRP 3:287–292
6. Shanmugamani R, Sadique M, Ramamoorthy B (2015) Detection and classification of surface defects of gun barrels using computer vision and machine learning. Measurement 60:222–230
7. Rajasekaran T, Palanikumar K, Vinayagam BK (2011) Application of fuzzy logic for modeling surface roughness in turning CFRP composites using CBN tool. Prod Eng 5(2): 191–199

8. Maity KP, Bagal DK (2015) Effect of process parameters on cut quality of stainless steel of plasma arc cutting using hybrid approach. Int J Adv Manuf Technol 78(1–4):161–175
9. Özek C, Çaydaş U, Ünal E (2012) A fuzzy model for predicting surface roughness in plasma arc cutting of AISI 4140 steel. Mater Manuf Process 27(1):95–102
10. Ananthakumar K, Rajamani D, Balasubramanian E, Davim JP (2019) Measurement and optimization of multi-response characteristics in plasma arc cutting of Monel 400™ using RSM and TOPSIS. Measurement 135:725–737
11. Dhanasekar B, Ramamoorthy B (2010) Restoration of blurred images for surface roughness evaluation using machine vision. Tribol Int 43(1–2):268–276

# Biomass Material Selection for Sustainable Environment by the Application of Multi-Objective Optimization on the Basis of Ratio Analysis (MOORA)

**C. Sowmya Dhanalakshmi, Manoj Mathew, and P. Madhu**

**Abstract** The objective of this study is to develop a method of sustainability assessment for ranking the prior order of biomass-based conversion technologies for the production of biofuel. Both tactical and environmental issues make biomass a vital energy source and part of sustainable energy policy. A multi-criteria decision-making process permits the decision makers to use multiple variables to assess its consequences on biofuel production. In this study, five criteria relevant to environmental and technological aspects have been taken for sustainability assessment. This proposed method considered two biomass-based technologies such as conventional pyrolysis and gasification, and biomass pyrolysis has been taken as the most sustainable process, since it gives three types of biofuels (bio-oil, char and biogas) and can be chosen for further development. This study suggests a multi-criteria decision-making framework for prioritization of biomass materials based on multi-objective optimization on the basis of ratio analysis (MOORA) method. During this evaluation, the performance of the reference point approach and full multiplicative MOORA method are also tested for this problem. The study concluded that all these applied methods are very simple to understand, and easy to implement and endow with almost exact rankings.

**Keywords** Biomass materials · Sustainable energy · Multi-criteria decision making · MOORA

---

C. S. Dhanalakshmi (✉) · P. Madhu
Department of Mechanical Engineering, SNS College of Technology, Coimbatore, Tamil Nadu, India
e-mail: dr.sowmi100782@gmail.com

P. Madhu
e-mail: pnpmadhu@gmail.com

M. Mathew
Department of Mechanical Engineering, Shri Shankaracharya Institute of Professional Management and Technology, Raipur, Chhattisgarh, India
e-mail: mathewmanojraipur@gmail.com

© The Editor(s) (if applicable) and The Author(s), under exclusive license to Springer Nature Singapore Pte Ltd. 2021
S. Mohan et al. (eds.), *Materials, Design, and Manufacturing for Sustainable Environment*, Lecture Notes in Mechanical Engineering,
https://doi.org/10.1007/978-981-15-9809-8_28

## 1 Introduction

In the past three decades, use of bioenergy systems, particularly for power production, liquid biofuels and heating, has grown rapidly. In addition, next to traditional biomass material sectors such as paper and wood industries, biomass is increasingly used as a feedstock for bio-based fuels that replace fossil fuels [1]. Biomass is the principle wellspring of sustainable power source, while half of the biomass vitality is being exposed to change into heat energy [2]. Biomass is the principle wellspring of sustainable power source, while half of the biomass vitality is being exposed to change into heat energy. The heat energy obtained from biomass constitutes the largest share, and it is the quickest developing energy blend component. For more than three decades, the biofuel production from farm wastes such as wood bark, stem, pods and leaves has been increasing predominantly [3]. There are many studies carried out to identify the use of various biomass materials for energy purpose. For the developing countries, superfluous energy is the main factor for their development and its growing financial system. On the other hand, the sustainability gives highly reliable and affordable energy which is also susceptible by many traders causing all kinds of environmental problems [4]. To speak about the ecological issues impending in the way of sustainable development, the clean and green energy resources are playing a vital role. Hence, for developing and under developing countries like India, to succeed on the way of development without obstructing the clean environmental systems, the development of sustainable and green energy sources can be proved to be more beneficial [5]. The effective utilization of renewable energy resources is also essential for developing nations. In many countries, the government introduced many policies to convert the available energy systems into non-polluting green sustainable energy systems. It also enhances the renewable energy usages by more than five times from 32 GW in 2014 to 175 GW in 2022 [6].

A biomass-based sustainability evaluation technology for the production of biofuels with multiple evaluation parameters is also a multi-criteria decision-making (MCDM) problem. Thus, a sufficient planning methodology with basic social, economic and environmental aspects is necessary to conquer the shortage of energy with a dream of sustainability development. MCDM is proved as an outstanding method among other tools for energy planning sector [7]. It deals with the problems associated with the compromise selection of the best solutions from the set of various alternatives. This technique has applied widely on education, investment, transport, agriculture, environment, health care, etc. [8]. To recognize the potential of biomass for contributing global energy generation, a full evaluation of its potential for sustainable development should be carried out. Previously, various literatures have worked to investigate the crop growth and processing [9]. The conversion efficiency of different types of biomass materials is based on the different biomass conversion technologies [10]. A huge amount of studies have been carried out to access the potential usage of biomass-based electric power generation for steering the high carbon and nitrogen emission [11]. Various reports are available related to different

MCDM methods to work out several material selection problems for various engineering applications. Some of the studies are utilizing various MCDM methods for decision making related to biomass sources. Yazdani-Chamzini et al. [12] propose an integrated COPRASAHP method by considering social, economic and environmental criteria for green energy project. The suitable biomass material for the boiler operation was predicted using TOPSIS multi-criteria model by Saelee et al. [13]. For that study, the authors compared the biomass materials such as wood chips, palm shells and wood pellets under the evaluation criteria of conversion efficiency, cost, operation simplicity, acidification potential and greenhouse gas emission. Madhu et al. [14] applied FAHP TOPSIS method for the selection of best biomass material. This study focused for yielding maximum condensable liquid during flash pyrolysis. The study is conducted to apt the best biomass material among five different alternatives by considering seven evaluation criteria. The study concluded that the hardwood was the best choice for pyrolysis yielding maximum bio-oil. Some of the researchers have developed some mathematical modeling approaches. Kabak and Dagdeviren [15] developed a mathematical model with a hybrid MCDM model for renewable energy sources based on benefits, opportunities, costs and risks. In every one of these techniques, the rankings of the chosen alternatives are affected by the criteria weights and normalization procedure taken up to build the elements of the decision matrix which are dimensionless and comparable [16]. Hence, separate normalization equations are mandatory to take care of the beneficial and non-beneficial criteria of the decision matrices. So, these processes are fairly difficult to understand and multi-faceted. Thus, the straightforward, logical and efficient approach is to be needed to the material selection issues. To the best of our knowledge, biomass material selection for the production of biofuel using MCDM tool is very limited. Many of the researchers focus directly on maximization of biofuel yield during various biomass conversion techniques, but normally they never consider the environmental outcomes.

This study focused on the sustainability of the biomass materials in terms of social, economic, energy and environmental contexts. The applications of three MCDM methods such as (i) MOORA, (ii) reference point approach and (iii) full multiplicative MOORA methods are illustrated and used for ranking the biomass materials. Ten biomass samples are selected as alternatives such as sawdust (SD), wheat straw (WS), rice straw (RS), sugarcane bagasse (SB), corn cop (CC), rise husk (RH), switch grass (SG), lemongrass (LG), palm shell (PS) and sunflower shell (SS) with its elemental compositions such as carbon (C), hydrogen (H), nitrogen (N), sulfur (S) and oxygen (O) that are considered as the evaluation parameters to choose the suitable material for pyrolysis process.

## 2 Materials

There are so many types of biomass materials available for the production of biofuels. It is an environmentally friendly, important energy resource, constituting 14% of the global primary energy. Dedicated energy crops are grown particularly for the production energy, whereas the residues are waste products from the agricultural fields. Ultimate biomass characteristics are low value, low energy, lower nutrient and easy compostable with smallest amount of contaminants. Essential properties of lignocellulosic biomass are considerable amount of volatile matters, fixed carbon with lower moisture and ash content [17]. The selected material used for this study was obtained from the Coimbatore-based agricultural field, which is in south part of India. Direct combustion, and thermochemical and biochemical processes are the various types of biomass conversion techniques, whereas thermochemical conversion processes are used to convert the low-grade biomass materials into energy-rich biofuels such as bio-oil, char and biogas. Among the other thermochemical conversion techniques, pyrolysis is the notable one for the production of liquid biofuels. It is the process of thermal destruction of biomass materials with oxygen-absent condition to produce condensable gases [18, 19]. This is the only conversion techniques yielding three types of biofuels (bio-oil, bio char and biogas); hence, it is focused for this study.

## 3 Methods and Materials

### 3.1 *Moora*

In this method of evaluation, a ratio is calculated in which every performance value of an alternative on an objective is compared to a value, which is the representative of all alternatives with respect to that objective. The value is calculated by taking the square root of the sum of square of all alternative with respect to that objective.

The steps in MOORA method are as follows [20].

Let $\left[x_{ij}\right]_{n\times m}$ be the evaluation-making matrix with "n" alternative and "m" objectives:

$$\left[x_{ij}\right]_{n\times m} = \begin{bmatrix} x_{11} & x_{12} & \dots x_{1m} \\ x_{21} & x_{22} & \dots x_{2m} \\ \vdots & \vdots & \vdots\vdots \\ x_{n1} & x_{n2} & \cdots x_{nm} \end{bmatrix} \tag{1}$$

where $x_{ij}$ indicates the performance value of *i*-th alternative on *j*-th objective, in which $i \in \{1, 2,\ldots, n\}$ and $j \in \{1, 2,\ldots, m\}$.

Next, the ratio is calculated using the equation

$$\overline{X}_{ij} = \frac{x_{ij}}{\sqrt{\sum_{j=1}^{m} x_{ij}^2}} \tag{2}$$

Finally, the optimized value (assessment value) $y_i$ is calculated by adding all the values in case of beneficial objectives and subtracted in case of non-beneficial objectives:

$$y_i = \sum_{i=1}^{g} \overline{X}_{ij} - \sum_{i=g+1}^{n} \overline{X}_{ij} \tag{3}$$

The final rank will be calculated using the $y_i$ value (1st rank will be given to highest $y_i$ value).

## 3.2 *Reference Point MOORA Method*

In reference point MOORA method, a maximal objective reference point [16] is calculated using the formula

$$P_i = \min_j \left( \max_i \left| r_i - \overline{X}_{ij} \right| \right) \tag{4}$$

where $r_i$ is the set of reference points and $\left(r_i - \overline{X}_{ij}\right)$ is the deviation of criterion value from the reference point. $r_i$ is maximum value for beneficial objective and minimum value for non-beneficial objective.

## 3.3 *Full Multiplicative MOORA Method*

In the full multiplicative form of MOORA (MULTIMOORA), the overall utility of alternative $U_i$ is calculated using the formula

$$U_i = \frac{\prod_{i=1}^{g} \overline{X}_{ij}}{\prod_{i=g+1}^{n} \overline{X}_{ij}} \tag{5}$$

# 4 Results and Discussion

## 4.1 *Biomass Material Selection Illustration*

The decision matrix is shown below with 10 alternatives, i.e., sawdust, wheat straw, rice straw, sugarcane bagasse, corn cop, rice husk, switch grass, lemongrass, palm

**Table 1** Data for biomass selection process

| S. No. | Material | Properties in wt% | | | | |
|---|---|---|---|---|---|---|
| | | Carbon | Hydrogen | Nitrogen | Sulfur | Oxygen |
| 1 | Sawdust | 48.52 | 6.39 | 0.13 | 0.32 | 44.65 |
| 2 | Wheat straw | 38.34 | 5.47 | 0.6 | 0.37 | 55.22 |
| 3 | Rice straw | 36.07 | 5.2 | 0.64 | 0.26 | 57.83 |
| 4 | Sugarcane bagasse | 44.86 | 5.87 | 0.24 | 0.06 | 48.97 |
| 5 | Corn cop | 42.1 | 5.9 | 0.5 | 0.48 | 51 |
| 6 | Rice husk | 47.4 | 6.5 | 0.8 | 0.26 | 46.24 |
| 7 | Switch grass | 46.08 | 5.44 | 0.36 | 0.05 | 47.38 |
| 8 | Lemongrass | 39.34 | 5.81 | 1.54 | 0.01 | 53.3 |
| 9 | Palm shell | 49.8 | 5.32 | 0.1 | 0.16 | 44.86 |
| 10 | Sunflower shell | 47.4 | 1.4 | 5.8 | 0.05 | 41.4 |

shell and sunflower shell, and 5 objectives, i.e., carbon, hydrogen, nitrogen, sulfur and oxygen. The heat value is one of the most important characteristics of a biomass because it indicates the total amount of energy that is available in it [21]. It is mostly a function of the elements of the biomass. The oxygen content present in the biomass not only affects the heating value of the biofuel product, but also affects the efficiency of the conversion process. So, the good biomass should have the minimum oxygen content [22]. The nitrogen and sulfur content present in the biomass should be minimum. The release of nitrogen and sulfur during thermochemical conversion affects the environment. Table 1 shows the various elemental properties (criteria) of the biomass materials. Among these criteria, carbon and hydrogen are considered as the beneficial properties, and nitrogen, sulfur and oxygen are the non-beneficial properties.

Next, the ratio is calculated by dividing the performance value of an each selected biomass with respect to objective with the square root of the sum of square of all alternative with respect to that objective, which is described in Table 2.

The assessment value $y_i$ is calculated using Eq. 3, and ranks are given to alternative based on those values. The ratio $\overline{X}_{ij}$ is used in the reference point MOORA method, which the deviations from the reference points are calculated and are shown in Table 3. This deviation is used to identify the maximal objective reference point $P_i$. Minimum the value of $P_i$, better the alternative. Then, the rank to the alternatives is provided based on $P_i$. The overall utility of alternative $U_i$ is calculated using Eq. 5. Based on $U_i$ value, the ranks are given to the alternatives. The $y_i$, $P_i$ and $U_i$ values and their corresponding rank are shown in Table 4.

This MOORA-based decision-making method provides a comparative ranking of the biomass materials as 4-7-9-8-1-6-3-2-5-10 when the assessment values are arranged in descending order. Sugarcane bagasse and switch grass were ranked first two by this MOORA method. This method predicted sunflower shell with ranking last. The reference point MOORA method calculated the performance score of the

**Table 2** Ratio (normalized decision matrix)

| S. No. | Material | Carbon | Hydrogen | Nitrogen | Sulfur | Oxygen |
|---|---|---|---|---|---|---|
| 1 | Sawdust | 0.3470 | 0.3671 | 0.0211 | 0.4003 | 0.2862 |
| 2 | Wheat straw | 0.2742 | 0.3143 | 0.0975 | 0.4628 | 0.3540 |
| 3 | Rice straw | 0.2579 | 0.2988 | 0.1040 | 0.3252 | 0.3707 |
| 4 | Sugarcane bagasse | 0.3208 | 0.3373 | 0.0390 | 0.0750 | 0.3139 |
| 5 | Corn cop | 0.3011 | 0.3390 | 0.0812 | 0.6004 | 0.3269 |
| 6 | Rice husk | 0.3390 | 0.3735 | 0.1300 | 0.3252 | 0.2964 |
| 7 | Switch grass | 0.3295 | 0.3126 | 0.0585 | 0.0625 | 0.3037 |
| 8 | Lemongrass | 0.2813 | 0.3338 | 0.2502 | 0.0125 | 0.3417 |
| 9 | Palm shell | 0.3561 | 0.3057 | 0.0162 | 0.2001 | 0.2876 |
| 10 | Sunflower shell | 0.3390 | 0.0804 | 0.9423 | 0.0625 | 0.2654 |

**Table 3** Deviations from the reference points

| S. No. | Material | Carbon | Hydrogen | Nitrogen | Sulfur | Oxygen |
|---|---|---|---|---|---|---|
| 1 | Sawdust | 0.0092 | 0.0063 | 0.0049 | 0.3877 | 0.0208 |
| 2 | Wheat straw | 0.0820 | 0.0592 | 0.0812 | 0.4503 | 0.0886 |
| 3 | Rice straw | 0.0982 | 0.0747 | 0.0877 | 0.3127 | 0.1053 |
| 4 | Sugarcane bagasse | 0.0353 | 0.0362 | 0.0227 | 0.0625 | 0.0485 |
| 5 | Corn cop | 0.0551 | 0.0345 | 0.0650 | 0.5879 | 0.0615 |
| 6 | Rice husk | 0.0172 | 0.0000 | 0.1137 | 0.3127 | 0.0310 |
| 7 | Switch grass | 0.0266 | 0.0609 | 0.0422 | 0.0500 | 0.0383 |
| 8 | Lemongrass | 0.0748 | 0.0396 | 0.2340 | 0.0000 | 0.0763 |
| 9 | Palm shell | 0.0000 | 0.0678 | 0.0000 | 0.1876 | 0.0222 |
| 10 | Sunflower shell | 0.0172 | 0.2930 | 0.9261 | 0.0500 | 0.0000 |

**Table 4** Ranks obtained from different MOORA methods

| S. No. | Material | $y_i$ | Rank | $P_i$ | Rank | $U_i$ | Rank |
|---|---|---|---|---|---|---|---|
| 1 | Sawdust | 0.0065 | 5 | 0.3877 | 7 | 52.6502 | 5 |
| 2 | Wheat straw | −0.3258 | 8 | 0.4503 | 8 | 5.3961 | 9 |
| 3 | Rice straw | −0.2432 | 7 | 0.3127 | 5 | 6.1480 | 8 |
| 4 | Sugarcane bagasse | 0.2301 | 1 | 0.0625 | 2 | 117.7872 | 1 |
| 5 | Corn cop | −0.3685 | 9 | 0.5879 | 9 | 6.4010 | 7 |
| 6 | Rice husk | −0.0392 | 6 | 0.3127 | 5 | 10.1042 | 6 |
| 7 | Switch grass | 0.2173 | 2 | 0.0609 | 1 | 92.7122 | 3 |
| 8 | Lemongrass | 0.0108 | 4 | 0.2340 | 4 | 87.8326 | 4 |
| 9 | Palm shell | 0.1579 | 3 | 0.1876 | 3 | 116.4273 | 2 |
| 10 | Sunflower shell | −0.8508 | 10 | 0.9261 | 10 | 1.7434 | 10 |

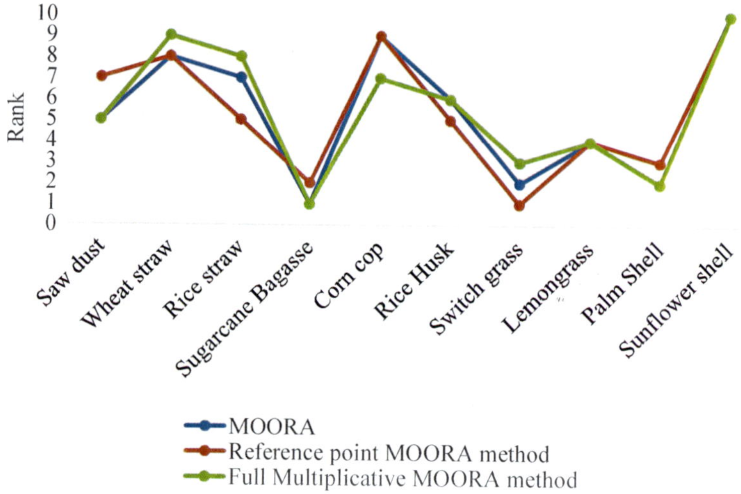


**Fig. 1** Rank obtained from different MOORA methods

selected biomass materials, and the order of ranking of the alternatives is 7-4-9-8-6-4-1-2-5-10. Table 4 also compares the ranking performance of these three methods. From the result, it can be shown that the predictions of these methods are moderately acceptable. Using MULTIMOORA technique and by Eq. (5), the effectiveness of all the selected biomass is found and ranked as 4-9-7-8-1-6-5-3-2-10. Figure 1 shows the rank obtained from different MOORA methods. From the figure, it can be seen that SB is dominating rank one in two MOORA-based MCDM methods out of three methods so it finalized sugarcane bagasse is the best alternative.

## 5 Conclusion

The production of biofuel from agricultural wastes faces many scientific, ecological and social challenges. Emission of polluting and greenhouse gas, and human health issues are the currently facing problems during biomass thermochemical conversion techniques. However, significant consideration must be given to diminishing the ecological contamination before sustainability can be achieved. This paper focused on three mathematical approaches related to biomass material selection for sustainable environment. These proposed methods are more convenient, precise and efficient tool for the selection of best biomass material among the other biomass materials. The results concluded that the obtained rankings by these techniques are almost same and accurate. The main advantage of the selected method used for this study is that they are not dependant on the adopted normalization procedure and also on the criteria weights. A simple ratio system is espoused to make the decision matrices dimensionless and comparable, and it does not depend on the type of the criterion. The results exemplified that sugarcane bagasse becomes the best choice for the biofuel production. This MOORA technique is giving significant results and also a bridge gap between the previous researches related to pyrolysis techniques. This is simple, well suitable and efficient tool for decision makers. Thus, the research can be extended to identify the suitable biomass materials for various biomass conversion techniques such as pyrolysis, gasification and torrefaction.

## References


1. Madhu P, Matheswaran MM, Periyanayagi G (2017) Optimization and characterization of bio-oil produced from cotton shell by flash pyrolysis using artificial neural network. Energy Sources Part A Recovery Utilization Environ Effects 39(23):2173–2180
2. Balezentiene L, Streimikiene D, Balezentis T (2013) Fuzzy decision support methodology for sustainable energy crop selection. Renew Sustain Energy Rev 17:83–93
3. Chandra R, Takeuchi H, Hasegawa T (2012) Methane production from lignocellulosic agricultural crop wastes: a review in context to second generation of biofuel production. Renew Sustain Energy Rev 16(3):1462–1476
4. Mateo JRSC (2012) Multi criteria analysis in the renewable energy industry. Springer Science & Business Media
5. Kumar A, Sah B, Singh AR, Deng Y, He X, Kumar P, Bansal RC (2017) A review of multi criteria decision making (MCDM) towards sustainable renewable energy development. Renew Sustain Energy Rev 69:596–609
6. Ministry of Power, Coal and New Renewable Energy, Governemnt of India " Ujwal Bharat, 2 year Achievements and Initiatives", 2016
7. Diaby V, Campbell K, Goeree R (2013) Multi-criteria decision analysis (MCDA) in health care: a bibliometric analysis. Oper Res Health Care 2(1–2):20–24
8. Hayashi K (2000) Multicriteria analysis for agricultural resource management: a critical survey and future perspectives. Eur J Oper Res 122(2):486–500
9. Elauria JC, Castro MLY, Racelis DA (2003) Sustainable biomass production for energy in the Philippines. Biomass Bioenergy 25:531–540
10. Madhu P, Kanagasabapathy H, Manickam IN (2016) Flash pyrolysis of palmyra palm (*Borassus flabellifer*) using an electrically heated fluidized bed reactor. Energy Sources Part A Recovery Utilization Environ Effects 38(12):1699–1705
11. Dornburg V, van Dam J, Faaij A (2007) Estimating GHG emission mitigation supply curves of large-scale biomass use on a country level. Biomass Bioenergy 31:46–65
12. Yazdani-Chamzini A, Fouladgar MM, Zavadskas EK, Moini SHH (2013) Selecting the optimal renewable energy using multi criteria decision making. J Business Econ Manage 14(5):957–978
13. Saelee S, Paweewan B, Tongpool R, Witoon T, Takada J, Manusboonpurmpool K (2014) Biomass type selection for boilers using TOPSIS multi-criteria model. Int J Environ Sci Dev 5(2):181–186
14. Madhu P, Kumar CN, Anojkumar L, Matheswaran M (2018) Selection of biomass materials for bio-oil yield: a hybrid multi-criteria decision making approach. Clean Technol Environ Policy 20(6):1377–1384
15. Kabak M, Dağdeviren M (2014) Prioritization of renewable energy sources for Turkey by using a hybrid MCDM methodology. Energy Convers Manage 79:25–33
16. Karande P, Chakraborty S (2012) Application of multi-objective optimization on the basis of ratio analysis (MOORA) method for materials selection. Mater Des 37:317–324
17. Sowmya Dhanalakshmi C, Madhu P (2019) Utilization possibilities of Albizia amara as a source of biomass energy for bio-oil in pyrolysis process. Energy Sources Part A Recovery Utilization Environ Effects 41(15):1908–1919

18. Madhu P, Manickam IN, Kanagasabapathy H (2015) Production and upgradation of cotton shell pyrolytic oil for biofuel from flash pyrolysis by fluidized bed reactor. Proc Natl Acad Sci India Sect A Phys Sci 85(3):457–462
19. Sowmya Dhanalakshmi C, Madhu P (2019) Biofuel production of neem wood bark (Azadirachta indica) through flash pyrolysis in a fluidized bed reactor and its chromatographic characterization. Energy Sources Part A Recovery Utilization Environ Effects. https://doi.org/10.1080/15567036.2019.1624893
20. Majumder H, Mishra SK, Sahu AR, Bavche AL, Valekar M, Padaseti BK (2019) Application of MOORA to optimize WEDM process parameters: a multi-criteria decision making approach. International conference on reliability, risk maintenance and engineering management. Springer, Singapore, pp 73–78
21. Dhanalakshmi CS, Madhu P (2019) Recycling of wood bark of Azadirachta indica for bio-oil and chemicals by flash pyrolysis. Indian J Ecol 46(2):347–353
22. Madhu P, Stephen Livingston T, Manickam IN (2017) Fixed bed pyrolysis of lemongrass (*Cymbopogon flexuosus*): bio-oil production and characterization. Energy Sources Part A Recovery Utilization Environ Effects 39(13):1359–1368

# A Multi-criteria Decision-Making Method to Analyze Service Quality Risks in Healthcare Industries

R. K. A. Bhalaji, S. Bathrinath, S. G. Ponnambalam, and S. Saravanasankar

**Abstract** Patient satisfaction in the service quality of orthopedic hospitals is considered as a prospective performance criterion in evaluating the medical services of a healthcare system. In this research, the most critical and crucial risk factors in service quality are identified by using relevant literature survey and inputs from hospital experts, these factors are tough to control. To solve this issue, the combined risk-based decision-making method for identifying and ranking the risk factors regarding service quality performance is proposed. The objective of this paper is to recognize the best one between five orthopedic hospitals in the service quality performance by using Interval-Valued Fuzzy Modified (IVFM) TOPSIS. Here, a contextual case is conducted in five south Indian orthopedic hospitals for verifying the efficiency of the suggested framework. To arrive at the best decision, Interval-valued fuzzy set theory is used in the IVFM TOPSIS for controlling subjectivity and ambiguity in fuzzy risk scores. The outcomes will help hospital managers implementing the performances of service quality.

**Keywords** IVFM TOPSIS · Orthopedic hospitals · Risk-based decision making

---

R. K. A. Bhalaji · S. Bathrinath (✉) · S. Saravanasankar
Department of Mechanical Engineering, Kalasalingam Academy of Research and Education, Krishnankoil, Tamil Nadu, India
e-mail: bathri@gmail.com

R. K. A. Bhalaji
e-mail: bhalaji28@gmail.com

S. Saravanasankar
e-mail: ssaravanasankar@klu.ac.in

S. G. Ponnambalam
School of Mechanical Engineering, VIT University, Vellore 632014, India
e-mail: sgponnambalam@yahoo.com

© The Editor(s) (if applicable) and The Author(s), under exclusive license to Springer Nature Singapore Pte Ltd. 2021

S. Mohan et al. (eds.), *Materials, Design, and Manufacturing for Sustainable Environment*, Lecture Notes in Mechanical Engineering,
https://doi.org/10.1007/978-981-15-9809-8_29

# 1 Introduction

Service quality is considered an important aspect in orthopedic hospitals it covers both satisfaction of patients and well-being. Recently, lifestyles have changed because of the increasing service quality of orthopedic hospitals. From the patient perspective, enhancing the service quality in the hospital is the primary concern and also it turned out to be a vital factor for hospitals in fulfilling the needs of the patients [1]. To assess the performance of inpatients in orthopedic hospitals will help to develop the outcome of existing framework of healthcare and also to improve the quality of service. Subsequently, many no. of inpatients will come to hospitals for their treatments and they are satisfied [2]. Hospitals can meet their probable loss of patients because of neglecting the importance of patients satisfaction and also delivering service quality [3]. For keeping the long-term behaviour of customer, satisfaction of the customer is viewed as a major factor [4]. Therefore, poor service quality in the orthopedic hospitals leads to loss of patients and behavioural intention. This paper aim is to recognize, rank, and analyze the risk factors of orthopedic hospital service quality. This will definitely help to improve service quality of orthopedic hospitals and also the mangers for comprehending the impacts of service quality, which directly relates satisfaction of patients and behavioural intention as well as continuing relations with their patients.

# 2 Relevant Literature

Relevant literature is sorted into three sub-segments namely: (i) Risk factors associated with service quality of hospitals (ii) Gap of literature.

## *2.1 Risk Factors Associated in Service Quality of Hospitals*

Analyzed the factors of service quality in hospital from the patient's perspective and also case study is conducted between both inpatients and outpatients in the hospital from Pakistan. The findings of this paper showed that high service quality to make loyalty and fulfillment between patients [5]. Meesala and Paul [6] assessed the parameters of service quality in Indian private hospitals. Responsiveness and trustworthiness affect the satisfaction of patients are identified from the results. Pai et al. [7] examined the parameters for quantifying service quality in Karnataka hospitals from the patient's view. The communication between patients and employees is the most influential one in the service quality process and it will surely assist the hospital managers in implementing the service quality. Westbrook et al. [8] assessed the factors for quantifying hospital service quality and to find out the critical one among the two hospitals in the service quality process. The consequences demonstrated that communication gaps, responsiveness, and poor housekeeping is the impacting factors and Hospital B is critical than Hospital Amin and Nasharuddin [9] investigated the criteria for service quality in the Malaysian hospital. The result suggested high service quality will encounter patient satisfaction and Assurance is the vital one in the service quality process.

## 2.2 *Gap of Literature*

To accomplish business quality and world market trade, the organization must be more skilled in service quality performance. Service quality is termed as an evaluation of how strong a conveyed service conforms to the expectations of clients. Business operators from service often assess the service quality delivered to their consumers to improve their service, to quickly identify problems, and to finer evaluate the satisfaction of the client. Service quality risk factors in different industrial streams are recognized in this literature review. While undertaking the service quality initiatives, organizations may face different hurdles. As an outcome, industry to industry the impact of specific risk factors may vary. Risk factors are not discussed sufficiently in the literature and also the ranking of orthopedic hospitals in the service quality process is missing. To overcome this issue, this paper looks to find the best one among five orthopedic hospitals in the service quality performance for the case instance. Therefore, this paper uses IVFM TOPSIS method for assessing the interrelationship between risk factors and for ranking orthopedic hospitals. Some highlights of this paper which are detailed below.

- Recognize risk factors of service quality in orthopedic hospitals from the relevant literature and also the input from hospital experts shown in Table 1.
- Propose a framework to evaluate the best one among five orthopedic hospitals in the service quality process with the help of MCDM technique.
- Verify the suggested framework with a case study from the five orthopedic hospitals. The acquired results are contrasted with existing literature.

**Table 1** Risk Factors for orthopedic hospital service quality

| Risk factors | References |
|---|---|
| Trustworthiness (RF1) | [6, 10] |
| Responsiveness (RF2) | [11, 12] |
| Guarantee (RF3) | [13, 14] |
| Tangibles (RF4) | [15] |
| Sympathy (RF5) | [16, 17] |

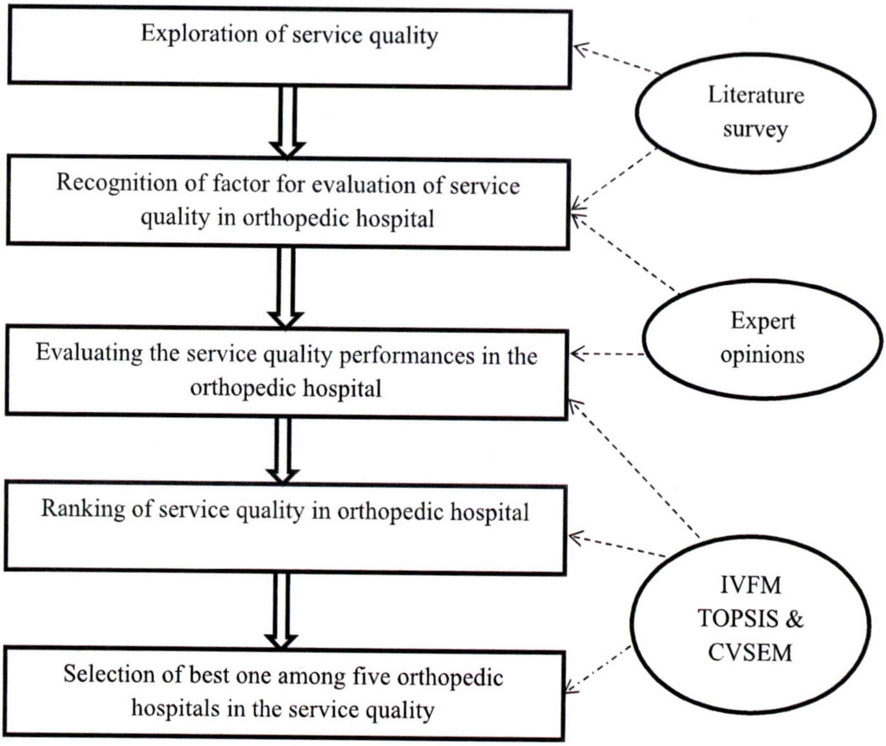


**Fig. 1** Framework used in this paper

## 3 Framework for Orthopedic Hospital Service Quality Assessment

See Fig. 1.

## 4 Methodology of IVFM TOPSIS

MCDM techniques are used to solve the problems of risk-based decision making with multiple risk factors. Yue [18] said that to find out the most influential risk factors from multiple factors is the goal of MCDM. Caputo et al. [19] said that to rank the risk factors regarding multiple factors by using different kinds of MCDM techniques like ANP, VIKOR, AHP and ELECTRE. Hwang and Yoon [20] initiated the method of TOPSIS and also it is one of the eminent MCDM techniques. Dependent on the fact that the selected alternative must have the smallest space from the optimistic-ideal solution (OIS) and the largest from the pessimistic-ideal solution (PIS) is the notion of this method. Methodology of TOPSIS was first created by Chen [21], for a problem of risk-based decision making in the fuzzy setting. The linguistic terms provided the rating of weight of every risk factor and alternative and it can be stated in TFNs (Triangular fuzzy numbers). Cornelis et al. [22] said that the conventional linguistic method is not clear, due to its performance in the normal fuzzy sets and also to signify the level of certainty in the form of interval. To overcome this issue, Vahdani et al. [23] were first introduced the method of IVFM TOPSIS for the rating of weights of every risk factor and every alternative and it can be stated in triangular IVFNs. For the improvement of risk-based decision-making, this methodology has been used. This paper used the fundamental notion of fuzzy decision matrix and it detailed below:

Here fuzzy decision matrix is $H = [y_{ij}]_{e\times f}$ which can be stated in a format of matrix as

$$H = \begin{bmatrix} y_{11} & y_{12} & \cdots & y_{1n} \\ y_{21} & y_{22} & \cdots & y_{2n} \\ \cdots & \cdots & \cdots & \cdots \\ y_{e1} & y_{e2} & \cdots & y_{ef} \end{bmatrix}$$

$$V = [v_1, v_2, \ldots, v_f],$$

Performance of alternatives can be quantified with $P_1, P_2, \ldots, P_e$ are potential alternatives, $R_{f1}, R_{f2}, \ldots, R_{fn}$ are factors, $y_{ij}$ is the alternative rating of $P_i$ regarding factor $R_j$, and $V_j$ is the weight of the factor $R_j$. Fuzzy numbers signified the both $y_{ij}, \forall i, j$ and $v_j, j = 1, 2, \ldots, f$. Then the phases of IVFM TOPSIS methodology are discussed below.

Phase 1: Formed a team of decision-makers $G$.

Phase 2: A set of regarding risk factors has to be delineated and explained.

Phase 3: Decision-makers are given the rating of every alternative regarding every risk factor.

Phase 4: Combine the alternative ratings against every individual factor $y_{ij}$ and chosen factor with fuzzy weights ($V_j$).

Suppose a team of decision has G decision-makers, $G = 1, 2, \ldots, g$. For every risk factor, the combined fuzzy weight and fuzzy score of alternatives can be calculated as follows.

$$y_{ij} = \frac{1}{G}\left[y_{iij}^1 + y_{ij}^2 + \cdots + y_{ij}^g\right], \quad i = 1, 2, \ldots, e, \quad j = 1, 2, \ldots, f \tag{1}$$

$$V_j = \frac{1}{G}\left[v_j^1 + v_j^2 + \cdots + v_j^g\right], \quad j = 1, 2, \ldots, f \tag{2}$$

Regarding $j$th factor, where $y_{ij}^g$ is the fuzzy score of $i$th alternative and $v_j^g$ is the fuzzy significance weight provided by the decision-maker $G$.

Phase 5: Calculate the matrix for normalized decision. To compute $f_{ij}$ and $f_{ij}^*$ for the normalization of vector as discussed below:

$$f_{ij} = \frac{y_{ij}}{\sqrt{\sum_{i=1}^{e} y_{ij}^2}}, \quad i = 1, 2, \ldots, e, \quad j = 1, 2, \ldots, f \tag{3}$$

$$f_{ij}^* = \left[ \left( \frac{p_{ij}}{r_j^+}, \frac{p'_{ij}}{r_j^+} \right); \frac{q_{ij}}{r_j^+}; \left( \frac{r'_{ij}}{r_j^+}, \frac{r_{ij}}{r_j^+} \right) \right], \quad j \in \Omega_q \tag{4}$$

$$f_{ij}^* = \left[ \left( \frac{p_j^-}{r_{ij}}, \frac{p_j^-}{r'_{ij}} \right); \frac{p_j^-}{q_{ij}}; \left( \frac{p_j^-}{p'_{ij}}, \frac{p_j^-}{p_{ij}} \right) \right], \quad j \in \Omega_r \tag{5}$$

$$r_j^+ = \underset{i}{\text{Max}}\, r_{ij}, \quad j \in \Omega_q$$

$$p_j^- = \underset{i}{\text{Min}}\, p_{ij}, \quad j \in \Omega_r$$

Here $\Omega_q$ shows the benefit factor and $\Omega_r$ shows the cost factor correspondingly. Hereafter, to acquire the matrix for normalization $F = [f_{ij}]_{e \times f}$. To keep the property within the ranges of normalized interval [0, 1] is previously mentioned in the method of normalization.

Phase 6: To build the matrix for weighted normalization, $Z = [z_{ij}]_{e \times f}$. To calculate the matrix for fuzzy weighted normalized decision by multiplying every matrix column by the weight of fuzzy $(v_j)$. Therefore,

$$Z_{ij} = v_j \times f_{ij}, \quad i = 1, 2, \ldots, e, \quad j = 1, 2, \ldots, f \tag{6}$$

The multiply operator can be used as per the description 1,

$$Z_{ij} = \left[ \left( v_{1j} \times f_{1ij}, v'_{1j}, v'_{1j} \times f'_{1ij} \right); v_{2j} \times f_{2ij}; \left( v'_{3j} \times f'_{3ij}, v_{3j} \times f_{3ij} \right) \right] \tag{7}$$

Phase 7: To define the OIS and PIS. The $P^*$ and $P^-$ values are described as

$$P^* = \{z_1^*, \ldots, z_f^*\} = \left\{ \left( \max_i z_{ij} | j \in \Omega_q \right), \left( \min_i z_{ij} | j \in \Omega_r \right) \right\}, \tag{8}$$

$$P^- = \{z_1^-, \ldots, z_f^i\} = \left\{ \left( \min_i z_{ij} | j \in \Omega_q \right), \left( \max_i z_{ij} | j \in \Omega_r \right) \right\}, \tag{9}$$

Here $\Omega_q$ shows the sets of benefit factors and $\Omega_r$ shows the sets of advantage factors correspondingly. A decision-maker who wants to have a most extreme value between the alternatives is the benefit factor and also decision-maker wants to have lowest value between alternatives is the cost factor. Clearly, OIs or the maximum desirable orthopedic hospital is denoted as $P^*$ and PIS or minimum desirable alternative is denoted as $P^-$.

Phase 8: To build the matrix for anti-ideal separation $(S^-)$ and ideal separation $(S^*)$ which are described as

$$S^* = \left[S^*_{ij}\right] = \begin{bmatrix} \left|z_{11} - z^*_1\right| & \left|z_{12} - z^*_2\right| & \ldots & \left|z_{1f} - z^*_f\right| \\ \left|z_{21} - z^*_1\right| & \left|z_{22} - z^*_2\right| & \ldots & \left|z_{2f} - z^*_f\right| \\ \ldots & \ldots & \ldots & \ldots \\ \left|z_{e1} - z^*_1\right| & \left|z_{e2} - z^*_1\right| & \ldots & \left|z_{ef} - z^*_f\right| \end{bmatrix} \tag{10}$$

$$S^- = \left[s^-_{ij}\right] = \begin{bmatrix} \left|z_{11} - z^-_1\right| & \left|z_{12} - z^-_2\right| & \ldots & \left|z_{1f} - z^-_f\right| \\ \left|z_{12} - z^-_1\right| & \left|z_{22} - z^-_2\right| & \ldots & \left|z_{2f} - z^-_f\right| \\ \ldots & \ldots & \ldots & \ldots \\ \left|z_{e1} - z^-_1\right| & \left|z_{e2} - z^-_1\right| & \ldots & \left|z_{ef} - z^-_f\right| \end{bmatrix} \tag{11}$$

Matrix for ideal separation ($S^*$) and matrix for anti-ideal $\left(S^-\right)$ is transformed into a matrix with accurate no's based on description 5 which are detailed below:

$$S^* = \begin{bmatrix} s^*_{11} & s^*_{12} & \ldots & s^*_{1f} \\ s^*_{21} & s^*_{22} & \ldots & s^*_{2f} \\ \ldots & \ldots & \ldots & \ldots \\ s^*_{e1} & s^*_{e2} & \ldots & s^*_{ef} \end{bmatrix}, \tag{12}$$

and

$$S^- = \begin{bmatrix} s^-_{11} & s^-_{12} & \ldots & s^-_{1f} \\ s^-_{21} & s^-_{22} & \ldots & s^-_{2f} \\ \ldots & \ldots & \ldots & \ldots \\ s^-_{e1} & s^-_{e2} & \ldots & s^-_{ef} \end{bmatrix}, \tag{13}$$

Phase 9: To compute the collective index (CI).

$$\alpha_i\left(S^*, S^-\right) = \left(\sum_{j-1(P)}^{K} \frac{s^*_{ij}}{s^-_{ij}}\right)^{\frac{1}{K}} + Y_{ij} \quad \forall i = 1, 2, \ldots, e, \tag{14}$$

For which $s^-_{ij} > 0$, where the aggregation of first one $\left(\sum P\right)$ denotes to all $j$ and for $s^-_{ij} = 0$, while $\left(Y_{ij}\right)$ denotes to all $j'$. Likewise, $K$ shows the no of chosen factor, and for which $s^-_{ij'} > 0$ and $v_j$ for $s^-_{ij} = 0$ to compute the value of $Y_{ij}$, such that $Y_{ij'} = \left(\max_{j'}\left(s^*_{ij}/s^-_{ij}\right)\right)^{1/\max_j v_j}$

$$\beta_i\left(S^*, S^-\right) = \left(\sum_{j-1}^{k} s^*_{ij}\right)^{1/k} + \left(\sum_{j=1(P)}^{k} \frac{1}{s^-_{ij}}\right)^{1/k} + B'_{ij} \quad \forall i = 1, 2, \ldots, e, \tag{15}$$

For which $s_{ij}^{-} > 0$, where the aggregation of second one $(\sum P)$ denotes to all $j$ and for $s_{ij}^{-} = 0$ while $B_{ij}$ denotes to all $j'$. Also, $K$ shows that no. of chosen factor and for which $s_{ij}^{-} > 0$ and $v_j$ for $s_{ij}^{-} = 0$ to compute the value $B_{ij}$ such that $B = ij'\left(\min_{j'}\left(s_{ij'}^{-}\right)\right)^{\max_j v_j}$.

CI is computed as:

$$\mathrm{CI}_i = \alpha_i + \beta_i \tag{16}$$

Phase 10: As per the order of sequence to be rank. Based on the CI value, to find out the best one among alternatives. Lowest CI value signifies the best one among alternatives $(\mathrm{AL})_i$.

## 5 Case Study

To conduct the case study by using the data gathered from a top-five orthopedic hospitals in India for verifying the suggested methodology. The aim of the paper is to choose the best one among five orthopedic hospitals in service quality performance. A team of experts like hospital managers and authorities was formed for the review and they were effectively related in the department of service quality in aforesaid orthopedic hospitals. The five experts are chosen for this review with over 5 years' experience in the field of service quality. Expert's profile has not been revealed because of obscurity reasons and they have been shortened as $DM_1$, $DM_2$, $DM_3$, $DM_4$ and $DM_5$. Experts have been asked to given their opinions about risk factors by using linguistic scale in a questionnaire. They are five risk factors related to service quality in the questionnaire and every risk factor has been defined by failure possibility and loss severity. The chosen specialists have participated in this review and their skill helped in the case study.

### 5.1 *Recognition of Risk Factor for Service Quality and Selection of Orthopedic Hospital*

Service quality is the process of comparison of both the performances and expectations in the process. An orthopedic hospital with good service quality will meet expectations from patients as well as competitive in finance. Therefore, this paper recognizes five major risk factors for service quality such as trustworthiness, responsiveness, guarantee, tangibles and sympathy from the literature survey as well as inputs from hospital experts. The service quality factors are gathered from five orthopedic hospitals such as $(OH_1)$, $(OH_2)$, $(OH_3)$, $(OH_4)$ and $(OH_5)$. The risk factors can be used for enhancing the service quality of all orthopedic hospitals because it improves productivity, satisfaction from patients as well as turnover. Based on these risk factors, to choose the best one among five orthopedic hospitals in the service quality performance can be evaluated.

## 5.2 *Gathering of Data*

Inputs are gathered individually from experts in the orthopedic hospital such as linguistic data on failure possibility, loss severity, and weights of the risk factor. The significance weights of every risk factor given by decision-makers are depicted in Table 4 and also Table 5 shows the failure possibility and loss severity relevant to every risk factor. Tables 2 and 3 denotes the fuzzy linguistic scale, here linguistic data has converted into suitable triangular IVFM. Dependent on the triangular IVFM, the significance weights of every risk factor are depicted (Table 6).

By using Eqs. 14 and 15, to calculate the matrix for perfect separation ($S^*$) and the matrix for anti-perfect separation $\left(S^-\right)$ correspondingly. The specific outcomes are provided as follows:

**Table 2** Descriptions of Linguistic scales for failure possibility and loss severity regarding risk factor

| TIVFN | Linguistic scales |
|---|---|
| [(8.5, 9.5); 10; (10, 10)] | Very high (VH) |
| [(5.5, 7.5); 9; (9.5, 10)] | High (H) |
| [(4.5, 5.5); 7; (8.0, 9.5)] | Medium high (MH) |
| [(2.5, 3.5); 5; (6.5, 7.5)] | Medium (M) |
| [(0, 1.5); 3; (4.5, 5.5)] | Medium low (ML) |
| [(0, 0.5); 1; (2.5, 3.5)] | Low (L) |
| [(0, 0); 0; (1, 1.5)] | Very low (VL) |

**Table 3** Descriptions of linguistic scales for significance risk factor weights

| TIVFN | Linguistic scales |
|---|---|
| [(0.85, 0.95); 1; (1, 1)] | Very High (VH) |
| [(0.55, 0.75); 0.9; (0.95, 1)] | High (H) |
| [(0.45, 0.55); 0.7; (0.80, 0.95)] | Medium-high (MH) |
| [(0.25, 0.35); 0.5; (0.65, 0.75)] | Medium (M) |
| [(0, 0.15); 0.3; (0.45, 0.55)] | Medium low (ML) |
| [(0, 0.05); 0.1; (0.25, 0.35)] | Low (L) |
| [(0, 0); 0; (0.1, 0.15)] | Very low (VL) |

**Table 4** Significance weights of every risk factor provided by DM's

| Factors | Decision-makers | | | | |
|---|---|---|---|---|---|
| | DM1 | DM2 | DM3 | DM4 | DM5 |
| Trustworthiness (RF1) | H | MH | H | M | H |
| Responsiveness (RF2) | MH | H | VH | H | MH |
| Guarantee (RF3) | H | M | H | MH | H |
| Tangibles (RF4) | MH | H | MH | H | M |
| Sympathy (RF5) | H | MH | H | M | H |

$$S^* = \begin{bmatrix} 0.574 & 0.262 & 0.94 & 0.692 & 0.94 \\ 0.15 & 0.544 & 0.502 & 0.392 & 0.321 \\ 0.316 & 0.94 & 0.772 & 0.538 & 0.165 \\ 0.152 & 0.845 & 0.344 & 0.93 & 0.401 \\ 0.94 & 0.659 & 0.472 & 0.702 & 0.452 \end{bmatrix}$$

$$S^- = \begin{bmatrix} 0.982 & 0.973 & 0.961 & 0.961 & 0.971 \\ 0.985 & 0.960 & 0.961 & 0.959 & 0.972 \\ 0.985 & 0.963 & 0.963 & 0.956 & 0.982 \\ 0.989 & 0.959 & 0.972 & 0.953 & 0.974 \\ 0.983 & 0.963 & 0.971 & 0.951 & 0.977 \end{bmatrix}$$

By using Eq. (16) to compute the $\alpha_i$, $\beta_i$, $\text{CI}_i$ values correspondingly and displayed in Table 7.

For instance, $\alpha_1 = [(0.574/0.982) + (0.262/0.973) + (0.94/0.961) + (0.692/0.961) + (0.94/0.971)]^{\frac{1}{5}} = 1.286\beta_1 = (0.574+0.262+0.94+0,692+0.94)^{\frac{1}{5}} + \{(1/0.982) + (1/0.973) + (1/0.961) + (1/0.961) + (1/0.971\}^{\frac{1}{5}} = 2.66$.

To calculate the $\text{CI}_i$ for orthopedic hospital ($\text{OH}_1$) is as follows:

$$\text{CI}_1 = \alpha_1 + \beta_1 = 1.286 + 2.666 = 3.952.$$

In the problem of service quality performance, the ranking order of five potential orthopedic hospitals is ($\text{OH}_2$), ($\text{OH}_4$), ($\text{OH}_3$), ($\text{OH}_5$) and ($\text{OH}_1$) based on Table 7 and depicted in Fig. 2. So, orthopedic hospital ($\text{OH}_2$) is the best one among five orthopedic hospitals in the service quality performance and also has a minimal risk value between the other orthopedic hospitals. From Table 7, ($\text{OH}_1$) has the highest risk value therefore the hospital administration should implement the performance of service quality based on the ISO standards such as ISO 10002:2018 and ISO 9001:2015.

**Table 5** Linguistic data for assessing fuzzy risk scores of various orthopedic hospitals against every risk factor allocated by DMs

| DMs | Orthopedic hospitals ($OH_i$) | Trustworthiness (RF1) | | Responsiveness (RF2) | | Guarantee (RF3) | | Tangibles (RF4) | | Sympathy (RF5) | |
|---|---|---|---|---|---|---|---|---|---|---|---|
| | | ($Se_{ij}$) | ($Pr_{ij}$) | ($Se_{ij}$) | ($Pr_{ij}$) | ($Se_{ij}$) | ($Pr_{ij}$) | ($Se_{ij}$) | ($Pr_{ij}$) | ($Se_{ij}$) | ($Pr_{ij}$) |
| DM1 | ($OH_1$) | MH | ML | M | L | M | ML | H | MH | ML | L |
| | ($OH_2$) | H | M | MH | M | MH | L | VH | H | M | ML |
| | ($OH_3$) | M | ML | M | ML | M | M | H | MH | M | L |
| | ($OH_4$) | H | MH | H | M | ML | ML | H | M | MH | M |
| | ($OH_5$) | MH | ML | M | ML | M | L | MH | M | M | ML |
| DM2 | ($OH_1$) | M | MH | MH | M | M | M | VH | H | M | M |
| | ($OH_2$) | MH | H | M | ML | M | ML | VH | MH | ML | ML |
| | ($OH_3$) | MH | ML | M | ML | ML | M | H | H | ML | ML |
| | ($OH_4$) | H | ML | H | L | MH | L | MH | H | L | L |
| | ($OH_5$) | M | M | MH | L | M | ML | H | MH | L | M |
| DM3 | ($OH_1$) | VH | M | ML | ML | MH | L | H | M | L | ML |
| | ($OH_2$) | H | MH | M | M | M | L | VH | MH | ML | ML |
| | ($OH_3$) | H | M | M | L | ML | M | MH | M | M | L |
| | ($OH_4$) | MH | H | MH | L | ML | ML | VH | MH | M | M |
| | ($OH_5$) | H | MH | M | ML | M | ML | H | MH | ML | M |
| DM4 | ($OH_1$) | MH | M | M | M | M | ML | H | MH | M | ML |
| | ($OH_2$) | H | MH | H | M | MH | L | H | M | MH | M |
| | ($OH_3$) | H | M | H | L | MH | L | VH | M | MH | M |
| | ($OH_4$) | VH | ML | MH | ML | ML | M | MH | MH | ML | ML |
| | ($OH_5$) | MH | M | M | L | M | L | VH | H | M | L |

(continued)

**Table 5** (continued)

| DMs | Orthopedic hospitals ($OH_i$) | Trustworthiness (RF1) | | Responsiveness (RF2) | | Guarantee (RF3) | | Tangibles (RF4) | | Sympathy (RF5) | |
|---|---|---|---|---|---|---|---|---|---|---|---|
| | | ($Se_{ij}$) | ($Pr_{ij}$) | ($Se_{ij}$) | ($Pr_{ij}$) | ($Se_{ij}$) | ($Pr_{ij}$) | ($Se_{ij}$) | ($Pr_{ij}$) | ($Se_{ij}$) | ($Pr_{ij}$) |
| DM5 | ($OH_1$) | H | M | MH | ML | ML | ML | MH | M | ML | L |
| | ($OH_2$) | MH | H | H | M | M | M | MH | M | L | ML |
| | ($OH_3$) | VH | MH | MH | M | MH | M | H | H | MH | ML |
| | ($OH_4$) | MH | ML | H | L | ML | L | VH | H | M | M |
| | ($OH_5$) | H | ML | H | ML | M | ML | H | MH | M | M |

**Table 6** Weighted normalized interval-valued fuzzy decision matrix

| Orthopedic hospitals | RF1 | RF2 | RF3 | RF4 | RF5 |
|---|---|---|---|---|---|
| ($OH_1$) | [(0.017, 0.033); 0.073; (0.173, 0.574)] | [(0.026, 0.045); 0.078; (0.143, 0.262)] | [(0.038, 0.068); 0.141; (0.309, 0.94)] | [(0.038, 0.069); 0.134; (0.289, 0.692)] | [(0.028,0.054); 0.114; (0.276,0.94)] |
| ($OH_2$) | [(0.014, 0.023); 0.039; (0.071, 0.150)] | [(0.039, 0.064); 0.135; (0.218, 0.544)] | [(0.038, 0.063); 0.115; (0.225, 0.502)] | [(0.040, 0.065); 0.109; (0.199, 0.392)] | [(0.027,0.044); 0.077; (0.148,0.321)] |
| ($OH_3$) | [(0.014, 0.027); 0.054; (0.120, 0.316)] | [(0.036, 0.064); 0.157; (0.284, 0.94)] | [(0.036, 0.067); 0.146; (0.321, 0.772)] | [(0.043, 0.07); 0.133; (0.258, 0.538)] | [(0.017,0.03); 0.048; (0.86,0.165)] |
| ($OH_4$) | [(0.011, 0.018); 0.033; (0.066, 0.152)] | [(0.040, 0.073); 0.149; (0.298, 0.845)] | [(0.027, 0.047); 0.085; (0.162, 0.344)] | [(0.046, 0.073); 0.143; (0.320, 0.93)] | [(0.025,0.043); 0.083; (0.162,0.401)] |
| ($OH_5$) | [(0.016, 0.032); 0.073; (0.194, 0.94)] | [(0.036, 0.068); 0.137; (0.289, 0.659)] | [(0.028, 0.050); 0.093; (0.193, 0.472)] | [(0.048, 0.087); 0.175; (0.330, 0.702)] | [(0.022,0.039); 0.072; (0.156,0.452)] |
| Optimistic-ideal solution | [(0, 0); 0; (0, 0)] | [(0, 0); 0; (0, 0)] | [(0,0); 0; (0,0)] | [(0,0); 0; (0,0)] | [(0,0); 0; (0,0)] |
| Pessimistic-ideal | [(1, 1); 1; (1, 1)] | [(1, 1); 1; (1, 1)] | [(1, 1); 1; (1, 1)] | [(1, 1); 1; (1, 1)] | [(1,1); 1; (1,1)] |

**Table 7** Values of $\alpha_i$, $\beta_i$ and $CI_i$ by the suggested method

| Orthopedic hospitals | $\alpha_i$ | $\beta_i$ | $CI_i$ | Rankings |
|---|---|---|---|---|
| ($OH_1$) | 1.286 | 2.666 | 3.952 | 5 |
| ($OH_2$) | 1.146 | 2.526 | 3.673 | 1 |
| ($OH_3$) | 1.231 | 2.610 | 3.842 | 3 |
| ($OH_4$) | 1.226 | 2.605 | 3.832 | 2 |
| ($OH_5$) | 1.271 | 2.652 | 3.924 | 4 |

## 6 Conclusion

In this paper, IVFM TOPSIS method is used to determine the best one between five orthopedic hospitals in the service quality process. The present case of risk

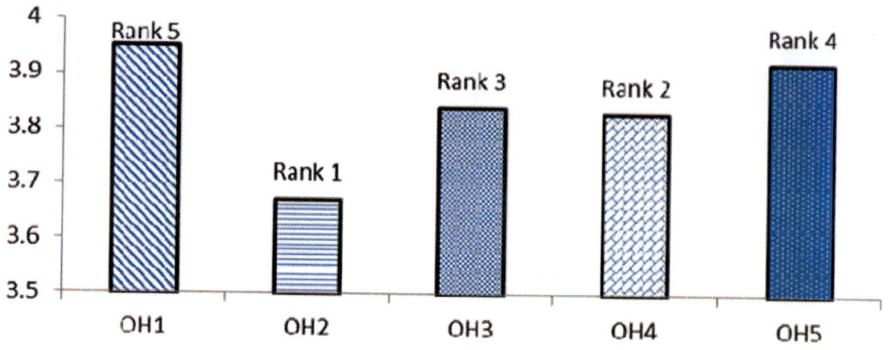


**Fig. 2** Ranking of orthopedic hospital service quality

management paves the tactic for enhancing the risk mitigation strategies towards the implementation of service quality. The outcomes of this paper depicted that orthopedic hospital ($OH_1$) is the best one in the process of service quality. Consequently, the organization needs to deliberate the other orthopedic hospitals which have a strong role in ascertaining the service quality.

- The findings of the paper shown that $OH_1$ is the best one among others and it useful for orthopedic hospital managers to predict the risks in service quality.
- Moreover, this finding assists to execute service quality processes efficiently by concentrating on enhancing skill sets to diminish the risks.
- By using the proposed MCDM method to examine the interaction between risks and also to rank the hospital in the process of service quality by giving valuable information about the risks engaged in developing an effective service quality process in a resilient and secured manner.

As it is evident that that suggested method is determined to be useful in the south Indian orthopedic hospitals, a suitable model can be constructed by an executive in operations of any organization by recognizing the exact risk factors with the help of the suggested MCDM method, for the service quality of any organization concerned outside or within India.

## References


1. Akdere M, Top M, Tekingündüz S (2020) Examining patient perceptions of service quality in Turkish hospitals: the SERVPERF model. Total Qual Manage Bus Excell 31(3–4):342–352
2. Buckė V, Ruževičius J, Buckus R (2020) Service quality management at Lithuanian healthcare institutions. Qual-Access Success 21:174
3. Pekkaya M, Pulat İmamoğlu Ö, Koca H (2019) Evaluation of healthcare service quality via Servqual scale: an application on a hospital. Int J Healthc Manage 12(4):340–347
4. Jacob SC, Manalel J, Minimol MC (2020) Service quality in the healthcare sector: do human resource management practices matter? Br J Healthc Manage 26(2):1–9

5. Fatima T, Malik SA, Shabbir A (2018) Hospital healthcare service quality, patient satisfaction and loyalty: an investigation in context of private healthcare systems. Int J Qual Reliab Manage 35(6):1195–1214
6. Meesala A, Paul J (2018) Service quality, consumer satisfaction and loyalty in hospitals: thinking for the future. J Retail Consum Serv 40:261–269
7. Pai YP, Chary ST, Pai RY (2018) Patient-perceived hospital service quality: an empirical assessment. Int J Health Care Qual Assur 31(1):76–91
8. Westbrook KW, Babakus E, Grant CC (2014) Measuring patient-perceived hospital service quality: validity and managerial usefulness of HCAHPS scales. Health Mark Q 31(2):97–114
9. Amin M, Zahora Nasharuddin S (2013) Hospital service quality and its effects on patient satisfaction and behavioural intention. Clin Gov Int J 18(3):238–254
10. Mohebifar R, Hasani H, Barikani A, Rafiei S (2016) Evaluating service quality from patients' perceptions: application of importance–performance analysis method. Osong Public Health Res Perspect 7(4):233–238
11. Nambisan P, Gustafson DH, Hawkins R, Pingree S (2016) Social support and responsiveness in online patient communities: impact on service quality perceptions. Health Expect 19(1):87–97
12. Kondasani RKR, Panda RK (2015) Customer perceived service quality, satisfaction and loyalty in Indian private healthcare. Int J Health Care Qual Assur 28(5):452–467
13. Shabbir A, Malik SA, Malik SA (2016) Measuring patients' healthcare service quality perceptions, satisfaction, and loyalty in public and private sector hospitals in Pakistan. Int J Qual Reliab Manage 33(5):538–557
14. Aliman NK, Mohamad WN (2016) Linking service quality, patients' satisfaction and behavioral intentions: an investigation on private healthcare in Malaysia. Procedia-Soc Behav Sci 224:141–148
15. Sadeh E, Garkaz M (2019) Interpretive structural modeling of quality factors in both medical and hospitality services in the medical tourism industry. J Travel Tour Market 36(2):253–267
16. Ameryoun A, Najafi S, Nejati-Zarnaqi B, Khalilifar SO, Ajam M, Ansarimoghadam A (2017) Factor selection for service quality evaluation: a hospital case study. Int J Health Care Qual Assur 30(1):58–66
17. Pramanik A (2016) Patients' perception of service quality of health care services in India: a comparative study on urban and rural hospitals. J Health Manage 18(2):205–217
18. Yue Z (2011) A method for group decision-making based on determining weights of decision makers using TOPSIS. Appl Math Model 35(4):1926–1936
19. Caputo AC, Pelagagge PM, Palumbo M (2011) Economic optimization of industrial safety measures using genetic algorithms. J Loss Prev Process Ind 24(5):541–551
20. Hwang CL, Yoon K (2012) Multiple attribute decision making: methods and applications a state-of-the-art survey, vol 186. Springer Science & Business Media
21. Chen CT (2000) Extensions of the TOPSIS for group decision-making under fuzzy environment. Fuzzy Sets Syst 114(1):1–9
22. Cornelis C, Deschrijver G, Kerre EE (2006) Advances and challenges in interval-valued fuzzy logic. Fuzzy Sets Syst 157(5):622–627
23. Vahdani B, Tavakkoli-Moghaddam R, Mousavi SM, Ghodratnama A (2013) Soft computing based on new interval-valued fuzzy modified multi-criteria decision-making method. Appl Soft Computing 13(1):165–172

# Experimental Investigation on Electrically Assisted Incremental Sheet Metal Forming of Ti–6Al–4V Alloy

**R. Mohanraj, S. Elangovan, and A. R. Shanmathy**

**Abstract** Electrically Assisted Incremental Sheet Metal Forming (EAISMF) is a die-less forming technique to deform the hard to form sheet metals into complex three-dimensional shapes by a series of small incremental local deformations. A lightweight alloy like Ti–6Al–4V titanium alloy has enormous applications in aerospace, automobile, and biomedical industries. Although Ti–6Al–4V alloy can be formed at high temperatures, the formability of the material achieved by EAISMF process has to be identified for industrial application. In the present research, the formability of Ti–6Al–4V alloy in EAISMF process was investigated by forming a truncated cone under consideration of different process parameters such as tool diameter, sheet thickness, temperature, and wall angle. Taguchi L9 orthogonal array was employed for an experimental investigation to identify the influencing process parameter on the formability of the material during EAISMF process. The result shows that temperature is a key influencing parameter on the formability of Ti–6Al–4V sheet metal, where an increase in temperature limits the force required for plastic deformation and improves the formability of sheet metal.

**Keywords** Electrical assisted forming · Ti–6Al–4V titanium alloy · Formability · Fracture strain

R. Mohanraj (✉) · S. Elangovan · A. R. Shanmathy
Department of Production Engineering, PSG College of Technology, Coimbatore, India
e-mail: mohanraj839@gmail.com

S. Elangovan
e-mail: elangosurya@gmail.com

A. R. Shanmathy
e-mail: shanmathyar1996@gmail.com

© The Editor(s) (if applicable) and The Author(s), under exclusive license to Springer Nature Singapore Pte Ltd. 2021

S. Mohan et al. (eds.), *Materials, Design, and Manufacturing for Sustainable Environment*, Lecture Notes in Mechanical Engineering,
https://doi.org/10.1007/978-981-15-9809-8_30

## 1 Introduction

Incremental Sheet Forming (ISF) was originated in 1978 by Mason, where the blacksmith's hammering, old forging, and metal spinning are the processes behind the development of the ISF process. Mason proposed that a single spherical roller tool would follow the surface with three-axis control thereby it is the simple way to create a shaped surface [1]. In 2003, a patent was filed by Okada and co-workers where a local heating device was used in the two-point incremental forming process to reduce the effect of spring back. In the conventional forming process, the manufacturing of dies and punches for various shapes and dimensions was a tedious process. Manufacturing of separate dies and punches with greater dimensional accuracy will consume more time and high cost [2]. Recently, electrical assistance for ISF has been developed in which the positive terminal of Direct Current (DC) power supply was connected to the forming tool whereas the negative terminal to the sheet metal [3]. Such type of process was called Electric Hot Incremental Forming (EHIF) and this technique was adopted in the present study with a slight difference where the DC power source was subjected to either side of sheet metal. The resistance to conducting electricity was the main phenomenon for the increase in temperature of sheet metal. The heat treatment of sheet metal reduces the force required for the forming process [4]. EAISMF technique was widely used to manufacture hard to form sheet metals into three-dimensional intricate shapes. This process enables localized plastic deformation with the help of forming tools. Frictional effects are established as there was a direct contact between the forming tool of certain rotational speed and sheet metal, which enhances the increase in temperature of sheet metal. EAISMF technique was suitable for automobile and aerospace industries as there was no necessity for mass production. The sheet metal of required dimension, geometry, surface finish, and high precision can be manufactured by employing the EAISMF process [5]. For bio-medical applications, biocompatibility of material with tissue or bone plays a vital role, where Ti–6Al–4V sheet metal poses poor shear strength and thus can be utilized for bone screws and plates. It was prone to surface wear properties and gets interlocked when it was in sliding contact with other metals. Surface treatments can improve the wear properties of titanium alloys. The replacement of human skull structure which was crucial to adapt for an individual were manufactured using incremental sheet metal forming [6].

## 2 Electrically Assisted Incremental Sheet Metal Forming

In electrically-assisted incremental sheet metal forming process, the positive and negative electrode terminal of the DC source was directly connected to sheet metal. The sheet was fixed in the fixture and it is placed between the insulating material to isolate the sheet and fixture. The current allowed to flow through the sheet, which rises the temperature due to high resistance of the sheet by following Joules law of

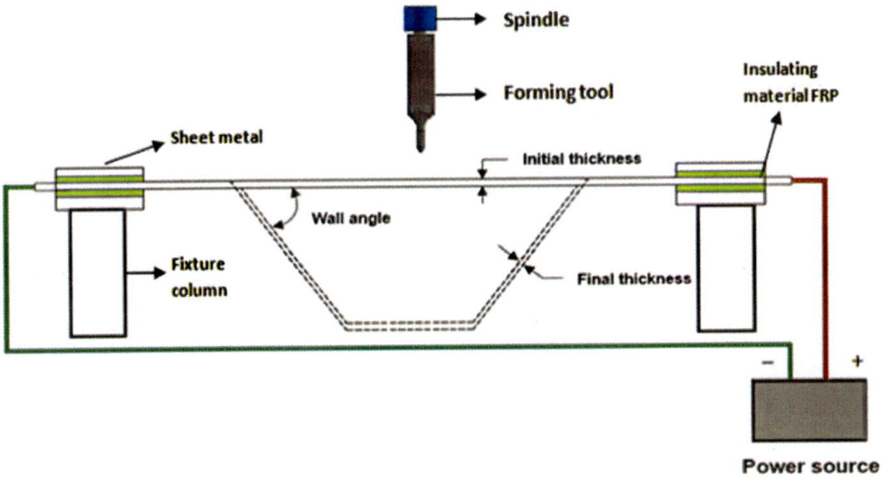


**Fig. 1** Schematic representation of EAISMF processes

heating. In this process, tool was coated with insulating material to avoid the flow of current and follows a path generated by Computer-Aided Manufacturing (CAM) software to produce a part. The generated heat is a key variable to increase the material temperature and consequently its ductility. The closed-circuit was electrically insulated from the milling machine by using ceramic as the base material for the EAISMF fixture.

EAISMF improves the forming limits and reduces the forming force compared to conventional forming methods. As electrons pass through the sheet material, it theories the aid dislocation motion in three ways as locally heating the area around dislocations, increasing the kinetic energy deposited on the dislocation creating an electron wind force, and increasing the number of electrons which breaks the bonds to form with ease. The forming tool has to be machined and processed based on requirements as per the application since the standardized forming tool was not available as drilling and milling tools. Figure 1 shows the schematic view of the working principle involved in the EAISMF process.

## 2.1 *Process Parameters of EAISMF*

The formability of the sheet metal was influenced by factors such as wall angle, tool radius, tool speed, sheet thickness, and scallop height. The sheet metal tends to fail when the wall angle increases due to limited stretching ability, which shows the wall angle has its influence on formability [7]. The speed of the tool was responsible for frictional effects on sheet metal during EAISMF process. Forming a tool of large diameter produces non-uniform deformation causing non-uniform thinning of sheet metal and tool with small diameter encourages localized deformation of sheet metal.

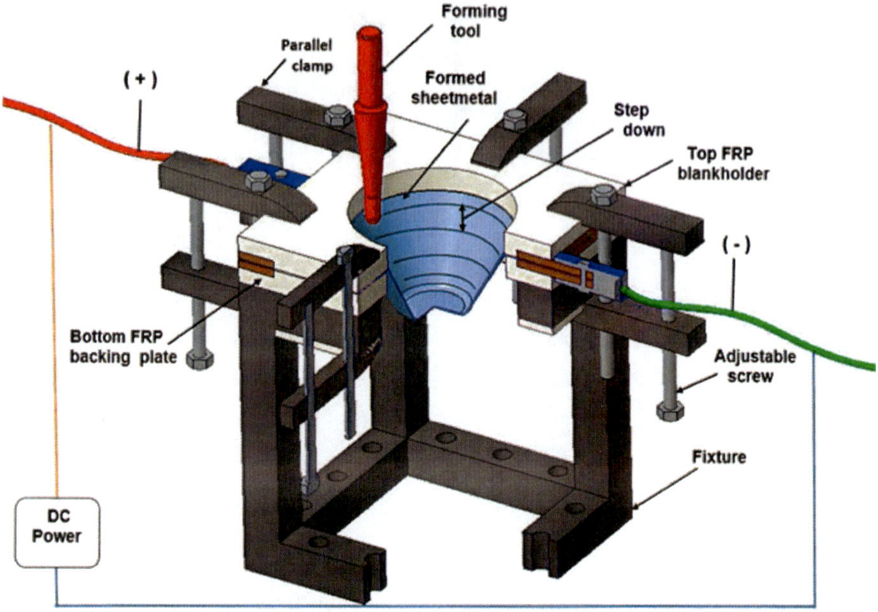


**Fig. 2** Experimental setup of EASIMF process

Formability is influenced by sheet metal thickness, where minimum sheet thickness has better formability with reduced thinning effect [8]. Incremental depth enables step-wise forming of sheet metal, where the surface quality of the finished part was influenced by the scallop height [9]. Figure 2 shows the experimental setup of EAISMF process.

# 3 Experimental Work

## *3.1 Methodology*

The experimental study was carried out using three-axis Computer Numerical Control (CNC) milling machines as shown in Fig. 3. The machine specification includes spindle speed of 9000 rpm, drive motor of 14 kW, and maximum stroke length of 835 mm × 510 mm × 510 mm in *x*, *y* and *z*-directions, respectively. CAM software was used to generate the tool path required to form the desired shape and fed into the CNC milling machine. Ti–6Al–4V titanium alloy sheet metal of different thicknesses with a dimension of 250 mm × 170 mm is clamped to the fixture provided with the insulating material to avoid current flow into the fixture. EAISMF process involves localized deformation of sheet metal with the effect of temperature as the electric current allowed to flow through sheet metal causes an increase in temperature

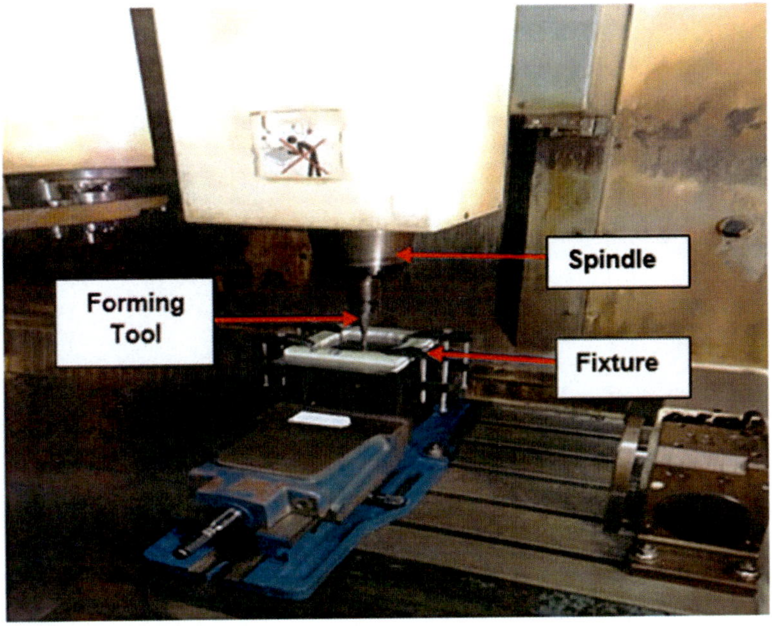


**Fig. 3** Experimental setup in CNC machine

of sheet metal due to electrical resistivity of material [10]. Increase in temperature at the localized zone of tool and sheet interface increases the ductility of the material at the contact zone [11]. Formability of the material was studied by forming truncated cone with different wall angles of 35°, 45°, and 55°.

## 3.2 *Design of Experiments*

Taguchi method is considered as a suitable experimental design tool for testing a greater number of parameter combinations for the purpose of optimizing a process and to determine the parameters that have the greatest influence on the output [12].

The process parameters considered for experimental study are tool diameter, sheet thickness, forming temperature, and wall angle with three levels as shown in Table 1.

**Table 1** Process parameters with their levels

| Parameters | Level 1 | Level 2 | Level 3 |
|---|---|---|---|
| Tool diameter (mm) | 8 | 10 | 12 |
| Sheet thickness (mm) | 0.5 | 1 | 1.5 |
| Temperature (°C) | 150 | 200 | 250 |
| Wall angle (°) | 35 | 45 | 55 |

**Table 2** Taguchi L9 experimental design

| S. No. | Tool diameter (mm) | Thickness of sheet metal (mm) | Temperature (°C) | Wall angle (°) |
|---|---|---|---|---|
| 1 | 8 | 0.5 | 150 | 35 |
| 2 | 8 | 1.0 | 200 | 45 |
| 3 | 8 | 1.5 | 250 | 55 |
| 4 | 10 | 0.5 | 200 | 55 |
| 5 | 10 | 1.0 | 250 | 35 |
| 6 | 10 | 1.5 | 150 | 45 |
| 7 | 12 | 0.5 | 250 | 45 |
| 8 | 12 | 1.0 | 150 | 55 |
| 9 | 12 | 1.5 | 200 | 35 |

Taguchi L9 orthogonal array was used to study the influence of process parameters on formability as shown in Table 2.

## 3.3 *Fixture and Forming Tools*

The fixture was designed and fabricated to clamp the sheet metal during the forming process as shown in Fig. 4. Improper clamping of sheet metal leads to non-uniform sheet metal thinning and even fracture due to concentrated stress developed by the forming tool [4]. Fixture plays a vital role in even distribution of strain along the cross-section of the sheet metal. In this study, Fibre-Reinforced Plastic (FRP) material was used as an insulator to prevent current flow to the CNC machine.

The forming tool was made up of EN8 material with hemispherical end of three different diameters such as 8 mm, 10 mm, and 12 mm. The forming tool design used for the experimental study was shown in Fig. 5.

## 3.4 *Preprocessing of Forming Tool and Sheet Metal*

The forming tool was heat-treated to improve the hardness and toughness property. The hardening process was carried out by heating the forming tool to temperature of 950 °C for 120 min and then oil quenched. Tempering process was carried out where the forming tool was heated to a temperature of 650 °C for 120 min and then it was air-cooled. The vertical muffle furnace heats the material to the desired temperature by conduction, convection, and blackbody radiation, where the operating temperature can be varied from 100 to 1200 °C. Figure 6 shows the heat treatment procedure followed for EN8 material before forming process.

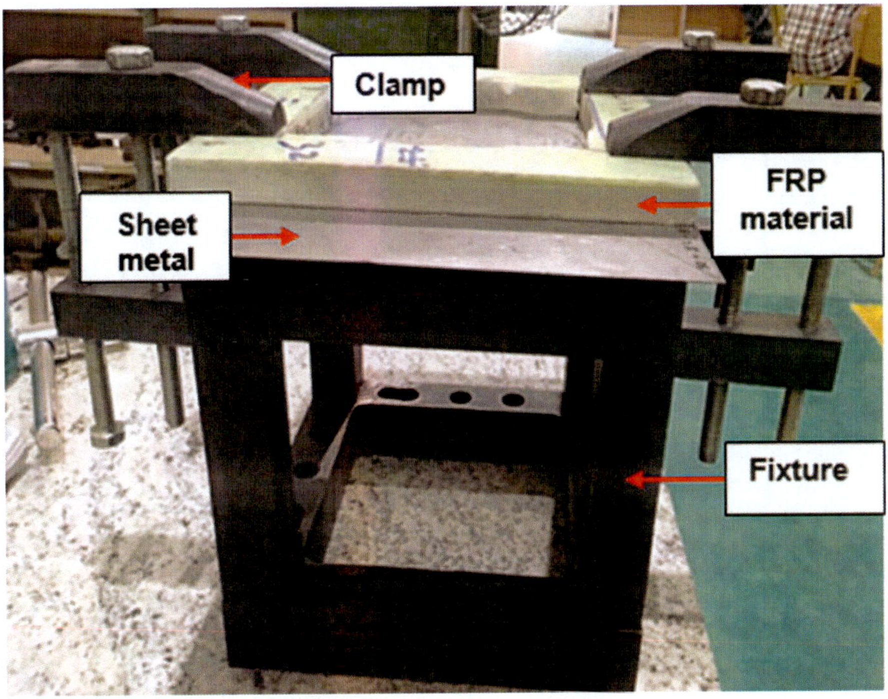


**Fig. 4** EAISMF fixture with FRP insulator material

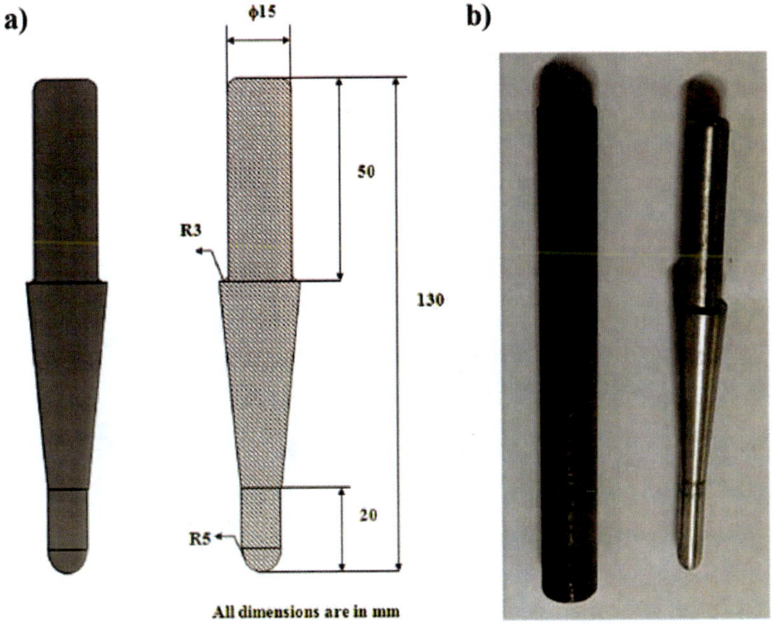


**Fig. 5** **a** Design of forming tool, **b** forming tool fabricated according to design

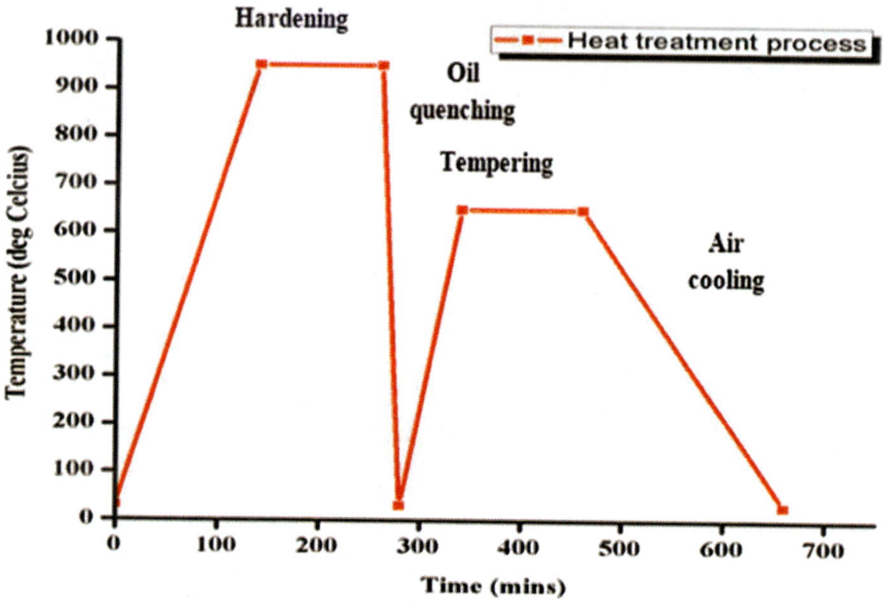


**Fig. 6** Heat treatment process cycle

The heat treatment process enhances the forming tool by increasing wear-resistant property and strength during fatigue loads [13]. The heat treatment process on EN 8 tool material was shown in Fig. 7.

Titanium sheet metal parts are used in aerospace, automobile, and biomedical applications are limited to simple shapes due to manufacturing difficulties in conventional incremental forming methods. Almost 50% of components used in aerospace applications are made up of Ti–6Al–4V titanium alloy [2]. Ti–6Al–4V sheet has poor

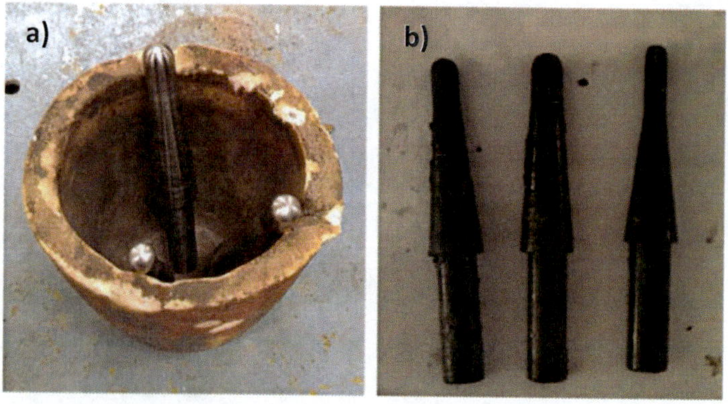


**Fig. 7** Forming tool, **a** before heat treatment, **b** after heat treatment

formability at room temperature and the ultimate tensile strength of the material is 895 MPa.

Circular grid analysis is a useful technique in diagnosing the forming severity of a stretched product. The circular grid pattern with 3 mm diameter is printed on sheet metal for determining major and minor strain as shown in Fig. 8.

At the end of EAISMF process, circular grid pattern deforms to an elliptical shape from which the formability limit of the material can be determined. The major and minor strain were calculated using Eqs. (1) and (2) [11]. Figure 9 shows the major and minor strains due to tension and compression.

$$\text{Major strain} = \frac{\text{major axis length} - \text{original circle diameter}}{\text{original circle diameter}} \times 100 \quad (1)$$

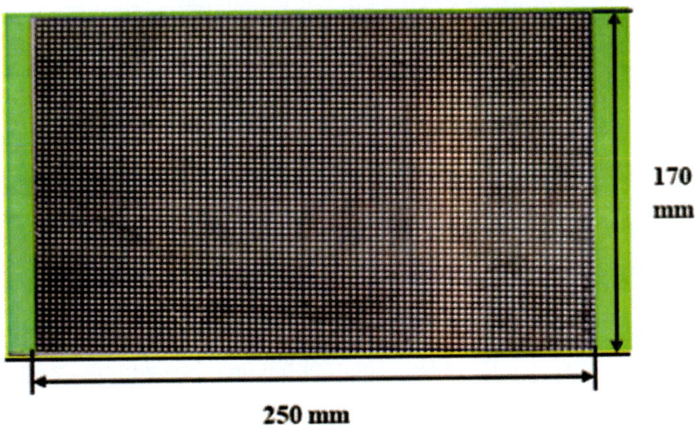


**Fig. 8** Circular grid pattern on sheet metal

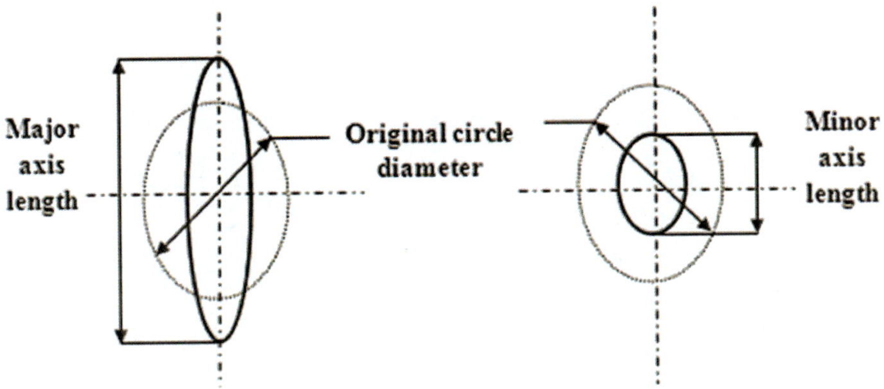


**Fig. 9** Major and minor axis length

$$\text{Minor strain} = \frac{\text{minor axis length} - \text{original circle diameter}}{\text{original circle diameter}} \times 100 \quad (2)$$

# 4 Results and Discussion

## 4.1 Forming Limit Diagram (FLD)

In EAISMF forming process, sheet metal flows according to the tool movement direction, where biaxial stretching deformation mode can be observed at the corners and plane strain stretching at the straight face of the component. Forming limit curve was assessed by determining major and minor strain of the formed component. During the EAISMF process, circular grid patterns deform into elliptical shape, where the major and minor strain values are calculated by measuring along the lengths of major and minor axes.

Ti–6Al–4V titanium alloy sheet metal was formed by using EAISMF process and the observations such as forming depth and time are determined for all experimental trials. The component formed up to fracture generation using EAISMF process is shown in Fig. 10.

Table 3 shows the forming time and forming depth obtained for the experiments conducted on Ti–6Al–4V sheet metal.

The deformation strain experienced by the sheet metal can be measured in terms of change in dimension to the original dimension. Fracture arises due to the stretching of sheet metal with the impact of forming tool beyond the forming limit. The major and minor strain measured using a traveling microscope is shown in Table 4. The

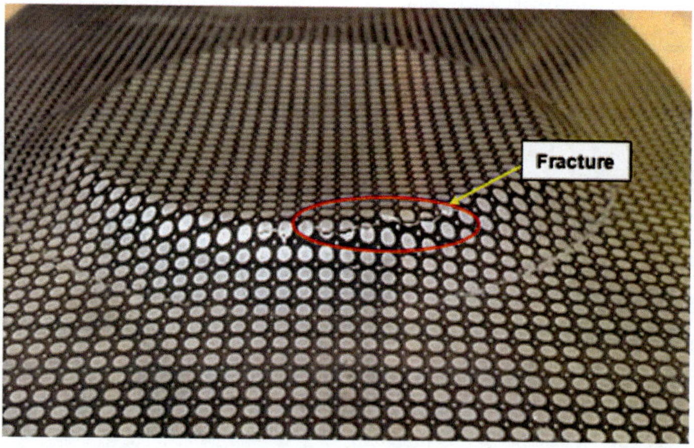


**Fig. 10** Fracture region on formed sheet metal

**Table 3** Forming depth and time

| S. No. | Tool diameter (mm) | Thickness of sheet metal (mm) | Temp (°C) | Wall angle (°) | Forming time (min: sec) | Forming depth (mm) |
|---|---|---|---|---|---|---|
| 1 | 8 | 0.5 | 150 | 35 | 13:42 | 9.79 |
| 2 | 8 | 1.0 | 200 | 45 | 10:23 | 7 |
| 3 | 8 | 1.5 | 250 | 55 | 03:42 | 2.92 |
| 4 | 10 | 0.5 | 150 | 35 | 12.53 | 9.20 |
| 5 | 10 | 1.0 | 200 | 45 | 11.89 | 8.40 |
| 6 | 10 | 1.5 | 250 | 55 | 13.21 | 10.01 |
| 7 | 12 | 0.5 | 250 | 45 | 06:00 | 4 |
| 8 | 12 | 1.0 | 150 | 55 | 16:10 | 11.99 |
| 9 | 12 | 1.5 | 200 | 35 | 14:00 | 10.5 |

**Table 4** Measured values of major and minor strain

| S. No. | Tool diameter (mm) | The thickness of sheet metal (mm) | Temp (°C) | Wall angle (°) | Major axis length (mm) | Minor axis length (mm) | Major strain | Minor strain |
|---|---|---|---|---|---|---|---|---|
| 1 | 8 | 0.5 | 150 | 35 | 2.45 | 2.15 | 0.225 | 0.075 |
| 2 | 8 | 1.0 | 200 | 45 | 2.67 | 2.18 | 0.335 | 0.090 |
| 3 | 8 | 1.5 | 250 | 55 | 2.21 | 2.05 | 0.105 | 0.025 |
| 4 | 10 | 0.5 | 150 | 35 | 2.31 | 2.11 | 0.155 | 0.055 |
| 5 | 10 | 1.0 | 200 | 45 | 2.99 | 2.02 | 0.495 | 0.010 |
| 6 | 10 | 1.5 | 250 | 55 | 2.77 | 2.16 | 0.385 | 0.080 |
| 7 | 12 | 0.5 | 250 | 45 | 2.95 | 2.03 | 0.475 | 0.015 |
| 8 | 12 | 1.0 | 150 | 55 | 3.39 | 2.74 | 0.695 | 0.370 |
| 9 | 12 | 1.5 | 200 | 35 | 2.25 | 2.15 | 0.125 | 0.050 |

difference in the formability of sheet metal was observed for the different combination of process parameters.

Figure 11 shows the forming limit diagram by plotting major and minor strain values in *Y*-axes and *X*-axes respectively. The forming limit curves are obtained by forming the sheet metal to the point of fracture. FLD plotted for all experimental trials was shown in Fig. 12. Thus the FLD of sheet metal traces different paths with the influence of process parameters combinations.

The region above the limiting curves in FLD diagram represents failure and below the curve represents safe deformations. The formability is higher for the parametric combination of tool diameter of 12 mm, sheet thickness of 1 mm, temperature of 150 °C, and wall angle of 55° compared to other experimental trials. This shows

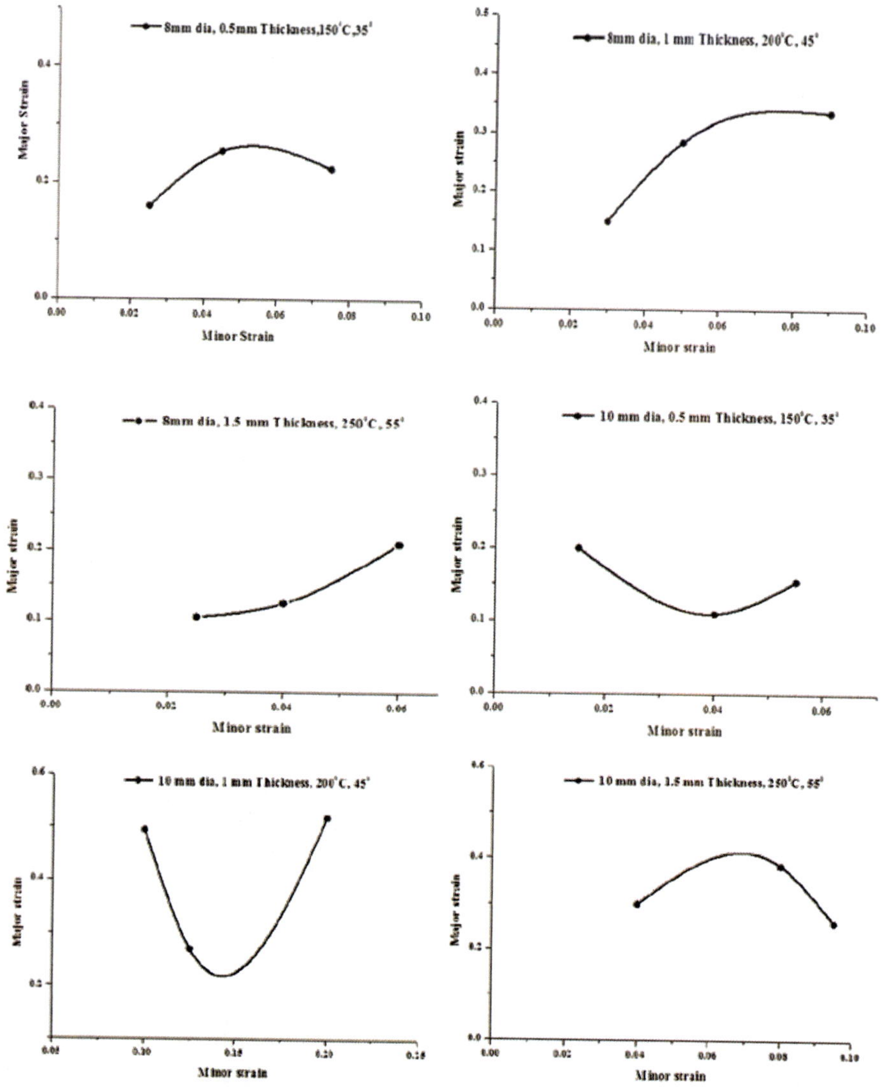


**Fig. 11** Forming limit diagrams (FLD)

an increase in temperature of sheet metal to improve the ductility property of sheet metal [14].

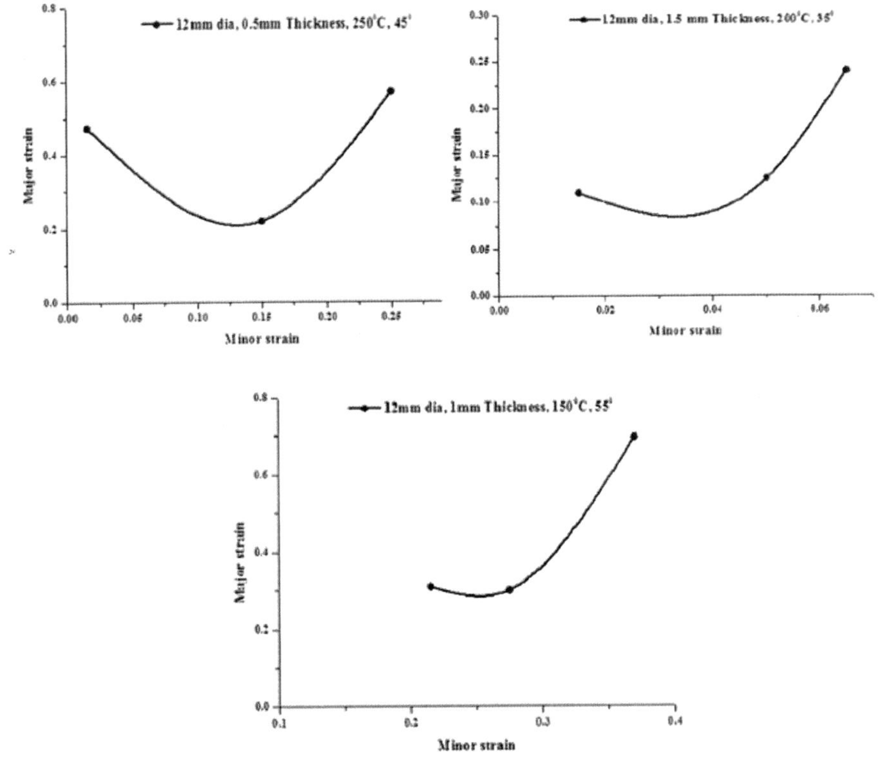


Fig. 11 (continued)

# 5 Conclusions

Formability of Ti–6Al–4V titanium alloy was determined by conducting experiments using EAISMF process. The forming limit of the material has been analyzed based on the effect of process parameters such as wall angle, incremental depth, thickness of sheet metal, tool diameter. Based on the experimental investigation, it was found that temperature was the major influencing factor and responsible for minimizing the force required for localized plastic deformation of sheet metal. From the experimental observation, the formability and forming depth for parameter combination of diameter tool of 12 mm, sheet metal thickness of 1 mm, temperature of 150 °C, and wall angle of 55° was higher compared to other combinations.

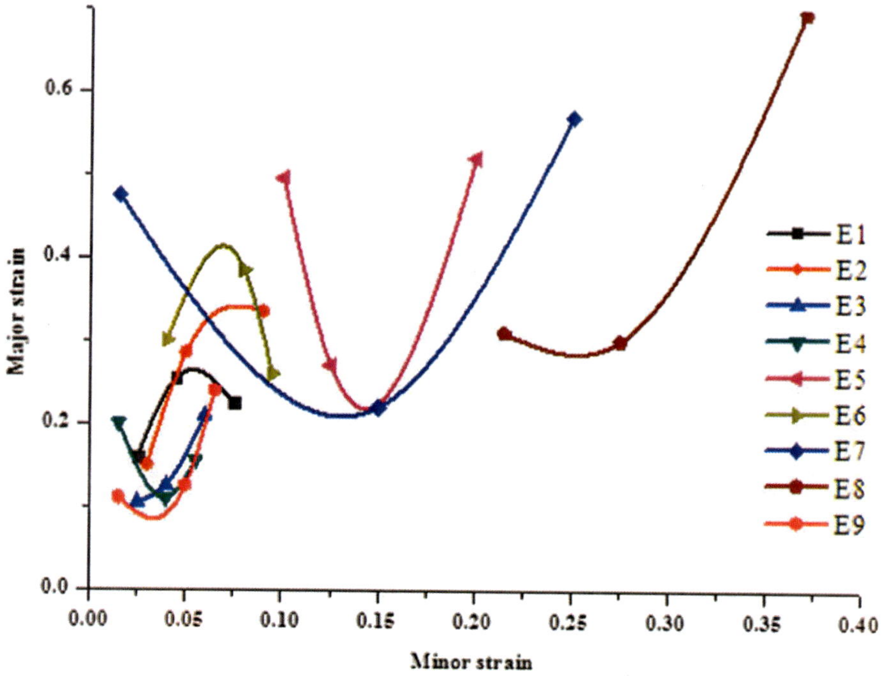


**Fig. 12** Comparison of forming limit curves for all experimental trials

## References


1. Emmens WC, Sebastiani G, van den BoogaardA H (2010) The technology of incremental sheet forming—a brief review of the history. J Mater Process Technol 210(8):981–997
2. Vahdani M, Mirnia MJ, Bakhshi-Jooybari M, Gorji H (2019) Electric hot incremental sheet forming of Ti-6Al-4V titanium, AA6061 aluminum, and DC01 steel sheets. Int J Adv Manuf Technol 103:1199
3. Honarpisheh M, Abdolhoseini MJ, Amini S (2016) Experimental and numerical investigation of the hot incremental forming of Ti-6Al-4V sheet using electrical current. Int J Adv Manuf Technol 83:2027–2037
4. Pacheco PAP, Silveira ME, Silva JA (2019) Heat distribution in electric hot incremental sheet forming. Int J Adv Manuf Technol 102:991
5. Li Z, Lu S, Zhang T, Zhang C, Mao Z (2018) Electric assistance hot incremental sheet forming: an integral heating design. Int J Adv Manuf Technol 96:3209–3215
6. Saidi B, Giraud Moreau L, Mhemed S, Cherouat A, Adragna PA, Nasri R (2019) Hot incremental forming of titanium human skull prosthesis by using cartridge heaters: a reverse engineering approach. Int J Adv Manuf Technol 101:873–880
7. Hagan E, Jeswiet J (2003) Effect of wall angle on Al 3003 strain hardening for parts formed by computer numerical control incremental forming. Proc IMechE Part B: J Eng Manuf 217(11):1571–1579
8. SilvaM B, SkjoedtM AG, BayN MAF (2008) Single-point incremental forming and formability-failure diagrams. J Strain Anal Eng 43(1):15–35
9. Aerens R, Eyckens P, Duflou JR, VanBael A (2010) Force prediction for single point incremental forming deduced from experiment and FEM observation. Int J Adv Manuf Technol 46:969–982

10. Fan G, Sun F, Meng X, Gao L, Tong G (2010) Electric hot incremental forming of Ti-6Al-4V titanium sheet. Int J Adv Manuf Technol 49:941–947
11. Amrut M, Ben S, Ismail S, Kocanda A (2017) Experimental investigations into the effects of SPIF forming conditions on surface roughness and formability by design of experiments. J Braz Soc Mech Sci Eng 39:3997–4010
12. Joshua JJ, Mears L (2013) Thermal response modeling of sheet metals in uniaxial tension during electrically assisted forming. J Manuf Sci Eng 135(2):021011–021022
13. Zhang X, He T, Miwa H, Nanbu T, Murakami R, Liu S, Cao J, Wang QJ (2019) A new approach for analyzing the temperature rise and heat partition at the interface of coated tool tip-sheet incremental forming systems. Int J Heat Mass Transf 129:1172–1183
14. Magnus CS (2017) Joule heating of the forming zone in incremental sheet metal forming: Part 1. Int J Adv Manuf Technol 91:1309–1319

# Investigation of Cutting Temperature on Machining Titanium Alloys Using Micro-textured Cutting Inserts

**N. Abishekraj, T. Gowtham, R. Bibeye Jahaziel, V. Krishnaraj, and B. Geetha Priyadarshini**

**Abstract** In the current investigation, it was focused mainly to compare and evaluate the analytical and experimental trials on the cutting temperature in machining the titanium alloy using micro-hole textured cutting inserts. Initially, the main input parameters such as cutting speed, depth of cut, and feed rate have been used to evaluate the performances. Box–Behnken technique of response surface methodology was used to perform machining experiments. Secondly, with the help of width of cut and thickness of the chip formed at each trial, the analytical calculation of cutting temperature at the two-deformation zone, namely primary and secondary of the orthogonal cutting, has been done. The efficacy of the input parameters on cutting temperatures has been also compared with the analytical calculation. Importance of the cutting parameters was studied with the help of mean effective plots. Results also revealed that the deviation of cutting temperature between the numerical and experimental investigation was only 13%. This depicts the analytical formulation which was much close relation to that of machined trials of cutting temperatures. Further, the cutting temperatures are found to be minimized at lower feed rates and cutting speed using the observation of mean effect plots.

N. Abishekraj · T. Gowtham · R. Bibeye Jahaziel · V. Krishnaraj (✉)
Department of Production Engineering, PSG College of Technology, Coimbatore, Tamil Nadu, India
e-mail: vkr.prod@psgtech.ac.in

N. Abishekraj
e-mail: abishekraj6464@gmail.com

T. Gowtham
e-mail: tgowtham636@gmail.com

R. Bibeye Jahaziel
e-mail: bij@psgias.ac.in

B. Geetha Priyadarshini
Nanotech Research Innovation and Incubation Centre, PSG Institute of Advanced Studies, Coimbatore, Tamil Nadu, India
e-mail: bgp@psgias.ac.in

© The Editor(s) (if applicable) and The Author(s), under exclusive license to Springer Nature Singapore Pte Ltd. 2021

S. Mohan et al. (eds.), *Materials, Design, and Manufacturing for Sustainable Environment*, Lecture Notes in Mechanical Engineering,
https://doi.org/10.1007/978-981-15-9809-8_31

**Keywords** Cutting temperature · Dry machining · Ti–6Al–4V · Textured cutting tools

## 1 Introduction

Machining is an intricate procedure which holds various parameters. Due to this assertion, the identification and rectification of problems in machining that would be counteracted were generally severe. The inserts, workpiece, and the machining conditions furnish the issues which affect them. There are various parameters which mark it as a source of heat generation in the machining operations. The deformation caused by friction and shear action is primarily due to the input parameters. These actions of work were simultaneously created in the form of heat and temperature fields. About 90% of work in the process machining was transferred into heat [1].

During the process of machining, the cutting tool or the workpiece would be influenced due to the dissipation of heat in the deformation zones which could tend to violate the geometry of the insert. The cutting temperature predominantly influences the characteristic attributes of the workpiece and cutting forces on the tools [1]. This generally influences wear debris on the rake and flank faces of the insert. It is considered as one of the significant consumable parameters in order to predict the wear occurring in the tool [2]. For this process, the projection of the cutting temperature is the derived goal [3]. The factors that reconcile the heat dissipation while machining was workpiece, friction between the tool–chip interface [4]. The measurement of cutting temperature in the cutting process was usually difficult. This was generally because of unnatural contact phenomena between the tool and workpiece in the secondary deformation zone [5]. This was used as the constricted approach in the machining process [6].

Xie et al. illustrated that the temperature occurred at the primary deformation zone while using micro-grooved textured tool was lower when compared with the conventional tool. The major assertion was the textures that allow the heat dissipation in higher manner, when the contact of the chip on the rake face was found to be less which tends the temperature not to rise above 500 °C while dry machining on titanium alloys. This also results in good convective heat transfer while the case of micro-grooved cutting inserts [7].

Jianxin et al. stated an increment in the cutting temperature causes due to the increase in cutting speed while machining steel. The elliptical grooves and wavy pattern resulted in higher temperature minimization to the linear and inclined micro-grooves. This was purely due to the fact that the wavy pattern induced higher surface area for heat dissipation in the cutting tool. This was also because this pattern produced between the tool–chip interface that infers in air aspiration, which provided to reduce the cutting temperature while machining [8].

Xing et al. reported that the pulse energy of laser irradiation on the surface of the cutting tool would majorly influence the increase in temperature during machining high grade steel. The impact of higher pulse energy of laser irradiated resulted in a significant reduction in temperature with the lower pulse energy in the plane tool. This could be inferred due to the fact that higher pulse energy would conclude in depth of the nano-textures on cutting tools. This could also minimize the contact length that produces greater heat radiation. This could result in making the cutting tool treated with high pulse energy in order to minimize the cutting temperature [9].

In similar work, Ze et al. [10] depicted that micro-texturing results in minimal cutting temperature on the rake than flank surface of the insert during metal cutting on titanium alloys. This was mainly due to the chip travelling being found to be more which results in higher heat generation during machining over the rake surface. Henceforth, the micro-grooves on the rake face minimize the area of contact length, which rather reduces the heat transfer in the secondary zone.

Bouchelaghem et al. [11] evaluated the influence of response variables on machining of steel using cubic boron nitride cutting tool. The observations depicted higher time for material removal tend to severe wear which would rather explode the heat generation on the secondary deformation area. Aouici et al. [12] investigated a similar research on the effect of input parameters on the heat generation phenomena while turning steel using CBN cutting tools. They used response surface methodology to evaluate the characterization between the response variables with the desired cutting temperature. They also concluded that the occurrence of higher temperature was mainly due to the explosion in all the process parameters.

Abhang and Hameedullah [13] evaluated the significance of minimum quantity lubrication (MQL) method in minimizing the cutting temperature during dry machining of alloy steel. The performances showed that the MQL can decline the secondary zone temperature of about 25% in correlation upon optimization of the response parameters and type of material used for cutting. Moreover, they observed that the decline trend towards the cutting temperature using this technique was large at lower levels of cutting variables and smaller at high levels of cutting variables.

Heigel et al. [14] performed the distribution of temperature at the secondary deformation zone in dry cutting by steel cutting inserts with the assist of infrared technique for various cutting speeds. Marcio and James [15] assessed the cutting temperatures taken experimentally by dry cutting and the calculated analytically.

Amritkar et al. [16] investigated the turning of alloy steel material at several cutting speeds and feed rate using commercial insert. In this, they used analysis of variance for finding the connection between the heat generated and the voltage generated. Moreover, the efficiency of the experiment while metal cutting of diverse workpiece materials in the grade of hardened alloy steel grade using commercial tungsten carbide insert. Their observations proved that this technique was providing much higher precision and better repeatability. The contribution of cutting temperature in the secondary deformation zone with textured could be henceforth the occurrence of minimal temperature near the aperture of the cutting tool. The coolant action was introduced between the contact surfaces which also provide good reduction in the cutting temperature. In addition, this would rather allow to control the cutting tool tip phenomena and also conform to lower tool wear and chip adhesion.

The current study focuses to calculate and investigate the conformation of cutting temperature analytical and experimental trials. The evaluation of the cutting variables has been analysed by employing a response surface methodology technique. Influence of the cutting parameters was studied with the help of mean effective plots. Finally, the cutting temperature measurements of the machining processes were analysed for better understanding of the effect of input parameters with textured cutting tools.

## 2 Experimental Set-Up

### *2.1 Materials and Methods*

The experiment was initiated by using a PINACHO turning lathe. The cutting temperature was measured with a FLIR infrared thermal camera having range limit of 200–1100 °C. It was technically derived by a point which is defined by the aperture of the cutting insert. The infrared thermography was used generally due to which it is a contactless technique. The tungsten carbide cutting tools utilized were Kennametal TCMT KC5025 AlTiN coated cutting tool. The tools used were micro-hole textured with pitch in the vertical direction which is about 100 m and is approximately 80 m in the horizontal direction. The depth of hole is about 50 m as shown in Fig. 1 modelled using SolidWorks design software.

The cutting temperatures of each of the machining trials were measured using the FLIR infrared thermal camera. The chip thickness and width of the chip were measured using a vernier caliper. A total of three readings for chip thickness and

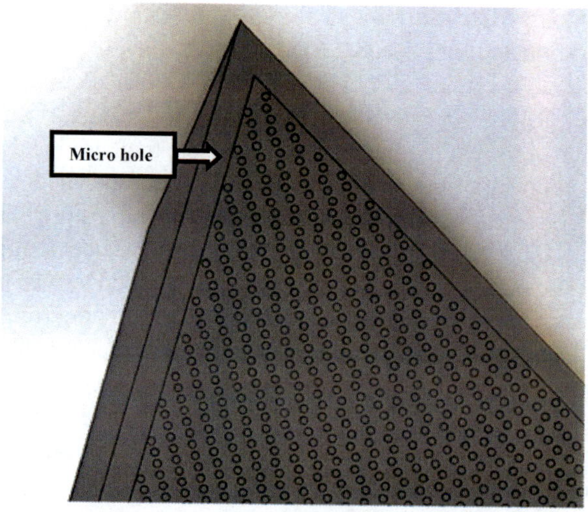


**Fig. 1** Modelled image of the micro-hole textured inserts

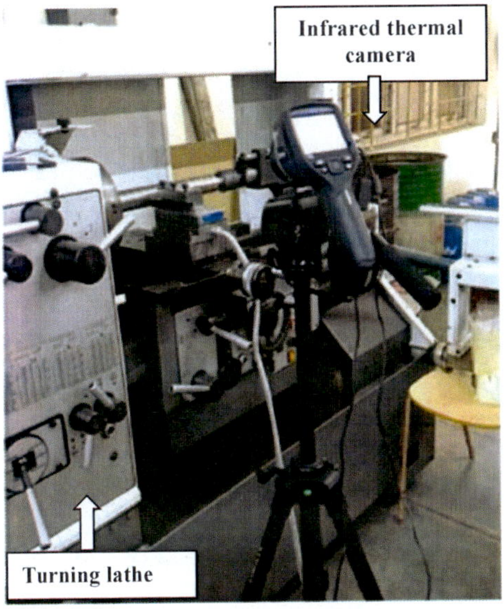


**Fig. 2** Experimental set-up for measurement of cutting temperatures

width of the chip were measured for each trial, and the average was taken as the final reading. The experimental set-up is used as shown in Fig. 2.

It was fixed to use the Box–Behnken design (RSM) technique to minimize count of the trial of experiments. In the current work, the main attributes selected were cutting speed and depth of cut and feed rate. The responses measured were the cutting temperatures at the tool–chip interface. It was very convenient of the repeatable experimental results with the last three trials which also help for the conformation trials.

## 3 Results and Discussion

### 3.1 Analytical Calculation

The total heat generation is given by the equation $P_m$ as follows:

$$P_m = F_c V_c \quad (1)$$

where

$V_c$ is the cutting velocity of the tool (m/s), and

$F_c$ is the cutting force of the work material (N).

Heat generation due to friction:

$$P_f = F_t V_c r_c \tag{2}$$

Heat generation due to shear:

$$P_s = P_m - P_f \tag{3}$$

Primary deformation zone temperature:

The primary deformation zone is caused due to the contact between the insert and the workpiece, and the temperature due to the shear of the material removal is calculated as,

$$\theta_s = \frac{(1 - T)P_s}{\rho c v a_w a_c} \tag{4}$$

where

- $c$ specific heat capacity of the work material (J/kg K),
- $k$ thermal conductivity of the work material (J/sm K),
- $\rho$ density of the work material (kg/m$^3$).

Secondary deformation zone temperature:

Since the secondary deformation zone is mainly caused due to the frictional force between the tool and the chip, the temperature due to the friction is calculated as,

$$\theta_f = \frac{P_f}{\rho c v a_w a_c} \tag{5}$$

Thermal number is calculated by using the formula:

$$R = \frac{\rho c v w}{k} \tag{6}$$

In order to find the actual temperature due to friction, the ratio get the width of the secondary deformation zone divided by the chip thickness is assumed to be 0.5 for the titanium alloy ($w_o$) under lubricating conditions which state that there would not be any difference in the temperature due to friction. The ratio of $\theta_m$ and $\theta_f$ is **1.** It was taken from the plot of the effect of secondary deformation on chip thickness [17].

Therefore,

$$\frac{\theta_m}{\theta_f} \cong 1 \tag{7}$$

Table 1 shows the analytical calculation for all the 15 experimental trials. This was measured with the help of width of the chip and chip thickness which were

**Table 1** Analytical calculation of cutting temperature using micro-textured cutting inserts

| Cutting velocity (m/s) | Chip thickness ratio | $P_m$ | $P_f$ | $P_s$ | Temp due to shear | Thermal number ($R$) | Temp due to friction | Max temp |
|---|---|---|---|---|---|---|---|---|
| 1 | 0.75 | 113.05 | 22.07 | 90.98 | 338.94 | 32.87 | 102.79 | 441.73 |
| 1.33 | 0.75 | 142.96 | 27.48 | 115.48 | 358.55 | 43.72 | 106.66 | 465.21 |
| 1 | 0.83 | 202.30 | 19.24 | 183.06 | 348.57 | 65.75 | 45.79 | 394.37 |
| 1.33 | 0.81 | 247.42 | 29.97 | 217.45 | 333.56 | 87.45 | 57.47 | 391.02 |
| 1 | 0.81 | 102.42 | 14.06 | 88.36 | 342.20 | 49.31 | 68.07 | 410.27 |
| 1.33 | 0.79 | 155.72 | 24.39 | 131.33 | 347.12 | 65.58 | 80.57 | 427.69 |
| 1 | 0.77 | 156.18 | 30.02 | 126.16 | 323.90 | 49.31 | 96.35 | 420.25 |
| 1.33 | 0.78 | 209.04 | 35.03 | 174.00 | 335.89 | 65.58 | 84.53 | 420.42 |
| 1.16 | 0.75 | 125.95 | 17.30 | 108.65 | 493.89 | 38.13 | 98.33 | 592.21 |
| 1.16 | 0.85 | 196.23 | 12.52 | 183.70 | 357.10 | 76.27 | 30.43 | 387.53 |
| 1.16 | 0.75 | 173.73 | 25.30 | 148.43 | 456.85 | 38.13 | 97.33 | 554.19 |
| 1.16 | 0.82 | 253.80 | 27.60 | 226.19 | 351.76 | 76.27 | 53.66 | 405.41 |
| 1.16 | 0.83 | 148.83 | 14.77 | 134.06 | 338.55 | 57.20 | 46.62 | 385.17 |
| 1.16 | 0.82 | 164.91 | 19.37 | 145.54 | 337.28 | 57.20 | 56.10 | 393.38 |
| 1.16 | 0.79 | 157.85 | 26.13 | 131.73 | 305.27 | 57.20 | 75.68 | 380.95 |

taken after dry machining on titanium alloys using surface textured cutting inserts. From that the chip thickness, ratio was determined for each trial, and using all the equations step by step, the result was achieved.

The maximum temperature is calculated as the sum of the temperature rise of the material passing through the secondary deformation and primary deformation zone.

$$\boldsymbol{\theta}_{max} = \boldsymbol{\theta}_s + \boldsymbol{\theta}_m \tag{8}$$

where

$\theta_m$ temperature increment in the secondary deformation.
$\theta_s$ temperature increment in the primary deformation zone.

Based on main effect plot shown in Fig. 3, it gives the input parameter such as lower feed rate, lower depth of cut, and moderate cutting velocity which provides optimum temperature as shown in 9th experimental trial. But the maximum deviation occurs in the 9th and 11th experiments, due to the convective heat transfer in that region, which was significantly reduced by usage of textured tools. From the analytical investigation, it infers that the optimum temperature occurs in the primary zone due to lower input parameter, but secondary deformation correlates as per standard dimensions resulting in maximum value. Also, there is deviation which exists in both the cases. The main reason was due to more heat dissipation at the primary and secondary zone.

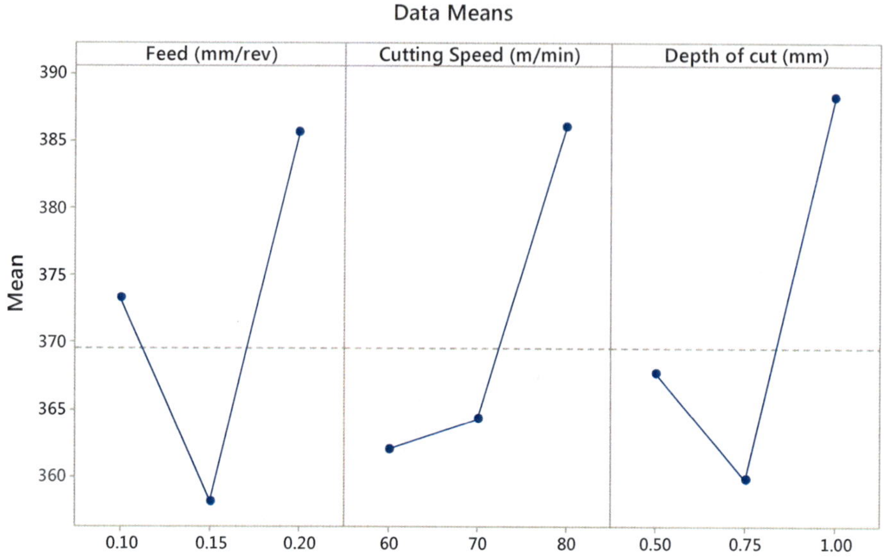


**Fig. 3** Main effect plots for the experimental cutting temperature

### 3.2 *Investigation of Analytical Cutting Temperature and Experimental Using Surface Textured Cutting Tools*

Experimental results are compared with numerical estimation and found close correlation which is presented in Fig. 4. From the observation, it is clear that there are many occurrences of fluctuations in the analytical calculations with an overall deviation of 13%. Experiments have shown that equations produce the exact plot, for instance they predict that higher the spindle speed which produces greater heat generation, but in tradition, it tends to the overvaluation in cutting temperature. The several shortening conjectures which were made in the primary deformation area infer that a higher inadequate treatment of temperature cannot be enumerated and that the equations given are at least accurate enough for the attempt to relate the calculated values [18].

## 4 Conclusion

Current research was aimed at calculating the cutting temperature experimentally with the infrared device by using the micro-hole textured cutting insert on titanium alloys. The main cutting variables used were cutting speed, feed rate, and depth of cut which were diagnosed with RSM design. The response parameter was cutting temperature and is compared with the analytically calculated results. On the basis of the response parameter analysis, the results were derived.

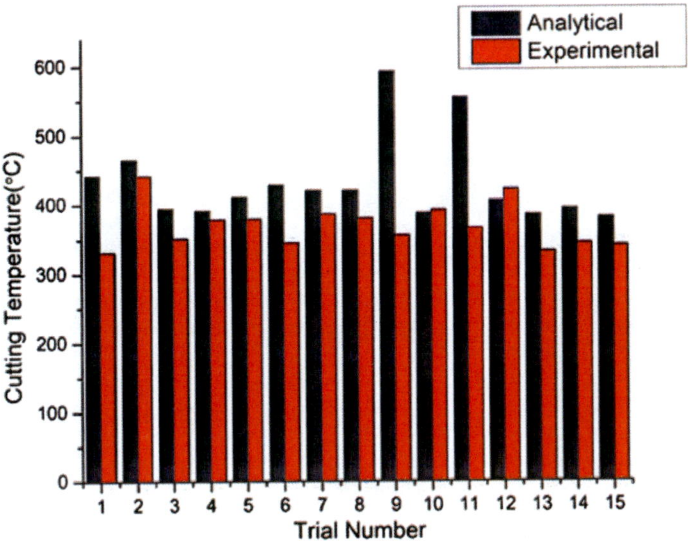


**Fig. 4** Deviation of numerical and experimental readings

- Main effect plots depict that machining at moderate cutting speed with lower feed rates and lower depth of cut gives lower cutting temperature while dry cutting on titanium alloys using surface textured tools.
- The analytical values and the experimental values are almost equal with the overall deviation of 13%.

**Acknowledgements** The authors are thankful to SERB, DST, New Delhi, for providing financial assistance to carry out this work [EMR/2016/007134].

## References

1. Aydin M, Karakuzu C, Uçar M, Cengiz A, Çavuşlu MA (2013) Prediction of surface roughness and cutting zone temperature in dry turning processes of AISI304 stainless steel using ANFIS with PSO learning. Int J Adv Manuf Technol 67:957–967
2. Andjelković BR, Djordjevi BR, Milovancević MD, Jovanović NR (2016) Modeling steady-state thermal defectoscopy of steel solids using two side testing. Therm. Sci. 20(5):S1333–S1343
3. Valiorgue F, Brosse A, Naisson P, Rech J, Hamdi H, Bergheau JM (2013) Emissivity calibration for temperature measurement using thermography in the context of machining. Appl Therm Eng 58:321–326
4. Li B, Zhu D, Pang J, Yang J (2011) Quadratic curve heat flux distribution model in the grinding zone. Int J Adv Manuf Technol 54:931–940
5. Childs T, Maekawa K, Obikawa T, Yamane Y (2000) Machining: theory and applications. Arnold, London
6. Ning J, Liang SY (2018) Evaluation of an analytical model in the prediction of machining temperature of AISI 1045 Steel and AISI 4340 Steel. J Manuf Mater Process 2:74

7. Xie J, Luo M-J, He J-L, Liu X-R, Tan T-W (2012) Micro-grinding of micro-groove array on tool rake surface for dry cutting of titanium alloy. Int J Precis Eng Manuf 13:1845–1852
8. Jianxin D, Ze W, Yunsong L, Ting Q, Jie C (2012) Performance of carbide tools with textured rake-face filled with solid lubricants in dry cutting processes. Int J Refract Metal Hard Mater 30:164–172
9. Xing Y, Deng J, Zhao J, Zhang G, Zhang K (2014) Cutting performance and wear mechanism of nanoscale and microscale textured Al2O3/TiC ceramic tools in dry cutting of hardened steel. Int J Refract Metal Hard Mater 43:46–58
10. Ze W, Jianxin D, Yang C, Youqiang X, Jun Z (2012) Performance of the self-lubricating textured tools in dry cutting of Ti-6Al-4V. Int J Adv Manuf Technol 62:943–951
11. Bouchelaghem H, Yallese MA, Mabrouki T, Amirat A, Rigal JF (2010) Experimental investigation and performance analyses of CBN insert in hard turning of cold work tool steel (D3). Int J Mach Sci Technol 14:471–501
12. Aouici H, Yallese MA, Chaoui K, Mabrouki T, Rigal JF (2012) Analysis of surface roughness and cutting force components in hard turning with CBN tool: prediction model and cutting conditions optimization. Measurement 45:344–353
13. Abhang BL, Hameedullah M (2010) Experimental investigation of minimum quantity lubricants in alloy steel turning. Int J Eng Sci Technol 2:3045–3053
14. Heigel JC, Whitenton E, Lane B, Donmez MA, Madhavan V, Moscoso-Kingsley W (2016) Infrared measurement of the temperature at the tool-chip interface while machining Ti-6Al-4V. J Mater Process Technol 243
15. Silva M, Wallbank J (1999) Cutting temperature: prediction and measurement methods—a review. J Mater Process Technol 88:195–202
16. Amritkar A, Prakash CH, Kulkarni AP (2012) Development of temperature measurement setup for machining. World J Sci Technol 2:15–19
17. Boothroyd GR, Winston A (1989) Fundamentals of machine and machine tools. Marcel Dekker Inc., Newyork
18. Fenton DRG, Oxley PLB (1968) Mechanics of orthogonal machining: allowing for the effects of strain rate and temperature on tool-chip friction. Proc Inst Mech Eng 83(1):417–438

# Experimental Investigation on the Effects of Welding Parameters in Tungsten Inert Gas Welding of Hastelloy C-276

**R. S. Nandha Kumar and J. Pradeep Kumar**

**Abstract** This research article reported the weldability aspects of the material namely Hastelloy C 276. The Tungsten Inert Gas (TIG) type of welding was employed on performing bead on plate welding using SS304 filler material. The weld bead geometry study had been conducted for the welded samples depicting front width, front height, and area in the welded zone. The experiment results reveal that lower the welding current, voltage, gas flow lower the width and height in the weld bead. The increase in the welding current, voltage, gas flow results in an increase in the width and height of the welded zone. The increase in the welding current, voltage, and gas flow results in an increase in the area of the weld bead.

**Keywords** TIG welding · Taguchi method · Weld bead geometry · Hastelloy · Bead on plate welding

## 1 Introduction

Welding process has developed in many ways for several applications. There is probably no industry that is not using the welding methods and the manufacturing process of several products. The research in the welding area has developed and hence new products are been developed as per the requirements of the industries. Devendranath et al. examined the structure and its properties of gas tungsten arc weldments in Monel 400 and Hastelloy C-276 which was carried out by multi-pass current. The microstructure analysis is carried out by the Scanning Electron Microscope equipment [1]. It reveals that there exists a zone which is unmixed at all sides of the material. The bend test reveals that the welded materials show higher value

R. S. Nandha Kumar (✉) · J. Pradeep Kumar
Department of Production Engineering, PSG College of Technology, Peelamedu, Coimbatore, Tamil Nadu, India
e-mail: nandhusaravanan1996@gmail.com

J. Pradeep Kumar
e-mail: jp.psgtech@gmail.com

© The Editor(s) (if applicable) and The Author(s), under exclusive license to Springer Nature Singapore Pte Ltd. 2021

S. Mohan et al. (eds.), *Materials, Design, and Manufacturing for Sustainable Environment*, Lecture Notes in Mechanical Engineering,
https://doi.org/10.1007/978-981-15-9809-8_32

in the process. The results show that the weld strength would be high in comparison with the base metal. Yasheng et al. examined the effects of temperature exposure in corrosion performance of welded specimens of nickel alloy. The corrosion test was conducted in eutectic conditions at various temperatures. The results indicate that weld joints show chromium depletion and Iron particles at a particular temperature and also the grains shows coarser by the annealing process at higher temperature [2].

Welding process yields stress in the materials. This strength is initiated by weld metal compression and also the compression in the heat-affected zone. The metal which is unheated forces restriction on over and also the weld metal cannot contract openly, and stress is developed within the welding joints. It occurs often for the most part known as remaining push, and for a few basic applications had to be evacuated by warm treatment in the entire manufacture. The metal joining is critical to the utilitarian abilities of the joint. Ahmed et al. influenced the tensile strength of the joint in TIG welding process on welding speed. The experiment is conducted on the material of the single V butt joint with various bevel height and the bevel angles. The specimen used for the process is Aluminium AA6351 alloy. The strength of the joints is checked by using Universal Testing Machine. The depth of the penetration of the weld bead decreases with an increase in the bevel height of the V butt joint [3].

Jigar et al. have optimized the effect in process parameters of mechanical and metallurgical properties in dissimilar metals weld by GTAW process. The metals used for the process is Stainless steel and Copper. The response surface methodology was used for the design of the experiment. The tensile analysis and hardness test are carried out which reveals the optimized parameter based on the experimentation conducted [4]. The reactions extremely can cause harm by the properties of a material. Many metals rapidly combine during molten. The proper bonding in the metal can be prevented by a layer of oxide.

The welding has better scope in several areas and hence it has a developing nature in various applications for several purposes. The process varies as per the welding and hence the material properties also vary in each welding. Hedayati et al. conducted the microstructure analysis and diffusion process during the bimetals heat treatment of Hastelloy/stainless steel. The bimetal samples were heat treated and the microstructural analysis is carried out for the material [5]. Savyasachi et al. examined the mechanical and metallurgical properties of different continuous GTA welds of Monel 400 and C-276. The tensile strength is performed to analyze the weld strength of the weld zone [6]. Figure 1 shows the schematic diagram of the weld bead geometry. It depicts the front width and front height of the weld bead. Front width is the distance between the two ends of the weld bead. The front height is the projection area of the weld bead.

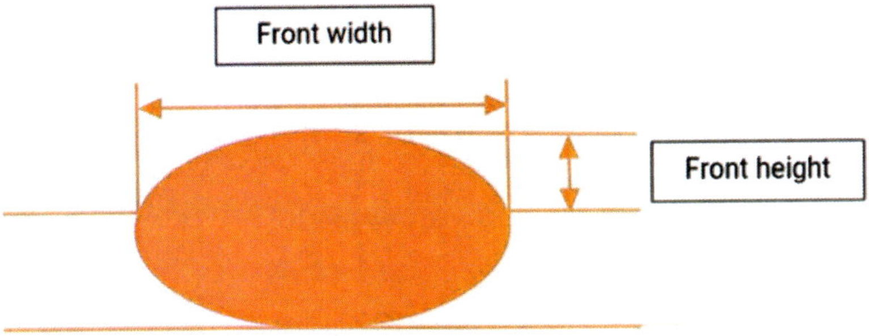


**Fig. 1** Schematic diagram of weld bead geometry

## 2 Methods and Materials

The material Hastelloy is generally a Nickel-Chromium-Molybdenum alloy. Hastelloy is a superalloy arranged basically from nickel and little rate of numerous other components. It is designed to have great stress and at high temperatures, the corrosion-resistance of Hastelloy is incomparable when it comes to execution. Other than the extraordinary resistance that the Hastelloy offers to all conduct of splitting or setting, parts made from this metal mix have an inclination to have ideal execution for a few chemical applications, which might oxidize the metals. Other components like chromium or molybdenum include in the profile of this combination, which demonstrates to be one of the hardest around. The making of Hastelloy may be a complex handle that includes the blending of numerous components like tungsten, carbon, titanium, aluminum, manganese, copper, cobalt, chromium, etc. into the fundamental fixing of the transition metal that's essentially a nickel-based substance. The dimensions of the workpiece are taken as 90 * 50 * 3 mm. The model of the workpiece is shown in Fig. 2. The chemical composition test is conducted for the Hastelloy C276 material. The chemical composition test was conducted in Optical Emission Spectrophotometer. The standard belonging to the test is ASTM E 3047. The test was conducted for the dimensions of 90 * 50 * 3 mm. The chemical composition test is shown in Table 1.

In this work, the filler material selected is SS305L. The filler material in the form of wire and the diameter of 1.2 mm is selected and process is carried out for the material. The reason for the selection of this filler material is given as follows: The Hastelloy material is a Nickel (Ni)-Chromium(Cr)-Molybdenum(Mb). Comparing to the chemical composition of Hastelloy C 276 it consists mainly of Nickel, chromium, and Molybdenum, and also it contains many other elements in less composition. The power source is an essential unit. The high current control is required for TIG welding. It is suitable for both AC and DC control. For the most part, DC current is utilized in stainless steel, Gentle Steel, Copper, Titanium, Nickel amalgam, etc. and AC current is utilized for aluminum, their combination, and magnesium. Control comprises a transformer, rectifier, and some controls for electronics. For the most part, 10–35 V is used at 5–300 for a proper process.

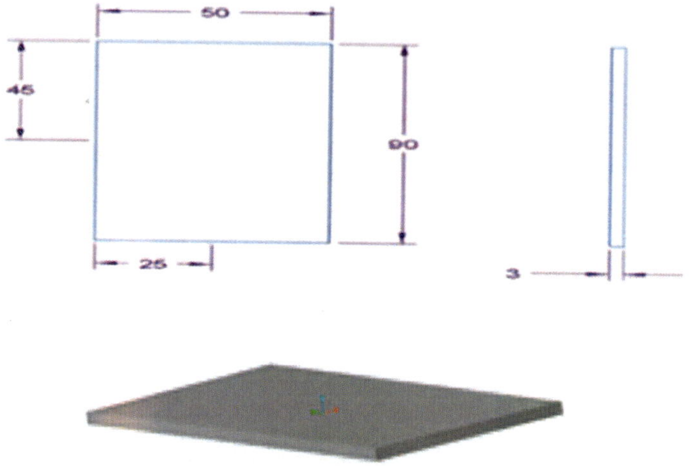


**Fig. 2** Model of the workpiece (all dimensions are in mm)

**Table 1** Chemical composition of Hastelloy C-276

| Element | Ni | Cr | Mb | Fe | W | Co | Va | Si | C | Cu | Mn |
|---|---|---|---|---|---|---|---|---|---|---|---|
| Percentage (%) | 55.69 | 15.20 | 17.76 | 5.70 | 3.83 | 0.30 | 0.20 | 0.12 | 0.031 | 0.082 | 0.42 |

The filler material SS305L consists approximately of more nickel, chromium, and molybdenum and with less carbon content and so it is selected for welding the Hastelloy plates. The filler material and the tungsten electrode are used for welding the plates and it produces heat flow for welding the plates. The mechanized TIG welding is used for welding the specimens. A suitable jig and fixture are used for holding the workpiece. The fixture is shown in Fig. 3. The parameters and their levels are shown in Table 2. The welded specimens are shown in Fig. 4. The methodology is shown in Fig. 4 [5].

The levels are set as per the trial experiments conducted. The design matrix is considered based on three factors and four levels. The maximum and the minimum limits are noted and the levels are fixed as per the design matrix (L16). The input parameters are welding current, voltage, and gas flow. The machine selected for welding the Hastelloy C-276 (nickel-based alloy) material is a TIG welding machine. The mechanized type of welding is performed for the material. The Design matrix (L16 orthogonal array) is shown in Table 3.

The welded specimens of Hastelloy C-276 are shown in Fig. 5. The weld bead geometry study is conducted on the welded specimens by using a machine vision system. The system inspects, identifies, and evaluates the images. In this work, the welded specimens are considered as the image. The front width, front height, and area are measured by using the machine vision system. The value of the front width and the front height is represented by pixels. For the conversion of the pixels into

**Fig. 3** Fixture for welding

**Table 2** Parameters and their levels

| Parameters | Level 1 | Level 2 | Level 3 | Level 4 |
|---|---|---|---|---|
| Welding current (A) | 100 | 120 | 140 | 160 |
| Voltage (V) | 10 | 15 | 20 | 25 |
| Gas flow (lit/min) | 15 | 14 | 12 | 13 |

**Fig. 4** Welded specimens of Hastelloy C-276

mm, a suitable slip gauge is used for the measurement. Figure 6 shows the weld bead geometry of the specimen. Figures 7 and 8 shows the slip gauge arrangement and machine vision setup.

**Table 3** Design matrix (L16 orthogonal array)

| S. No. | Welding current (A) | Voltage (V) | Gas flow (lit/min) |
|---|---|---|---|
| 1 | 100 | 10 | 15 |
| 2 | 120 | 10 | 14 |
| 3 | 140 | 10 | 12 |
| 4 | 160 | 10 | 13 |
| 5 | 100 | 15 | 14 |
| 6 | 120 | 15 | 12 |
| 7 | 140 | 15 | 15 |
| 8 | 160 | 15 | 13 |
| 9 | 100 | 20 | 12 |
| 10 | 120 | 20 | 13 |
| 11 | 140 | 20 | 15 |
| 12 | 160 | 20 | 14 |
| 13 | 100 | 25 | 13 |
| 14 | 120 | 25 | 15 |
| 15 | 140 | 25 | 14 |
| 16 | 160 | 25 | 12 |

# 3 Results and Discussion

Based on the experimentation conducted, the effects of front width, front height, and the gas flow are discussed based on the input parameters.

## *3.1 Effect of Welding Parameter on Front Width*

The front width is measured by using machine vision system for all the runs as per the design of experiments. The front width of the weld bead geometry shows that the level of 160 A welding current, 25 V voltage, and the 15 lit/min gas flow has the highest front width compared to all other runs in the design of experiments. The values of weld bead geometry are shown in Table 4. The main effects plot for the mean is shown in Fig. 9.

## *3.2 Effect of Welding Parameter on Front Height*

The front height is also measured by using a machine vision system with the help of a slip gauge arrangement. The front height shows that the 100 A welding current,

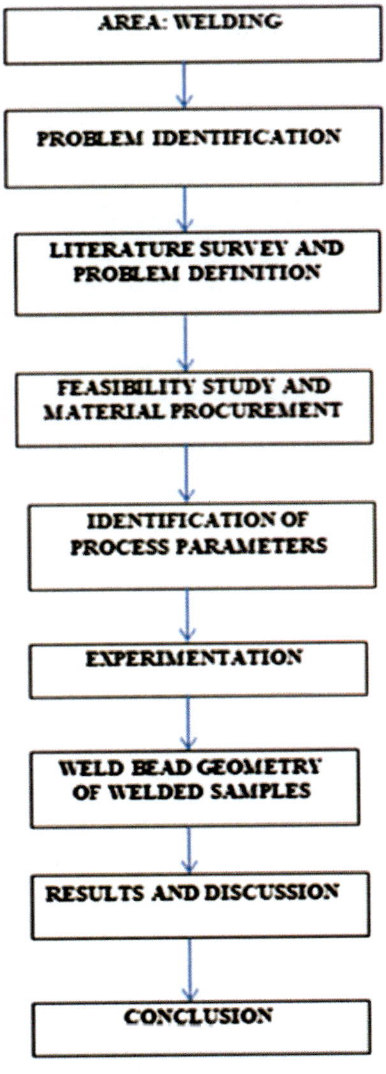


**Fig. 5** Methodology of the TIG welding process

10 V voltage, and 10 lit/min gas flow shows the minimum front height compared to other runs as per the design of experiments. It also shows that 160 A welding current, 25 V voltage, and 15 lit/min gas flow shows the highest front height. It shows that the more the welding current, voltage, and gas flow, higher the front height in the weld bead geometry. The main effects plot for the mean is shown in Fig. 10.

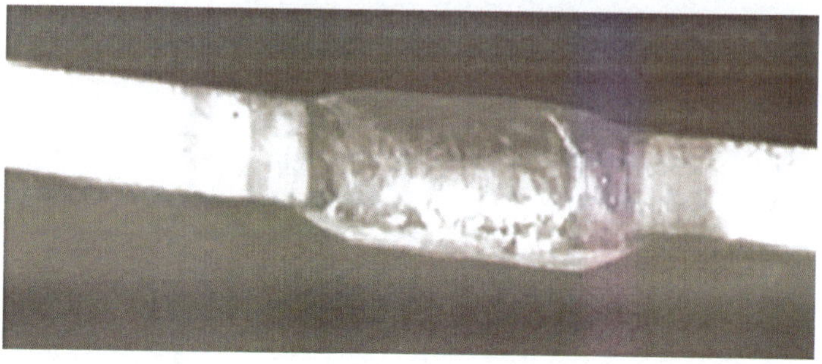

**Fig. 6** Weld bead geometry of the specimen

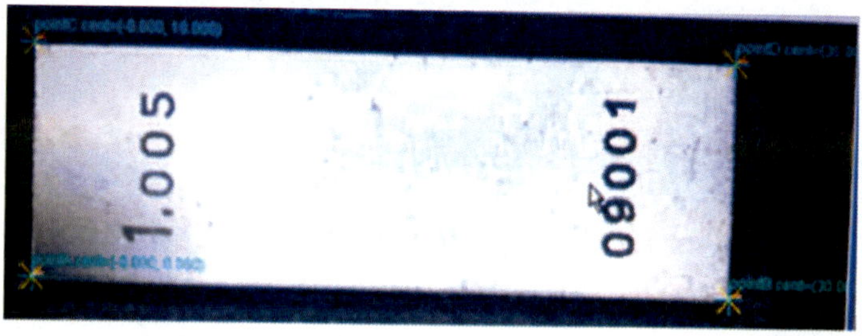


**Fig. 7** Slip gauge arrangement

**Fig. 8** Machine vision setup

**Table 4** Output parameters of weld bead geometry

| S. No. | Front width (mm) | Front height (mm) | Area ($mm^2$) |
|---|---|---|---|
| 1 | 13.413 | 0.914 | 6.706 |
| 2 | 11.540 | 0.785 | 5.77 |
| 3 | 16.622 | 0.592 | 8.31 |
| 4 | 16.449 | 0.359 | 8.22 |
| 5 | 15.050 | 0.358 | 7.52 |
| 6 | 13.866 | 0.889 | 6.93 |
| 7 | 14.992 | 0.706 | 7.49 |
| 8 | 14.009 | 0.214 | 7.00 |
| 9 | 13.870 | 0.329 | 6.93 |
| 10 | 14.069 | 0.718 | 7.03 |
| 11 | 12.909 | 0.698 | 6.45 |
| 12 | 17.753 | 0.465 | 8.87 |
| 13 | 14.150 | 0.155 | 7.07 |
| 14 | 18.709 | 0.392 | 9.35 |
| 15 | 15.132 | 0.197 | 7.56 |
| 16 | 16.118 | 0.950 | 8.05 |

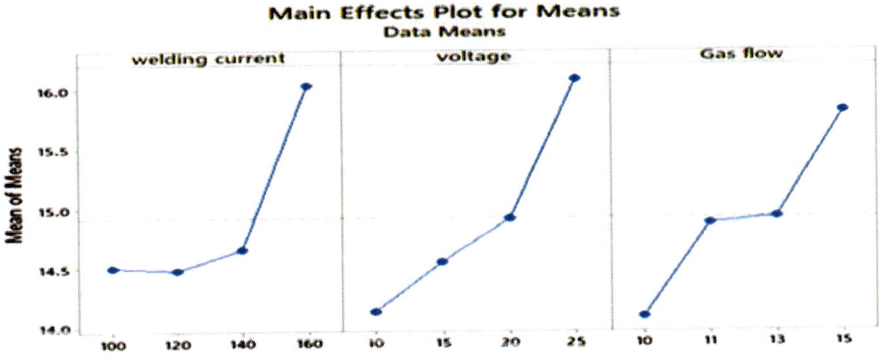


**Fig. 9** Main effects plot for a mean of front width

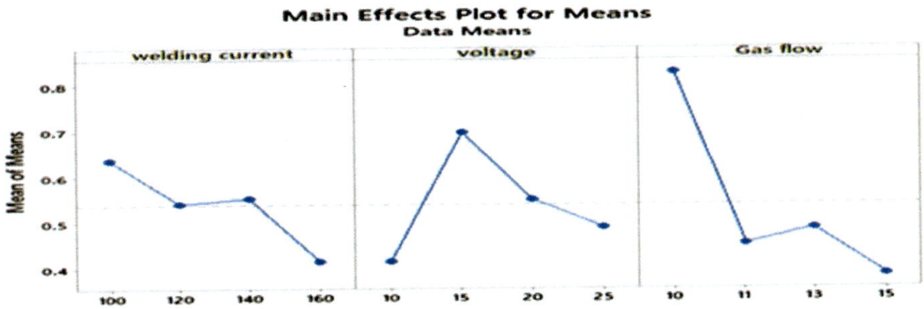


**Fig. 10** Main effects plot for mean for front height

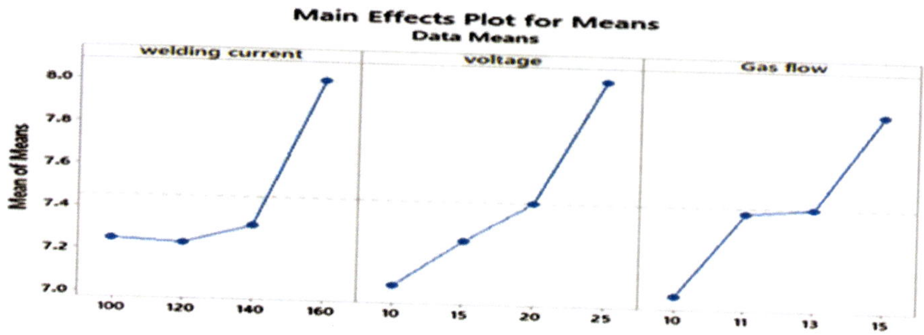


Fig. 11 Main effects plot for mean for area

### 3.3 Effect of Welding Parameter on Area

The area of the weld bead geometry is determined by using machine vision system. The welding parameters 160 A welding current, 25 V voltage, and 15 lit/min gas flow shows the highest area compared to other runs. The reason is that the welding current and the voltage are high compared to other levels of the experiment. The main effects plot for the mean of the area is shown in Fig. 11. The areas of the welded specimens are in an unbounded region and so the machine vision system is applied for the accurate measurement of the area in the weld bead geometry.

## 4 Conclusion

The experimental study was conducted on the process parameters such as welding current, voltage, and gas flow. The study on weld bead geometry is conducted for Hastelloy C-276 material where Tungsten Inert gas (TIG) welding is done by Taguchi L16 orthogonal array. The conclusion from the study is as follows:

- The influence of welding current, voltage, and gas flow on the front width is higher for the higher levels since the input parameters levels are higher compared to other levels of parameters.
- The influence of welding current, voltage, and gas flow on the front height is higher for the medium level of input parameters since the levels are at a moderate level compared to other levels.
- The influence of welding current, voltage, and gas flow on the area is higher for the high level of input parameters since there is a wide range of areas due to higher welding current.

**Acknowledgements** Sincere thanks to management and staff for giving me moral support and guidance throughout my project work and also for giving me the technical support to improve my skills and for their guidance.

## References


1. Ramkumar D, Joshi V, Pandit S, Agrawal M, Kumar OSM (2014) Investigations on the microstructure and mechanical properties of multi pass pulsed current gas tungsten arc weldments of Monel 400 and Hastelloy C-276, 64:775–782
2. Zhu Y, Hou J, Yu G, Qiu J, Chen S, Zhou X (2016) Effect of exposing temperature on corrosion performance of weld joints of Ni-Mo-Cr alloy, 18:269–275
3. Hussain AK, Lateef A, Javed M, Pramesh T (2015) Influence of welding speed on Tensile strength of welded joint in TIG welding process. 1(3)
4. Parasiya JR, Sathavara K, Trevedi PT (2015) Effect of process parameters on mechanical and metallurgical properties of dissimilar metal welds using GTAW process. 2(6)
5. Hedayati O, Korei N, Adeli M, Etminanbakhsh M (2017) Microstructural evolution and interfacial diffusion during heat treatment of Hastelloy/stainless steel bimetals. 712:172–178
6. Pandit S, Joshi V, Agrawal M, Manikandan M, Ramkumar D (2017) Investigations on mechanical and metallurgical properties of dissimilar continuous GTA welds of Monel 400 and C-276. 75:61–65

# Optimization of Particle Size of Teak Wood Saw Powder Using Taguchi Method

**P. K. Palani, K. Chithambaram, and B. Rajeswari**

**Abstract** Teak is a tropical wood species and placed in the flowering plant family Lamiaceae. It has a high degree of natural durability, moderately hard, excellent decay resistance, and good dimensional stability. The major applications of the teak wood are exterior and interior, indoor and outdoor furniture, ship decks. The major issue in wooden doors is that it absorbs moisture over a period of time and results in condensation where the stability gets decreased. Also during the high temperature season, the wood material undergoes expansion. But the wooden materials made from the micro-particles provide a greater strength than made from the normal size particles. In this project, teak wood saw particles are collected as a waste material from wooden design door industry and the size of the particles is reduced in ball mill machine. Optimization of these wood particles involving the various factors was carried out by using Taguchi method. Then the micro-wood particles are finally obtained by using sieve. The results indicate that the samples obtained from the factors having speed 150 rpm, 18 numbers of balls, and a period of 2 h provides the optimum size.

**Keywords** Windows · Optimization · Ball milling · Sieve analysis · Taguchi method and parameters

---

P. K. Palani · K. Chithambaram (✉) · B. Rajeswari
Department of Mechanical Engineering, Government College of Technology, Coimbatore, Tamil Nadu, India
e-mail: chithambaram97@gmail.com

P. K. Palani
e-mail: pkpalaniku@gmail.com

B. Rajeswari
e-mail: rajes.aravind@gmail.com

© The Editor(s) (if applicable) and The Author(s), under exclusive license to Springer Nature Singapore Pte Ltd. 2021

S. Mohan et al. (eds.), *Materials, Design, and Manufacturing for Sustainable Environment*, Lecture Notes in Mechanical Engineering,
https://doi.org/10.1007/978-981-15-9809-8_33

# 1 Introduction

Teak is tropical hard wood tree species which is widely used in construction of wooden doors and windows. It has high degree of natural durability, moderately hard, excellent decay resistance, and good dimensional stability. The major issue in wooden doors is that it absorbs moisture over a period of time [1, 2] and results in condensation where the stability gets decreased [3]. During high-temperature season, it undergoes expansion [4]. The main objective is to obtain the minimum particle size of teak wood saw by optimization using Taguchi method where the sample is milled by ball mill and analyzed using sieve [5] (Fig. 1).

## 1.1 Taguchi Method

Taguchi method is a quality control method which is primarily used to design methods for the improvement in the quality of goods and development at its initial stage. This method has greater influence in sorting out the defects and failures so as to improve the quality. Many industrial fields use this method for their optimizing need. It provides a greater performance on the quality by governing the process at its design stage which is an offline function [6].

## 1.2 Planetary Ball Mill

The planetary ball mill is a grinding machine used to pulverize the hard ceramic materials to fine particles. The milling operation is based upon the planetary motion

**Fig. 1** Teak wood saw

**Fig. 2** Planetary ball mill

of the jar give the maximum grinding energy to the grinding media available in the grinding jar and makes the material to very fine particles [7] (Fig. 2).

## *1.3 Sieve*

Sieve analysis is the technique used to eliminate the particles that are larger in size in the group of complex sized particles. These particles are differentiated by the varying hole size in the sieves. The smaller size particles are obtained after successive sieving of particles [7] (Fig. 3).

## *1.4 Research Gap*

Based on the literature review, it was found that limited work has been carried out in reducing the particle size of teak wood saw powder materials that are made from smaller particles provides greater strength and stability ball mill makes a greater impact [8] in reducing the particle size in a greater level wood saw particles remained as a waste can be recycled and reused [9].

**Fig. 3** Sieve

## 2 Design of Experiments

A systematic experimental design that performs the task to determine the factors that provide the expected outcome. Initially, the independent variables should be given as input to obtain the output by relation with those variables. It also involves the design of conducting the experiment in a optimal condition so this information is needed to manage process inputs in order to optimize the output [6] in which they are given in Tables 1 and 2.

**Table 1** Varying process parameters

| Parameters | Values | Unit |
|---|---|---|
| No. of balls | 14, 18 | – |
| Speed | 125, 150 | RPM |
| Milling hour | 1, 2 | Hr |

**Table 2** Design of experiment

| Runs | No. of Balls | Speed (RPM) | Time (h) |
|---|---|---|---|
| 1 | 14 | 125 | 1 |
| 2 | 14 | 125 | 2 |
| 3 | 14 | 150 | 1 |
| 4 | 14 | 150 | 2 |
| 5 | 18 | 125 | 1 |
| 6 | 18 | 125 | 2 |
| 7 | 18 | 150 | 1 |
| 8 | 18 | 150 | 2 |

## 2.1 *Taguchi Analysis*

Taguchi analysis makes the design to minimize the variation instead of performing the entire data values [6]. The analysis involves in a two successive path. The first path involves on identifying the minimal variability, and the second path concentrates on performing that minimal parameter task. The notation of Taguchi design for this experiment is given as Level $(A)$ $(A \wedge B)$, and it informs the following:

$$\text{Level}(A) = \text{number of runs}(A \wedge B)$$

$A =$ number of levels for every single factor,

$B =$ exponent factors. It follows L8 Orthogonal array.

### 2.1.1 Signal-to-Noise Ratio

Here Table 3 shows the ratio in which in provides information of essential data.

**Table 3** Signal-to-noise ratio

| Signal-to-noise ratio | Experiment goal | Essential data |
|---|---|---|
| Larger size is better | Highest response | Positive value |
| Smaller size is better | Least response | Non negative value with target value zero |

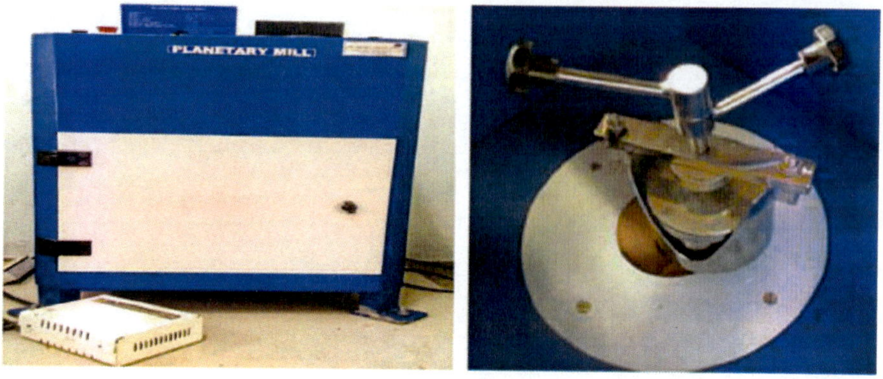


Fig. 4 Ball mill apparatus

# 3 Experimental Work

## 3.1 Blending

Introduce the powders (both matrix and reinforcement) into the ball mill vase. Balls were introduced into the vial, and the vial is closed. Parameters such as milling time and milling speed are set. After the time is completed, the vial is removed. The balls are removed first, and then the milled powders are collected and the next set of samples was introduced. The process is continued for eight samples (Fig. 4).

### 3.1.1 Process Parameters for Milling

Milling Hour—(1, 2) hours.

Rotational Speed—125, 150 RPM.

No. of Balls—14, 18.

Ball diameter—10 mm.

Ball Material—Tungsten Carbide.

Milling Vial Material—Tungsten carbide (Figs. 5, 6 and 7).

## 3.2 Sieve

Sieve often involves the elimination of bigger particles. Sieve is commonly available in various sizes. For this analysis over the teak wood saw powder, the least available sieve of size about less than 75 microns is taken. The individual sieves that are placed

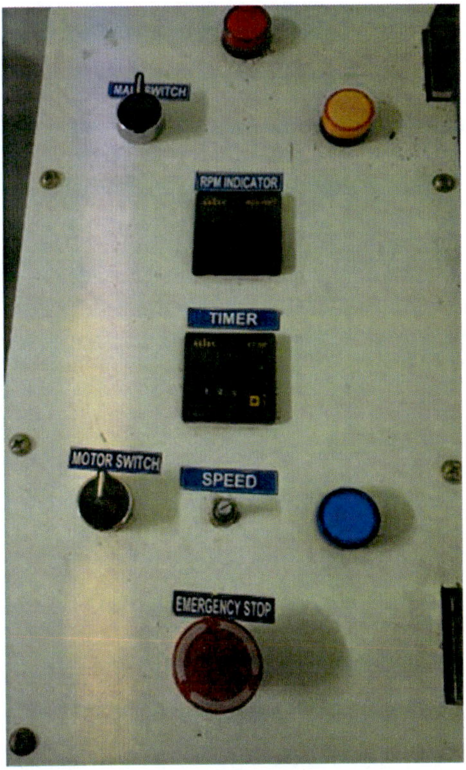


**Fig. 5** Setting of parameters

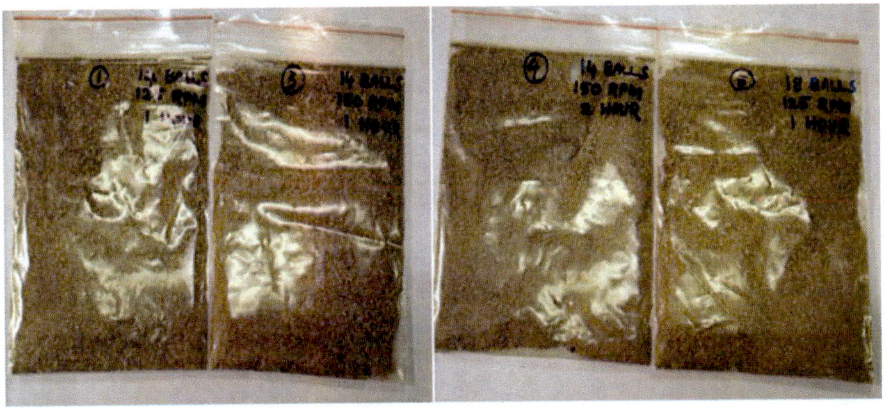

**Fig. 6** Weighed sample powders

**Fig. 7** Milled sample powders

separately in a sequential order of their size. The maximum sieve size in this analysis is above 600 mm, and the least sieve size in this analysis is less than 75 microns (Fig. 8).

### 3.2.1 Sieve Analysis

In every individual sieves, the sieve analysis is carried out to eliminate the particles of greater size with respective to the others. Figure 10 shows the teak wood saw sample which is obtained after the sieve analysis in respective of 75 micron particle size. These provide the different amounts of weight for its varying parameters. In these samples, sample 6 shows higher amount of sieve particles (Fig. 9).

**Fig. 8** Individual sieves

# 4 Results and Discussion

## *4.1 Particle Size Analysis*

These are samples that are collected after the sieve analysis. These collected samples are in the range of less than 75 microns and in different grams of weight. The sample with the higher grams in weight shows the optimum run in the planetary ball mill. Table 4 shows the result of weight of sample after sieve and SNRA1 ratio (Fig. 11 and Tables 5 and 6).

**Fig. 9** Sieving the sample

### *4.2 Optimal Values*

In all the response tables, milling hour has the highest delta rank which indicates it has largest effect (Table 7).

## 5 Conclusion

The results indicate that milling hour has more effect on the signal-to-noise ratio followed by the speed of rotation. Taguchi method was used to optimize the process parameters to obtain the minimum particle size of teak wood saw powders. The optimal parameters obtained are 2 h of milling, 18 number of balls, and speed at 150 RPM.

## 6 Future Scope

It is clear that the particle size can be reduced by varying the process parameters. Finalizing the optimal parameters for the optimization of particle size of teak wood saw powder need some more analysis and discussion. Since some more parameters

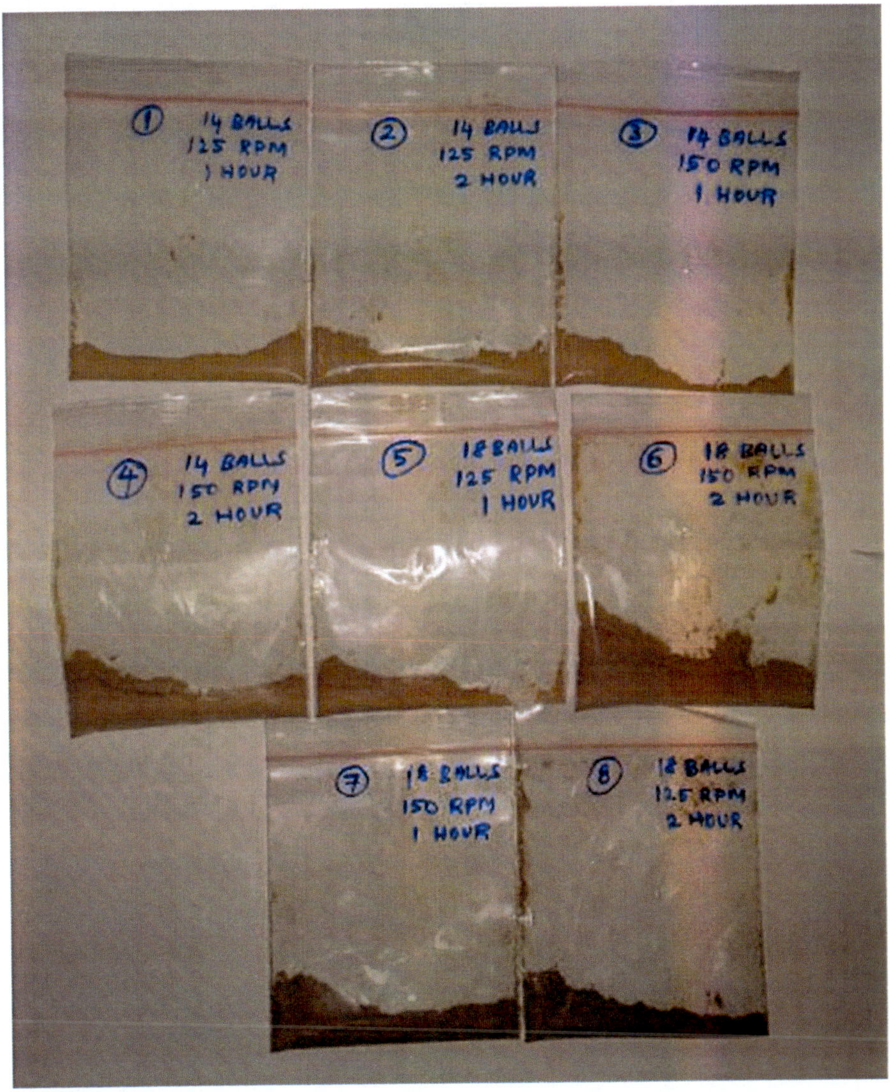

**Fig. 10** Sieved sample

such as varying the sample weight, ball diameter and milling atmosphere can affect the particle size [10]. In the present work, the higher size reduction parameter has been evaluated. Further, it can be extended to study the properties of materials that are made from these wood particle composites [9].

**Table 4** Analysis result

| Runs | No of Balls | Speed (RPM) | Time (Hr) | Grams after Sieve (g) | SNRA1 |
|---|---|---|---|---|---|
| 1. | 14 | 125 | 1 | 0.8 | -1.93820 |
| 2. | 14 | 125 | 2 | 1.5 | 3.52183 |
| 3. | 14 | 150 | 1 | 1.1 | 0.82785 |
| 4. | 14 | 150 | 2 | 2.0 | 6.02060 |
| 5. | 18 | 125 | 1 | 1.0 | 0.00000 |
| 6. | 18 | 125 | 2 | 1.7 | 4.60898 |
| 7. | 18 | 150 | 1 | 1.1 | 0.82785 |
| 8. | 18 | 150 | 2 | 3.0 | 9.54243 |

Colour indication is to make some difference between the optimal value and the normal values. To show the values unique i.e, the highest and lowest

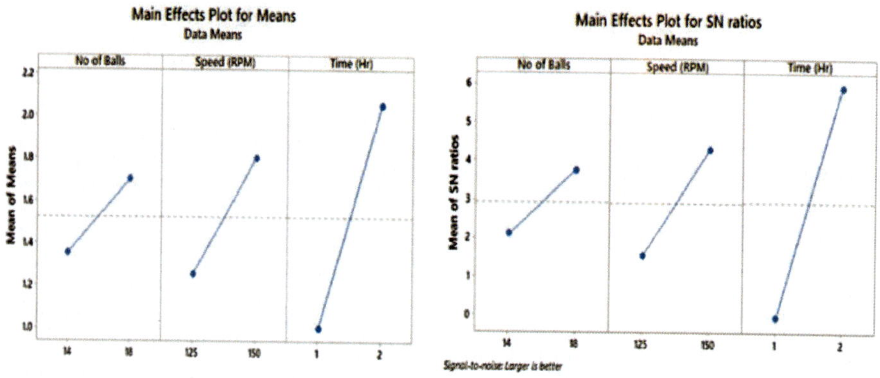


**Fig. 11** Graphically plotted data

**Table 5** Response to signal-to-noise ratio

| Level | No. of balls | Speed (RPM) | Time (h) |
|---|---|---|---|
| 1 | 2.10802 | 1.54815 | −0.07062 |
| 2 | 3.74481 | 4.30468 | 5.92346 |
| Delta | 1.63679 | 2.75653 | 5.99408 |
| Rank | 3 | 2 | 1 |

**Table 6** Response to mean

| Level | No. of balls | Speed (RPM) | Time (h) |
|---|---|---|---|
| 1 | 1.350 | 1.250 | 1.000 |
| 2 | 1.700 | 1.800 | 2.050 |
| Delta | 0.350 | 0.550 | 1.050 |
| Rank | 3 | 2 | 1 |

**Table 7** Optimal values

| No. of balls | Milling hour (h) | Speed (RPM) |
|---|---|---|
| 18 | 2 | 150 |

## References


1. Nevander J (1987) Closing remarks. In: Proceedings of windows in building design and maintenance, gothenburg, sweden, sweden council of building research, pp 133–134
2. Ortiz O, Pasqualino JC, Diez G, Castells F (2010) The environmental impact of the construction phase: an application to composite walls from a life cycle perspective. Resour Conserv Recycl 54:832–840
3. Kalita K, Sharma U (2019) Studies on change of strain developed in different wood samples due to change in relative humidity. Sens Bio-sens Res 22
4. Sanders CH (1982) Thermal insulation and condensation, building research establishment news, no 55
5. Minne E, Wingrove K, Crittenden JC (2015) Influence of climate on the environmental and economic life cycle assessments of window options in the United States. Energy Build 102:293–306
6. Ng'andwe P (2003) Timber certification, optimization and value added wood products—a case study for Zambia. University of Wales
7. Fu S-Y, Feng X-Q, Lauke B, Mai Y-W (2008) Effects of particle size, particle/matrix interface adhesion and particle loading on mechanical properties of particulate-polymer composites. Compos Part B Eng 39:933–961
8. Salazar J, Sowlati T (2008) A review of life cycle assessment of windows. For Prod J 58:91–96
9. Craighill AL, Powell JC (1996) Life cycle assessment and economic evaluation of recycling: a case study. Resour Conserv Recycl 17:75–96
10. Srinivasababu N (2019) Edge crack effect on tensile behaviour of diversified wood particulate composites. Wood head publishing series, pp 109–131

# Optimization of Machining Parameters in Drilling Ti–6Al–4V Using User's Preference Rating-Based TOPSIS

**S. Samsudeensadham, A. Mohan, R. ArunRamnath, and R. Keshav Thilak**

**Abstract** Titanium alloys have gained prominence in aerospace and automotive industries due to higher strength to weight ratio and lesser weight. Titanium alloys employed in diverse applications cannot be produced as near net shape components and require holes and other machining processes for structural assembly. Drilling the most conventional machining process is adopted for hole making in titanium alloys. Titanium alloy Ti–6Al–4V is termed as hard to machine materials. In this experimental investigation, drilling experiments are carried out on Ti–6Al–4V with solid carbide drill bits as tool material of 8 mm diameter. Drilling experiments with 16 different trials are carried out by Taguchi's experimental design. Speed and feed rate are the two input parameters considered ranging from 20 to 50 rpm and 0.05 to 0.2 m/rev. The response attributes measured from experimental investigations are surface roughness, burr thickness, hole diameter and circularity. With an objective of enhancing the drilling process and achieve the desired machining characteristics, machining parameters are optimized based on TOPSIS model. From preference ranking among the alternatives, experimental run 4 is determined as the optimized solution. Optimal machining parameter combinations are obtained at speed 20 rpm and feed 0.2 m/rev.

**Keywords** Titanium alloy · Drilling · Surface roughness · Burr thickness · TOPSIS

S. Samsudeensadham · A. Mohan · R. ArunRamnath (✉) · R. Keshav Thilak
Department of Mechanical Engineering, PSG College of Technology, Peelamedu, Coimbatore, Tamil Nadu, India
e-mail: arunraamnath.89@gmail.com

S. Samsudeensadham
e-mail: samsudeensadham@gmail.com

A. Mohan
e-mail: mohan3mmm@gmail.com

R. Keshav Thilak
e-mail: thilakr26@gmail.com

© The Editor(s) (if applicable) and The Author(s), under exclusive license to Springer Nature Singapore Pte Ltd. 2021

S. Mohan et al. (eds.), *Materials, Design, and Manufacturing for Sustainable Environment*, Lecture Notes in Mechanical Engineering,
https://doi.org/10.1007/978-981-15-9809-8_34

# 1 Introduction

Titanium alloys are preferred for high performance aerospace applications because of its higher strength to weight ratio and lightweight. Also the low modulus of elasticity and thermal conductivity nature made these alloys as difficult to cut materials [1]. Drilling titanium alloys with a suitable tool material and ideal machining parameter combination is a critical and recent research topic in manufacturing process which has gained higher importance. Research investigations on machining titanium alloy predominantly focus on the enhanced surface integrity with minimum surface roughness values [2]. Drilling studies on titanium alloy and quality of the drilled components are evaluated in terms of parameters such as circularity, burr thickness, surface roughness and the hole diameter. Titanium alloys are found in aerospace applications with the need of higher reliability, and hence, the surface roughness and other surface damages have to be minimized. The study on drilling of titanium alloys is carried out with the view of minimized surface roughness, reduced burr thickness and enhanced hole quality [3, 4]. Surface roughness plays a critical role in measure of quality in machining process. Authors have investigated the impact of speed and feed rate on temperature and surface roughness while machining the titanium alloy grade 5. It is reported that speed is the most influential parameter for temperature and surface roughness, and this could be attributed to wiping action during end milling of titanium alloy [5, 6]. Research problems involving trade-off among attributes are often being solved by employing decision-making models. Sarbjit Singh [7] and co-authors applied grey relation analysis in the selection of optimum machining parameter combination in drilling metal matrix composites. Many other decision-making models such as TOPSIS, AHP, VIKOR, ENTROPY and other hybrid decision-making models are employed by several researchers depending on the applications [8–11]. Arun Ramnath [12] machined epoxy granite a hard-to-cut material and determined the optimum machining parameter combination in end milling based on TOPSIS model. Three attributes with contrasting characteristics are considered, and the optimum solution is determined. In this research investigation, four different attributes, namely surface roughness, burr thickness, circularity and hole diameter, are considered, and the machining parameters are optimized. In this research study, optimal solutions are determined by TOPSIS a decision-making model. Selection of optimum machining combination is a critical and demanding task among limited experimental trials.

# 2 Experimental Procedure

## *2.1 Materials and Methods*

Titanium alloy (Ti–6Al–4V) with heat treatment processed (annealed-grade 5) material was preferred for drilling with dimensions of 250 mm (length) × 50 mm (breadth)

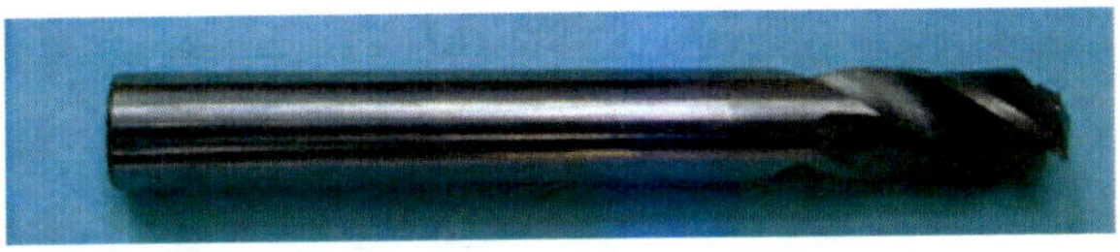

**Fig. 1** Solid carbide drilling tool

**Table 1** Design of experiments

| Experiments | Cutting speed (m/min) | Feed (mm/rev) |
|---|---|---|
| 1–4 | 20 | 0.05, 0.10, 0.15, 0.20 |
| 5–8 | 30 | |
| 9–12 | 40 | |
| 13–16 | 50 | |

× 8 mm (thickness). In order to remove the oxide layer formation in the workpiece material, a grinding process was performed using hydraulic surface grinding machine. All the experiments were carried out using VMC VF-30 with speed of 6000 rpm. Solid carbide drill tool had been selected for the drilling process with the diameter of 8 mm and point angle of 120° as shown in Fig. 1. This tool has a shank and flute length of 50 mm and 25 mm, respectively.

In this experimental investigation, full factorial design (2 factors and 4 levels) had been chosen to design the experiments and conduct the same. Table 1 shows the design of experiments. After conducting the experiments, the responses such as circularity, hole diameter error, surface roughness and burr thickness were measured.

A coordinate measuring machine (CMM) made by CONTURA G2—Zeiss as shown in Fig. 2 with a working range of 700 mm ($X$), 1000 mm ($Y$) and 600 mm ($Z$) and a maximum measuring points of 500 no.'s to measure the circularity and hole diameter. Calypso software was integrated with the CMM machine to acquire and analyse the data. Surface roughness was measured using handheld compact type SURFTEST SJ-210 Series surface roughness tester as shown in Fig. 3. TOPSIS method has been applied by numerous authors in solving user preference rating of optimization problems.

**Stage 1**: TOPSIS considers equal weights of each attribute in a decision matrix and is derived as:

$$R_x = \begin{bmatrix} y_{11} & y_{12} & \cdots\cdots & y_{1n} \\ y_{21} & y_{22} & \cdots\cdots & y_{2n} \\ \vdots & \vdots & \ddots\ddots & \vdots \\ \vdots & \vdots & \ddots\ddots & \vdots \\ y_{m1} & y_{m2} & \cdots\cdots & y_{mn} \end{bmatrix} \quad (1)$$

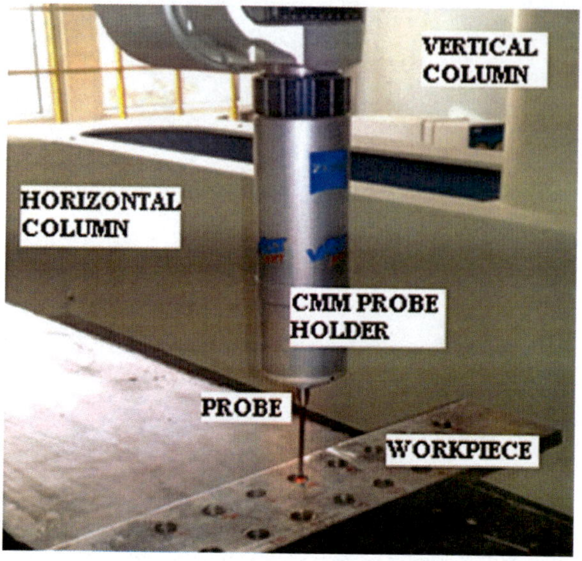


**Fig. 2** Circularity measurement set-up

**Fig. 3** Surface roughness measurement set-up

where $y_{ij}$ denotes responses of the subsequent experimental runs.

**Stage 2**: Normalized decision matrix of the attributes is computed based on the expression below:

$$q_{ij} = \frac{y_{ij}}{\sqrt{\sum_{j=1}^{n} y_{ij}^2}}, \quad j = 1, 2, \ldots, n \tag{2}$$

**Stage 3**: Weighted normalized matrix of the attributes is determined as:

$$I = w_j q_{ij} \quad \text{where} \sum_{j=1}^{n} Wj = 1 \tag{3}$$

**Stage 4**: Positive ideal solution ($I^+$) and negative ideal solution ($I^-$) of each attribute were determined as per the expressions below:

$$I^+ = \left\{ \left( \sum_i^{\max} i_{ij} | j \in J \right), \left( \sum_i^{\max} i_{ij} | j \in J | i = 1, 2, \ldots, m \right) \right\} \tag{4}$$

$$I^+ = \{ i_1^+, i_2^+, i_3^+, \ldots, i_n^+ \}$$

$$I^- = \left\{ \left( \sum_i^{\min} i_{ij} | j \in J \right), \left( \sum_i^{\max} i_{ij} | j \in J | i = 1, 2, \ldots, m \right) \right\}$$

$$I^- = \{ i_1^-, i_2^-, i_3^-, \ldots, i_n^- \} \tag{5}$$

**Stage 5**: Alternative distances ($D_i{}^+$) and ($D_i{}^-$) of ideal solution are found from the below expression:

$$D_i^+ = \sqrt{\sum\nolimits_{j=1}^{n} (i_{ij} - i_j^+)^2}, \quad i = 1, 2, \ldots, m \tag{6}$$

$$D_i^- = \sqrt{\sum\nolimits_{j=1}^{n} (i_{ij} - i_j^-)^2}, \quad i = 1, 2, \ldots, m \tag{7}$$

**Stage 6**: Preference values of every alternative are estimated from Eq. (8):

$$P_i = \frac{D_i^-}{D_i^+ + D_i^-}, \quad i = 1, 2, \ldots, m \tag{8}$$

# 3 Results and Discussions

## *3.1 Investigation of Machining Parameters on Drilling Responses*

Circularity is defined as the degree of roundness that resulting from the various combinations of machining parameters such as speed and feed while drilling of titanium alloy. In this study, it was measured by a CMM. Also the same measuring set-up was used for acquiring the hole diameter data to analyse the quality of drilled hole. Figures 4 and 5 show the three-dimensional surface plot of effects of drilling parameters on circularity and diameter error. Considering the hole diameter, there is a fall in hold diameter error at lower speeds from 20 to 30 m/min. When cutting speed increases, there is sudden rise in diameter error which leads up to 8.2 mm which could be the reason of increased temperature effects during high speed conditions. After reaching the maximum diameter error at 40 m/min speed, there is radical drop in error due to stability of the spindle speed with the workpiece that reaching the plastic deformation condition. While considering the circularity plot in Fig. 5, it is clearly understood that there is a contribution of both cutting speed and feed rate. The lower conditions of both machining parameters lead to increase in circularity error due to an abrupt uncontrolled movement action of drill bit. Comparing these geometrical characteristics, the feed rate is influential parameter only on circularity.

Surface roughness and burr thickness are considered as important quality characteristics in the assembly of bolted joints on aircraft structures. The roughness plot as shown in Fig. 6 represents the nonlinear surface characteristics in variation with cutting speeds, but the feed rate has produced consistent progression in surface roughness. Burr thickness was measured for all the 16 set of holes which produces different forms of burrs. The plot as shown in Fig. 7 represents that there is a linear increment in burr thickness with regard to increase in cutting speed especially at the speed range of 30–50 m/min. It seems cutting speed has substantial effect on burr thickness compared to feed rate. This could be because of thermal expansion of workpiece material due to high temperature during high speed conditions.

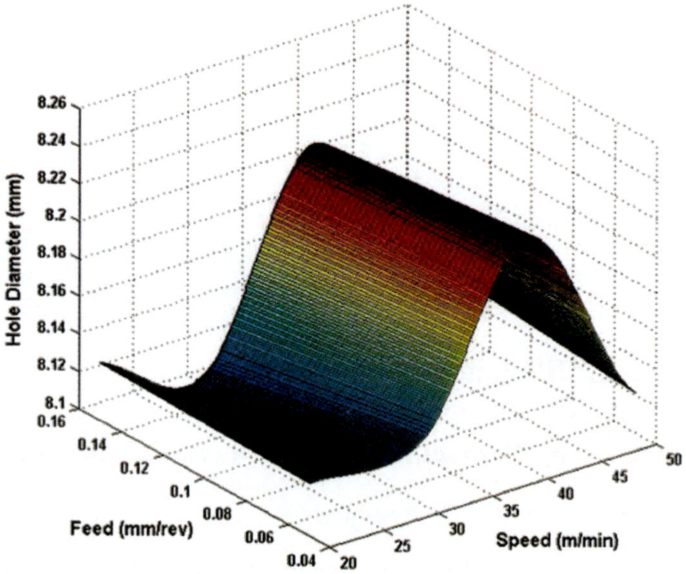


**Fig. 4** Effect of machining parameters on hole diameter

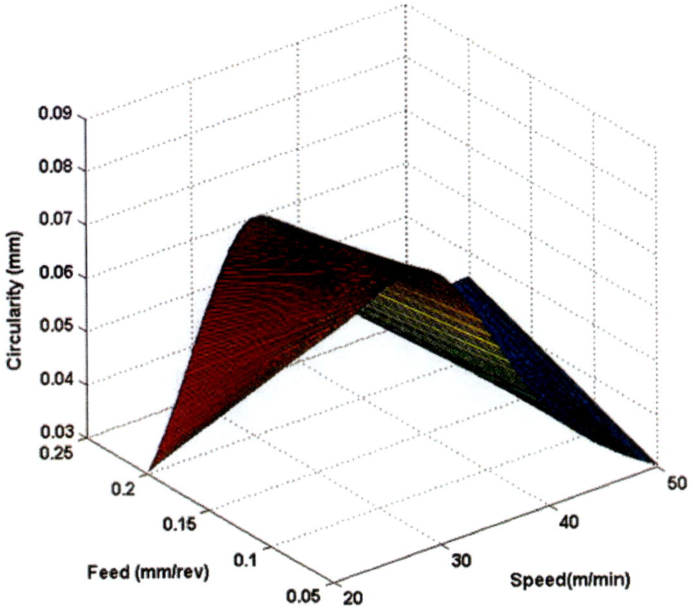


**Fig. 5** Effect of machining parameters on circularity

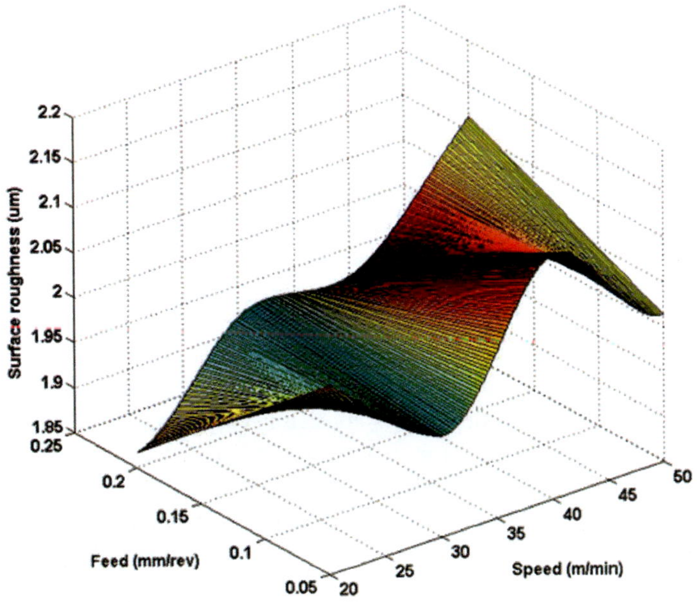


**Fig. 6** Effect of machining parameters on surface roughness

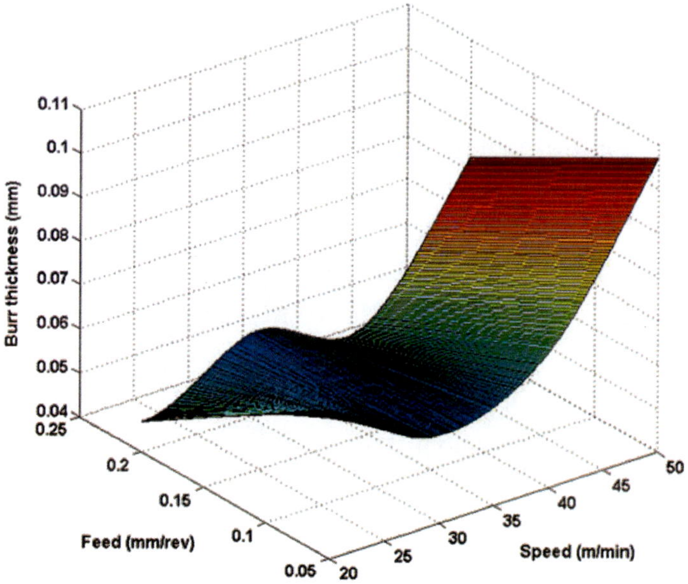


**Fig. 7** Surface effect of machining parameters on burr thickness

## 3.2 *Optimization of Machining Parameters*

The measured values of hole diameter, circularity, surface roughness and burr thickness are listed in Table 2. In order to obtain an optimal solution, consisting of multiple attributes S/N ratio is no longer valid and hence not applied in this research investigation. A unique solution with machining parameter combination is determined with the aid of TOPSIS a decision-making model. A unique solution cannot be obtained by Taguchi method due to the trade-off among attributes, and only single-objective optimization problems can be solved. Optimization of drilling of titanium alloy consisting multiple attributes like hole diameter, circularity, surface roughness and burr thickness was performed using TOPSIS.

The initial stage in determination of optimal solution is formulation of decision matrix. The formulated decision matrix consisting four different attributes is shown in Table 2. Further stage in optimization is development of normalized decision matrix. Attributes are in varying scale of units and normalized by vector approach. Normally in TOPSIS method, the response values are vector normalized. The normalized decision matrixes of the measured responses are depicted in Table 3. In TOPSIS method, uniform weights are distributed among all the attributes. In this research problem with four different attributes, the weights of each attribute are considered as 0.25. Based on the weights of attributes, the weighted normalized decision matrix is developed. From the weighted decision matrix, ideal solutions are identified. Four different attributes considered are non-beneficial, and hence, minimum values are

**Table 2** Decision matrix of attributes

| Experiment run | Hole diameter (mm) | Circularity (mm) | Surface roughness (μm) | Burr thickness (mm) |
|---|---|---|---|---|
| 1 | 8.128 | 0.0773 | 2.06 | 0.074 |
| 2 | 8.1654 | 0.0888 | 2.197 | 0.086 |
| 3 | 8.1308 | 0.0364 | 1.912 | 0.046 |
| 4 | 8.0977 | 0.0305 | 1.867 | 0.047 |
| 5 | 8.1304 | 0.0805 | 1.962 | 0.060 |
| 6 | 8.1999 | 0.098 | 1.973 | 0.058 |
| 7 | 8.1092 | 0.0539 | 1.858 | 0.033 |
| 8 | 8.2793 | 0.071 | 1.984 | 0.059 |
| 9 | 8.225 | 0.048 | 2.121 | 0.071 |
| 10 | 8.211 | 0.0401 | 1.989 | 0.084 |
| 11 | 8.2112 | 0.046 | 1.875 | 0.045 |
| 12 | 8.1745 | 0.0454 | 1.987 | 0.052 |
| 13 | 8.1212 | 0.0306 | 2.014 | 0.107 |
| 14 | 8.1944 | 0.0345 | 2.121 | 0.122 |
| 15 | 8.1316 | 0.0464 | 1.997 | 0.084 |
| 16 | 8.0848 | 0.0455 | 2.115 | 0.083 |

**Table 3** Normalized decision matrix of attributes

| Experiment run | Hole diameter | Circularity | Surface roughness | Burr thickness |
|---|---|---|---|---|
| 1 | 0.2489 | 0.3309 | 0.2569 | 0.2526 |
| 2 | 0.2500 | 0.3801 | 0.2740 | 0.2936 |
| 3 | 0.2490 | 0.1558 | 0.2384 | 0.1570 |
| 4 | 0.2480 | 0.1305 | 0.2328 | 0.1604 |
| 5 | 0.2490 | 0.3446 | 0.2447 | 0.2048 |
| 6 | 0.2511 | 0.4195 | 0.2460 | 0.1980 |
| 7 | 0.2483 | 0.2307 | 0.2317 | 0.1126 |
| 8 | 0.2535 | 0.3039 | 0.2474 | 0.2014 |
| 9 | 0.2518 | 0.2054 | 0.2645 | 0.2424 |
| 10 | 0.2514 | 0.1716 | 0.2480 | 0.2867 |
| 11 | 0.2515 | 0.1969 | 0.2338 | 0.1536 |
| 12 | 0.2503 | 0.1943 | 0.2478 | 0.1775 |
| 13 | 0.2487 | 0.1309 | 0.2512 | 0.3653 |
| 14 | 0.2509 | 0.1476 | 0.2645 | 0.4165 |
| 15 | 0.2490 | 0.1986 | 0.24908 | 0.2867 |
| 16 | 0.2476 | 0.1947 | 0.2638 | 0.2833 |

identified as the positive ideal solution. Positive and negative ideal solutions of each attribute were identified and are presented in Table 4.

From the computed values of ideal solutions, the separation distances of alternatives are evaluated. The separation distances of each alternative and the corresponding preference values calculated are listed in Table 5. From the preference values estimated, the optimal solutions are determined in drilling titanium alloys. The optimal machining parameter combinations of alternatives are arranged in descending order and are as follows: 4 > 3 > 11 > 7 > 12 > 9 > 10 > 16 > 15 > 8 > 13 > 5 > 14 > 1 > 6 > 2. The optimal process parameter combination is obtained for the experimental run 4. It is preferred to choose the experimental run 4 to achieve the lower magnitudes of surface roughness, burr thickness, hole diameter and circularity. The optimal solution obtained in experimental alternative 4 has a machining parameter combination with cutting conditions such as speed of 20 rpm and feed rate of 0.2 m/min.

**Table 4** Ideal solutions of attributes

| Attributes | Positive ideal solution | Negative ideal solution |
|---|---|---|
| Hole diameter | 0.0619 | 0.0633 |
| Circularity | 0.0326 | 0.1048 |
| Surface roughness | 0.0579 | 0.0685 |
| Burr thickness | 0.0281 | 0.1041 |

**Table 5** Separation distance and preference values of alternatives

| Experimental run | Separation distance | | Preference value |
|---|---|---|---|
| | Positive ideal solution | Negative ideal solution | |
| 1 | 0.06146 | 0.04674 | 0.4319 |
| 2 | 0.07783 | 0.03223 | 0.2928 |
| 3 | 0.01292 | 0.09289 | 0.8778 |
| 4 | 0.01200 | 0.09703 | 0.8899 |
| 5 | 0.05836 | 0.05660 | 0.4923 |
| 6 | 0.07539 | 0.05505 | 0.4220 |
| 7 | 0.02500 | 0.09009 | 0.7827 |
| 8 | 0.04883 | 0.06143 | 0.5571 |
| 9 | 0.03839 | 0.06899 | 0.6424 |
| 10 | 0.04496 | 0.07018 | 0.6095 |
| 11 | 0.019560 | 0.08666 | 0.8158 |
| 12 | 0.02305 | 0.08240 | 0.7814 |
| 13 | 0.06339 | 0.07345 | 0.5367 |
| 14 | 0.07618 | 0.06794 | 0.4714 |
| 15 | 0.04697 | 0.06433 | 0.5779 |
| 16 | 0.04632 | 0.06537 | 0.5852 |

## 4 Conclusions

In this investigation, selection of optimal machining parameter combination in drilling titanium alloy is discussed by employing TOPSIS decision-making model. The conclusions and research findings in this experimental investigation are as follows:

- Considering the circularity results, it is clearly understood that there is a contribution of both cutting speed and feed rate.
- The lower condition of both cutting conditions leads to increase in circularity error, and this can be attributed to an uncontrolled movement action of drill bit.
- Experimental results reveal that there exists linear increment in burr thickness with respect to the increase in cutting speed especially at the speed range of 30–50 m/min.
- Optimal machining parameter combination found by TOPSIS method is speed of 20 rpm and feed rate of 0.15 m/min.
- Different solutions are obtained by employing single-objective optimization methods such as S/N ratio. Unique solutions are determined from TOPSIS and hence employed in solving multi-objective optimization research studies.

## References

1. Boyer (1996) An overview on the use of titanium in the aerospace industry. Mater Sci Eng A 213:103–114
2. Cantero RRJL, Tardı'o MM, Canteli JA, Marcos M, Migue'lez MH (2005) Dry drilling of alloy Ti–6Al–4V. Int J Mach Tools Manuf 45:1246–1255
3. Li R, Hegde P, Shih AJ (2007) High-throughput drilling of titanium alloys. Int J Mach Tools Manuf 47:63–74
4. Azizi MW, Belhadi S, Yallese MA, Mabrouk T (2012) Surface roughness and cutting forces modeling for optimization of machining condition in finish hard turning of AISI 52100 steel. J Mech Sci Technol 26(12):4105–4114
5. Samsudeensadham S, Krishnaraj V (2014) An analysis on temperature & surface roughness during end milling of Ti-6Al-4V alloy. In: International mechanical engineering congress, Applied mechanics and materials. Transtech Publications, vols 592–594 pp 38–42
6. Vijayan K, Sadham S, Sangeetha S, Palaniyandi K, Zitoune R (2014) Study on cutting forces and surface finish during end milling of titanium alloy. In: ASME 2014 international mechanical engineering congress and exposition, volume 2A: Advanced Manufacturing, Montreal, Quebec, Canada, 14–20
7. Singh S, Singh I, Dvivedi A (2013) Multi objective optimization in drilling of Al6063/10% SiC metal matrix composite based on grey relational analysis. Proc Inst Mech Eng Part B: J Eng Manuf 227(12):1767–1776
8. Thirumalai R, Senthilkumaar JS (2013) Multi-criteria decision making in the selection of machining parameters for Inconel 718. J Mech Sci Technol 27(4):1109–1116
9. Pawade RS, Joshi SS (2011) Multi-objective optimization of surface roughness and cutting forces in high-speed turning of Inconel 718 using Taguchi grey relational analysis (TGRA). Int J Adv Manuf Technol 56(1–4):47–62

10. Ahmed LS, Pradeep Kumar M (2016) Multiresponse optimization of cryogenic drilling on Ti-6Al-4V alloy using topsis method. J Mech Sci Technol 30(4):1835–1841
11. Rajmohan T, Palanikumar K, Kathirvel M (2012) Optimization of machining parameters in drilling hybrid aluminium metal matrix composites. Trans Nonferrous Met Soc China 22(6):1286–1297
12. Arun Ramnath R et al (2018) Optimization of machining parameters of composites using multi-attribute decision-making techniques: a review. J Reinf Plast Compos 37(2):77–89

# Multi-objective Optimization of CNC Turning Parameters of Grey Cast Iron Using Response Surface Methodology and Genetic Algorithm

**S. R. Devadasan, S. T. Kiruba Shankaran, A. K. Deepak Raj, R. Narain Krishna, and S. Hariharan**

**Abstract** Carbon emission is one of the factors that have recently gained more negative environmental impact. In this project, optimization has been done to obtain the optimum values of the machining parameters- speed, feed, and depth of cut in computer numerical control turning of Grey Cast Iron material to reduce carbon emissions. Other objective functions are processing time, which characterizes the production rate, and surface finish, which characterizes material quality. The optimization has been done using genetic algorithm and response surface methodology.

**Keywords** Optimization · Carbon emission · Genetic algorithm · RSM

## 1 Introduction

Machining is a technique with which an unfinished material is converted to the desired form by various material removal processes. Machining has been right there from the Stone Age, where people made sharp weapons out of stones. Before two centuries, a person who built and repaired machines were referred to as a machinist.

S. R. Devadasan · S. T. Kiruba Shankaran · A. K. Deepak Raj (✉) · R. Narain Krishna · S. Hariharan
Department of Production Engineering, PSG College of Technology, Peelamedu, Coimbatore, Tamilnadu, India
e-mail: deepakkrishnaraj1998@gmail.com

S. R. Devadasan
e-mail: devadasan_srd@yahoo.com

S. T. Kiruba Shankaran
e-mail: kirubashankaran09@gmail.com

R. Narain Krishna
e-mail: narainkrishna98@gmail.com

S. Hariharan
e-mail: harishudt1998@gmail.com

© The Editor(s) (if applicable) and The Author(s), under exclusive license to Springer Nature Singapore Pte Ltd. 2021

S. Mohan et al. (eds.), *Materials, Design, and Manufacturing for Sustainable Environment*, Lecture Notes in Mechanical Engineering,
https://doi.org/10.1007/978-981-15-9809-8_35

He mostly manual processes such as hand carving, hand forging, etc. In the past two centuries, technological advancements were made in the past two centuries and machining has got a new dimension altogether.

The terms 'machine tool' and 'machining' were coined in the middle of the 1900s. The early 2000s as well as the late 1900s were called the machine age. This period occurred between the first and the Second World War. The machine age saw huge improvements such as

- Invention of IC engines
- Mass production concept
- Gigantic production machineries
- Development of modern war machines.

Such evolutions meant more modern technologies and faster and precise machines. Machining took a huge drift from manual material removal to the usage of tools and machines. This period mostly consisted of the usage of conventional machining tools. Authors Brynjolfsson and McAfee [1] call the period after the machine age as second machine age where digital manufacturing came into existence.

Today, we are in the age of smart machining, where machines need minimal or no human intervention and are capable of completing tasks by themselves. Technologies like the internet of things, big data, etc., have taken such smart machines from the dream into reality.

In such a technically advanced era, one of the main issues is an environmental threat. A lot of resources, both renewable and non-renewable are being continuously exhausted from the environment. In return, poisonous gases, dust, waste, toxic substances, etc., are being dumped into the atmosphere. If this situation persists, the successive generations will not have enough resources to survive.

This is where the concept of sustainability came into the picture. Sustainability is a concept of making the world a fit place to live for future generations too. Sustainability includes concepts such as optimal utilization of resources, reducing the amount of toxic substances emitted into the atmosphere, etc.

### *1.1 Optimization*

Optimization can be defined as the process of finding the best possible outcome under constrained situations and conditions. Any engineering process has a critical component of decision making in it. The decisions have to be made such that the effort is reduced or the profit or benefit is maximized. Mathematical optimization also has a similar meaning. Mathematical optimization tries to optimize the input parameters such that the output parameter or function is either minimized or maximized.

Optimization has its roots placed in operations research. Operations research concepts evolved during the Second World War, where countries were suffering from lack of adequate resources. In such a crisis situation, people had to use the available resources in the best possible way. Hence, they mathematically formulated the real-world situations and found out the most optimal way of utilizing the resources.

### *1.2 Methods of Optimization*

There are four main categories of optimization [2]. These are

- Mathematical programming techniques, wherein the optimum value can be obtained under a given set of conditions. Examples include calculus methods, nonlinear programming methods, geometric programming techniques, etc.
- Statistical methods are used where an accurate representation model of a physical situation is required. Examples of this method are regression analysis, design of experiments, etc.
- Stochastic processes like reliability theory and queuing theory are used to analyze processes having known probabilistic distributions.
- Non-traditional processes are used to solve complex engineering problems. Examples include genetic algorithm, particle swarm optimization, etc.

### *1.3 Non-traditional Optimization*

Using traditional optimization techniques it is possible to solve simple optimization problems. Some of the traditional optimization techniques are as follows:

- Steepest descent
- Conjugate gradient
- Quasi-Newton
- Graphical technique
- Simplex method and its variants
- Etc.

These traditional methods have some drawbacks

- The solution highly depends on the starting random value. The obtained solution may or may not be a globally optimal one.
- There is every possibility that the solutions obtained by gradient-based techniques may give the local minima as the solution.
- No one optimization method that can be used to solve all optimization problems.

To overcome these drawbacks and to solve complex engineering problems, non-traditional optimization methods came into the picture. Some of the non-traditional optimization methods are as follows:

- Genetic algorithm
- Response surface methodology

- Simulated annealing method
- Particle swarm optimization technique
- Ant colony optimization technique
- Etc.

### 1.3.1 Genetic Algorithm Technique

Genetic algorithm is a technique derived from biology. Genetic algorithm has lesser chances of getting stuck at local minima or maxima and hence overcomes the disadvantage of traditional optimization techniques. Moreover, it can also solve problems having discontinuous objective functions and also problems with integer or real variables or a combination of both. The main disadvantage of this method is that it is slow in computation.

### 1.3.2 Response Surface Method

RSM is a non-traditional technique. It usually creates a response surface of the output variable for all possible combinations of the input variables. RSM is preferred because it gives the optimum value with the minimum number of experiments.

## *1.4 Need for the Current Study*

Sustainability, today, has become a buzz word and the world has realized the importance of keeping resources for the future. In such a period, emission of carbon into the atmosphere from a predominant sector like the industrial sector has very undesirable after effects. Though it is not possible to completely eradicate carbon emissions, it is possible to reduce carbon emissions, which is the scope of this work.

## *1.5 Problem Statement*

Right from the invention of stone, machining and material removal have been a prime task. From there, machining has grown to a large extent and has taken different forms. As of today, the machining industry is heading towards more environmental consciousness. Many aspects that have an environmental impact like water footprint, biological oxygen demand, emissions, etc., are being worked upon.

Carbon emission is one of the prime environmental problems experienced when performing machining at high speeds. When performing machining at high speeds, operators are exposed to a certain level of carbon. Such exposure causes many health hazards to the operators like Loss of concentration, Increased heart rate, Oxygen deprivation, Coma, Death, and Changes in body metabolism.

Carbon emissions can occur in a machine through many possible ways. Emissions may occur from the tool, workpiece, or the cutting fluid. As operators are exposed to this carbon on a frequent basis, there is a heavy chance of the operator being affected by the long term ill effects of carbon. In some cases, this might even lead to life threats.

## *1.6 Scope*

- The scope limits of the project include the following.
- Machining parameters are restricted to rotational speed, machine feed, and depth of cut as these three are the important contributing factors that have effects on all the three output parameters [3].
- The study has been done for the material Grey Cast Iron grade 20 alone.
- Machining has been restricted to CNC turning alone.

## *1.7 List of Possible Solutions*

The issue of carbon emissions can be overcome by many alternative solutions apart from optimization of machining parameters. Some of the possible alternative solutions include

- Arranging a closed machining setup
- Use of masks or covers to reduce the exposure to carbon
- Using dry machining to reduce coolant levels
- Revise the coolant formulation to reduce carbon levels
- Revise the material composition to reduce carbon levels
- Considering alternate tools that have lesser emission levels.

Of all these solutions, optimization solution has been selected because of its effectiveness without affecting the machining output much.

## *1.8 Objectives*

The objective of the current study is to find out the optimum machining parameters of CNC turning process of Grey cast iron workpiece that gives the minimum carbon emissions at a minimal processing time, without affecting the surface roughness. To achieve this, two processes, Genetic Algorithm, and RSM are used. The results of both methods are to be compared.

## 1.9 Methodology

The following methodology was adopted to complete the project (Fig. 1).

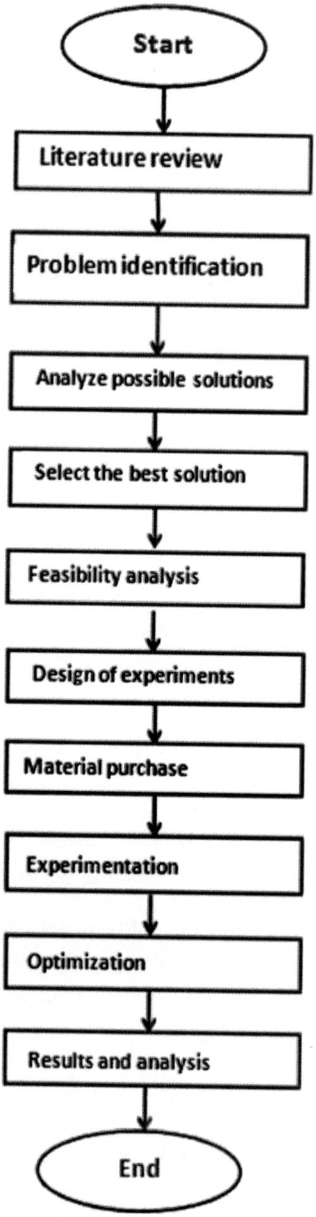


**Fig. 1** Methodology

### 1.10 Literature Review

Several authors have done specific works in optimization in various industries to solve various optimization problems. Most of these works ended up in success and the authors were able to achieve the objective that they have set. Before starting an optimization project, it becomes essential to make a brief literature review on the various literature that carried out optimization projects successfully.

Rajesh [3] has performed the experimental analysis considering how rotational speed, feed rate, depth of cut, and nose radius affect CNC turning of 7075 Al alloy with 15% of Silicon Carbide by weight. The objectives of the experiment were to minimize power consumption and maximize tool life. Response Surface Methodology (RSM) was used to attain this goal. The optimum levels of factors were identified and significant contributors were determined by using Analysis of Variance (ANOVA). Through desirability analysis, the result showed the reduction of power consumption by 13.55% and an increase in tool life by 22.12%.

Zhigang et al. [4] proposed an optimization model of machining factors in CNC turning of C45 steel. Process cost and cutting fluid consumption were set as the two objective functions and they were affected by 4 variables namely cutting fluid flow, cutting depth, speed, and feed rate. The method adopted to solve the problem was hybrid genetic algorithm programmed in Matlab7. The optimum parameters have been identified and a 17% decrease in fluid consumption was obtained in the simulation results.

Nitesh et al. [5] proposed a systematic procedure to calculate the carbon emissions of CNC machine tools. The process for optimization is CNC turning of SS 1.4542. The objectives were minimizing carbon emission, processing time, and surface roughness. The parameters that affect the objectives were depth of cut, speed, and feed. RSM and ANOVA were the methods used to solve the optimization model. Experimental trials were done to validate the model and optimum cutting parameters have been identified.

Zhi et al. [6] developed a numerical model and technique to find the optimized cutting parameters that reduce carbon emission costs. The optimization variables were taken as rotational speed and feed rate. The objectives were carbon emissions and machining hours. The optimization model was the turning of C45 steel. Non-dominated sorting genetic algorithm (NSGA-2) was used to perform the optimization. The optimal cutting parameters were identified.

Murat et al. [7] used DoE to analyze the effect of the prime cutting parameters. The parameters taken for the study were cooling condition, rotational speed, feed rate, and depth of cut. The objective was to minimize arithmetic average roughness (Ra) and average maximum height of the profile (Rz) when turning o AISI 1050 steel. Experimental trials were done under dry machining and wet cooling. Taguchi's L16 array was used and ANOVA analysis was performed to determine the optimal combination.

Mozammel [8] proposed a mathematical model of cutting energy and surface roughness (Ra) in the end milling of hardened AISI 4140 steel under the Minimum

Quantity Lubrication(MQL). The parameters taken for the model were speed, feed rate, and flow rate of lubricant. RSM, Taguchi, and ANOVA analysis were performed to identify the optimal combination.

Vijay et al. [9] used CNC turning operation to machine EN 19 steel material. The process parameters were spindle speed, feed rate, depth of cut, and lubricant. The objective was to minimize surface roughness and maximize material removal rate. Taguchi's L18 mixed type orthogonal array and ANOVA were used to identify significant parameters.

Salem et al. [10] performed optimization of machining parameters of dry CNC turning of SS 316. The objectives were set as minimizing tool wear, surface roughness, and power consumption. The factors are taken that affect the objectives were speed, feed, and depth of cut. RSM was used in the optimization of the machining parameters. The optimal levels of parameters were found and a confirmation test was conducted. The results show that there was a reduction in power consumption by 14.94%, tool wear by 13.98%, and surface roughness by 4.71%.

Paramjit et al. [11] carried out an analysis to get the optimized machining parameters for the CNC rough turning of EN 353 alloy steel. The objective was to focus on the optimization of energy efficiency, active energy consumed by the machine, and power factor. The input process variables were cutting speed, feed rate, depth of cut, and nose radius. Taguchi's L27 array was used to optimize the response parameters using MINITAB software and Analysis of variance was used to determine the significant factors.

Carmita [12] conducted an experimental study to perform optimization of cutting parameters during turning of AISI 6061 T6. The objective was to minimize energy consumption (EC), surface roughness (SR), and maximize material removal rate(MRR). Taguchi orthogonal array was used to optimize the parameters and ANOVA was used to find the effects of input variables such as cutting speed, feed rate, and depth of cut on the objective. The optimal and significant factors were identified.

Congbo et al. [13] found a method for complex optimization of cutting parameters in CNC milling of AISI 1018 steel. The objectives were set as energy efficiency and minimizing processing time. The input variables were speed, feed, and depth of cut, and width of cut. Taguchi, RSM, and Multi-Objective Particle Swarm Optimization (MOPSO) were the methods integrated to find the optimal parameters which minimize processing time and maximize energy efficiency.

Issam et al. [14] optimized the cutting parameters in the turning of a composite with matrix, Poly Ether Ether Ketone, and reinforcement, carbon fibers. The optimization objective was to get minimum power consumption and optimal surface finish. Grey relational theory and Taguchi approach were applied to find the optimized cutting parameters. The results showed that cutting speed and depth of cut were the most important parameters.

Munish et al. [15] found an optimization model in turning titanium alloy under nano-fluid based Minimum Quantity Lubrication (MQL). The objectives were the tool wear, cutting temperature, cutting forces, and surface finish which was influenced by four input variables namely, speed, feed, depth of cut, and type of cutting fluid.

RSM integrated with MOPSO and Bacterial Foraging Optimization was utilized to conduct experiments and identify the optimal parameters. The results were analyzed by comparing with the desirability function approach technique.

Jihong et al. [16] proposed a multi-objective optimization method to find optimum cutting parameters in the milling of C45 steel. The objectives were set as achieving the best surface roughness, MRR, and cutting energy. The factors considered were rotational speed, feed rate, depth of cut, and width of cut. RSM, Taguchi, and grey relational analysis were the methods used to find the optimal milling parameters and the conclusions showed that the width of cut is the highly influential parameter.

Kuram et al. [17] presented a model to perform optimization of the parameters in the milling of AISI 304 Austenitic stainless steel. The objectives were set as tool wear and cutting force components. The machining parameters were taken as cutting speed, feed rate, depth of cut, and type of cutting fluids. The L9 orthogonal array was used for the experiment. Regression analysis, S/N ratio, and ANOVA analysis were performed to identify the significant factors.

Sweety et al. [18] focused on optimizing the machining parameters of electric discharge machining of composites containing metal matrix based on hybrid Aluminium added with Silicon carbide and B4C as reinforcements. The objectives were power consumption and surface roughness. The process parameters used were pulse current, pulse duty factor, pulse on time, and gap voltage. Genetic Algorithm (GA) and ANOVA analysis were done to determine the optimal and significant factors. Pulse current was the most significant contributor.

Uttam et al. [19] conducted experiments to attain the optimal parameters in EDM in the machining of high carbon steel. The objectives were to optimize MRR, tool wear rate, surface roughness, and radial overcut. The input factors were current, pulse on time, and voltage. L9 orthogonal array was utilized for the experiments. ANOVA analysis and multiple linear regression analysis were used to determine the significant input parameters.

Raman et al. [20] performed experiments to optimize parameters in the rough turning of EN 353 alloy steel with multiple layers coated tungsten carbide insert. The input factors were depth of cut, feed, speed, and nose radius. The objectives were MRR, surface roughness, and energy consumption. Taguchi L27 orthogonal array was utilized for DOE by using Minitab software. The optimal parameters have been identified to meet the objectives.

Zhaohui et al. [21] found out a multiple objective optimization models in the grinding of steel 45. Optimal carbon emissions and minimum processing time were set as the optimization objectives. The input grinding parameters were grinding depth, wheel speed, and feed. GA and Taguchi orthogonal array were performed to ensure the feasibility of the result and the optimal grinding parameters were ensured to reduce process time and increase carbon efficiency.

Carmita [22] conducted an experimental analysis to find the optimized cutting parameters during rough turning of AISI 6061 T6 aluminium. The objective was to minimize energy consumption (EC), surface roughness (SR), and maximize material removal rate (MRR). Central Composite Design and RSM were used to obtain the regression model and ANOVA was utilized to find the influence of input variables such as cutting speed, feed rate, and depth of cut on the objective. Feed rate was significant to SR and feed rate and depth of cut was significant to energy consumption. It was found that energy consumption was reduced by 14.41% and surface roughness by 36.47%.

Nan et al. [23] proposed a methodology to identify the optimal cutting parameters in CNC finish turning of C45 E4. The optimization objectives were SR, energy consumption, and energy efficiency. The input factors taken for the model were speed, feed, depth of cut, and tool wear. Taguchi, RSM, and NSGA-3 were the methods used to construct and solve the problem.

Kuldip et al. [24] invented a predictive and optimization model for the turning of AISI 1045 steel with a tungsten carbide tool. This is based on an integrated RSM and GA approach. The objectives of the model were set as energy consumption. The input process variables taken for the study were speed, feed, and depth of cut, Taguchi was used to design experiments and ANOVA analysis was used to find the important contributors.

Sivaiah et al. [25] focused on optimizing cutting conditions in turning of 17-4 precipitation hardened stainless steel. Cutting speed, feed rate, depth of cut, and cutting environment were selected as input process parameters. Surface roughness (Ra) and tool flank wear were taken as optimization objectives. Taguchi orthogonal array was used for the experimental design plan and to identify the optimal parameters. ANOVA was used to predict the important factors and it was concluded that cutting speed was the most important factor.

Mozammel et al. [26] presented an optimization model for Minimum Quantity Lubrication (MQL) turning of 60 Rc steel. The average surface roughness parameter (Ra) was the objective of the model with the input process parameters being spindle speed, feed rate, depth of cut, and time gap between MQCL pulsing. The least-square support vector machine method and interior point method were employed for prediction and optimization purposes respectively.

Reza et al. [27] carried out an optimization attempt in Ultrasonic burnishing of Al 6061 T6. The objectives were surface roughness, arc height, and hardness. The parameters were feed, vibration amplitude, burnishing depth, and step over. Adaptive network-based fuzzy inference system and particle swarm optimization were the methods used to carry out optimization process.

Laura et al. [28] conducted experiments to optimize the input parameters for milling of AISI 1018 steel. The input factors were cutting speed, feed rate, and depth of cut and nano-particle concentration in the cutting fluid. The objectives were to achieve the best spindle load, cutting insert radius, and surface roughness (Ra). ANOVA analysis was performed to identify the significant factors from input parameters that have effects on all the objectives simultaneously.

Deepak et al. [29] proposed an optimization model to optimize cutting parameters and fluid application parameters in turning of oil-hardened nonshrinkable steel with minimum cutting fluid application. The optimization objective was surface roughness. The controllable input factors were pressure at fluid injector and composition of cutting fluid. Taguchi technique was used to obtain the optimum set of parameters.

Mozammel et al. [30] presented an optimization model for the hard turning of AISI 1060 steel. The tool wear, MRR, and surface roughness were the objectives of the model with the input process parameters being speed, feed rate, and depth of cut. Taguchi orthogonal array was used to design experiments and S/N ratio was used for optimization. Results showed that the cutting speed, feed rate, and depth of cut influenced SR, MRR, and tool wear, respectively.

Congbo et al. [31] made a quantitative model calculate carbon emissions that occur during a machining process. This work has been inferred by many authors in their work for the calculation of carbon emissions.

Yusuf et al. [32] optimized the machining of aluminium alloys using CNC machines. The objectives were surface roughness and carbon emissions. The input parameters ere tool material, workpiece material, speed, and depth of cut. RSM technique was used to obtain the optimized parameters. Minitab software was used to generate the results.

From the above literature studied, a research gap has been identified. Grey cast iron is one of the major materials that has been used and is more frequently machined using CNC machines. But no optimization works have been done on optimization of the machining parameters of computer numerical control machining of grey cast iron with the objective of reducing carbon emissions. Hence this area has been selected as the area of research.

## 2 Design of Experiments

### *2.1 Output Parameters and Input Parameters*

Three output factors were considered to perform optimization.

- Carbon emission
- Surface roughness
- Processing time.

Carbon emission was considered so as to optimize the process from an environmental perspective. The surface roughness parameter was considered so as not to sacrifice the quality of the workpiece and processing time was considered so as to not sacrifice productivity [5].

Three input factors were taken into account. These were

- Rotational speed
- Feed
- Depth of cut.

Surface roughness and processing time affect the machining parameters directly, whereas carbon emission too has a relationship with machining parameters [3].

**Table 1** Factors and levels

| S. No | Factors | Levels | | | Unit |
|---|---|---|---|---|---|
| | | 1 | 2 | 3 | |
| 1 | Speed | 1.0 | 1.5 | 2.0 | m/s |
| 2 | Feed | 0.2 | 0.5 | 0.8 | mm/rev |
| 3 | Depth of cut | 0.5 | 1.0 | 1.5 | mm |

### 2.2 *Selection of Levels*

The three factors were considered at three levels [33] each. These are shown (Table 1).

### 2.3 *The Central Composite Design*

Two methods were selected for performing optimization. For the RSM, the DoE was created using Minitab software. For RSM, the CCD method was selected to perform the DoE (Table 2) as it includes the centre points, edge points as well as the circumscribed points, which gives a high degree of accuracy. CCD (Fig. 2) helps to find the effects of quadratic terms but doesn't give the effect of interaction terms. The formula for calculating the number of points in a CCD is given by Eq. 1.

$$\text{No. of points} = 2^k + 2k + n \quad (1)$$

where $k$ is the degrees of freedom and $n$ is the number of central runs.

In our case, the degrees of freedom are 3 and the number of central runs is 6. Hence we get a total number of experiments as 20.

## 3 Experimental Details

### 3.1 *Insert*

The recommended insert [] CNMA120408 was purchased (Fig. 3) from a local metal mart.

The detailed expansion for the selected insert designation is provided below.

- C—Insert shape—Rhombic 80°.
- N—Relief angle—0°.
- M—Tolerance class—tolerance of nose height = ±0.08 to ±0.18 mm, tolerance of inscribed circle = ±0.05 to ±0.15 mm, tolerance of thickness = ±0.13 mm.
- A—Chip breaker and clamping systems—cylindrical hole with no chip breaker.

**Table 2** Design of experiments

| S. No | Speed (m/s) | Feed (mm/rev) | DOC (mm) |
|---|---|---|---|
| 1 | 1 | 0.8 | 0.5 |
| 2 | 1.5 | 0.5 | 1 |
| 3 | 1.5 | 0.5 | 1 |
| 4 | 1 | 0.2 | 1.5 |
| 5 | 1.5 | 0.5 | 1 |
| 6 | 2 | 0.2 | 1.5 |
| 7 | 2 | 0.8 | 0.5 |
| 8 | 1 | 0.2 | 0.5 |
| 9 | 2 | 0.8 | 1.5 |
| 10 | 2 | 0.2 | 0.5 |
| 11 | 1.5 | 0.5 | 1 |
| 12 | 1 | 0.8 | 1.5 |
| 13 | 1.5 | 0.5 | 1.841 |
| 14 | 2.341 | 0.5 | 1 |
| 15 | 1.5 | 0.5 | 1 |
| 16 | 1.5 | 0.05 | 1 |
| 17 | 1.5 | 1.0046 | 1 |
| 18 | 1.5 | 0.5 | 1 |
| 19 | 1.5 | 0.5 | 0.159 |
| 20 | 0.659 | 0.5 | 1 |

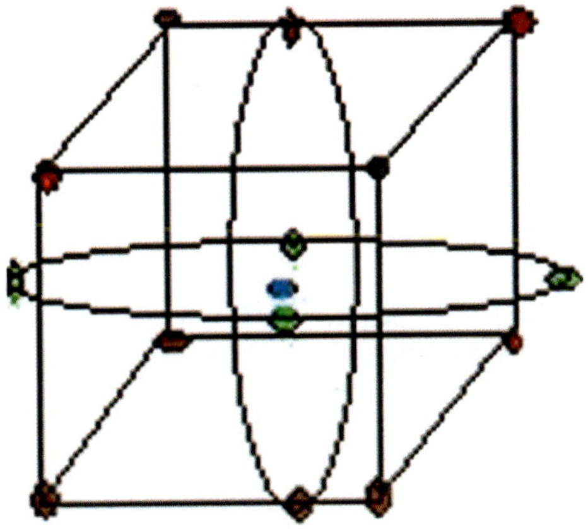

**Fig. 2** CCD

**Fig. 3** Insert

- 12—Diameter of inscribed circle = 12 mm.
- 04—Insert thickness = 4.76 mm.
- 08—Nose radius = 0.8 mm.

## 3.2 *Material*

Grey cast iron shafts of diameter 25 mm and length 100 mm were purchased. They were rough turned to diameter 22 mm using lathe machine.

## 3.3 *Specifications of the Computer Numerical Control Machine*

The features of the CNC lathe (Fig. 4) are as follows.

- Model: Horizontal turning center-Smarturn
- Manufacturer: LMW Limited
- Max. turning diameter: 200 mm
- Max. turning length: 262 mm
- Spindle speed: 4500 rpm
- Cross travel X: 105 mm
- Longitudinal travel Z: 320 mm
- Rapid traverse X/Z: 20 m/min
- Weight (approx.): 2300 kg
- Controller: Fanuc.

**Fig. 4** CNC machine

### *3.4 Experimentation*

After the completion of material purchase and insert purchase, experimentation process was started. The CNC lathe machine available at production engineering laboratory, PSG College of Technology was utilized to perform the experiments. The dimensions before and after turning are shown below (Figs. 5 and 6).

## 4 Calculation of Output Parameters

The output parameters were found out in the following ways.

### *4.1 Calculation of Processing Time*

Processing time was calculated using the stopwatch while performing the experiments.

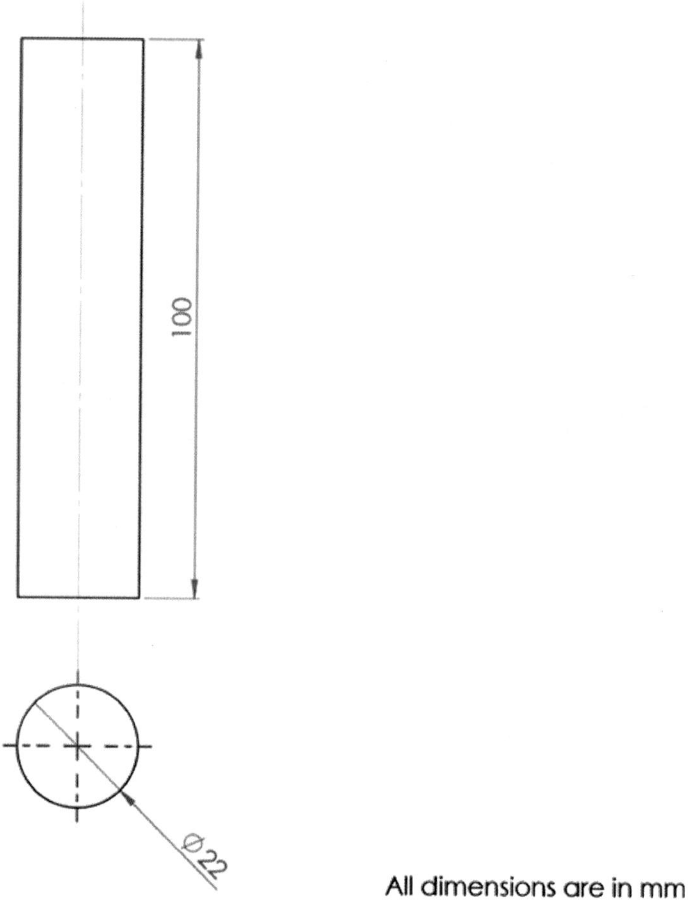


**Fig. 5** Dimensions before experiment

## 4.2 Calculation of Carbon Emission

The overall carbon emission can be found out using the formula [32].

$$CE_{tot} = CE_{el} + CE_{tool} + CE_{mat} + CE_{chip}$$

where

$CE_{tot}$ total carbon emissions in $kgCO_2$
$CE_{el}$ carbon emissions due to electricity in $kgCO_2$
$CE_{tool}$ carbon emissions due to cutting tool in $kgCO_2$
$CE_{mat}$ carbon emissions due to production of raw material in $kgCO_2$
$CE_{chip}$ carbon emissions due to metal removal in $kgCO_2$.

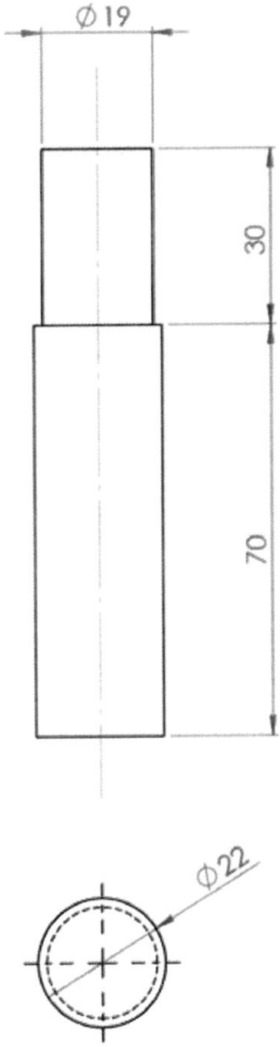


**Fig. 6** Dimensions after experiment

$$\mathrm{CE_{el}} = \mathrm{CEF_{elec}} \times \mathrm{EC_{mc}}$$

where

$\mathrm{CEF_{elec}}$ Carbon emission factor for producing electricity

$$\text{in India in} \frac{\mathrm{kgCO_2}}{\mathrm{kWh}} = 1.41 \frac{\mathrm{kgCO_2}}{\mathrm{kWh}}$$

$\mathrm{EC_{mc}}$ Energy consumed during machining in kWh

$$\mathrm{CE_{tool}} = \mathrm{CEF_{tool}} \times W_{\mathrm{tool}} \times \left(\frac{t_{\mathrm{c}}}{T_{\mathrm{tool}}}\right)$$

where

$CEF_{tool}$ Carbon emission factor of cutting tool in $\frac{kgCO_2}{kg} = 29.6\frac{kgCO_2}{kg}$
$W_{tool}$ Mass of tool in kg
$t_c$ Cutting time in min
$T_{tool}$ tool life of cutting tool in min.

$$CE_{mat} = CEF_m \times M_{chip}$$

where

$CEF_{mat}$ carbon emission factor of material in $\frac{kgCO_2}{kg} = 2.22\,kgCO_2/kg$
$M_{chip}$ Mass of chip in kg.

$$CE_{chip} = CEF_{chip} \times M_{chip}$$

where

$CEF_{chip}$ carbon emission factor of chip in $\frac{kgCO_2}{kg} = 0.361\frac{kgCO_2}{kg}$.

The carbon emission factors are referred to from literature [5, 31]. Energy consumption was measured from an electric meter (Fig. 7).

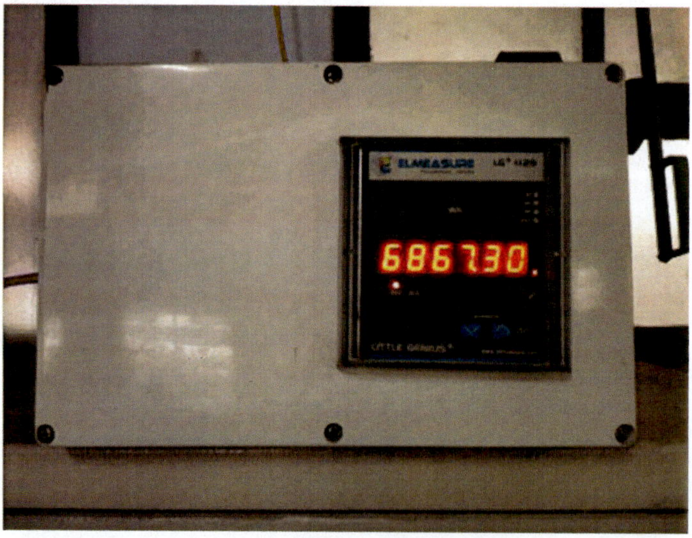


**Fig. 7** Electric meter

### 4.3 *Calculation of Surface Roughness*

Surface roughness was calculated for the turned workpieces using surface roughness tester (Fig. 8). The specifications of the tester are provided.

- Manufacturer: TIME GROUP INC/MMS
- Model: TR100
- Tracking length: 6 mm
- Tracking speed: 1 mm/s
- Accuracy: ±5 μm.

## 5 Results and Discussions

After completion of experiments and gathering all the required data, a calculation model was created using MS Excel, which would automatically calculate the required values from the input values (Tables 3 and 4).

**Fig. 8** Roughness tester setup

**Table 3** Output parameters

| S. No | Speed (m/s) | Feed (mm/rev) | DOC (mm) | Idle time (min) | Cutting time (min) | Total time (min) | Mass of the chip (kg) | Energy consumption (kWh) | Ra (μm) |
|---|---|---|---|---|---|---|---|---|---|
| 1 | 1 | 0.8 | 0.5 | 0.05 | 0.1997 | 0.2497 | 0.023 | 0.03 | 3.873 |
| 2 | 1.5 | 0.5 | 1 | 0.041667 | 0.15 | 0.191667 | 0.024 | 0.02 | 3.497 |
| 3 | 1.5 | 0.5 | 1 | 0.1 | 0.3 | 0.4 | 0.02 | 0.02 | 2.564 |
| 4 | 1 | 0.2 | 1.5 | 0.08333 | 0.3333 | 0.416667 | 0.019 | 0.03 | 3.884 |
| 5 | 1.5 | 0.5 | 1 | 0.1 | 0.35 | 0.45 | 0.016 | 0.03 | 2.392 |
| 6 | 2 | 0.2 | 1.5 | 0.05 | 0.1666 | 0.216667 | 0.02 | 0.02 | 1.904 |
| 7 | 2 | 0.8 | 0.5 | 0.06667 | 0.1666 | 0.2333 | 0.025 | 0.02 | 3.851 |
| 8 | 1 | 0.2 | 0.5 | 0.25 | 0.9 | 1.15 | 0.024 | 0.06 | 2.391 |
| 9 | 2 | 0.8 | 1.5 | 0.0333 | 0.0666 | 0.1 | 0.023 | 0.01 | 2.956 |
| 10 | 2 | 0.2 | 0.5 | 0.085667 | 0.342833 | 0.4285 | 0.022 | 0.042 | 2.580 |
| 11 | 1.5 | 0.5 | 1 | 0.05 | 0.15 | 0.2 | 0.022 | 0.04 | 3.272 |
| 12 | 1 | 0.8 | 1.5 | 0.025 | 0.0666 | 0.091667 | 0.023 | 0.02 | 4.030 |
| 13 | 1.5 | 0.5 | 1.841 | 0.041667 | 0.1 | 0.141667 | 0.024 | 0.02 | 3.197 |
| 14 | 2.341 | 0.5 | 1 | 0.05 | 0.2 | 0.25 | 0.022 | 0.02 | 3.102 |
| 15 | 1.5 | 0.5 | 1 | 0.042833 | 0.1285 | 0.171333 | 0.019 | 0.008 | 3.098 |
| 16 | 1.5 | 0.05 | 1 | 0.157 | 1.2428 | 1.399833 | 0.02 | 0.09 | 3.257 |
| 17 | 1.5 | 1.0046 | 1 | 0.025 | 0.05 | 0.075 | 0.021 | 0.01 | 4.268 |
| 18 | 1.5 | 0.5 | 1 | 0.041667 | 0.15 | 0.191667 | 0.021 | 0.02 | 3.739 |
| 19 | 1.5 | 0.5 | 0.159 | 0.333 | 1.05 | 1.3833 | 0.024 | 0.06 | 3.267 |
| 20 | 0.659 | 0.5 | 1 | 0.041667 | 0.3 | 0.341667 | 0.023 | 0.03 | 3.247 |

**Table 4** Output parameters—2

| S. No | CE electric ($kgCO_2$) | CE tool ($kgCO_2$) | CE material ($kgCO_2$) | CE chip ($kgCO_2$) | CE total ($kgCO_2$) |
|---|---|---|---|---|---|
| 1 | 0.0423 | 7.71315E−05 | 0.05106 | 0.008303 | 0.101740131 |
| 2 | 0.0282 | 0.000313804 | 0.05328 | 0.008664 | 0.090457804 |
| 3 | 0.0282 | 0.000627608 | 0.0444 | 0.00722 | 0.080447608 |
| 4 | 0.0423 | 0.000128746 | 0.04218 | 0.006859 | 0.091467746 |
| 5 | 0.0423 | 0.000732209 | 0.03552 | 0.005776 | 0.084328209 |
| 6 | 0.0282 | 0.001156096 | 0.0444 | 0.00722 | 0.080976096 |
| 7 | 0.0282 | 0.001156096 | 0.0555 | 0.009025 | 0.093881096 |
| 8 | 0.0846 | 0.000347613 | 0.05328 | 0.008664 | 0.146891613 |
| 9 | 0.0141 | 0.000462438 | 0.05106 | 0.008303 | 0.073925438 |
| 10 | 0.05922 | 0.002378089 | 0.04884 | 0.007942 | 0.118380089 |
| 11 | 0.0564 | 0.000313804 | 0.04884 | 0.007942 | 0.113495804 |
| 12 | 0.0282 | 2.57491E−05 | 0.05106 | 0.008303 | 0.087588749 |
| 13 | 0.0282 | 0.000209203 | 0.05328 | 0.008664 | 0.090353203 |
| 14 | 0.0282 | 0.002673351 | 0.04884 | 0.007942 | 0.087655351 |
| 15 | 0.01128 | 0.000268825 | 0.04218 | 0.006859 | 0.060587825 |
| 16 | 0.1269 | 0.002600039 | 0.0444 | 0.00722 | 0.181120039 |
| 17 | 0.0141 | 0.000104601 | 0.04662 | 0.007581 | 0.068405601 |
| 18 | 0.0282 | 0.000313804 | 0.04662 | 0.007581 | 0.082714804 |
| 19 | 0.0846 | 0.002196627 | 0.05328 | 0.008664 | 0.148740627 |
| 20 | 0.0423 | 2.03859E−05 | 0.05106 | 0.008303 | 0.101683386 |

## *5.1 Results*

Initially, surface plots and contour plots were created for combinations of two input parameters against an output parameter. Then, surface plots were created for the same.

The contour and surface plots of speed and feed against carbon emissions (Figs. 9 and 10) show that as feed is being increased carbon emission gets reduced and at a point, it changes direction. Speed doesn't have a significant impact.

The contour and surface plots of speed and depth of cut against carbon emissions (Figs. 11 and 12) show that as the depth of cut is being increased carbon emission gets reduced and at a point, it changes direction. From the contour plot, at an increased rotational speed and depth of cut, carbon emissions tend to be lesser.

The contour and surface plots of machine feed and depth of cut against carbon emissions (Figs. 13 and 14) shows that at a medium to high range of both feeds as well as depth of cut, carbon emissions tend to touch the lowest point.

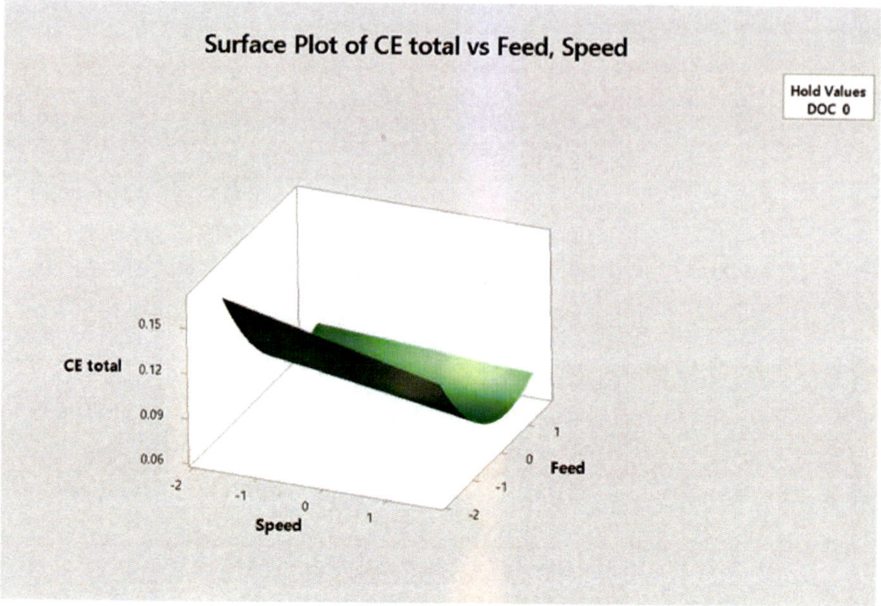


**Fig. 9** Surface plot of rotational speed, feed against carbon emission

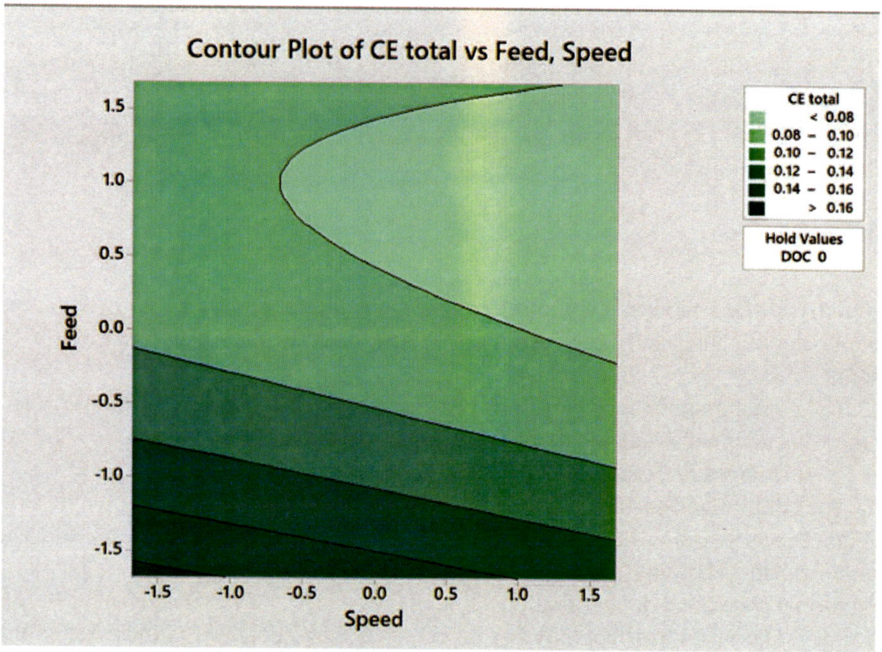


**Fig. 10** Contour plot of speed, feed against carbon emission

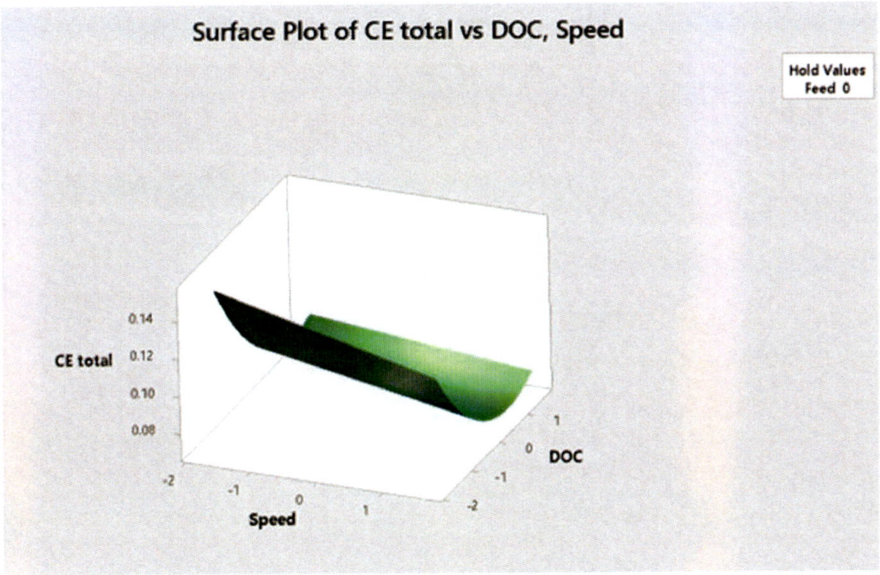


**Fig. 11** Surface plot of speed, depth of cut versus carbon emission

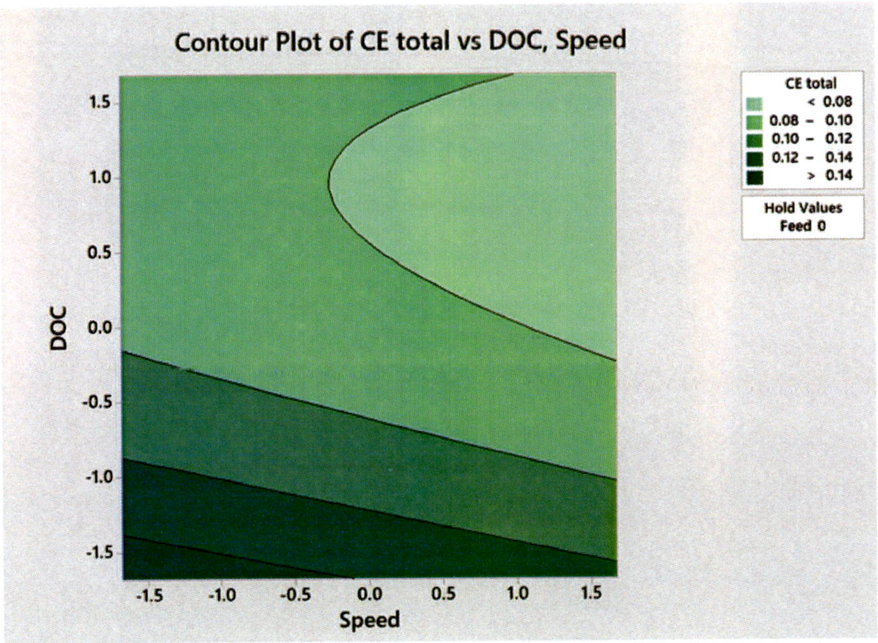


**Fig. 12** Contour plot of rotational speed, depth of cut versus carbon emission

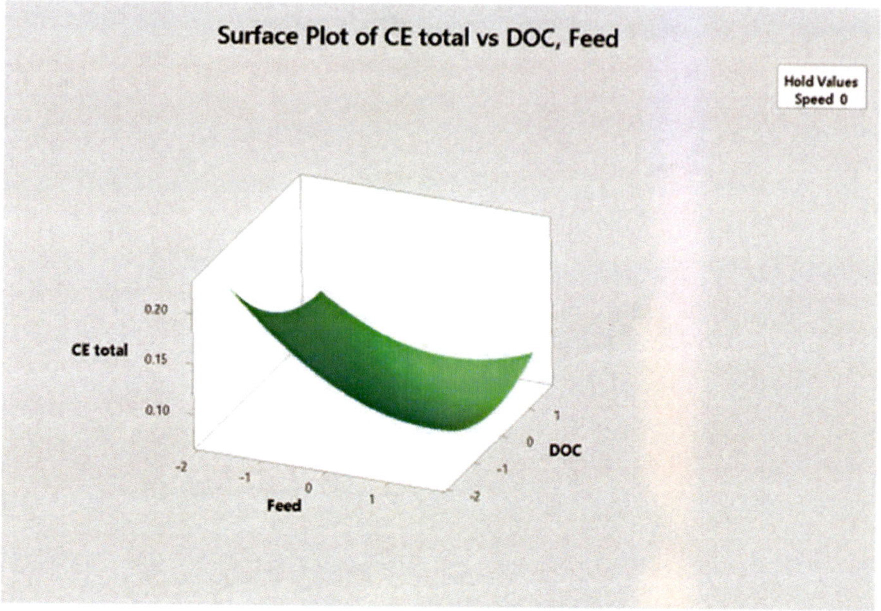


**Fig. 13** Surface plot of depth of cut, feed versus carbon emission

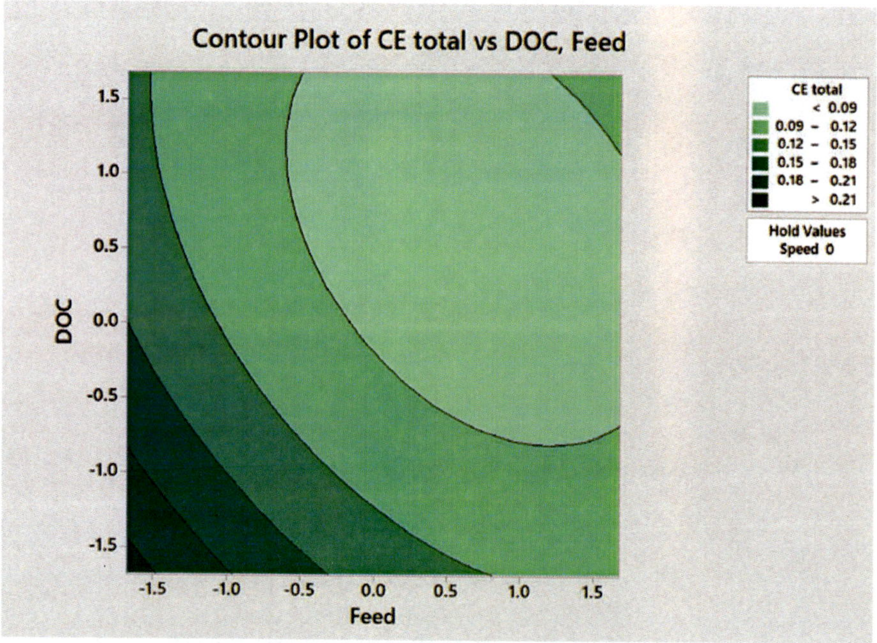


**Fig. 14** Contour plot of depth of cut, feed versus carbon emission

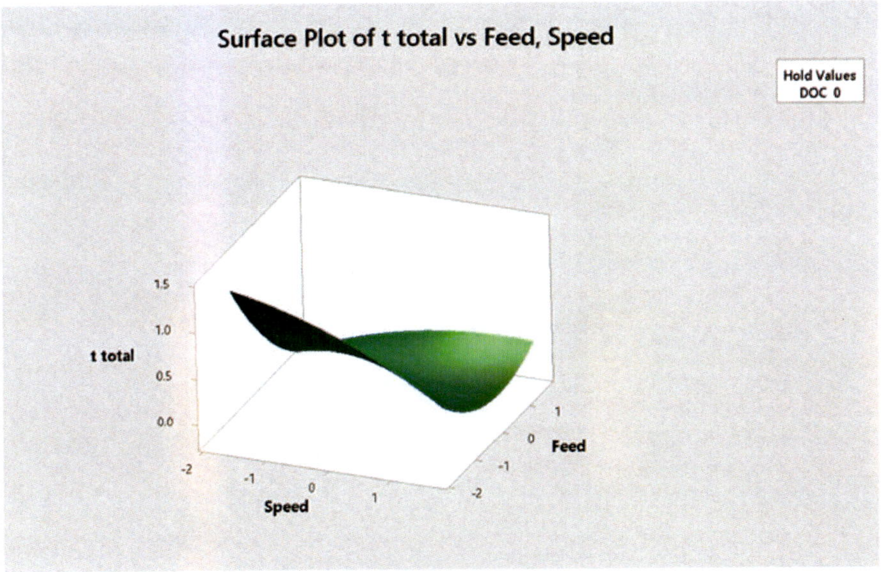


**Fig. 15** Surface plot of speed, feed versus processing time

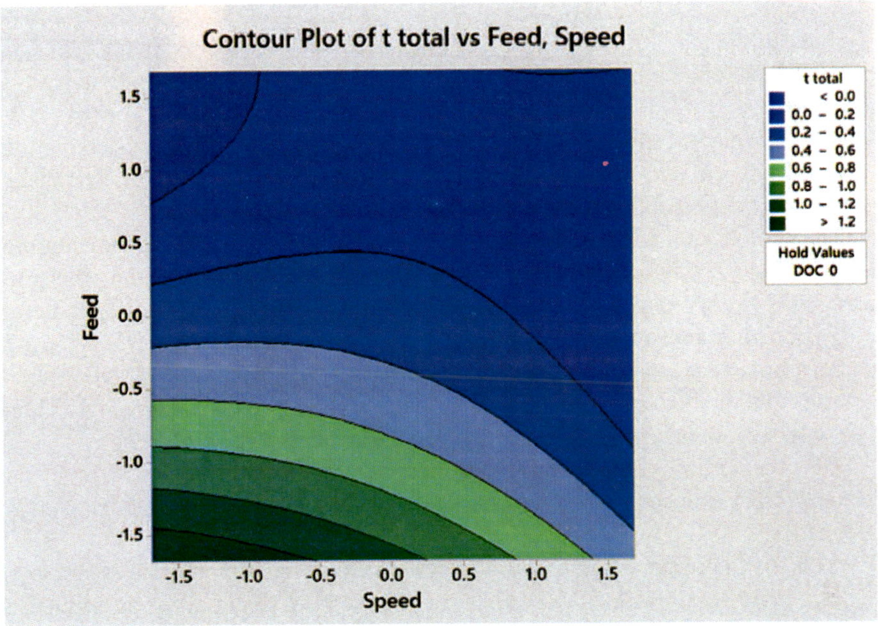


**Fig. 16** Contour plot of speed, feed versus processing time

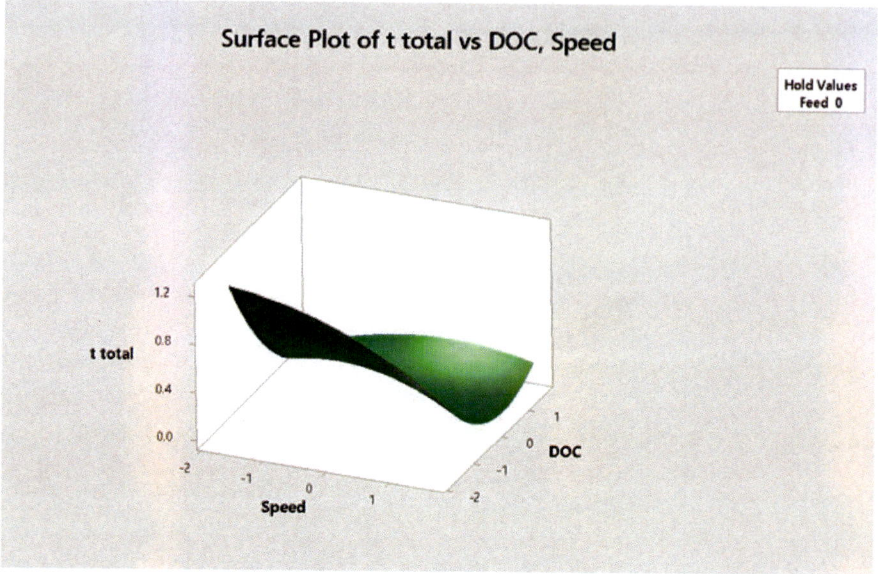


**Fig. 17** Surface plot of speed, depth of cut versus processing time

The contour and surface plots of feed and speed against processing time (Figs. 15 and 16) show that with the increase in speed and feed, the processing time tends to get lower.

The contour and surface plots of depth of cut and speed against processing time (Figs. 17 and 18) show that at higher depth of cut, the processing time is reduced. Speed has a very feeble effect on processing time.

The contour and surface plots of depth of cut and feed against processing time (Figs. 19 and 20) show that with increase in feed and depth of cut, the processing time tends to get reduced and this attains lowermost value at a particular point.

The contour and surface plots of rotational speed and feed against Ra (Figs. 21 and 22) show that at a minimum feed, the roughness value tends to get reduced.

The contour and surface plots of speed and depth of cut against Ra (Figs. 23 and 24) show that at a high speed, the roughness value tends to get reduced.

The contour and surface plots of feed and depth of cut against Ra (Figs. 25 and 26) show that as feed and depth of cut decreases, the roughness value tends to get reduced.

Analysis of variance (ANOVA) was done for both carbon emissions (Table 6), surface roughness (Table 7) as well as processing time (Table 5) using Minitab to find out the contribution of each factor.

From Table 5, it is found that feed and depth of cut are the major contributing parameters and the squared terms of feed and depth of cut also have a considerable effect. Speed has a minimal contribution. *R* square value obtained for the above table was 83.2%. Higher the *R* square value, higher the accounted variation.

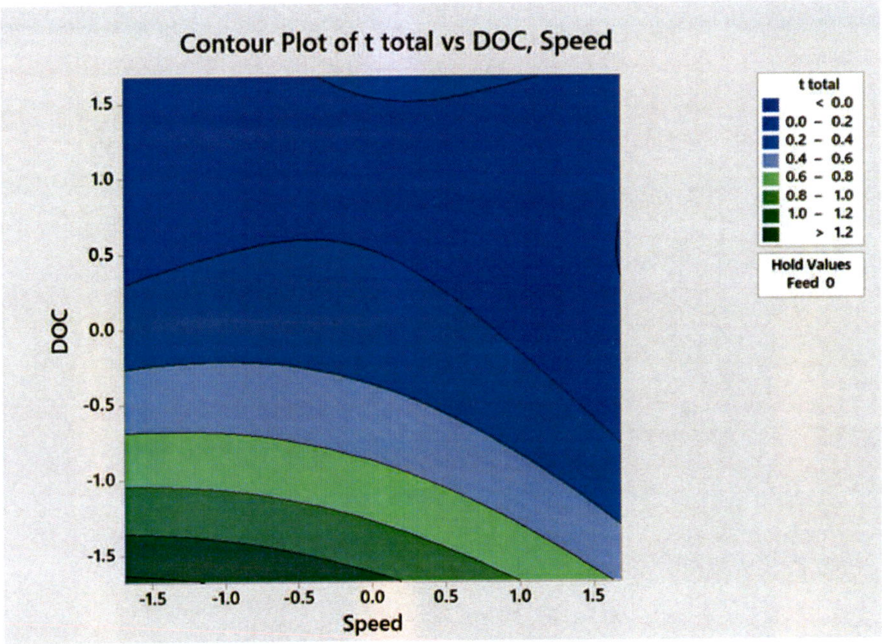


**Fig. 18** Contour plot of speed, depth of cut against processing time

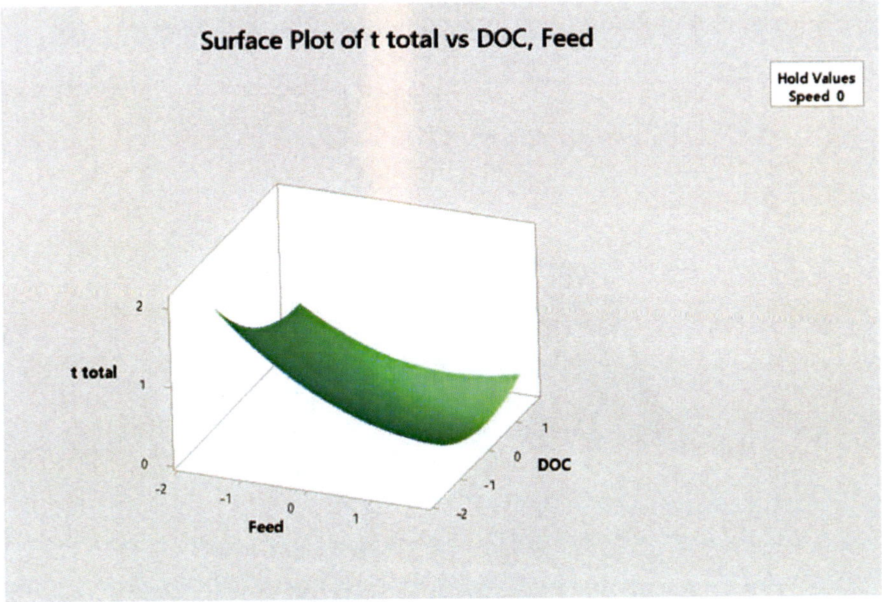


**Fig. 19** Surface plot of depth of cut, feed versus processing time

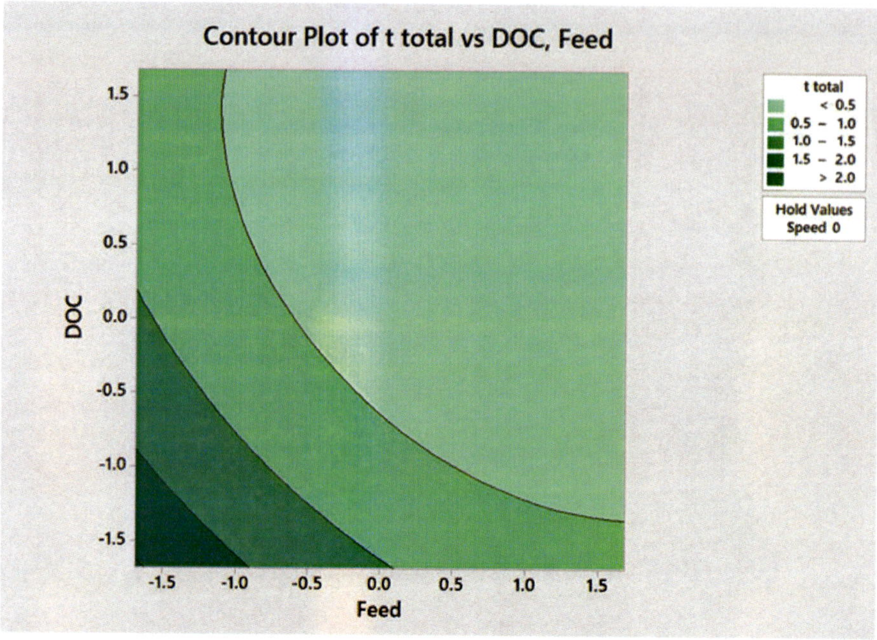


**Fig. 20** Contour plot of depth of cut, feed against processing time

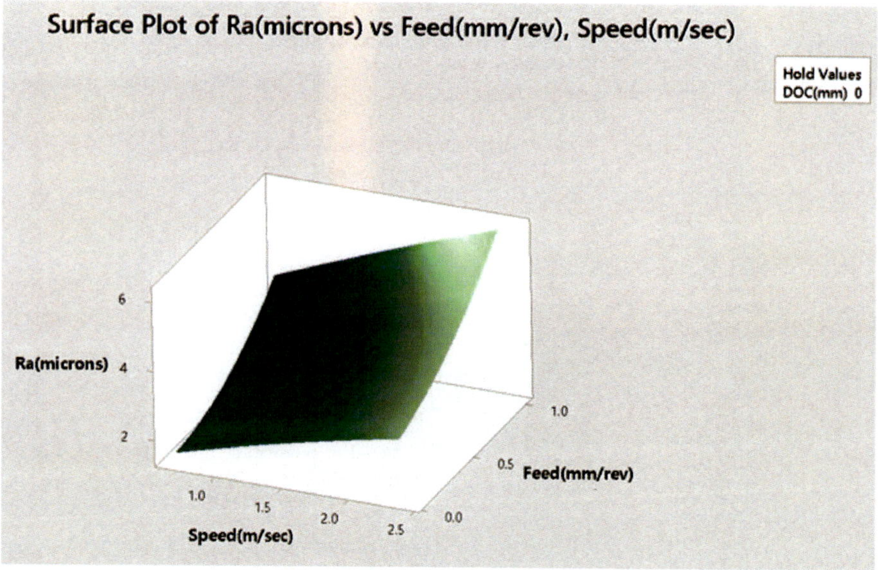


**Fig. 21** Surface plot of speed, feed versus roughness

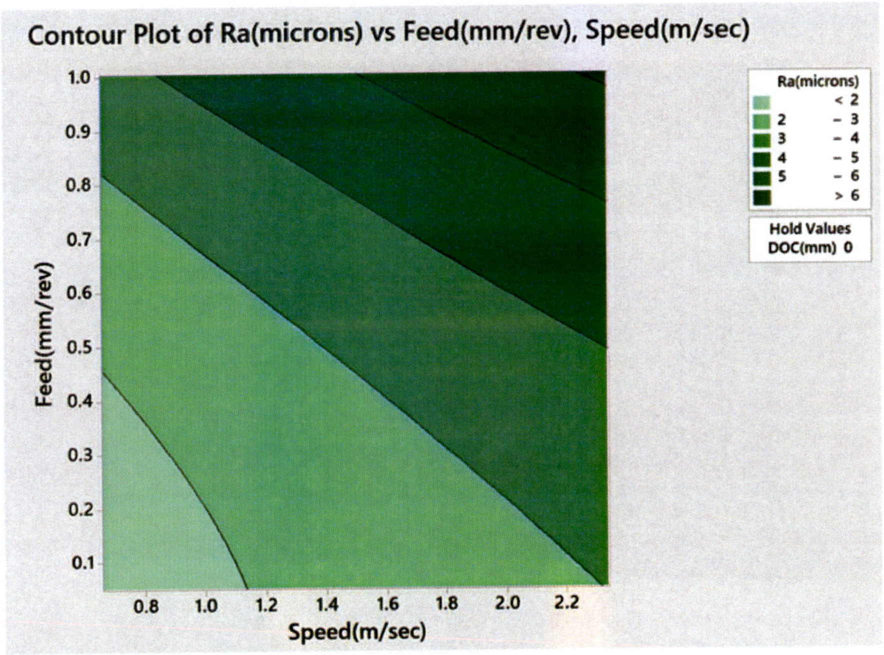


**Fig. 22** Contour plot of speed, feed versus roughness

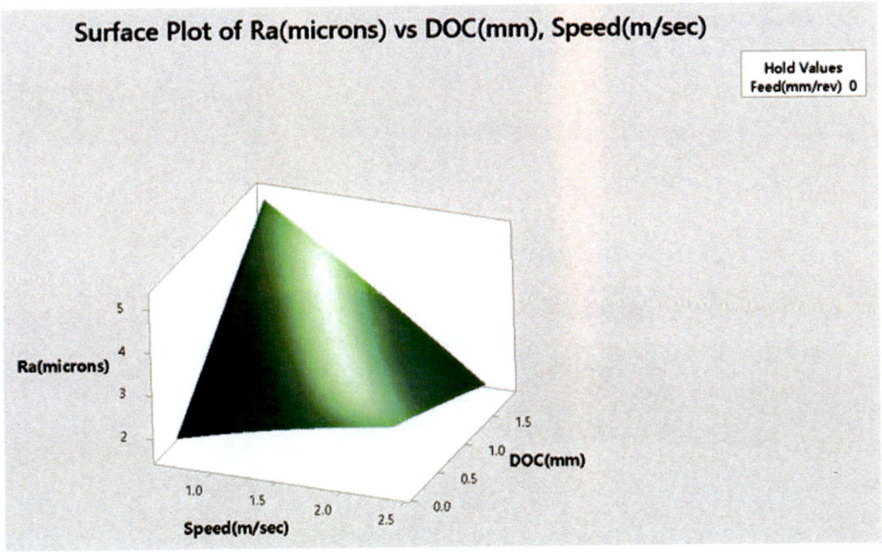


**Fig. 23** Surface plot of rotational speed, depth of cut against roughness

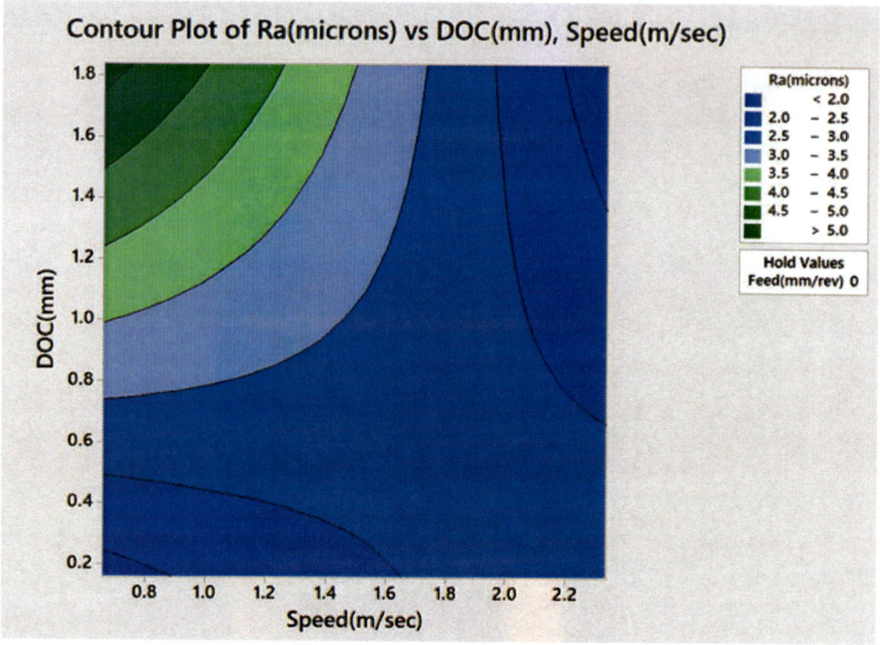


**Fig. 24** Contour plot of speed, depth of cut against roughness

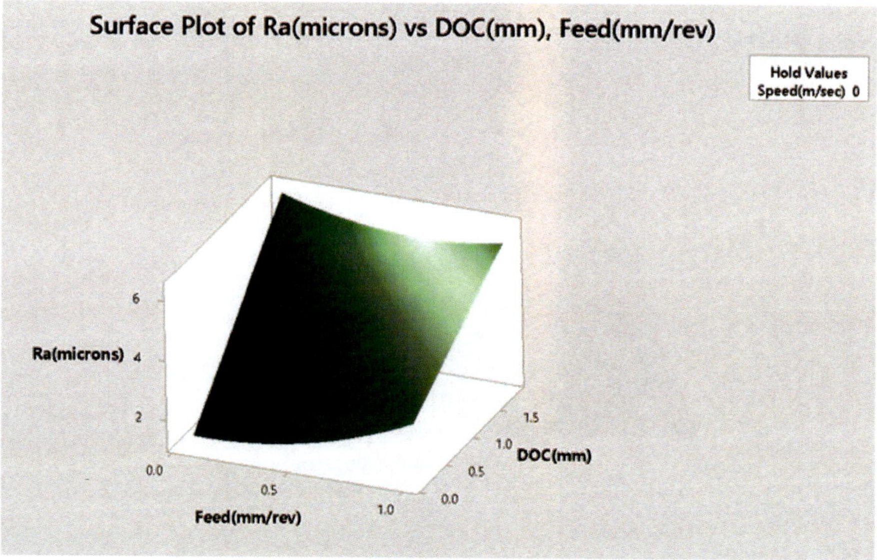


**Fig. 25** Surface plot of feed, depth of cut versus roughness

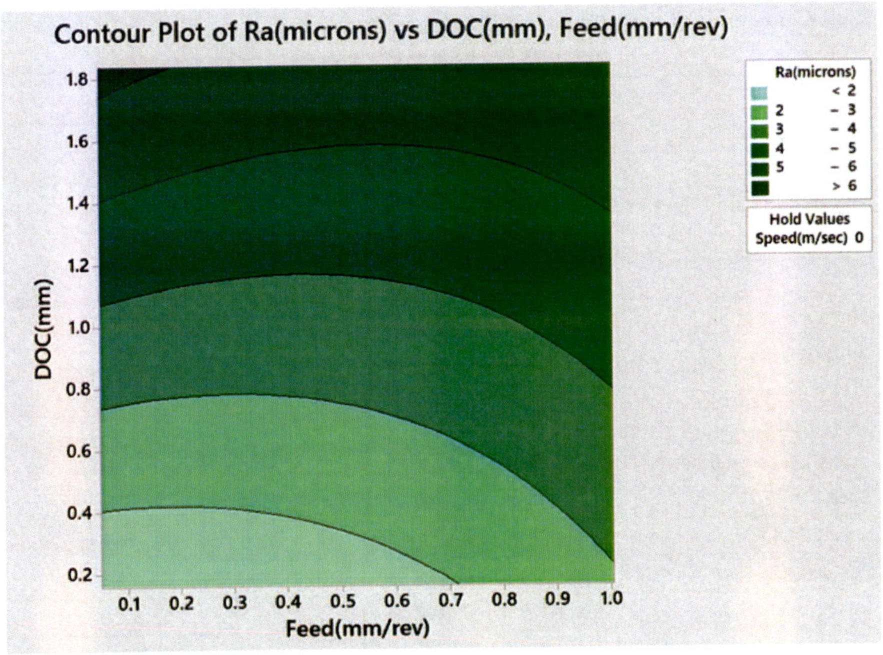


**Fig. 26** Contour plot of feed, depth of cut versus roughness

**Table 5** Anova—processing time

| Source | dof | Adj SS | Adj MS | *F*-value | *P*-value | % contribution |
|---|---|---|---|---|---|---|
| Model | 10 | 2.64379 | 0.26438 | 4.45 | 0.017 | |
| Blocks | 1 | 0.09788 | 0.09788 | 1.65 | 0.231 | |
| Linear | 3 | 1.93357 | 0.64452 | 10.84 | 0.002 | |
| Speed | 1 | 0.08599 | 0.08599 | 1.45 | 0.260 | 2.7 |
| Feed | 1 | 1.03813 | 1.03813 | 17.46 | 0.002 | 32.65 |
| DOC | 1 | 0.80945 | 0.80945 | 13.62 | 0.005 | 25.46 |
| Square | 3 | 0.40677 | 0.13559 | 2.28 | 0.148 | |
| Speed * Speed | 1 | 0.03136 | 0.03136 | 0.53 | 0.486 | 0.98 |
| Feed * Feed | 1 | 0.17250 | 0.17250 | 2.90 | 0.123 | 5.42 |
| DOC * DOC | 1 | 0.20158 | 0.20158 | 3.39 | 0.099 | 6.34 |
| Two way interaction | 3 | 0.19503 | 0.06501 | 1.09 | 0.401 | |
| Speed * Feed | 1 | 0.10430 | 0.10430 | 1.75 | 0.218 | 3.28 |
| Speed * DOC | 1 | 0.03729 | 0.03729 | 0.63 | 0.449 | 1.17 |
| Feed * DOC | 1 | 0.05343 | 0.05343 | 0.9 | 0.368 | 1.68 |
| Error | 9 | 0.53498 | 0.05944 | | | 16.8 |
| Total | 19 | 3.17877 | | | | 100 |

**Table 6** Anova—carbon emissions

| Source | dof | Adj SS | Adj MS | $F$-value | $P$-value | % contribution |
|---|---|---|---|---|---|---|
| Model | 10 | 0.012757 | 0.001276 | 2.85 | 0.065 | |
| Blocks | 1 | 0.000115 | 0.000115 | 0.26 | 0.624 | |
| Linear | 3 | 0.009573 | 0.003191 | 7.12 | 0.009 | |
| Speed | 1 | 0.000518 | 0.000518 | 1.16 | 0.310 | 3.08 |
| Feed | 1 | 0.005344 | 0.005344 | 11.93 | 0.007 | 31.8 |
| DOC | 1 | 0.003711 | 0.003711 | 8.28 | 0.018 | 22.1 |
| Square | 3 | 0.002540 | 0.000847 | 1.89 | 0.202 | |
| Speed * Speed | 1 | 0.0000 | 0.00 | 0 | 0.999 | 0 |
| Feed * Feed | 1 | 0.001627 | 0.001627 | 3.63 | 0.089 | 9.69 |
| DOC * DOC | 1 | 0.001111 | 0.001111 | 2.48 | 0.150 | 6.61 |
| Two way interaction | 3 | 0.000488 | 0.000163 | 0.36 | 0.781 | |
| Speed * Feed | 1 | 0.000038 | 0.000038 | 0.09 | 0.777 | 0.022 |
| Speed * DOC | 1 | 0.000019 | 0.000019 | 0.04 | 0.843 | 0.011 |
| Feed * DOC | 1 | 0.000431 | 0.000431 | 0.96 | 0.352 | 2.56 |
| Error | 9 | 0.004033 | 0.000448 | | | 24.02 |
| Total | 19 | 0.016790 | | | | 100 |

From Table 6, it is found that feed and depth of cut are the major contributing parameters and the interaction terms have very minimal effect. $R$ square value obtained for the above table was 75.98%.

From Table 7, it is found that interaction of speed and depth of cut is the major contributing factor. R square value obtained for the above table was 66.23%.

Regression equations were obtained for both carbon emissions as well as processing time using Minitab.

$$\begin{aligned}\text{CE total} &= 0.08648 - 0.00616\ \text{Speed} - 0.01978\ \text{Feed} - 0.01648 \\ &\text{DOC} - 0.00001\ \text{Speed} * \text{Speed} + 0.01063\ \text{Feed} * \text{Feed} + 0.00878 \\ &\text{DOC} * \text{DOC} + 0.00219\ \text{Speed} * \text{Feed} + 0.00153\ \text{Speed} * \text{DOC} \\ &+ 0.00734\ \text{Feed} * \text{DOC}\end{aligned}$$

$$\begin{aligned}t\ \text{total} &= 0.295 - 0.0793\ \text{Speed} - 0.2757\ \text{Feed} - 0.2434 \\ &\text{DOC} - 0.0467\ \text{Speed} * \text{Speed} + 0.1094\ \text{Feed} * \text{Feed} + 0.1183 \\ &\text{DOC} * \text{DOC} + 0.1142\ \text{Speed} * \text{Feed} + 0.0683 \\ &\text{Speed} * \text{DOC} + 0.0817\ \text{Feed} * \text{DOC}\end{aligned}$$

$$\begin{aligned}\text{Ra(microns)} &= 0.78 + 1.17\ \text{Speed} - 0.22\ \text{Feed} + 3.10 \\ &\text{DOC} - 0.101\ \text{Speed} * \text{Speed} + 2.02\ \text{Feed} * \text{Feed} - 0.020\end{aligned}$$

**Table 7** Anova—surface roughness

| Source | dof | Adj SS | Adj MS | *F*-value | *P*-value | % contribution |
|---|---|---|---|---|---|---|
| Model | 10 | 3.46282 | 0.34628 | 1.76 | 0.063 | |
| Blocks | 1 | 0.41698 | 0.41698 | 2.13 | 0.179 | 7.9 |
| Linear | 3 | 0.95835 | 0.31945 | 1.63 | 0.251 | |
| Speed | 1 | 0.09964 | 0.09964 | 0.51 | 0.494 | 1.9 |
| Feed | 1 | 0.00159 | 0.00159 | 0.01 | 0.930 | 0.03 |
| DOC | 1 | 0.90470 | 0.90470 | 4.61 | 0.060 | 17.3 |
| Square | 3 | 0.42800 | 0.14267 | 0.73 | 0.561 | |
| Speed * Speed | 1 | 0.00923 | 0.00923 | 0.05 | 0.833 | 0.176 |
| Feed * Feed | 1 | 0.40448 | 0.40448 | 2.06 | 0.185 | 7.7 |
| DOC * DOC | 1 | 0.00035 | 0.00035 | 0.00 | 0.967 | 0.0066 |
| Two way interaction | 3 | 1.65949 | 0.55316 | 2.82 | 0.100 | |
| Speed * Feed | 1 | 0.06038 | 0.06038 | 0.31 | 0.593 | 1.15 |
| Speed * DOC | 1 | 1.29686 | 1.29686 | 6.61 | 0.030 | 24.8 |
| Feed * DOC | 1 | 0.30225 | 0.30225 | 1.54 | 0.246 | 5.78 |
| Error | 9 | 1.76597 | 0.19622 | | | 33.77 |
| Total | 19 | 5.22879 | | | | 100 |

$$\text{DOC} * \text{DOC} + 0.58\ \text{Speed} * \text{Feed} - 1.610\ \text{Speed} * \text{DOC}$$
$$- 1.30\ \text{Feed} * \text{DOC}$$

where,

CE total Total carbon emission
$t$ total Processing time
Ra Surface roughness in microns.

## 6 Conclusion

It is found that feed and depth of cut play a major role in determining carbon emissions, surface roughness as well as processing time. Hence, selecting these two values will be a critical task while performing machining. By using optimizer in Minitab, the following conclusions were obtained to minimize carbon emissions, surface roughness, and processing time.

- Speed: 2.341 m/s
- Feed: 0.532121 mm/rev
- DOC: 1.60015 mm.

**Table 8** Pareto solutions

| S. No | Speed (m/s) | Feed (mm/rev) | DOC (mm) | CE ($kgCO_2$) | Time (min) | Ra (μm) |
|---|---|---|---|---|---|---|
| 1 | 1.9970 | 0.2196 | 1.3259 | 0.0710 | 0.0351 | 2.4508 |
| 2 | 1.9984 | 0.2247 | 1.1977 | 0.0697 | 0.0085 | 2.5085 |
| 3 | 1.9999 | 0.3041 | 1.4785 | 0.0731 | 0.0828 | 2.3832 |
| 4 | 1.9999 | 0.2080 | 1.4618 | 0.0728 | 0.0660 | 2.3856 |
| 5 | 1.9999 | 0.2115 | 1.2562 | 0.0703 | 0.0187 | 2.4776 |
| 6 | 1.9996 | 0.2615 | 1.1627 | 0.0693 | 0.0054 | 2.5385 |
| 7 | 1.9962 | 0.2294 | 1.4998 | 0.0733 | 0.0791 | 2.3714 |
| 8 | 2.0000 | 0.2437 | 1.4996 | 0.0733 | 0.0803 | 2.3651 |

By using the above parameters, it is possible to achieve minimal processing time, surface roughness in addition to minimal carbon emissions. The value of minimal output factors were

- CE = 0.070 $kgCO_2$
- SR = 1.90254 μm
- PT = 0.035 min.

The same problem was solved using the multi objective genetic algorithm toolbox in Matlab. The Pareto set is shown in Table 8.

All these Pareto solutions are optimal in their own way. It is up to the operations manager or the concerned decision-maker to select the optimal parameter from the list according to their situation and machine.

## 7 Scope

The scope of the current work is limited to CNC turning of Grey Cast Iron. The same work can also be done for machining of other materials as well as other machining operations too. This work can also be extended by considering other factors like coolant type, coolant level, various tool materials, and various workpiece materials as input factors.

The same optimization can also be accomplished for material removal processing of non-traditional materials including composite materials, which form a vast area to be explored. The same can also be done for traditional machining processes like lathe machining, etc. Other objective functions like power consumption can also be considered.

**Acknowledgements** We have taken sincere efforts in this project. However, it would not have been possible without the support of the people listed below:

We wish to express gratitude to Dr. Prakasan K, Principal In-charge, PSG College of Technology for providing an opportunity and necessary facilities in carrying out this project work.

We wish to express gratitude to Dr. Senthilkumar M, Professor and Head In-charge, Department of Production Engineering, PSG College of Technology, for his constant source of support, encouragement, and motivation throughout our work.

We express heartfelt thanks to the project guide Dr. Devadasan S R, Professor, Department of Production Engineering for his guidance, suggestions, constant support, and encouragement given to us throughout this project.

On specific, we are highly intended to thank our faculty committee, Dr. Krishnaraj V, Associate Professor, Dr. Somasundara Vinoth K, Assistant Professor, Dr. Madhan Mohan G, Associate professor (CAS) and Mr. Rajesh R, Assistant professor, Department of Production Engineering, PSG College of Technology for aiding their help and support on this project.

We would like to thank Mr. Anand K, Assistant professor (Sr. Gr), and Mr. Krishnakumar N, Assistant professor, Department of Production Engineering, PSG College of Technology for their assistance in this project.

We would also like to extend our gratitude to the technicians from MaxByte, Coimbatore, for helping in gathering data required for the project.

We would like to extend our gratitude to the technicians in the production engineering laboratory at PSG College of Technology.

We would like to thank the faculties of the department of Metallurgical Engineering, PSG College of technology for extending technical help.

We would like to thank the faculties of the department of Mechanical engineering, Ramakrishna Engineering College for extending help.

## References

1. Brynjolfsson E, McAfee A (2016) The second machine age: work, progress, and prosperity in a time of brilliant technologies. W. W. Norton & Company, New York
2. Rao SS (2020) Engineering optimization theory and practice. Wiley, Hoboken, NJ, USA
3. Bhushan RK (2013) Optimization of cutting parameters for minimizing power consumption and maximizing tool life during machining of Al alloy SiC particle composites. J Clean Prod 39:242–254. https://doi.org/10.1016/j.jclepro.2012.08.008
4. Jiang Z, Zhou F, Zhang H, Wang Y, Sutherland JW (2015) Optimization of machining parameters considering minimum cutting fluid consumption. J Clean Prod 108:183–191. https://doi.org/10.1016/j.jclepro.2015.06.007
5. Sihag N, Sangwan KS (2018) Development of a multi-criteria optimization model for minimizing carbon emissions and processing time during machining. Procedia CIRP 69:300–305. https://doi.org/10.1016/j.procir.2017.11.060
6. Liu Z-J, Sun D-P, Lin C-X, Zhao X-Q, Yang Y (2016) Multi-objective optimization of the operating conditions in a cutting process based on low carbon emission costs. J Clean Prod 124:266–275. https://doi.org/10.1016/j.jclepro.2016.02.087
7. Sarıkaya M, Güllü A (2014) Taguchi design and response surface methodology based analysis of machining parameters in CNC turning under MQL. J Clean Prod 65:604–616. https://doi.org/10.1016/j.jclepro.2013.08.040
8. Mia M (2018) Mathematical modeling and optimization of MQL assisted end milling characteristics based on RSM and Taguchi method. Measurement 121:249–260. https://doi.org/10.1016/j.measurement.2018.02.017
9. Kumar MV, Kumar BK, Rudresha N (2018) Optimization of machining parameters in CNC turning of stainless steel (EN19) By TAGUCHI'S orthogonal array experiments. Mater Today: Proc 5(5):11395–11407. https://doi.org/10.1016/j.matpr.2018.02.107
10. Bagaber SA, Yusoff AR (2017) Multi-objective optimization of cutting parameters to minimize power consumption in dry turning of stainless steel 316. J Clean Prod 157:30–46. https://doi.org/10.1016/j.jclepro.2017.03.231

11. Bilga PS, Singh S, Kumar R (2016) Optimization of energy consumption response parameters for turning operation using Taguchi method. J Clean Prod 137:1406–1417. https://doi.org/10.1016/j.jclepro.2016.07.220
12. Camposeco-Negrete C (2013) Optimization of cutting parameters for minimizing energy consumption in turning of AISI 6061 T6 using Taguchi methodology and ANOVA. J Clean Prod 53:195–203. https://doi.org/10.1016/j.jclepro.2013.03.049
13. Li C, Xiao Q, Tang Y, Li L (2016) A method integrating Taguchi, RSM and MOPSO to CNC machining parameters optimization for energy saving. J Clean Prod 135:263–275. https://doi.org/10.1016/j.jclepro.2016.06.097
14. Hanafi I, Khamlichi A, Cabrera FM, Almansa E, Jabbouri A (2012) Optimization of cutting conditions for sustainable machining of PEEK-CF30 using TiN tools. J Clean Prod 33:1–9. https://doi.org/10.1016/j.jclepro.2012.05.005
15. Gupta MK, Sood P, Sharma VS (2016) Optimization of machining parameters and cutting fluids during nano-fluid based minimum quantity lubrication turning of titanium alloy by using evolutionary techniques. J Clean Prod 135:1276–1288. https://doi.org/10.1016/j.jclepro.2016.06.184
16. Yan J, Li L (2013) Multi-objective optimization of milling parameters—the trade-offs between energy, production rate and cutting quality. J Clean Prod 52:462–471. https://doi.org/10.1016/j.jclepro.2013.02.030
17. Kuram E, Simsek BT, Ozcelik B, Demirbas E, Askin S (2010) Optimization of the cutting fluids and parameters using Taguchi and ANOVA in milling
18. Mahanta S, Chandrasekaran M, Samanta S (2018) GA based optimization for the production of quality jobs with minimum power consumption in EDM of hybrid MMCs. Mater Today: Proc 5(2):7788–7796. https://doi.org/10.1016/j.matpr.2017.11.457
19. Mohanty UK, Rana J, Sharma A (2017) Multi-objective optimization of electro-discharge machining (EDM) parameter for sustainable machining. Mater Today: Proc 4(8):9147–9157. https://doi.org/10.1016/j.matpr.2017.07.271
20. Kumar R, Bilga PS, Singh S (2017) Multi objective optimization using different methods of assigning weights to energy consumption responses, surface roughness and material removal rate during rough turning operation. J Clean Prod 164:45–57. https://doi.org/10.1016/j.jclepro.2017.06.077
21. Deng Z, Lv L, Li S, Wan L, Liu W, Yan C, Zhang H (2016) Study on the model of high efficiency and low carbon for grinding parameters optimization and its application. J Clean Prod 137:1672–1681. https://doi.org/10.1016/j.jclepro.2016.03.022
22. Camposeco-Negrete C (2015) Optimization of cutting parameters using response surface method for minimizing energy consumption and maximizing cutting quality in turning of AISI 6061 T6 aluminum. J Clean Prod 91:109–117. https://doi.org/10.1016/j.jclepro.2014.12.017
23. Xie N, Zhou J, Zheng B (2018) Selection of optimum turning parameters based on cooperative optimization of minimum energy consumption and high surface quality. Procedia CIRP 72:1469–1474. https://doi.org/10.1016/j.procir.2018.03.099
24. Sangwan KS, Kant G (2017) Optimization of machining parameters for improving energy efficiency using integrated response surface methodology and genetic algorithm approach. Procedia CIRP 61:517–522. https://doi.org/10.1016/j.procir.2016.11.162
25. Sivaiah P, Chakradhar D (2019) Modeling and optimization of sustainable manufacturing process in machining of 17-4 PH stainless steel. Measurement 134:142–152. https://doi.org/10.1016/j.measurement.2018.10.067
26. Mia M, Morshed M, Kharshiduzzaman M, Razi MH, Mostafa MR, Rahman SM, … Kamal AM (2018) Prediction and optimization of surface roughness in minimum quantity coolant lubrication applied turning of high hardness steel. Measurement 118:43–51. https://doi.org/10.1016/j.measurement.2018.01.012
27. Teimouri R, Amini S (2019) A comprehensive optimization of ultrasonic burnishing process regarding energy efficiency and workpiece quality. Surf Coat Technol 375:229–242. https://doi.org/10.1016/j.surfcoat.2019.07.038

28. Peña-Parás L, Maldonado-Cortés D, Rodríguez-Villalobos M, Romero-Cantú AG, Montemayor OE, Herrera M, Hugler W (2019) Optimization of milling parameters of 1018 steel and nanoparticle additive concentration in cutting fluids for enhancing multi-response characteristics. Wear 426–427:877–886. https://doi.org/10.1016/j.wear.2019.01.078
29. Johnson RDJ, Wins KLD, Raj A, Beatrice BA (2014) Optimization of cutting parameters and fluid application parameters during turning of OHNS steel. Procedia Eng 97:172–177. https://doi.org/10.1016/j.proeng.2014.12.239
30. Mia M, Dey PR, Hossain MS, Arafat MT, Asaduzzaman M, Ullah MS, Zobaer ST (2018) Taguchi S/N based optimization of machining parameters for surface roughness, tool wear and material removal rate in hard turning under MQL cutting condition. Measurement 122:380–391. https://doi.org/10.1016/j.measurement.2018.02.016
31. Li C, Tang Y, Cui L, Li P (2013) A quantitative approach to analyze carbon emissions of CNC-based machining systems. J Intell Manuf 26(5):911–922. https://doi.org/10.1007/s10845-013-0812-4
32. Ic YT, Güler ES, Cabbaroğlu C, Yüksel ED, Sağlam HM (2018) Optimisation of cutting parameters for minimizing carbon emission and maximising cutting quality in turning process. Int J Prod Res 56(11):4035–4055. https://doi.org/10.1080/00207543.2018.1442949
33. Tech PSG (n.d.) Design data book. Coimbatore-12: Kalaikathir

# Comparison of Dispatching Rules in a Flow Shop Scheduling Problem Using Simulation

M. Suryaprakash, R. Sridhar, R. Jayachitra, and M. Gomathi Prabha

**Abstract** Just-in-time delivery is one of the major concerns in today's business environment. In the manufacturing world, the scheduling problems are extensively solved by implementing the dispatching rules. The major aim of this paper is to adopt the best production scheduling rule to the valve manufacturing company. The priority sequencing rules include shortest processing time, largest processing time, first come first serve, earliest due date, etc. Simulation software Arena 10.0 is used to compare the performance measures such as work in progress (WIP), number of products waiting and average machine utilization. The results from the Arena simulation provide the best dispatching rule for production scheduling.

**Keywords** Optimization · FCFS · SPT · LPT · Production scheduling

## 1 Introduction

Production scheduling is the process of syncing, arranging, controlling and optimizing work and workloads in a production plant. Production scheduling is used to allocate men, machine and materials in a manufacturing plant. The purpose of scheduling is to reduce the cost and time of production and to maximize the efficiency of processes and minimize the production cost. In this paper, the case company

M. Suryaprakash (✉) · R. Sridhar · R. Jayachitra · M. Gomathi Prabha
Department of Mechanical Engineering, PSG College of Technology, Peelamedu, Coimbatore, Tamil Nadu, India
e-mail: suryaprakash80497@gmail.com

R. Sridhar
e-mail: sridhar.sulur@gmail.com

R. Jayachitra
e-mail: jayachitrapsg@gmail.com

M. Gomathi Prabha
e-mail: prabhamohan.psg@gmail.com

© The Editor(s) (if applicable) and The Author(s), under exclusive license to Springer Nature Singapore Pte Ltd. 2021
S. Mohan et al. (eds.), *Materials, Design, and Manufacturing for Sustainable Environment*, Lecture Notes in Mechanical Engineering,
https://doi.org/10.1007/978-981-15-9809-8_36

which is considered for study follows FCFS dispatching rules for scheduling. This results in increase in operator waiting time in the assembly station. The major aim of this paper is to adopt the best scheduling rules to reduce the above-mentioned shortcoming.

### 1.1 Dispatching Rules

Priority sequencing rules are the most frequently applied heuristics for solving flow shop and job shop scheduling problems in practice because of their low time complexity and ease of implementation. The priority function is a function that assigns a value to the waiting jobs and based on the priority rule used the job to be processed further. The scalar value allocated to the waiting jobs is called an attribute. For SPT, the waiting job with smaller attribute value will be processed first. Similarly, the job with the longest processing time is processed first in LPT rule. Table 1 shows some of the most commonly used rules. Table 2 shows the dispatching rules and queue priority set-up considered for this paper.

## 2 Literature Review

Trucker et al. [1] explain about the dynamic job shop scheduling using real-time data with simulation using arena. The paper provides the dispatching rules based on the job such as shortest processing time (SPT), first come first serve (FCFS), longest processing time (LPT) and earliest due date (EDD).

**Table 1** Dispatching rules

| Rule | Description |
|---|---|
| First come first serve (FCFS) | A routing management that automatically executes queued requests and proceeds by the sequence of their arrival |
| Shortest processing time (SPT) | The job with the shortest total processing time in the queue |
| Longest processing time (LPT) | The job with the longest processing time in the queue |
| Earliest due date (EDD) | The job which possesses the earliest due date in the queue will be executed first |

**Table 2** Dispatching rules, attribute value, queue priority

| Dispatching rule | Attribute value | Queue priority |
|---|---|---|
| FCFS | Zero (not used) | First in first out |
| SPT | Total processing time | Lowest attribute value |
| LPT | Total processing time | Highest attribute value |

Kumar et al. [2] explain about the simulation optimization of a flexible manufacturing system using arena for effective scheduling of machines which improves the total system utilization which was performed using certain dispatching rules.

Kaban et al. [3] simulate the scheduling problem using dispatching rules in job shop scheduling and the fact that routing and processing times are dependent on group of items machined or processed.

Windsor et al. [4] describe how simulation can be applied to model a job shop scheduling problem (JSSP) using Arena simulation software.

Kassa et al. [5] describe the importance of JSSP and in its applications over wide area. This paper aims at reducing makespan and improving the performance of the manufacturing system.

Nasiri [6] explains a multi-response optimization methodology based on software simulation for production scheduling as a non-preemptive open shop scheduling (OSS) with stochastic ready times is represented to reduce the mean waiting time of the jobs.

Yang et al. [7] explain the effects of priority sequencing rules and resources allocations on certain performance measures. The outcome shows that shortest processing time (SPT) and earliest due date (EDD) are much more superior to first in first out (FIFO).

de Oliveira Teixeira et al. [8] aim to determine which scheduling rule to be applied in order to minimize makespan. The first simulation model was random and for successive modelling, dispatching rules and hybrid rules through weighted sum was used.

Chiang et al. [9] address the job shop scheduling problem with due date-based objectives including the tardy rate, mean tardiness and maximum tardiness. Eighteen priority sequencing rules are selected, and their design concepts and features are discussed.

Hicks et al. [10] explore the influence of data update period and the minimum set-up, machining and transfer times under stochastic capacity conditions.

## 2.1 Problem Background

Flow shops are an important part in the world of manufacturing. Scheduling is the process of organizing the resources for production through varying lots and routings.

The problems addressed in this paper are as follows: operators are waiting for the components in the assembly station because of poor scheduling. The scheduling was manual and does not follow any technique.

## 3 Methodology

The following flow chart shown in Fig. 1 has adopted to solve the above-described problems in the flow shop.

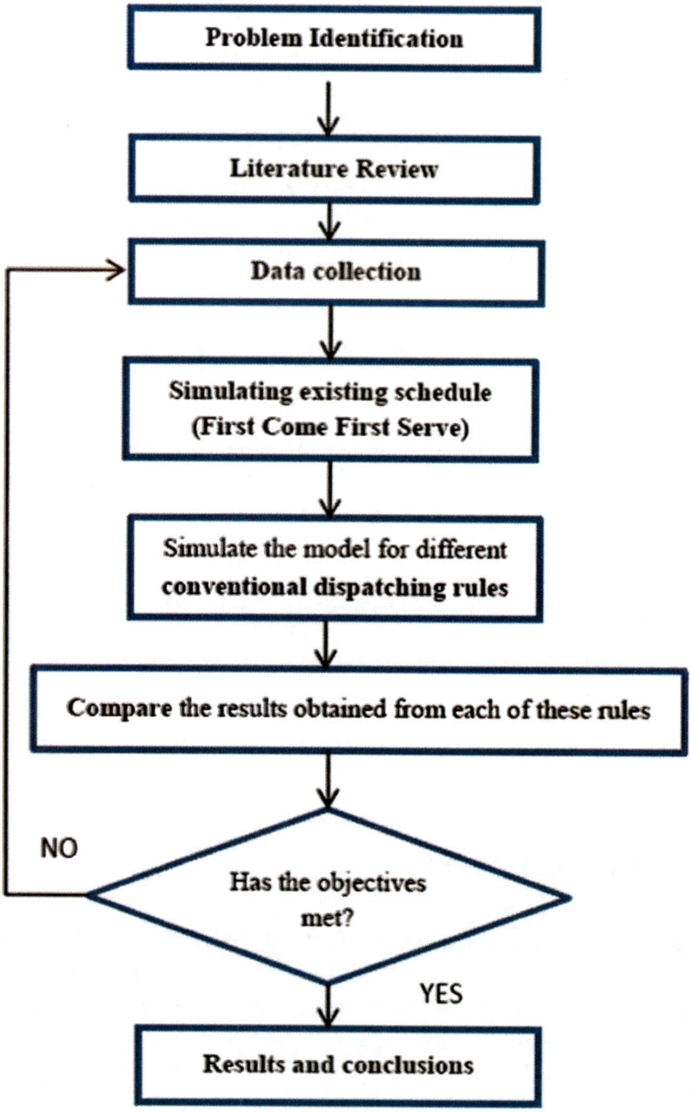


**Fig. 1** Methodology flow chart

# 4 Performance Measurement

Basically, there are 21 performance measures and the most important three features are considered for this project.

- **Average work in progress**: The goal of the project is to minimize the total work in progress inventory between the stations. This results in the reduction of waiting time for the successive operations.
- **Average number of parts waiting in the queue**: The goal is to minimize average number of parts waiting in the queue. This specifies the number of parts waiting in the queue.
- **Average machine utilization**: The aim is to improve the average utilization of machines. This is the measure of how intensively a machine is engaged.

# 5 Simulation Model Creation

In this paper, four model types are considered. Input data file for model creation consists of the following information:

- **Job processing sequence**: The routing of parts in a simulation depends upon the entity process flow. The sequences are provided to accomplish the processes in a certain order. Figure 2 depicts the process flow for both the valves.
- **Process data**: Part process plans with machine and processing time information. Table 3 shows the processing time and corresponding machine plans for gate valve and globe valve, and the rest of the data is shown in Appendix.

## *5.1 Simulation Assumptions*

Simulation made with arena follows certain assumptions, and those assumptions are listed below:

- Processing time is defined by their set-up times and machining time which is processed independently on machine.
- Loading and unloading time on the machine is considered negligible.
- Time for routing was assumed to be negligible.
- Part rejection rate was assumed to be negligible.
- Only machining operations such as turning and finishing are modelled for a job. Additional operations like grinding, welding, final inspection, etc., have not been considered for simulation.

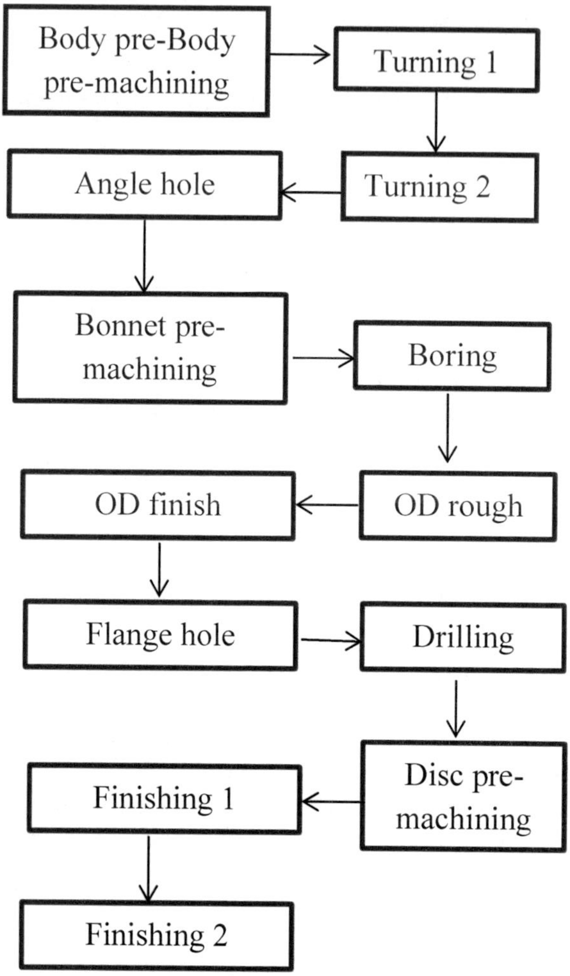


**Fig. 2** Process flow

## 6 Simulation Using Arena

The simulation software, Arena v10.0, is used to simulate the existing model (FCFS) and other two dispatching rules, SPT and LPT. Then, the results were compared with each other and a conclusion was made. The time study performed for collecting the process time only includes job set-up and machining time. All the other activity times such as loading, unloading, cleaning and routing time are not taken into consideration for study. All the parts were taken into consideration for study irrespective of its customer requirements. Operations such as welding, grinding, burr removal and inspection are not considered for simulation.

**Table 3** Process times

| Size and class | Components | Process time globe valve (mins) | Process time gate valve (mins) |
|---|---|---|---|
| DN15CL1500 (0.5 inch) | *Body* | | |
| | Pre-machining | 16 | 20 |
| | Finishing | | |
| | Turning 1 | 28 | 32 |
| | Turning 2 | 12 | 16 |
| | Angle hole | 8 | 12 |
| | *Bonnet* | | |
| | Boring | 30 | 30.5 |
| | Finishing | | |
| | OD roughing | 18 | 18 |
| | OD finishing | 22 | 22 |
| | Flange hole | 6 | 6 |
| | Gland flange | | |
| | Drilling | 3 | 3.5 |
| | *Disc* | | |
| | Pre-machining | 4 | 5 |
| | Finishing | | |
| | Operation 1 | 7 | 7.5 |
| | Operation 2 | 10 | 10 |

The products considered for study are globe valve and gate valve. Figures 3 and 4 show the globe valve and gate valve, respectively.

## 6.1 Existing Model Simulation

The simulation has been carried out for the existing model. The case company follows first come first serve (FCFS) basis. FCFS is a scheduling algorithm and network routing mechanism which executes waiting requests and processes automatically in the order of their arrival. The part routing (sequence of part arrival) and attribute value for the FCFS model are shown in Table 4.

From the simulation, the following results are obtained, the average WIP between stations was found out to be 4.6175, total number of parts waiting was found to be 4 and the average machine utilization was found to be 35%. In the component name, '15, 20, 40, 50' specifies the size of the valve, 12.7 mm, 19.05 mm, 38.1 mm and 50.8 mm, respectively. The numbers '5 and 2' represent globe valve and gate valve, respectively. The terms '1500 and 4500' represent the pressure rating for the valves.

**Fig. 3** Globe valve

**Fig. 4** Gate valve

**Table 4** Sequence of part arrival and attribute value for FCFS model

| Sequence of part arrival | Component name | Attribute value |
|---|---|---|
| Sequence 1 | DN155CL1500 | 0 |
| Sequence 2 | DN155CL4500 | 0 |
| Sequence 3 | DN205CL1500 | 0 |
| Sequence 4 | DN205CL4500 | 0 |
| Sequence 5 | DN405CL1500 | 0 |
| Sequence 6 | DN405CL4500 | 0 |
| Sequence 7 | DN505CL1500 | 0 |
| Sequence 8 | DN505CL4500 | 0 |
| Sequence 9 | DN152CL1500 | 0 |
| Sequence 10 | DN152CL4500 | 0 |
| Sequence 11 | DN202CL1500 | 0 |
| Sequence 12 | DN202CL4500 | 0 |
| Sequence 13 | DN402CL1500 | 0 |
| Sequence 14 | DN402CL4500 | 0 |
| Sequence 15 | DN502CL1500 | 0 |
| Sequence 16 | DN502CL4500 | 0 |

### 6.2 *Shortest Processing Time*

The shortest processing time (SPT) is a priority sequencing rule which orders the job in the order of increasing processing times. The lowest attribute value is considered for the SPT simulation; i.e. the job with the shortest processing time is considered for processing on a machine. Table 5 shows the sequence for SPT and its corresponding attribute value.

From the simulation, the following results are obtained, the average WIP between stations was found out to be 1.1986, total number of parts waiting was found to be 1 and the average machine utilization was found to be 49%.

### 6.3 *Longest Processing Time*

The longest processing time is a priority sequencing rule, which orders the job in the order of decreasing processing times. Whenever a machine is free, the job with the largest processing time gets executed on that machine. The highest attribute value is considered for the LPT simulation. Table 6 shows the sequence for LPT and its corresponding attribute value.

**Table 5** Sequence of part arrival and attribute value for SPT model

| Sequence of part arrival | Component name | Attribute value |
|---|---|---|
| Sequence 1 | DN155CL1500 | 164 |
| Sequence 2 | DN155CL4500 | 164 |
| Sequence 3 | DN205CL1500 | 169.5 |
| Sequence 4 | DN205CL4500 | 169.5 |
| Sequence 5 | DN152CL1500 | 182.5 |
| Sequence 6 | DN152CL4500 | 183.5 |
| Sequence 7 | DN202CL1500 | 187 |
| Sequence 8 | DN202CL4500 | 188.5 |
| Sequence 9 | DN405CL1500 | 243.5 |
| Sequence 10 | DN405CL4500 | 243.5 |
| Sequence 11 | DN402CL1500 | 258.5 |
| Sequence 12 | DN402CL4500 | 258.5 |
| Sequence 13 | DN505CL1500 | 293 |
| Sequence 14 | DN505CL4500 | 293 |
| Sequence 15 | DN502CL1500 | 302 |
| Sequence 16 | DN502CL4500 | 302 |

**Table 6** Sequence of part arrival and attribute value for LPT model

| Sequence of part arrival | Component name | Attribute value |
|---|---|---|
| Sequence 1 | DN502CL4500 | 302 |
| Sequence 2 | DN502CL1500 | 302 |
| Sequence 3 | DN505CL4500 | 293 |
| Sequence 4 | DN505CL1500 | 293 |
| Sequence 5 | DN402CL4500 | 258.5 |
| Sequence 6 | DN402CL1500 | 258.5 |
| Sequence 7 | DN405CL4500 | 243.5 |
| Sequence 8 | DN405CL1500 | 243.5 |
| Sequence 9 | DN202CL4500 | 188.5 |
| Sequence 10 | DN202CL1500 | 187 |
| Sequence 11 | DN152CL4500 | 183.5 |
| Sequence 12 | DN152CL1500 | 182.5 |
| Sequence 13 | DN205CL4500 | 169.5 |
| Sequence 14 | DN205CL1500 | 169.5 |
| Sequence 15 | DN155CL4500 | 164 |
| Sequence 16 | DN155CL1500 | 164 |

From the simulation, the following results are obtained, the average WIP between stations was found out to be 78.7613, total number of parts waiting was found to be 78 and the average machine utilization was found to be 49%

## 6.4 Comparison of Results

The results obtained from each simulation model have been compared with each other, and the best dispatching rule was decided based on that comparison. The performance measures such as WIP, numbers waiting and average machine utilization are compared for each dispatching rule, and they are shown in Figs. 5, 6 and 7, respectively.

The above pie chart represents the number of parts waiting between workstations. This also makes sure that SPT model has the lowest number of components waiting between stations.

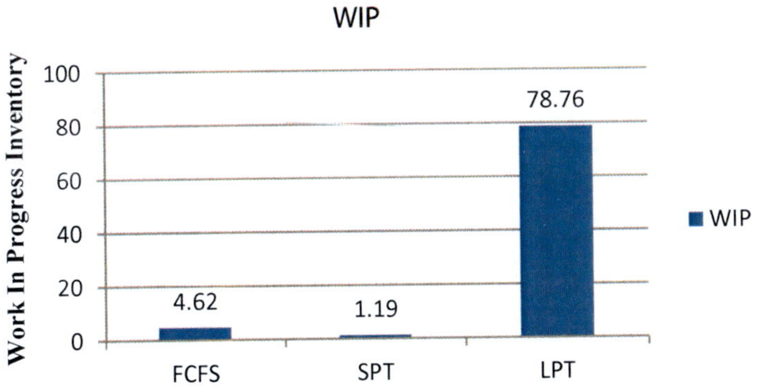


**Fig. 5** Comparison of WIP

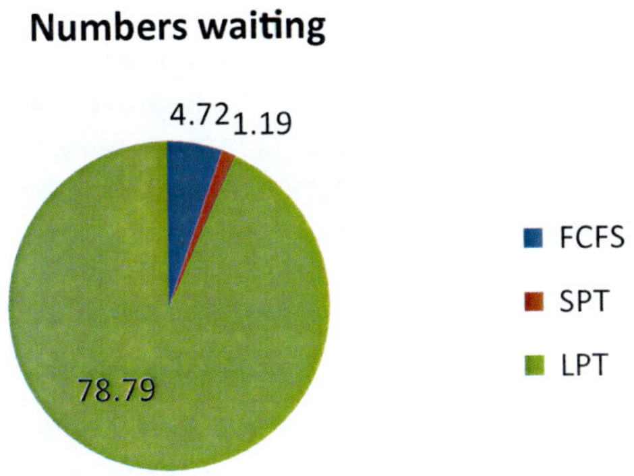


**Fig. 6** Comparison of numbers waiting

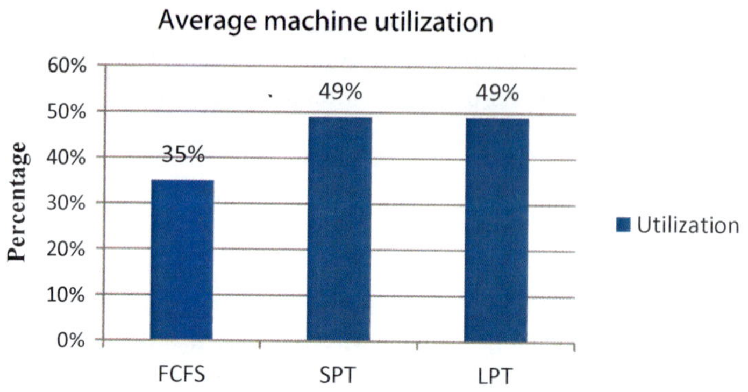


**Fig. 7** Comparison of average machine utilization

The above graph shows that FCFS model has the least machine utilization when compared relatively with SPT and LPT models. Even though LPT model has equal machine utilization, but the other two performance measures perform better than LPT model.

## 6.5 *Waiting Time*

Based on the data collection by brainstorming, it is found that the operators are waiting in the assembly station for 14 and 17 min per batch. Batch size and demand data are shown in Table 7.

The case company works for 26 days per month. The total available time was found out to be 480 min. The available production time excluding breaks and lunch was 420 min per shift. There are four batches per day, and each batch arrives to the assembly station at an interval of 2 h. So, the waiting time for the existing model was 56 min and 68 min for gate valve and globe valve, respectively. So, the total available time for assembly was 296 min. After implementing SPT rule, the waiting time for gate valve and globe valve was reduced by 3.5 min and 4.25 min per batch. This means that the total available assembly time was improved by 31 min. Waiting time per day and net available time for assembly are shown in Figs. 8 and 9, respectively.

From the above comparison, it is inferred that the net available time per day is improved by 31 min after scheduling using SPT priority sequencing rule.

**Table 7** Demand data and batch size

| Valve | Demand per month (nos) | Demand per day (nos) | Batch size (nos) |
|---|---|---|---|
| Gate valve | 800 | 32 | 8 |
| Globe valve | 900 | 36 | 9 |

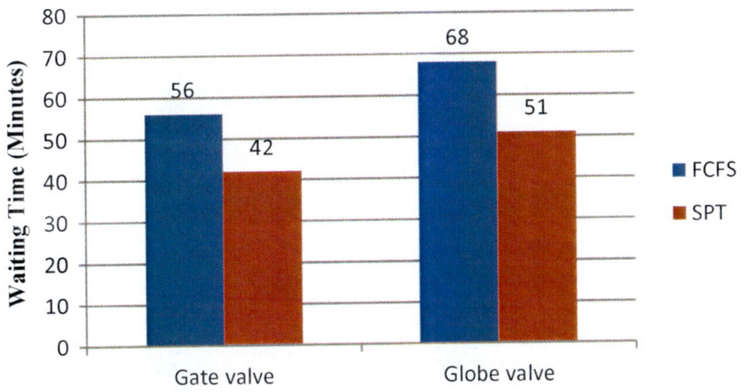


**Fig. 8** Waiting time in assembly station per day

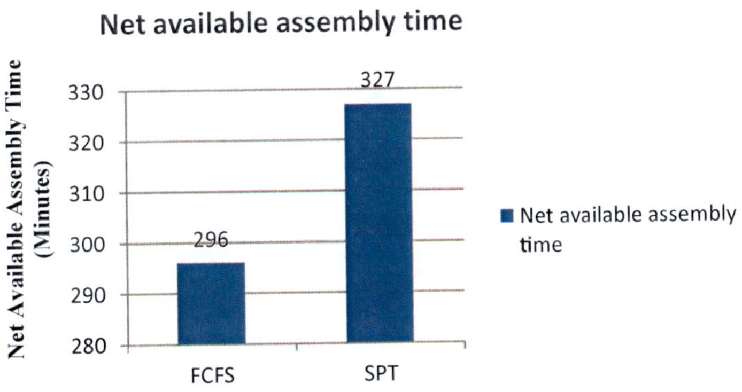


**Fig. 9** Net available assembly time per day

## 6.6 *Takt Time*

Takt time is defined as the rate at which the products are produced or assembled in order to meet the customer demand. Mathematical representation of takt time is given in Eq. 1.

$$\text{Takt time} = \frac{\text{Net available time}}{\text{customer demand}} \tag{1}$$

The takt time of assembly operation for gate valve and globe valve was found out with the data given in Table 7.

Takt time for gate valve = 13.5 min

Takt time for globe valve = 12 min

| | A | B | C | D | E | F | G |
|---|---|---|---|---|---|---|---|
| 11 | | Master Formula | | 80 | | 0.00 | |
| 12 | Task | Activity | VA | Rating | Average Time | Cycle 1 | Cycle 2 |
| 13 | Time Study- Gate valve assembly | | | | | | |
| 14 | | Open benchwise | | 80 | 0.500 | 0.500 | 0.500 |
| 15 | | Place body | | 80 | 0.500 | 0.500 | 0.500 |
| 16 | | Tighten the benchwise | | 80 | 0.500 | 0.500 | 0.500 |
| 17 | | Reach for airgun | N | 80 | 0.500 | 0.500 | 0.500 |
| 18 | | Take the air gun | R | 80 | 0.500 | 0.500 | 0.500 |
| 19 | | Clean the body | R | 80 | 0.500 | 0.500 | 0.500 |
| 20 | | Grab seat ring mandrel | | 80 | 1.100 | 1.000 | 1.200 |
| 21 | | Fix the ring mandrel to the body | | 80 | 0.500 | 0.500 | 0.500 |
| 22 | | Grab seat ring | | 80 | 0.750 | 1.000 | 0.500 |
| 23 | | Place seat ring over the body | R | 80 | 0.500 | 0.500 | 0.500 |
| 24 | | Grab disc | | 80 | 0.500 | 0.500 | 0.500 |
| 25 | | Place the disc | | 80 | 1.050 | 1.000 | 1.100 |
| 26 | | Lapping | | 80 | 1.050 | 1.000 | 1.100 |
| 27 | | Match disc | R | 80 | 0.500 | 0.500 | 0.500 |
| 28 | | Check for grovee match | R | 80 | 1.000 | 1.000 | 1.000 |
| 29 | | Remove the mandrel from the body | | 80 | 1.000 | 1.000 | 1.000 |
| 30 | | Place the gasket over the body | | 80 | 1.000 | 1.000 | 1.000 |
| 31 | | Place the bonnet over the gasket | | 80 | 1.000 | 1.000 | 1.000 |
| 32 | | Grab bolts | | 80 | 0.900 | 1.000 | 0.800 |
| 33 | | Tighten body and bonnet | | 80 | 1.000 | 1.000 | 1.000 |
| 34 | | Place the handwheel over the valve | | 80 | 1.000 | 1.000 | 1.000 |
| 35 | | Grab the nut | | 80 | 1.000 | 1.000 | 1.000 |
| 36 | | Tighten the nut | | 80 | 0.900 | 1.000 | 0.800 |
| 37 | | Open benchwise and remove the valve | | 80 | 0.500 | 0.500 | 0.500 |

**Fig. 10** Input details of gate valve assembly for time study

In order to check the assembly time in the assembly station, time study was performed using Timer Pro software. The input details and output of the time study for gate valve and globe valve are shown in Figs. 10, 11, 12 and 13, respectively.

The total time taken for assembling a gate valve was found to be 18 min if the production follows FCFS model. But, the actual takt time for assembling a gate valve is 13.5 min. After scheduling the production using SPT model, the assembly time was reduced to 16.1 min.

The total time taken for assembling a globe valve was found to be 16.4 min if the production follows FCFS model. But, the actual takt time for assembling a gate valve is 12 min. After scheduling the production using SPT model, the assembly time was reduced to 15 min. The comparison of takt time before and after optimizing the scheduling is shown in Fig. 14.

**Time Study Summary**

Copyright © 2006, ACS Inc. All rights reserved.

| | |
|---|---|
| Task: | Gate valve assembly |
| Department: | Quality |
| Shift: | 01 |
| M/C No | Assembly station |
| Manager | XXX |
| Engineer | YYY |
| Skill | 80% |
| Study | ManualTime_Raw |

Send to Balance

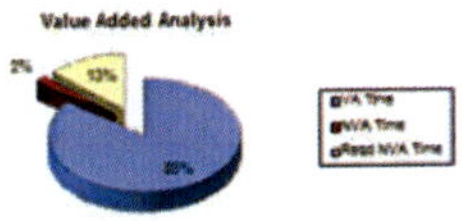


| Element Description | Total Time | Leveled Time | Average Rating | Min Time | MaxTime | Count | Average Time | VA Indicator | VA Time | NVA Time | Reqd NVA Time | Avoid Time | Unavoid Time |
|---|---|---|---|---|---|---|---|---|---|---|---|---|---|
| <total> | 18.000 | .152 | 1 | .500 | 1.200 | 24 | .750 | | .129 | .002 | .020 | .000 | .000 |
| Open benchwise | .500 | .002 | 1 | .500 | .500 | 1 | .500 | | .002 | .000 | .000 | .000 | .000 |
| Place body | .500 | .002 | 1 | .500 | .500 | 1 | .500 | | .002 | .000 | .000 | .000 | .000 |
| Tighten the benchwise | .500 | .002 | 1 | .500 | .500 | 1 | .500 | | .002 | .000 | .000 | .000 | .000 |
| Reach for airgun | .500 | .002 | 1 | .500 | .500 | 1 | .500 | N | .000 | .002 | .000 | .000 | .000 |
| Take the air gun | .500 | .002 | 1 | .500 | .500 | 1 | .500 | R | .000 | .000 | .002 | .000 | .000 |
| Clean the body | .500 | .002 | 1 | .500 | .500 | 1 | .500 | R | .000 | .000 | .002 | .000 | .000 |
| Grab seat ring mandrel | 1.200 | .013 | 1 | 1.200 | 1.200 | 1 | 1.200 | | .013 | .000 | .000 | .000 | .000 |
| Fix the ring mandrel to the body | .500 | .002 | 1 | .500 | .500 | 1 | .500 | | .002 | .000 | .000 | .000 | .000 |
| Grab seat ring | .500 | .004 | 1 | .500 | .500 | 1 | .500 | | .004 | .000 | .000 | .000 | .000 |
| Place seat ring over the body | .500 | .002 | 1 | .500 | .500 | 1 | .500 | R | .000 | .000 | .002 | .000 | .000 |
| Grab disc | .500 | .002 | 1 | .500 | .500 | 1 | .500 | | .002 | .000 | .000 | .000 | .000 |
| Place the disc | 1.100 | .012 | 1 | 1.100 | 1.100 | 1 | 1.100 | | .012 | .000 | .000 | .000 | .000 |
| Lapping | 1.100 | .012 | 1 | 1.100 | 1.100 | 1 | 1.100 | | .012 | .000 | .000 | .000 | .000 |
| Match disc | .500 | .002 | 1 | .500 | .500 | 1 | .500 | R | .000 | .000 | .002 | .000 | .000 |
| Check for grovee match | 1.000 | .010 | 1 | 1.000 | 1.000 | 1 | 1.000 | R | .000 | .000 | .010 | .000 | .000 |
| Remove the mandrel from the body | 1.000 | .010 | 1 | 1.000 | 1.000 | 1 | 1.000 | | .010 | .000 | .000 | .000 | .000 |
| Place the gasket over the body | 1.000 | .010 | 1 | 1.000 | 1.000 | 1 | 1.000 | | .010 | .000 | .000 | .000 | .000 |
| Place the bonnet over the gasket | 1.000 | .010 | 1 | 1.000 | 1.000 | 1 | 1.000 | | .010 | .000 | .000 | .000 | .000 |
| Grab bolts | .800 | .007 | 1 | .800 | .800 | 1 | .800 | | .007 | .000 | .000 | .000 | .000 |
| Tighten body and bonnet | 1.000 | .010 | 1 | 1.000 | 1.000 | 1 | 1.000 | | .010 | .000 | .000 | .000 | .000 |
| Place the handwheel over the valve | 1.000 | .010 | 1 | 1.000 | 1.000 | 1 | 1.000 | | .010 | .000 | .000 | .000 | .000 |
| Grab the nut | 1.000 | .010 | 1 | 1.000 | 1.000 | 1 | 1.000 | | .010 | .000 | .000 | .000 | .000 |
| Tighten the nut | .800 | .007 | 1 | .800 | .800 | 1 | .800 | | .007 | .000 | .000 | .000 | .000 |
| Open benchwise and remove the valve | .500 | .002 | 1 | .500 | .500 | 1 | .500 | | .002 | .000 | .000 | .000 | .000 |

**Fig. 11** Output of time study for gate valve assembly

| | Task | Activity | VA | Rating | Average Time | Cycle 1 | Cycle 2 |
|---|---|---|---|---|---|---|---|
| 11 | | Master Formula | | 80 | | 0.00 | |
| 13 | Time Study- Globe valve assembly | | | | | | |
| 14 | | Open benchwise | | 80 | 0.008 | 0.008 | 0.008 |
| 15 | | Place body | | 80 | 0.008 | 0.008 | 0.007 |
| 16 | | Tighten the benchwise | | 80 | 0.008 | 0.008 | 0.008 |
| 17 | | Reach for airgun | N | 80 | 0.008 | 0.008 | 0.008 |
| 18 | | Take the air gun | R | 80 | 0.008 | 0.008 | 0.008 |
| 19 | | Clean the body | R | 80 | 0.008 | 0.008 | 0.008 |
| 20 | | Grab seat ring mandrel | | 80 | 0.008 | 0.008 | 0.008 |
| 21 | | Fix the ring mandrel to the body | | 80 | 0.035 | 0.036 | 0.033 |
| 22 | | Grab seat ring | | 80 | 0.008 | 0.008 | 0.008 |
| 23 | | Place seat ring over the body | R | 80 | 0.017 | 0.017 | 0.017 |
| 24 | | Grab disc | | 80 | 0.008 | 0.008 | 0.008 |
| 25 | | Place the disc | | 80 | 0.017 | 0.017 | 0.016 |
| 26 | | Check for grovee match | R | 80 | 0.033 | 0.033 | 0.033 |
| 27 | | Remove the mandrel from the body | | 80 | 0.018 | 0.018 | 0.018 |
| 28 | | Place the gasket over the body | | 80 | 0.017 | 0.017 | 0.017 |
| 29 | | Place the bonnet over the gasket | | 80 | 0.053 | 0.055 | 0.050 |
| 30 | | Grab bolts | | 80 | 0.017 | 0.017 | 0.017 |
| 31 | | Tighten body and bonnet | | 80 | 0.036 | 0.036 | 0.036 |
| 32 | | Place the handwheel over the valve | | 80 | 0.017 | 0.017 | 0.017 |
| 33 | | Grab the nut | | 80 | 0.017 | 0.017 | 0.017 |
| 34 | | Tighten the nut | | 80 | 0.033 | 0.033 | 0.033 |
| 35 | | Open benchwise and remove the valve | | 80 | 0.017 | 0.017 | 0.017 |

**Fig. 12** Input details of globe valve assembly for time study

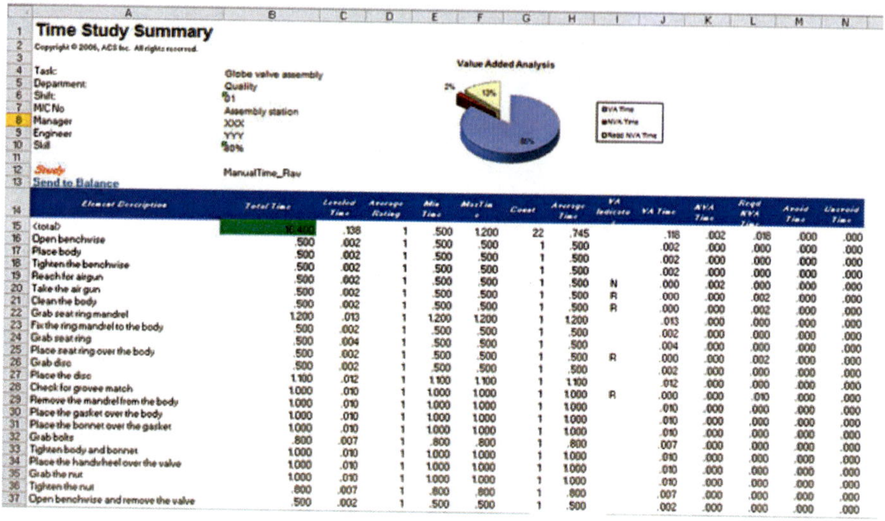


| | Element Description | Total Time | Leveled Time | Average Rating | Min Time | MaxTime | Count | Average Time | VA Indicator | VA Time | NVA Time | Reqd NVA Time | Avoid Time | Unavoid Time |
|---|---|---|---|---|---|---|---|---|---|---|---|---|---|---|
| 15 | <total> | 16.400 | .138 | 1 | .500 | 1.200 | 22 | .745 | | .118 | .002 | .018 | .000 | .000 |
| 16 | Open benchwise | .500 | .002 | 1 | .500 | .500 | 1 | .500 | | .002 | .000 | .000 | .000 | .000 |
| 17 | Place body | .500 | .002 | 1 | .500 | .500 | 1 | .500 | | .002 | .000 | .000 | .000 | .000 |
| 18 | Tighten the benchwise | .500 | .002 | 1 | .500 | .500 | 1 | .500 | | .002 | .000 | .000 | .000 | .000 |
| 19 | Reach for airgun | .500 | .002 | 1 | .500 | .500 | 1 | .500 | N | .000 | .002 | .000 | .000 | .000 |
| 20 | Take the air gun | .500 | .002 | 1 | .500 | .500 | 1 | .500 | R | .000 | .000 | .002 | .000 | .000 |
| 21 | Clean the body | .500 | .002 | 1 | .500 | .500 | 1 | .500 | R | .000 | .000 | .002 | .000 | .000 |
| 22 | Grab seat ring mandrel | 1.200 | .013 | 1 | 1.200 | 1.200 | 1 | 1.200 | | .013 | .000 | .000 | .000 | .000 |
| 23 | Fix the ring mandrel to the body | .500 | .002 | 1 | .500 | .500 | 1 | .500 | | .002 | .000 | .000 | .000 | .000 |
| 24 | Grab seat ring | .500 | .004 | 1 | .500 | .500 | 1 | .500 | | .004 | .000 | .000 | .000 | .000 |
| 25 | Place seat ring over the body | .500 | .002 | 1 | .500 | .500 | 1 | .500 | R | .000 | .000 | .002 | .000 | .000 |
| 26 | Grab disc | .500 | .002 | 1 | .500 | .500 | 1 | .500 | | .002 | .000 | .000 | .000 | .000 |
| 27 | Place the disc | 1.100 | .012 | 1 | 1.100 | 1.100 | 1 | 1.100 | | .012 | .000 | .000 | .000 | .000 |
| 28 | Check for grovee match | 1.000 | .010 | 1 | 1.000 | 1.000 | 1 | 1.000 | R | .000 | .000 | .010 | .000 | .000 |
| 29 | Remove the mandrel from the body | 1.000 | .010 | 1 | 1.000 | 1.000 | 1 | 1.000 | | .010 | .000 | .000 | .000 | .000 |
| 30 | Place the gasket over the body | 1.000 | .010 | 1 | 1.000 | 1.000 | 1 | 1.000 | | .010 | .000 | .000 | .000 | .000 |
| 31 | Place the bonnet over the gasket | 1.000 | .010 | 1 | 1.000 | 1.000 | 1 | 1.000 | | .010 | .000 | .000 | .000 | .000 |
| 32 | Grab bolts | .800 | .007 | 1 | .800 | .800 | 1 | .800 | | .007 | .000 | .000 | .000 | .000 |
| 33 | Tighten body and bonnet | 1.000 | .010 | 1 | 1.000 | 1.000 | 1 | 1.000 | | .010 | .000 | .000 | .000 | .000 |
| 34 | Place the handwheel over the valve | 1.000 | .010 | 1 | 1.000 | 1.000 | 1 | 1.000 | | .010 | .000 | .000 | .000 | .000 |
| 35 | Grab the nut | 1.000 | .010 | 1 | 1.000 | 1.000 | 1 | 1.000 | | .010 | .000 | .000 | .000 | .000 |
| 36 | Tighten the nut | .800 | .007 | 1 | .800 | .800 | 1 | .800 | | .007 | .000 | .000 | .000 | .000 |
| 37 | Open benchwise and remove the valve | .500 | .002 | 1 | .500 | .500 | 1 | .500 | | .002 | .000 | .000 | .000 | .000 |

**Fig. 13** Output of time study for globe valve assembly

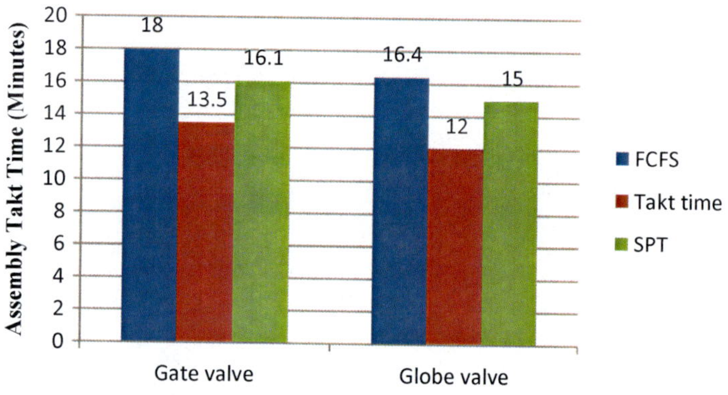


**Fig. 14** Comparison of takt time before and after optimizing schedule

## 7 Results

After the experimentation of production scheduling using FCFS, SPT and LPT, it is found that shortest processing time rule provides effective scheduling. The average WIP for SPT was 1.19 components which was considerably less when compared to other two sequencing rules. Similarly, average numbers waiting according to SPT rule were found to be one component, which proves the production follows single piece flow and the average machine utilization was found to be 49% for SPT which is in level with LPT. The comparison of performance measures for different dispatching rules is shown in Table 8. The waiting time has reduced from 56 to 42 min for gate

**Table 8** Comparison of performance measures

| Performance measure | FCFS | SPT | LPT |
|---|---|---|---|
| Average WIP | 4.62 | 1.19 | 78.86 |
| Average numbers waiting | 4 | 1 | 78 |
| Average machine utilization (%) | 35 | 49 | 49 |

valve and to 51 min from 68 min for globe valve after applying SPT rule. Similarly, the net available time has improved by 31 min.

## 8 Conclusions

Production scheduling is an important factor in helping the organization by providing accurate, real-time schedules and just-in-time delivery. Effective scheduling reduces the waiting time in the production line and improves the time of delivery to the customers.

The establishment of well-structured priority sequencing rules could improve the firm's overall growth and performance. In this project, priority sequencing rules or dispatching rules were used to reduce the waiting time and to achieve the takt time in a flow shop. After collecting data by means of brainstorming and time study, the experimentation was carried out to optimize the schedule using dispatching rules by simulation using Arena. The simulation resulted in obtaining best sequencing rule which provides the optimized schedule.

## Appendix

| Size and class | Components | Process time globe valve (mins) | Process time gate valve (mins) |
|---|---|---|---|
| DN15CL4500 (0.5 inch) | *Body* | | |
| | Pre-machining | 16 | 20 |
| | Finishing | | |
| | Turning 1 | 28 | 32 |
| | Turning 2 | 12 | 16 |
| | Angle hole | 8 | 12 |
| | *Bonnet* | | |
| | Boring | 30 | 30.5 |
| | Finishing | | |

(continued)

(continued)

| Size and class | Components | Process time globe valve (mins) | Process time gate valve (mins) |
|---|---|---|---|
| | OD roughing | 18 | 18 |
| | OD finishing | 22 | 22 |
| | Flange hole | 6 | 6 |
| | *Gland flange* | | |
| | Drilling | 3 | 3.5 |
| | *Disc* | | |
| | Pre-machining | 4 | 5 |
| | Finishing | | |
| | Operation 1 | 7 | 7.5 |
| | Operation 2 | 10 | 11 |
| DN20CL1500 (0.75 inch) | *Body* | | |
| | Pre-machining | 16 | 20 |
| | Finishing | | |
| | Turning 1 | 28 | 32 |
| | Turning 2 | 12 | 16 |
| | Angle hole | 8 | 12 |
| | *Bonnet* | | |
| | Boring | 31 | 31 |
| | Finishing | | |
| | OD roughing | 19 | 19 |
| | OD finishing | 22 | 22 |
| | Flange hole | 7 | 7 |
| | *Gland flange* | | |
| | Drilling | 4 | 4 |
| | *Disc* | | |
| | Pre-machining | 4 | 5 |
| | Finishing | | |
| | Operation 1 | 7.5 | 8 |
| | Operation 2 | 11 | 11 |
| DN20CL4500 (0.75 inch) | *Body* | | |
| | Pre-machining | 16 | 20 |
| | Finishing | | |
| | Turning 1 | 28 | 32 |
| | Turning 2 | 12 | 16 |
| | Angle hole | 8 | 12 |
| | *Bonnet* | | |

(continued)

(continued)

| Size and class | Components | Process time globe valve (mins) | Process time gate valve (mins) |
|---|---|---|---|
| | Boring | 31 | 31 |
| | Finishing | | |
| | OD roughing | 19 | 19.5 |
| | OD finishing | 22 | 23 |
| | Flange hole | 7 | 7 |
| | *Gland flange* | | |
| | Drilling | 4 | 4 |
| | *Disc* | | |
| | Pre-machining | 4 | 5 |
| | Finishing | | |
| | Operation 1 | 7.5 | 8 |
| | Operation 2 | 11 | 11 |
| DN40CL1500 (1.5 inch) | *Body* | | |
| | Pre-machining | 40 | 44 |
| | Finishing | | |
| | Turning 1 | 38 | 42 |
| | Turning 2 | 36 | 40 |
| | Angle hole | 16 | 18 |
| | *Bonnet* | | |
| | Boring | 32 | 32 |
| | Finishing | | |
| | OD roughing | 20 | 20 |
| | OD finishing | 23 | 23 |
| | Flange hole | 8 | 8 |
| | Gland flange | | |
| | Drilling | 6 | 6 |
| | *Disc* | | |
| | Pre-machining | 4 | 5 |
| | Finishing | | |
| | Operation 1 | 8.5 | 8.5 |
| | Operation 2 | 12 | 12 |
| DN40CL4500 (1.5 inch) | *Body* | | |
| | Pre-machining | 40 | 44 |
| | Finishing | | |
| | Turning 1 | 38 | 42 |
| | Turning 2 | 36 | 40 |

(continued)

(continued)

| Size and class | Components | Process time globe valve (mins) | Process time gate valve (mins) |
|---|---|---|---|
| | Angle hole | 16 | 18 |
| | *Bonnet* | | |
| | Boring | 32 | 32 |
| | Finishing | | |
| | OD roughing | 20 | 20 |
| | OD finishing | 23 | 23 |
| | Flange hole | 8 | 8 |
| | *Gland flange* | | |
| | Drilling | 6 | 6 |
| | *Disc* | | |
| | Pre-machining | 4 | 5 |
| | Finishing | | |
| | Operation 1 | 8.5 | 8.5 |
| | Operation 2 | 12 | 12 |
| DN50CL1500 (2 inch) | *Body* | | |
| | Pre-machining | 50 | 52 |
| | Finishing | | |
| | Turning 1 | 50 | 52 |
| | Turning 2 | 60 | 62 |
| | Angle hole | 18 | 20 |
| | *Bonnet* | | |
| | Boring | 32.5 | 32.5 |
| | Finishing | | |
| | OD roughing | 20 | 20 |
| | OD finishing | 23 | 23 |
| | Flange hole | 8 | 8 |
| | Gland flange | | |
| | Drilling | 7 | 7 |
| | *Disc* | | |
| | Pre-machining | 4 | 5 |
| | Finishing | | |
| | Operation 1 | 8.5 | 8.5 |
| | Operation 2 | 12 | 12 |
| DN50CL4500 (2 inch) | *Body* | | |
| | Pre-machining | 50 | 52 |
| | Finishing | | |

(continued)

(continued)

| Size and class | Components | Process time globe valve (mins) | Process time gate valve (mins) |
|---|---|---|---|
| | Turning 1 | 50 | 52 |
| | Turning 2 | 60 | 62 |
| | Angle hole | 18 | 20 |
| | *Bonnet* | | |
| | Boring | 32.5 | 32.5 |
| | Finishing | | |
| | OD roughing | 20 | 20 |
| | OD finishing | 23 | 23 |
| | Flange hole | 8 | 8 |
| | *Gland flange* | | |
| | Drilling | 7 | 7 |
| | *Disc* | | |
| | Pre-machining | 4 | 5 |
| | Finishing | | |
| | Operation 1 | 8.5 | 8.5 |
| | Operation 2 | 12 | 12 |

## References


1. Türker AK, Aktepe A, FiratInal A, Ersoz OO, Das GS (2019) A decision support system for dynamic job-shop scheduling using real-time data with simulation. Mathematics 7:278. https://doi.org/10.3390/math7030278
2. Kumar G, Bisoniya TS (2015) The simulation optimization of a flexible manufacturing system with arena. Int J Res Technol (IJERT) 4(09). ISSN: 2278-0181
3. Kaban AK, Othman Z, Rohmah DS (2012) Comparison of dispatching rules in job-shop scheduling problem using simulation: a case study. Int J Simul Mod 11 3:129–140. ISSN: 1726-4529
4. Windsor AM, Brigoren B (2013) Mathematical modelling and a meta-heuristic for flexible job shop scheduling. Int J Prod Res 51(20)
5. Kassa A, Balasundaram (2016) A simulation modeling approach for job shop scheduling problems: case of metal industry. Int J Sci Res (IJSR) ISSN: 2319-7064 Index Copernicus Value
6. Nasiri NM, Yazdanparast R, Jolai F (2017) A simulation optimisation approach for real-time scheduling in an open shop environment using a composite dispatching rule. Int J Comp Integr Manufact. https://doi.org/10.1080/0951192X.2017.1307452
7. Yang C-L, Hsieh C-C (2014) A production scheduling simulation model for improving production efficiency. Cogent Eng 1(1)
8. de Oliveira Teixeira F, Oliveira L, Varela LR (2014) Comparative analysis of scheduling rules through arena for parallel machines. 978-1-4799-5937-2114/$31. 00 © IEEE

9. Chiang TC, Fu LC (2007) Using dispatching rules for job shop scheduling with due date-based objectives. Int J Prod Res 45(14):3245–3262. https://doi.org/10.1080/00207540600786715
10. Hicks C, Pongcharoen P (2006) Dispatching rules for production scheduling in the capital goods industry. Int J Prod Econ 104:154–163. https://doi.org/10.1016/j.ijpe.2005.07.005

# Prioritizing the Challenges for Lean and Industry 4.0 Integration Using Fuzzy TOPSIS

**Vigneshvaran R. and S. Vinodh**

**Abstract** There exists a vital need for integrating lean and Industry 4.0 to withstand competitiveness. Lean focuses on increased value addition, and Industry 4.0 makes product and process smarter. As there exists need to develop smart products through smart processes, lean concepts are to be integrated with Industry 4.0. In order to enable this integration, challenges need to be analysed. It is vital for an organization to identify the challenges and need to employ remedial measures in order to become a successful firm. In this viewpoint, this paper presents the analysis of challenges of integrated lean and Industry 4.0 using multi-criteria decision-making (MCDM) method fuzzy Technique for Order Performance by Similarity to Ideal Solution (TOPSIS).

**Keywords** Lean · Industry 4.0 · Challenges · Multi-criteria decision-making · TOPSIS

## 1 Introduction

Manufacturing industries have been witnessing four Industrial Revolutions out of which the present focus is on Industry 4.0 (I4.0) [1]. I4.0 focuses on integration of nine technologies such as CPS, IoT, augmented and virtual reality, big data and analytics, simulation, autonomous robots, additive manufacturing, cloud computing and system integration [2]. Lean integration with I4.0 is essential to facilitate smart network of machines, product, process and ICT in the overall value chain to have a smart and flexible manufacturing [3]. Industry 4.0 technologies mainly aim to ease the production process and to address the future challenges in manufacturing

Vigneshvaran R. · S. Vinodh (✉)
Department of Production Engineering, National Institute of Technology, Tiruchirappalli, Tamil Nadu, India
e-mail: vinodh_sekar82@yahoo.com

Vigneshvaran R.
e-mail: vickyvignesh2427@gmail.com

© The Editor(s) (if applicable) and The Author(s), under exclusive license to Springer Nature Singapore Pte Ltd. 2021

S. Mohan et al. (eds.), *Materials, Design, and Manufacturing for Sustainable Environment*, Lecture Notes in Mechanical Engineering,
https://doi.org/10.1007/978-981-15-9809-8_37

sector, they are rising population growth, and increasing need of customized product requires decrease in production lead time, time to market and deliver and additional challenges such as globalized market and volatile nature of customer requirements [4]. As there exists potential for integration of lean and I4.0, the challenges need to be understood and analysed. In this context, this paper presents the analysis of challenges of lean and I4.0 using fuzzy Technique for Order Performance by Similarity to Ideal Solution (TOPSIS). The prioritized challenges have to be focused for effective implementation.

## 2 Literature Review

### 2.1 *Review on Lean Challenges*

Bhasin [5] explored the prominent elements which influenced the effectuation of lean in small, medium and large enterprises and were listed based on outcomes from survey and confirmed with case studies. Fifteen barriers were listed in which size of the organization and time of lean journey of each firm were also taken into consideration. Sixty-eight organizations participated in the survey, and among them seven participated in case studies. Organizational culture, insufficient time and improper communication means were some of the notable challenges depicted in this study.

Kumar and Kumar [6] explained the influencing parameters which restrict the effectuation of lean manufacturing in Indian industry and identified seven primarily imputed factors that affect lean manufacturing effectuation such as management-related barriers, resource-influenced barriers, knowledge-related barriers, conflicts, employee-related barriers, financial barriers and past experience, and 25 barriers were found. In this study, medium- and large-sized companies were evaluated. Top rated barriers were management was not focused, insufficient support and lack of capital fund.

Belhadi et al. [7] explored the undermining factors to lean implementation in SMEs in the context of practical, theoretical, financial and organizational sides, found top five barriers along with its solutions and prioritized the solutions using fuzzy AHP-TOPSIS approach. In this study, 20 barriers were grouped in five categories such as strategic, technical, cultural, market-related and knowledge-related barriers and 17 solutions were provided. They mentioned that study does not consider the interdependence between barriers and remedial measures.

AlManei et al. [8] discussed on prominent lean implementation framework challenges in the context of SMEs. The challenges were identified by virtue of literature review, and this study reviewed lean implementation approaches. Key enablers and inhibitors were also identified and depicted in force field analysis structure.

Bakke and Johansen [9] investigated about lean implementation methods in organization. In this study, the main challenges in implementing lean in large organization were identified as management support as most essential followed by employee training.

## 2.2 Review on Industry 4.0 Challenges

Glass et al. [10] examined and identified the barriers to I4.0. Fifteen barriers were obtained from the statistical analysis. A survey of 253 companies was considered. In this study, 79% of respondents expressed interest in support for enhancement in efficiency of their production process. The study depicted that not all challenges discussed in literature existed in industry.

Orzes et al. [11] investigated on barriers and challenges in implementation of I4.0 by small and medium-size industries. Small focus group studies in four countries, namely USA, Italy, Austria and Thailand, were set up, and systematic literature review was carried out. Identified barriers were grouped in six categories such as financial, cultural, resources, legal, technical and implementation. In this study, high costs, uncertain return on investment and long implementation process were found out to be the primary impediments to I4.0 implementation in SMEs.

Agostini and Filippini [1] studied about organizational and managerial challenges of Italian manufacturing companies in addition to Industry 4.0 technological challenges. They found compelling number of organizational and managerial challenges statistically with $t$-test and concluded that overcoming these challenges offers implementation of I4.0 technologies.

Muller [12] identified barriers to I4.0 implementation and causes for those barriers through interview of about 41 representatives. In this study, workers' perception on Industry 4.0 and their ways of assessing disputes associated with I4.0 were provided and suggested longer vision, clear strategy and goals should be set properly at initial stage for successful implementation of I4.0.

Raj et al. [13] explored the context of barriers in the deployment of I4.0 technologies in manufacturing sector of both developed and developing countries. In this study, 15 barriers were identified and analysed through Grey Decision Making Trial and Evaluation Laboratory (DEMATEL) approach and it stated that improvement in standard and regulation would promote I4.0 technological implementation in developing countries and digitalized framework were needed to elevate the implementation of these features in the developed country.

Rauch et al. [14] explored the requirements and barriers for smart manufacturing in SMEs. Fourteen barriers were found, and they are grouped in six categories, namely cultural, implementation, people, resource management, security and strategy. In this study, managerial action recommendations were made for successful implementation of I4.0 technologies in SMEs.

### 2.3 Review on Interdependencies of Lean and Industry 4.0

Mayr et al. [15] explored and analysed I4.0 technology interaction with lean tools. In this study, most I4.0 techniques like human–computer interaction (HCI), real-time computing, digital twin, digital object memory and many more were related to primary and secondary lean tools. The result depicted that real-time computing was mapped with almost all lean tools like VSM, JIT, Kanban, Heijunka and everything.

Dombrowski et al. [16] examined the interdependencies of lean and I4.0. In this study, machine-to-machine communication (M2M), cyber-physical system (CPS), Internet of things (IoT) and many more are projected as I4.0 systems and big data, RFID, cloud computing and so on are mentioned as I4.0 technologies that support lean tools in order to have an efficient process and organization.

From the literature study, it is found that there exists good potential for integrating lean and I4.0 tools and techniques. In order to facilitate this integration, challenges for integration of lean and I4.0 are to be analysed. In this context, this article focuses on analysis of challenges for integrating lean and I4.0 and it is clearly evident that there is lack of prioritizing or ranking of major challenges or impediments of integrated lean and I4.0 approach which formed the scope of this study.

## 3 Fuzzy TOPSIS Methodology

TOPSIS is a realistic and effective multi-criteria decision-making (MCDM) technique. Two types of criteria are used in this methodology; they are benefit criteria and cost criteria; two types of ideal solution are obtained in this method, namely positive ideal solution and negative ideal solution. The benefit criteria should lie closer to the positive ideal solution and away from negative ideal solution. The methodology could be referred from Mittal and Sangwan [4], Wang and Lee [17], Mahdavi et al. [18], Singh et al. [18], Aswathi et al. [19], and Sun [20].

The ambiguity and biasedness are managed with linguistic term guidelines by TFN. The ambiguity in mapping of human judgement on challenges can be surmounted using fuzzy TOPSIS [20]. The MCDM approach revolves around alternatives and more than one criterion to establish a priority ranking of challenges [17].

A triangular fuzzy number (TFN) represents as a triplet $a = (a_1, a_2, a_3)$. This method is used widely as it is very simple in computation. The membership function $\mu_a(x)$ of TFN could be refereed from Mittal and Sangwan [4].

Assume there are m challenges (impediments) $A_i (i = 1, 2, \ldots, m)$ to be assessed against n selection criteria $c_j (j = 1, 2, \ldots, n)$.

The goal is to rank the challenges with reference to selection criteria based on their relative importance. Steps are:

Step 1: Assignment of ratings to the criteria and alternatives.
The associated relationship ratings for each challenge with reference to each criterion will be given by the decision-maker $D_k$ separately, and also the criterion weight will also be given by the $D_k$.

Step 2: Compute aggregate fuzzy ratings for criteria and the alternatives.
The fuzzy rating of all decision-maker is denoted as TFN.
$R_k = (a_k, b_k, c_k), k = 1, 2, \ldots, K$, and then the aggregate fuzzy rating is given by $R = (a, b, c), k = 1, 2, \ldots, K$, where

$$a = \min_k\{a_k\}, \quad b = \frac{1}{K}\sum_{k=1}^{K} b_k \quad \text{and } c = \max_k\{c_k\} \tag{1}$$

If the fuzzy rating and importance weight of the $k$th decision-maker are $x_{ijk} = (a_{ijk}, b_{ijk}, c_{ijk})$ and $w_{ijk} = (w_{jk1}, w_{jk2}, w_{jk3})$, $(i = 1, 2, \ldots, m)$ and$(j = 1, 2, \ldots, n)$, respectively, then the aggregate fuzzy rating $(x_{ijk})$ of alternatives with reference to each criterion is given by $x_{ij} = (a_{ij}, b_{ij}, c_{ij})$ where

$$a_{ij} = \min_k\{a_{ijk}\} \tag{2.1}$$

$$b_{ij} = \frac{1}{K}\sum_{k=1}^{K} b_{ijk} \tag{2.2}$$

$$c_{ij} = \max_k\{c_{ijk}\} \tag{2.3}$$

The aggregated fuzzy weights $(w_{ij})$ of each criterion are computed as $w_j = (w_{j1}, w_{j2}, w_{j3})$ where

$$w_{j1} = \min_k\{w_{jk1}\} \tag{3.1}$$

$$w_{j2} = \frac{1}{K}\sum_{k=1}^{K} w_{jk2} \tag{3.2}$$

$$w_{j3} = \max_k\{w_{jk3} \tag{3.3}$$

Step 3: Compute the fuzzy decision matrix.
The combined fuzzy decision matrix is given as:

$$C_1 \quad C_2 \quad \ldots \quad C_n$$

$$D = \begin{matrix} A_1 \\ A_2 \\ A_3 \end{matrix} \begin{bmatrix} x_{11} & x_{12} & \cdots & x_{1n} \\ x_{21} & x_{22} & \cdots & x_{2n} \\ x_{m1} & x_{m2} & \cdots & x_{mn} \end{bmatrix} \tag{4}$$

Step 4: Normalization of combined fuzzy decision matrix (R).
It is given by:

$$R = \left[r_{ij}\right]_{m \times n} \tag{5}$$

where

$$r_{ij} = \left(\frac{a_{ij}}{c_j^*}, \frac{b_{ij}}{c_j^*}, \frac{c_{ij}}{c_j^*}\right) \text{ and } c_j^* = \max_i c_{ij} \tag{6}$$

(benefit criteria)

$$r_{ij} = \left(\frac{a_j^-}{c_{ij}}, \frac{a_j^-}{b_{ij}}, \frac{a_j^-}{a_{ij}}\right) \text{ and } a_j^- = \min_i a_{ij} \tag{7}$$

(cost criteria)

Step 5: Calculate the weighted normalized matrix V.
It is calculated by the product of $(w_j)$ and $r_{ij}$.
where

$$V = \left[v_{ij}\right]_{m \times n}, i = (1, 2, \ldots, m); \; j = (1, 2, \ldots, n) \tag{8}$$

and

$$v_{ij} = r_{ij}(.)w_j \tag{9}$$

Step 6: Calculate the fuzzy positive ideal solution (FPIS) and fuzzy negative ideal solution (FNIS).
The FPIS and FNIS of the challenges are computed as follows:

$$A^* = \left(v_1^*, v_2^*, \ldots, v_n^*\right) \tag{10}$$

where $v_j^* = \max_i \left\{v_{ij3}\right\}$

$$A^- = (v_1, v_2, .., v_n) \tag{11}$$

where $v_j^- = \min_i \left\{v_{ij1}\right\}$

Step 7: Calculate the range between challenges and to FPIS and FNIS, respectively. The span $(d_i^*, d_i^-)$ of weighted challenges from FPIS and FNIS is calculated as follows:

$$d_i^* = \sum_{j=1}^{n} d_v\left(v_{ij}, v_j^*\right) \tag{12}$$

$$d_i^- = \sum_{j=1}^{n} d_v\left(v_{ij}, v_j^-\right) \tag{13}$$

Step 8: Closeness coefficient ($\mathrm{CC}_i$) of challenges.
The closeness coefficient ($\mathrm{CC}_i$) denotes the range between FPIS ($A^*$) and the FNIS $(A^-)$ and is computed as:

$$\mathrm{CC}_i = \frac{d_i^-}{d_i^- + d_i^*} \tag{14}$$

Step 9: Hierarchical order of challenges.
The hierarchical order of challenges can be assessed based on their closeness coefficient ($\mathrm{CC}_i$). The challenges or impediments with higher value of closeness coefficient are considered as the most important challenge, it can be ranked as "first", and all other challenges are ranked in descending order with respect to their closeness coefficient.
In order to deal with uncertainty in a better manner, grey theory could be integrated with TOPSIS in future studies.

## 4 Case Study

The challenges for the case study are obtained from the literature study. Analysis of challenges of lean and I4.0 integration shows that it involves all level of hierarchical structure in organization which includes top-level management, middle-level management as well as workers, and all these challenges are discussed in the context of selected criteria. The inputs for these challenges and criteria weight are obtained from industrial practitioners, academicians and experts in the relevant field. Challenges to integration of lean and Industry 4.0 approach are depicted in Table 1.

In this study, three decision-makers' opinion is taken into consideration. The experts' opinion on impediments and criteria weights is given by the decision-maker individually in linguistic terms, and they need to be converted into fuzzy number. Criteria weight further needs to be converted into aggregated criteria weight. The linguistic term and their relevant fuzzy number are depicted in table below.

**Table 1** Challenges of integrated lean and Industry 4.0

| S. No. | Challenges | Study |
|---|---|---|
| 1 | Organizational culture | Bhasin [5], Shawan et al. [21] |
| 2 | Lack of time | Bhasin [5], Kumar and Kumar [6], Belhadi et al. [7] |
| 3 | Lack of communication | Bhasin [5], Kumar and Kumar [6], Bakke et al. [9] |
| 4 | Management support | Kumar and Kumar [6], Belhadi et al. [7], Yadav et al. [22], AlManei et al. [8], Orzes et al. [11], Müller [12] |
| 5 | Idea innovation | Kumar and Kumar [6] |
| 6 | Organizational structure | Dorah et al. [23], AlManei et al. [8], Müller [12]. |
| 7 | Digital strategy | Glass et al. [10], Müller [12], Raj et al. [13] |
| 8 | Technical integration | Orzes et al. [11] |
| 9 | Cooperation | Müller [12], Rauch et al. [14] |
| 10 | Financial resources | Orzes et al. [11], Müller JM [12], Raj et al. [13] |
| 11 | Employee motivation and training | Agostini and Filippini [1], Bakke and Johansen [9], Rauch et al. [14] |
| 12 | Complexity | Glass et al. [10] |

Notations used in this study are follows:

$A_i$—number of challenges (12 challenges were used in our study)
$c_j$—number of criteria (6 criteria were used in our study)
$D_k$.—number of decision-maker (three decision-makers in our study).

Table 2 depicts the linguistic terms for criteria and challenges and its associated membership function in fuzzy numbers. Decision-maker responses regarding the challenges with respect to criteria are provided in the form of linguistic terminologies in Table 3 along with criteria weight, and these weightages are from the decision-makers' view point described in linguistic terms. Then, it is converted into fuzzy

**Table 2** Linguistic term and corresponding fuzzy number

| Assessments of challenges | | Assessment of criteria | |
|---|---|---|---|
| Terminologies | Fuzzy number | Terminologies | Fuzzy number |
| Very low (VL) | (1, 1, 3) | Not important (NI) | (1, 1, 3) |
| Low (L) | (1, 3, 5) | Less important (LI) | (1, 3, 5) |
| Medium (M) | (3, 5, 7) | Fairly important (FI) | (3, 5, 7) |
| High (H) | (5, 7, 9) | Important (I) | (5, 7, 9) |
| Very high (VH) | (7, 9, 9) | Very important (VI) | (7, 9, 9) |

**Table 3** Expert opinions and the criteria weight of impediments in linguistic terms

| | C1 | | | C2 | | | C3 | | | C4 | | | C5 | | | C6 | | |
|---|---|---|---|---|---|---|---|---|---|---|---|---|---|---|---|---|---|---|
| | DM1 | DM2 | DM3 | DM1 | DM2 | DM3 | DM1 | DM2 | DM3 | DM1 | DM2 | DM3 | DM1 | DM2 | DM3 | DM1 | DM2 | DM3 |
| C.W | VI | I | VI | I | I | I | VI | I | VI | VI | I | VI | I | VI | VI | VI | I | VI |
| I1 | H | VH | H | H | M | H | H | M | H | H | VH | H | H | H | H | H | H | H |
| I2 | H | H | VH | VH | H | H | M | H | H | H | H | H | H | H | VH | H | H | H |
| I3 | H | M | H | H | M | H | H | M | H | H | M | H | M | M | M | M | M | H |
| I4 | H | H | M | M | H | M | M | M | H | M | H | M | M | M | M | M | M | M |
| I5 | H | M | H | M | M | M | H | M | H | H | M | H | H | M | M | H | M | H |
| I6 | M | M | H | M | H | M | H | H | H | M | M | H | H | H | H | H | H | M |
| I7 | H | M | H | M | M | M | M | H | M | H | M | H | M | M | H | M | H | H |
| I8 | H | M | H | H | M | H | VH | H | VH | M | H | M | M | L | M | M | M | H |
| I9 | H | M | H | M | M | M | M | H | M | M | H | M | M | L | M | M | H | M |
| I10 | M | M | H | H | M | H | M | M | M | H | H | H | H | M | H | M | H | M |
| I11 | H | H | M | H | H | M | M | M | M | M | H | M | M | H | M | H | M | H |
| I12 | M | H | M | M | H | H | H | M | H | M | M | H | M | H | H | H | H | M |

numbers as depicted in Table 4. In Table 4, challenges associated with fuzzy numbers as per decision-maker opinion are depicted in the form of triangular fuzzy number and then it will be converted into fuzzy numbers based on step 2 in methodology using Eqs. 2.1–2.3.

Table 5 depicts the combined decision matrix of challenges with reference to criteria of lean and I4.0 integration as explained in steps 2 and 3 of fuzzy TOPSIS methodology.

Table 6 depicts the normalized fuzzy decision matrix. The normalization is computed using the equation mentioned in step 4 of methodology, and Table 7 depicts the weighted normalized fuzzy decision matrix as per step 5. The criteria weights are multiplied with challenges.

Table 8 depicts the fuzzy positive and negative ideal solutions of challenges, and Table 9 depicts the distance from fuzzy positive ideal solution to challenges with reference to each criterion. Challenge value of I1 and I2 depicts very closer distance from positive ideal solution, and I6, I10, I12 are very farthest to ideal positive solution in terms of first criterion.

Table 10 depicts the distance from fuzzy negative ideal solution to challenges with reference to each criterion. Challenge value of I1 and I2 depicts farthest distance from negative ideal solution, and I6, I10, I12 are very closer to ideal negative solution in terms of first criterion. Table 11 depicts the fuzzy positive solution distances for each criterion, and it is obtained by summing all criteria values with reference to challenges. The lowest value corresponds to digital strategy challenge (I1), and this indicates that it lies closer to positive ideal solution.

Table 12 depicts the fuzzy negative solution distances for each criterion, and it is obtained by summing all criteria values with reference to challenges. The highest value corresponds to digital strategy challenge (I1), and this indicates that it lies far away from negative ideal solution. The closeness coefficient index is depicted in Table 13 and is ranked correspondingly. Challenge with highest closeness index ranks top.

## 4.1 Result

The computation results depict that the impediment digital strategy has highest closeness coefficient ($CC_i$) and ranked first with respect to all twelve challenges. Technical integration and lack of idea innovation rank second and third, and these three are identified as the top three challenges for integration of lean and I4.0. The study enabled the identification of prioritized challenges which would facilitate the deployment of integrated lean and I4.0. Tools and techniques of integrated lean and I4.0 such as dynamic VSM (DVSM), Kaizen integrated with IoT (Kaizen 4.0) and total productive maintenance integrated with augmented and virtual reality (TPM 4.0) could be better focused.

**Table 4** Decision-maker opinion in fuzzy number

| | C1 | | | C2 | | | C3 | | | C4 | | | C5 | | | C6 | | |
|---|---|---|---|---|---|---|---|---|---|---|---|---|---|---|---|---|---|---|
| | DM1 | DM2 | DM3 | DM1 | DM2 | DM3 | DM1 | DM2 | DM3 | DM1 | DM2 | DM3 | DM1 | DM2 | DM3 | DM1 | DM2 | DM3 |
| C.W | 7, 9, 9 | 5, 7, 9 | 7, 9, 9 | 5, 7, 9 | 5, 7, 9 | 5, 7, 9 | 7, 9, 9 | 5, 7, 9 | 7, 9, 9 | 7, 9, 9 | 5, 7, 9 | 7, 9, 9 | 5, 7, 9 | 7, 9, 9 | 7, 9, 9 | 7, 9, 9 | 5, 7, 9 | 7, 9, 9 |
| I1 | 5, 7, 9 | 7, 9, 9 | 5, 7, 9 | 5, 7, 9 | 3, 5, 7 | 5, 7, 9 | 5, 7, 9 | 3, 5, 7 | 5, 7, 9 | 5, 7, 9 | 7, 9, 9 | 5, 7, 9 | 5, 7, 9 | 5, 7, 9 | 5, 7, 9 | 5, 7, 9 | 5, 7, 9 | 5, 7, 9 |
| I2 | 5, 7, 9 | 5, 7, 9 | 7, 9, 9 | 7, 9, 9 | 5, 7, 9 | 5, 7, 9 | 3, 5, 7 | 5, 7, 9 | 5, 7, 9 | 5, 7, 9 | 5, 7, 9 | 5, 7, 9 | 5, 7, 9 | 5, 7, 9 | 7, 9, 9 | 5, 7, 9 | 5, 7, 9 | 5, 7, 9 |
| I3 | 5, 7, 9 | 3, 5, 7 | 5, 7, 9 | 5, 7, 9 | 3, 5, 7 | 5, 7, 9 | 5, 7, 9 | 3, 5, 7 | 5, 7, 9 | 5, 7, 9 | 3, 5, 7 | 5, 7, 9 | 3, 5, 7 | 3, 5, 7 | 3, 5, 7 | 3, 5, 7 | 3, 5, 7 | 5, 7, 9 |
| I4 | 5, 7, 9 | 5, 7, 9 | 3, 5, 7 | 3, 5, 7 | 5, 7, 9 | 3, 5, 7 | 3, 5, 7 | 3, 5, 7 | 5, 7, 9 | 3, 5, 7 | 5, 7, 9 | 3, 5, 7 | 3, 5, 7 | 3, 5, 7 | 3, 5, 7 | 3, 5, 7 | 3, 5, 7 | 3, 5, 7 |
| I5 | 5, 7, 9 | 3, 5, 7 | 5, 7, 9 | 3, 5, 7 | 3, 5, 7 | 3, 5, 7 | 5, 7, 9 | 3, 5, 7 | 5, 7, 9 | 5, 7, 9 | 3, 5, 7 | 5, 7, 9 | 5, 7, 9 | 3, 5, 7 | 3, 5, 7 | 5, 7, 9 | 3, 5, 7 | 5, 7, 9 |
| I6 | 3, 5, 7 | 3, 5, 7 | 5, 7, 9 | 3, 5, 7 | 5, 7, 9 | 3, 5, 7 | 5, 7, 9 | 5, 7, 9 | 5, 7, 9 | 3, 5, 7 | 3, 5, 7 | 5, 7, 9 | 5, 7, 9 | 5, 7, 9 | 5, 7, 9 | 5, 7, 9 | 5, 7, 9 | 3, 5, 7 |
| I7 | 5, 7, 9 | 3, 5, 7 | 5, 7, 9 | 3, 5, 7 | 3, 5, 7 | 3, 5, 7 | 3, 5, 7 | 5, 7, 9 | 3, 5, 7 | 5, 7, 9 | 3, 5, 7 | 5, 7, 9 | 3, 5, 7 | 3, 5, 7 | 5, 7, 9 | 3, 5, 7 | 5, 7, 9 | 5, 7, 9 |
| I8 | 5, 7, 9 | 3, 5, 7 | 5, 7, 9 | 5, 7, 9 | 3, 5, 7 | 5, 7, 9 | 7, 9, 9 | 5, 7, 9 | 7, 9, 9 | 3, 5, 7 | 5, 7, 9 | 3, 5, 7 | 3, 5, 7 | 1, 3, 5 | 3, 5, 7 | 3, 5, 7 | 3, 5, 7 | 5, 7, 9 |
| I9 | 5, 7, 9 | 3, 5, 7 | 5, 7, 9 | 3, 5, 7 | 3, 5, 7 | 3, 5, 7 | 3, 5, 7 | 5, 7, 9 | 3, 5, 7 | 3, 5, 7 | 5, 7, 9 | 3, 5, 7 | 3, 5, 7 | 1, 3, 5 | 3, 5, 7 | 3, 5, 7 | 5, 7, 9 | 3, 5, 7 |
| I10 | 3, 5, 7 | 3, 5, 7 | 5, 7, 9 | 5, 7, 9 | 3, 5, 7 | 5, 7, 9 | 3, 5, 7 | 3, 5, 7 | 3, 5, 7 | 5, 7, 9 | 5, 7, 9 | 5, 7, 9 | 5, 7, 9 | 3, 5, 7 | 5, 7, 9 | 3, 5, 7 | 5, 7, 9 | 3, 5, 7 |
| I11 | 5, 7, 9 | 5, 7, 9 | 3, 5, 7 | 5, 7, 9 | 5, 7, 9 | 3, 5, 7 | 3, 5, 7 | 3, 5, 7 | 3, 5, 7 | 3, 5, 7 | 5, 7, 9 | 3, 5, 7 | 3, 5, 7 | 5, 7, 9 | 3, 5, 7 | 5, 7, 9 | 3, 5, 7 | 5, 7, 9 |
| I12 | 3, 5, 7 | 5, 7, 9 | 3, 5, 7 | 3, 5, 7 | 5, 7, 9 | 5, 7, 9 | 5, 7, 9 | 3, 5, 7 | 5, 7, 9 | 3, 5, 7 | 3, 5, 7 | 5, 7, 9 | 3, 5, 7 | 5, 7, 9 | 5, 7, 9 | 5, 7, 9 | 5, 7, 9 | 3, 5, 7 |

**Table 5** Combined decision matrix

| | C1 | C2 | C3 | C4 | C5 | C6 |
|---|---|---|---|---|---|---|
| A.C.W | 5, 8.33, 9 | 5, 7, 9 | 5, 8.33, 9 | 5, 8.33, 9 | 5, 8.33, 9 | 5, 8.33, 9 |
| I1 | 5, 7.67, 9 | 3, 6.33, 9 | 3, 6.33, 9 | 5, 7.67, 9 | 5, 7, 9 | 5, 7, 9 |
| I2 | 5, 7.67, 9 | 5, 7.67, 9 | 3, 6.33, 9 | 5, 7, 9 | 5, 7.67, 9 | 5, 7, 9 |
| I3 | 3, 6.33, 9 | 3, 6.33, 9 | 3, 6.33, 9 | 3, 6.33, 9 | 3, 5, 7 | 3, 5.67, 9 |
| I4 | 3, 6.33, 9 | 3, 5.67, 9 | 3, 5.67, 9 | 3, 5.67, 9 | 3, 5, 7 | 3, 5, 7 |
| I5 | 3, 6.33, 9 | 3, 5, 7 | 3, 6.33, 9 | 3, 6.33, 9 | 3, 5.67, 9 | 3, 6.33, 9 |
| I6 | 3, 5.67, 9 | 3, 5.67, 9 | 5, 7, 9 | 3, 5.67, 9 | 5, 7, 9 | 3, 6.33, 9 |
| I7 | 3, 6.33, 9 | 3, 5, 7 | 3, 5.67, 9 | 3, 6.33, 9 | 3, 5.67, 9 | 3, 6.33, 9 |
| I8 | 3, 6.33, 9 | 3, 6.33, 9 | 5, 8.33, 9 | 3, 5.67, 9 | 1, 4.33, 7 | 3, 5.67, 9 |
| I9 | 3, 6.33, 9 | 3, 5, 7 | 3, 5.67, 9 | 3, 5.67, 9 | 1, 4.33, 7 | 3, 5.67, 9 |
| I10 | 3, 5.67, 9 | 3, 6.33, 9 | 3, 5, 7 | 5, 7, 9 | 3, 6.33, 9 | 3, 5.67, 9 |
| I11 | 3, 6.33, 9 | 3, 6.33, 9 | 3, 5, 7 | 3, 5.67, 9 | 3, 5.67, 9 | 3, 6.33, 9 |
| I12 | 3, 5.67, 9 | 3, 6.33, 9 | 3, 6.33, 9 | 3, 5.67, 9 | 3, 6.33, 9 | 3, 6.33, 9 |

## 5 Conclusion

This study is focused on analysis of integration of lean and I4.0. Twelve challenges have been focused. The prioritization is formulated as MCDM problem. Fuzzy TOPSIS is applied as solution methodology. The top prioritized challenges are digital strategy, technical integration and lack of idea innovation. Remedial measures must be taken continuously for sustainable competition of any organization. The practitioners have to focus on prioritized challenges to effectively deploy the integrated concept.

In future, other challenges could be considered. Also, integration of I4.0 with other manufacturing strategies also could be considered and before generalizing these challenges some empirical case studies need to be carried out. Also, implementation of lean and I4.0 technologies could be focused for improvement in work effectiveness.

**Table 6** Normalized fuzzy decision matrix

| | C1 | C2 | C3 | C4 | C5 | C6 |
|---|---|---|---|---|---|---|
| A.C.W | 5, 8.33, 9 | 5, 7, 9 | 5, 8.33, 9 | 5, 8.33, 9 | 5, 8.33, 9 | 5, 8.33, 9 |
| I1 | 0.5556, 0.852, 1 | 0.333, 0.473, 1 | 0.333, 0.473, 1 | 0.5556, 0.852, 1 | 0.556, 0.778, 1 | 0.556, 0.778, 1 |
| I2 | 0.5556, 0.852, 1 | 0.333, 0.391, 0.6 | 0.333, 0.473, 1 | 0.556, 0.778, 1 | 0.5556, 0.852, 1 | 0.556, 0.778, 1 |
| I3 | 0.333, 0.7036, 1 | 0.333, 0.473, 1 | 0.333, 0.473, 1 | 0.333, 0.7036, 1 | 0.333, 0.556, 0.778 | 0.333, 0.63, 1 |
| I4 | 0.333, 0.7036, 1 | 0.333, 0.529, 1 | 0.333, 0.529, 1 | 0.333, 0.63, 1 | 0.333, 0.556, 0.778 | 0.333, 0.556, 0.778 |
| I5 | 0.333, 0.7036, 1 | 0.428, 0.6, 1 | 0.333, 0.473, 1 | 0.333, 0.7036, 1 | 0.333, 0.63, 1 | 0.333, 0.7036, 1 |
| I6 | 0.333, 0.63, 1 | 0.333, 0.529, 1 | 0.333, 0.428, 0.6 | 0.333, 0.63, 1 | 0.556, 0.778, 1 | 0.333, 0.7036, 1 |
| I7 | 0.333, 0.7036, 1 | 0.428, 0.6, 1 | 0.333, 0.529, 1 | 0.333, 0.7036, 1 | 0.333, 0.63, 1 | 0.333, 0.7036, 1 |
| I8 | 0.333, 0.7036, 1 | 0.333, 0.473, 1 | 0.333, 0.36, 0.6 | 0.333, 0.63, 1 | 0.111, 0.481, 0.778 | 0.333, 0.63, 1 |
| I9 | 0.333, 0.7036, 1 | 0.428, 0.6, 1 | 0.333, 0.529, 1 | 0.333, 0.63, 1 | 0.111, 0.481, 0.778 | 0.333, 0.63, 1 |
| I10 | 0.333, 0.63, 1 | 0.333, 0.473, 1 | 0.428, 0.6, 1 | 0.556, 0.778, 1 | 0.333, 0.7036, 1 | 0.333, 0.63, 1 |
| I11 | 0.333, 0.7036, 1 | 0.333, 0.473, 1 | 0.428, 0.6, 1 | 0.333, 0.63, 1 | 0.333, 0.63, 1 | 0.333, 0.7036, 1 |
| I12 | 0.333, 0.63, 1 | 0.333, 0.473, 1 | 0.333, 0.473, 1 | 0.333, 0.63, 1 | 0.333, 0.7036, 1 | 0.333, 0.7036, 1 |

**Table 7** Weighted normalized fuzzy decision matrix

| | C1 | | | C2 | | | C3 | | | C4 | | | C5 | | | C6 | | |
|---|---|---|---|---|---|---|---|---|---|---|---|---|---|---|---|---|---|---|
| I1 | 2.778 | 7.09716 | 9 | 1.665 | 3.311 | 9 | 1.665 | 3.94009 | 9 | 2.778 | 7.09716 | 9 | 2.78 | 6.48074 | 9 | 2.78 | 6.48074 | 9 |
| I2 | 2.778 | 7.09716 | 9 | 1.665 | 2.737 | 5.4 | 1.665 | 3.94009 | 9 | 2.78 | 6.48074 | 9 | 2.778 | 7.09716 | 9 | 2.78 | 6.48074 | 9 |
| I3 | 1.665 | 5.860988 | 9 | 1.665 | 3.311 | 9 | 1.665 | 3.94009 | 9 | 1.665 | 5.85599 | 9 | 1.665 | 4.63148 | 7.002 | 1.665 | 5.2479 | 9 |
| I4 | 1.665 | 5.860988 | 9 | 1.665 | 3.703 | 9 | 1.665 | 4.40657 | 9 | 1.665 | 5.2479 | 9 | 1.665 | 4.63148 | 7.002 | 1.665 | 4.63148 | 7.002 |
| I5 | 1.665 | 5.860988 | 9 | 2.14 | 4.2 | 9 | 1.665 | 3.94009 | 9 | 1.665 | 5.85599 | 9 | 1.665 | 5.2479 | 9 | 1.665 | 5.85599 | 9 |
| I6 | 1.665 | 5.2479 | 9 | 1.665 | 3.703 | 9 | 1.665 | 3.56524 | 5.4 | 1.665 | 5.2479 | 9 | 2.78 | 6.48074 | 9 | 1.665 | 5.85599 | 9 |
| I7 | 1.665 | 5.860988 | 9 | 2.14 | 4.2 | 9 | 1.665 | 4.40657 | 9 | 1.665 | 5.85599 | 9 | 1.665 | 5.2479 | 9 | 1.665 | 5.85599 | 9 |
| I8 | 1.665 | 5.860988 | 9 | 1.665 | 3.311 | 9 | 1.665 | 2.9988 | 5.4 | 1.665 | 5.2479 | 9 | 0.555 | 4.00673 | 7.002 | 1.665 | 5.2479 | 9 |
| I9 | 1.665 | 5.860988 | 9 | 2.14 | 4.2 | 9 | 1.665 | 4.40657 | 9 | 1.665 | 5.2479 | 9 | 0.555 | 4.00673 | 7.002 | 1.665 | 5.2479 | 9 |
| I10 | 1.665 | 5.2479 | 9 | 1.665 | 3.311 | 9 | 2.14 | 4.998 | 9 | 2.78 | 6.48074 | 9 | 1.665 | 5.85599 | 9 | 1.665 | 5.2479 | 9 |
| I11 | 1.665 | 5.860988 | 9 | 1.665 | 3.311 | 9 | 2.14 | 4.998 | 9 | 1.665 | 5.2479 | 9 | 1.665 | 5.2479 | 9 | 1.665 | 5.85599 | 9 |
| I12 | 1.665 | 5.2479 | 9 | 1.665 | 3.311 | 9 | 1.665 | 3.94009 | 9 | 1.665 | 5.2479 | 9 | 1.665 | 5.85599 | 9 | 1.665 | 5.85599 | 9 |

**Table 8** Fuzzy positive and negative ideal solution

| | C1 | | | C2 | | | C3 | | | C4 | | | C5 | | | C6 | | |
|---|---|---|---|---|---|---|---|---|---|---|---|---|---|---|---|---|---|---|
| I1 | 2.778 | 7.09716 | 9 | 1.665 | 3.311 | 9 | 1.665 | 3.94009 | 9 | 2.778 | 7.09716 | 9 | 2.78 | 6.48074 | 9 | 2.78 | 6.48074 | 9 |
| I2 | 2.778 | 7.09716 | 9 | 1.665 | 2.737 | 5.4 | 1.665 | 3.94009 | 9 | 2.78 | 6.48074 | 9 | 2.778 | 7.09716 | 9 | 2.78 | 6.48074 | 9 |
| I3 | 1.665 | 5.860988 | 9 | 1.665 | 3.311 | 9 | 1.665 | 3.94009 | 9 | 1.665 | 5.85599 | 9 | 1.665 | 4.63148 | 7.002 | 1.665 | 5.2479 | 9 |
| I4 | 1.665 | 5.860988 | 9 | 1.665 | 3.703 | 9 | 1.665 | 4.40657 | 9 | 1.665 | 5.2479 | 9 | 1.665 | 4.63148 | 7.002 | 1.665 | 4.63148 | 7.002 |
| I5 | 1.665 | 5.860988 | 9 | 2.14 | 4.2 | 9 | 1.665 | 3.94009 | 9 | 1.665 | 5.85599 | 9 | 1.665 | 5.2479 | 9 | 1.665 | 5.85599 | 9 |
| I6 | 1.665 | 5.2479 | 9 | 1.665 | 3.703 | 9 | 1.665 | 3.56524 | 5.4 | 1.665 | 5.2479 | 9 | 2.78 | 6.48074 | 9 | 1.665 | 5.85599 | 9 |
| I7 | 1.665 | 5.860988 | 9 | 2.14 | 4.2 | 9 | 1.665 | 4.40657 | 9 | 1.665 | 5.85599 | 9 | 1.665 | 5.2479 | 9 | 1.665 | 5.85599 | 9 |
| I8 | 1.665 | 5.860988 | 9 | 1.665 | 3.311 | 9 | 1.665 | 2.9988 | 5.4 | 1.665 | 5.2479 | 9 | 0.555 | 4.00673 | 7.002 | 1.665 | 5.2479 | 9 |
| I9 | 1.665 | 5.860988 | 9 | 2.14 | 4.2 | 9 | 1.665 | 4.40657 | 9 | 1.665 | 5.2479 | 9 | 0.555 | 4.00673 | 7.002 | 1.665 | 5.2479 | 9 |
| I10 | 1.665 | 5.2479 | 9 | 1.665 | 3.311 | 9 | 2.14 | 4.998 | 9 | 2.78 | 6.48074 | 9 | 1.665 | 5.85599 | 9 | 1.665 | 5.2479 | 9 |
| I11 | 1.665 | 5.860988 | 9 | 1.665 | 3.311 | 9 | 2.14 | 4.998 | 9 | 1.665 | 5.2479 | 9 | 1.665 | 5.2479 | 9 | 1.665 | 5.85599 | 9 |
| I12 | 1.665 | 5.2479 | 9 | 1.665 | 3.311 | 9 | 1.665 | 3.94009 | 9 | 1.665 | 5.2479 | 9 | 1.665 | 5.85599 | 9 | 1.665 | 5.85599 | 9 |
| A* | 2.778 | 7.09716 | 9 | 2.14 | 4.2 | 9 | 2.14 | 4.998 | 9 | 2.778 | 7.09716 | 9 | 2.778 | 7.09716 | 9 | 2.78 | 6.48074 | 9 |
| $A^-$ | 1.665 | 5.2479 | 9 | 1.665 | 2.737 | 5.4 | 1.665 | 2.9988 | 5.4 | 1.665 | 5.2479 | 9 | 0.555 | 4.00673 | 7.002 | 1.665 | 4.63148 | 7.002 |

**Table 9** Distance from fuzzy positive ideal solution (FPIS)

| | C1 | C2 | C3 | C4 | C5 | C6 |
|---|---|---|---|---|---|---|
| I1 | 0.000 | 0.336 | 0.387 | 0.000 | 0.205 | 0.000 |
| I2 | 0.000 | 1.305 | 0.387 | 0.205 | 0.000 | 0.000 |
| I3 | 0.554 | 0.336 | 0.387 | 0.556 | 1.121 | 0.554 |
| I4 | 0.554 | 0.229 | 0.253 | 0.719 | 1.121 | 0.981 |
| I5 | 3.456 | 0.000 | 0.387 | 0.556 | 0.719 | 0.426 |
| I6 | 0.719 | 0.229 | 1.301 | 0.719 | 0.205 | 0.426 |
| I7 | 0.554 | 0.000 | 0.253 | 0.556 | 0.719 | 0.426 |
| I8 | 0.554 | 0.336 | 1.382 | 0.719 | 1.433 | 0.554 |
| I9 | 0.554 | 0.000 | 0.253 | 0.719 | 1.433 | 0.554 |
| I10 | 0.719 | 0.336 | 0.000 | 0.205 | 0.556 | 0.554 |
| I11 | 0.554 | 0.336 | 0.000 | 0.719 | 0.719 | 0.426 |
| I12 | 0.719 | 0.336 | 0.387 | 0.719 | 0.556 | 0.426 |

**Table 10** Distance from fuzzy negative ideal solution (FNIS)

| | C1 | C2 | C3 | C4 | C5 | C6 |
|---|---|---|---|---|---|---|
| I1 | 0.719 | 1.215 | 1.240 | 0.964 | 1.294 | 0.981 |
| I2 | 0.719 | 0.000 | 1.240 | 0.847 | 1.433 | 0.981 |
| I3 | 0.204 | 1.215 | 1.240 | 0.422 | 0.425 | 0.697 |
| I4 | 0.204 | 1.242 | 1.288 | 0.370 | 0.425 | 0.000 |
| I5 | 0.204 | 1.305 | 1.240 | 0.422 | 0.867 | 0.781 |
| I6 | 0.000 | 1.242 | 0.189 | 0.742 | 1.294 | 0.781 |
| I7 | 0.204 | 1.305 | 1.288 | 0.422 | 0.867 | 0.781 |
| I8 | 0.204 | 1.215 | 0.000 | 0.000 | 0.000 | 0.697 |
| I9 | 0.204 | 1.305 | 1.288 | 0.000 | 0.000 | 0.697 |
| I10 | 0.000 | 1.215 | 1.382 | 0.553 | 0.980 | 0.697 |
| I11 | 0.204 | 1.215 | 1.382 | 0.370 | 0.867 | 0.781 |
| I12 | 0.000 | 1.215 | 1.240 | 0.370 | 0.980 | 0.781 |

**Table 11** Fuzzy positive ideal solution

| | C1 | C2 | C3 | C4 | C5 | C6 | $d_i^*$ |
|---|---|---|---|---|---|---|---|
| I1 | 0.000 | 0.336 | 0.387 | 0.000 | 0.205 | 0.000 | 0.928 |
| I2 | 0.000 | 1.305 | 0.387 | 0.205 | 0.000 | 0.000 | 1.897 |
| I3 | 0.554 | 0.336 | 0.387 | 0.556 | 1.121 | 0.554 | 3.508 |
| I4 | 0.554 | 0.229 | 0.253 | 0.719 | 1.121 | 0.981 | 3.858 |
| I5 | 3.456 | 0.000 | 0.387 | 0.556 | 0.719 | 0.426 | 5.544 |
| I6 | 0.719 | 0.229 | 1.301 | 0.719 | 0.205 | 0.426 | 3.601 |
| I7 | 0.554 | 0.000 | 0.253 | 0.556 | 0.719 | 0.426 | 2.509 |
| I8 | 0.554 | 0.336 | 1.382 | 0.719 | 1.433 | 0.554 | 4.979 |
| I9 | 0.554 | 0.000 | 0.253 | 0.719 | 1.433 | 0.554 | 3.514 |
| I10 | 0.719 | 0.336 | 0.000 | 0.205 | 0.556 | 0.554 | 2.371 |
| I11 | 0.554 | 0.336 | 0.000 | 0.719 | 0.719 | 0.426 | 2.755 |
| I12 | 0.719 | 0.336 | 0.387 | 0.719 | 0.556 | 0.426 | 3.143 |

**Table 12** Fuzzy negative ideal solution

| | C1 | C2 | C3 | C4 | C5 | C6 | $d_i^-$ |
|---|---|---|---|---|---|---|---|
| I1 | 0.719 | 1.215 | 1.240 | 0.964 | 1.294 | 0.981 | 6.414 |
| I2 | 0.719 | 0.000 | 1.240 | 0.847 | 1.433 | 0.981 | 5.221 |
| I3 | 0.204 | 1.215 | 1.240 | 0.422 | 0.425 | 0.697 | 4.203 |
| I4 | 0.204 | 1.242 | 1.288 | 0.370 | 0.425 | 0.000 | 3.530 |
| I5 | 0.204 | 1.305 | 1.240 | 0.422 | 0.867 | 0.781 | 4.820 |
| I6 | 0.000 | 1.242 | 0.189 | 0.742 | 1.294 | 0.781 | 4.248 |
| I7 | 0.204 | 1.305 | 1.288 | 0.422 | 0.867 | 0.781 | 4.868 |
| I8 | 0.204 | 1.215 | 0.000 | 0.000 | 0.000 | 0.697 | 2.116 |
| I9 | 0.204 | 1.305 | 1.288 | 0.000 | 0.000 | 0.697 | 3.495 |
| I10 | 0.000 | 1.215 | 1.382 | 0.553 | 0.980 | 0.697 | 4.827 |
| I11 | 0.204 | 1.215 | 1.382 | 0.370 | 0.867 | 0.781 | 4.819 |
| I12 | 0.000 | 1.215 | 1.240 | 0.370 | 0.980 | 0.781 | 4.587 |

**Table 13** Closeness coefficients and rank

| | $d_i^*$ | $d_i^-$ | $(CC_i)$ | Rank |
|---|---|---|---|---|
| I1 | 0.928 | 6.414 | 0.873598 | 1 |
| I2 | 1.897 | 5.221 | 0.733491 | 2 |
| I3 | 3.508 | 4.203 | 0.545096 | 7 |
| I4 | 3.858 | 3.530 | 0.477819 | 10 |
| I5 | 5.544 | 4.820 | 0.465068 | 11 |
| I6 | 3.601 | 4.248 | 0.541217 | 8 |
| I7 | 2.509 | 4.868 | 0.659922 | 4 |
| I8 | 4.979 | 2.116 | 0.298294 | 12 |
| I9 | 3.514 | 3.495 | 0.49863 | 9 |
| I10 | 2.371 | 4.827 | 0.670624 | 3 |
| I11 | 2.755 | 4.819 | 0.636239 | 5 |
| I12 | 3.143 | 4.587 | 0.59337 | 6 |

# References


1. Agostini L, Filippini R (2019) Organizational and managerial challenges in the path toward Industry 4.0. Eur J Innov Manag
2. Vaidya S, Ambad P, Bhosle S (2018) Industry 4.0–a glimpse. Procedia Manuf 20:233–238
3. Mrugalska B, Wyrwicka MK (2016) Towards lean production in industry 4.0
4. Mittal VK, Sangwan KS (2014) Fuzzy TOPSIS method for ranking barriers to environmentally conscious manufacturing implementation: government, industry and expert perspectives. Int J Environ Technol Manage 17(1):57–82
5. Bhasin S (2012) Prominent obstacles to lean. Int J Prod Perform Manag
6. Kumar R, Kumar V (2014) Barriers in implementation of lean manufacturing system in Indian industry: a survey. Int J Latest Trends Eng Technol 4(2):243–251
7. Belhadi A, Touriki F E (2017) Prioritizing the solutions of lean implementation in SMEs to overcome its barriers. J Manuf Technol Manag
8. AlManei M, Salonitis K, Xu Y (2017) Lean implementation frameworks: the challenges for SMEs. Procedia CIRP 63:750–755
9. Bakke AL, Johansen A (2019) Implementing of lean–challenges and lessons learned. Procedia Comput Sci 164:373–380
10. Glass R, Meissner A, Gebauer C, Stürmer S, Metternich J (2018) Identifying the barriers to Industrie 4.0 Procedia CIRP 72:985–988
11. Orzes G, Rauch E, Bednar S, Poklemba R (2018) Industry 4.0 implementation barriers in small and medium sized enterprises: a focus group study. In: 2018 IEEE international conference on industrial engineering and engineering management (IEEM). IEEE, pp 1348–1352
12. Müller JM (2019) Assessing the barriers to Industry 4.0 implementation from a workers' perspective. IFAC-PapersOnLine 52(13):2189–2194
13. Raj A, Dwivedi G, Sharma A, de Sousa Jabbour ABL, Rajak S (2019) Barriers to the adoption of industry 4.0 technologies in the manufacturing sector: an inter-country comparative perspective. Int J Prod Econ 107546
14. Rauch E, Dallasega P, Unterhofer M (2019) Requirements and barriers for introducing smart manufacturing in small and medium-sized enterprises. IEEE Eng Manage Rev 47(3):87–94
15. Mayr A, Weigelt M, Kühl A, Grimm S, Erll A, Potzel M, Franke J (2018) Lean 4.0-A conceptual conjunction of lean management and Industry 4.0. Procedia CIRP 72(1):622–628

16. Dombrowski U, Richter T, Krenkel P (2017) Interdependencies of Industrie 4.0 & Lean production systems: a use cases analysis. Procedia Manuf 11:1061–1068
17. Wang TC, Lee HD (2009) Developing a fuzzy TOPSIS approach based on subjective weights and objective weights. Expert Syst Appl 36(5):8980–8985
18. Mahdavi I, Mahdavi-Amiri N, Heidarzade A, Nourifar R (2008) Designing a model of fuzzy TOPSIS in multiple criteria decision making. Appl Math Comput 206(2):607–617
19. Awasthi A, Chauhan SS, Omrani H (2011) Application of fuzzy TOPSIS in evaluating sustainable transportation systems. Expert Syst Appl 38(10):12270–12280
20. Sun CC (2010) A performance evaluation model by integrating fuzzy AHP and fuzzy TOPSIS methods. Expert Syst Appl 37(12):7745–7754
21. Sahwan MA, Ab Rahman MN, Deros B M (2012) Barriers to implement lean manufacturing in Malaysian automotive industry. Jurnal Teknologi 59(2)
22. Yadav V, Jain R, Mittal ML, Panwar A, Sharma MK (2019) An appraisal on barriers to implement lean in SMEs. J Manuf Technol Manag
23. Dora M, Kumar M, Gellynck X (2016) Determinants and barriers to lean implementation in food-processing SMEs–a multiple case analysis. Prod Plan Control 27(1):1–23

# Hardfacing of Ni-Based Alloys on Medium Carbon Steel to Improve Turbine Blade Properties

**P. Karuppuswamy, C. Bhagyanathan, S. Sathish, and D. Elangovan**

**Abstract** Hardfacing involves depositing a required superior material over a substrate for improving its surface properties like wear and corrosion to function properly in the given working conditions. For example, normally, carbide forming elements that have excellent tribological properties are fused over the steel substrates for the improvement in performance or life. The cited research work described in this article has been undertaken to investigate the suitability of the substrate and the coating material to be used for high temperature applications such as turbine blade which has to work against the corrosive and erosive working environment. The failure of the turbine blades is predominantly due to combustion by-products exhausted from the combustion chamber attributed to high temperature and velocity. In this research work, medium carbon steel EN24 substrate was hardfaced with the coating material nickel-based alloy Inconel 625 selected for its toughness, high corrosion and creep resistance using MIG welding process. The experiments were conducted as a means to improve the surface properties of the turbine blades through hardfacing. The results of the experiments indicated that the properties of Inconel 625 can be comfortably used as turbine blades.

**Keywords** Ni-based alloys · Hardfacing · Medium carbon steel · Turbine blade

P. Karuppuswamy (✉) · C. Bhagyanathan · S. Sathish
Department of Mechanical Engineering, Sri Ramakrishna Engineering College, Coimbatore, Tamil Nadu, India
e-mail: pkaruppuswamy_cbe@rediffmail.com

C. Bhagyanathan
e-mail: bhagyanathan@srec.ac.in

S. Sathish
e-mail: sathish.selvaraj@srec.ac.in

D. Elangovan
Department of Mechanical Engineering, PSG Institute of Technology and Applied Research, Coimbatore, Tamil Nadu, India
e-mail: delangovan_cbe@rediffmail.com

© The Editor(s) (if applicable) and The Author(s), under exclusive license to Springer Nature Singapore Pte Ltd. 2021

S. Mohan et al. (eds.), *Materials, Design, and Manufacturing for Sustainable Environment*, Lecture Notes in Mechanical Engineering,
https://doi.org/10.1007/978-981-15-9809-8_38

## 1 Introduction

Materials subjected to high temperature applications such as turbines and aero engines warranted the development of superalloys. Superalloys exhibit excellent mechanical properties at high temperature with surface stability [1]. Among the three categories of superalloys, the nickel-based alloys are predominantly used for turbine blade applications [2]. The research paper describes the feasibility study undertaken for hardfacing of Inconel 625 alloy on medium carbon steel to be used for high temperature applications such as turbine blades. Hardfacing is one of the techniques used to deposit the required alloy to improve the surface properties like resistance against wear, tear, abrasion and corrosion [3]. The hardfaced part behaves like a laminate with the substrate and improves the surface properties. The superior creep and stress rupture resistance of nickel superalloys are suitable for engines to operate at higher temperatures for producing greater thrust.

Globally, steel finds more than 85% of applications, and obviously, the substrate material was selected from the steel family for experimentation. Though medium carbon steels are possessing good welding characteristics, they are subjected to pre- and post-heating for stress relieving [4]. The substrates can be hardfaced with alloys using different welding processes ranging from the traditional oxy-acetylene gas welding to more sophisticated PVD, thermal barrier coatings [5–7].

## 2 Materials and Methods

The combination of hardfaced medium carbon steel substrate with Inconel 625, a nickel-chromium alloy, improves tribological properties. The good formability and weldability of Inconel 625 were found to be suitable for being coated with the selected substrate material. Hardfacing at low temperature is preferred to control the grain development [8]. Work hardening of Inconel 625 is not a complex one on account of its ductility, and therefore, moderate annealing methods are employed while subjected to welding. Superior properties like weldability and surface properties of the nickel-based superalloys such as Inconel 625 made it suitable for aeronautical, chemical, petrochemical and marine applications and could sustain while exposed to aggressive environments. Inconel 625 is made out of alloying elements, namely chromium, molybdenum and niobium [9] whose chemical composition is provided in Table 1.

Other than iron and carbon, steel consists of manganese, chromium, nickel, etc. and non-metals such as carbon, silicon, phosphorus and sulphur that influence its properties [10–12]. Preheating of the substrate is not required if the percentage of carbon does not exceed 0.15, and it is essential if the percentage of carbon exceeds 0.15 for hardfacing. The parameters of preheating depend on the thickness of parts being hardfaced [13, 14]. Precautions are to be taken while hardfacing the steels

**Table 1** Chemical composition of Inconel 625

| Cr | Fe | Mo | Nb + Ta | C | Mn | Si | P | S | Al | Ti | Co | Ni |
|---|---|---|---|---|---|---|---|---|---|---|---|---|
| 20.0–23.0 | 5.0 | 8.0–10.0 | 3.15–4.15 | 0.10 max | 0.50 max | 0.50 max | 0.015 max | 0.015 max | 0.40 max | 0.40 max | 1.0 max | Balance |

**Table 2** Chemical composition of EN 24

| C | Si | Mn | P | S | Mo | Cr | Ni |
|---|---|---|---|---|---|---|---|
| 0.36–0.44 | 0.10–0.35 | 0.45–0.70 | 0.035 (min) | 0.04 (min) | 0.20–0.35 | 1.00–1.40 | 1.30–1.70 |

that consist of 0.25–0.50% carbon and 0.60–1.65% manganese [15–17]. Chemical composition of EN24 is provided in Table 2.

EN24 has wide application in the field of automotive, aerospace components manufacturing [18–20].

## 3 Experimentation

Experiment on hardfacing of Inconel 625 over medium carbon steel was carried out using MIG welding process. The base material EN24 was procured as a block of dimension 50 × 50 × 50 mm, and five blocks of the mentioned dimension were hardfaced using MIG welding machine with different welding parameters [21–24]. Inconel 625 was deposited in multiple layers over the substrate to the height of 2 mm, and another 8 passes were made over the surface to increase the bead height to 4 mm ensuring better dilution of Inconel 625 over the base material [25, 26]. The input parameters of the MIG welding process standardized based on the literature survey are as given below:

- Shielding gas—100% pure ARGON,
- Gas flow rate—5LPM,
- Number of weld layers—2,
- Thickness of the weld—4 mm,
- Diameter of the filler wire—1.2 mm.

Voltage level was set as 16, 20, 21.2, 21.2 and 22 V for each of the five specimens, and feed rate was set as 4.6 m/min for all the specimens with varying voltage. The weld speed was maintained at constant pace of 2.5 mm/s for better dilution. The parameters for the experiment were monitored thoroughly until the process was completed for its soundness [27, 28]. These parameters played an important role for the process to be completed with uniform deposition without any defect. The resulted hardfaced specimen is shown in Fig. 1.

## 4 Results and Discussion

The specimens were tested for wear and corrosion to study the suitability of the coating for the intended high temperature application. The wear testing was carried

**Fig. 1** Overlay of Inconel 625 over EN24

out using pin-on-disc wear testing apparatus, and the corrosion testing was carried using salt spray chamber.

## 4.1 Wear Testing

Wear rate of the material plays a vital role in material selection as it causes loss of the material over a period which leads to the deterioration in mechanical properties. Mechanical erosion and chemical corrosion affect the properties such as fatigue and creep. Loss of important properties degrades the functionality of the material and becomes inappropriate for the intended application.

Wear test was conducted in pin-on-disc apparatus to assess the wear rate of the material or specimen in which friction would be the predominant factor causing deterioration of the life of the material or coating [29]. The specimens having dimensions $10 \times 10 \times 50$ mm were fixed in the holder set with a sliding distance of 94 mm. A load of 5 kg was applied on the specimens. A tungsten carbide plate was used to make contact with specimen surface where Inconel 625 was deposited [30, 31]. The specimen fixed in the collet was placed in the holder, and the time was set as 25 min. The process was started, and the values of the parameters such as wear, friction force and the coefficient of friction (COF) were noted at the end for each specimen. The values obtained were plotted in graph as shown in Figs. 2 and 3. The experiments done on the pin-on-disc indicated that wear rate was 547 microns, and the coefficient of friction was 13.9. The results indicated that Inconel 625 would be an appropriate coating material with excellent wear resistant characteristics for high temperature applications.

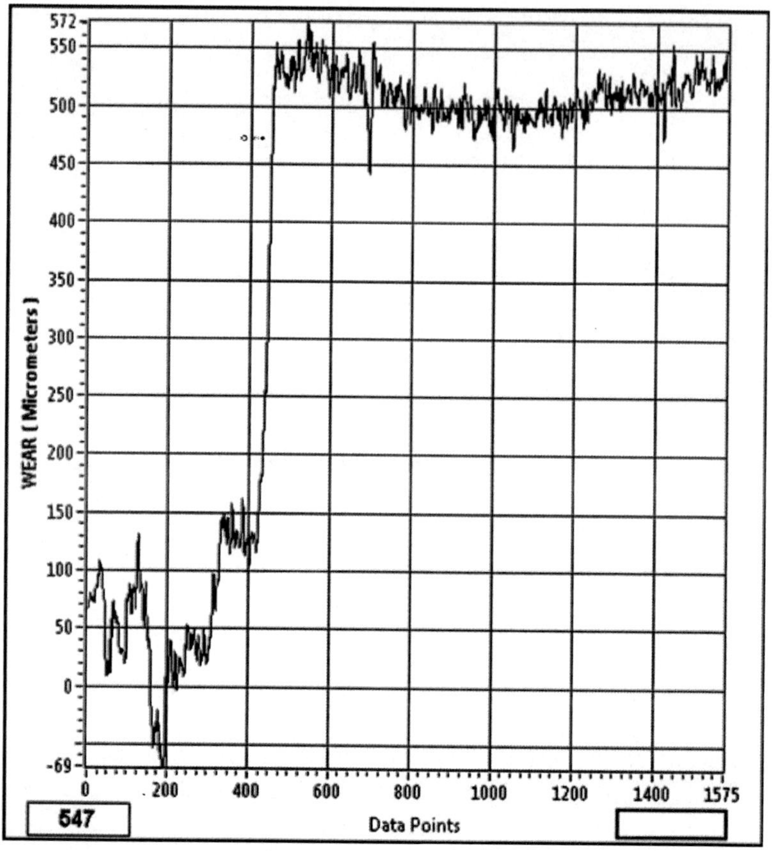


**Fig. 2** Wear rate—Inconel 625 coated on EN24

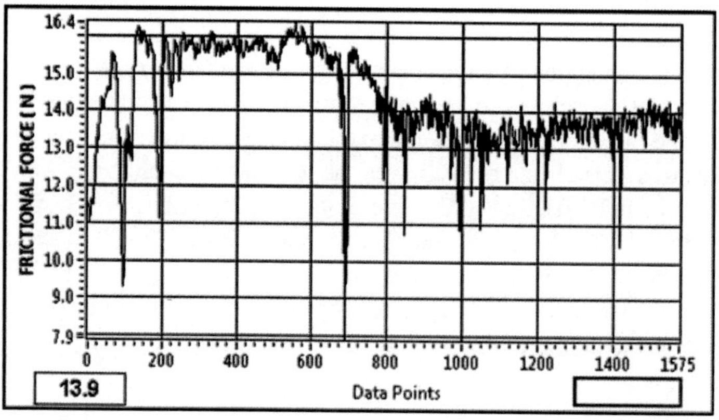


**Fig. 3** Co-efficient of friction—Inconel 625 coated on EN24

## 4.2 Corrosion Testing

Corrosion deteriorates the performance of materials and surface coatings at exponential rate and needs to be assessed through reliable means. Salt spray testing is very common to evaluate corrosion of the material [19]. The cited test involves imposition of an accelerated corrosion attack on the test material intentionally to assess the response to corrosion attack. The salt spray chamber is having an enclosure or a tank in which test specimen is held to expose it with corrosion environment accelerated intentionally. Artificial mist or fog was stimulated with the help of an atomizer. Liquid corrosion solution was mixed with pressurized air to spray on the specimen in the mist form. The very commonly used corrosive liquid is water mixed with 5% sodium chloride. The specimens were masked suitably, and the exposed surface was cleaned with tap water. This cleaned specimen was then hanged in the chamber with the help of a plastic tray. The specimen was placed at an angle of 15–30°. The set-up was then placed inside the chamber which was maintained at temperatures between 35 and 37 °C. The temperature during the procedure was maintained between 25 and 30 °C. The relative humidity during test was maintained between 45 and 75% RH. The samples were inspected for every 24 h, and the results were tabulated. The test was carried out for 50 h. Corrosion test results were obtained using weight loss method. It is clear that before conducting the test, the specimen was at the state where no corrosion was inhibited. It was kept inside the chamber under the corrosive environment created intentionally with the decided parameters. Figures 4 and 5 show the test specimen before and after the corrosion test.

After conducting the test, it was noted that the weight loss was not occurred significantly as the weight loss was in the fourth digit of the decimal and began to corrode after 50 h of exposure. Also, it is clear that the uncoated side of the specimen corroded severely whereas the coated side exposed was not corroded significantly. After comparing the weights of the specimen before and after conducting the test, it was concluded that the coating material Inconel 625 would be an appropriate material that could be used for high temperature applications like turbine blades and aero engines.

**Fig. 4** Specimen—before corrosion test

**Fig. 5** Specimen—after corrosion test

## 5 Conclusion

Hardfacing of EN24 with Inconel 625 was carried out using MIG welding process with an objective of improving the surface properties such as resistance against wear and corrosion. Hardfacing of candidate alloys can be applied to substrates through different welding processes ranging from the traditional oxyacetylene gas welding to highly sophisticate metal inert gas welding. Hardfacing technique is one of the most feasible and cost-efficient. Diffusion of Inconel on EN24 after heat treatment really improved the wear and corrosion properties of the EN24 substrate. The turbine blades can be operated at higher temperatures for an extended period if made out of nickel-based superalloys that exhibit exceptional creep and stress rupture resistance. It is very difficult to produce a material which will be equivalent to single crystal Inconel alloy. Even if produced, processing will not be easier and economically not viable. It is evident that the method and material proposed can be used as an alternate to Inconel which would exhibit 80% of the similar properties.

## References

1. Pokluda J, Kianicová M (2010) Assessment of performance capability of turbine blades with protective coatings after overheating events. Eng Fail Anal 17(6):1389–1396. 24
2. Barbosa C, Nascimento JL, Caminha IMV, Abud IC (2005) Microstructural aspects of the failure analysis of nickel base superalloys components. Eng Fail Anal 12(3):348–361
3. Pradeep GRC, Ramesh A, Durga Prasad B (2010) A review paper on hardfacing processes and materials. Int J Eng Sci Technol 2(11):6507–6510. 3
4. Wang X, Han F, Liu X, Qu S, Zou Z (2008) Microstructure and wear properties of the Fe–Ti–V–Mo–C hardfacing alloy. Wear 265(5–6):583–589. 5
5. Wang X, Han F, Liu X, Qu S, Zou Z (2008) Microstructure and wear properties of the Fe–Ti–V–Mo–C hardfacing alloy. Wear 265(5–6):583–589. 10

6. Wang XH, Zou ZD, Qu SY, Song SL (2005) Microstructure and wear properties of Fe-based hardfacing coating reinforced by TiC particles. J Mater Process Technol 168(1):89–94. 17
7. Kai W, Hui P, Hongbo G, Shengkai G (2012) Effect of sintering on thermal conductivity and thermal barrier effects of thermal barrier coatings. Chin J Aeronaut 25(5):811–816. 23
8. Hou JS, Guo JT, Zhou LZ, Yuan C, Ye HQ (2004) Microstructure and mechanical properties of cast Ni-base superalloy K44. Mater Sci Eng: A 374(1–2):327–334. 6
9. Chang CM, Lin CM, Hsieh CC, Chen JH, Wu W (2009) Micro-structural characteristics of Fe–40 wt% Cr–xC hard facing alloys with [1.0–4.0 wt%] carbon content. J Alloy Compd 487(1–2):83–89. 20
10. Wang XH, Han F, Liu XM, Qu SY, Zou ZD (2008) Effect of molybdenum on the microstructure and wear resistance of Fe-based hardfacing coatings. Mater Sci Eng: A 489(1–2):193–200. 11
11. Wang XH, Han F, Qu SY (2008) Microstructure of the Fe-based hardfacing layers reinforced by TiC–VC–Mo2C particles. Surf Coat Technol 202(8):1502–1509. 12
12. Chang CM, Chen YC, Wu W (2010) Microstructural and abrasive characteristics of high carbon Fe–Cr–C hard facing alloy. Tribol Int 43(5–6):929–934
13. Ünlü BS, Atik E (2010) Investigation of tribological properties of boronized Fe-based SAE 1020 and TS-DDK 40 journal bearings at high loads. Mater Des (1980–2015) 31(5):2690–2696. 13
14. Chawla V, Sidhu BS, Puri D, Prakash S (2008) Performance of plasma sprayed nanostructured and conventional coatings. J Aust Ceram Soc 44(2):56–62. 26
15. Buytoz S (2006) Microstructural properties of SiC based hardfacing on low alloy steel. Surf Coat Technol 200(12–13):3734–3742. 16
16. Bulloch JH, Henderson JL (1991) Some considerations of wear and hard facing materials. Int J Press Vessel Pip 46(3):251–267. 22
17. Peng H, Wang L, Guo L, Miao W, Guo H, Gong S (2012) Degradation of EB-PVD thermal barrier coatings caused by CMAS deposits. Prog Nat Sci: Mater Int 22(5):461–467. 25
18. Mouritz AP (2012) Introduction to aerospace materials. Elsevier. 4
19. Morad AM, Shash YM (2014) Nickel base superalloys used for aero engine turbine blades. In: Proceedings of the 16th international AMME conference, vol 27, p 29. 1
20. Sims CT, Stoloff NS, Hagel WC, II S (1987) High temperature materials for aerospace and industrial power. A Wiley-Interscience Publication John Wiley and Sons, New York. 7
21. Wang X, Han F, Liu X, Qu S, Zou Z (2008) Microstructure and wear properties of the Fe–Ti–V–Mo–C hardfacing alloy. Wear 265(5–6):583–589. 10.5
22. Pradeep GRC, Ramesh A, Prasad BD (2013) Comparative study of hard facing of AISI 1020 steel by three different welding processes. Glob J Res Eng. 29
23. https://super-metals.com/alloy/inconel-625/. 8
24. https://www.azom.com/article.aspx?ArticleID=4459. 9
25. Lin YC, Chang KY (2010) Elucidating the microstructure and wear behavior of tungsten carbide multi-pass cladding on AISI 1050 steel. J Mater Process Technol 210(2):219–225. 18
26. Němeček S, Fidler L, Fišerová P (2014) Corrosion resistance of laser clads of Inconel 625 and Metco 41C. Phys Procedia 56:294–300. 28
27. Atamert S, Bhadeshia HKDH (1990) Microstructure and stability of Fe Cr C hardfacing alloys. Mater Sci Eng: A 130(1):101–111. 15
28. Mostafaei A, Hilla C, Stevens EL, Nandwana P, Elliott AM, Chmielus M (2018) Comparison of characterization methods for differently atomized nickel-based alloy 625 powders. Powder Technol 333:180–192. 30
29. Okechukwu C, Dahunsi OA, Oke PK, Oladele IO, Dauda M (2017) Review on hardfacing as method of improving the service life of critical components subjected to wear in service. Niger J Technol 36(4):1095–1103. 27
30. Katsich C, Badisch E, Roy M, Heath GR, Franek F (2009). Erosive wear of hardfaced Fe–Cr–C alloys at elevated temperature. Wear 267(11):1856–1864. 14
31. Karuppuswamy P, Bhagyanathan C, Narendhar C, Kirupakaran RN, Arun Kumar SD (2017) The effect of carburization of ASTM A36 steel substrate in improving wear properties of plasma sprayed WC-12%Co coating. Surf Rev Lett 24(05):1750068

# Development of Supply Chain Risk Management Strategies for Mitigating Loss Prevention in Manufacturing Organizations

**D. Elangovan, G. Sundararaj, S. R. Devadasan, P. Karuppuswamy, and R. Vishnupriyan**

**Abstract** Customer demand requirements are changing rapidly in the current industrial scenario. Changing customer needs necessitates the adoption of appropriate modern tools, equipment, and other needed delivery methods. These efforts are to be integrated by considering the demand requirements of customers, the needed supply of raw material, availability of sub-assembly, manufacturing systems management, delivery of products based on the due date requirements of the customers, and after-sales services. These activities are to be integrated with dynamic operable strategies to face the challenges in the pierce competitive market environment. In order to amalgamate these activities dynamically, supply chain management integrated with risk management strategies are initiated. Risk mitigation through the identification of risk, assessment of risk, prioritization of risk, and control of risk are effective in many of the manufacturing organizations. But, the risk control measure just pinpoints

D. Elangovan (✉) · R. Vishnupriyan
Department of Mechanical Engineering, PSG Institute of Technology and Applied Research, Coimbatore, Tamil Nadu, India
e-mail: delangovan_cbe@rediffmail.com

R. Vishnupriyan
e-mail: rvishnupriyanmech@gmail.com

G. Sundararaj
Department of Mechanical Engineering, PSG College of Technology, Coimbatore, Tamil Nadu, India
e-mail: psgtechgsraj@yahoo.com

S. R. Devadasan
Department of Production Engineering, PSG College of Technology, Coimbatore, Tamil Nadu, India
e-mail: devadasan_srd@yahoo.com

P. Karuppuswamy
Department of Mechanical Engineering, Sri Ramakrishna Engineering College, Coimbatore, Tamil Nadu, India
e-mail: pkaruppuswamy_cbe@rediffmail.com

© The Editor(s) (if applicable) and The Author(s), under exclusive license to Springer Nature Singapore Pte Ltd. 2021

S. Mohan et al. (eds.), *Materials, Design, and Manufacturing for Sustainable Environment*, Lecture Notes in Mechanical Engineering,
https://doi.org/10.1007/978-981-15-9809-8_39

the loss reduction at that point of time only and the development of feasible strategies for risk mitigation. Then, moving towards routine production activities will be usually resumed. The loss of producing elements in the supply chain flows was not analyzed in detail to prevent them in the future. In order to tackle these issues in a cursory manner in manufacturing organizations, the development, and implementation of strategies by adopting the integrated supply chain risk management principles are essential for the mitigation of loss prevention efforts. For moving towards the continual improvement of manufacturing organizations with appropriate outcomes, it is essential to integrate these types of risk prevention measures.

**Keywords** Suppliers · Customers · Supply chain risk management · Production · Loss prevention · Manufacturing organizations

## 1 Introduction

Every manufacturing organization must focus its flows of supply chain activities in a phased manner. These activities and efforts must be appropriately integrated along the entire supply chain flow by adopting the needed cooperation, coordination, continuous monitoring, and follow-up strategies. These strategies will be influenced by the reduction of probability of occurrence and the frequency of consequence of the likelihood of risk or loss producing events along with the flows of the entire supply chain. Hence systematic continuous monitoring and follow-up activities along the entire supply chain flows are vital to stay against the competition. Thus, the organizations are focusing on implementing new techniques and strategies to achieve the overall success of the entire supply chain system. Varieties of similar products with more attractive price tags are introduced from various competitors from time-to-time. Thus, organizations are necessary to focus on feasible alternative strategies in the entire internal and external supply chain flows. This will further helpful to meet customer expectations, achieve customer acceptations, and enhance customer retention [1–3]. These strategies will be significantly influencing the cycle time reduction, time delay reduction, delivery performance improvement, focus on customer needs, product customization, response to customer feedback, overall time reduction, and overall cost reduction [4–6]. Hence strategies for mitigating loss prevention are highly essential to stay against competitive business success with appropriate outcomes.

## 2 Concept of Risk Management

The probability or chance or likelihood of occurrence of loss producing event is called as a risk. It is also a probability of occurrence of an adverse event in the process or system. The risk events are having the potential chance or probability for causing an undesirable change in the process or system. Through risk mitigation and development of appropriate risk management strategies, the risk level of the system or process is assessed, managed, and controlled to an acceptable level. Thus, risk mitigation strategies will provide the basic means to self-audit a firm's current exposures and practices. Risk management process involves risk identification, risk evaluation, and risk control.

## 2.1 Risk Identification

All possible risk events that may significantly affect the process or system, which are ranging from high impact to low impact with probabilities are identified. Risk-causing events can be listed, summarized, and analyzed by flow charts, cost data, brainstorming with employees, consulting with experts, and historical data analysis.

## 2.2 Risk Evaluation

It is used to analyze the risk events that are potential hazards of the process or system. The identified risk events will be consolidated, categorized, listed, and summarized. The probabilities of occurrence of the risk events are then assessed based on the collected details. The consequences based particularly on the impact of cost and their associated severities of the risk events are determined. Among them, risks with high frequency and high probability of occurrence indicate the greatest impact on the process or system. Thus, it should be given immediate priority and predominant attention. After identifying the relevant consequences and their significance, appropriate risk planning measures are done to reduce or to eliminate the likelihood of risk events. Finally, appropriate decisions are taken by analyzing the results of the assessment as the corrective measures to reduce or eliminate the risk of loss producing events one by one based on the identified priority level.

## 2.3 Risk Control

This phase develops sets and implements suitable or appropriate feasible strategies, needed policies, modified procedures, fine-tuned goals, and improved responsibility standards for risk mitigation cum risk control measures. The significant potential risk control measures are implemented with avoidance, retention, transfer, and reduction of appropriate risk control strategies. The risk control data are then compiled and documented for the database and further review analysis.

## 3 Concept of Supply Chain Management

Supply chain management integrates all the activities needed for the transformation and delivery of products as per the customer needs. It is integrated with all the necessary material, information, and financial flows. A supply chain network facilitates the relationship and inter-relationship of various supply chain activities and its associated flows. It includes the after-sales service also by properly using the available resources and facilities. It exists in the case of either manufacturing or service organizations. The nature and integrity of the supply chain flows will vary from one organization to another organization.

The information-based revolution, customer need-based demands (quality, cost, delivery, and service) and appropriate inter-organizational relationships are the three major developments in the flows of emerging supply chain management. Among these three, the inter-organizational relationship is a highly essential one for implementing effective flows in the supply chain management to face the challenges in the pierce competition.

## 4 Concept of Supply Chain Risk Management

The supply chain risks need to be much more investigated by both academia and research. The issue of supply chain risk handling or sharing is an emerging research agenda. This is needed to be focused especially on those organizations with longer supply chains. This is to face the challenges in the uncertain demand and supply along with the entire flows of the supply chain network. From a theoretical and research perspective the flows of the supply chain network will include a number of real-time issues. The analysis of flows of supply chain networks, the identification of risk event types, and the risk handling stages have recently found importance in the emerging research agenda.

The analysis of flows of supply chain networks focuses on the flows of the network of activities in the organizations. These complex relationships are useful in implementing better buyer-seller supply chain network flows. This concept can be furthermore expanded externally for two or three or many organizations for more effectiveness. The identification of risk event types can be varied for organizations with different kinds and scope. The impact of the risk can be quantified by the product of frequency or probability of occurrence and its associated cost related consequence. The frequency or probability of occurrence may be defined by the levels of highly probable, probable, occasional, and rare. The associated cost consequence may be defined by the levels of high, substantial, medium, and low. The segregation of these types of risk levels is described by the concerned organization based on the nature, type, and impact of the risk events that are occurred. The risk handling stage discusses the risk mitigation analysis by adopting the different ways of managing risks by implementing appropriate strategies like risk avoidance, reduction, transfer, and share, as the case that may be applicable.

Supply chain risks are unpredictable events with the uncertainty of loss producing events that will influence the achievement of the business objectives. The influence of flows of any link in the supply chain will affect the dynamic operability of the entire organization. To effectively achieve the business objectives, organizations must focus on the supply chain entities and its associated partners. Thus, it becomes essential to understand the uncertainties at any point along the supply chain to ensure that the organizational objectives are being attained from time-to-time in a phased manner. Supply chain risk management is a synergetic process and structured framework of the entire supply chain, which is used to integrate the feasible strategies, facilities, processes, human resources, technology, and knowledge. Figure 1 illustrates the basic processes of supply chain risk management in the extremities of suppliers and customers of the entire flow of the supply chain of the organization.

Thus, it is essential to mitigate, identify, evaluate, control, and monitor the supply chain risk events for achieving business success. There are new uncertainties that are arising from the recent competitive emerging revolutionary trends. These are to be furthermore addressed and investigated. However, there is a lack in real industrial situations in practice for handling supply chain risk management, and there is more scope available for doing research improvements [7].

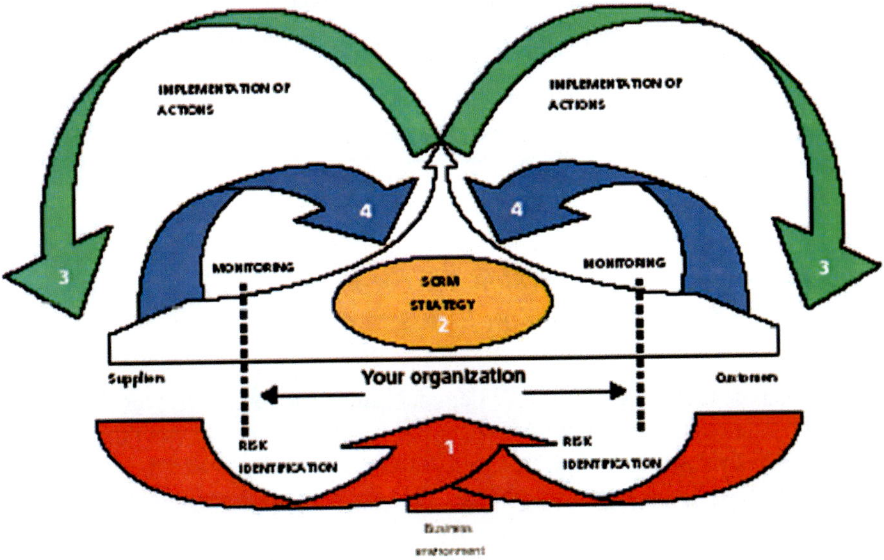


Fig. 1 Supply chain risk management process

# 5 Review of Literature

## 5.1 Supply Chain Risk Management: Need and Industrial Importance

Industries are recently focusing on each and every entity of supply chain flows and its inter-relationships in a cursory manner for staying in the pierce competition. Thus, close monitoring and subsequent follow-ups of the entire flows of supply chain integration are essential for the reduction of the occurrence of likelihood of risk events as well as its elimination in the future. The implementation measures in practice by the organizations play a key challenge due to the trade-off observed in real-time between the loss incurred and opportunity for improvement [8, 9]. The various real-time factors in the organizations by considering the reasons, root causes, and its subsequent strategies are inevitably analyzed in real-time practice for the risk mitigation and its subsequent loss reduction. The formulated feasible strategies, its alignment with corporate strategy, and its follow-ups with learning are used to reduce the likelihood of occurrence of risk producing events [10, 11]. The implemented strategies are to be reviewed, rephrased, and revised for better effectiveness in the organizations [12–14]. Numerical based solutions are needed for the detailed analysis including the cumulative effect of loss accountability with respect to time to achieve stable solutions, which are suitable for the pierce competitive environment [15, 16].

In this paper, the development of appropriate preventive supply chain risk management strategies for producing and supply of precision manufacturing components by using computer numerical control machine tools was investigated to reduce the likelihood of occurrence of risk issues related to time-delay for facing the pierce competition.

## 5.2 Investigation and Development of Pilot Preventive Supply Chain Risk Management Strategies

Global market focuses on customer demand requirements and timely delivery of products. The major challenges are changing customer expectations rapidly in the current industrial scenario. Changing customer needs necessitates the adoption of modern tools, equipment, and other needed delivery methods by analyzing the various practical constraints.

Thus, the relationship and inter-relationship of the flows of the supply chain activities including after-sales service are to be properly integrated along the entire supply chain with feasible operable strategies. Hence highly integrated follow-ups are needed. From time-to-time outcomes are to be extracted in a phased manner. In order to integrate these efforts, supply chain risk management strategies are initiated in many organizations. These efforts are carried out through risk mitigation.

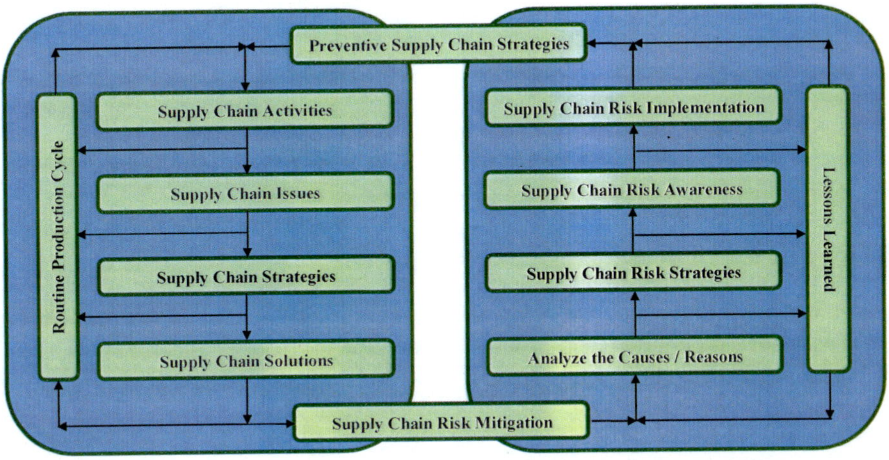


**Fig. 2** The generic salient features of the developed framework for the investigation of the preventive supply chain risk management strategies

This is achieved through identification, assessment, prioritization, and control of the occurrence of likelihood of risk events. These efforts are effective in many of the manufacturing organizations. But, the risk control measures just pinpoint the loss reduction at that point in time and development of strategies for risk mitigation. Then, moving towards routine production activities will be resumed. The loss producing elements in the supply chains were not analyzed in detail to prevent them in the future [7]. Hence the development and initiation of preventive supply chain risk management strategies for initiating and mitigating loss prevention efforts in the manufacturing organizations are needed to investigate these issues further in a cursory manner. For moving towards the continual improvement of manufacturing organizations with appropriate outcomes, it is essential to integrate these types of risk prevention measures. The generic salient features of the developed framework for the investigation of the preventive supply chain risk management strategies were shown in Fig. 2.

## 6 Details of the Profile of the Case Study Organization

A batch production case study organization producing and supply of precision manufacturing components by using computer numerical control machine tools are taken into consideration. The organization's product profile includes top flanges, bottom flanges, shafts, cylinder, ball valves, carriers, cases, and housing products. The production and supplying of major precision components of the case organization by using computer numerical control machine tools are shown in Fig. 3.

**Fig. 3** Major precision components produced of the case organization

These products serve for general engineering applications. The product design of the components, its associated geometry, needed material, and required production processes are well defined in the routine production cycle.

## 7 Investigations with Risk Analysis

The production and supply details of the batch production case study organization producing and supply of precision manufacturing components by using computer numerical control machine tools are taken into consideration for further investigation. In the routine production activities, the supply chain of the case study organization has executed by considering the customers' demand requirements, required spaces or sizes, needed ultimate load carrying capacity, and preferred service life. It was observed that there are some time delay issues observed in the conducted plot investigation during the period of April 2019 to September 2019. The observed issues related to time-delays for production and supply of precision manufacturing components, its occurrence frequency risk categories and associated ratings with risk description, its cost consequence risk categories and associated ratings with risk description and risk rating matrix developed for the observed time-delays are studied during the pilot investigation period by using the concept of risk ranking are illustrated in the Tables from 1, 2, 3 to 4, respectively.

By considering the occurrence frequency and associated cost implications of the time delay related risk events, it was found that the *Supply Chain: Internal executions* (Table 1; *Serial Number 2*) as a predominant time-delay in the case study organization during the pilot investigation period.

### *7.1 Development of Appropriate Preventive Supply Chain Risk Management Pilot Strategies and Its Implementation*

The reasons and root causes for the time delay due to supply chain: internal executions are further analyzed in detail. This is carried out to reduce and eliminate the risk event occurrence and its associated cost implications.

**Table 1** Observed issues related to time-delays for production and supply of precision manufacturing components

| Serial Number | Description of Observed Time-Delay in the Supply Chain | Occurrence Frequency |
|---|---|---|
| 1 | Raw materials, parts and standard accessories related | 55 |
| 2 | Supply chain: Internal executions | 100 |
| 3 | Supply chain: External executions | 35 |
| 4 | Financial inward and outward transactions | 25 |
| 5 | Service and after sales service | 45 |
| | Total / Annum (Observed) | 250 |

**Table 2** Occurrence frequency risk categories and associated ratings with risk description

| Risk category | Occurrence frequency range | Associated ratings | Risk description |
|---|---|---|---|
| E | >80 | 4 | Highly probable |
| F | 61–80 | 3 | Probable |
| G | 41–60 | 2 | Occasional |
| H | ≤40 | 1 | Rare |

**Table 3** Cost consequence risk categories and associated ratings with risk description

| Risk category | Cost consequence range (INR) | Associated ratings | Risk description |
|---|---|---|---|
| A | >40,000 | 4 | High |
| B | 20,001–40,000 | 3 | Substantial |
| C | 10,001–20,000 | 2 | Medium |
| D | ≤10,000 | 1 | Low |

The major reasons and root causes are analyzed by cost data, brainstorming with employees, consulting with experts, and historical data analysis. The development of appropriate preventive supply chain risk management pilot strategies and its implementation are narrated below:

- The reduction of unnecessary movements is carried out by indicating and training the workers appropriately. Further monitoring and control measures also carried for risk control measures.

**Table 4** Risk rating matrix developed for the observed time-delay issues (Sl. No. 1–5: indicated with relevant colours, as per Table 1)

| | | Versus | Occurrence Frequency Risk Categories and Associated Ratings | | | |
|---|---|---|---|---|---|---|
| | | | H | G | F | E |
| | | | 1 | 2 | 3 | 4 |
| Cost Consequence Risk Categories and Associated Ratings | A | 4 | | | | 2 |
| | B | 3 | 4 | | | |
| | C | 2 | | 1 | | |
| | D | 1 | 3 | 5 | | |

- Reductions in overproduction of few components which are not having the customer demand are also pin-pointed. The concept of ensuring proper and required resources at right place is trained and educated. Needed quantity estimates are derived based on the requirements of the customer demand with appropriate quality, as a risk control measure.
- Reduction in inventory is carried out by adopting a supplier database and two bin levels as well as streamlined mode of communications. Needed time estimates are derived based on the indicated customer due dates, as a risk control measure.
- Reduction in wastes is carried out by better training and material inspections. Close monitoring of inward material and follow-up is carried out as a risk control measure.
- Reduction in waiting time and idle time are created by better training and maintenance activities. Preventive maintenance and follow-up activities are initiated for further reduction as well as risk control measures.
- Reduction of underproduction is indicated by referring to the customer demand requirements related data. Needed quantity estimates and time estimates are derived based on the requirements of the customer demand with required quality and derived based on the indicated customer due dates respectively, as risk control measures.

- Slight layout modification is also incorporated for minimizing the time, motion, and operator fatigue. Further modification in layout is suggested for further follow-up as risk control measure.
- Reduction in processing waste is carried out by a suitable and correct quantity of material selection. Needed quantity estimates are derived based on the requirements of the materials with relevant specifications, as a risk control measure.
- Reduction in transportation time is carried out by better and improved ways of carrying out jobs. Accurate estimates of internal processing follow-up are risk control measures.

Table 5 indicates the actions taken for risk reduction with Preventive Supply Chain Risk Managements Pilot Strategies for the observed prioritized predominant time-delay issue (*Supply chain: Internal executions*).

The developed preventive supply chain risk managements pilot strategies are implemented during the trial implementation period from October 2019 to December 2019, the predominant time delay observed (*Supply chain: Internal executions*) are reduced to a frequency level of 83 on an average basis. This indicates that there is a reduction of time-delay in the flow of *Supply chain: Internal executions* by 17%.

**Table 5** Actions taken for risk reduction with preventive supply chain risk managements pilot strategies for the observed prioritized predominant time-delay issue (*supply chain: internal executions*)

| Actions taken for risk reduction | Preventive supply chain risk managements pilot strategies |
|---|---|
| Reduction of unnecessary movements | Time, waste of movements, delay and awareness program on self-monitoring and follow-ups |
| Reduction in overproduction | Just-in-time concepts, needed quantity estimates are derived for the requirements of the customer demand with appropriate quality |
| Reduction in inventory | Needed time estimates are derived for the indicated customer due dates |
| Reduction in defects | Inward material inspection and follow-ups |
| Reduction in waiting time and idle time | Preventive maintenance and follow-ups |
| Reduction of under production | Needed quantity estimates and time estimates are derived for the requirements of the customer demand with required quality and indicated customer due dates respectively |
| Slight layout modification | Convenience to minimize the processing time and movements with follow-ups |
| Reduction in processing waste | Accurate estimates of material requirements, manufacturing resources, tools and follow-ups |
| Reduction in transportation time | Accurate estimates of internal processing and follow-ups |

### 7.2 Advantages and Benefits of the Preventive Supply Chain Risk Management Strategies

The developed preventive supply chain risk management pilot strategies will provide many benefits to the case study organization under consideration. Few of the advantages and benefits of these developed pilot strategies are listed below:

- Anticipative response to face the pierce competitive market.
- Reduction in the uncertainty level of the supply chain flows by reducing the occurrence of the likelihood of risk events in the supply chain activities.
- Integrity on the entire supply chain entities and activities with smooth consistent flows.
- Improved supply chain inter-organizational relationships for better understanding and problem-solving interactions including after-sales service
- A systematic approach in considering and analyzing the strength, weaknesses, opportunities, and threats in the supply chain flows of the organization.
- Identification and development of appropriate strategies by adopting the concept of risk prevention measures.

### 7.3 Future Scope of Work

The entire supply chain flows and its visibility from end to end are needed to be focused to achieve significant outcomes in the business organizations. Thus, it is essential to trace and track the uncertainties in the flows of supply chain from time-to-time to ensure that the organizational objectives with outcomes are adequately achieved. The supply chain risk management facilitates the integration of relationship and inter-relationship of various activities including after-sales service by properly selecting and using the available resources and facilities towards continuous improvement. Future works may include more realistic data collection with software integrity as well as qualitative cum quantitative analysis with streamlined model for predictive strategies. This will enhance the utility of supply chain risk management practices towards continuous monitoring and improvement.

## 8 Conclusions

In this paper, the investigation on the development and trial implementation of preventive supply chain risk management pilot strategies for a batch production case study organization was carried out. These strategies were helpful for timely production and supply of precision manufacturing components by using computer numerical control machine tools of the case study organization. The pilot preventive supply chain risk management strategies for the case organization were investigated to reducing the identified predominant time-delay of *Supply chain: Internal executions* for achieving competitive advantage. The predominant time-delay related causes and suitable preventive supply chain risk management pilot strategies for loss reduction were developed and implemented for a trial period of investigation. There is a reduction in time–delay in the *Supply chain: Internal executions* by 17%. These are used to achieve the indicated customer due to date requirements and improved inter-organizational relationship, reduction in time-delay, cost reduction, and effective after sales service. Future works were also indicated with a requirement of a modified streamlined model with software integrity as well as qualitative cum quantitative analysis. This paper finally highlights the importance of loss prevention for the success of an industry in the fierce competition for continuous improvement. This will significantly enhance the utility of available effective use of resources with supply chain risk management practices in the manufacturing organizations with appropriate outcomes towards continuous monitoring and improvement of business success.

## References

1. Kuei C, Madu CN, Lin C (2001) The relationship between supply chain quality management practices and organizational performance. Int J Qual Reliab Manag 18(8):864–872. https://doi.org/10.1108/EUM0000000006031
2. Rungtusanatham M, Salvador F, Forza C, Choi TY (2003) Supply-chain linkages and operational performance. Int J Oper Prod Manag 23(9):1084–1099. https://doi.org/10.1108/01443570310491783
3. Jüttner U (2005) Supply chain risk management. Int J Logist Manag 16(1):120–141. https://doi.org/10.1108/09574090510617385
4. Nie X, Zhang H M (2005) Delay-function-based link models: their properties and computational issues. Transp Res Part B: Methodol 39(8):729–751. https://doi.org/10.1016/j.trb.2004.10.002
5. Wook Kim S (2006) Effects of supply chain management practices, integration and competition capability on performance. Supply Chain Manag: An Int J 11(3):241–248. https://doi.org/10.1108/13598540610662149
6. Manuj I, Mentzer J T (2008) Global supply chain risk management strategies. Int J Phys Distrib Logist Manag 38(3):192–223. https://doi.org/10.1108/09600030810866986
7. Elangovan D, Sundararaj G, Devadasan SR, Karuppuswamy P (2011) Development of futuristic supply chain risk management pilot strategies for achieving loss reduction in manufacturing organizations. World J Entrep, Manag Sustain Dev 6(1/2):39–51. https://doi.org/10.1108/20425961201000004
8. Wong C Y, Boon-itt S, Wong CWY (2011) The contingency effects of environmental uncertainty on the relationship between supply chain integration and operational performance. J Oper Manag 29(6):604–615. https://doi.org/10.1016/j.jom.2011.01.003
9. Sodhi MS, Son B-G, Tang CS (2011) Researchers' perspectives on supply chain risk management. Prod Oper Manag 21(1):1–13. https://doi.org/10.1111/j.1937-5956.2011.01251.x
10. Tang O, Nurmaya Musa S (2011) Identifying risk issues and research advancements in supply chain risk management. Int J Prod Econ 133(1):25–34. https://doi.org/10.1016/j.ijpe.2010.06.013
11. Maruf H (2013) Sustainable supply chain management practices and operational performance. Am J Ind Bus Manag 3(1):42–48. https://doi.org/10.4236/ajibm.2013.31006

12. Chae B (Kevin), Olson D, Sheu C (2013) The impact of supply chain analytics on operational performance: a resource-based view. Int J Prod Res 52(16):4695–4710. https://doi.org/10.1080/00207543.2013.861616
13. Ellram LM, Cooper MC (2014) Supply chain management: it's all about the journey, not the destination. J Supply Chain Manag 50(1):8–20. https://doi.org/10.1111/jscm.12043
14. Yu W, Chavez R, Feng M, Wiengarten F (2014) Integrated green supply chain management and operational performance. Supply Chain Manag: An Int J 19(5/6):683–696. https://doi.org/10.1108/SCM-07-2013-0225
15. Ho W, Zheng T, Yildiz H, Talluri S (2015) Supply chain risk management: a literature review. Int J Prod Res 53(16):5031–5069. https://doi.org/10.1080/00207543.2015.1030467
16. Prakash S, Soni G, Rathore APS (2017) A critical analysis of supply chain risk management content: a structured literature review. J Adv Manag Res 14(1):69–90. https://doi.org/10.1108/jamr-10-2015-0073

# Effective Supply Chain Management by Using the Data-Based Decision Support System

**D. Ramesh Kumar, D. Elangovan, S. R. Devadasan, and B. Gokulakrishnan**

**Abstract** Managing supply chain becomes very much difficult due to the interaction between various stakeholders. Hence, it is essential to make strategic decisions in time. From time to time, the status of in-progress performance of effective supply chain management is varying due to its nature of interrelated and interactive upstream and downstream flows. Thus, it is necessary to make a database from the point of suppliers to customers to integrate the process flow and also reduce the operational gap. The manufacturing processes involved, fluctuation in demand, costs related to transportation, costs related to inventory and costs related to decisions, are the significant operational aspects, but not limited to these alone. For carrying out this integration, it is necessary to update, consolidate and categorize the database for deriving feasible decisions. Data-based decision support plays a critical role in encompassing the routine production and triggers to meet the customer requirements. The decisions in time will be quite useful by considering the database and orienting towards the effective, smooth and successful supply chain management.

**Keywords** Suppliers · Customers · Supply chain management · Production · Decision support system

D. Ramesh Kumar (✉) · B. Gokulakrishnan
Department of Mechanical Engineering, Adithya Institute of Technology, Coimbatore 641107, Tamil Nadu, India
e-mail: dramesh184@gmail.com

B. Gokulakrishnan
e-mail: scmgokul@gmail.com

D. Elangovan
Department of Mechanical Engineering, PSG Institute of Technology and Applied Research, Coimbatore 641062, Tamil Nadu, India
e-mail: delangovan_cbe@rediffmail.com

S. R. Devadasan
Department Production Engineering, PSG College of Technology, Coimbatore 641004, Tamil Nadu, India
e-mail: devadasan_srd@yahoo.com

© The Editor(s) (if applicable) and The Author(s), under exclusive license to Springer Nature Singapore Pte Ltd. 2021

S. Mohan et al. (eds.), *Materials, Design, and Manufacturing for Sustainable Environment*, Lecture Notes in Mechanical Engineering,
https://doi.org/10.1007/978-981-15-9809-8_40

## 1 Introduction

The change of market scenario has emerged from producer centric to consumer centric with lot of focuses on producing the product as well rendering the service. This is particularly having impact on mass production and batch production of products that are similar in manufacturing aspects. Larger the quantity of products produced, total costs associated will be spread along these products, thereby reducing the cost of product and increasing profits. While the market was flooded with products, little attention may the given to quality of the products. This gave birth to the concept of total quality management five decades ago to solve all the product quality-related issues. Once the concept of total quality management was established, the quality of the products was made a must include characteristic of products. Since there were more quantity of high-quality goods with little or no variety, customer started to seek variety. When the high-quality goods were delivered to the customers, the need for more variety of products skyrocketed after a decade. This is when all the enterprises started transitioning from product centric to customer, and for this to be made possible the inventories had to be managed optimally and cycle times of production had to be minimized. Managing high product variety became more challenging and complex with diverse requirements demanded by the consumers. The ability of being flexible to market changes in terms of time to deliver and modify design was given significant importance. Hence, the processes involved inside the enterprise and supplier management had to be developed flawlessly and was later incorporated to be a deep-rooted belief in the concepts such as just-in-time and lean manufacturing. Globalization made a vast impact on all the companies and so did the new developments in sophisticated technology and adopting processes to overall world economy. For the next decade, most of the enterprises paid significant attention on developing their core competencies through various approaches, innovating for better results, reducing costs by outsourcing processes and proactive association. This is when the concept of supply chain management was established.

The decisions related to global supply chain network design are obscure due to various cases. One such problem includes hundreds and thousands of data and parameters related to decisions which defines location, size, number, etc., of facilities, products and possible modes of transportation and so on [1]. Supply chain management is a finely constructed network of organizations designed to satisfy the consumer. The core concept of supply chain management (SCM) is to manage the sourcing, transporting and warehousing of raw materials, spare parts and finished product throughout the firm and its market links so as to improve profit margin and customer satisfaction [2]. The various stakeholders in a business are linked to fulfil customer's desires through a constructed and organized web called supply chain. Supply chain management deals with using several modern tools in order to add value to the product. The objective of SCM is to maintain optimal levels of supply and the demand of a product to fulfil the customers' need. The flow in the supply chain must be optimized to achieve effectiveness and flexibility. The types of flow are information, material and data which flows on either side of the value chain (Fig. 1).

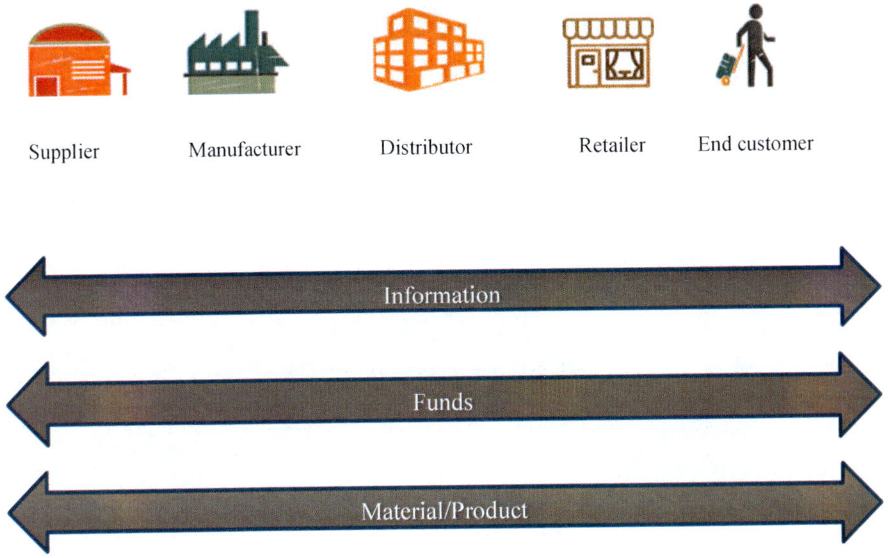


**Fig. 1** Supply chain

There are different links in business like supply, production plant, distribution centre, retail and consumer. The person who finally uses the product is known as the end customer. The supply chain has to be designed based on a lot of parameters.

There are number of interactive elements, which makes supply chain becomes complex. The nature of the product is categorized as perishable and non-perishable goods. Perishable goods are those products or goods whose value deteriorates as time passes. There are several companies that have developed best practices in the industries to improve their business and have served as an example over the course of history. One such problem is to maintain inventories containing outdated goods. If the goods or products' values change with trend like a fashion item, it is also considered as perishable goods. Non-perishable goods are those goods whose value stays same even as time passes.

The type of supply chain varies considering the source of raw material. The decisions such as mode of transportation and cost associated with procurement vary based on source location of supply. In order to produce a product, the basic element required is the supply which enables business. Sourcing of raw material is a fundamental process which enables business [3]. B2B or B2C business type also affects the supply chain decisions. Companies identified that SCM is the method to reach their organizational goals and objectives. The supply chain management deals with not only the logistics but inventory planning, demand forecasting, operational planning. These decisions are mostly made through pure experience and intuition, but the company cannot always depend on their intuition. But due to unstable market conditions, a decision support system is essential to provide faster and reliable solutions. In current market with unpredictable trends and uncertainties, the best way to improve and evolve is to boost the responsiveness of the system to these changes.

**Fig. 2** Supply chain management flow

The decision time has to be as low as possible while ensuring the quality of the decision. Time required for decision-making is influenced by the data and information collected. These data include information about production plant, capacity and data related to forecasting [4]. Time cannot be compromised in business as it comes with a price. Increased time causes delay, and delay is not encouraged in business by the customers. In current market with continuous changes in trend, decisions have to be made quickly. Communication is a very important factor to achieve faster decision-making. Modern technologies have facilitated the possibility of making decisions faster without compromising the quality of the decision. Hence, new software tools and DSS are constructed to facilitate managers with making critical decisions as shown in Fig. 2.

## 2 Literature Review

Reich et al. [1] used mixed-integer linear programming to construct a framework to provide solution for problems related to strategic global supply chain network design. The objective is to provide a feasible solution for decision-makers.

Vishnu et al. [2] have summarized the present methods available in creating and administrating risks in supply chain to facilitate decision-makers to provide better solutions to tackle uncertainties.

Ross [3] has provided a review paper on the supply chain management including history, problems faced in business and means to effectively transport products through supply chain.

Zheng et al. [4] have presented a paper on strategic production networks. This paper focuses on the practices and tools used to facilitate decision-makers.

Karaesmen et al. [5] presented a paper that addresses the perishable products whose value deteriorates over time.

Issam et al. [6] have presented a paper based on transportation management system to provide better routing and mapping that formulates the supply chain.

## 3 Methodology

The methodology adopted in the research work is shown in Fig. 3.

The first step is to acquire the required data to perform the analysis and obtain faster and cost-effective solution. These data could be possible routes and availability of transportation so as to arrive at optimum logistics service. In order to evaluate,

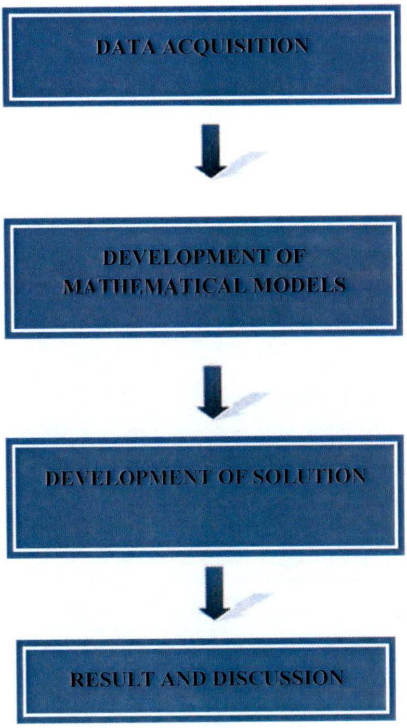


**Fig. 3** Methodology adopted

there is a need of a tool, and mathematical models are one for those tools. These mathematical models are usually made up of programming methods, models and algorithms. Model is developed, and the data are then fed into the model for critical evaluation. This model produces a feasible and faster solution by analysing all the possible solutions. Finally, the solution is implemented and discussed.

# 4 Decision Support System

The DSS gathers the information, sorts it and analyses it to provide feasible solutions for management, operations and planning in making better decisions. The data required by the DSS are of raw data, documents and personal knowledge. These data are collected from employees, management, executives and business models. The information being used in DSS has to be accurate, accessible and available for it to be effective [7]. A few desirable features of the decision support system are simplicity, flexible and strong in construction [8]. The salient features of a decision support system are shown in Fig. 4.

## 4.1 *Elements of DSS*

The elements of DSS include database for input, tools for analysing the problem and means to present the solution [9]. The elements of DSS are shown in Fig. 5.

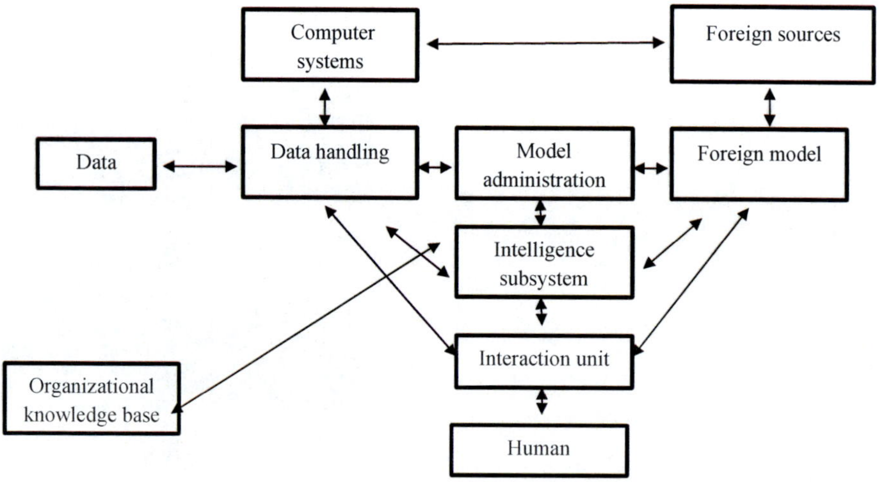


**Fig. 4** Salient features of decision support system

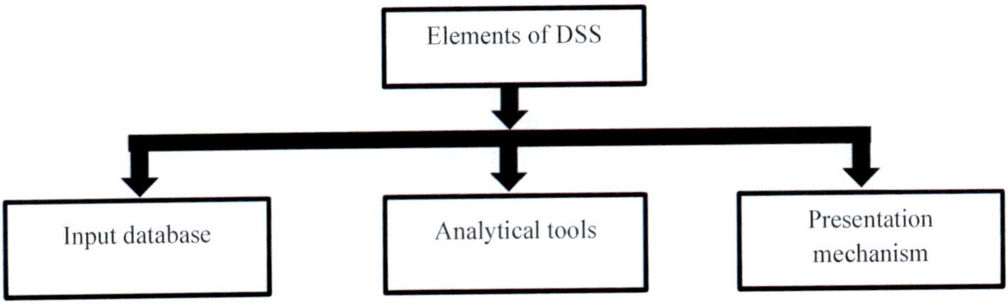


**Fig. 5** Elements of DSS

The input database contains the different data from internal and external source stored in it. The analytical tools are used to analyse the data given using mathematical models, algorithms and sophisticated mathematical approaches. The analytical tools provide the solutions to the unstructured and semi-structured problems. Results obtained from the analysis are in the form of tables, lists and large information which would make it difficult for the decision-maker to understand. To tackle this problem, presentation tools such as geographic information system, graphs and Gantt charts are used. The various decision support systems used as mentioned in [9] are

- Design for logistic network
- Plan for operations
- Planning for transportation
- Planning for requirement of material
- Operation execution system
- Inventory management.

## 5 Logistic Network Design

The investigated logistic network design of a case study industry of making chain sprocket is shown in Fig. 6.

Logistic network design is a DSS that facilitates in making decisions about transportation involved in a production of a product. The required information such as size, number and location of warehouses, customers, suppliers and distribution centres is fed to the DSS. The various constraints like the desired service level, cost, facilities required are mentioned. The logistics network provides a number of optimum solutions. The decision-maker should pick the suitable solution for the logistic network and implement to increase the efficiency of the current network design. In this case, we provided an optimum logistics network for a chain sprocket manufacturing company with low-cost logistic network across its various stakeholders. The logistic network design makes it easier for the manufacturer to reach customers and reduces the overall lead time.

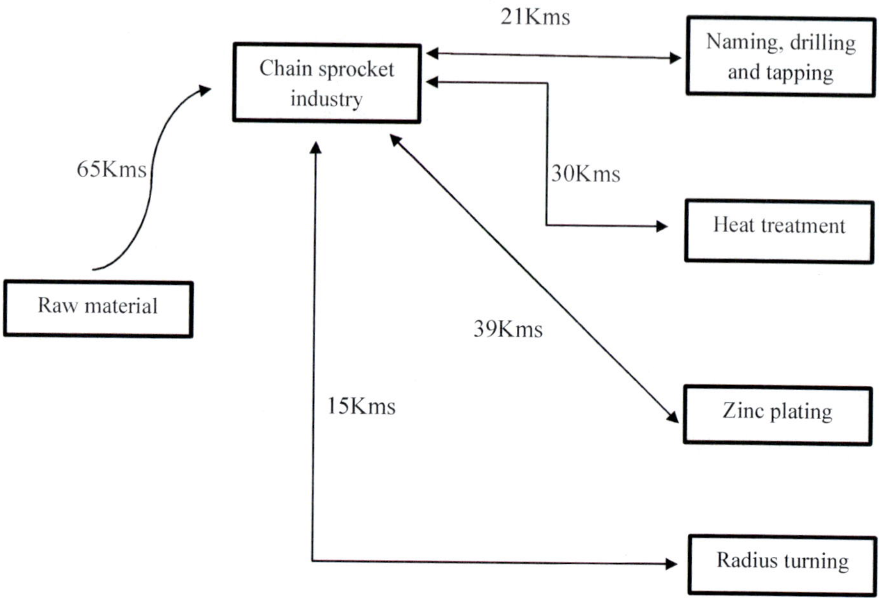


**Fig. 6** Schematic diagram of the logistic network design

## 6 Discussion

The goal of this paper is to provide a synopsis of supply chain management, its processes and usage of decision support system for faster and reliable decision-making. From the literature review, it is clear that usage of decision support system is an essential factor to survive the current unstable market conditions and sustain on the long run. Due to high level of changing market trends and uncertainties, a data-driven approach is required to obtain a beneficial and practical solution in SCM [5]. The data acquired from various tools like data mining are fed into mathematical models to develop a faster and feasible solution to empower businesses. In a complex supply network, the person responsible to make decision is handling large amount of data appropriately for achieving business advantage [10]. Incorporating logistic network design in a chain sprocket manufacturing organization, the production rate increases and reduces cycle time and overall lead time which is very beneficial in sustaining the global market.

## 7 Conclusions

Supply chain is complex network which could be interlinked for multiple businesses and industry. Optimizing the effectiveness and response of a supply chain is a challenging task without a decision support system. The decision of one business will directly or indirectly affect the other business linked to it. Effective management of supply web or chain is the process which improves various traits of products and service provided by a business. Even though there are a lot of review papers in supply chain management, this paper provides an easily understandable overview of effective supply chain management through data-based decision support system. All the companies around the globe are linked as a result of globalization, and the risks have become high and complex [2]. In future, supply chain can be made much more effective and responsive through data science. Studying the behaviour and trends of data to provide solutions to markets is already being developed. The chain sprocket organization has to produce 300,000 products in 25 days, and it is highly unlikely to make it possible without logistic network design. With the help of logistic network design, it is possible to produce 300,000 products in 22 days. Since the products are delivered in time, it results in increased customer satisfaction. Hence, using DSS like logistic network design is very essential in current competitive industry in order to provide quality goods with desired service level.

## References


1. Reich J, Kinra A, Kotzab H (2019) Strategic decision-making in global supply chain network design—how a decision support system combining MILP and AHP on a Pareto front can alleviate decision-making. IFAC-PapersOnLine 52(13):2285–2290. https://doi.org/10.1016/j.ifacol.2019.11.546
2. Vishnu CR, Sridharan R (2019) Supply chain risk management: models and methods, 32–65. https://doi.org/10.1504/IJMDM.2019.096689
3. Ross DF (2015) Introduction to supply chain management. Distribution Planning and Control, 3–43. https://doi.org/10.1007/978-1-4899-7578-2_1
4. Zheng L, Frank P-D (2002) Strategic production networks. Springer, Berlin. https://doi.org/10.1007/978-3-540-24812-5
5. Karaesmen IZ, Scheller-Wolf A, Deniz B (2010) Managing perishable and aging inventories: review and future research directions. In: Planning production and inventories in the extended enterprise. International Series in Operations Research & Management Science, pp 393–436. https://doi.org/10.1007/978-1-4419-6485-4_15
6. Issam et al (2018) Transportation management and decision support systems within the supply chain management framework, pp 338–342. https://doi.org/10.1007/978-3-030-02131-3_30
7. Twafik M et al (2017). Decision support system in supply chain management: literature review. Eur J Logist Purchas Supply Chain Manag 5(5):40–51. ISSN 2054-0949
8. Little DC (1970) Models and managers: the concept of a Decision Calculus. Manage Sci 16(8):466–485
9. Bozarth CC, Handfield RB (2008) Introduction to operations and supply chain management. Pearson Education Inc., New Jersey
10. Manuj I, Sahin F (2011) A model of supply chain and supply chain decision-making complexity. Int J Phys Dist Log Manage 41(5). https://doi.org/10.1108/09600031111113884

# Design and Analysis of Flow Field Patterns in Bipolar Plates of a Proton Exchange Membrane Fuel Cell

T. Prem Kumar, P. Sai Subramanian, and G. Naresh

**Abstract** In this study, an attempt has been made to study the effect of different flow field patterns in bipolar plates on the pressure drop and flow distribution of reactants in the proton exchange membrane fuel cell (PEMFC). Various patterns were designed in SolidWorks and analysed using computational fluid dynamics (CFD) in Ansys Fluent software to analyse the patterns created. The pressure drop and the flow distribution were studied and analysed. Results have proven that both serpentine and the parallel pattern are more suitable than any other pattern for improving the efficacy of PEMFC. It is imperative to note that serpentine model provides a uniform flow distribution at the cost of relatively increase in pressure drop, and on the other hand, in the parallel model the pressure drop is reduced at the cost of non-uniform flow distribution of reactants.

**Keywords** PEM fuel cells · Flow distribution · Pressure drop · Serpentine fillet

## 1 Introduction

In the current era, the need for using renewable energy is increasing due to the depletion of non-renewable resources and its associated environmental pollution. Non-renewable resources like coal and fossil fuels are used extensively to meet the energy demands. Though they serve as the best resource to meet energy demand, they will be depleted overtime and cause significant level of pollution too. The issues like global warming and ozone depletion are the main reasons for the development

T. Prem Kumar (✉) · P. Sai Subramanian · G. Naresh
Department of Mechanical Engineering, PSG Institute of Technology and Applied Research, Neelambur, Coimbatore, India
e-mail: prem50567@gmail.com

P. Sai Subramanian
e-mail: saisubbu2000@gmail.com

G. Naresh
e-mail: nareshgunasegaran@gmail.com

© The Editor(s) (if applicable) and The Author(s), under exclusive license to Springer Nature Singapore Pte Ltd. 2021

S. Mohan et al. (eds.), *Materials, Design, and Manufacturing for Sustainable Environment*, Lecture Notes in Mechanical Engineering,
https://doi.org/10.1007/978-981-15-9809-8_41

of new alternate source of energy that is renewable and pollution-free. One such renewable energy is the usage of hydrogen. One of the methods to retrieve energy from that is the concept of fuel cell which is pollution-free. Hydrogen and oxygen can be used as inputs to generate electricity in PEMFC which is one type of the fuel cells. In the presence of catalyst like carbon-supported platinum at anode side, the hydrogen is ionized and forms $H^+$ ions and electrons. The electrons produced cannot be passed through semi-permeable electrolyte which is electron insulator, and hence, it is induced to move through external circuit to generate current and reach the cathode side. $H^+$ ions produced at anode side can penetrate through the semi-permeable membrane which is proton conductive to the cathode side. In the presence of catalyst like carbon-supported platinum at the cathode side, hydrogen ions, electrons and oxygen combined and form water which is exothermic reaction. Among the many fuel cells that have been developed, the proton exchange membrane fuel cells (PEMFC) have served as the best because of its low operating temperature point and its ability to be coupled as modules to form stack.

In the past decade, researches have focussed their attention on aspects like geometry of the flow field channel [1–4]. Among the critical factors, PEM fuel cell's bipolar plate pattern geometry has gained more importance as they influence the homogenous distribution of gases all over the catalyst surface and thereby the overall stack performance [1]. Two important factors that the different patterns on the plate affect are pressure drop and flow distribution of reactants. These eventually affect the performance of the fuel cell. To obtain a good efficiency from the fuel cell, an excellent flow distribution at optimal pressure drop is required.

### 1.1 Effect of Flow Distribution on Cell Performance

One of the main contributors to the cell performance is the flow distribution of reactants in the flow field plates. The flow channels are responsible for supplying reactants to the membrane as well as carry the by-product water that is obtained after the reaction [5].

A good flow distribution channel provides a good flow distribution of reactants in the flow field plates; however, a non-uniform distribution of flow channel results in degradation of fuel cell. Non-uniform distribution leads to waterlogging at some places in the flow channel. Areas in which water gets logged or areas where the reactants are under supplied are often referred to as dead spots. The remaining area becomes dry spots where water is not present. These hotspots cause mechanical degradation of membrane. Non-uniform distribution also leads less supply of the reactants, and hence, the output, i.e. current density, will be very less.

### 1.2 *Effect of Pressure Drop on Cell Performance*

The geometry of flow field not only decides the flow distribution of reactants, but it also determines the total pressure losses that occur across the flow field and the local pressure distribution over the active area. The power consumption of the auxiliary system required to operate a fuel cell directly corresponds to the total pressure drop across the flow field plate. This auxiliary system power requirement can be considered as parasitic loss to system performance, and hence, it is desirable to reduce auxiliary power as much as possible [5].

When comparing the performance of various flow fields to one another, all performance variation is commonly attributed to reactant distribution only, and the pressure effect is often neglected. In this study, the pressure drop is also taken into consideration. Although it seems to be relatively less for single flow fuel cell, it will consider big impact during scaling up of the fuel cell, i.e. fuel stack. The pressure drop can be obtained using the pressure contour from Ansys software, and the results will be validated using experimentation.

## 2 Methodology

The flow field channel geometry was designed in SolidWorks. In the flow field channel, a square cross-section of 9 $mm^2$ was incorporated. Ansys Fluent was used to carry out computational fluid dynamics (CFD) analysis for the designed models after importing the geometry from the SolidWorks.

### 2.1 *Design Works*

The different flow field design patterns studied are ring type, parallel type, pin type and serpentine (fillet corners and sharp corners as shown in Figs. 1, 2, 3, 4 and 5).

**Fig. 1** Ring type

**Fig. 2** Parallel type

**Fig. 3** Pin type

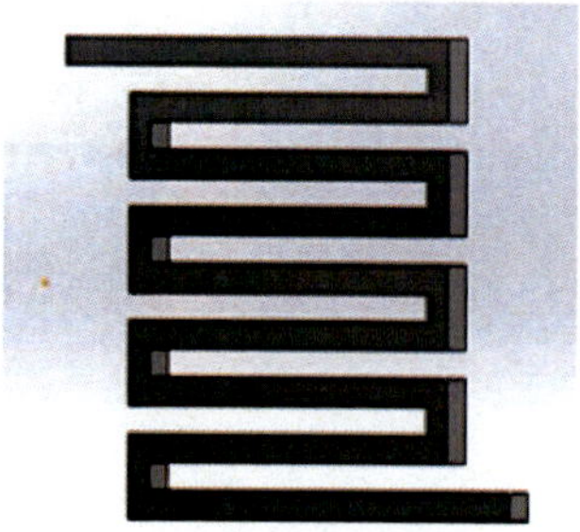

**Fig. 4** Serpentine (fillet corners)

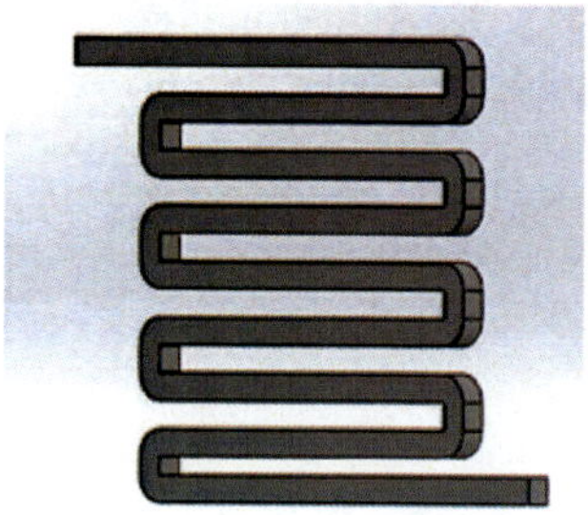

**Fig. 5** Serpentine (sharp corners)

# 3 Results and Discussion

## 3.1 *Serpentine Model (Sharp Corners)*

From Figs. 6 and 7, it is inferred that the flow distribution in serpentine model is excellent, but the pressure drop is around 2800 Pa. This increased pressure drop is likely to affect the performance of the cell. But the good flow distribution in the channel is very helpful for providing a good active site for the gas and water to flow, and hence, output current density will be high. Better flow distribution helps in the flow of water generated and prevents flooding in the channel too.

### 3.1.1 Modification in Serpentine Model

Serpentine Model (Fillet)

Several modifications were done in the serpentine model such as changing shapes of the cross-sectional area of channel to improve the flow distribution of water and gases in the channel by many researchers. However, in the present study, a novel approach of converting sharp corners into fillets has been undertaken. The modified

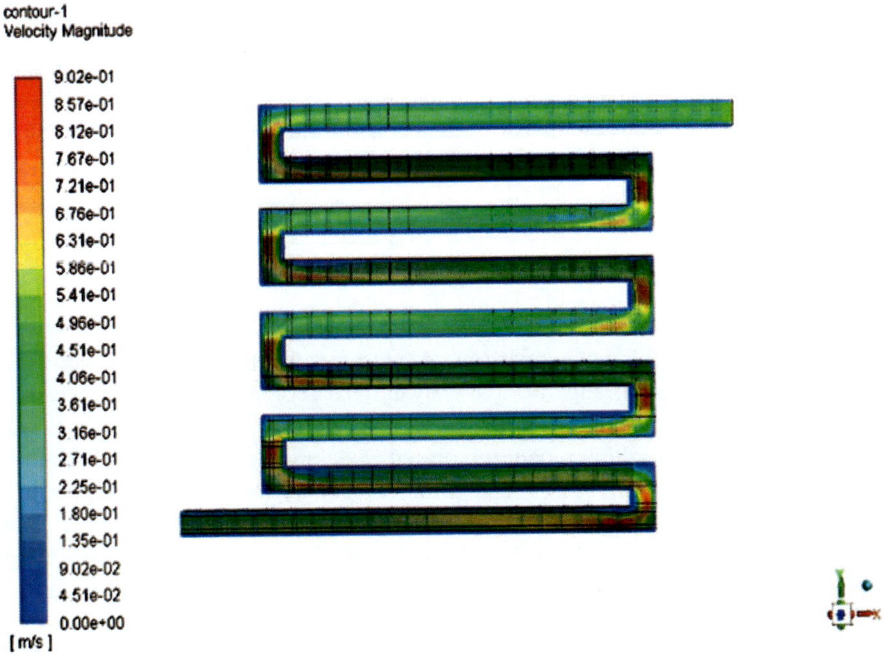


**Fig. 6** Velocity contour (flow distribution)

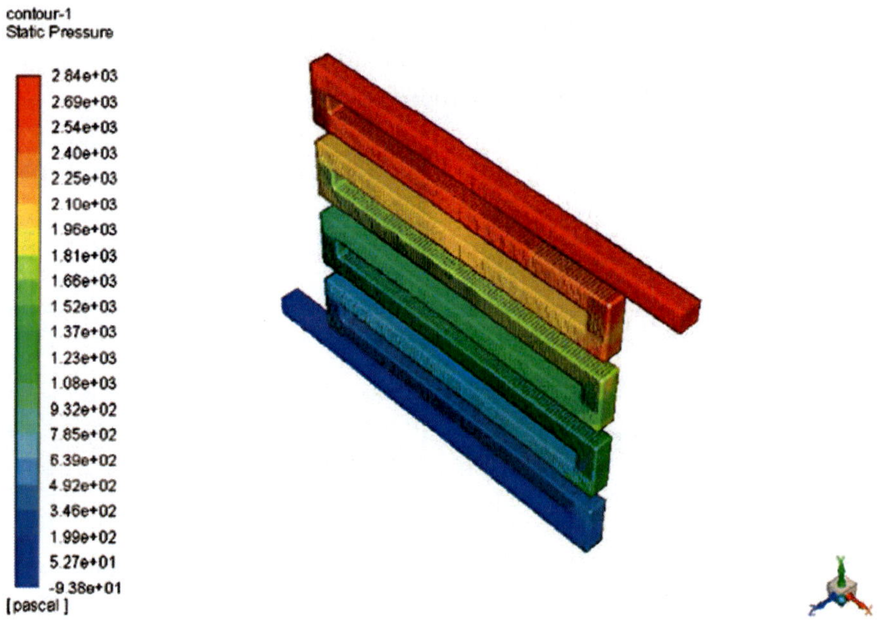


**Fig. 7** Pressure contour (pressure drop)

serpentine model with fillets was designed and analysed. The results are presented in Figs. 8 and 9.

The pressure drop in the modified serpentine was found to be 2220 Pa. The flow distribution of the water is very uniform in nature like the conventional serpentine model. The pressure drop obtained in modified serpentine is less than the conventional serpentine model by 580 Pa. This reduction in pressure drop was obtained without disturbing the flow distribution.

### Ring Type

In this modification, the rectangular row of the serpentine model was converted to series of circular rings arranged in the conventional serpentine model. The results are shown in Figs. 10 and 11.

The pressure drop in this model was found to be around 2900 Pa, and the flow distribution in the channel was found to be slightly uniform in nature. The flow distribution is better than the conventional parallel model, but the pressure drop obtained is significantly high, and the active area for the reactants to flow is less compared to the conventional serpentine model, and hence, in overall this pattern has not given a desirable output.

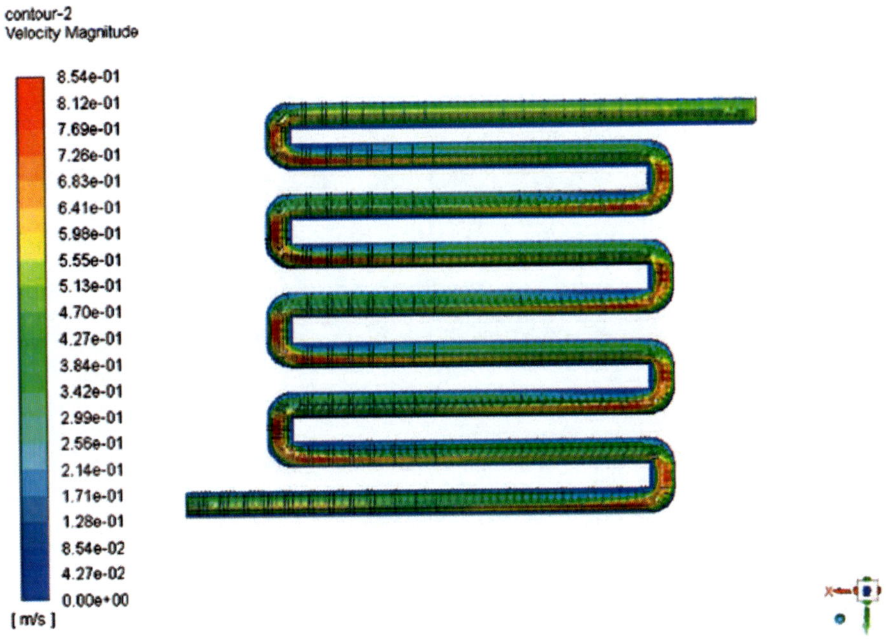


**Fig. 8** Velocity contour (flow distribution)

## *3.2 Parallel Model*

The parallel model proposed by Pollegri and Spaziante was designed and analysed [6]. From the parallel model simulation results shown in Figs. 12 and 13, it can be inferred that the pressure drop is very minimal compared to the serpentine model, and the pressure drop is around 346 Pa. This minimal pressure drop is very helpful in reduction of auxiliary power consumption, and hence, the performance of the cell will be good, but this reduced pressure drop compared to serpentine model comes at the cost of non-uniform flow distribution of the reactants. This also leads to non-uniform distribution of the water in the flow channel results in waterlogging in the channels which ultimately reduces the performance of fuel cell.

### 3.2.1 Modifications in Parallel Model

Tapered Parallel Flow

Li and Sabi [3] have proposed in modifying the parallel model with decreasing width of the channel from inlet to outlet. However, the drop in pressure and flow properties has not been extensively reported by them. In the present study, this modification in the channel has been studied and analysed in detail in Figs. 14 and 15.

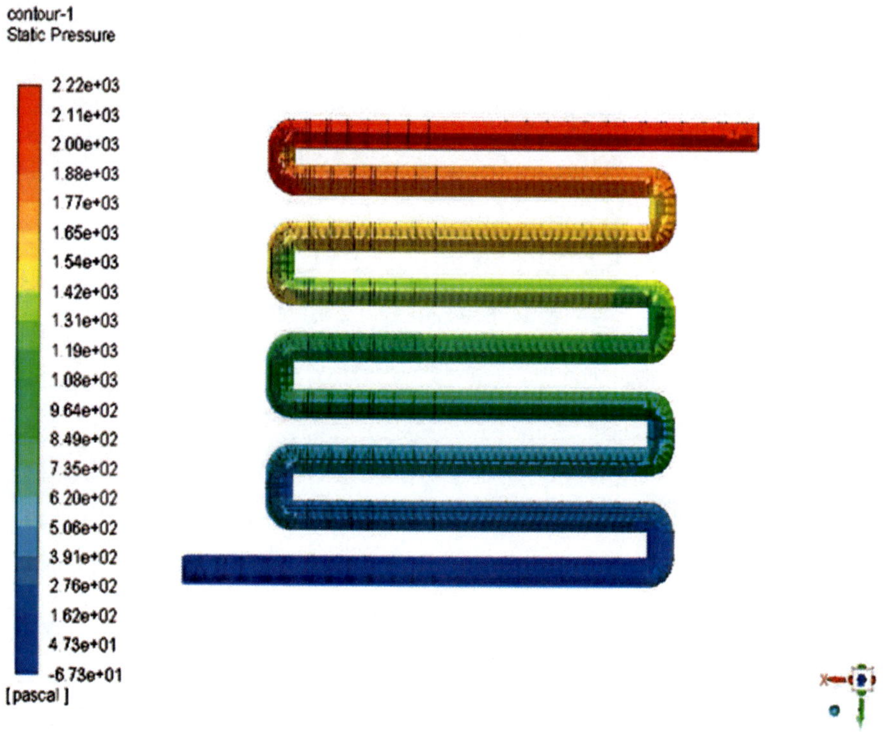


**Fig. 9** Pressure contour (pressure drop)

From Fig. 15, it is clear that the pressure drop is 359 Pa which is almost equal to the normal parallel model. From the velocity contour presented in Fig. 14, it can be inferred that the flow distribution is uneven like the normal parallel model, and from the above pressure contour presented in Fig. 15, it is inferred that this modification in width of the channel (decreasing from inlet to outlet) did not yield a desirable impact on reducing the pressure drop. From this, it could be suggested that this model would be ineffective.

### Pin Flow

The pin type channel proposed by Reiser and Sawyer [7] was designed and analysed. The results obtained are presented in Figs. 16 and 17.

From the above velocity and pressure contours presented in Figs. 16 and 17, respectively, it can be inferred that the pressure drop in the pin type flow is minimal and measures around 350 Pa. The flow distribution is uneven like the parallel type. This indicates that the results are exactly similar to the results obtained for the parallel model, and this modification has not yielded the desirable output.

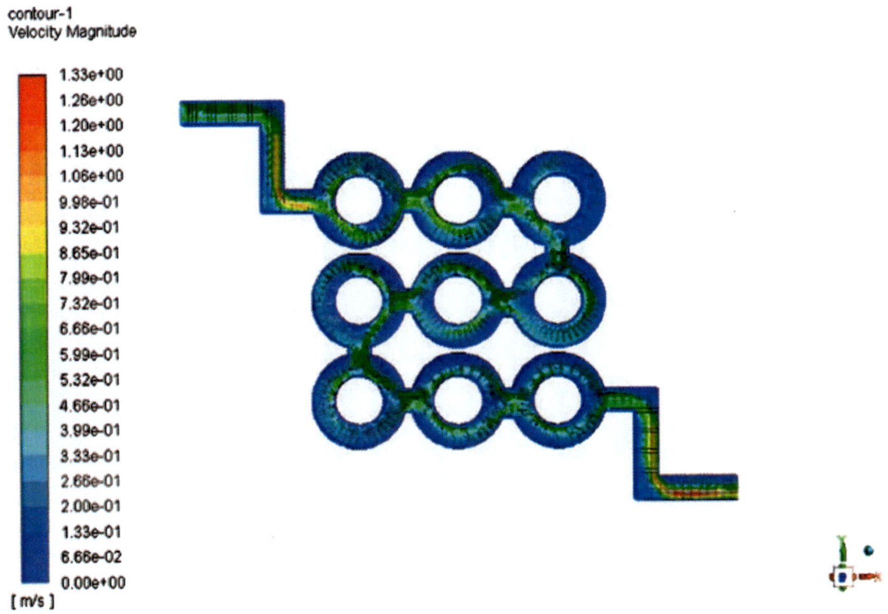


**Fig. 10** Velocity contour (flow distribution)

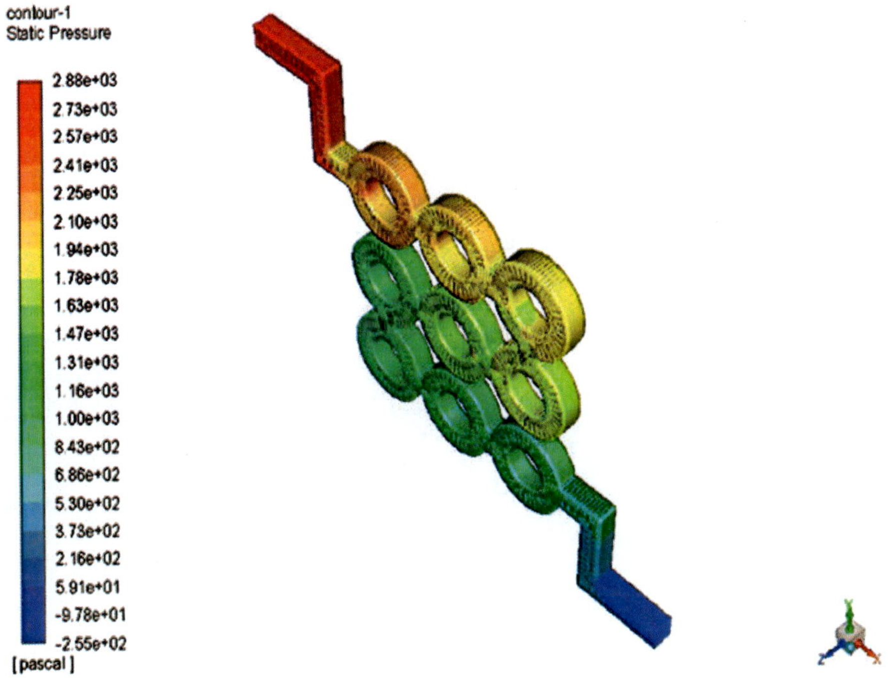


**Fig. 11** Pressure contour (pressure drop)

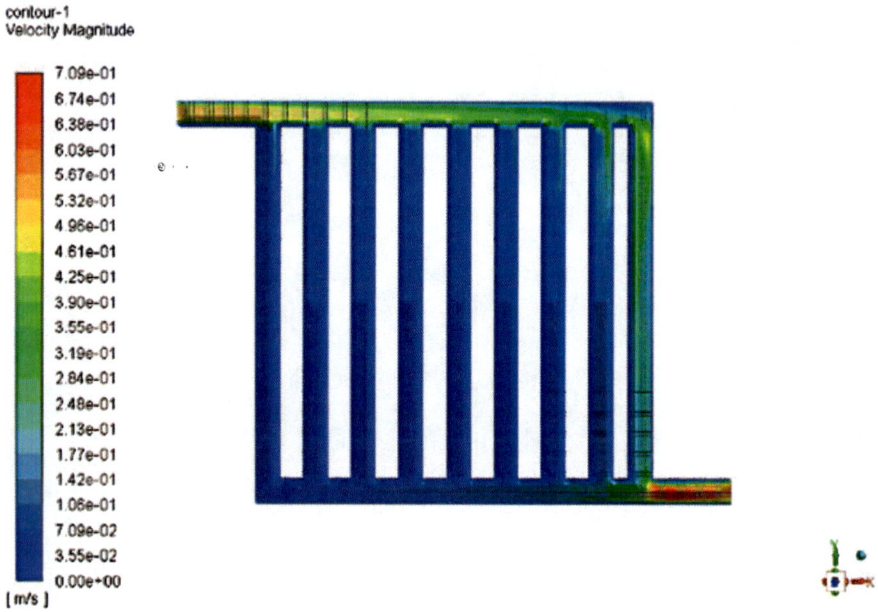


**Fig. 12** Velocity contour (flow distribution)

**Fig. 13** Pressure contour (pressure drop)

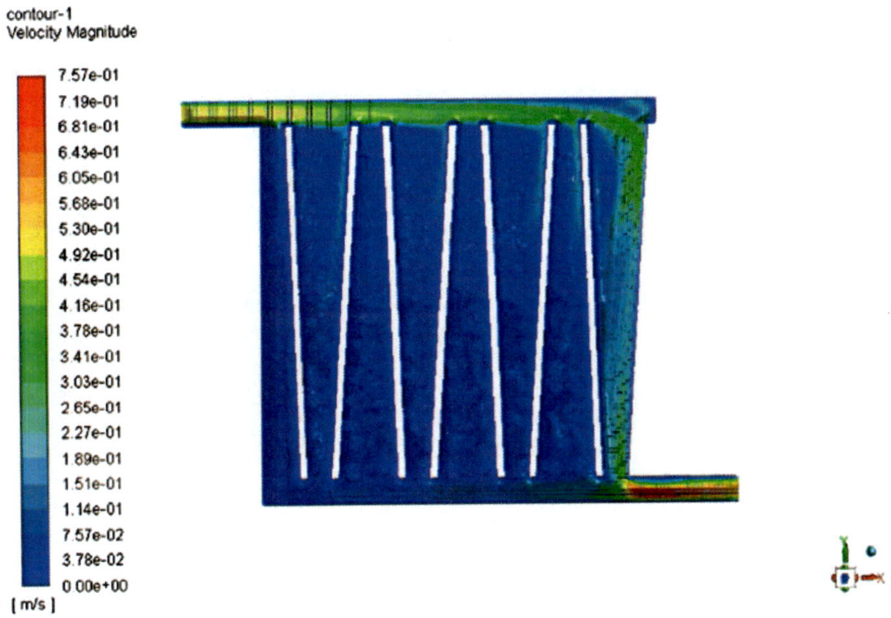


**Fig. 14** Velocity contour (flow distribution)

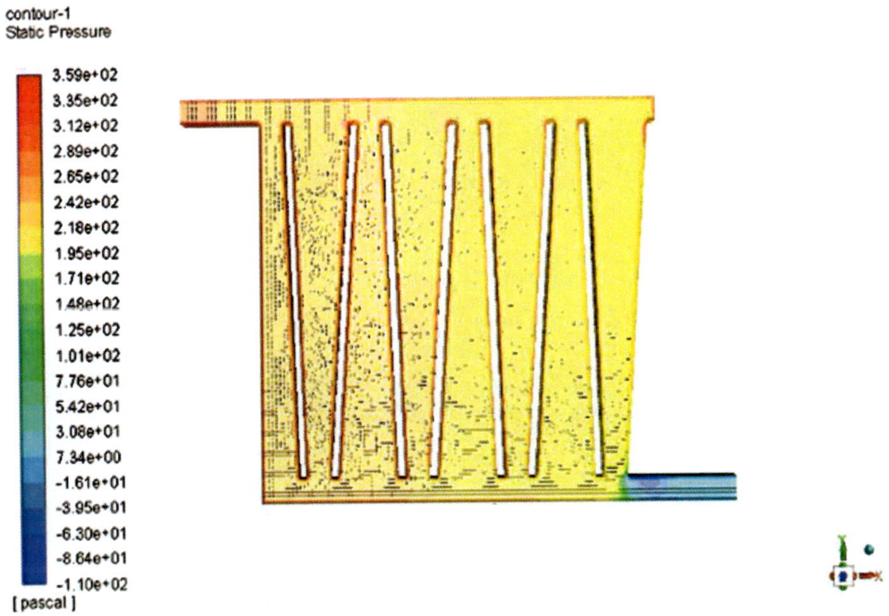


**Fig. 15** Pressure contour (pressure drop)

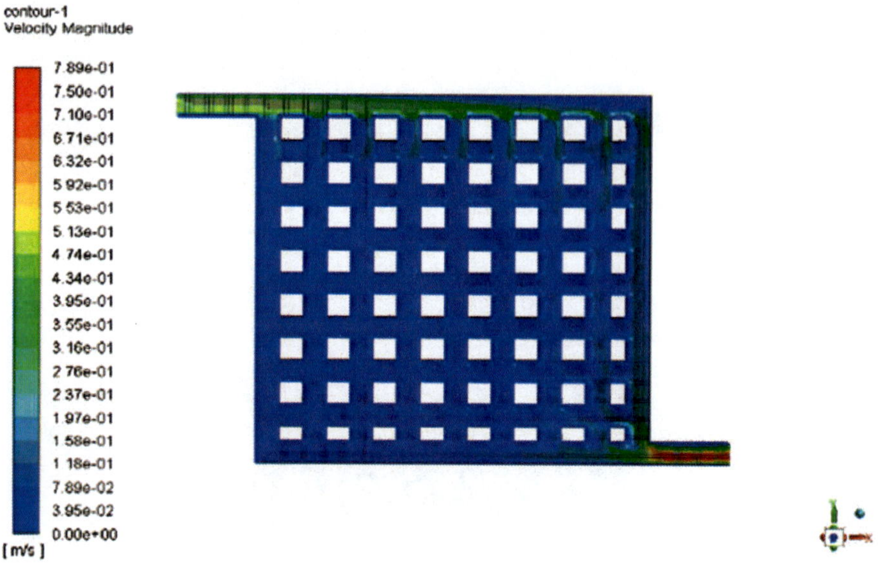


**Fig. 16** Velocity contour (flow distribution)

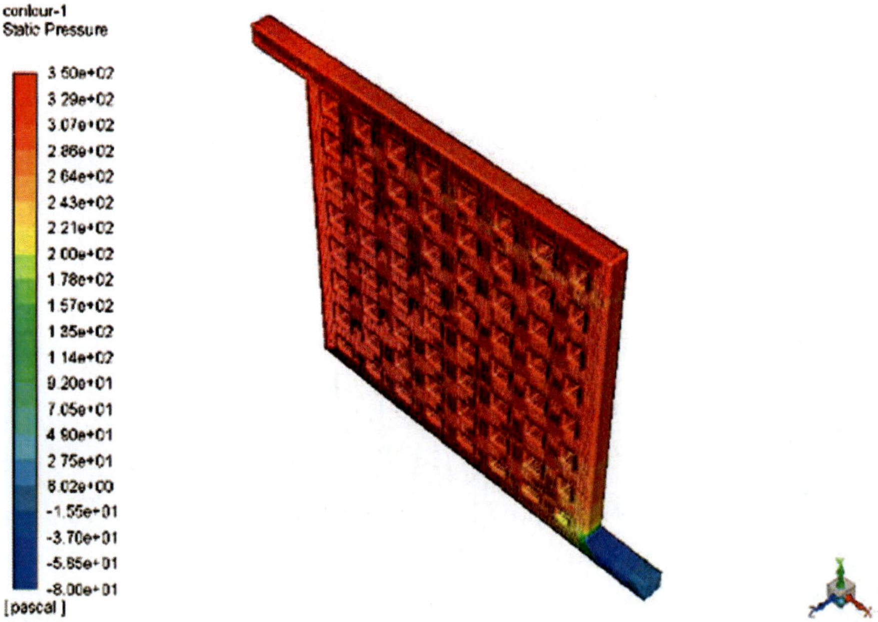


**Fig. 17** Pressure contour (pressure drop)

## 4 Conclusion

Results of the present study suggest that the parallel pattern was found to have the least pressure drop among the other flow field design patterns while the serpentine pattern provides good uniform flow distribution. In the serpentine pattern, replacing the sharp corners with fillets proved to reduce the pressure drop by 580 Pa without disturbing the flow distribution, thereby increasing the overall performance of the PEMFC. However, the findings of the present study need to be experimentally validated to confirm the results obtained from computational software.

## References

1. Manso AP, Marzo FF, Barranco J, Garikano X, Mujika MG (2012) Influence of geometric parameters of the flow fields on the performance of a PEM fuel cell. A review. Int. J. Hydrogen Energy 37(20):15256–15287
2. Arvay A, French J, Wang JC, Peng XH, Kannan AM (2013) Nature inspired flow field designs for proton exchange membrane fuel cell. Int J Hydrogen Energy 38(9):3717–3726
3. Li X, Sabir I (2005) Review of bipolar plates in PEM fuel cells: flow-field designs. Int J Hydrogen Energy 30(4):359–371
4. Sathishkumar S, Praveenkumar S, Vetrivel A, Vidhyasri E (2016) Design and analysis of flow field geometry designs of proton exchange membrane (PEM) fuel cell. Int J Eng Trends Technol (IJETT) 42(3):114–125
5. Wang J (2011) Flow distribution and pressure drop in different layout configurations with z-type arrangement. Energy Sci Technol 2(2):1–12
6. Pollegri A, Spaziante PM (1980) Bipolar Separator for Electrochemical Cells and Method of Preparation Thereof US Patent No. 4197178
7. Reiser CA, Sawyer RD (1988) Solid polymer electrolyte fuel cell stack water management system. US Patent No. 4769297

# CFD Analysis of NACA 4412 Aerofoil Considering Ground Effect

**S. Suraj, S. S. Nivedha Sri, P. Ragul Krishna, S. Jagatheaswaran, and P. Manoj Kumar**

**Abstract** The presence of ground can affect the lift and drag forces during the take-off and landing of an aircraft. A two-dimensional, steady-state, turbulent CFD model for flow over a NACA4412 aerofoil is developed for studying the ground effect on the aerofoil. The Spalart–Allmaras model was used to solve for the turbulent flow. The scope of the study is limited for three different angle of attack (0°, 5° and 10°) and for the *h*/*c* ratios of 0.5, 1.0, and 2.0 and unbounded flow.

**Keywords** NACA 4412 · Ground effect · Drag force · Lift forces

## 1 Introduction

Ground effect is considered to be one of the important parameters in aerodynamics that helps in determining the lift and drag coefficient values of an aeroplane during landing and takeoff.

Lao and Wong [1] reported a numerical analysis of WIG effect for an unmanned aerial vehicle operated at low altitudes and found an enhancement in the ratio of lift coefficient to drag coefficient but affecting the stability. Jonathan David and Chaim

S. Suraj · S. S. Nivedha Sri · P. Ragul Krishna · S. Jagatheaswaran · P. Manoj Kumar (✉)
Department of Mechanical Engineering, PSG Institute of Technology and Applied Research, Coimbatore, Tamil Nadu, India
e-mail: manoj@psgitech.ac.in

S. Suraj
e-mail: surajshank28@gmail.com

S. S. Nivedha Sri
e-mail: s.nivedhasri@gmail.com

P. Ragul Krishna
e-mail: ragulparam@gmail.com

S. Jagatheaswaran
e-mail: jagatheaswaran.sj@gmail.com

© The Editor(s) (if applicable) and The Author(s), under exclusive license to Springer Nature Singapore Pte Ltd. 2021

S. Mohan et al. (eds.), *Materials, Design, and Manufacturing for Sustainable Environment*, Lecture Notes in Mechanical Engineering,
https://doi.org/10.1007/978-981-15-9809-8_42

Zerihan [2] investigates the ground effect experimentally and computationally. They also examined the applications of Gurmey flaps. Evaluation of the turbulence models for the simulation of the flow over a NACA 0012 aerofoil by Eleni et al. [3] analyzed 2D subsonic flow using Spalart–Allmaras, realizable k-ω, and k-ε SST turbulence models and validated using experimental data. Tofa et al. [4] developed a compound wing with new configuration in UTM to improve the wing aerodynamic performance. Allmaras et al. [5] presented a modified Spalart–Allmaras (S–A) turbulence model that targets toward situations of under-resolved grids and unphysical transient states. They also presented a new analytic solution to Spalart–Allmaras for law of the wall velocity. Kolomenskiy et al. [6] presents the numerical study of an insect during takeoff and concluded effect of ground during natural voluntary takeoff is negligible except for modified takeoff's. Agarwal et al. [7] performed numerical simulation of landing to investigate the influence of dynamic ground effect. Study by Kevadiya and Vaidya [8] focuses on low wind power density blade designs, since aerodynamic aerofoils have crucial influence on aerodynamic efficiency of wind turbine. Sezer [9] studied the variation of pressure distribution, lift, and drag force of NACA 4412 at different angle of attacks using ANSYS Fluent and indicated that Spalart–Allmaras model almost matches in the cases where angle of attack is varied.

This paper studies the lift and drag factors of an aeroplane inside the ground effect region through which lift and drag forces can be found for different angle of attacks. ANSYS Fluent has been chosen for solving and finding the lift and drag forces along with the change in pressure and velocity. Design foil is used to find the boundary layer separation.

## 2 Modeling Details

Every commercial flight wing has a unique NACA profile. NACA 4412 profile suits wide range of non-commercial flights and hence is taken into consideration for the present analysis (Fig. 1).

### *2.1 Model Domain*

Figure 2 shows the computational domain for the present study. In order to study the effect of the ground on the aerodynamic lift and drag on the NACA 4412 profile, the profile is considered at a height 'h' from the ground. '*c*' is the chord length of the NACA 4412 profile. The computational studies are conducted to evaluate the effect

**Fig. 1** NACA 4412 profile

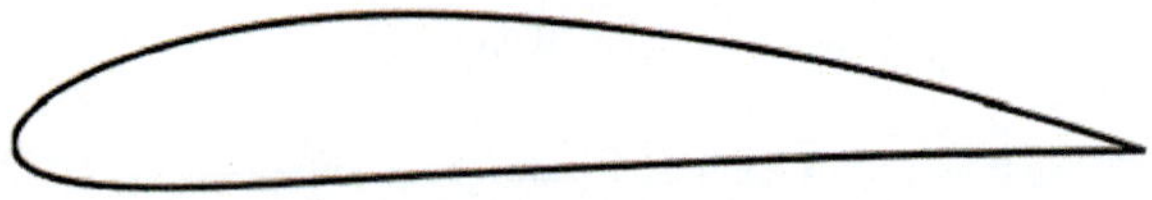

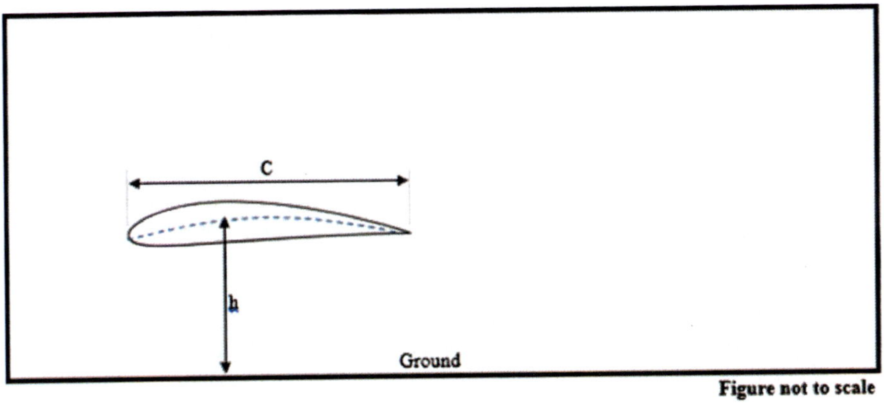


**Fig. 2** Computational geometry

of ground on the lift and drag forces for the chosen NACA 4412 profile for different '*h*/*c*' ratios and for different angle of attack.

## 2.2 *Mesh Details*

The two-dimensional mesh generated for the computational study is shown in Fig. 3. In order to capture the flow dynamics near the aerofoil, very fine mesh is used. The mesh size near the aerofoil is fine to capture the velocity gradient in the viscous sub-layer of the turbulent boundary layer. Grid-independent study is conducted, and the optimized mesh has 40,318 elements.

The left-side boundary and the top boundaries are given velocity inlet boundary conditions, bottom boundary represents the ground, the right side boundary is given the pressure outlet condition, and wall boundary condition for the aerofoil. Typical flight takeoff speed is in the range of 250 to 285 km/h, and hence, an inlet velocity of 80 m/s is given as the input. The Reynolds number corresponding to the velocity of 80 m/s is around $5.6 \times 10^6$ by taking chord length of the aerofoil as the characteristic dimension for the calculation, and hence, the flow is in the turbulent region.

## 2.3 *Governing Equations*

ANSYS Fluent software was used to solve the governing equations and to obtain the results. The steady-state, incompressible flow continuity and momentum equations solved by the software are as follows:

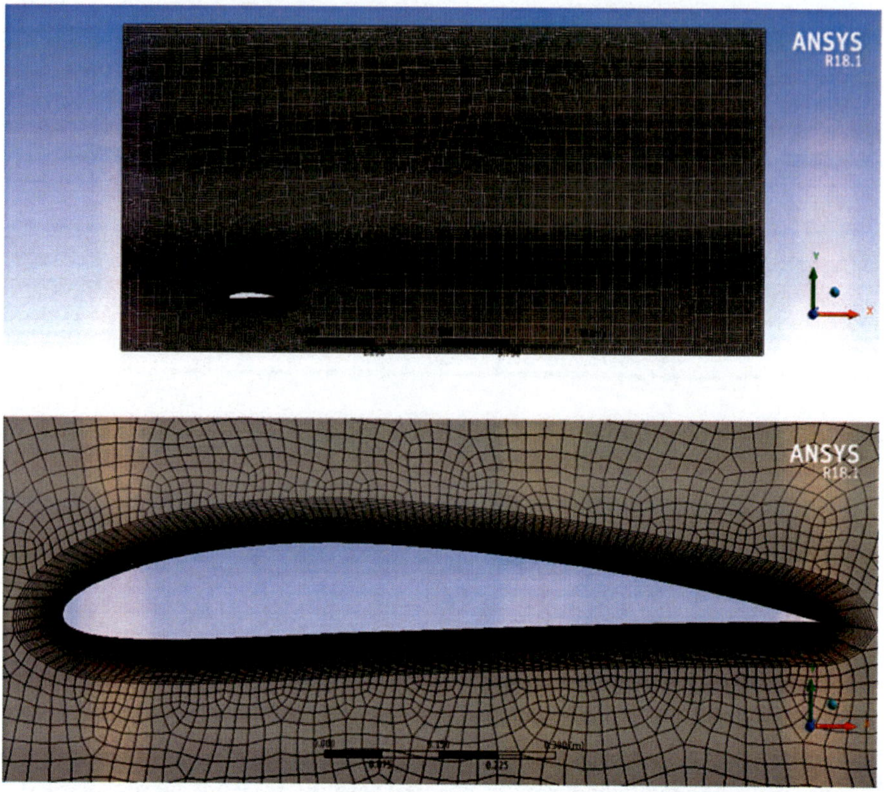


**Fig. 3** Mesh generated for the computational study

*Continuity Equation:*

$$\nabla \cdot \vec{V} = 0$$

*Momentum Equation:*

$$\rho\left(\vec{V} \cdot \nabla\right)\vec{V} = -\nabla P + \mu \nabla^2 \vec{V}$$

As the flow is turbulent, the **Spalart–Allmaras** turbulence model is used as it is proved to be very suitable for aerodynamic flow simulations.

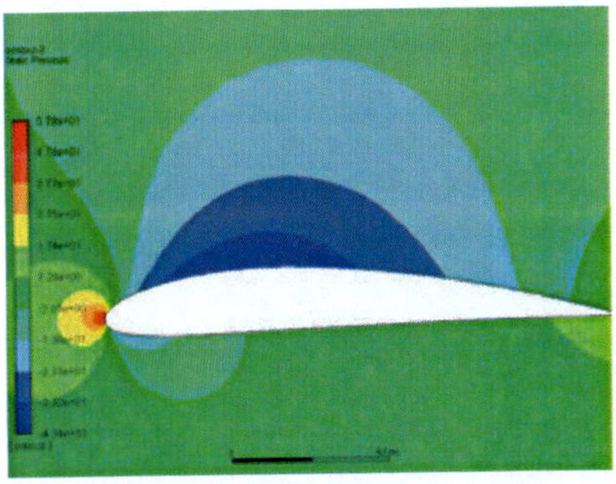

a) Present model

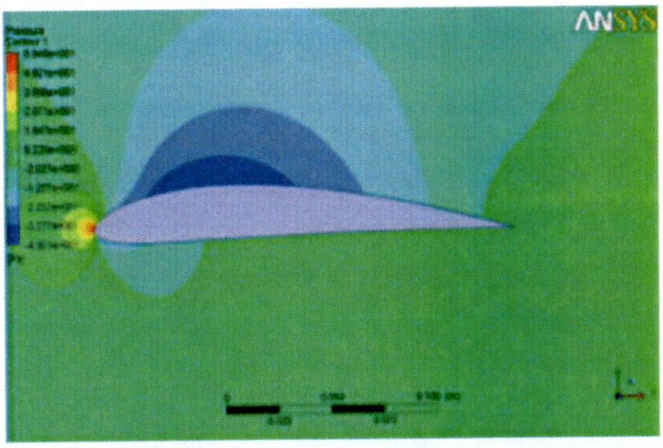

b) Halima and Djilali (2018)

**Fig. 4** Drag coefficient, $C_D$ ($v = 10$ m/s, $\alpha = 0°$)

### 2.4 *Model Validation*

Figure 4 shows the pressure contours over a NACA 4412 profile from the present study and that reported by Halima and Djilali (2018) for an air flow velocity of 10 m/s and for zero degrees angle of attack. The results matches exactly, and hence, the present model is validated.

## 3 Results and Discussion

Figure 5 shows the pressure and velocity contours for the flow over NACA 4412 aerofoil for different angle of attack (0°, 5° and 10°) and for an $h/c$ ratio of 1. At lower angle of attack, the lift force is mainly due to the asymmetric nature of the aerofoil. Air has to flow at higher velocity over the aerofoil as it has to move through a larger distance due to the asymmetric nature of the aerofoil. This leads to a decrease in pressure over the aerofoil, thus developing a lift force in the upward direction. At higher angle of attack, there is a pressure build up at the bottom of the aerofoil as the air flow is obstructed it. Also it can be seen that there is flow separation happening toward the trail edge of the aerofoil for higher angle of attack. Figure 6 shows the pressure and velocity contours for different $h/c$ ratios, for an angle of attack of 5°.

The variation of pressure coefficient, $c_p$, over the aerofoil is show in Fig. 7. The negative values correspond to the top surface of the aerofoil where the velocities are high, leading to lowering of pressure.

Figure 8 shows the variation of coefficient of drag $C_D$ and the drag force for different angle of attack and for different $h/c$ values. The $C_D$ value and the drag force increase with angle of attack for all $h/c$ ratios. This is expected because the projected area increases which in turn increases the pressure drag at higher angle of attacks. The presence of the ground decreases the drag, and the effect is more predominant at higher angle of attack.

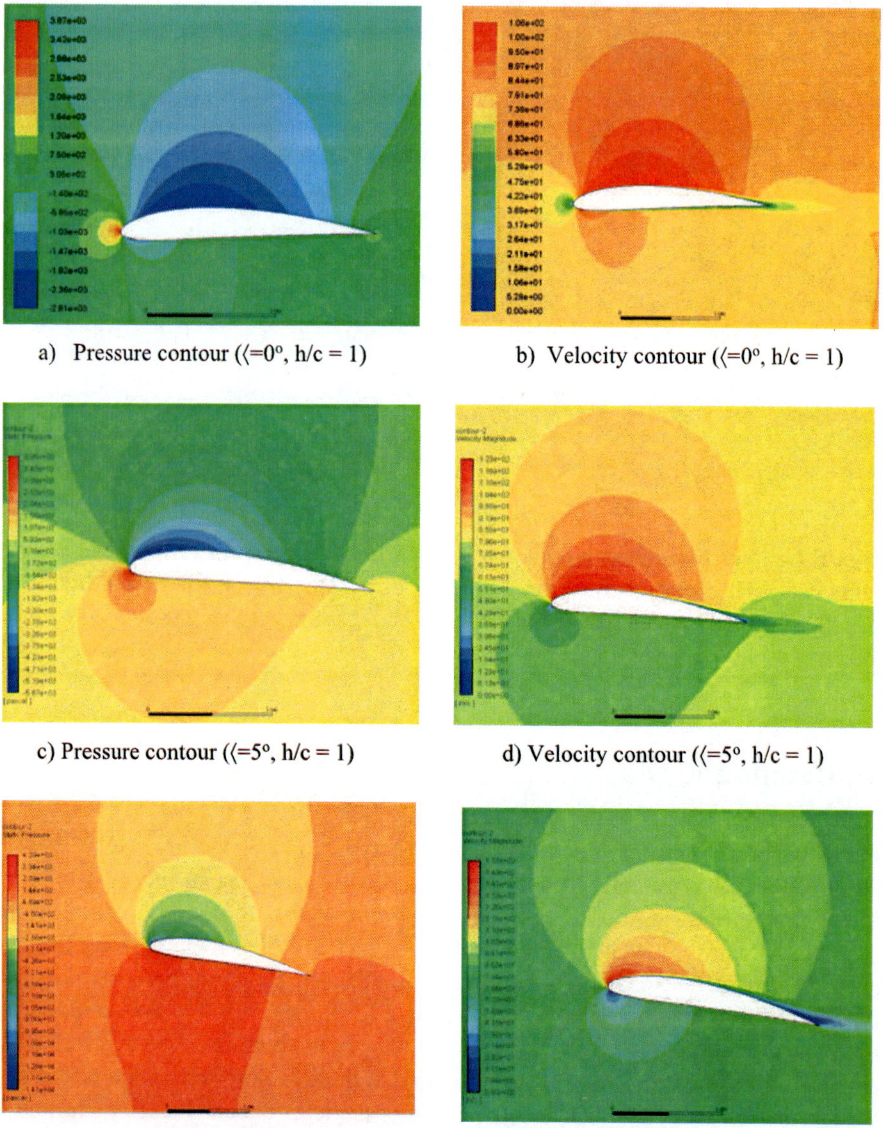

a) Pressure contour (∠=0°, h/c = 1)
b) Velocity contour (∠=0°, h/c = 1)
c) Pressure contour (∠=5°, h/c = 1)
d) Velocity contour (∠=5°, h/c = 1)
e) Pressure contour (∠=10°, h/c = 1)
f) Velocity contour (∠=10°, h/c = 1)

**Fig. 5** Pressure and velocity contours for different angle of attack (∠)

Figure 9 shows the variation of lift coefficient and the lift force with angle of attack and for different *h*/*c* ratios. The lift coefficient and the lift force increases with angle of attack for all the *h*/*c* ratios. As the angle of attack increases, the pressure build up at the bottom of the aerofoil has a component in the vertical direction with aids in the lift of the aerofoil. The effect of the ground does not seem to have much effect on the lift coefficient and lift force of the aerofoil.

a) Pressure contour (⟨=5°, h/c = 0.5)

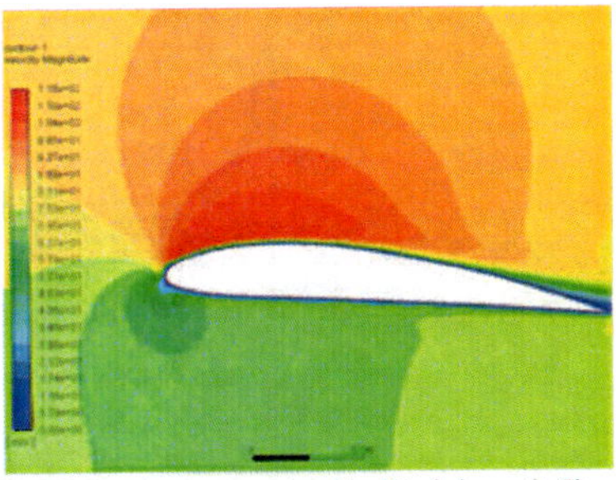

b) Velocity contour (⟨=5°, h/c = 0.5)

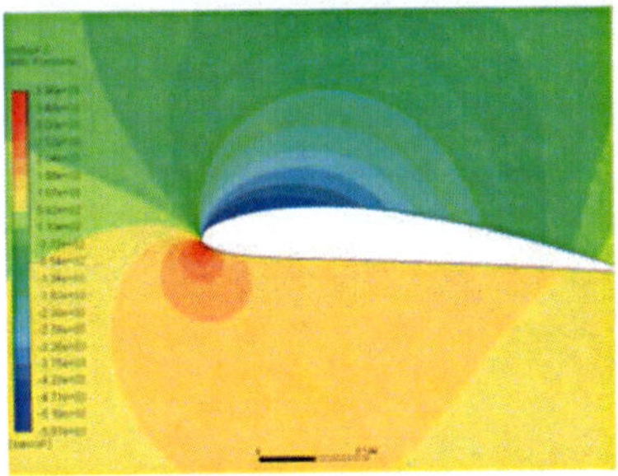

c) Pressure contour (⟨=5°, h/c = 1.0)

d) Velocity contour (⟨=5°, h/c = 1.0)

e) Pressure contour (⟨=5°, h/c = 2.0)

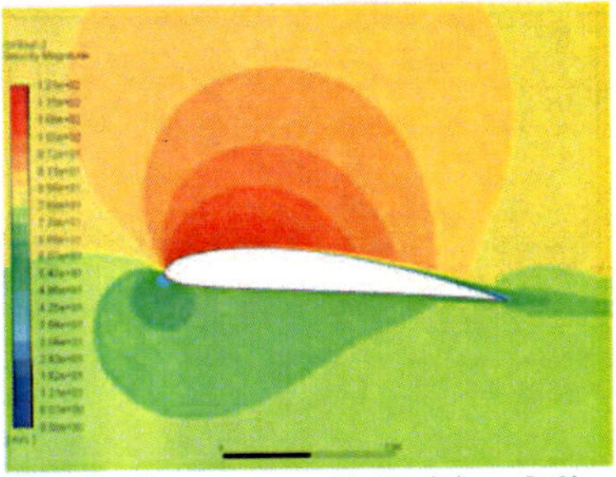

f) Velocity contour (⟨=5°, h/c = 2.0)

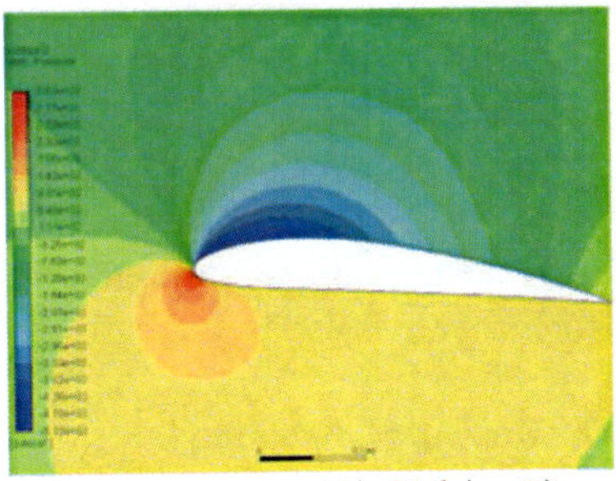

g) Pressure contour (⟨=5°, h/c =□)

h) Pressure contour (⟨=5°, h/c = □)

**Fig. 6** Pressure and velocity contours for different h/c ratios ($\angle = 5°$)

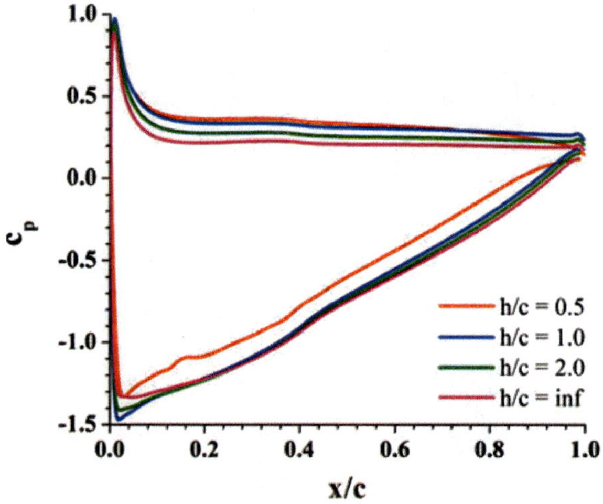


**Fig. 7** Pressure coefficient variation over the aerofoil

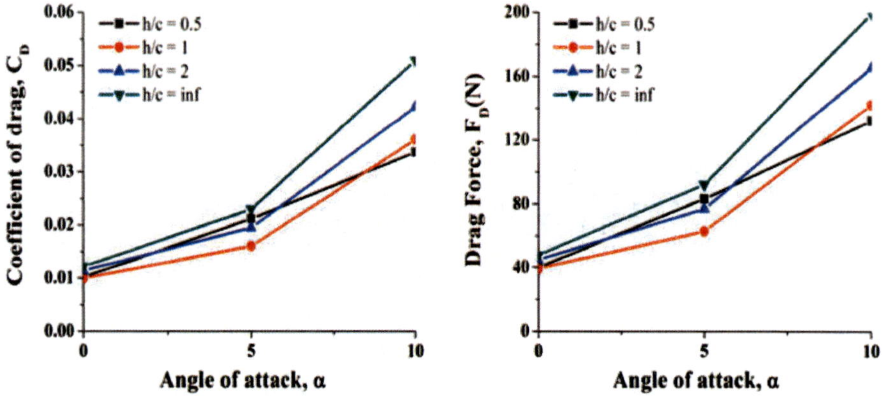


**Fig. 8** Variation of coefficient of drag and drag force

# 4 Conclusions

The ground effect on the lift and drag forces of a NACA 4412 aerofoil was studied using CFD software. The forces were evaluated at three different angle of attack and for various h/c ratios. The lift and drag forces were found to increase with the angle of attack. The presence of ground has significant effect on the drag force on the aerofoil, whereas the lift force is not much effected by the ground.

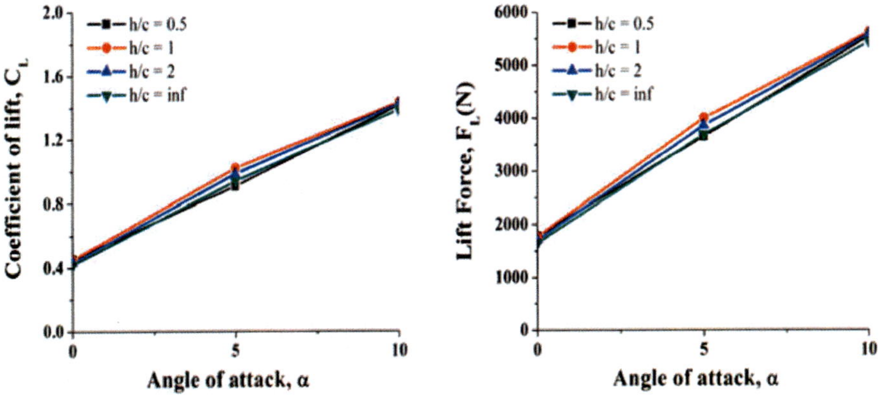


**Fig. 9** Variation of coefficient of lift and lift force

# References


1. Lao CT, Wong ETT (2018) CFD simulation of a wing-in-ground-effect UAV. IOP Conference Series: Materials Science and Engineering, Malaysia, 21–22 Nov 2017
2. Zerihan J (2001) An investigation into the aerodynamics of wings in ground effect. Doctoral dissertation, University of Southampton
3. Eleni DC, Athanasios TI, Dionissios MP (2012) Evaluation of the turbulence models for the simulation of the flow over a National Advisory Committee for Aeronautics (NACA) 0012 airfoil. J Mech Eng Res 4(3):100–111
4. Tofa MM, Maimun A, Ahmed YM, Jamei S, Priyanto A (2014) Experimental investigation of a wing-in-ground effect craft. Sci World J 2014:1–7
5. Allmaras SR, Johnson FT (2012) Modifications and clarifications for the implementation of the Spalart-Allmaras turbulence model. In: Seventh international conference on computational fluid dynamics (ICCFD7), Hawaii, 9–13 July
6. Kolomenskiy D, Maeda M, Engels T, Liu H, Schneider K, Nave JC (2016) Aerodynamic ground effect in fruit fly sized insect takeoff. PLoS ONE 11:e0152072
7. Qu Q, Jia X, Wang W, Liu P, Agarwal RK (2014) Numerical study of the aerodynamics of a NACA 4412 airfoil in dynamic ground effect. Aerosp Sci Technol 38:56–63
8. Ahmed MR, Takasaki T, Kohama Y (2007) Aerodynamics of a NACA4412 airfoil in ground effect. AIAA J 45(1):37–47
9. Sezer F (2016) Computational fluid dynamics analysis of NACA 4412 Airfoil. Project Report, Trakya University, Turkey

# Numerical Model and Simulation of Photovoltaic Cell Heat Transfer Performance Integrated with PCM

**Vinay Sai Kumar Devarapu and S. Ranganathan**

**Abstract** The performance of PV panel decreases with increasing temperature which was tried to reduce its temperature by integrating the phase changing materials and finding its heat transfer performance in numerical and simulation analysis which was carried out using ANSYS and MATLAB software. The results have been validated with the simulation and mathematical model values obtained assuming through different environmental conditions. The PCM integrated photovoltaic cell solar panel along with fins shows better performance and enhanced life duration. The efficiency of the solar photovoltaic cell integrated with PCM is increased by 18%. The heat removal from the PCM integrated photovoltaic cell solar panel is more effective in summer climates, and the electricity production increases by 8.9%.

**Keywords** Photovoltaic cell · Solar panel · PCM · Heat transfer · Numeric model and simulation · MATLAB · ANSYS

## Nomenclature

| | |
|---|---|
| $C_p$ | specific heat of solids (J/kg K), |
| $K$ | thermal conductivity (W/mK), |
| $\alpha$ | function of absorptivity and reflectivity properties of glass, laminating material and PV material, |
| $G_{pv}$ | solar irradiance, |
| $\eta$ | real efficiency of PV module, |
| $A$ | area of surface, |
| $H$ | convective heat transfer co efficient, |
| $\xi_{gl}$ | glass emissivity, |

V. S. K. Devarapu · S. Ranganathan (✉)
Department of Mechanical Engineering, GMR Institute of Technology, GMR Nagar, Rajam, Andhra Pradesh 532127, India
e-mail: ranganathan.s@gmrit.edu.in

© The Editor(s) (if applicable) and The Author(s), under exclusive license to Springer Nature Singapore Pte Ltd. 2021

S. Mohan et al. (eds.), *Materials, Design, and Manufacturing for Sustainable Environment*, Lecture Notes in Mechanical Engineering,
https://doi.org/10.1007/978-981-15-9809-8_43

$F$ PV surface view factor,
$T^4_{\text{surface}}$ temperature of surface (front),
$T^4_{\text{sky}}$ temperature of the ambient air,
$Re$ Reynolds number,
$Pr$ Prandtl number: $\frac{\mu C_p}{K}$,
$Gr$ Grashof's number: $\frac{L^3 \beta g \Delta t}{\upsilon^2}$
$\beta$ $\frac{1}{T}$,
$L$ characteristic length,
$g$ acceleration due to gravity,
$h_{fg}$, latent heat of PCM (solidification or liquefaction),
$X_{\text{old}}$ quality of PCM before time step,
$X_{\text{new}}$ quality of PCM after each time step,
$Q_{\text{added}}$ quantity of energy added,
$E$ heat absorbed by the PCM,
$E_0$ heat required for starting the melting of PCM,
$E_1$ latent heat capacity of the PCM,
$T$ current PCM temperature,
$T_s$ low temperature (solid),
$T_l$ high temperature (liquid),
$C_s$ specific heat of PCM in solid,
$C_l$ specific heat of PCM in liquid,
$P_{\text{out}}$ power output (W),
$\eta$ efficiency of PV conversion,
$\eta_{ref}$ efficiency provided by manufacture,
$\beta_{\text{ref}}$ temperature coefficient of PV efficiency,
$T_{\text{surface}}$ temperature of PV surface,
$T_{\text{ref}}$ reference temperature at the standard test conditions,
$V_{\text{oc}}$ open circuit voltage (V),
$I_{\text{sc}}$ short circuit current (A)

# 1 Introduction

As the world takes every single step towards the developing side of future, there is always something to worry about. Both the population growth and industrial development have led to the massive usage of energy resources. The main input is electricity for industrial, economic, social and domestic life to carry on, which is being produced by depletion of fossil fuels producing pollutants as a result of it. In order to reduce the exploitation of relic energies and to reduce emission of pollutants by consumption electricity, the best alternative for producing the electricity, efficiently and effectively is by usage of solar cells. A solar cell or photovoltaic cell uses a mechanism that converts sun radiation falling on it into electricity by photoelectric effect which was discovered by Heinrich Rudolf Hertz.

The conversion of sunlight to useful energy reached huge advancement through technology in recent years, but huge financing and small conversion efficiency were the principal problems facing by solar photovoltaic cells. It has been found out that conversion of electricity from sunlight by PV cells is greatly depended on its surface temperature.

The effectiveness of the solar cell conversion is greatly depended upon the model of PV cell utilised, outdoor provisions plus functional parameters. The operating temperature of the cell/module/panel has its effect on both electrical efficiency and power output of the panel [1]. According to [2], for every rise of temperature equal to 1 K, the power output of solar panel decreases only by 0.4%. Although solar cells and widely used for power generation, most of them have 15–20% conversion efficiency [3].

In order to solve these problems and improve the energy conversion of PV cells, cooling methods or techniques have been employed, namely active cooling and passive cooling. Active cooling includes usage of pumps or fans for cooling, and this method includes forced air and water cooling, nanofluid cooling and refrigerant cooling. Passive cooling includes passing of natural air and water through tubes, thermoelectric and phase change material cooling methods. Shukla [4] reported that the cooling methods are proven more worthy and advantage for getting the maximum power output of the solar panel. Among all the cooling methods for solar panel present, PCM was proven more capable of reducing the temperature because of its properties.

## *1.1 Phase Changing Materials (PCMs)*

Phase change materials (PCMs) are those that absorb heat at low or desirable temperatures and give out heat energy at almost constant temperatures during its phase transition period. These depend upon the latent heat storage capacity. Desirable properties for PCM are:

- Melting points should be at room temperature,
- High latent heat of fusion values,
- Less coefficient of expansion,
- Must possess low vapour pressure at the temperature of use.

These properties of the PCM made them utilise as a heat storage and cooling medium for electrical devices such as PV panels. The scientific interest in using PCM as a cooling and thermal management of various electrical applications is grown rapidly in recent years.

## 2 Investigations

The first-ever experimentation and investigation of using the PCM integrated with PV panel were done in the year 1978, which shown that the PCM cooling effect has lowered the temperature of PV and increased its conductivity as well as rate of heat transfer from panel to material [5].

Stropnik [6] conducted experiment in open-air climatic surroundings for one week and simulated the experimentation using TRANSYS software. Based on his experiment results, it is shown that there is a drop of 35.6 °C using PCM as a cooling method for the PV panel and an increase in its average electrical energy output by 9.2% by experimentation. Simulated results are given increased output electrical power by 4.5–8.7% and efficiency by 0.5–1%.

Kant [7] has developed a mathematical model considering all the parameters of the heat transfer conditions for PV-PCM. Simulation model is done, and analysis is carried out using COMSOL Multiphysics 5.0 and is compared with other researcher results. The simulated results have shown that PCM installed at back of PV has lowered a maximum temperature of 6 °C on equated with conventional PV, and the productivity of PV panel has increased by 5% comparing with other performance parameters such as wind velocity, panel tilt angle and radiations.

Hasan [8] has conducted experiment on yearly bases and evaluated effects in both summer and winter seasons using PCM (paraffin wax). In the month of April and June, the maximum drop of temperature of PV achieved was high. On taking the average of yearly basis, the PV showed a drop of 10.5 °C and average power yield of 5.9%.

Using foam-stable paraffin wax (RT 28) as a PCM, the PV panel [9] showed a decrease of 4.7 °C than the convectional PV which was kept under 50 °C about 3.33 h. The output current was maximised which is 0.51 V about the maximum voltage of conventional voltage of PV panel through experimental analysis.

On conducting experiment under sunlight with aluminium sheet as TCE, Rajvikram [10] has shown that there is an average increase in its efficiency by 24.4% and decrease of 10.35 °C temperature by using aluminium as a TCE and 2% increase in its total electrical productivity.

There are other methods of refining the conductivity of PV-PCM by using nano-materials inside the PCM as heat transfer catalyst. Both the PCM (RT 55) and PCM (RT 55) with $Al_2O_3$ nano-materials were compared [11] in his paper for the enhancement of heat transfer and efficiency of PV panel which tested experimentally under same atmospheric conditions and found that PCM and with nano-materials drop the temperature of the PV panel by 8.1 °C and 10.6 °C and showed increase in the PV efficiency by 5.7% and 13.2%, respectively. Therefore, proving that PCM with nano-materials can improve the thermal regulation of the PCM much higher.

Similarly, integration of cooling procedures is used in improving concentrated PV panel efficiency. Nada [12] has developed a model where pure water having nano-material (CuO) of 0.4% concentration is used as a heat transfer fluid integrated directly on the PV panel back, which is then passed on to the container having PCM

(RT 35HC) which gave a reduction of 60% in average temperature and increase in system efficiency of 224% and power output of 12% increment.

In a similar manner, two PCMs with varying thermal conductivities of 0.19 and 2.4 were used to determine that the effective heat transfer is carried between these two, which were experimented under similar conditions. The PCM with higher thermal conductivity gave lower temperature than the other. Therefore, PCMs with higher thermal conductivities were preferred and are more appropriate for PV module application.

## 3 Numerical Modelling

### 3.1 PV Panel Temperature

The governing equation of the heat transfer model, fluid model, power generation and momentum equation is derived to analyse the thermal removal calculations.

The radiation heat transfer is governed by Eq. (1), and the total solar radiation is given by Eq. (2).

Basic governing equation in 1D is:

$$\rho C_p \frac{\mathrm{d}T}{\mathrm{d}t} = K \frac{\mathrm{d}T}{\mathrm{d}x} \tag{1}$$

$$\text{Total solar radiationsinking in PV surface } = \alpha G_{pv} \eta A \tag{2}$$

$$\text{Radiation heat loss from surface to ambient air } = \xi_{\mathrm{gl}} F \sigma \left( T_{\mathrm{surface}}^4 - T_{\mathrm{sky}}^4 \right) \tag{3}$$

The heat transfer through convection is governed by convective heat loss given by.

$$\text{Convective heat loss from surface to ambient air } = HA(T_{\mathrm{surface}} - T_{\mathrm{ambient}}) \tag{4}$$

Therefore,

The total radiation falling on PV surface is given by

$$Q_{\mathrm{rad}} = \alpha G_{pv} \eta + \xi_{\mathrm{surface}} F \sigma \left( T_{\mathrm{sky}}^4 - T_{\mathrm{surface}}^4 \right) \left( \mathrm{W/m^2} \right) \tag{5}$$

Then, total heat lost from exterior of the plate of PV panel is given by

$$\{Q_{total}\}\rho C_p \frac{\mathrm{d}T}{\mathrm{d}t} = \alpha G_{pv}\eta A + \xi_{gl} F\sigma\left(T_{\text{surface}}^4 - T_{\text{sky}}^4\right) + HA(T_{\text{surface}} - T_{\text{ambient}}) \tag{6}$$

Similarly, the net heat lost on the back of the panel is given by

$$\rho C_p \frac{\mathrm{d}T}{\mathrm{d}t} = \xi_{\text{material used}} F\sigma\left(T_{\text{back}}^4 - T_{\text{ground}}^4\right) + HA\left(T_{\text{back}} - T_{\text{ground}}\right) \tag{7}$$

Since wind flow plays a major role in heat conduction, this parameter is also considered for accurate results.

Considering vertical plate empirical co-relation for.

Free convection

$$h_{\text{free}} = \frac{K}{L}(0.825) + \left[\frac{0.387(\text{Gr}\cdot\text{Pr})^{\frac{1}{6}}}{\left[1 + \left(\frac{0.492}{\text{Pr}}\right)^{\frac{9}{16}}\right]^{\frac{8}{27}}}\right]^2 \tag{8}$$

Forced convection,

$$h_{\text{force}} = \frac{K}{L} 0.664(\text{Re})^{0.5}(\text{Pr})^{0.33} \tag{9}$$

## 3.2 *Temperature of PCM*

The PCM possesses two stages, one is solid phase and other is liquid phase, and both of the phases possess temperature which additional equations are employed.

The energy required for melting the totally solid PCM is

$$Q_{\text{solidification}} = Q_{\text{liquefaction}} = mh_{fg} \tag{10}$$

where

$Q_{\text{solidification}} = Q_{\text{liquefaction}}$, energy required for PCM to convert into liquid.

Change in PCM accompanying by adding or subtracting the energy

$$(X_{\text{old}} - X_{\text{new}})mh_{fg} = Q_{\text{added}} \tag{11}$$

$$X_{\text{new}} = X_{\text{old}} - \frac{Q_{\text{added}}}{mh_{fg}}$$

The temperature of PCM found out established on quality

$$T_{\mathrm{PCM}} = \left(T_{\text{liquefacation}} - T_{\text{solidification}}\right)X_{\text{new}} + T_{\text{solidification}} \tag{12}$$

Variations along 2D with respect to time are given by

$$\rho_p c p_p \frac{\partial T_p}{\partial t} = \nabla \cdot \left(K_p \nabla T_p\right) - \frac{\partial}{\partial x}\left(\rho_p c p_p U_x T_p\right) - \frac{\partial}{\partial y}\left(\rho_p c p_p U_y T_p\right) \tag{13}$$

PCM heat diffusion in all the phases is given as

$$\rho_p c p_p \frac{\partial T_p}{\partial t} + \nabla \cdot \left(K_p \nabla T_p\right) + \frac{\partial}{\partial x}\left(\rho_p c p_p U_x T_p\right) + \frac{\partial}{\partial y}\left(\rho_p c p_p U_y T_p\right) = 0 \tag{14}$$

Since PCM possesses both sensible heat and latent heat, phase change temperature can be divided as three main categories, as

$$T = \begin{cases} E/c_s + T_0 & E < E_0, \quad T < T_l \\ T_s + \left(\frac{E-E_0}{E_l}\right)(T_l - T_s) & E_0 < E < E_0 + E_l, \quad T_s < T < T_l \\ T_l + \frac{E-E_0-E_l}{c_l} & E \geq E_0 + E_l, \quad T > T_l \end{cases}$$

## 3.3 Power Output and Efficiency

The output power of PV panel generated is given by

$$P_{\text{out}} = \eta \cdot A \cdot G_{pv} \tag{15}$$

The relation between the decrease in PV and its temperature rise is represented in

$$\eta = \eta_{\text{ref}}\{1 - \beta_{\text{ref}}(T_{\text{surface}} - T_{\text{ref}})\} \tag{16}$$

Electrical efficiency of the PV panel is

$$\eta_{\text{electrical }pv} = \frac{p_{\max}}{p_i} = \frac{I_{\max} * V_{\max}}{\text{incident solar radiation} * \text{area of solar cell}} = \frac{V_{\text{oc}} * I_{\text{sc}} * \text{FF}}{I * A_c} \tag{17}$$

FF = fill factor, given by

$$\text{FF} = \frac{P_{\max}}{V_{\text{oc}} * I_{\text{sc}}} = \frac{V_{\max} * I_{\max}}{V_{\text{oc}} * I_{\text{sc}}}$$

## 4 Simulation Analysis

Design, simulation and analysis are done by using ANSYS 2019 R3 software. Under steady-state thermal analysis, the design of the panel is done. A basic mono-crystalline PV panel of 20 watts' capacity is considered whose properties and parameters remain same by every manufacture. In order to effectively cool any heat source, fins were provided at the outer surface such that their heat transfer rate would be maximised. Based on that idea, fins were also designed on rear of PV panel such that the rate of heat transfer from PV panel to the PCM would be at a faster rate.

**Panel Parameters**

Capacity: 20 watts
Material: silicon solar panel, aluminium
Type of panel: mono-crystalline
Number of cells: 36
Output voltage: 12
Solar power output: 20.

The actual physical model and cross-section of the PV panel made by aluminium casing and container with PCM and fins are designed using ANSYS 2019 software are shown Figs. 1 and 2 with internal fins arrangement. The fins are integrated with the PV-PCM internally with the copper plates, and the plates are 2 mm thickness and filled with RT35. The fins are arranged in parallel with regular intervals of 70 mm and consist of six fins. Proper insulation was made in the PV-PCM arrangement. The material specification, thermal and physical properties are given in Table 1.

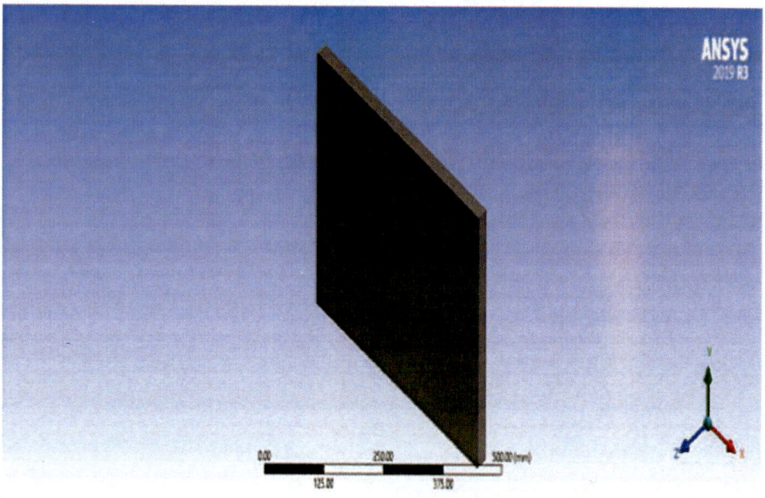


**Fig. 1** Model designed

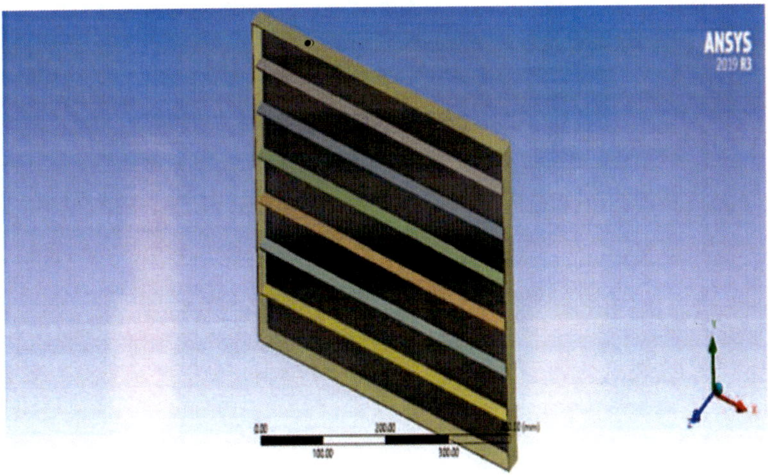


**Fig. 2** Model designed with fins

**Table 1** Properties of PV panel

| | Density (kg/m$^3$) | Thickness (mm) | Specific heat (J/kg k) | Thermal conductivity (w/cm k) |
|---|---|---|---|---|
| Glass | 3000 | 3 | 500 | 1.8 |
| EvA | 960 | 0.5 | 2100 | 0.35 |
| Silicon cells | 2300 | 0.3 | 680 | 148 |
| Tedlar | 1200 | 0.1 | 1250 | 0.2 |
| Aluminium | 2675 | 4 | 900 | 211 |

Under steady-state thermal analysis, taking convection of temperature 35 °C and with default radiation as a source of heat input on surface was carried out in steady-state thermal analysis. Figure 3 shows the heat generated result at peak hour of the

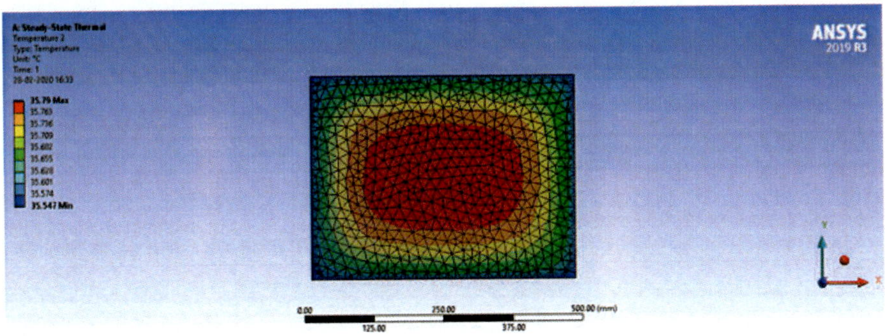


**Fig. 3** Heat generated on the glass

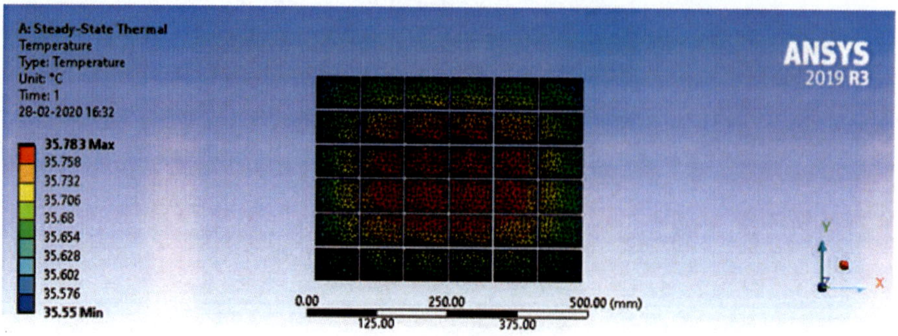


**Fig. 4** Heat generated on silicon cells

luminous day on the glass surface, and Fig. 4 shows the thermal distribution of the silicon cells by convention.

After the design of fins, copper material is used as fin materials because of its high thermal conductivity, and the cavity was filled with PCM (RT35), and heat spawned is shown in Figs. 5, 6 and 7. These simulation images prove that the heat generated will be conducted to the fins and thereby to PCM lowering the temperature of the PV panel. And on increasing the number of fins to 6, the temperature distribution was even and proven optimal for the thermal regulation, thereby ensuring the prolonged life of the panel.

Depending upon the temperature given out through ANSYS results, power and current generated curves can be found out by utilising MATLAB R2018a software, and simulation was carried out. The solar panel considered was default in MATLAB. Two circuits were considered one with same temperature but different solar radiation levels and other with different temperatures but same solar radiation. The first circuit diagram is shown in Fig. 8 having different radiation with same temperature levels. Figure 9 shows circuit diagram of different temperature and radiations. The graph shown in Fig. 10 shows the result of same temperature but different solar radiation.

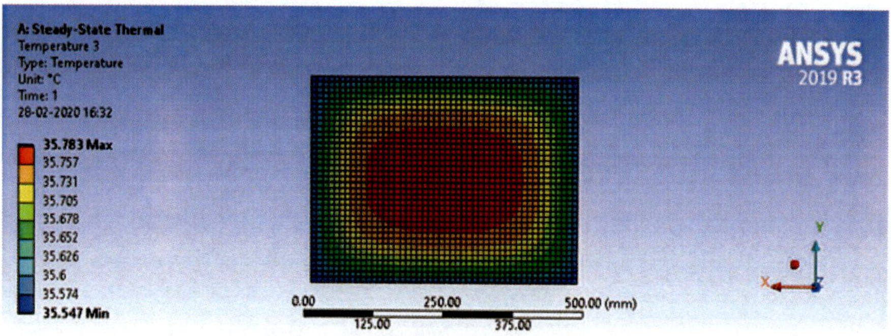


**Fig. 5** Thermal distribution of panel with PCM

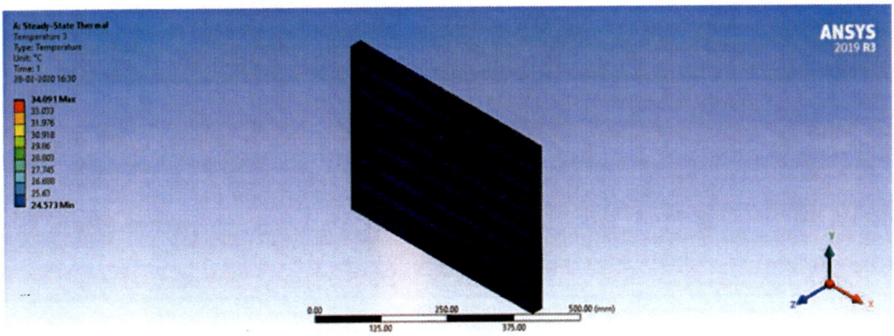


Fig. 6 Directional of heat flux (fins)

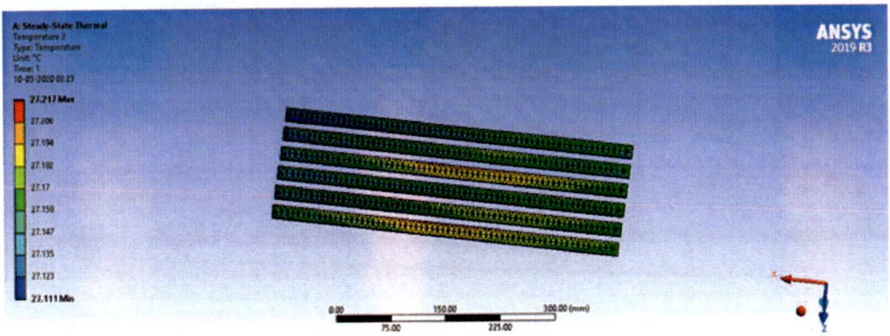


Fig. 7 Heat transfer from fins to PCM

This shows that there is always drop of power when compared to that of manufacturer result.

## 5 Conclusions

It is found out that parameters such as materials of fins, their arrangement, solar irradiance levels and ambient temperature contribute major part in transferring heat to PCM and show variations in temperature reduction by 6–9 °C when compared with numerical and ANSYS results. The numerical model is taken into consideration by some major assumptions which gave accurate results. The ANSYS simulation results prove that most of the heat is being accumulated at the centre of the PV panel, showing that there is unequal melting of the PCM.

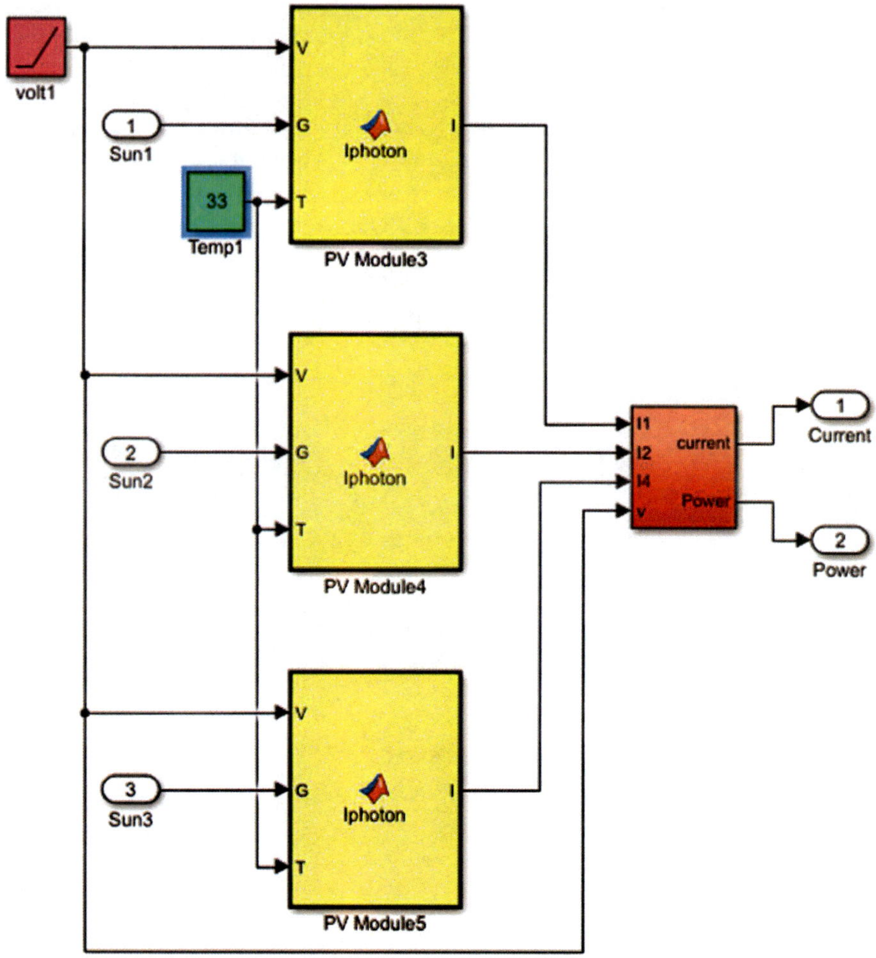


**Fig. 8** Circuit diagram with temperature

## 6 Future Work

- Simulation in ANSYS can be done more accurately by taking the actual ambient conditions.
- Geometry of the fins can be altered for better heat distribution for the PCM.
- Better material selection for fins and PCM can be compared with each other that might suitable for any atmospheric conditions.

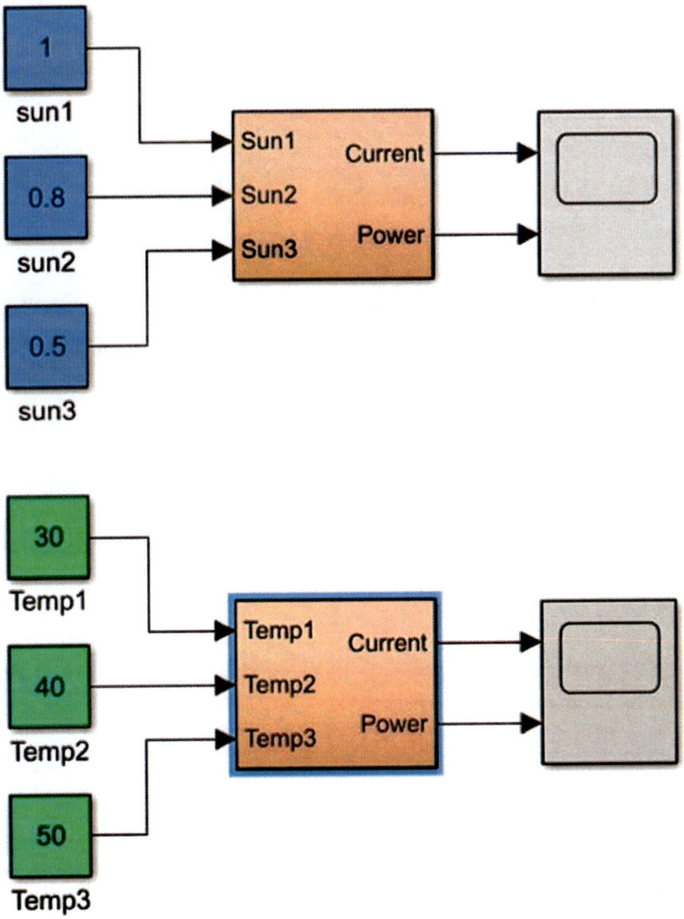


**Fig. 9** Two different circuit diagrams with different temperature and radiation

**Fig. 10** Power output and current curves generated by MATLAB

# References


1. Skoplaki E (2009) On the temperature dependence of photovoltaic module electrical performance: a review of efficiency/power correlations. Sol Energy 83(2009):614–624
2. Radziemska E (2003) The effect of temperature on the power drop in crystalline silicon solar cells. Renewable Energy 28(2003):1–12
3. Ma T (2015) Using phase change materials in photovoltaic systems for thermal regulation and electrical efficiency improvement: a review and outlook. Renew Sustain Energy Rev 43(2015):1273–1284
4. Shukla A (2017) Cooling methodologies of photovoltaic module for enhancing electrical efficiency: a review. Sol Energy Mater Sol Cells 160(2017):275–286
5. Japs E (2016) Experimental study of phase change materials for photovoltaic modules: energy performance and economic yield for the EPEX spot market. Sol Energy 140:51–59
6. Stropnik R (2016) Increasing the efficiency of PV panel with the use of PCM ". 10. Renewable Energy 97:671–679
7. Karunesh Kant, 2016," Heat transfer studies of photovoltaic panel coupled with phase change Material " Solar Energy 140 (2016) 151–161.
8. Hasan A (2017) Yearly energy performance of a photovoltaic-phase change material (PV-PCM) system in hot climate. Sol Energy 146:417–429
9. Luo Z (2017) Numerical and experimental study on temperature control of solar panels with form-stable paraffin/expanded graphite composite PCM. Energy Convers Manage 149:416–423
10. Rajvikram M (2019) Experimental investigation on the abasement of operating temperature in solar photovoltaic panel using PCM and aluminium. Sol Energy 188(2019):327–338
11. Nada SA (2018) Improving the thermal regulation and efficiency enhancement of PCM-Integrated PV modules using nano particles. Energy Convers Manage 166(2018):735–743
12. Nasef HA (2019) Integrative passive and active cooling system using PCM and nanoflfluid for thermal regulation of concentrated photovoltaic solar cells. Energy Convers Manage 199:112065

# Study on Performance Enhancement of SPV Panel Incorporating a Nanocomposite PCM as Thermal Regulator

**P. Manoj Kumar, G. Mukesh, S. Naresh, D. Mohana Nitthilan, and R. Kishore Kumar**

**Abstract** The conversion of solar energy into electricity is becoming popular as the cost of solar electricity is continuously declining due to the government policy decisions in encouraging green electricity. However, the efficiency of the panel is a notable factor which is always lying around 16%. Because of getting direct solar rays, the surface temperature of the PV panel is varying significantly throughout the day which plays a crucial role in their performance. Hence, the proper thermal management of PV panel is required to attain the improved performance of the panel in terms of their efficiency. This present work made an attempt to improve the efficiency of the SPV panel by properly regulating their absorbed thermal energy using a nanocomposite phase change material (NCPCM). NCPCM was obtained by diffusing lower mass % of nano-$SiO_2$ particles within paraffin matrix (1.0% mass). Two PV panels of similar capacity (30 $W_p$) and configuration have been used during the experimentation. The first panel, without any modification, was named as SPV 1, and the second panel was named as SPV 2 which has been integrated with the NCPCM for thermal energy regulation. Both the PV panels were investigated during the clear solar days between 7.00 a.m. and 5.00 p.m. The results showed that the incorporation of NCPCM significantly reduced the panel surface temperature and improved electrical efficiency.

---

P. Manoj Kumar (✉) · G. Mukesh · S. Naresh · D. Mohana Nitthilan · R. Kishore Kumar
Department of Mechanical Engineering, KPR Institute of Engineering and Technology, Coimbatore, Tamil Nadu, India
e-mail: pasupathimanojkumar@gmail.com

G. Mukesh
e-mail: mukeshgokul01@gmail.com

S. Naresh
e-mail: naresh121999@gmail.com

D. Mohana Nitthilan
e-mail: nitthinitthinitthi@gmail.com

R. Kishore Kumar
e-mail: kishore.7475.rk@gmail.com

© The Editor(s) (if applicable) and The Author(s), under exclusive license to Springer Nature Singapore Pte Ltd. 2021

S. Mohan et al. (eds.), *Materials, Design, and Manufacturing for Sustainable Environment*, Lecture Notes in Mechanical Engineering,
https://doi.org/10.1007/978-981-15-9809-8_44

**Keywords** SPV panel thermal regulation · Nanocomposite PCM · SPV surface temperature · Electrical efficiency

## 1 Introduction

The energy conservation and eco-friendly energy generation are the need of the hour worldwide. The majority of the world energy requirement is fulfilled by the conventional fuels such as coal, petroleum oils and gases. They become the prime reason for greenhouse emission, global warming and air pollution [1, 2]. Hence, the countries are thriving toward generation of energy through the non-conventional sources such as energy from sun, wind energy and tidal energy, to avoid further damage to the environment [3, 4]. Among the different form of energy generation, sun is considered to be the predominant energy source as it is available in greater quantity in worldwide at no cost, no emission and environment-friendly [5, 6].

The solar PV panels are used to transform solar radiation energy into electrical power which can be applied for different needs such as domestic, commercial and industrial purposes. Due to the technological developments and government policies, the cost of the PV panels is persistently decreasing every year, and hence, solar PV conversion becomes the most viable solution for eco-friendly energy generation [7]. Nonetheless, the SPV efficiency is a limiting factor in this kind of power production. The available solar PV panel's efficiency is varying between 15 and 20%, nominally [8, 9]. The surface temperature of the panel is considered to be a key parameter, which directly impacts on the SPV efficiency. It was referred from the literature that the SPV surface temperature is inversely proportional to the panel electrical efficiency [10, 11]. It was acknowledged that the increase of panel surface temperature by 1 °C would result in reduction of panel efficiency by 0.65% [12]. The researchers showed that the PV panel efficiency can drop by 10% during the day time for the temperature rise between 12 and 20 °C [13, 14]. Hence, the thermal regulation of the PV panel is considered and studied by the number of researchers.

The circulation of water behind the SPV panel would decrease the surface temperature appropriately; further, it could enhance the electrical efficiency to 13%. However, it requires a dedicated system to circulate the water continuously [15]. Xiang et al. [16] incorporated soil-generator hybrid method to the thermal management of SPV panel and acknowledged the augment in efficiency of around 6.85%. Bahaidarah [17] analyzed the jet impingement cooling for solar PV panels and concluded that this arrangement would lessen the PV surface temperature by 33.1 °C.

Among the different proposed methods of PV cooling, the incorporation of suitable phase change material (PCM) on the rear side of SPV was strongly recommended by the recent review papers for the cost-effective and efficient thermal regulation of PV modules [18, 19]. The paraffin wax is one of the widely used PCMs because of its superior qualities such as less supercooling, great latent heat, non-harmful, wide melting range, easy and cheap availability. However, they are inherently poor in thermal conductivity which limits their applications as solar PV thermal regulations [20]. Huang et al. [21] incorporated the internal fins within the PCM tank to improve rate of heat transfer within PCM for PV thermal management. The dispersion of small volume percentage of $TiO_2$ nanoparticles in paraffin matrix upgraded thermal conductivity of the paraffin matrix to a considerable level [22]. Lin and Al-Kayiem [23] mixed copper nanoparticles of 0.5, 1.0, 1.5 and 2.0 wt. % in paraffin wax for attaining greater thermal characteristics of paraffin. They determined that the nano-cu particles considerably boosted the thermal conductivity and other thermal properties in proportional to their wt.%. Mohamed et al. [24] used nano-alumina as the nano-additive in paraffin and concluded that the rate of heat conduction inside the paraffin was greatly enriched with the nano-alumina.

The literatures proved that the PCM can intensify the thermal regulation of SPV panels, and further, it was clearly revealed that the nanoparticles with high conductivity would improve the heat conduction inside the paraffin. Nevertheless, the literatures scarcely reported the application of nanocomposite PCM for the thermal regulation of SPV panels even though a huge potential is available with this kind of mechanism. Sardarabadi et al. [25] piloted an investigation on PV/T system with a water/silica ($SiO_2$) nanofluid and acknowledged the improvement in the system efficiency by 7.9%. In this present work, an effort has been made to study the influence of incorporating nano-$SiO_2$ doped NCPCM on surface panel temperature and further the electrical efficiency of the SPV panels.

The main objective of this work is to prepare the nanocomposite PCM by diffusing lower mass fraction of nano-silica particles within paraffin to prepare the NCPCM and assess their thermo-physical properties. Further, it is planned to integrate the NCPCM with the solar PV panel to study the variation in panel surface temperature and electrical efficiency in the presence of NCPCM.

## 2 Materials and Methods

### *2.1 Synthesizing Nanocomposite Phase Change Material (NCPCM)*

Nanocomposite PCM was prepared by dissolving 1.0% mass of silica ($SiO_2$) with paraffin wax. The lower mass percentage of nanoparticles was chosen with reference to Arshad et al. [26]. Paraffin wax of commercial grade and nano-$SiO_2$ particles of mean diameter 25 nm were commercially acquired from Sun Chemicals, Coimbatore. The NCPCM was prepared in two-step method [27, 28]. The paraffin was heated up to 80 °C and placed on the magnetic stirrer, where silica nanoparticles were poured in drop by drop and stirred simultaneously for 2 h to ensure the homogeneous mixture of NCPCM.

## 2.2 Evaluating Thermo-physical Properties of NCPCM

Thermal storage characteristics of paraffin wax and NCPCM samples were characterized with the help of differential scanning calorimetry (DSC) and then with a thermal properties analyzer to evaluate the thermal conductivity, latent heat, melting point, solidification point and degree of supercooling of paraffin and NCPCM. DSC 6000 was used to measure latent capacity, fusion point, crystallization point and degree of supercooling of the paraffin and NCPCM, whereas KD2 Pro thermal properties analyzer was engaged to measure thermal conductivity.

Figure 1 shows the DSC curve of paraffin and NCPCM. There are two distinguished peaks at during heating and cooling of the paraffin/NCPM. The small peak shows that there is phase change of solid to solid transition, and the larger peak describes the latent heat of phase transition between crystal and liquid phases. The similar trend was acknowledged by Lin and Al-Kayiem [22]. Thermal conductivity of paraffin was measured as 0.18 W/mK and 0.22 W/mK for the NCPCM which is 22.22% higher the pure paraffin wax.

The obtained properties are consolidated and given in Table 1. The fusion and crystallization temperatures of the base PCM were suitably modified in the presence of nano-$SiO_2$ particles. The degree of supercooling of paraffin wax was substantially reduced by 1.5 °C with the help of nanoparticles.

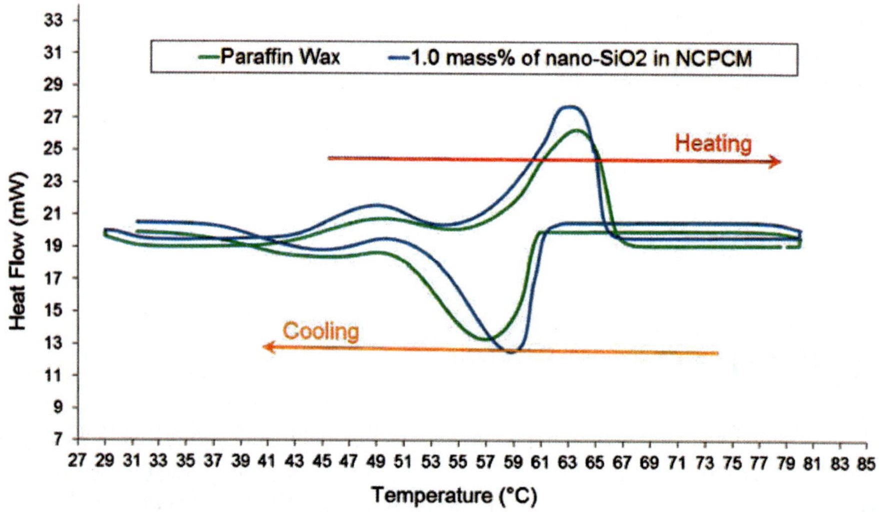


**Fig. 1** DSC curve of paraffin and NCPCM

**Table 1** Thermo-physical properties of constituents

| Properties | Paraffin (PCM) | Nano-$SiO_2$ particles | NCPCM |
|---|---|---|---|
| Temp. of fusion (°C) | 63.7 | 1715 | 62.7 |
| Solidifying temp. (°C) | 57.0 | – | 57.5 |
| Degree of supercooling (°C) | 6.7 | – | 5.2 |
| Latent heat capacity (J/kg) | 140,000.0 | – | 135,000 |
| Thermal conductivity (W/m°C) | 0.18 | 1.50 | 0.22 |

## 2.3 *Fabrication of NCPCM Integrated Solar PV Panel*

The defined numbers of solar photovoltaic cells are connected series to form the solar PV (SPV) panels. Each solar cell can create the potential difference of 0.5 V. Hence, the required voltage potential can be obtained by suitable arrangement of the cells. In this work, two SPVs (SPV1 and SPV2) of same configuration were procured as with the specifications of nominal power 30 $W_p$ and maximum voltage rating 18 V.

Among the different used panels, SPV1 is a panel without NCPCM and SPV2 is the panel integrated with NCPCM. Both SPV1 and SPV2 are having same physical structure except backpacking of NCPCM. As shown in Fig. 2, each cell of the SPV1 consists of a semiconductor PV cell covered by thin layer of ethyl vinyl acetate (EVA) on either rims. Further, top layer is having a transparent glass to protect the cell from environmental deterioration, and bottom portion is covered with a back support sheet followed by the aluminum frame.

The SPV2 is similar to SPV1; except, it is packed with NCPCM of 2.5 kg at the rear end as presented in Fig. 3. The encapsulation was done with the help of box structure made up of aluminum, in order to ensure the airtight covering of the NCPCM.

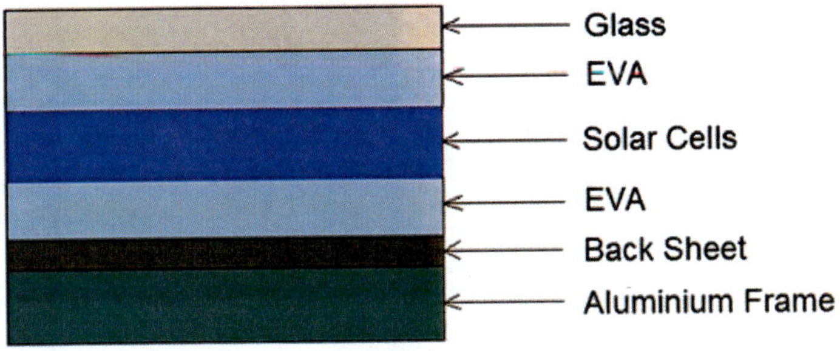


**Fig. 2** Schematic diagram of SPV1

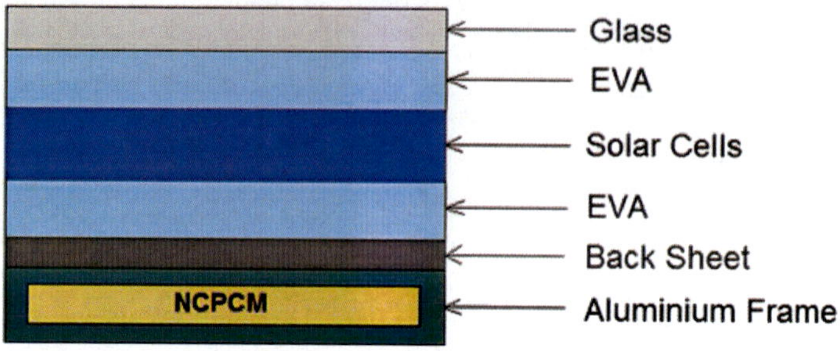


**Fig. 3** Schematic diagram of SPV2

## 3 Experimental Work

The two PV panels SPV1 and SPV2 both were experimented simultaneously on an average for three clear solar days to ensure the correctness and repeatability of the results. The high precision thermocouples were employed to measure the temperature of panel surface and atmospheric temperature for both cases. Three such thermocouples were placed arbitrarily on each panel to read out the panel temperature, and their average was used for further analysis. One more thermocouple was used to observe the atmospheric temperature, and all these thermocouples were coupled with a digital indicator for the ease of taking measurements. A multi-meter was utilized to observe SPV's instantaneous current and voltage, and TES-1333 model solar power meter was employed to note down the solar radiation. The deviation in hourly solar irradiance during the analysis is shown in Fig. 4. The crowning radiation was observed in the mid of the day between 1 p.m. and 2 p.m., and it was slowly declined toward the dusk.

After taking the necessary readings, the required parameters are calculated with the help of following equations [8].

The input solar irradiation fallen on the PV panel is given by

$$P_{\text{in}} = S_{\text{in}} \cdot A_p \tag{1}$$

The electrical power output can be deliberated as

$$P_{\text{out}} = V \cdot I \tag{2}$$

The electrical efficiency of the SPV can be projected as

$$\eta_{\text{elec}} = P_{\text{out}} / P_{\text{in}} \tag{3}$$

where '$S_{\text{in}}$' is the instantaneous solar insolation in watt per sq. meter, '$A_{\text{p}}$' is the effective area of SPV in sq. meter, '$V$' refers SPV voltage in volt, and '$I$' refers SPV current in ampere.

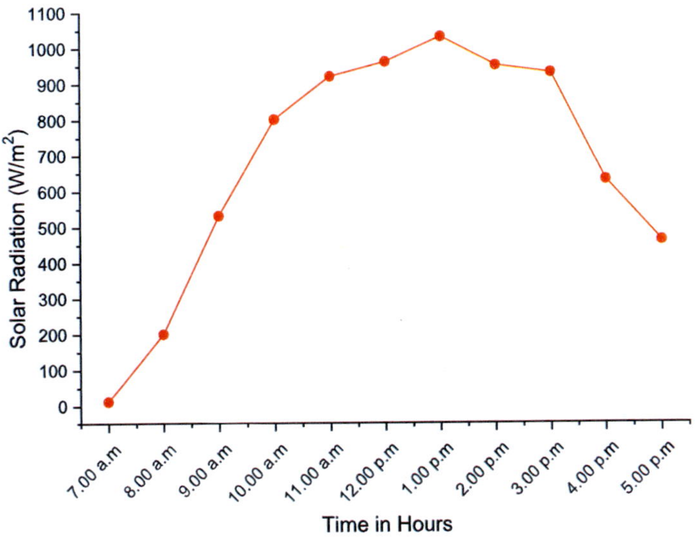


**Fig. 4** Deviation of solar radiation during the analysis

## 4 Result and Discussion

The analysis was carried out between 7 a.m. and 5 p.m. during November 2018. The panel surface temperature, atmospheric temperature, multi-meter readings and solar radiation data were observed in the interval of every one hour. Then, the collected data were carefully analyzed to determine the electrical efficiencies of both panels in the similar working conditions. The interpreted results are further discussed in subsequent divisions.

### *4.1 Temperature Trends in SPV Panels*

Figure 5 shows the variation in SPV surface temperatures and the atmospheric temperature at different hours in the day of investigation. It could be perceived that external temperature of the panels was almost equal to the atmospheric temperature in the morning. Then, their temperature increased with the received solar radiation and reached the maximum value between 1 p.m. and 2 p.m. It is predicted that the surface temperature of SPV1 (without NCPCM) increased up to 71 °C by 2 p.m. which is 24 °C higher than SPV2 (with NCPCM). Hence, it can be evidenced that the integration of NCPCM substantially depressed the panel surface temperature by absorbing a great quantity of heat energy.

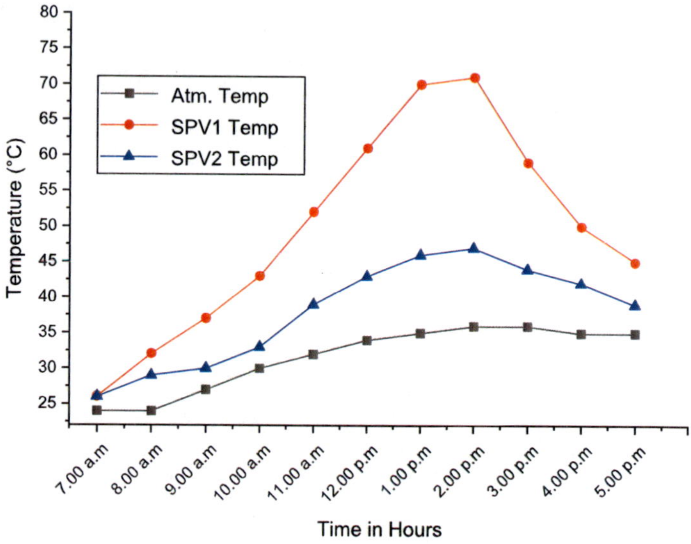


**Fig. 5** Trends of atmospheric and panel surface temperature for SPV1 and SPV2

## 4.2 *Efficiency Trend of the SPV Panels*

The efficiency of SPV1 and SPV2 panels were deliberated with the help of Eqs. (1), (2) and (3), and their values are plotted as shown in Fig. 6. It is conceived that the efficiency of both the panels was similar at the beginning of the experiment.

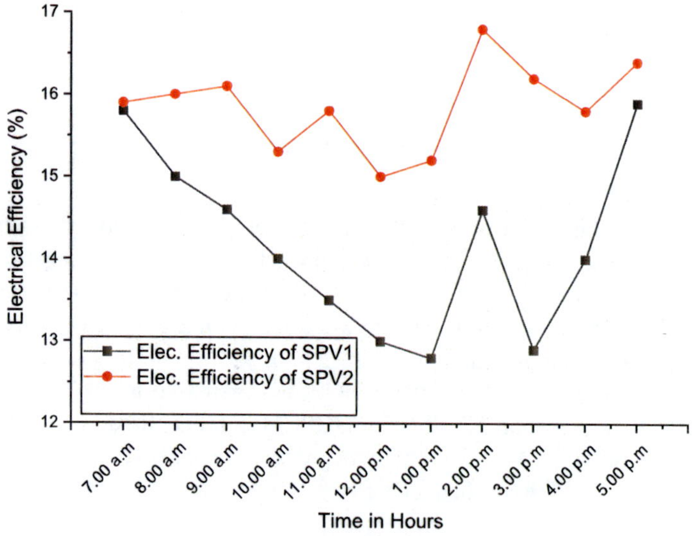


**Fig. 6** Electrical efficiency trends for SPV1 and SPV2 panels

Nevertheless, the efficiency of SPV1 (without NCPCM) declined drastically with time as the temperature of panel was increased continuously. Also, the variation in efficiency of the panel was non-uniform and unpredictable due to the increased surface temperature. The maximum efficiency recorded in SPV1 was 15.9% in the evening, whereas the efficiency of SPV2 was not substantially changed and maintained within the short range between 15 and 16% on an average. The maximum efficiency observed with SPV2 was 16.8% which is 1.2% more compared to the efficiency of SPV1 measuring at the same time and 0.9% more compared to the peak efficiency value of SPV1. These results proved that the incorporation of NCPCM improved thermal regulation of the SPV which in turn ensured the increased the life of SPV and enhanced the electrical efficiency.

## 5 Conclusion

Improvement in performance of the Solar PV panel had been analyzed through the means of thermal regulation of the panel. In this regard, a nanocomposite phase change material (NCPCM) with the improved properties was synthesized by disbanding nano-silica particles within paraffin matrix. Further, its thermo-physical properties were characterized with DSC and thermal properties analyzer. Then, two SPVs with same configuration were commercially purchased for the experimentation. One SPV (SPV1) was tested without any modification, and the second SPV (SPV2) was tested simultaneously, after integrating it with NCPCM. The inferences are summarized below:

1. DSC images proved that incorporating silica ($SiO_2$) nanoparticles considerably reduced the supercooling of the paraffin wax without much affecting its latent heat capacity.
2. NCPCM's thermal conductivity was enriched by 22.22% with the presence of $SiO_2$ nanoparticles.
3. The integration of NCPCM maintained the panel surface temperature within the short range and reduced the peak surface temperature of the SPV by 24 °C. Further, it improved the peak electrical efficiency by 5.66%, comparatively.

Hence, it is recommended to use NCPCM for maintaining the surface temperature of SPV at the lowest possible level and to improve the performance of the SPV.

## References

1. Lenzen M (1999) Greenhouse gas analysis of solar-thermal electricity generation. Sol Energy 65(6):353–368
2. Zhou G, Chung W, Zhang Y (2014) Carbon dioxide emissions and energy efficiency analysis of China's regional thermal electricity generation. J Clean Prod 83:173–184

3. Demirbas A (2006) Electricity generation via unconventional methods. Energ Explor Exploit 24(1–2):131–138
4. Zhao Y, Tang KK, Wang LL (2013) Do renewable electricity policies promote renewable electricity generation? Evidence from panel data. Energ Policy 62:887–897
5. Adam AD, Apaydin G (2016) Grid connected solar photovoltaic system as a tool for greenhouse gas emission reduction in Turkey. Renew Sust Energ Rev 53:1086–1091
6. Kumar PM, Mylsamy K, Prakash KB, Nithish M, Anandkumar R (2020) Investigating thermal properties of Nanoparticle Dispersed Paraffin (NDP) as phase change material for thermal energy storage. Mater Today-Proc. https://doi.org/10.1016/j.matpr.2020.02.800
7. Lacerda JS, van den Bergh JC (2016) Diversity in solar photovoltaic energy: Implications for innovation and policy. Renew Sust Energ Rev 54:331–340
8. Zhou S, Wang Y, Zhou Y, Clarke LE, Edmonds JA (2018) Roles of wind and solar energy in China's power sector: implications of intermittency constraints. Appl Energ 213:22–30
9. Kumar PM, Anandkumar R, Sudarvizhi D, Prakash KB, Mylsamy K (2019) Experimental investigations on thermal management and performance improvement of solar PV panel using a phase change material. AIP Conf Proc AIP Publishing 2128(1):020023
10. Du D, Darkwa J, Kokogiannakis G (2013) Thermal management systems for photovoltaics (PV) installations: a critical review. Sol Energy 97:238–254
11. Chauhan A, Tyagi VV, Anand S (2018) Futuristic approach for thermal management in solar PV/thermal systems with possible applications. Energ Convers Manage 163:314–354
12. Hasan A, McCormack SJ, Huang MJ, Norton B (2010) Evaluation of phase change materials for thermal regulation enhancement of building integrated photovoltaics. Sol Energy 84(9):1601–1612
13. Tripanagnostopoulos Y, Nousia TH, Souliotis M, Yianoulis P (2002) Hybrid photovoltaic/thermal solar systems. Sol Energy 72(3):217–234
14. Fraisse G, Ménézo C, Johannes K (2007) Energy performance of water hybrid PV/T collectors applied to combisystems of Direct Solar Floor type. Sol Energy 81(11):1426–1438
15. Krauter S (2004) Increased electrical yield via water flow over the front of photovoltaic panels. Sol Energ Mat Sol C 82(1–2):131–137
16. Xiang B, Cao X, Yuan Y, Sun L, Wu H, Haghighat F (2018) A novel hybrid energy system combined with solar-road and soil-regenerator: dynamic model and operational performance. Energ Convers Manage 156:376–387
17. Bahaidarah HM (2016) Experimental performance evaluation and modeling of jet impingement cooling for thermal management of photovoltaics. Sol Energy 135:605–617
18. Preet S (2018) Water and phase change material based photovoltaic thermal management systems: a review. Renew Sust Energ Rev 82:791–807
19. Manikandan S, Selvam C, Poddar N, Pranjyal K, Lamba R, Kaushik SC (2019) Thermal management of low concentrated photovoltaic module with phase change material. J Clean Prod 219:359–367
20. Manoj Kumar P, Mylsamy K, Saravanakumar PT (2019) Experimental investigations on thermal properties of nano-$SiO_2$/paraffin phase change material (PCM) for solar thermal energy storage applications. Energ Source Part A, 1–14
21. Huang MJ, Eames PC, Norton B, Hewitt NJ (2011) Natural convection in an internally finned phase change material heat sink for the thermal management of photovoltaics. Sol Energ Mat Sol C 95(7):1598–1603
22. Kumar PM, Saravanakumar PT, Mylsamy K, Kishore P, Prakash KB (2019) Study on thermal conductivity of the candle making wax (CMW) using nano-$TiO_2$ particles for thermal energy storage applications. AIP Conf Proc AIP Publishing 2128(1):020027
23. Lin SC, Al-Kayiem HH (2016) Evaluation of copper nanoparticles–Paraffin wax compositions for solar thermal energy storage. Sol Energ 132:267–278
24. Mohamed NH, Soliman FS, El Maghraby H, Moustfa YM (2017) Thermal conductivity enhancement of treated petroleum waxes, as phase change material, by α nano alumina: energy storage. Renew Sust Energ Rev 70:1052–1058

25. Sardarabadi M, Passandideh-Fard M, Heris SZ (2014) Experimental investigation of the effects of silica/water nanofluid on PV/T (photovoltaic thermal units). Energy 66:264–272
26. Arshad A, Jabbal M, Yan Y (2020) Preparation and characteristics evaluation of mono and hybrid nano-enhanced phase change materials (NePCMs) for thermal management of microelectronics. Energ Convers Manage 205:112444
27. Kumar PM, Anandkumar R, Mylsamy K, Prakash KB (2020) Experimental investigation on thermal conductivity of nanoparticle dispersed paraffin (NDP). Mater Today-Proc. https://doi.org/10.1016/j.matpr.2020.02.798
28. Kumar PM, Mylsamy K (2019) Experimental investigation of solar water heater integrated with a nanocomposite phase change material. J Therm Anal Calorim 136(1):121–132

# Modal Analysis of Pipe Line Under Fluid-Structure Interaction by Simulation and Experiment

**A. Tamil Chandran, T. Suthakar, K. R. Balasubramanian, S. Rammohan, and Jacob Chandapillai**

**Abstract** In this paper, modal analysis of a test loop filled with and without water was studied experimentally and numerically. For experimental modal analysis, a section of a test loop is considered and the considered piping system is divided in to fourteen elements with fifteen nodal points. At each nodal point, response was measured in two mutually perpendicular directions using vibration sensors. For excitation, Impact hammer with a force transducer is used. Hammer input force and response from the accelerometers are taken to an LMS PIMENTO vibration analyzer. Frequency Response Function between the response and force are plotted and further analyzed. First four mode shapes in each two mutually perpendicular directions and a radial mode shape are plotted. Amplitude ratio between the pipe with water and without water are studied and compared. For numerical analysis, the commercially available Finite Element Package ANSYS is used. The pipe is modeled using the geometry creation option available in WORKBENCH. For meshing, three dimensional solid elements was used and analyzed using the standard model analysis solver available with FEM package. Fluid loading is simulated using the special element available in ANSYS. Analysis was performed in both conditions. The reduction in frequency, due to fluid loading is studied and observed. Natural frequencies obtained by both

A. Tamil Chandran (✉) · S. Rammohan · J. Chandapillai
Fluid Control Research Institute (FCRI), Palakkad, Kerala 678623, India
e-mail: tamilchandran@fcriindia.com

S. Rammohan
e-mail: s.rammohan@fcriindia.com

J. Chandapillai
e-mail: jacobchandapillai@fcriindia.com

T. Suthakar · K. R. Balasubramanian
Department of Mechanical Engineering, National Institute of Technology Tiruchirapalli, Tiruchirapalli, Tamil Nadu, India
e-mail: suthakar@nitt.edu

K. R. Balasubramanian
e-mail: krbala@nitt.edu

© The Editor(s) (if applicable) and The Author(s), under exclusive license to Springer Nature Singapore Pte Ltd. 2021
S. Mohan et al. (eds.), *Materials, Design, and Manufacturing for Sustainable Environment*, Lecture Notes in Mechanical Engineering,
https://doi.org/10.1007/978-981-15-9809-8_45

FEM and experimental modal analysis (EMA) are studied and compared. Scope for further study is discussed.

**Keyword** Modal analysis · Fluid loading · Mode shapes · Natural frequency · Fluid–structure interaction

## Nomenclature

| | |
|---|---|
| $m$ | mass of element |
| $c$ | damping of element |
| $k$ | stiffness of element |
| $F(t)$ | force |
| $[M]$ | mass matrix |
| $[C]$ | damping matrix |
| $[k]$ | stiffness matrix |
| $\lambda$ | eigen value |
| X | eigenvector |
| $\rho$ | density |
| $\rho_0$ | acoustic density |
| '$p_0$' | atmospheric pressure |
| '$p$' | acoustic pressure |
| '$a$' | acceleration |
| $\ddot{x}(t)$ | acceleration |
| $\dot{x}(t)$ | velocity |
| $x(t)$ | displacement |
| '$\omega$' | natural frequency |
| '$c$' | speed of sound |
| $\gamma$ | heat capacity |
| $P$ | absolute pressure |

## Subscripts

| | |
|---|---|
| $s$ | structure element |
| $e$ | element |
| $f$ | fluid element |
| $fs$ | fluid Solid interaction |

## 1 Introduction

Dynamic properties of a mechanical system, such as natural frequency, damping and mode shapes can be determined using modal analysis. Methods used for modal analysis are mainly classified in to experimental (EMA) and analytical modal analysis. EMA can be used to determine its natural frequency and damping using frequency domain/time domain methods. In EMA, responses are recorded by vibration sensors installed at different locations. In analytical modal analysis, finite element analysis (FEA) is one of the widely used methods to get natural frequency and mode shapes.

When studying about modal analysis of structures coupled with fluid, it is to understand about fluid–structure interaction mechanisms and are friction coupling, Poisson coupling and junction coupling [1]. In fluid–structure interaction, mass of the fluid influenced is significantly in the system natural frequency. Viscosity and density plays a major role in its damping. It was well documented in a series of successive review articles of Wiggert [2, 3], Tijsseling [12], and Li [4].

This paper concentrates analytical modal analysis and experimental validation of a pipe filled with and without water. Analytical work includes the modeling, establishing fluid structural coupling and analysis to extract frequencies in different mode shapes.

## 2 Literature Review

In prediction of pressure wave propagation in fluids, pressure and velocity are the unknown parameters. Considering this, during 1800 to 1900, momentum and continuity equations were used for study the propagation of pressure waves in pipe without considering the pipe wall. Korteweg and De Vries [5] has studied about the wave motion in a rectangular canal. Based on Korteweg's development in sound wave, an equation for flexible tube filled with water and submerged in water was derived by Lamb [6] for its first mode. He considered the pipe as an elastic membrane by including the effect of longitudinal stresses in the pipe wall. Gronwall [7] has studied about longitudinal vibrations of partially filled an elastic tube with liquid.

Jacobi [8] has studied about propagation of axis symmetric waves in a fluid. He has demonstrated the importance of including elastic effect of structure in fluid–structure interaction analysis. Skalak [1] has studied by considering the pipe as an elastic membrane by include the inertia and axial stress. Wiggert [2], extended the study of Skalak with analytical and experimental study for a elbow with different boundary conditions. Williams [9] studied the effect of longitudinal elastic strain of pipe walls and found it altering the effects of the system. Tijsseling [10] solved the four-equation model, developed by Skalak, for time-dependent boundary condition. Walker [11] has developed a six-equation model by including equations for radial inertia forces. Tijsseling [12] derived one-dimensional four equations model for thick walled pipe to validate thin wall assumption for thick pipe also. By using four-equation model, Lee [13] has developed a spectral element model and studies were conducted on a pipe line with simple support. Further, it was modified by You [14] including pipe material elastic anisotropy. With this he did an experiment and was able to accurately predict the valve closure time of a system, than the four-equation model. From the literature review conducted, it is observed that for straight pipe filled with incompressible fluids, all the four-equation model, six-equation model and eight-equation model are accurate and efficient.

# 3 Theory–Numerical Analysis

## 3.1 Structural Element Formulations

When a system is subjected to external disturbance, it starts to vibrate and by considering the mechanical properties of the system, equation of motion for a multi-degree of freedom system in matrix form is

$$[M_s]\ddot{x}(t) + [C_s]\dot{x}(t) + [k_s]x(t) = F(t) \tag{1}$$

By considering the effect of external force is negligible, i.e., $F(t) = 0$ and neglecting damping ratio, in Eq. (1) and it can be written as,

$$[M_s]\ddot{x}(t) + [k_s]x(t) = 0 \tag{2}$$

When the system is subjected to a simple harmonic motion of $x(t) = x_0\sin(\omega t)$ and $\ddot{x}(t) = -x_0\omega^2 \sin(\omega t)$, Eq. (2) can be rewritten as

$$-[M_s]\omega^2 + [k_s] = 0 \tag{3}$$

or

$$\omega^2 = [M_s]^{-1}[k_s] = \lambda \tag{4}$$

From Eq. (4), the eigenvector can be calculated as

$$[A - I\lambda]X = 0$$

where $A = [M_s]^{-1}[k_s]$ and $I$ is diagnal unit matrix,
$X$—is known as eigenvector or mode shape.

## 3.2 Fluid Element Formulations

Finite element formulations for fluid domain need to be considered as an acoustic medium. For acoustic medium, the wave equation can be derived from the continuity and momentum equation. For simplification purpose, linearized form of the equations is used.

Continuity and momentum equation for a fluid having dimensions of d$y$ and d$z$, the linear form is as follows.

Continuity equation is

$$\frac{\partial s}{\partial t} + \nabla \cdot u = 0 \tag{5}$$

where

$$s = \frac{\rho - \rho_0}{\rho_0} \tag{6}$$

The momentum equation for inviscid and incompressible fluid is

$$\rho \frac{\mathrm{d}u}{\mathrm{d}t} + \nabla p = 0 \tag{7}$$

Substituting the acoustic density for the total density and $S \ll 1$, we get

$$\rho_0 \frac{\partial u}{\partial t} = -\nabla p \tag{8}$$

### 3.2.1 Acoustic Equation

Acoustic equation for the element can be formed from the continuity and momentum equation.

The partial derivative of the continuity equation is multiplied with the acoustic density to give

$$\nabla \rho_0 \frac{\partial u}{\partial t} = -\rho_0 \frac{\partial^2 s}{\partial t^2} \tag{9}$$

Substituting the momentum equation in Eq. (9) will give

$$\nabla^2 p = \rho_0 \frac{\partial^2 s}{\partial t^2} \tag{10}$$

Bulk modulus of a fluid

$$B = \frac{P - P_0}{(\mathrm{d}V/V)} = \frac{P - P_0}{((\rho - \rho_0)/\rho_0)} \tag{11}$$

where acoustic pressure $p = P - P_0$.

Substituting Eq. (6) in Eq. (11), we get

$$B = \frac{P - P_0}{s} \tag{12}$$

for an adiabatic process in gases, $PV$ = constant and acoustic compressions are adiabatic,

The equation of state can be written $\frac{P}{P_0} = \left(\frac{\rho}{\rho_0}\right)^{\gamma}$

From the above, bulk modulus

$$B = \gamma P \tag{13}$$

From Eqs. (12) and (13), we get

$$s = \frac{p}{p_0 \gamma} \tag{14}$$

When considering Eq. (11), Eq. (8) becomes,

$$\nabla^2 p = \frac{1}{C^2} \frac{\partial^2 p}{\partial t^2} \tag{15}$$

where $C = \sqrt{\frac{p_0 \gamma}{\rho_0}}$,

Equation (15) is the acoustic wave equation, described for water pressure, which is known as Helmholtz's equation [16].

Matrix form of Helmholtz's equation is

$$\{L\}^T (\{L\} P) = \frac{1}{C^2} \frac{\partial^2 p}{\partial t^2} \tag{16}$$

where $\{L\}$ is matrix operator.

### 3.3 *Fluid–Structure Interaction Formulation*

After discretizing Eq. (16) using Galerkin procedure to obtain the element matrix with fluid–structure interaction as given by Liang (2007), the following equation is obtained.

$$\left[M_f\right]\left\{\ddot{P}\right\}+\left[C_f\right]\left\{\dot{P}\right\}+\left[K_f\right]\{P\}=\left\{F_{sf}\right\} \tag{17}$$

where $\left[M_f\right]$, $\left[C_f\right]$, $\left[K_f\right]$ are equivalent fluid matrices and $\left\{F_{sf}\right\}$ is the fluid force produced by fluid–structure interaction.

where

$$\left[M_f\right]=\frac{1}{c^2}\int\limits_V \{N_P\}\{N_P\}^T \mathrm{d}V$$

$$\left[C_f\right]=\frac{\beta}{c}\int\limits_S \{N_P\}\{N_P\}^T \mathrm{d}S$$

$$\left[K_f\right]=\int\limits_V [B]^T[B]\mathrm{d}V$$

$$\left\{F_{sf}\right\}=-\rho_0\int\limits_S \left\{N_p\right\}\{n\}^T\{N_u\}^T \mathrm{d}(S)\{\ddot{x}\} \tag{18}$$

$$\left\{F_{sf}\right\}=-\rho_0[R]^T\{\ddot{x}\}$$

where $[R]=\int\limits_S \{N_u\}\left\{N_p\right\}^T\{n\}\mathrm{d}(S)$

Likewise, the fluid pressure load acting on the structure can be obtained by integrating the pressure over the area of surface.

Fluid pressure load vector at the interface $\left\{F_{fs}\right\}=\int\limits_S\{N_u\}P\{n\}d(S)$

$$\left\{F_{fs}\right\}=\int\limits_S \{N_u\}\left\{N_p\right\}^T\{n\}\mathrm{d}(S)\{p\}=[R]\{p\} \tag{19}$$

Substituting fluid force component on the structural system Eq. (1), we get

$$[M_s]\ddot{x}+[C_s]\dot{x}+[k_s]x=F_s+F_{fs} \tag{20}$$

The combined equation can be get by assembly Eq. (20) and (17) as

$$\begin{bmatrix} [M_s] & [0] \\ \left[M_{fs}\right] & \left[M_f\right] \end{bmatrix}\begin{Bmatrix} \{\ddot{x}\} \\ \{\ddot{p}\} \end{Bmatrix}+\begin{bmatrix} [C_s] & [0] \\ [0] & \left[C_f\right] \end{bmatrix}\begin{Bmatrix} \{\dot{x}\} \\ \{\dot{p}\} \end{Bmatrix}+\begin{bmatrix} [k_s] & \left[k_{fs}\right] \\ [0] & \left[k_f\right] \end{bmatrix}\begin{Bmatrix} \{x\} \\ \{p\} \end{Bmatrix}=\begin{Bmatrix} \{F_s\} \\ \{0\} \end{Bmatrix} \tag{21}$$

where the equivalent coupling mass and coupling matrix are $\left[M_{fs}\right]=\rho_0[R]^T$, $\left[k_{fs}\right]=-[R]$, respectively.

## 4 Problem Formulation

A stainless steel pipe having a length of ($L$) 5800 mm, internal diameter ($d_i$) = 150 mm, and external diameter ($d_o$) = 164 mm is used for analysis. The part of the loop used for study is having three piping support, and it is rigidly clamped in the ground and connected with the pipe using 'U' clamps. The system was analyzed and studied for the natural frequency and mode shape with two test conditions, such as with and without water being filled.

## 5 Finite Element Formulation

Finite element analysis was performed by using computer code ANSYS. The piping was modeled as a three-dimensional solid element with constant thickness. The material of the piping is assumed to be homogeneous, isotropic, and linearly elastic. The pipe wall is modeled by the three-dimensional 20-nodded solid element SOLID186 from the ANSYS element library. The geometry of SOLID186 [15] is as given in Fig. 1.

ANSYS fluid element HSFLD242 is adopted to impose the fluid load in the pipe, and the construction of the element is shown in Fig. 2.

The FE model generated using the ANSYS workbench is shown in Fig. 3.

The material properties of the structural elements considered are as follows

Young's modulus $E = 18e + 10$ Pa,
Poisson's ratio $\upsilon = 0.3$,
Mass density $\rho_s = 7850$ (kg/m$^3$).

**Fig. 1** Three dimensional solid element

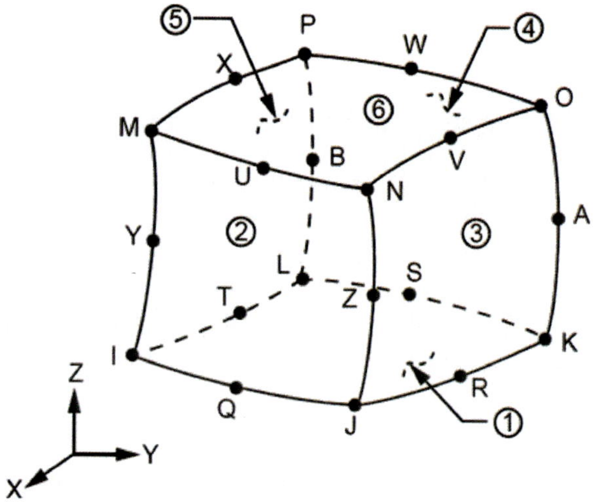

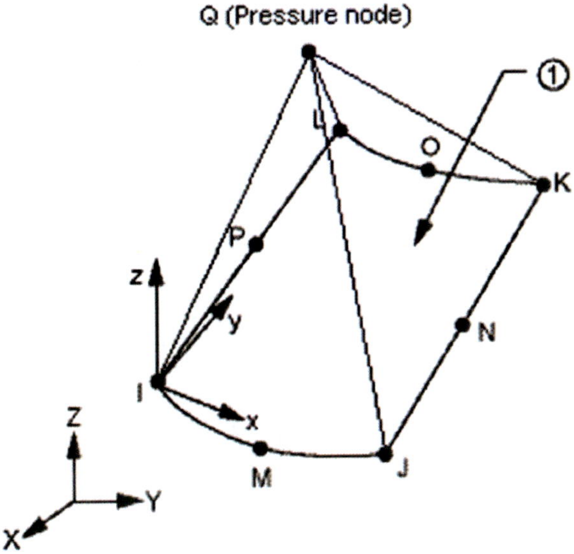


**Fig. 2** Fluid loading element

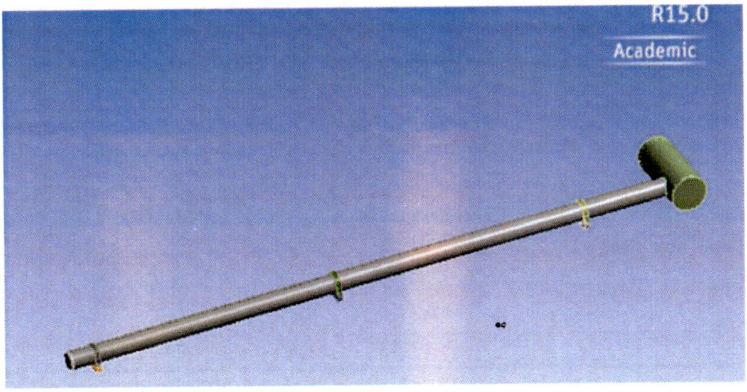


**Fig. 3** Finite element model

Water, as the contained fluid, has a density of $\rho_f = 997$ (kg/m$^3$). The sound speed in water is 1486 m/s, which is equivalent to a bulk modulus of elasticity of 2.2 GPa.

For analysis, Block Lanczos solver is recommended by ANSYS for general application, and hence, this method is used for the present work. The eigenvalues and eigenvectors are extracted for both the conditions.

## 6 Experimental Setup

The piping system used for this study is one of the test loops available at a research laboratory and is as shown in Figs. 4 and 5. It has the capability to fill the test loop with water or empty for other test condition. In this study, the instrumentation used for conducting experimental modal analysis consists of a vibration analyzer, two accelerometers and an impact hammer. The LMS makes PIMENTO vibration analyzer having the capability to perform EMA is used. It can take both force input and response of the accelerometers. It is having the basic capabilities for impact testing.

Pre-trigger delay ensures that the entire signal is captured for analysis. The windowing functions are applied to the signals after they are sampled, but before the FFT is applied to them in the analyzer. The force window is used to remove noise from the impulse force signal. The exponential window is applied to the impulse response signal to reduce leakage. In EMA, spectrum averaging was employed for frequency response function (FRF). 3 to 5 impacts were employed per measurement for FRF. The accelerometers used for the measurement are having flat frequency response curve within ±3% deviation from the nominal sensitivity for the frequency range of 2 Hz to 10 kHz. The impact hammer used can excite the system with flat amplitude up to 2 kHz. As discussed in the literature review, a number of modal analysis techniques are used. In this study, single input multiple output (SIMO) roving hammer roving accelerometer method was used.

Since accelerometers are used for vibration measurement, the acceleration signal is directly processed without converting it into velocity or displacement. From the measured signals, FRF between acceleration and force is calculated and is known as accelerance

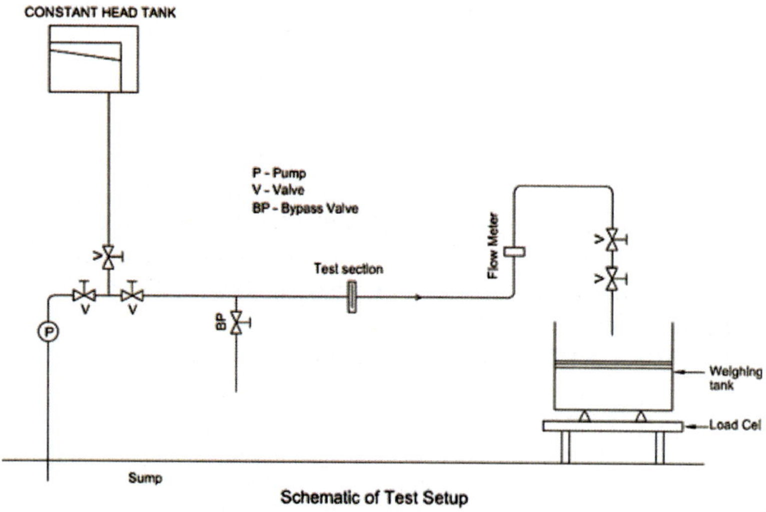


**Fig. 4** Schematic of test setup

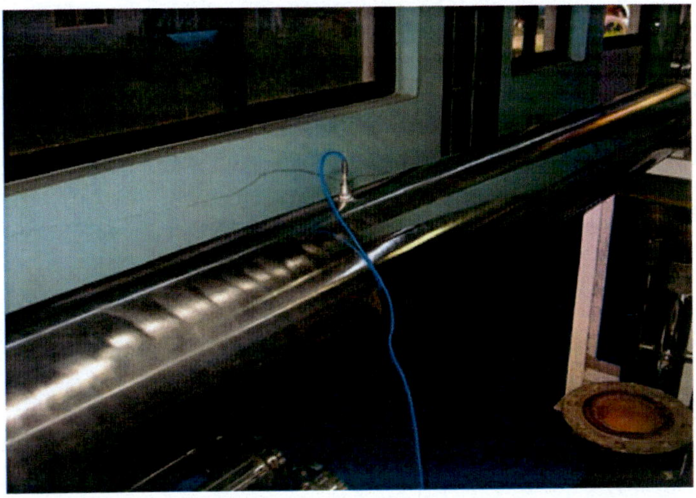

**Fig. 5** Test loop

$$\text{Accelerance} = \frac{\text{Response of accelerometer}}{\text{Force input of hammer}}$$

In the FRF of accelerance, the amplitude graph is having a peak amplitude corresponding to the FFT of the response spectrum. This confirms the natural frequency obtained in the response spectrum.

### 6.1 Uncertainty in Measurement

Major uncertainty contributions in frequency measurement by this method are from vibration sensor, force hammer used for excitation purpose, and vibration analyzer. The uncertainty in the frequency measurement was estimated as 4.2% with a confidence level of 95%.

## 7 Result

### 7.1 Finite Element Analysis

Finite element analysis was done for two conditions, i.e., one for pipe filled with fluid and the other for pipe without fluid. For both the cases, the piping is modeled using the ANSYS Workbench and is discretized using three-dimensional solid elements. It is discretized using 45,511 Elements having 155,386 nodes. The fluid element was modeled using fluid element HSFLD242. First four natural frequencies in vertical and horizontal directions and one in radial direction were extracted, and mode shapes are obtained. The obtained natural frequency in both conditions is as given in Table 1. Typical mode shape obtained in the analysis for its radial mode is as given in Figs. 6 and 7

### 7.2 *Experimental Modal Analysis*

Experimental modal analysis was also done for two conditions: one for pipe filled with fluid and other for pipe without fluid. For both the cases, the piping system is discretized in to 14 elements, and measurement was done in vertical and horizontal direction of the pipe. During the excitation, measurement at two directions (excited direction and perpendicular to excitation) was done simultaneously in the excited location. Cross-correlation between the excitation force and response in two mutually perpendicular locations at each location are studied to understand the influence of excitation direction in response.

The natural frequencies obtained in the direction of excitation, in both conditions (with and without water filling), are as given in Table 2.

Real part of FRF in all the locations for the direction of excitation is as given in Figs. 8, 9, 10, 11, 12, 13, 14 and 15.

First four mode shapes in both vertical and horizontal directions and one radial mode is as shown in Figs. 16 and 17.

## 8 Analysis and Comparison

As discussed in literature, due to the fluid loading effect in the piping system, reduction in frequency is always observed in both experiment and numerical analyses. The percentage of reduction in frequency was studied in both by numerical analysis and experiment. The numerical/FEA results of the study are as given in Table 3. Experimental results indicating the reduction in frequency due to fluid loading is as given in Table 4. Comparison between the experimental and numerical analysis, without fluid loading is as given in Table 5 and with fluid loading is as given in Table 6.

From the results, we see that, in FEA an average of 14.9% of reduction in frequency was observed in horizontal direction with a standard deviation of 5.6, and for vertical direction, it is 19.9% and 1.1, respectively. From the EMA, an average of 14.6% of reduction in frequency was observed in horizontal direction with standard deviation of 4.8, and for vertical direction, it is 22.4% and 4.4, respectively. Both the methods indicate that, reduction in frequency is more in vertical direction than for horizontal direction with lower standard deviation.

Vibration amplitude measured in the direction of excitation is normally more when comparing to that of perpendicular direction. But in radial direction, irrespective of

**Table 1** Natural frequency by finite element analysis

| Direction | | Horizontal | | | | Vertical | | | | Radial Mode |
|---|---|---|---|---|---|---|---|---|---|---|
| Mode number | | 1 | 2 | 3 | 4 | 1 | 2 | 3 | 4 | |
| | | Natural Frequency in Hz | | | | | | | | |
| Condition | Without water filled | 29.2 | 45.0 | 71.6 | 151.3 | 78.5 | 97.7 | 188.1 | 270.3 | 534.9 |
| | With water filled | 23.0 | 41.4 | 59.5 | 131.2 | 62.8 | 77.1 | 153.4 | 216.5 | 430.6 |

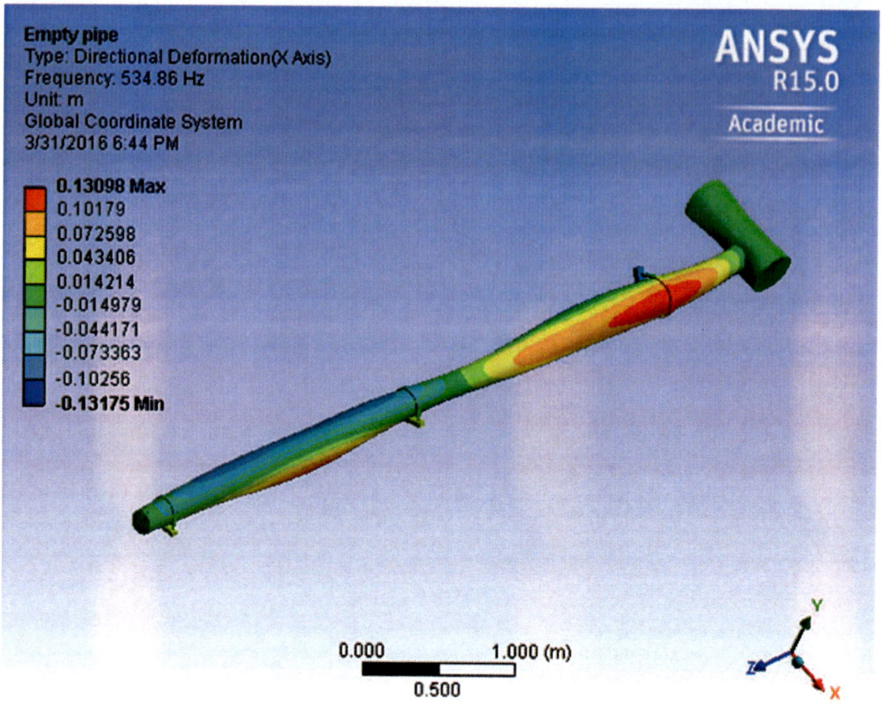


**Fig. 6** Radial mode, Direction: Horizontal

direction of excitation, at most same amplitude, is observed in vertical and horizontal directions.

From the experimental results, it is observed that the amplitude ratio in horizontal direction is reduced in water-filled condition. In vertical direction, reduction in amplitude is observed except third mode. In radial mode, significant reduction in amplitude is observed in water-filled condition.

## 9 Conclusion

In this paper, modal analysis was done, using finite element analysis and by experiment. The effects of fluid loading in the piping system are analyzed. In water-filled condition, reduction in frequency is observed in all modes. From the experiments and numerical analysis conducted, it is observed that the effect of fluid loading is higher in vertical direction than that of horizontal direction. It is also observed that the amplitude ratio in horizontal direction is reduced in water-filled condition. In vertical direction, reduction in amplitude is observed except in third mode. Compared to horizontal modes, the amplitude ratio is higher in vertical modes. In the radial mode,

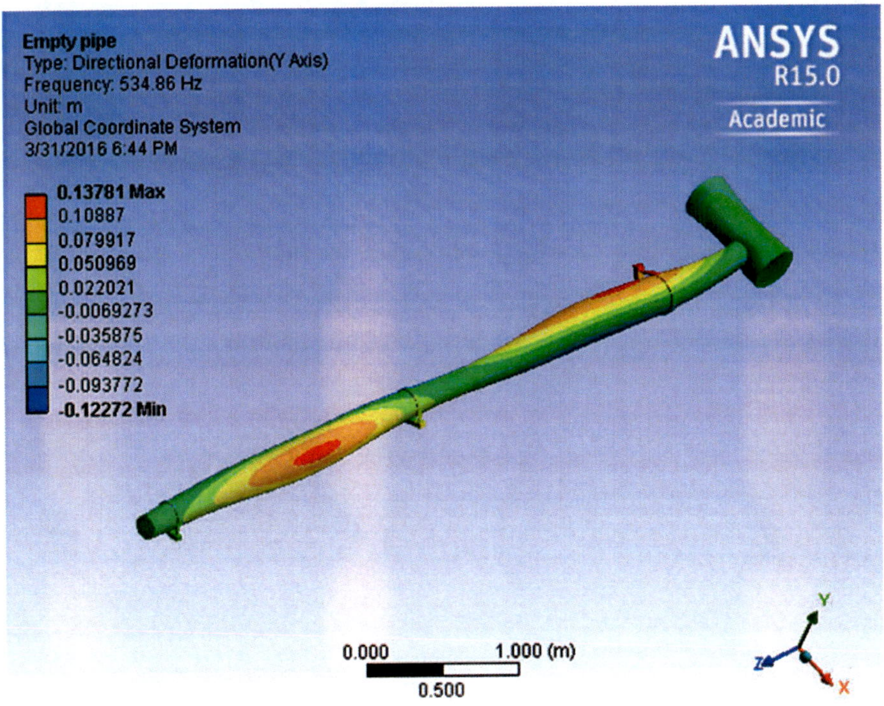


**Fig. 7** Radial mode, Direction: Vertical

significant reduction in amplitude is observed in water-filled condition. Irrespective of direction of excitation, there is no significant change in amplitude in radial mode. The radial mode, the amplitude ratio is significantly higher than that of other two directions. In radial mode, the FRF amplitude is approximately 80 times of other vertical modes and 160 times of other horizontal modes in pipe without fluid-filled condition. With water-filled condition, it is approximately 40 times of other vertical mode and 80 times of other horizontal modes.

## 10 Scope for Further Work

Irrespective of the direction of excitation, no significant change in amplitude is observed in radial mode. When the piping is filled with water, approximately fifty percentage reductions in amplitude are observed. It leads to further study in the area to operational modal analysis to identify the effect of excitation force in radial amplitude. The effect of fluid loading in amplitude reduction also can be studied in higher radial modes.

**Table 2** Natural frequency by experimental modal analysis

| Direction | | Horizontal | | | | Vertical | | | | Radial Mode |
|---|---|---|---|---|---|---|---|---|---|---|
| Mode | | 1 | 2 | 3 | 4 | 1 | 2 | 3 | 4 | |
| | | Natural Frequency in Hz | | | | | | | | |
| Condition | Without water filled | 27.5 | 49.1 | 71.7 | 146.5 | 82.4 | 99.7 | 193.1 | 252.5 | 573 |
| | With water filled | 22.1 | 45.0 | 61.8 | 122.3 | 64.8 | 75.5 | 140.5 | 210.0 | 458 |

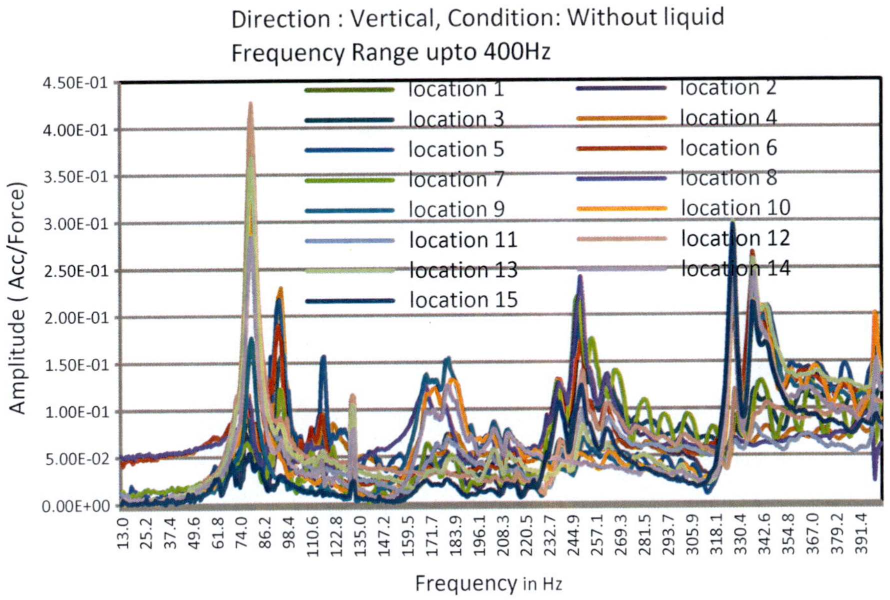


**Fig. 8** Amplitude ratio—frequency range of 10–400 Hz (Vertical direction, without liquid filled)

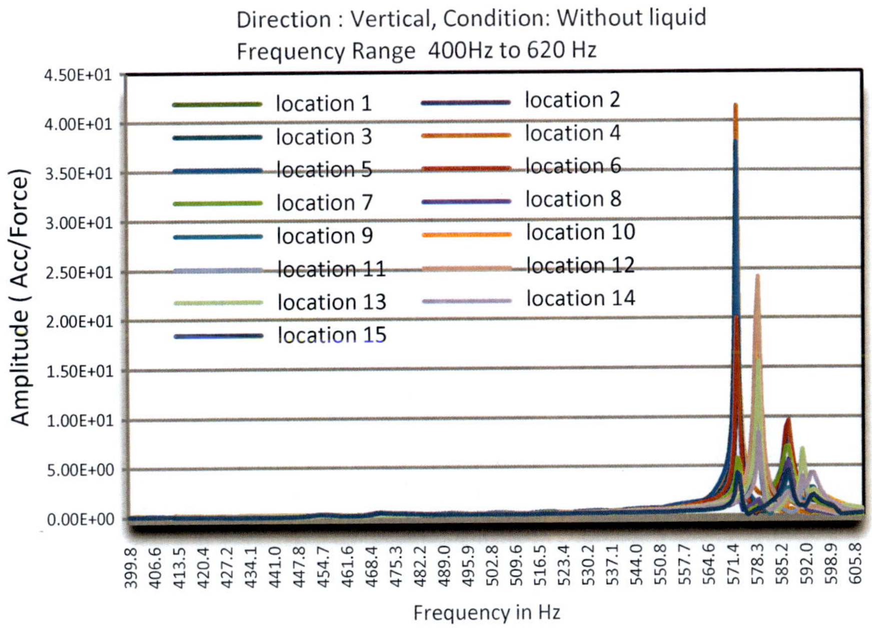


**Fig. 9** Amplitude ratio—frequency range of 400–620 Hz (Vertical direction, without liquid filled)

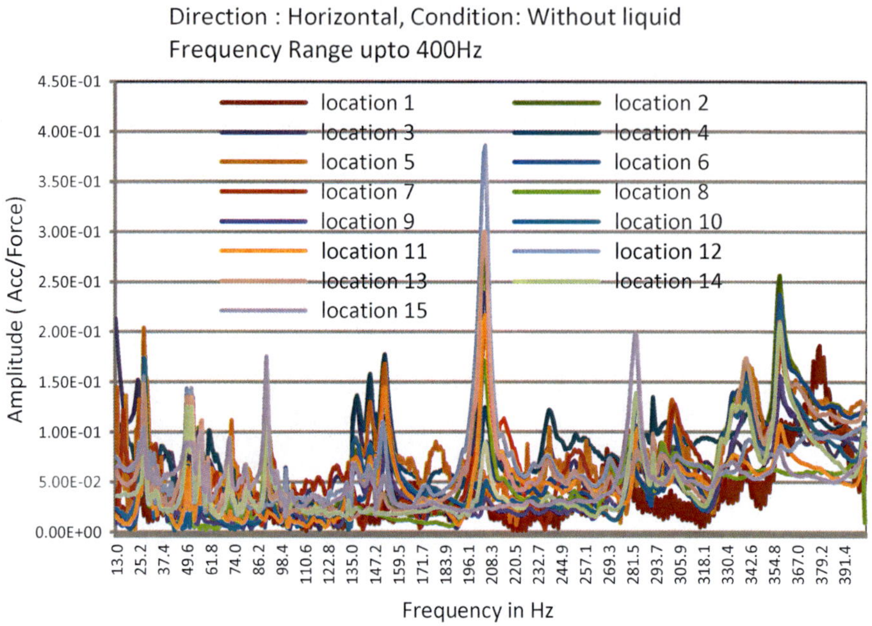


**Fig. 10** Amplitude ratio—frequency range of 10–400 Hz (Horizontal direction, without liquid filled)

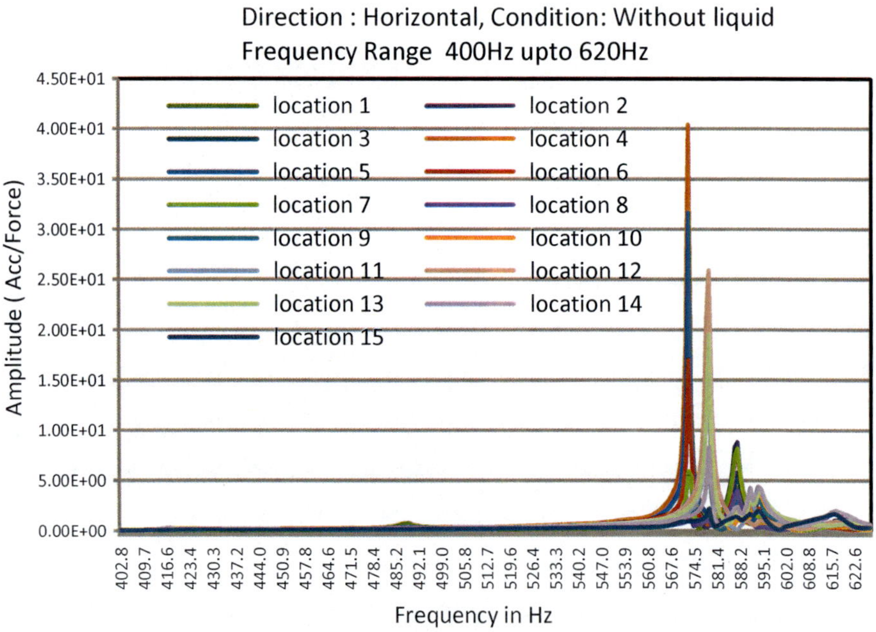


**Fig. 11** Amplitude ratio—frequency range of 400–620 Hz (Horizontal direction, without liquid filled)

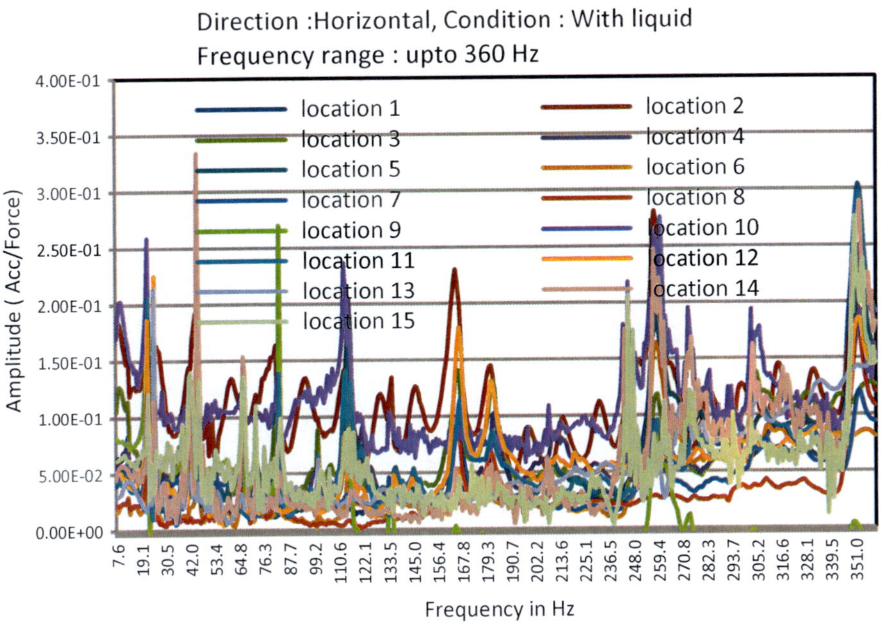


**Fig. 12** Amplitude ratio—frequency range of 10–360 Hz (Horizontal direction, with liquid filled)

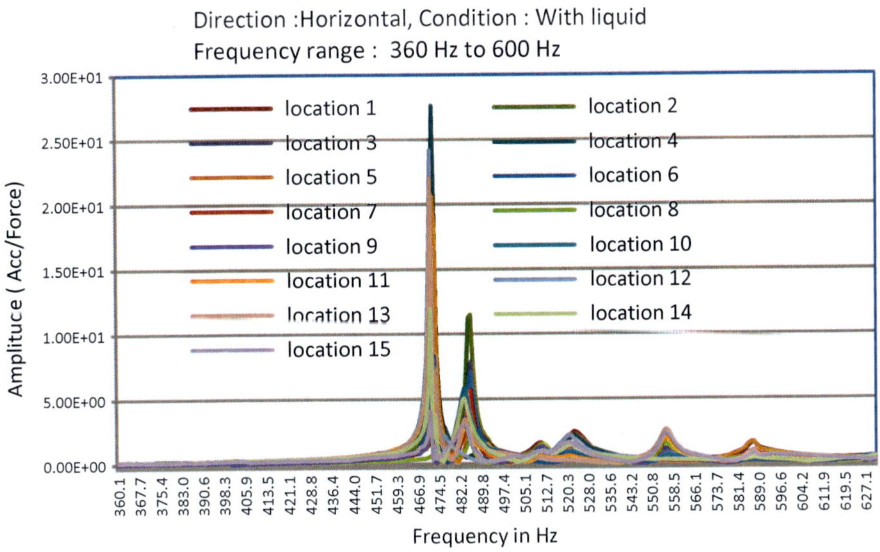


**Fig. 13** Amplitude ratio—frequency range of 360–600 Hz (Horizontal direction, with liquid filled)

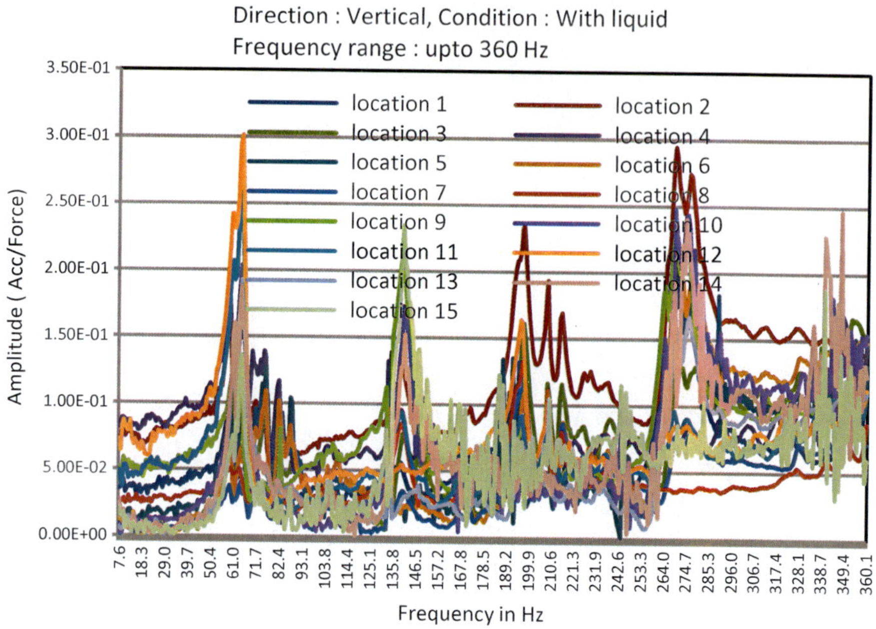


**Fig. 14** Amplitude ratio—frequency range of 10–360 Hz (Vertical direction, with liquid filled)

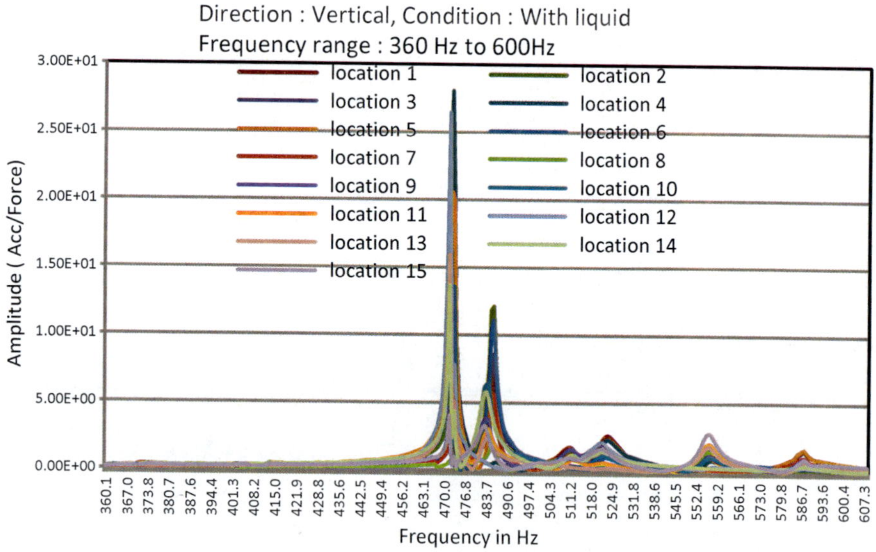


**Fig. 15** Amplitude ratio—frequency range of 360–600 Hz (Vertical direction, with liquid filled)

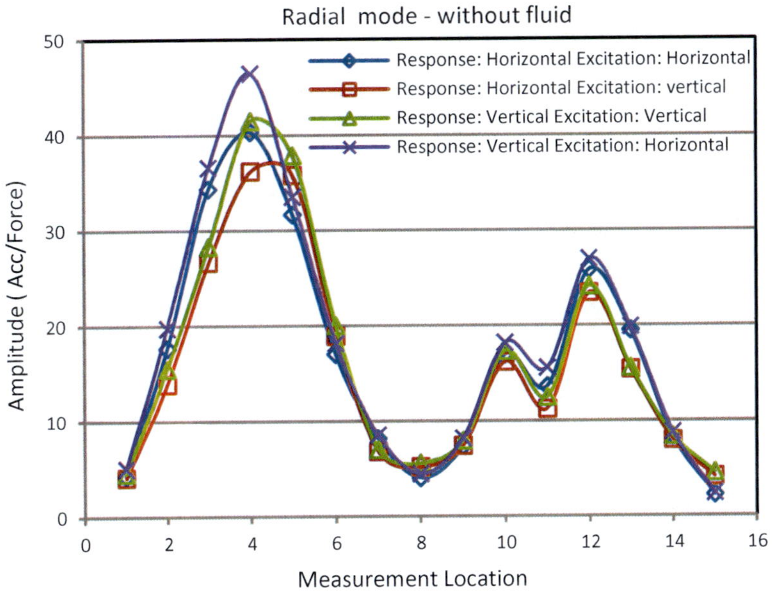


**Fig. 16** Radial mode shape (with out fluid)

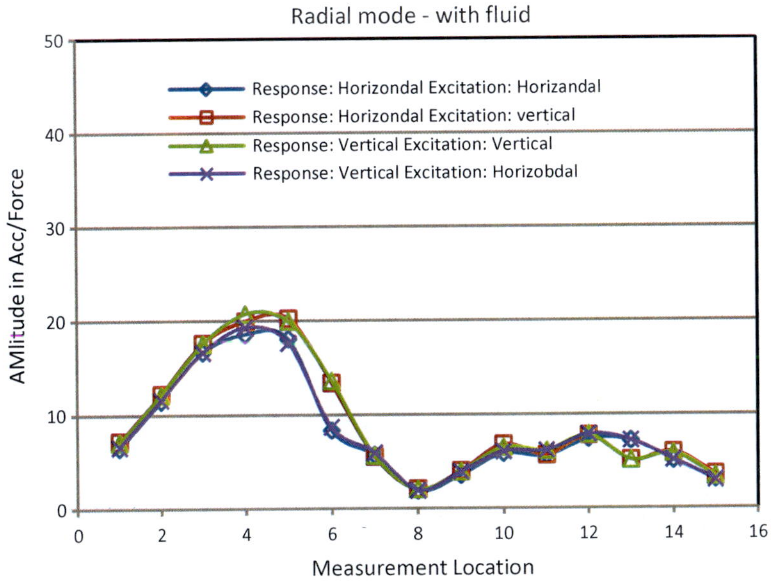


**Fig. 17** Radial mode shape (with fluid)

**Table 3** Finite element analysis results

| Direction | | Horizontal | | | | Vertical | | | | Radial Mode |
|---|---|---|---|---|---|---|---|---|---|---|
| Mode | | 1 | 2 | 3 | 4 | 1 | 2 | 3 | 4 | |
| | | Natural Frequency in Hz | | | | | | | | |
| Condition | Without water filled | 29.2 | 45.0 | 71.6 | 151.3 | 78.5 | 97.7 | 188.1 | 270.3 | 534.9 |
| | With water filled | 23.0 | 41.4 | 59.5 | 131.2 | 62.8 | 77.1 | 153.4 | 216.5 | 430.6 |
| Percentage of reduction in natural frequency | | 21.2 | 8.0 | 16.9 | 13.3 | 20.0 | 21.1 | 18.4 | 19.9 | 19.5 |

**Table 4** Experimental results

| Direction | | Horizontal | | | | Vertical | | | | Radial Mode |
|---|---|---|---|---|---|---|---|---|---|---|
| Mode | | 1 | 2 | 3 | 4 | 1 | 2 | 3 | 4 | |
| | | Natural Frequency in Hz | | | | | | | | |
| Condition | Without water filled | 27.5 | 49.1 | 71.7 | 146.5 | 82.4 | 99.7 | 193.1 | 252.5 | 573 |
| | With water filled | 22.1 | 45 | 61.8 | 122.3 | 64.8 | 75.5 | 140.5 | 210 | 458 |
| Percentage of reduction in natural frequency | | 19.6 | 8.4 | 13.8 | 16.5 | 21.4 | 24.3 | 27.2 | 16.8 | 20.1 |

**Table 5** Comparison between analysis and experimental results (Empty pipe)

| Direction | Horizontal | | | | Vertical | | | | Radial Mode |
|---|---|---|---|---|---|---|---|---|---|
| Mode | 1 | 2 | 3 | 4 | 1 | 2 | 3 | 4 | |
| | Natural Frequency in Hz | | | | | | | | |
| FEM | 29.2 | 45 | 71.6 | 151.3 | 78.5 | 97.7 | 188.1 | 270.3 | 534.9 |
| Experiment | 27.5 | 49.1 | 71.7 | 146.5 | 82.4 | 99.7 | 193.1 | 252.5 | 573 |
| Percentage of variation | 6.2 | 8.4 | 0.1 | 3.3 | 4.7 | 2.0 | 2.6 | 7.0 | 6.6 |

**Table 6** Comparison between analysis and experimental results (With fluid filled)

| Direction | | Horizontal | | | | Vertical | | | | Radial Mode |
|---|---|---|---|---|---|---|---|---|---|---|
| Mode | | 1 | 2 | 3 | 4 | 1 | 2 | 3 | 4 | |
| | | Natural Frequency in Hz | | | | | | | | |
| | FEM | 23 | 41.4 | 59.5 | 131.2 | 62.8 | 77.1 | 153.4 | 216.5 | 430.6 |
| | Experiment | 22.[illegible] | 45 | 61.8 | 122.3 | 64.8 | 75.5 | 140.5 | 210 | 458 |
| Percentage of variation | | 4.1 | 8.0 | 3.7 | 7.3 | 3.1 | 2.1 | 9.2 | 3.1 | 6.0 |

## References

1. Skalak R (1956) An extension of the theory of water hammer. Trans ASME 78:105–116
2. Wiggert DC, Otwell RS, Hatfield FJ (1985) The effect of elbow restraint on pressure transients. J Fluids Eng 107:402–406
3. Wiggert DC, Hatfield FJ, Stuckenbruck S (1987) Analysis of liquid and structural transients in piping by the method of characteristics. J Fluids Eng 109:161–165
4. Li S, Karney BW, Liu G (2015) FSI research in pipe line systems—a review of the literature. J Fluids Struct 57:277–297
5. Korteweg DJ, De Vries G (1895) On the change of form of long waves advancing in a rectangular canal, and on a new type of long stationary waves. Philos Mag Ser 55 39(240):422–443
6. Lamb H (1898) On the velocity of sound in a tube, as affected by the elasticity of the walls. Memoirs of the Manchester Literary and Philosophical Society, Manchester, UK 42(9):1–16
7. Gronwall TH (1927) The longitudinal vibrations of a liquid contained in a tube with elastic walls. Phys Rev 30:71–83
8. Jacobi WJ (1949) Propagation of sound waves along liquid cylinders. J Acoust Soc Am 21(2):120–127
9. Williams DJ (1977) Water hammer in non-rigid pipes: precursor waves and mechanical damping. IMechE J Mech Eng Sci 19:237–242
10. Tijsseling AS (2003) Exact solution of linear hyperbolic four-equation system in axial liquid-pipe vibration. J Fluids Struct 18:179–196
11. Walker JS, Phillips JW (1977) Pulse propagation in fluid-filled tubes. J Appl Mech 44:31–35
12. Tijsseling AS (2007) Water hammer with fluid–structure interaction in thick-walled pipes. Comput Struct 85:844–851
13. Lee U, Oh H (2003) The spectral element model for pipelines conveying internal steady flow. Eng Struct 25:1045–1055
14. You JH, Inaba K (2013) Fluid–structure interaction in water-filled thin pipes of anisotropic composite materials. J Fluids Struct 36:162–173
15. ANSYS Mechanical APDL Element Reference, Release 15.0" (2013), ANSYS, Inc.
16. Som SK, Biswas G, Chakraborty S (2011) Introduction to fluid mechanics and fluid machines, 3rd edn. Tata McGrawHill Education

# Flow Estimation Using Cross-Flow-Induced Vibration

A. Tamil Chandran, T. Suthakar, K. R. Balasubramanian, S. Rammohan, and Jacob Chandapillai

**Abstract** Flow passing around an object can create vertices behind it and static and dynamic pressure variation across it. Due to the dynamic pressure variation, the object/beam will be getting excited and vibrate. The major excitation frequency due to flow depends upon the flow velocity. The frequency range of excitation depends upon the type of flow regime (i.e., laminar or turbulent). When flow passes through, the beam is not only subjected to induce vibration, it will also damp the vibration and is also proportional to the flow velocity. In this paper, vibration induced on cantilever beams subjected to cross-flow is studied by numerical methods and by experiments. Vibration response of the cantilever beam at the free end is measured, using underwater accelerometer, and reported as waterfall spectrum and overall rms values. Recorded data are further analyzed in power spectral density (PSD) scale and amplitude at its fundamental frequency. It was done in three geometric configurations of beams with thirteen flow rates at each. For numerical simulation, ANSYS fluid and structural analysis package are used and results are reported. Close correlation in vibration levels is observed between the experiment and predicted results. Experimental values are plotted with polynomial regression curve for various flow velocities and acceleration levels. Obtained $R^2$ values are better than 0.99 for two samples and for one sample it is in the order of 0.95. In all the analysis methods, i.e., PSD, amplitude at fundamental frequency and overall rms value, same order of, $R^2$ values are obtained and repeated.

A. Tamil Chandran (✉) · K. R. Balasubramanian · S. Rammohan · J. Chandapillai
Fluid Control Research Institute, Palakkad, Kerala 678623, India
e-mail: tamilchandran@fcriindia.com

S. Rammohan
e-mail: s.rammohan@fcriindia.com

J. Chandapillai
e-mail: director@fcriindia.com

T. Suthakar
Department of Mechanical Engineering, National Institute of Technology-Tiruchirapalli, Tiruchirapalli, Tamil Nadu, India
e-mail: suthakar@nitt.edu

© The Editor(s) (if applicable) and The Author(s), under exclusive license to Springer Nature Singapore Pte Ltd. 2021

S. Mohan et al. (eds.), *Materials, Design, and Manufacturing for Sustainable Environment*, Lecture Notes in Mechanical Engineering,
https://doi.org/10.1007/978-981-15-9809-8_46

**Keywords** Flow measurement · Flow-induced vibration · Cross-flow

## Nomenclature

| | |
|---|---|
| $\rho$ | fluid density |
| $F_I$ | Inertia force |
| $m$ | mass matrix |
| $u$ | fluid velocity |
| $F_D$ | drag force |
| $K$ | stiffness matrix |
| $p$ | fluid pressure |
| $C_d$ | drag co efficient |
| $c$ | damping matrix |
| $p_s$ | static pressure |
| $q$ | distributed load |
| $\ddot{x}(t)$ | acceleration |
| $p_t$ | total pressure |
| $A$ | surface area |
| $\dot{x}(t)$ | velocity |
| $p'$ | dynamic pressure |
| $\omega$ | natural frequency |
| $x(t)$ | displacement |
| $M$ | bending moment |
| $\omega_d$ | damped natural frequency |

## Subscript

| | |
|---|---|
| $I$ | moment of inertia |
| $\xi$ | damping ratio |
| $s$ | structure |
| $E$ | Young's modulus |

## 1 Introduction

Flow measurement is essential to quantify the amount of flow passing through a section or through open channels. These are required for accounting the quantity of fluid used/consumed, custody transfer, determination of turbine efficiency, etc. Flow meter is the instrument to quantify the flow passing through it. Boiten [1] has discussed about various flow measurement structures are available for open channel flow measurement and its field of application. This has mainly done in rivers, streams, canals, partially filled pipes, etc. Closed conduit flow measurements are mainly carryout in industries and where the line is subjected to pressure. Flow measurement in closed conduit or open channel can be determined by various measurement techniques, such as volume flow rate measurement, mass flow rate measurement, flow velocity measurement, or other quantities related to these. In general, flow meters can be classified as differential pressure flow meters, velocity flow meter, and mass flow meters. Differential pressure devices include flow nozzles, orifices, Venturi tubes, etc. It works in the principle of difference in pressure across flow element. In this, pressure drop across the flow element is proportional to the square of the flow rate [2].

Direct velocity flow meter operates linearly with respect to the volume flow pass through it. Since there is no square root relationship, error in measurement is limited and it can be used for wide range. Wide variety of flow meters operates in the principle of velocity measurement. Velocity meter includes differential pressure flow meter and direct velocity flow meter. Some of velocity measurement flow meters are turbine flow meter, vortex shedding flow meter, electromagnetic flow meter, and ultrasonic flow meter. In this, turbine flow meter is having a small turbine and is directly exposed to the flow. It will not work properly with slurry flow, and on the other hand, electromagnetic flow meter can be used with corrosive fluids and slurries. It can measure forward as well as reverse flow with equal accuracy. It operates in the principle of electromagnetic induction. In this case, the flowing liquid will serves as a conductor.

In mass flow meter, mass of the flow passing through the flow element is measured directly. Two major types of mass flow meter are thermal mass flow meter and Coriolis flow meter. In this, thermal mass flow meter operates independent of flow parameters such as density and pressure. It is having, a heated sensing element and the flow stream conduct heat from the sensing element. Conducted heat is directly proportional to the mass flow rate of the fluid. Another type of mass flow meter is Coriolis flow meter. It uses the Coriolis principle to quantify the amount of fluid passing through it. In Coriolis flow meter, fluid to be measured runs through a U-shaped tube that is excited by an angular harmonic oscillation with known frequency, phase, and amplitude [3]. When flow through the tube, due to the Coriolis forces, the tubes will deform and generate an additional vibration component in the pipe. This additional component causes a phase shift on the tubes, and this will be used to determine the mass flow rate.

Most of the above flow meters are mainly used in industrial purpose and in installed condition. But for field efficiency testing of hydroturbine, it is not a case. For field efficiency testing, international standards ASME PTC 18 performance test code and IEC 60041 deal with various test methods used for flow measurement. Some of the methods suggested in standard are propeller current meter method, tracer method, pressure time method, ultrasonic method, thermodynamic method, and index method. In all these methods, flow meter need not be installed permanently. Out of the above, current meter method is frequently used in efficiency tests of the small water turbines and consists of the use of propeller hydrometric current meters located in the hydrometric section in the open channel or closed section in order to determine velocity field components in this section. The flow rate calculation is based on proper integration of the velocity field over the area of the measuring section. In pressure time method, the principle used is the relationship between the change of pressure in a section of the penstock and change of velocity in that section. Tracer method of flow measurement is suitable only where there are relatively long lengths of penstock available or where additional mixing of the tracer may be obtained by inclusion of machine in the measuring length. Radioactive and non-radioactive tracers can be used. Thermodynamic method uses the principle of conservation of energy to determine the efficiency of a hydraulic turbine. It will be done by measuring the flow variables such as pressure, temperature, velocity, and level and the thermodynamic properties of water. This method is preferably used to assess high-head turbines.

To overcome the practical difficulties of existing flow measurement devices and for utilizing the vibration measurement techniques and instrumentation, researchers have focused their research in flow meter development using flow-induced vibration. In this series, Pittard [4, 5], has studied the dynamic pressure fluctuation in the pipe numerically by using RANS and LES methods. In RANS method, pressure fluctuations are averaged out and its effect on piping vibration are eliminated. When using LES method, the pressure fluctuation in pipe flow is modeled similar to its original condition. The obtained dynamic pressure fluctuation was used as source of excitation for vibration simulation. Evans [6] has extended the work of Pittard and performed experiments to demonstrate the findings. It was also shown that the standard deviation of pipe vibration is proportional to the pressure fluctuations in the fluid. Evans (2004) has proved the work of Pittard by experiment and the vibration amplitudes are proportional to the flow velocity. Medeiros [7] extended the work done by Evans (2004) for quantifying the flow through pipe using accelerometers. It was found that the vibration measured is proportional to the flow velocity. Kim [8] has measured the flow velocities using accelerometers on steel pipe. By providing external excitation, he was able to measure small flow velocities in the subcritical region and correlate it with the flow transported through the pipe. Phase difference between the external excitation and vibrations measured at various locations was used for flow determination. Researches were done in the area of multiphase flow measurement using flow-induced vibration. In this, considerable amplitude of vibrations is created due to mixed medium. In the area of multiphase flow, the percentage of medium of flow can be determined without much difficulty from the measured vibration. Antonio Lopes Gama et al. [9] have studied the flow-induced vibration of 'U'-type bend pipe with two-phase flow. It was observed that the overall vibration amplitudes are proportional to the flow velocity.

# 2 Theory

## 2.1 Flow Analysis

Fluid flow analysis through pipe can be done using Newtons laws of motion, conservation of mass, energy, and momentum equations. From the conservation of momentum, one-dimensional fluid flow equation for inviscous fluid is as in Eq. (1) [10]

$$\rho u \frac{\mathrm{d}u}{\mathrm{d}x} = -\frac{\mathrm{d}p}{\mathrm{d}x} \tag{1}$$

where

$\rho$ fluid density
$u$ fluid velocity
$p$ fluid pressure

$$\rho u \frac{\mathrm{d}u}{\mathrm{d}x} + \frac{\mathrm{d}p}{\mathrm{d}x} = 0 \tag{2}$$

$$\frac{\mathrm{d}}{\mathrm{d}x}\left(\frac{\rho u^2}{2}\right) + \frac{\mathrm{d}p}{\mathrm{d}x} = 0 \tag{3}$$

$$\frac{\mathrm{d}}{\mathrm{d}x}\left(p + \frac{\rho u^2}{2}\right) = 0 \tag{4}$$

Integrating Eq. (4) gives

$$p_s + \frac{\rho u^2}{2} = \text{constant} = p_t \tag{5}$$

where

$p_s$ static pressure
$p_t$ total pressure
$\frac{\rho u^2}{2}$ dynamic pressure ($p'$)

When flow passes through a pipe, static pressure acts in all three directions. If there is no recirculation and flow conduit is uniform, flow will be moving in only one direction. So in pipe flow, velocity components in other two directions will be zero. It leads to the dynamic pressure acts only in flow direction. This dynamic pressure will be acting as an excitation force for the cantilever beam in cross-flow. In this, force will be in square of the flow velocity. So, vibration in cantilever beam is proportional to dynamic pressure and flow velocity.

In structural analysis, when a beam is subjected to bending, one side of this is in compression and the other side is tension.

Bending moment equation for beam is as given in Eq. (6) [11]

$$M = EI\frac{d^2y}{dx^2} \tag{6}$$

Differentiating the B. M Eq. (6) we get

$$\frac{d^2M}{dx^2} = EI\frac{d^4y}{dx^4} \tag{7}$$

The Euler–Bernoulli equation describes the relationship between the beam's deflection and the applied load and is [11]

$$EI\frac{d^4y}{dx^4} = q \tag{8}$$

and it can be rewritten as

$$\frac{d^2}{dx^2}\left(EI\frac{d^2y}{dx^2}\right) = q \tag{9}$$

where

'$q$' distributed load on the specimen.

When flow passes through the pipe and a beam is placed on perpendicular to the flow path, fluid pressure act as a load on the beam. As in Eq. (5), total pressure is sum of static and dynamic pressure. During dynamic analysis, effect of static pressure can be neglected and the load acting on the beam which can give dynamic response is dynamic pressure $p'$. With this, Eq. (9) becomes

$$\frac{d^2}{dx^2}\left(EI\frac{d^2y}{dx^2}\right) = p' \tag{10}$$

From Eqs. (7) & (10)

$$\frac{d^2M}{dx^2} = EI\frac{d^4y}{dx^4} = p' \tag{11}$$

Governing equation for forced transverse vibration of beam is obtained as below. This is the well-known Euler–Bernoulli equation [11]

$$EI\frac{\partial^4 y}{\partial x^4} + \rho A\frac{\partial^2 y}{\partial t^2} = f(x, t) \quad (12)$$

For the free vibration case, i.e., $f(x, t) = 0$, the equation of motion (12) becomes

$$EI\frac{\partial^4 y}{\partial x^4} + \rho A\frac{\partial^2 y}{\partial t^2} = 0 \quad (13)$$

where

$\frac{\partial^2 y}{\partial t^2}$ acceleration.

Equation (13) can be written as

$$\frac{EI}{\rho A}\frac{\partial^4 y}{\partial x^4} = -\frac{\partial^2 y}{\partial t^2} \quad (14)$$

From Eqs. (11) & (14), we get

$$\frac{\partial^2 y}{\partial t^2} = \frac{p'}{\rho A} \quad (15)$$

From Eq. (15), it can be observed that acceleration is proportional to the dynamic pressure action on the beam.

## 2.2 *Vibration Analysis*

Vibration on a system is initiated, when mechanical energy is imparted to it. Equation of motion for a single degree of freedom system having force excitation can be written as [11]

$$m\ddot{x}(t) + c\dot{x}(t) + kx(t) = F\sin(\omega t) \quad (16)$$

In this, $F\sin(\omega t)$ is force harmonic excitation.

When the system is excited by external harmonic force, the steady-state displacement of the mass will follow the path described by '$x = X\sin(\omega t - \phi)$' where $\phi$ is the phase lag caused in the response due to the system damping.

For simplification of mathematical calculation, Eq. (16) can be rewritten by neglecting the damping term i.e., [12],

$$m\ddot{x}(t) + kx(t) = 0 \quad (17)$$

When the system is subjected to a simple harmonic motion of $x(t) = x_0\sin(\omega t)$ and $\ddot{x}(t) = -x_0\omega^2\sin(\omega t)$ substituting in Eq. (17) it will become

$$-m\omega^2 + k = 0 \tag{18}$$

From Eq. (18)

$$\omega^2 = m^{-1}k \tag{19}$$

where '$\omega$' is natural frequency.

By considering the damping ratio, damped natural frequency can be written as

$$\omega_d = \sqrt{1-\xi^2}\omega$$

By solving Eq. (16) for harmonic excitation, maximum system response will be [13] as in Eq. (20)

$$X = \frac{F}{\sqrt{\left(k - m\omega^2\right)^2 + (c\omega)^2}} \tag{20}$$

In flow-induced vibration, the fluid force acts on the structure by fluid structure interaction. Considering the fluid properties, Eq. (16) can be rewritten as

$$(m_s + m_f)\ddot{x}(t) + \left(c_s + c_f\right)\dot{x}(t) + \left(k_s + k_f\right)x(t) = F\sin(\omega t) + F_f \tag{21}$$

Considering the excitation force on the structure is only due to fluid force and in comparison with fluid damping, structural damping is negligible, and Eq. (21) can be simplified and rereturn as,

$$(m_s + m_f)\ddot{x}(t) + c_f\dot{x}(t) + \left(k_s + k_f\right)x(t) = F_f \tag{22}$$

Undamped natural frequency of the system is

$$\omega_n = \sqrt{\frac{k_s + k_f}{m_s + m_f}} \tag{23}$$

But most of the cases, added stiffness effect due to fluid is negligible. For that case, Eq. (23) can be simplified and written as below.

$$\omega_n = \sqrt{\frac{k_s}{m_s + m_f}} \tag{24}$$

Damping ratio due to fluid loading is

$$\xi = \frac{C_f}{2\omega_n(m_s + m_f)} \tag{25}$$

## 2.3 Damping

All mechanical systems have some forms of energy dissipation and are known as damping. When the level of dissipation is small, damping is neglected [12]. Due to the presence of increased damping, natural frequency will slightly reduce [11]. There are three primary mechanisms of damping in a system, internal damping, structural damping, and fluid damping.

In this, fluid damping arises from the mechanical energy dissipation resulting from drag forces and associated dynamic interactions of viscous fluid. When study is about flow-induced vibration, fluid damping plays an important role.

Fluid damping is also known as hydrodynamic damping. Morison equation expresses hydrodynamic force as a function of relative fluid–structure velocity and acceleration [9].

Morison equation for a fixed body in an oscillatory flow is,

$$F(t) = F_I + F_D \tag{26}$$

where

$F(t)$ is total force on the object.

$F_I$—Inertia force, $F_D$—drag force, $F_I = \rho C_m V a$.

where

$\rho$—density of fluid,

$C_m = 1 + C_a$ is the inertia coefficient and $C_a$ is the added mass coefficient

V—Volume of body,

$a$—flow acceleration

and

$$F_D = \frac{1}{2}\rho C_d A u|u| \tag{27}$$

where

A Frontal area of body in the flow direction
$C_d$ drag coefficient
$u$ flow velocity
$|u|$ sign of flow velocity.

Drag force, which can generate damping on the beam, is

$$F_D = C_f u \tag{28}$$

Equation (27) can be rewritten as

**Table 1** Test specimen details

| Specimen identification | Physical dimensions in mm | | |
|---|---|---|---|
| | Length | Width | Thickness |
| Sample 1 | 110.0 | 10.0 | 1.4 |
| Sample 2 | 110.0 | 10.0 | 2.0 |
| Sample 3 | 110.0 | 10.0 | 2.0 |

$$C_f = \frac{F_D}{u} \tag{29}$$

Substituting Eq. (29) on Eq. (25), we get the fluid damping and is proportional to the flow velocity.

## 3 Test Specimen

To study the relationship between flow rate and vibration levels, a set of experiments and numerical analysis were undertaken. Test specimen considered for this flow correlation study is a cantilever beam with various lengths and thicknesses. For holding the cantilever beam and for assembling in the test line, 6 inch, 150 class flanges is used.

The cantilever beam is welded with the flange to hold it. The test specimen is made up of SS316 material. Its major physical properties are Young's modulus of 190 GPa, density of 7870 kg/m$^3$, Poisson's ratio of 0.27, and speed of sound is 5790 m/s. For this study, two sizes of cantilever beams are used. Width and thickness of both the beams are 10.0 mm and 1.4 mm, respectively. Lengths of the beams are 110.0 mm and 60.0 mm. Sample lengths are selected in such a way to study the thickness effect and intrusion length. In this, sample 1 and 2 will pass beyond the pipe centerline and sample 3 is little short of the center line. Blockage ratio of the samples used is from 3.4% to 6.2%. Sample dimensional details are as in Table 1. Photograph of a sample is in Fig. 1.

## 4 Modal Analysis

Preliminary stage for structural analysis is to study about its dynamic characteristics. For this, modal analysis is employed to determine the mode shapes and natural frequencies. Natural frequency of the test specimen can be determined by various methods, such as frequency response method, impact hammer method and finite element analysis, and analytical methods. As in Eq. (19), the natural frequency mainly depends upon the stiffness and mass. When the structure immersed in fluid, mass of

**Fig. 1** Photograph of the test specimen

the fluid displaced by the structure ($m_f$) will act as mass loading on the structure. In the case of structure in air, the mass of the fluid displaced is negligible when compared to mass of the structure. In analytical method to determine natural frequency, Eq. (24) can be used without any difficulty for structures having uniformly distributed loads. For the case, where the loads are non-uniform and having discrete masses, finite element analysis method is one of the appropriate methods. In this case, while the mass of sensor is considered as discrete mass, it is required to do finite element analysis for natural frequency determination.

For the present study, FEM/FVM software ANSYS15 has the capability to model the solid and fluid part and has the provision to set the required properties to perform modal analysis [14]. The test specimen is modeled as a three-dimensional solid element to obtain the modal characteristics.

To include the fluid structure interaction, the acoustic material available with ANSYS is used for the fluid part and interface in fluid/structure interaction problems [15].

The FE model generated using the ANSYS is shown in Fig. 2 and its first mode is as in Fig. 3.

Experiments were conducted to determine the natural frequency and were done using frequency response method. Test instruments required to perform vibration testing for frequency response method is a vibration exciter with controller and software, control and response measurement vibration sensors. The test facility used can simulate vibration for the frequency range of 5–2000 Hz and the acceleration of 900 $m/s^2$ with maximum displacement of 50 mm pk-pk. It can take a payload of 450 kg for testing and can generate up to 20 KN for testing. Obtained natural frequencies by analysis and experiment are in Table 2 and more details about the testing and analysis are presented in reference [16, 17]. Further, the mass loading effect of the sensor was also studied and reported [16, 17] and is in Table 3. Since the sensor used is the only available sensor having lowest weight with underwater capability, it was used for this study purpose.

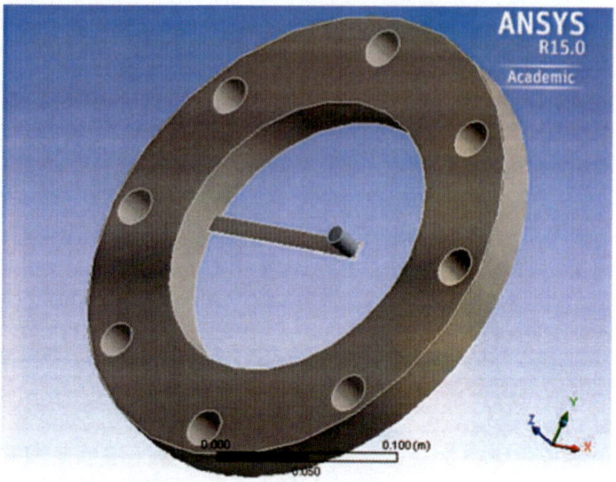


**Fig. 2** Solid model

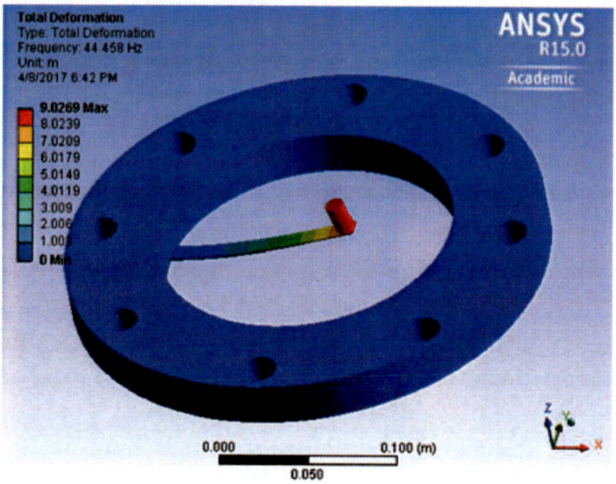


**Fig. 3** Mode shape with sensor at free end

**Table 2** First natural frequency in Hz

| | Numerical analysis | | Experiment | |
|---|---|---|---|---|
| | Air | Water | Air | Water |
| Sample 1 | 44.4 | 41.8 | 44.1 | 40.0 |
| Sample 2 | 72.1 | 68.1 | 71.5 | 65.6 |
| Sample 3 | 191.6 | 185.2 | 191.6 | 179.0 |

**Table 3** First natural frequency in Hz

| | Without mass loading of sensor | | With mass loading | |
|---|---|---|---|---|
| | Air | Water | Air | Water |
| Sample 1 | 92.3 | 75.6 | 44.4 | 41.8 |
| Sample 2 | 131.7 | 112.3 | 72.1 | 68.1 |
| Sample 3 | 443.9 | 383.0 | 191.6 | 185.2 |

## 5 Flow Measurement and Analysis Methods

Flow passing around the structure can create a vortex behind it and make static and dynamic pressure on the surface. Static pressure around it will make the specimen to bend and give static deflection and the dynamic pressure around it functions as an excitation source and it predominantly excites in flow velocity and natural frequencies of the structure. As in Eqs. (15) and (23), the dynamic pressure fluctuation influence in amplitude and natural frequency of the structure. Out of these two methods, a suitable method is needed to be selected to predict the flow rate.

### 5.1 *Frequency Method*

From Eq. (23), it is very well understood that, the change in stiffness made by the flow velocity will influence its fundamental natural frequency. However, the changes in natural frequency are typically very small for turbulent flow and in its subcritical flow rates. Therefore, the sensitivity of such a technique for relatively low flow rates would be poor. For higher flow rates, shift in natural frequency is significantly larger. Robert P. Evans [6] has observed an insignificant shift in frequency due to change in flow rate. For a small change in frequency, a large change in flow rate would make the system sensitivity low. So, the frequency method can be used only with advanced instrumentation, which can track the frequency accurately. But most of the vibration measurement instrumentation is giving very less focus for tracking frequency (in second decimal) so it is inappropriate to use this method.

### 5.2 *Amplitude Method*

From Eq. (15), it is clear that, acceleration output is proportional to the dynamic pressure generated in the system. The dynamic pressure is proportional to square of flow velocity. Hence, increase in acceleration is significant with increase in flow velocity. Evans [6] and Medeiros [7] used vibration amplitude method for predicting flow rate. Both of them used the standard deviation of vibration amplitude for predicting flow velocity. So, it is appropriate to use this method to predict flow velocity and this method is used in this paper.

# 6 Experimental Test Facility

## 6.1 *Test Loop*

Fluid Control Research Institute (FCRI), India, is having various test facilities for flow control device testing and calibration. Major facilities at FCRI are in the medium of water, air, oil, and in two-phase flow. Three different test loops are available to test in water and its capability varies from 200 to 15,000 $m^3$/h. One of the test loop used for this study is having capability up to 300 $m^3$/h. This loop consists of three different parts. They are reservoir, test loop, and reference measurement system. Flow reservoir consists of a overhead tank having head of 13 m, underground storage tank, and pumping system to pump the water from storage tank to over head tank. It can supply a flow rate of up to 300 $m^3$/h by gravity feed with a pressure head of 13 m. To increase the line pressure, pump will be used to directly pump the water through test loop. The test loop consists of three pipe sizes varying from ½ inch to 6 inch diameter and it has provision to change the pipe diameter with in 6 inch. Flow passing through the test loop is taken to the weighing system to weigh the water pass though the device under test or calibration. The flow is simultaneously measured by electromagnetic flow meter connected in the test loop. There a four range of flow meters are used to measure the flow rate and flow meter will be selected based on the required flow rate. Schematic of test loop and photograph of test loop are as given in Fig. 4a, b, respectively. Flow through the test loop is controlled using a flow control valve and bypass available in the loop. When the flow is directly taken from the gravity fed tank, bypass system will not be used to control the flow rate. Flow rates and its respective velocity of flow through the pipe used for study are as given in Table 4.

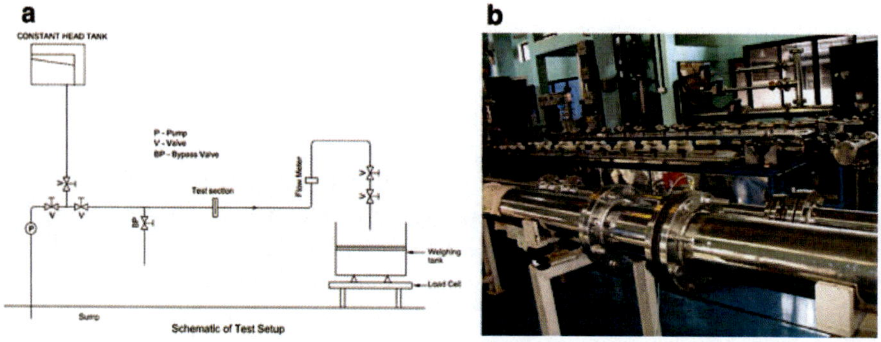


**Fig. 4** Schematic of test loop. Test loop with specimen in installed condition

**Table 4** Flow rate and velocity

| Flow rate in $m^3/h$ | 30 | 40 | 50 | 60 | 80 | 100 | 120 | 140 | 160 | 180 | 200 | 220 | 240 |
|---|---|---|---|---|---|---|---|---|---|---|---|---|---|
| Flow velocity in m/s | 0.45 | 0.60 | 0.75 | 0.89 | 1.19 | 1.49 | 1.79 | 2.09 | 2.39 | 2.68 | 2.98 | 3.28 | 3.58 |

### 6.2 *Instrumentation*

For this study, instrumentation used for vibration measurement and analysis are LMS make PIMENTO vibration analyzer, signal conditioner, and hermetically sealed underwater accelerometer. PIMENTO analyzer can be used to record various dynamic signals, with the input frequency range up to 20 kHz. It is suitable for noise and vibration measurement, analysis, and post-processing. It can take both ICP and voltage inputs. Some of the capabilities available with the analyzer are modules for modal analysis by impact hammer method, spectrum analysis, octave analysis, sound power calculation, and time recording. In this, time data recording and analysis modules are used. It has the capability to set the required sampling speed. Sampling rate needs to be selected based on the frequency range of interest. For this study, set sampling speed for measurement is 3125 samples/s. Accelerometer used for this study is having flat response up to 5 kHz and the deviation in sensitivity is less than 3% from its nominal sensitivity. Vibration sensor used is an AP Technology, USA, makes underwater accelerometer. The underwater sensor is connected at the tip of the cantilever and the cable is taken through the hole made on the flange and connected to the analyzer through the charge signal conditioner.

### 6.3 *Uncertainty in Flow Measurement*

Major uncertainty contributions in flow measurement by this method are from vibration sensor, vibration analyzer, and from flow measuring device used. The expanded uncertainty for the measurement of flow was estimated as 3.7% with a confidence level of 95%.

## 7 Numerical Analysis

For numerical analysis, the cantilever beam including the vibration sensor is modeled as structural part and the liquid around it is as fluid part. Both the parts are modeled using ANSYS Workbench. Geometry created and used for analysis is as shown in Fig. 5. To study the dynamic characteristics of the cantilever beam, the excitation

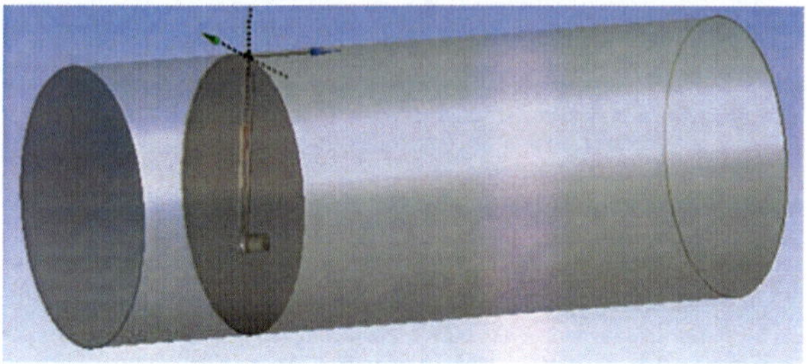

**Fig. 5** Flow and structural simulation geometry

force need to be known. As in Eq. (15), in fluid structure interaction studies, excitation force is the dynamic pressure force transferred from fluid part to the beam by fluid structure coupling. The dynamic pressure across the beam surface is varying throughout the length and width. To get the varying dynamic pressure over the surface area, the flow areas need to be analyzed using computational fluid dynamics (CFD) techniques. The commercially available CFD analysis software ANSYS Fluent is used to get the dynamic pressure distribution over the surface. It was done at discrete flow velocities and required dynamic pressure distributions over surface area are recorded.

To get the frequency response of the structure, 'Harmonic Response' module available in ANSYS is used. By including the fluid damping and fluid structure interaction, it can solve the fluid structure combined Eq. (22). There are three methods available in this module for analysis. They are full method, reduced method, and modal superposition method. One of the features of full method is it has the capability of handling unsymmetric matrices, which occur in fluid–structure interaction problems or rotor dynamics. Reduced method and modal superposition methods cannot handle unsymmetric matrices but it has the capability to take into consideration the effects of prestressing within a harmonic analysis. To include the fluid structure interaction (unsymmetric) and get the best result, full method is used for analysis in the study.

From the analysis, acceleration, velocity, and displacement at the tip of beam, i.e., where the sensor is mounted, are recorded for various flow velocities and is given in Figs. 6, 7 and 8, respectively. Damping determined as per Eq. (25) is used as one of the input parameters for various flow velocities.

## 8 Experimental Results

Details of experimental setup available to simulate the flow condition and instrumentation used for vibration measurement are discussed in detail in Sect. 6. At various flow velocities, vibration signals in acceleration parameter are recorded in time mode. At each flow velocity, data are recorded for the duration of 30 s with a sampling speed of 3125 samples/s. During measurement, constant flow velocity is maintained within less than 0.1% fluctuation in flow rate. Recorded signals are further analyzed and

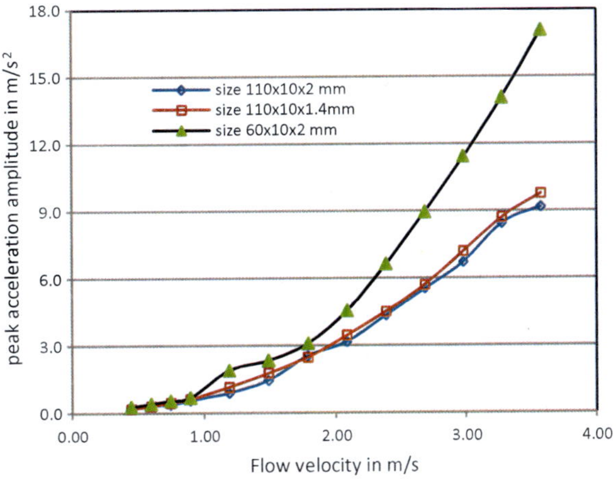


**Fig. 6** Vibration at tip of the beam in acceleration

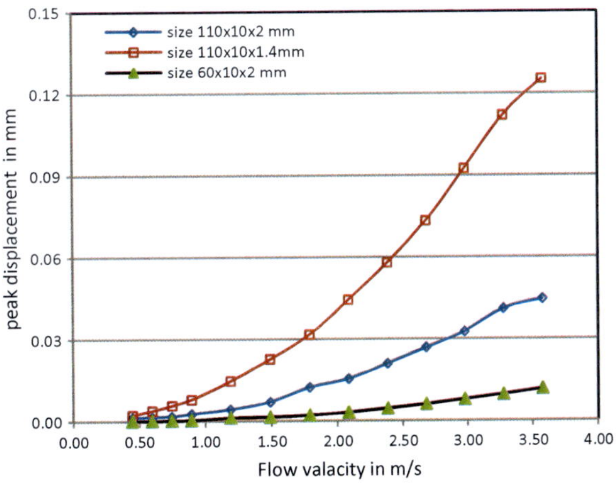


**Fig. 7** Vibration at tip of the beam in displacement

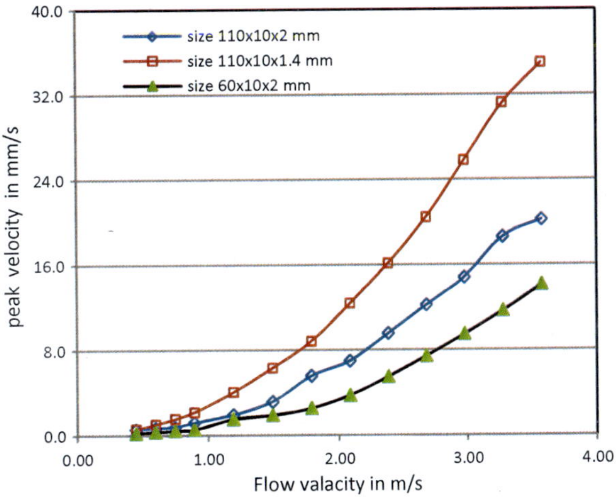


**Fig. 8** Vibration at tip of the beam in velocity

water fall diagram of the acceleration levels at various frequencies and flow velocity are reported. Further it was analyzed and reported in the form of power spectral density (PSD) vs flow rate and frequency. Vibration spectrum for all three samples is as given in Figs. 9, 10 and 11 and PSD spectrums are as given in Figs. 12, 13 and 14.

For all the cases, spectrum analysis was done and peak acceleration levels at its first fundamental frequency for all three samples are as given in Fig. 15. Recorded peak accelerations are further analyzed and velocity and displacement are determined by using standard conversion formulas.

Conversion formulas are

$$\text{Velocity} = \frac{\text{Acceleration}}{(2 * \pi * \text{frequency})}$$

$$\text{Displacement} = \frac{\text{Velocity}}{(2 * \pi * \text{frequency})}$$

Determined velocity and displacement plots are as given in Figs. 16 and 17, respectively. In comparison between acceleration, velocity, and displacement plots, it was observed that velocity and displacement amplitudes are significantly higher for the sample having lesser natural frequency than the other.

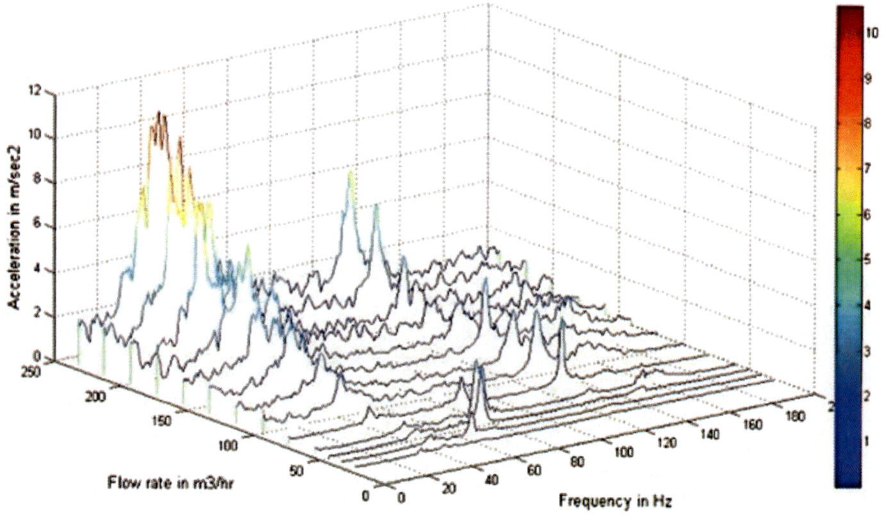


**Fig. 9** Acceleration levels for sample size 110 × 10 × 1.4 mm

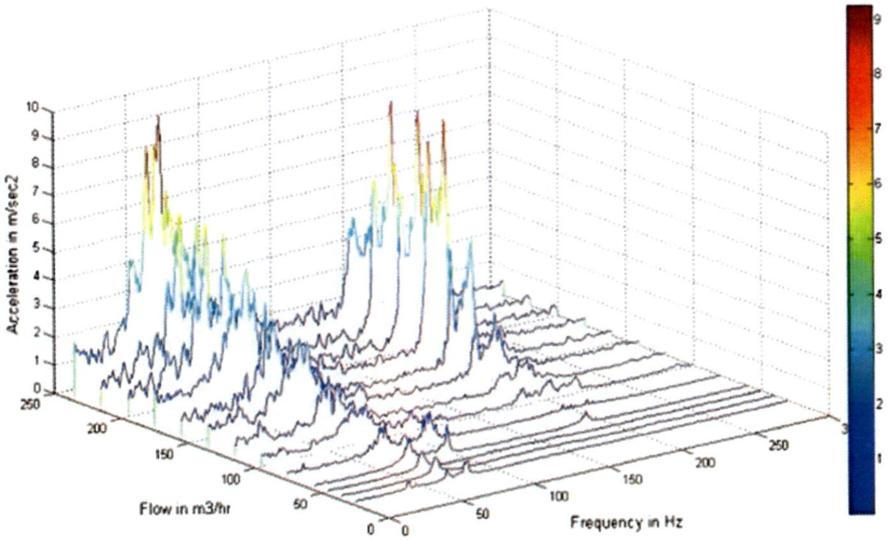


**Fig. 10** Acceleration levels for sample size 110 × 10 × 2 mm

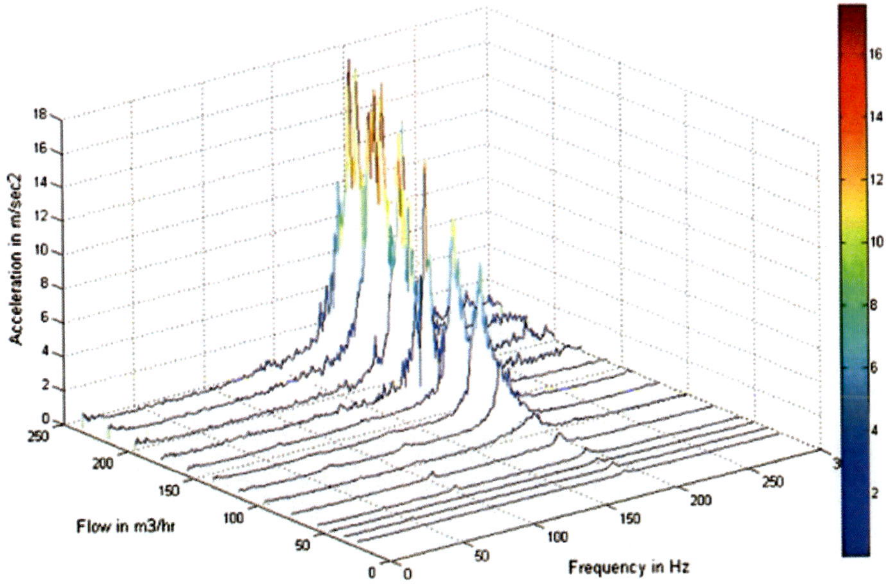


**Fig. 11** Acceleration levels for sample size 60 × 10 × 2 mm

## 9 Result and Discussion

From the water fall plots of acceleration spectrum shown in Figs. 9, 10 and 11, it is observed that the vibration amplitudes are significant in its first natural frequency.

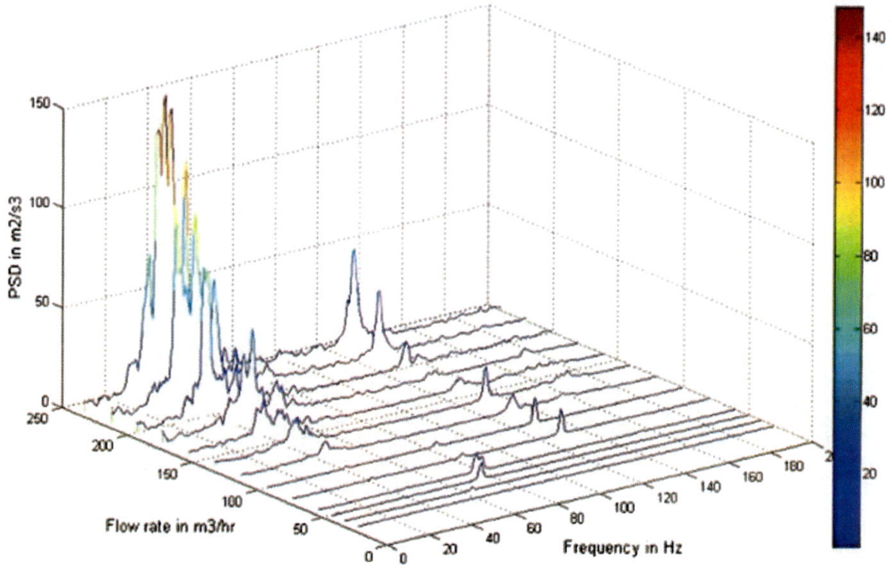


Fig. 12 PSD levels for sample size 110 × 10 × 1.4 mm

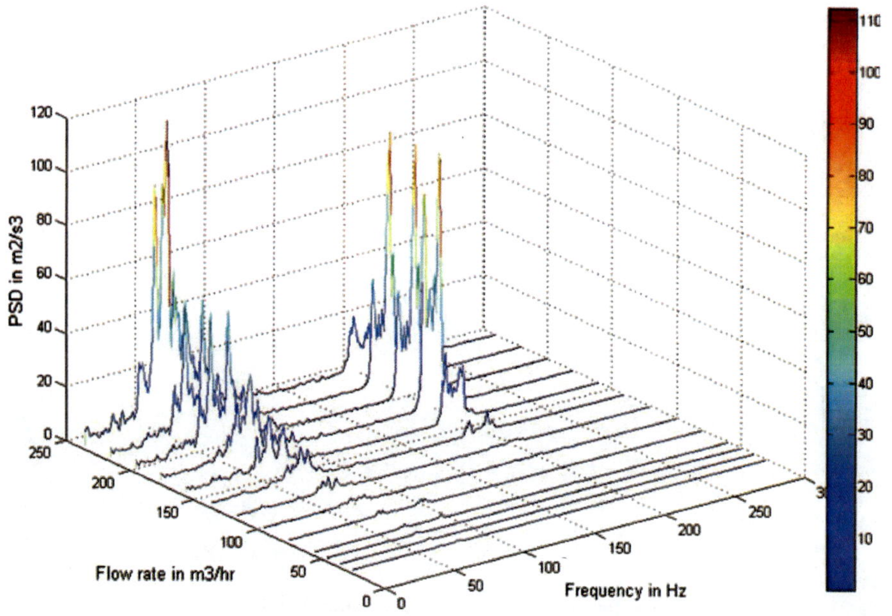


Fig. 13 PSD levels for sample size 110 × 10 × 2 mm

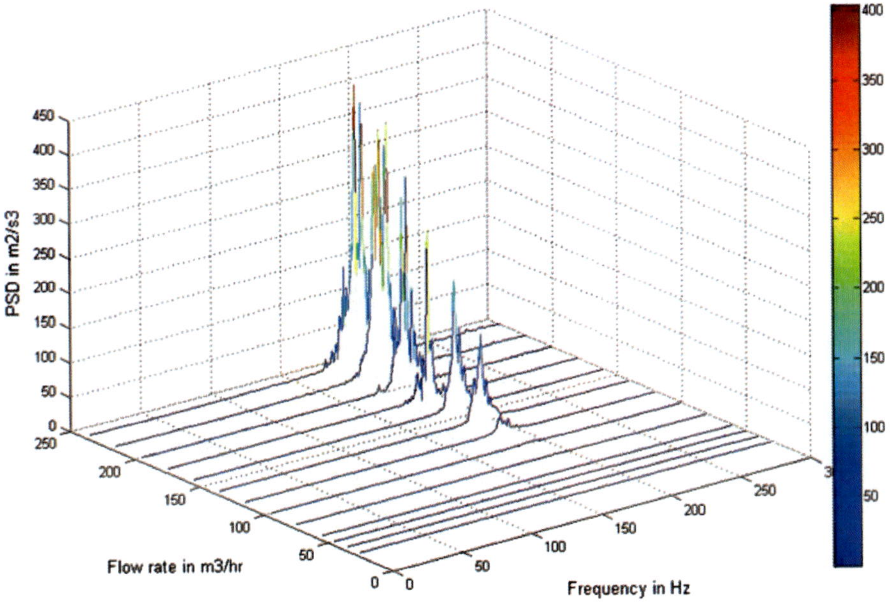


**Fig. 14** PSD levels for sample size 60 × 10 × 2 mm

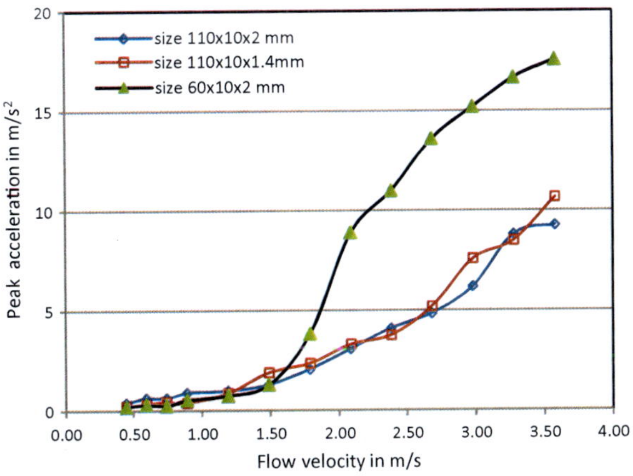


**Fig. 15** Acceleration at tip of the beam in first mode

When flow rates are increased, other than first natural frequency, vibration levels are observed in other frequencies also. At low flow velocity, apart from first natural frequency, predominant peak amplitude is observed at another frequency also. By plotting the water fall diagram in power spectral density scale, only distinct peaks are observed and are as shown in Figs. 12, 13 and 14. PSD plots will give more weightage for higher peaks and lesser weightage for surrounding small peak frequencies. Hence, by plotting the water fall diagram in PSD scale, it is easy to identify the peak amplitudes by suppressing the small amplitudes in broad band frequencies range.

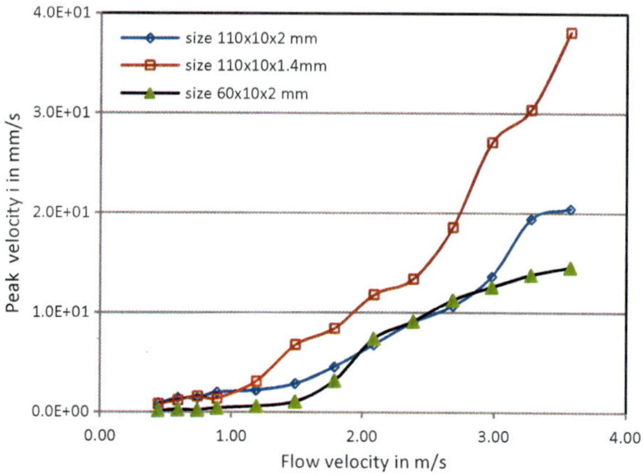


**Fig. 16** Velocity at tip of the beam in first mode

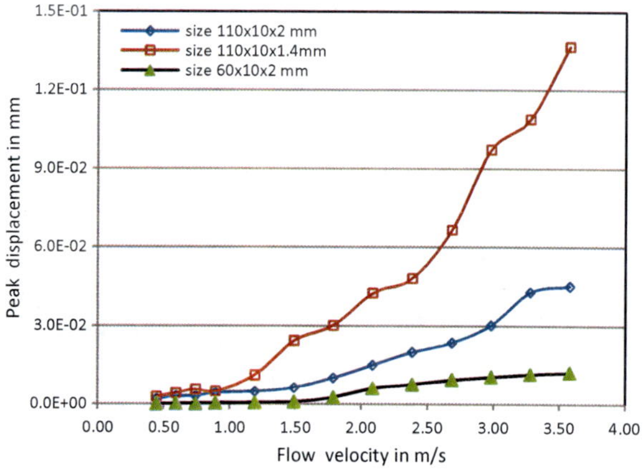


**Fig. 17** Velocity at tip of the beam in first mode

Comparison between the experimental results and predicted acceleration at its first natural frequency is plotted and is as given in Figs. 18, 19 and 20. Irrespective of beam thickness, both predicted and experimental results are closely matching for the beam having length of 110 mm, whereas for the beam having length of 60 mm, wide variation between the predicted and measured acceleration levels.

Experimental data are further analyzed for co-relating the flow with vibration levels. Measured data are fitted using polynomial regression and $R^2$ value also reported in Figs. 21, 22 and 23. In the regression equations, '$x$', represents the flow velocity in m/s and '$y$' corresponds to vibration levels. For all three samples, vibration levels such as acceleration, velocity, and displacement in its first mode and its best fit line are as in Figs. 21, 22 and 23, respectively. The PSD plot is as in Fig. 24. From the best fit line, $R^2$ values of 110 mm long 2 mm thick beam are 0.990 and for 1.4 mm thick beam it is 0.993. For 60 mm long beam, it has dropped to 0.957 and is in line with the deviation observed between predicted and measured values. $R^2$ values from the PSD plot are 0.931, 0.979, and 0.975, respectively. These curves

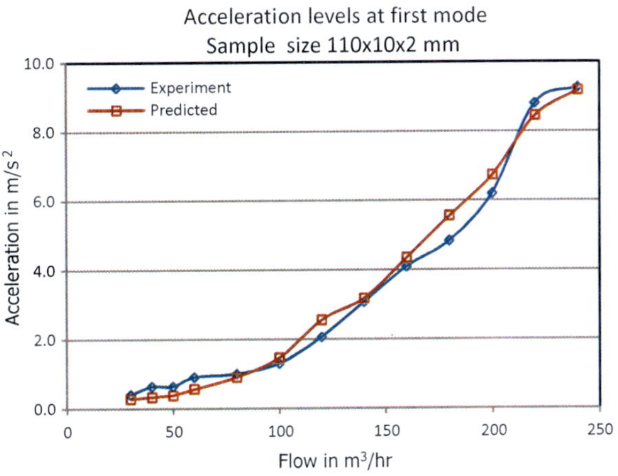


**Fig. 18** Experimental and predicted acceleration levels at first mode

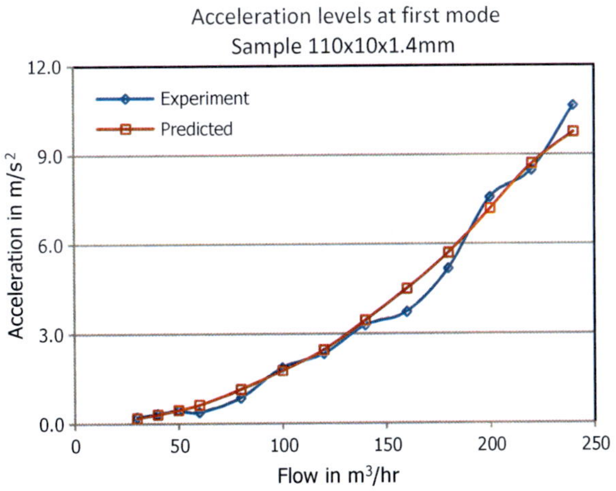


**Fig. 19** Experimental and predicted acceleration levels at first mode

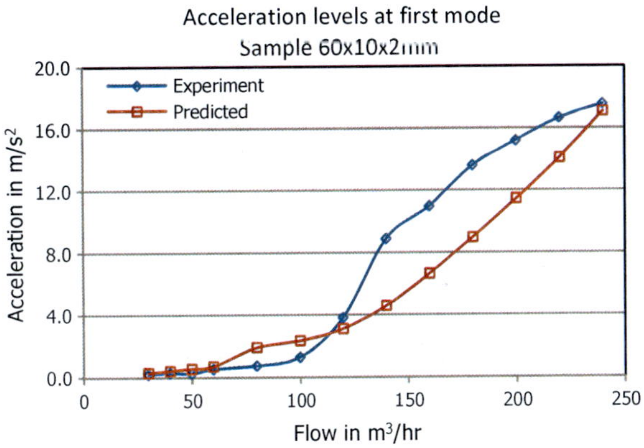


**Fig. 20** Experimental and predicted acceleration levels at first mode

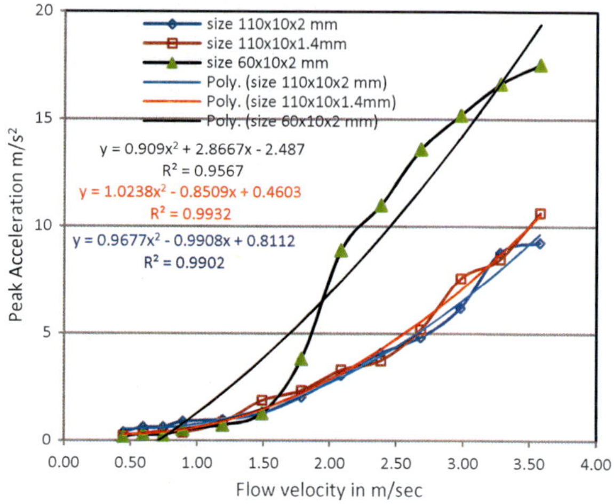


**Fig. 21** Maximum acceleration at first mode versus flow velocity

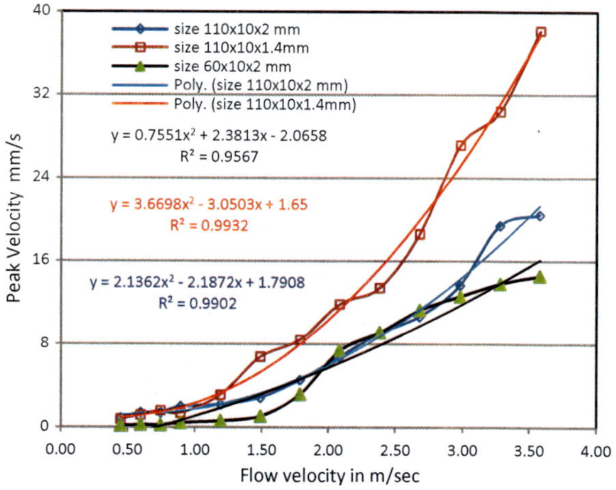


**Fig. 22** Maximum velocity at first mode versus flow velocity

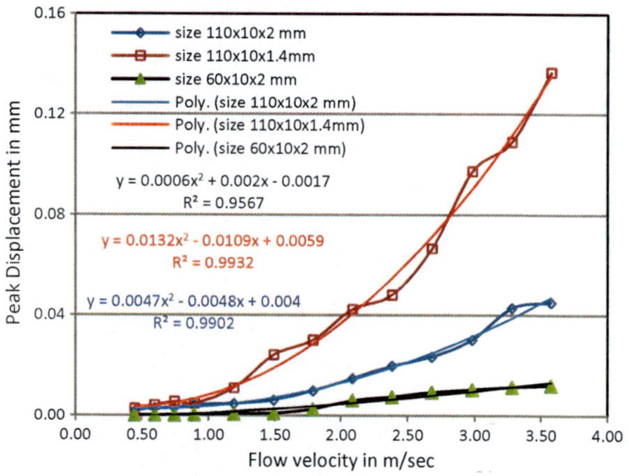


**Fig. 23** Maximum displacement at first mode versus flow velocity

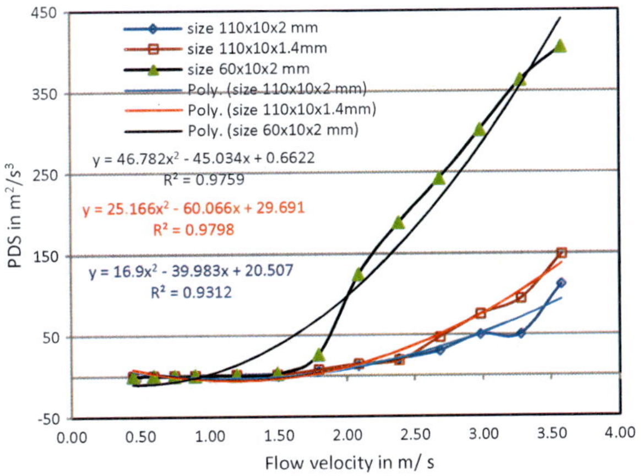


**Fig. 24** Maximum PSD at first mode versus flow velocity

clearly illustrate the strong correlation between vibration levels at first mode and flow velocity.

As indicated in water fall plots, when flow velocity is increased, vibration signals are becoming broad band and to understand the characteristics of it, overall rms acceleration levels are plotted. In first case, overall values are plotted from 5 Hz to 1.3 times of first natural frequency of the respective samples. Overall plot is in Fig. 25. From the best fit line of overall rms plots, $R^2$ values for 110 mm long 2 mm and 1.4 mm thick beams are 0.995 and for 60 mm long beam it drops to 0.962. For all the cases, it gives better representation than picking the peak value. Further, the analysis was extended up to 1.3 times of second peak amplitude. Since there is no second peak is observed in the 60 mm long beam results, it is not considered for this analysis. For 110 mm long and 2 mm thick beam, the $R^2$ value is 0.977 and for 1.4 mm thick beams it is 0.984.

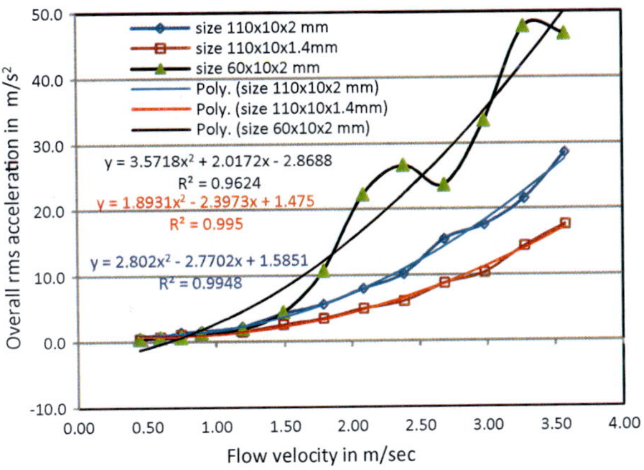


**Fig. 25** Overall rms acceleration, including first mode versus flow velocity

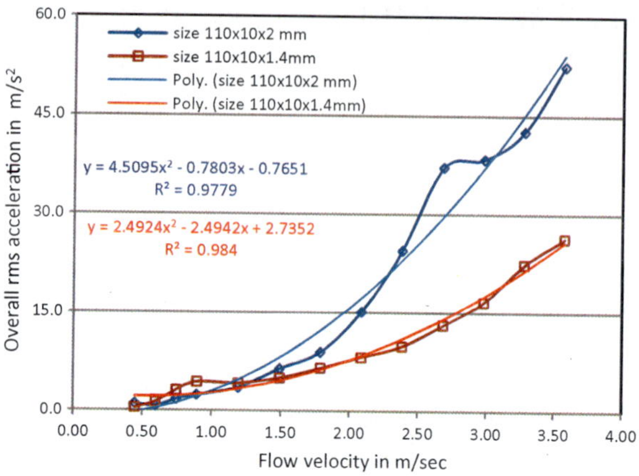


**Fig. 26** Overall rms acceleration, including first and second mode versus flow velocity

All the above tests and analysis results show a significant relationship between vibration level and flow velocity. Only constrain observed is optimization of geometrical length. Length of the beam extended beyond the center line of the pipe is providing better relationship than the shorter beam. It is providing sufficient scope for development of a flow sensor based on the vibration measurement approach presented in this paper (Fig. 26).

## 10 Conclusion

It can be concluded that vibration levels generated by a cantilever beam placed in cross-flow direction are strongly correlated with the mass flow rate passing through it. Based on measured vibration levels and predicted values using ANSYS, it can also be concluded that a flow rate measurement technique, based on the vibration levels at its first natural frequency and overall rms value around it, is having significant potential than a frequency shift-based technique. However at low flow rates, other than its first natural frequency, another predominant peak is also observed and it needs further study in this area. It was also shown that beam thickness and length has an effect on vibration levels and in flow versus vibration level relationship. Length of the beam extended beyond the center line of the pipe is providing better relationship than the shorter beam. Based on the studies conducted, it can also be concluded that, vibration amplitude-based flow measurement technique has potential for development, but it required further study with various fluids, pipe size, and with various size and shape of test specimens.

**Acknowledgements** The authors would like to acknowledge the support provided by Mr. P.Surendran, Deputy Director (retd) and Mr. U. Muthu kumar, Senior Research Engineer of CWM of M/s Fluid Control Research Institute, for conducting experimental studies.

## References

1. Boiten W (1993) Flow measuring structures—Publication no. 478. Delft hydraulics
2. Chattopadhyay P (2006) Flow meters and flow measurement. Asian Books Private Limited, New Delhi, India
3. Pursley WC, Mass flow measurement—Lecture note 16, The basic principles and practice of flow measurement. National Engineering Laboratory, Glagow
4. Thurlow PM, Blotter JB (2003) Numerical modeling of LES based turbulent-Flow induced vibration. In: ASME international mechanical engineering congress & exposition, Washington, pp. 1–8
5. Thurlow PM (2003) Large eddy simulation based turbulent flow-induced vibration of fully developed pipe flow. All Theses and Dissertations. Paper 103
6. Evans RP, Blotter JD, Stephens AG (2004) Flow rate measurements using flow-induced pipe vibration. ASME J Fluids Eng 126:280–285
7. Medeiros KAR, Barbosa CRH, de Oliveira EC (2015) Flow measurement by piezoelectric accelerometers: application in the oil industry. Pet Sci Technol 33(13–14):1402–1409
8. Jendrzejczyk. J.A, Chen.S.S, (1985).Experiments on tubes conveying fluid. Thin-Walled Structures, Volume 3, Issue 2, 1985, Pages 109–134.
9. Gama AL, dos Santos Ferreira LR, Walter Filho PHA (2009) Experimental study on the measurement of two phase flow rate using pipe vibration. In: Proceedings of COBEM 2009 20th international congress of mechanical engineering, ABCM
10. White FM (2016) Fluid mechanics, 8th edn. McGraw Hill Publication
11. de Silva CW (2000) Vibration—fundamentals and practice. CRC Press LLC
12. Caughey MTK, O'Kelly EJ (1961) Effect of damping on the natural frequencies of linear dynamic systems. J Acoust Soc Am 33
13. Lieven NAJ (2002) Forced response—encyclopedia of vibration, vol 2. Academic Press
14. ANSYS mechanical APDL element Reference, Release 15.0 (2013). ANSYS, Inc.
15. ANSYS mechanical APDL acoustic analysis Guide, Release 15.0 (2013). ANSYS, Inc
16. Tamil Chandran A, Suthakar T, Chandapillai J (2018) Modal analysis of flexible beam with fluid structure interaction. In: 2nd international conference on advances in dynamics vibration and control (ICADVC 2018), NIT Durgapur, pp 494–502
17. Tamil Chandran A, Suthakar T, Chandapillai J (2018) Modal analysis of flexible beam with fluid structure interaction. Int. J. Mech. Prod. Eng. Res. Dev. (IJMPERD) , Special Issue, 351–361. ISSN (P): 2249-6890; ISSN (E): 2249-8001

# Effect of Dodecagon Shape Frustum Concentrator and Internal Fins in 2 in 1 Box-Type Solar Cooker

**T. Prem kumar, V. Hariharan, and S. Manojkumar**

**Abstract** This paper aims to combine the applications of cooking and hot air generation in single box-type solar cookers to reduce environmental pollution and to save energy, cost and space. Because of the combined effect, a 33.42% improvement in efficiency is obtained. An innovative dodecagon shape of frustum concentrator is introduced which increases efficiency by 6.3% compared to conventional mode. Different types of pan (i.e. receiver) materials like Mild steel and Aluminium are analyzed. The usage of Aluminium leads to a 4.27% increase in total efficiency compared to Mild steel pans. The effect of internal fins in Aluminium pans is studied which leads to 3.59% increment in the efficiency of cookers compared to the aluminium pan without fins. The combined 2 in 1 effect leads to a reduction of 272 kg of $CO_2$ emission per year and the simple payback period of the cooker is 10 months.

**Keywords** Box-type solar cooker · Environmental pollution · Efficiency · Dodecagon shape · Frustum concentrator · Internal fins · $CO_2$ emission · Payback

## 1 Introduction

Demand for source of energy is never-ending process as energy is the source of all day-to-day activities and for sustainable growth. One of the major sources of energy is fossil fuel which includes crude oil and its distilled components, coal, natural gas, LPG, etc. Because of its energy density, availability, reliability, and simplicity

T. Prem kumar (✉) · V. Hariharan · S. Manojkumar
Department of Mechanical Engineering, PSG Institute of Technology and Applied Research, Coimbatore 641062, India
e-mail: prem50567@gmail.com

V. Hariharan
e-mail: harry181998@gmail.com

S. Manojkumar
e-mail: 18me130@psgitech.ac.in

© The Editor(s) (if applicable) and The Author(s), under exclusive license to Springer Nature Singapore Pte Ltd. 2021

S. Mohan et al. (eds.), *Materials, Design, and Manufacturing for Sustainable Environment*, Lecture Notes in Mechanical Engineering,
https://doi.org/10.1007/978-981-15-9809-8_47

in storage made it as more dependable. But the reduction in the availability of its reserves and the environmental pollution caused by the emission of flue gasses lead to alternative sources which are renewable energy. Renewable energy is pollution-free and sustainable in nature. Day-to-day household activities demand a huge amount of energy where the replacement of fossil fuels with renewable can bring a significant reduction in pollution and sustainable development.

One of the significant renewable energies is solar energy, particularly in tropical regions. It is being used for many applications in day-to-day activities [1–3]. Solar radiation can be harnessed to provide heat to a variety of applications such as cooking, Vapour absorption type air conditioning, and electricity generation either in a direct manner (by steam production) or direct manner (Solar photovoltaic). Cooking is one of the applications where solar radiation heat energy can be utilized. Solar dish type cookers [4] and box-type cookers [5–7] are the two types of cookers that are available in the market. In spite of its less efficiency, box-type solar cooker is one of the significant types of solar cooker because of its simplicity and portability. The inner walls of the box and the bottom face have to be coated with high absorptivity paint to increase absorption of heat. The combined usage of solar energy with fossil fuels to cook the food will be more reliable and effective [8, 9]. Reflecting materials like stainless steel, mirrors, or highly polished aluminium are also used around the cooker to increase the rate of heat transfer through multiple reflections. In order to increase the efficiency of box-type cookers, outer booster reflectors are used [10]. A cooker with internal reflector and outer reflectors [11–13] improves the thermal performance of the cooker significantly. The reflecting mirror [14] is used to track the sun resulted in an increase in efficiency by 6%. The temperature of the cooker was increased by 15–22% with the presence of reflectors [5].

A lot of innovations are made in box-type cookers to improve its efficiency [15, 16]. Hybrid solar box cooker was designed [17] with additional heat from the halogen lamps apart from the solar radiation. Hollow copper balls were also imparted inside the cooker to improve efficiency. Fins were also imparted in the absorber plates for the improvement in the efficiency of the cooker [18–20].

Hot air can be used in day-to-day lives for drying wet hands, drying wet clothes, drying wet utensils, and to obtain dried foods like dried fish, dried fruits, dried coconut, etc. Solar air dryers can be used in the drying applications mentioned above either directly or indirectly [21–27].

In this research apart from outer booster reflector, additional dodecagon shape frustum concentrator and internal finned absorbers are used to increase efficiency. The reduction in volume due to the presence of internal fins is compensated by the buffer volume given in the pan at the design stage itself. Apart from this the conventional box-type solar cooker and solar drier are combined in a single set up (2 in 1) and hence both cooked food and hot air can be obtained in a single set up.

## 2 Experimental Setup

The 2 in 1 box-type solar cooker with aperture area 0.8 $m^2$ is fabricated with materials like plywood, aluminium reflective sheets (0.2 mm), booster reflector, concentrator (dodecagon shape frustum), cooking vessel (aluminium and mild steel), and top glass cover as shown in Fig. 1. The material specification and dimensions are mentioned in Table 1 and the modeling of parts of 2 in 1 box-type solar cooker is shown in Fig. 2.

**Fig. 1** Two (2) in one (1) box-type cooker

**Table 1** Parts and dimensions of 2 in 1 box-type solar cooker

| Parts/dimensions | Values/specification |
|---|---|
| Reflecting mirror | 4 mm thickness<br>900 mm × 900 mm |
| Top glass cover | 4 mm thickness<br>900 mm × 900 mm |
| Concentrating mirror | 4 mm thickness |
| Aluminium reflecting sheet | 0.2 mm thickness<br>Reflectivity 95% |
| 230 V AC fan | Internal diameter = 65 mm<br>Outer diameter = 120 mm |
| Black painted pans (stainless steel, aluminium and aluminium internal finned) | Radius = 100 mm<br>Height = 100 mm<br>Fin (2 mm thickness) |
| Plywood | (12 and 8 mm thickness) |

Fig. 2 Modeling of parts of 2 in 1 box-type solar cooker

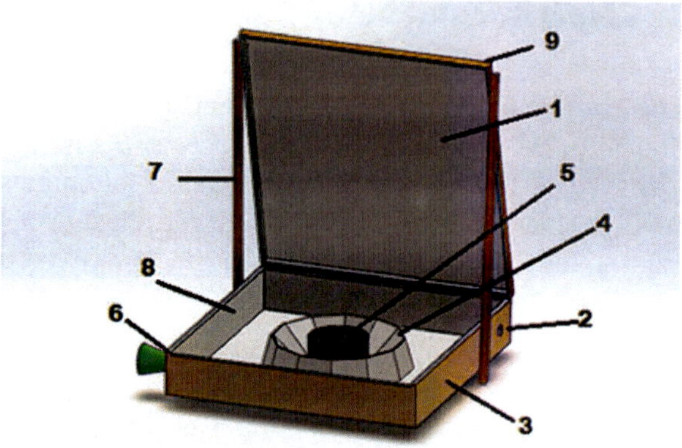


1. Booster reflector
2. Suction airport
3. Side wooden box cover
4. Dodecagon shape frustum concentrator
5. Pan (Mild steel or Aluminium or Aluminium with internal fins)
6. Induced Draft fan along with delivery duct
7. Support for Top cover (Booster reflector)
8. Top Glass cover
9. Top Wooden cover to support Booster reflector.

Apart from cooking the food, the provisions have been made to heat the air inside the box for the effective usage of the box as well as to increase the efficiency. Induced draft fan is used to induce the air from the atmosphere into the solar cooker in order to retrieve the heat at outer compartment, i.e., outside the central dodecagon concentrator and hence the hot air can be obtained.

One mild steel pan (without internal fin) and two Aluminium receiver pans (with and without internal fins) are fabricated to analyze the effect of internal fins and the material of pan as shown in Figs. 3, 4, and 5.

Fig. 3 Mild steel pan

**Fig. 4** Aluminium pan

**Fig. 5** Aluminium receiver pan with internal fins

The total global solar radiation is measured using class Second Class Pyranometer, Model No.: LP PYRA 03, Make Delta Ohm. The solar radiation and the temperatures measured by k type thermocouples are recorded simultaneously using the data logger which is eight channels Universal input process data logger of Model No.: Logger-01. The air velocity is measured using vane type anemometer. The accuracy of the instruments is mentioned in Table 2. Water heating is considered here as a food load for simplicity [17] and ease of calculation. Food cooking is also observed for edible noodles and found satisfactory with proper hygiene nature.

**Table 2** Instruments and its accuracy

| Instruments/sensors | Accuracy/sensitivity |
|---|---|
| Pyranometer (Model No.: LP PYRA 03) | 5–15 $\mu V/W/m^2$ |
| Vane type anemometer (Lutron AM-4201 digital anemometer) | $\pm(2\% + 0.1$ m/s) |

## 3 Efficiency Calculation

The useful heat rate absorbed by water ($\dot{Q}_{\text{cook}}$) in kilo Watt is given by

$$\dot{Q}_{\text{cook}} = \frac{m}{t} C_p.(T_f - T_i) \tag{1}$$

where $m$ is the mass of water taken in pan measured in kg, $t$ is the time taken for heating up to '$n$' seconds, $C_{\text{p}}$ is the specific heat capacity of water at constant pressure in kJ/kg K, $T_{\text{f}}$ is the final temperature of the water in °C at the end of time period $t$, $T_{\text{i}}$ is the initial temperature of the water in °C during the beginning of the experiment.

The average input heat rate given by sun ($\dot{Q}_{\text{in}}$) in kilo Watt over the time '$t$' up to '$n$' seconds is given as

$$\dot{Q}_{\text{in}} = Aa * I_{g_t} \tag{2}$$

where $Aa$ is the aperture area of the cooker in m$^2$, $I_{g_t}$ is the average instantaneous global radiation in kW/m$^2$ over the time period '$t$' up to '$n$' seconds.

The efficiency of the cooker ($\eta_{\text{cook}}$) contributed by cooking effect (here water) is given by

$$\eta_{\text{cook}} = \frac{\dot{Q}_{\text{cook}}}{\dot{Q}_{\text{in}}} * 100 \tag{3}$$

$\dot{Q}_{\text{air}}$ is average heat rate in kilo Watt carried by air over the time period '$t$' up to '$n$' seconds is given by

$$\dot{Q}_{\text{air}} = \dot{m}_{\text{air}} * C_{p_{\text{air}}} * \frac{\sum_{t=1}^{n} \left\{ \left[ \int_{T_{\text{air inlet}_t}}^{T_{\text{air Outlet}_t}} \mathrm{d}T \right] \right\}}{n} \tag{4}$$

where $\dot{m}_{\text{air}}$ mass flow rate of air sucked by induced AC fan in kg/s, $C_{p_{\text{air}}}$ is the specific heat capacity of air at constant pressure in kJ/kg K, $T_{\text{air Outlet}_t}$ and $T_{\text{air inlet}_t}$ are the air temperatures at Box cooker outlet and inlet, respectively in °C at '$t$th' second. Time period '$t$' varies from 1 to $n$ seconds with total intervals of '$n$' with the recess time of 1 s using the data logger.

The efficiency of the cooker ($\eta_{\text{air}}$) contributed by air is given by

$$\eta_{\text{air}} = \frac{\dot{Q}_{\text{air}}}{\dot{Q}_{\text{in}}} * 100 \tag{5}$$

The total efficiency ($\eta_{\text{total}}$) of the cooker (including cooking effect and heating the air) is given by

$$\eta_{\text{total}} = \eta_{\text{cook}} + \eta_{\text{air}} \tag{6}$$

# 4 Results and Discussion

## 4.1 *Effect of Dodecagon Shape Frustum Concentrator*

Apart from outer booster mirror, a separate concentrator of dodecagon shape (12 sides, side angle at 45° with respect to base) frustum is attached inside the cooker to concentrate the solar radiation towards the pan. Hence more solar radiation can be projected towards the pan and hence heat to the surrounding will be relatively less compared to conventional methods. The modeling diagram and actual picture are shown in Figs. 6 and 7, respectively.

The experimental setup is connected and the booster reflector is arranged such that the solar radiations are reflected towards the aperture area of solar cooker. The

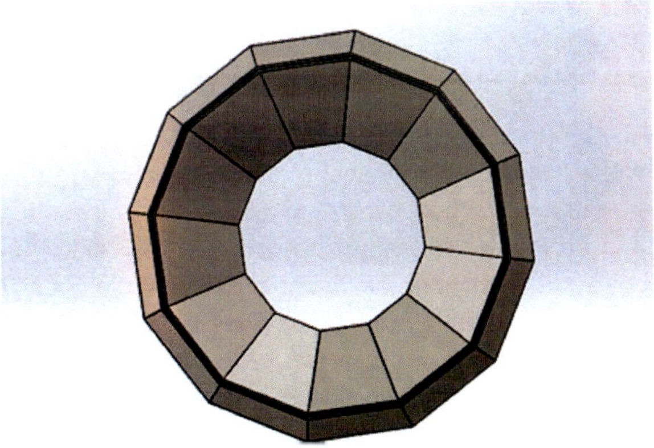

**Fig. 6** Modeling of concentrator

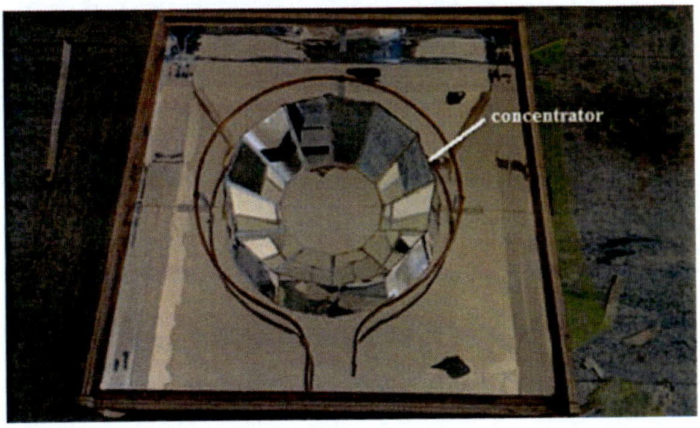


**Fig. 7** Concentrator

**Table 3** Effect of concentrator

| Condition | Initial temperature $T_i$ (°C) | Final temperature $T_f$ (°C) | Cooking heat rate $\dot{Q}_{cook}$ (kJ/s) | Input heat rate $\dot{Q}_{in}$ (kJ/s) | Efficiency $\eta$ (%) | Average solar radiation (kW/m$^2$) |
|---|---|---|---|---|---|---|
| With concentrator | 24 | 56.7 | 0.130 | 0.752 | 17.3 | 0.929 |
| Without concentrator | 23 | 43.5 | 0.081 | 0.742 | 11.0 | 0.917 |

temperature of the water is continuously recorded and stored along with total global radiation for the interval of 2100s. At first, the experiments were done using Mild steel receiver pan without concentrator in simple box-type solar cooker. Then the procedure is repeated with a concentrator in the cooker. The readings are tabulated in Table 3.

From the table, it is clear that the usage of concentrator gives more benefits and effective. Due to the introduction of Dodecagon shape frustum concentrator, there is a 6.3% rise in efficiency compared to setup where the concentrator was absent.

## 4.2 *Effect of Combined Cooking and Air Heating*

Apart from conventional cooking at the pan, set up has been made to heat air too as by-product. Induced Draft fan is fixed at side wooden cover box to suck air through inlet air suction port which is kept at opposite side wooden box cover. The induced air collected the heat in the chamber between the concentrator and sides of the wooden box and heated. The result of the combined experiment is shown in Table 4. By comparing Table 4 and Table 3 it is clear that the combined 2 in 1 usage increases the efficiency mainly because of continuous removal of heat at the outer chamber

**Table 4** Experiment on MS (both hot water and air generation)

| Contributing part | Ratio between mass to time (or) mass flow rate $\dot{m}$ (kg/s) | Initial temperature $T_i$ (°C) | Final temperature $T_f$ (°C) | Cooking heat rate $\dot{Q}_{cook}$ or air heat rate $\dot{Q}_{air}$ (kJ/s) | Input heat rate $\dot{Q}_{in}$ (kJ/s) | Efficiency $\eta$ (%) | Average solar radiation (kW/m$^2$) |
|---|---|---|---|---|---|---|---|
| Water | 2 | 22.5 | 47.7 | 0.100 | 0.686 | 14.7 | 0.848 |
| Air | 0.0298 | 38.2 | 46.5 | 0.248 | | 36.5 | |
| Combined | | | | 0.348 | 0.686 | 51.2 | 0.848 |

**Table 5** Experiment on Al pan (both hot water and hot air generation)

| Contributing part | Ratio between mass to time (or) Mass flow rate $\dot{m}$ (kg/s) | $T_i$ (°C) | $T_f$ (°C) | $Q$ (kJ/s) | $Q_{in}$ (kJ/s) | $\eta$ (%) | Solar radiation (kW/m$^2$) |
|---|---|---|---|---|---|---|---|
| Water | 2 | 22.4 | 50.2 | 0.110 | 0.651 | 17.4 | 0.804 |
| Air | 0.0298 | 38.2 | 46.5 | 0.248 | | 38.0 | |
| Combined | | | | 0.358 | 0.651 | 55.4 | 0.804 |

by air. The efficiency contributed by water is slightly reduced by 2.6% compared to conventional-type single-use. But this reduction in efficiency is insignificant by obtaining hot air as by-product which increases the combined efficiency by 36.5%. Hence the total efficiency obtained is 50.72% against the efficiency of 17.3% which is obtained when the cooker is used for single cooking usage.

## 4.3 *Effect of Material of Pan*

The mild steel receiver pan is replaced with aluminium receiver pan to improve the efficiency of the cooker. The efficiency of the cooker with Aluminium receiver pan is 54.99% whereas the contribution of water heating is 17.10% and the contributed efficiency of air heating is 38.56%. The improvement in water heating is 2.7% more when compared to mild steel receiver pan. So the usage of aluminium receiver pan is more effective. One of the main reasons for rise inefficiency is the high thermal conductivity of Aluminium which is 205 W/m K against the mild steel value of 50.2 W/m K [28]. Because of the better thermal conductivity, the heat conduction will be more which leads to an increase in heat absorption by water. At the same time, mild Steel pans are easy to weld whereas Aluminium welding demands a certain cautious procedure with respect to temperature of welding source and method of welding to avoid defect in welding. Mild steel pans are cheaper compared to aluminium but it has to be coated to avoid rusting and for hygienic food usage. The observed reading of the experiment using aluminium pan (both hot water and air generation) is mentioned in Table 5.

## 4.4 *Effect of Internal Fins in Pan*

The aluminium receiver pan is replaced with an internally finned aluminium receiver pan to study the effect of fins on the efficiency of the cooker. The measured efficiency of the cooker using aluminium pan with internal fins is 58.58% whereas the water heating contribution is 20.08% and the efficiency contribution by air heating is 38.50%. The improvement in water heating is 2.6% more when compared to

**Table 6** Experiment on Al pan with internal fins (both hot water and hot air generation)

| Contributing part | Ratio between mass to time (or) mass flow rate $\dot{m}$ (kg/s) | $T_1$ (°C) | $T_2$ (°C) | $t$ (s) | $Q$ (kJ/s) | $Q_{in}$ (kJ/s) | $\eta$ (%) | Solar radiation (kW/m$^2$) |
|---|---|---|---|---|---|---|---|---|
| Water | 2 | 18.5 | 55.1 | 2100 | 0.145 | 0.732 | 20.0 | 0.904 |
| Air | 0.0298 | 38.2 | 47.5 | 2100 | 0.278 | | 38.5 | |
| Combined | | | | | 0.423 | 0.732 | 58.5 | 0.904 |

aluminium receiver pan without fins. The increased surface area which is in contact with water increased the heat rate and hence increased the heat absorption or reduced the cooking time period. Hence the usage of fins in pan is high in advantage and effective. The observed reading of the experiment on an aluminium pan with internal fins is tabulated in Table 6.

## 5 Economic and Environmental Benefit

If the 2 in 1 solar box cooker is not available, then the cooking has to be done by LPG (Liquefied Petroleum Gas) stove and hot air has to be obtained by electric air heaters. The cost of 2 in 1 box-type solar cooker is INR (Indian National Rupee) 6127. The net amount which has to be spent excess of cost of LPG stove and electric air heater is INR 2577. The equivalent per day LPG consumption and electricity consumption would be 0.088 kg of LPG and 0.567 units (kWh) of electricity respectively. With the requirement consideration of 4 cooking hours per day and 2 hot air demand hours per day, the total running cost will be INR 3013. The fan electricity consumption considering demand hours is 0.08 units per day. Hence considering the net cost of solar cooker and running cost of LPG and electric heater leads to simple payback period of 10.3 months which is attractive. With the consideration of 3.336 kg of $CO_2$ emission per kg of LPG consumption and 0.93 kg of $CO_2$ per unit of electricity consumption, the total reduction in $CO_2$ emission per year is 271.96 kg of $CO_2$.

## 6 Conclusion

The 2 in 1 (both cooking and hot air generation) effect with innovative dodecagon shape frustum concentrator along with internally finned aluminium pan is more desirable which leads to 47.5% improvement in efficiency compared to conventional box-type solar cooker. The net effect leads to 272 kg of $CO_2$ per year along with simple payback period of 10 months. The dimensions and nature of the fins have to be studied more in the future. The internal fins have to be compared with external fins for the possibility of optimization and improvement. The Mild steel pans are used for experimental purposes whereas actual cooking application demands either coated Mild steel pans or Stainless steel pans which are slightly costlier than Aluminium pans. Hence the usage of Aluminium pans with internal fins is highly effective.

**Acknowledgements** The authors would like to thank the management of the PSG Institute of Technology and applied research for the resources provided to conduct the experiments. They would like to extend their thanks to their family and friends for financial support.

## References

1. Sekhar TVR, Prakash R, Nandan G, Muthuraman M (2018) Pressure drop characteristics and efficiency enhancement by using $TiO_2$-$H_2O$ nanofluid in a sustainable solar thermal energy collector. Int J Environ Sustain Dev 17(2/3):273–294
2. Shafiqul Islam M, Rakibul Hasan M (2016) Study of performance evaluation, economic and environmental impact of 1.68 kWp DC operated submersible centrifugal solar pump with auto tracker using low cost DAS. Int J Environ Sustain Dev 15(2):146–158
3. King J, Slesser M (1995) Can the world make the transition to a sustainable economy driven by solar energy? Int J Environ Pollut 5(2):14–29
4. Hassan K (2004) Experimental determination of energy and exergy efficiency of the solar parabolic-cooker. Sol Energy 77(1):67–71
5. Negi BS, Purohit I (2005) Experimental investigation of a box type solar cooker employing a non-tracking concentrator. Energy Convers Manage 46(4):577–604
6. Purohit I (2010) Testing of solar cookers and evaluation of instrumentation error. Renew Energy 35(9):2053–2064
7. Kumar S (2005) Estimation of design parameters for thermal performance evaluation of box-type solar cooker. Renew Energy 30(7):1117–1126
8. Prasanna UR, Umanand L (2011) Modeling and design of a solar thermal system for hybrid cooking application. Appl Energy 88(5):1740–1755
9. Nandwani SS (2007) Design, construction and study of a hybrid solar food processor in the climate of Costa Rica. Renew Energy 32(3):427–441
10. Sagade AA, Samdarshi SK, Panja PS (2018) Experimental determination of effective concentration ratio for solar box cookers using thermal tests. Sol Energy 159(1):984–991
11. Ibrahim SMA, Elreidy MK (1995) The performance of a solar cooker in Egypt. Renew Energy 6(8):1041–1050
12. El-Sebah AA (1997) Thermal performance of a box-type solar cooker with outer-inner reflectors. Sol Energy 22(10):1011–1021
13. Kahsay MB, Paintin J, Mustefa A, Haileselassie A, Tesfay M, Geberay B (2014) Theoretical and experimental comparison of box solar cookers with and without internal reflector. Energy Procedia 57:1613–1622
14. Nahar NM (2001) Design, development and testing of a double reflector hot box solar cooker with a transparent insulation material. Renew Energy 23:167–179
15. Buddhi D, Sahoo LK (1997) Solar cooker with latent heat storage: design and experimental testing. Energy Convers Manage 35(5):493–498
16. Cuce E, Cuce PM (2013) A comprehensive review on solar cookers. Appl Energy 102:1399–1421
17. Saxena A, Agarwal N (2018) New hybrid solar cooker with air duct. Sol Energy 159(1):628–637

18. Harmim A, Belhamel M (2010) Experimental investigation of a box-type solar cooker with a finned absorber plate. Energy 35(9):3799–3802
19. El-Sebaii AA, Ramadan MRI (2015) Effect of fin configuration parameters on single basin solar still performance. Desalination 365(1):15–24
20. Harmim A (2008) Double exposure solar cooker with finned vessel. Sol Energy 82(4):287–289
21. Anil Kumar S, Sridhar K, Vinod Kumar G (2018) Heat transfer analysis of solar air heating system for different tilt angles. Appl Solar Energy 54(1):17–22
22. Abbasov ES, Umurzakova MA, Boltoboeva MP (2016) Efficiency of solar air heaters. Appl Solar Energy 52(2):97–99
23. Tyagi RK, Ranjan R, Kishore K (2014) Performance studies on flat plate solar air heater subjected to various flow patterns. Appl Solar Energy 50(2):98–102
24. Omojaro AP, Aldabbagh LBY (2010) Experimental performance of single and double pass solar air heater with fins and steel wire mesh as absorber. Appl Energy 87:3759–3765
25. Saxena A, Agarwal N, Srivastava G (2013) Design and performance of a solar air heater with long term heat storage. Int J Heat Mass Transfer 60:8–16
26. Orozbaev MT (2007) On determining the thermo technical characteristics of the flat-plate solar air heating collectors. Appl Solar Energy 43(1):56–57
27. Prem Kumar T, Naveen C, Premalatha M (2019) Performance analysis of 2 in 1 Parabolic trough collector for both hot water and hot air production for domestic household applications. Appl Solar Energy 55(6):397–403
28. Kothandaraman CP, Subramanyan S (2018) Heat and mass transfer data book, 9th edn. New Age International Publisher, New Delhi

# Studies on the Improved Design in the Heat-Setting Platen Used in Textile Industry

**K. Sivananda Devi**

**Abstract** The paper discusses the design and development of the platen used in a special-purpose machine by ABC textile industry in Coimbatore. The objective is to attain uniform heat-setting temperature on both top and bottom platens of the machine. This research work is therefore important as it addresses the real need raised by the industry to have better heat-setting process in order to optimize and develop a better-quality end-product. Heat setting changed the structure and gave the fibers an unyielding property. Elongation of the fabric varies with different temperatures. Hence, it is very much essential to maintain precise and uniform temperature in heat setting. The author proposed design modifications on the heating element configuration and the material of the platen to achieve uniform temperature and improved heat transfer. Numerical validation and experimentation were carried out to ensure the design modifications and the attainment of uniform temperature with improved heat transfer in the top and bottom platen across different zones in the platen. The effect of changing the design on thermal characteristics, results and the benefits achieved by the company are discussed.

**Keywords** Thermal effect · Elongation · Heat setting · Bottom platen · Aluminum

## 1 Introduction

This research is significant for textile and fabric manufacturing industries as it is very difficult to get uniform temperature in the heat-setting process. Heat setting stabilizes the structure of the yarn by application of heat with the accurate temperature that gives optimized strength for fabric and avoids the chances for distortion [1]. Due to heat setting, the structural pattern of the fabric is improved. This research work is therefore important as it addresses the real need raised by the industry to have

K. Sivananda Devi (✉)
Department of Mechanical Engineering, PSG Institute of Technology and Applied Research, Coimbatore, Tamil Nadu, India
e-mail: ksd@psgitech.ac.in

© The Editor(s) (if applicable) and The Author(s), under exclusive license to Springer Nature Singapore Pte Ltd. 2021

S. Mohan et al. (eds.), *Materials, Design, and Manufacturing for Sustainable Environment*, Lecture Notes in Mechanical Engineering,
https://doi.org/10.1007/978-981-15-9809-8_48

better heat-setting process in order to optimize and develop a better-quality end-product. This investigative research was conducted on a special machine that is used for finishing the fabric. This study optimizes the heat-setting process by changing the design of the platen used in rotary curing press. In order to maintain anonymity, the company is referred to here as ABC. ABC Company belongs to small enterprise textile fabric manufacturing company located in Coimbatore, Tamil Nadu.

During the thermal process of heat setting, the fabrics go inside a dry heating chamber. In this heating chamber, the heat is applied by various methods [2]. Temperature and pressure are the parameters that affect the internal structural arrangement of the yarns in the fabric to get optimized strength [3]. Heat setting improved the structural arrangement of the fibers and gives unyielding property in fabrics [4]. Application of temperature will adjust the fibers and increase the firmness in yarns [5]. Shrinkage and elasticity of the fabric are directly dependent on the temperature of the heat-setting process [6, 7]. Applying temperature improved the elasticity of the fabrics that are user-friendly by modifying the amorphous interior of the elastic yarns [8]. The required fabric properties like elasticity, stretch and recovery will be obtained by adjusting the temperature precisely [5]. Heat setting also improves the stability, percentage of dye adherence and anti-crease properties of fabrics [9]. If the width of the fabric is less when compared to the finished width, the fabrics must be heated in expanded condition [10]. For elastic fabrics, heat setting was done in the elongated form which improved the elastic performances of fabrics [11]. Changing any one of the parameters such as temperature or machine speed in heat setting will affect the elasticity and strength of the fabric [12]. It was confirmed that application of heat created rigidity, permanent and stable condition for fibers [13]. One of the researchers said that it is very much essential to study the consequence of heat setting and maintain uniform temperature throughout the area of contact of the platen and the fabric [14].

Heat-setting process will be carried out after calendaring and steam curing of the fabric. Rotary curing press is used for heat-setting process that improves the physical properties such as stiffness and elongation of the fabric in textile industries. Elongation is the ability of the fabric to be stretched, extended or lengthened. Moreover, elongation of the fabric varies with different temperatures. This study was carried out to achieve uniform temperature in the top and bottom platen across different zones in the platen. Design changes in the configuration of heating elements and material change in the platen were introduced. Numerical analysis using ANSYS and experimental setup is done to validate and verify the design for uniform temperature and even heating of the fabric. This press operates in the setting temperature range of 160–280 °C with the speed ranging from 20 to 60 m/min and a pressure of 40 daN/cm. It is used for all groups of fabric mass ranged from 150 to 500 g/m$^2$.

The structure of the existing platen is shown in Fig. 1. Both the top and bottom platens in the press are made of rectangular plates of dimensions 1250 × 850 mm$^2$. The top plate is made up of mild steel plate of 40 mm thick fastened to the bottom aluminum plate of 6-mm thickness. There are five heating elements made of tungsten that are inserted inside the mild steel plate. These heating elements are arranged axially that span the entire length of the platen. A 12.5 kW heater is used for heating

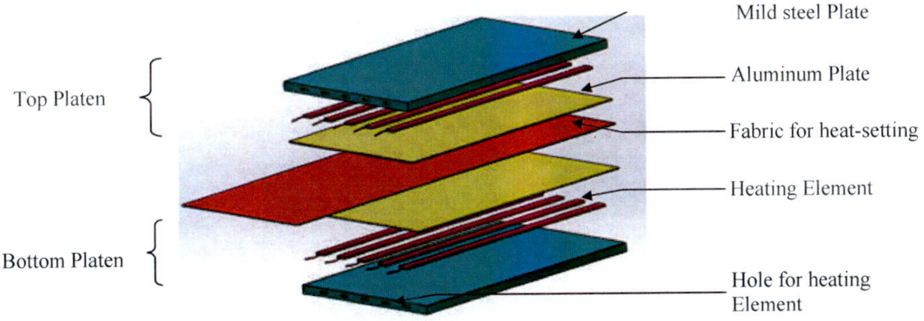


**Fig. 1** Exploded view of the heating platen

each element. The aluminum plate is directly in contact with the fabric. Three thermocouples were used to measure the temperature at the locations as shown in Fig. 2. The total weight of the existing heating platen is 400 kg.

## *1.1 Following Problems Are Faced with the Existing Platen Design*

**Problem 1** In the existing press when the fabric passes through the platen, the temperature obtained is different at various zones across the width and length of the platen. The configuration of different heating zones used for investigating the existing scenario (vertical heating element arrangement) is shown in Fig. 3. FLIR CM275 thermal imaging camera is used to measure the temperature differences across the zones for both top and bottom platens to identify the potential problem. The temperatures thus obtained are shown in Table 1.

The bottom platen temperatures are higher compared to top platen temperatures as shown in Table 1 because of the natural convection effects. Temperature at different zones of the bottom platen varies between minimum of 150 °C and maximum of 170 °C, whereas the temperature at the different zones of the top platen varies between 145 °C (minimum) and 161 °C (maximum). However, the temperature range for the top plate is 20 °C and for the bottom plate is found to be 16 °C. It is clear from the real-time analysis that there will not be uniform heat setting in the fabric as there is a variation in the temperature of top platen and bottom platen. For example at zone 5, the variation is found to be 11 °C leading to non-uniform heat setting in the fabric.

**Problem 2** Another problem faced by the existing press is the arrangement of heating elements. As the fabric passes through the existing platen, failure of one heating element leads to the absence of heat setting throughout the entire length of the fabric.

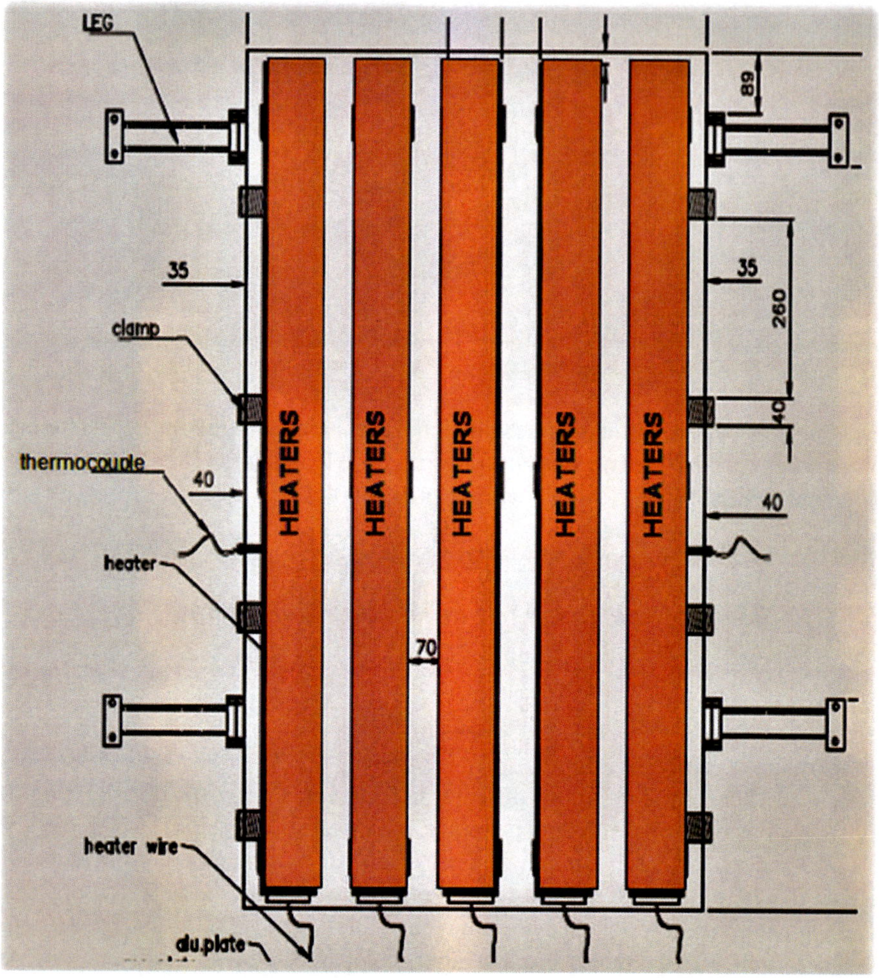


**Fig. 2** Heating element arrangement in the existing setups

## 2 Proposed Modifications and Its Implications

In this study, in order to maintain uniform temperature for the top and bottom platen and to reduce the temperature variation between the zones, certain design changes are proposed. Numerical analysis and experiments were carried out to validate and verify the design modifications.

**Modification 1** The existing vertical heating element arrangement is replaced with a new horizontal heating element arrangement. The heating elements are placed along the width (850 mm) of the plate instead of the length (1250 mm) of the plate. In the proposed design, the quantity of heating elements is found to be 12 numbers to cover the entire span of the platen. This configuration also ensures heating the entire

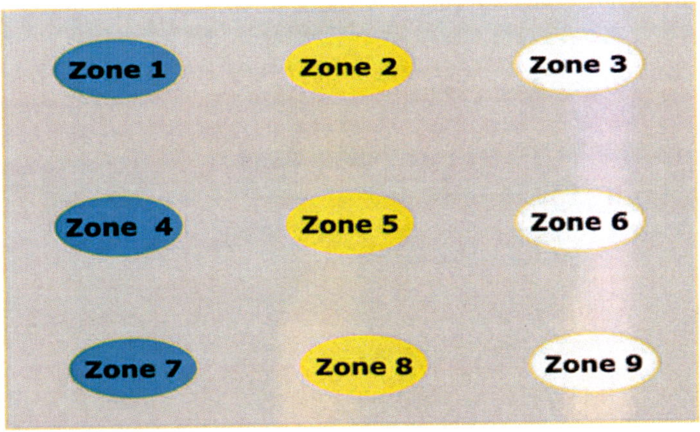


**Fig. 3** Heating zone configuration

**Table 1** Observation of temperature in different zones in the existing platen (vertical heating element arrangement)

| Temperature Observation (°C) | | | |
|---|---|---|---|
| Zone | FLIR – Measurement | | |
| | Bottom | Top | Δ T(°C) |
| 1 | 151 | 150 | 1 |
| 2 | 165 | 161 | 4 |
| 3 | 150 | 145 | 5 |
| 4 | 151 | 148 | 3 |
| 5 | 170 | 159 | 11 |
| 6 | 156 | 154 | 2 |
| 7 | 154 | 149 | 5 |
| 8 | 166 | 156 | 10 |
| 9 | 158 | 157 | 1 |
| Min | 150 | 145 | 1 |
| Max | 170 | 161 | 11 |
| Range | 20 | 16 | 10 |

length of the fabric as it passes through the platen, even if one heating element fails the fabric will be heated up by the previous and subsequent elements so that there will not be any portion of fabric left out without heat setting as shown in Fig. 4.

This arrangement ensures a less variation in temperature at various zones. These heating elements also make sure the surface is heated quickly, effectively and evenly.

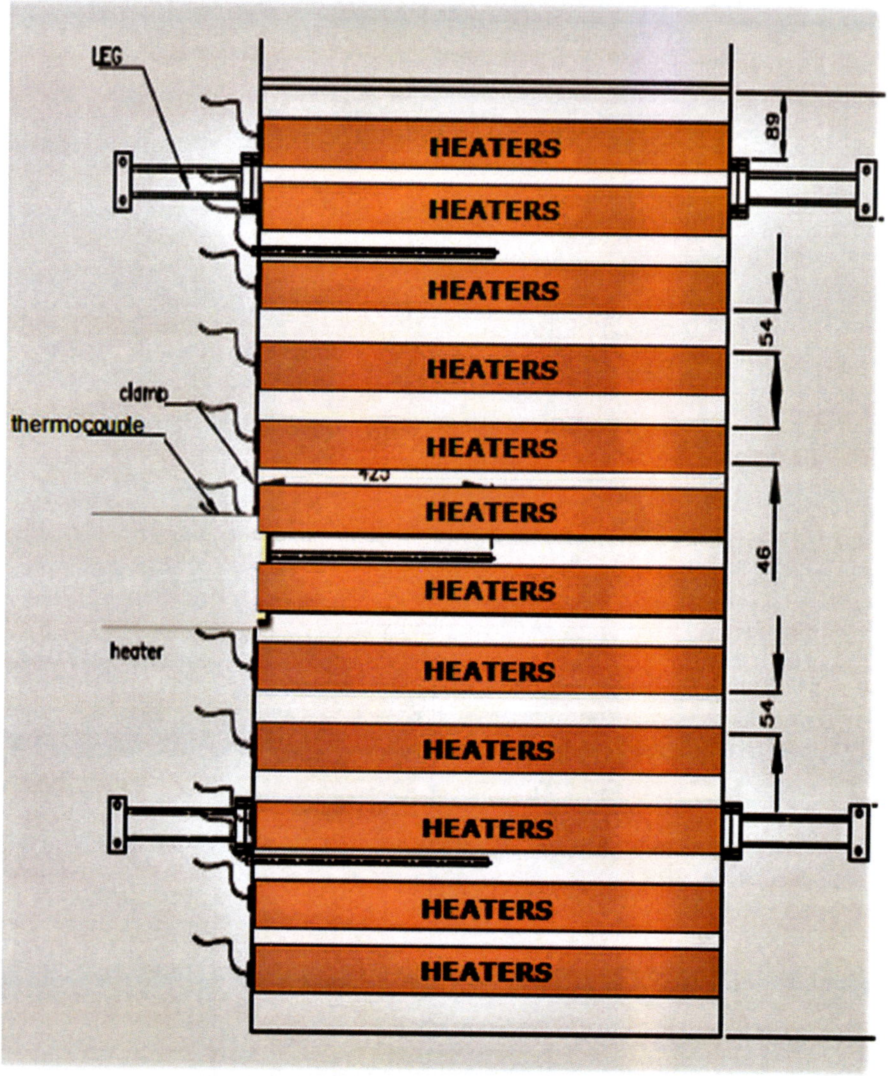


**Fig. 4** Heating element arrangement in the proposed setup

Placing the heating elements horizontally instead of the previous vertical arrangement ensures better heating. Thus, all the zones are heated uniformly, thus heating the entire fabric uniformly.

Heat transfer '*Q*' is an independent parameter. Thermal conductivity '*k*' is a transport property of the medium through which heat transfer takes place. It is a function of the material and the temperature. This can be expressed as follows:

$$\frac{\partial}{\partial x}\left(k\frac{\partial T}{\partial x}\right)+\frac{\partial}{\partial y}\left(k\frac{\partial T}{\partial y}\right)+\frac{\partial}{\partial z}\left(k\frac{\partial T}{\partial z}\right)+\dot{e}_{\text{gen}}=\rho c\frac{\partial T}{\partial t} \tag{1}$$

Equation 1 is the general three-dimensional heat conduction equation in unsteady state with thermal energy generation.

Since the system that the paper is addressing to is a one-dimensional steady-state heat conduction system with no heat generation, the equation is simplified to Eq. 2:

$$\frac{\partial^2 T}{\partial x^2}=0 \tag{2}$$

The heat transfer as given by Fourier's law of heat conduction is

$$Q=-kA\frac{T_2-T_1}{L}=-\frac{T_2-T_1}{R} \tag{3}$$

Considering the case of the two plates, mild steel and aluminum, the heat transfer rate for the existing platen setup can be calculated by the following equation

$$Q=-(T_{\text{Al}}-T_{\text{Ms}})/R_1+R_2=-kA(T_{\text{Al}}-T_{\text{Ms}})/L \tag{4}$$

where

$Q$ is the heat transfer through the plate [W]
$k$ is the thermal conductivity of the material [$\text{Wm}^{-1}\ \text{K}^{-1}$]
$L$ is the plate thickness [mm]
$A$ is the plate area [$\text{mm}^2$]
$T_{\text{Al}}$ = Temperature of the aluminum plate (°C)
$T_{\text{Ms}}$ = Temperature of the mild steel plate (°C)
Thermal resistance of plate 1 ($R_1$) = $L_1/k_1A_1$
Thermal resistance of plate 2 ($R_2$) = $L_2/k_2A_2$.

The thermal resistance '$R$' increases as '$L$' increases, as '$A$' decreases, and as '$k$' decreases. Moreover, '$Q$' is directly proportional to the area '$A$' and inversely proportional to '$L$'. In this system, the thickness of the plate cannot be changed in the machine and the thickness of 46 mm has to be retained as it is. As we have increased the number of heating elements to 12, the area '$A$' increases further leading to an improved heat transfer. As the heat transfer is high, the platen heats up fast leading to faster heating of the platen in the machine.

$$\begin{aligned}\text{Heating element surface area} &= \text{Number of heating elements}\\ &\quad\times \text{Surface area of contact of one element}\end{aligned} \tag{5}$$

Since the heating element surface area is increased after 12 horizontal heating elements are placed, the heating takes place much faster compared to the existing 5 vertical heating element setups. Therefore, the plate gets heated faster and hence

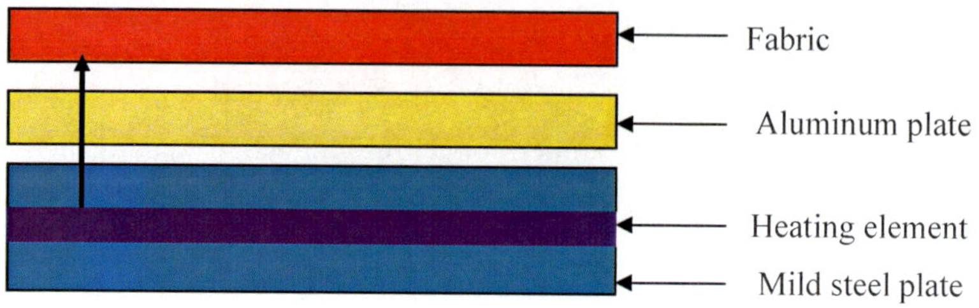


**Fig. 5** Representation of heat transfer in the existing setup

the fabric. This decreases the time taken for heating and therefore increases the production rate.

**Modification 2** The existing top plate is made of mild steel, the bottom plate is made of aluminum, and heat from the heating element passes through a layer of aluminum and a mild steel layer as shown in Fig. 5 in the existing setup.

To have an improved heat transfer in the setup, a material of low thermal resistivity is incorporated in the proposed design. So, it is proposed to use aluminum plate of 46 mm thick by retaining the same area of $1250 \times 850\ \text{mm}^2$ for both top and bottom platens. The proposed changes were incorporated, and the arrangements of heating elements are as shown in Fig. 6. As mild steel is replaced with aluminum, the better thermal conductivity of aluminum heats up the fabric very fast and time taken to heat the fabric reduces.

In the existing vertical heating element arrangement, a 12.5 kW heater is used for heating each element. Since the new horizontal heating element arrangement consists of 12 heating elements, the wattage of the heating elements can be reduced to obtain the same temperature. Hence, a heater with 5.2 kW rating is replaced as a heat source for the heating elements. Therefore, the power consumption remains the same before and after modifications.

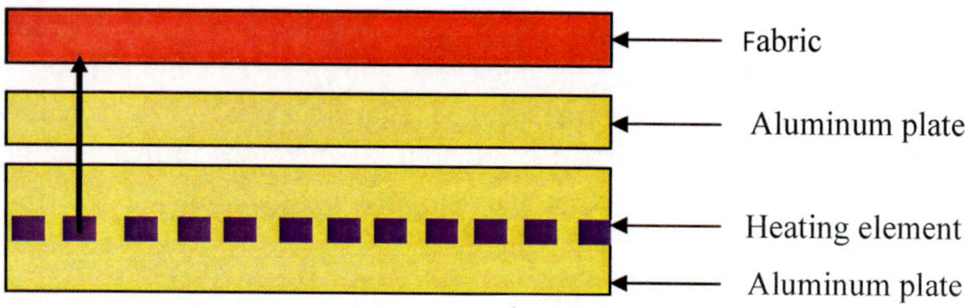


**Fig. 6** Representation of heat transfer in the proposed setup

## 3 Numerical Verification

ANSYS is a general-purpose finite element analysis (FEA) software package. The finite element analysis is a numerical method for solving problems of engineering. This type of analysis is typically used for the design and optimization of a system which is complex to solve by hand and to validate the design changes before setting up the actual experimental setup. ANSYS 16.0 is used in this paper for steady-state thermal analysis for the existing heating element configuration and the proposed modifications (Figs. 7 and 8; Tables 2 and 3).

For the existing vertical heating element configuration, the temperature is high at Zone 5 and is of 173 °C. And the temperature across various zones varies from 152 to 173 °C and the range is 21 °C, whereas in the proposed design the temperature

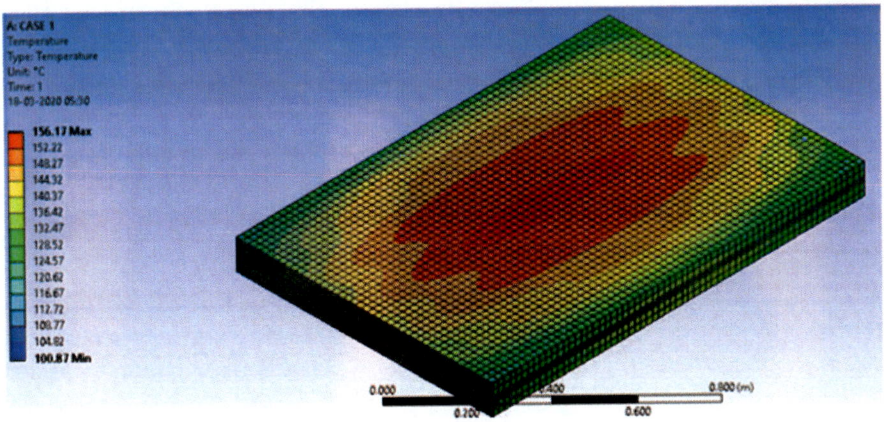


**Fig. 7** Temperature distribution in vertical heating element arrangement

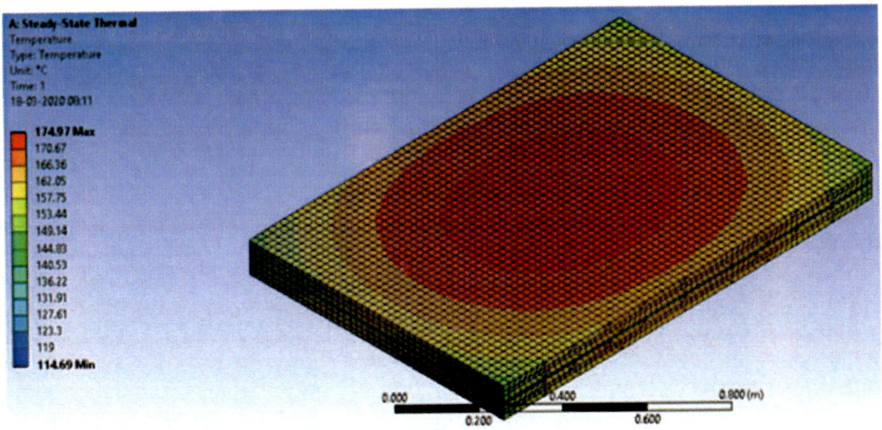


**Fig. 8** Temperature distribution in horizontal heating element arrangement

**Table 2** Temperature values in vertical heating element arrangement from ANSYS

| Zone | Bottom plate temperature (°C) |
|---|---|
| 1 | 152 |
| 2 | 167 |
| 3 | 152 |
| 4 | 153 |
| 5 | 173 |
| 6 | 160 |
| 7 | 159 |
| 8 | 168 |
| 9 | 159 |

**Table 3** Temperature values in horizontal heating element arrangement from Ansys

| Zone | Bottom plate temperature (°C) |
|---|---|
| 1 | 172 |
| 2 | 175 |
| 3 | 174 |
| 4 | 173 |
| 5 | 177 |
| 6 | 173 |
| 7 | 174 |
| 8 | 175 |
| 9 | 172 |

at Zone 5 is 177 °C and the temperature across various zones varies from 172 to 177 °C in the range of 5 °C. The theoretical calculations of temperature are higher when compared to the observed experimental values due to assumptions in material properties from standard library that will be different from actual material of the plates, which are used in real time. The numerical analysis gave the expected results, and the proposed modifications are validated before going for experimental setup.

## 4 Results and Discussion

Figures 9 and 10 show the FLIR images of different heating zones in the top and bottom platen. The thermal imaging camera is used to measure the temperature differences across different zones for both top and bottom platens, and the observed temperatures are given in Table 4. In the existing platen, the temperature in the top and bottom platen varied considerably. In the case of 5 vertical heating elements, failure

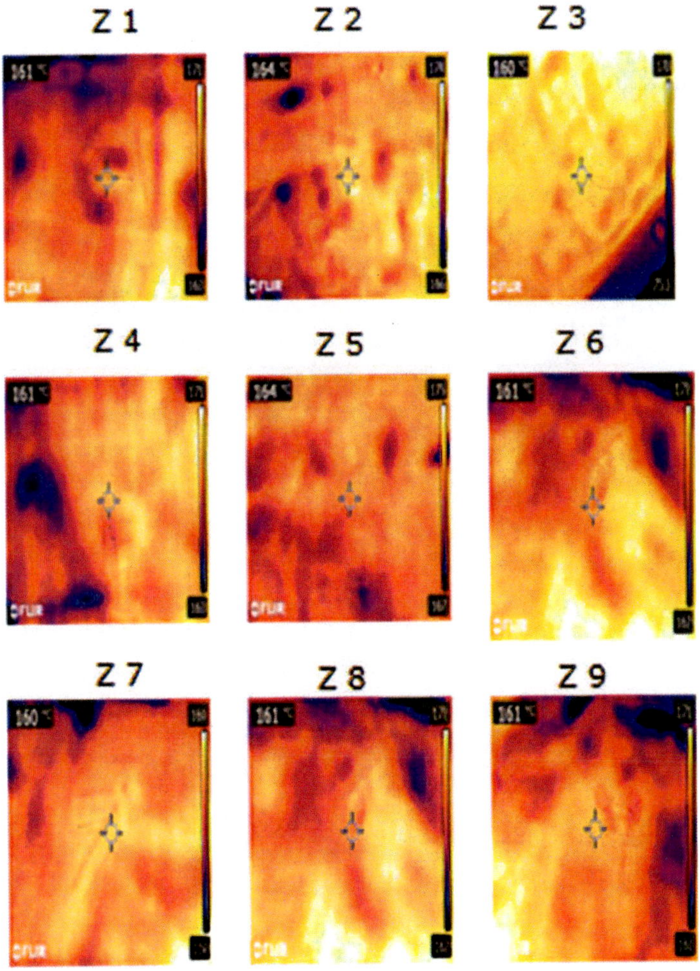


**Fig. 9** FLIR image—top platen

of one heating element leaves a portion of the fabric passing through it unheated. Moreover because of the placement of 5 heating elements along the length of the motion of the fabric, gap between the heating elements exists. So, different zones are heated up at different temperatures resulting in uneven heat setting in the fabric. At times when the element fails, heat setting will not happen throughout the length of the fabric, whereas in case of 12 horizontal heating elements, if one of the heating elements fails, the previous and subsequent heating elements will aid the heating process and the entire portion of the fabric gets heated. Therefore, the difference in temperature between various zones is reduced and uniform heating is achieved.

From Table 5, it is evident that the variations in temperature at different zones are reduced from 20 to 4 °C at the bottom platen. Similarly, the temperature variations at top platen are reduced from 16 to 5 °C at different zones that are investigated before

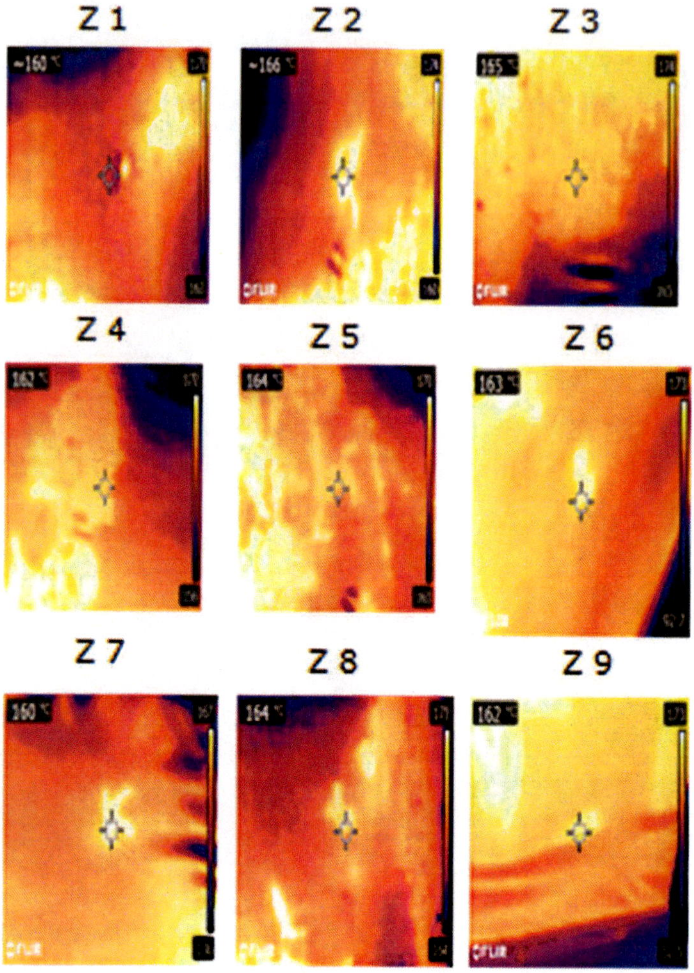


**Fig. 10** FLIR image—bottom platen

and after the change in design. The thermal conductivity at room temperature for Al is 200 W/m K and that of mild steel is 50 W/m K. Therefore, time taken to reach the desired temperature of 173 °C (approximately) is less with aluminum compared to mild steel. Since the power consumption remains the same, the running cost of the machine remains the same before and after the modifications.

## 5 Conclusion

Heat setting in textile industry is a thermal process to improve physical attributes of the fabric for subsequent processes by stabilizing or fixing of the fiber thread. The

**Table 4** Observation of temperature in different zones in the experimental platen

| Temperature Observation (°C) | | | |
|---|---|---|---|
| Zone | FLIR – Measurement | | |
| | Bottom | Top | Δ T (°C) |
| 1 | 171 | 170 | 2 |
| 2 | 173 | 171 | 2 |
| 3 | 173 | 170 | 3 |
| 4 | 172 | 171 | 1 |
| 5 | 175 | 175 | 0 |
| 6 | 171 | 170 | 1 |
| 7 | 173 | 172 | 1 |
| 8 | 174 | 171 | 3 |
| 9 | 171 | 170 | 1 |
| Min | 171 | 170 | 0 |
| Max | 175 | 175 | 3 |
| Range | 4 | 5 | 3 |

**Table 5** Comparison of temperature in top and bottom platen

| Platen type | Bottom | | | Top | | |
|---|---|---|---|---|---|---|
| | Min temperature (°C) | Max temperature (°C) | $\Delta T$ (°C) | Min temperature (°C) | Max temperature (°C) | $\Delta T$ (°C) |
| Old | 150 | 170 | 20 | 145 | 161 | 16 |
| New | 171 | 175 | 4 | 170 | 175 | 5 |

design modifications, analysis and experimentation conducted in this paper are to minimize the variation in the heat-setting process. The efficiency of this heat-setting process is ensured by maintaining uniform temperature in the process. The effect on thermal characteristics of changing the configuration of the heating elements and material of the platen has been studied. The following conclusions are drawn from the study.

- The change in the arrangement of the heating elements from the existing vertical arrangement to the proposed horizontal arrangement has ensured that the entire fabric is heated. No portion of the fabric is left unheated if a heating element fails as the previous and subsequent heating elements would heat the fabric. Thus, the entire span of fabric is heated.
- Placing heating elements horizontally in the proposed setup has the decreased variation in temperature at different zones. Hence, uniform heating is achieved across all the zones. Thus, the fabric is heated to almost the same temperature (173 °C) throughout.

- The surface area '$A$' of the heating elements is increased in the proposed horizontal arrangement of heating elements. This increases the rate of heat transfer, thereby reducing the time taken to achieve the desired temperature to heat the fabric.
- Thermal conductivity of the plate is the function of material being used. In order to achieve the required temperature with improved heat transfer, mild steel is replaced with aluminum which has greater thermal conductivity. Therefore, heat transfer is improved as and the desired temperature is achieved quickly.
- The size of the platen remains the same as $1250 \times 850 \times 46\ mm^3$. Even though the size remained same, due to the change in material from a dissimilar material combination to aluminum, the weight of both the top and bottom platens is reduced by 68% and is at 150 kg (each) from 400 kg.
- As 5 heating elements are replaced with 12 heating elements, the heater wattage of 12.5 kW is reduced to 5.2 kW in the proposed design. The power consumption thus remains the same in the new design.
- Uniform temperature is achieved by arranging the proposed 12 heating elements along the width instead of the 5 heating elements along the length. The proposed configuration of heating elements has significantly reduced the temperature variation between different zones in the platen. In the bottom plate, the variation is reduced by 80% from 20 to 4 °C, whereas the variation in top plate is reduced by 68.75% from 16 to 5 °C. The study recommends, in further research experiments, to use different fabrics and analyze the effect of heat setting.

## References


1. Hasani H, Avinc O, Khoddami A (2017) Effects of different production processing stages on mechanical and surface characteristics of polylactic acid and PET fibre fabrics. Indian J Fibre Text Res (IJFTR) 42(1):31–37
2. Pervez MN, Talukder ME, Shafiq F, Hasan KMF, Taher MA, Meraz MM, Lin L (2018) Effect of heat-setting on UV protection and antibacterial properties of cotton/spandex fabric. IOP Conf Ser: Mater Sci Eng 284(1):012010
3. Senthilkumar P, Suganthi T (2019) Influence of tuck stitch in course direction on thermal comfort characteristics of layered knitted fabrics. Indian J Fibre Text Res (IJFTR) 44(2):163–172
4. Asayesh A, Mirgoli F, Gholamhosseini A (2018) An investigation into the effect of fabric structure on the compressional properties of woven fabrics. J Text Inst 109(1):32–38
5. Jankoska M, Demboski G (2018) Influence of structure variation and finishing on woven fabric thermal properties. Fibres Text East Europe 26(1):127
6. Herath CN, Kang BC (2008) Dimensional stability of core spun cotton/spandex single jersey fabrics under relaxation. Text Res J 78(3):209–216. https://doi.org/10.1177/0040517507082958
7. Wei J, Xu S, Liu H, Zheng L, Qian Y (2015) Simplified model for predicting fabric thermal resistance according to its microstructural parameters. Fibres Text East Europe 4(112):57–60
8. Pervez MN, Talukder ME, Datta MK, Mia MS, Zaman A, Khan MMR, Cai Y, Lin L (2017) The influence of annealing process on crystallinity and structural properties of cotton/spandex fabric. In: MATEC web of conferences, vol 130 (EDP Sciences, 2017), p 02001

9. Herath CN, Choon Kang B (2007) Dimensional characteristics of core spun cotton–spandex 1 9 1 rib knitted fabrics in laundering. Int J Cloth Sci Technol 19(1):43–58. https://doi.org/10.1108/09556220710717044
10. Nassif NAA (2018) Influencing sheath type and core draft ratios on the physical properties of stretchable yarns. Int J ChemTech Res 11(05):500–508
11. Safdar F, Hussain T, Nazir A, Iqbal K (2014) Improving dimensional stability of cotton knits through resin finishi ng. J Eng Fabr Fibers (JEFF) 9(3):28–35
12. Mozafary V, Payvandy P, Rezaeian M (2018) A novel approach for simulation of curling behavior of knitted fabric based on mass spring model. J Text Inst 1–22
13. Sadek R, El-Hossini AM, Eldeeb AS, Yassen AA (2012) Effect of lycra extension percent on single jersey knitted fabric properties. J Eng Fibers Fabr 7(2):11–16
14. Wang Y, Zhang P, Zhang Y (2014) Experimental investigation the dynamic pressure attenuation of elastic fabric for compression garment. Text Res J 84(6):572–582. https://doi.org/10.1177/0040517513503726

# Drag Reduction in the Sedan Car by Implementing Diffuser to Improve the Fuel Efficiency

**S. Ajith Balaa, S. Aravind, N. Kowshik Santhakumar, and S. Saravana Kumar**

**Abstract** This paper focuses on studying the effect of diffuser in improving the fuel economy by reducing drag force and also improving the vehicle stability. Due to complex geometry (presence of axles, exhaust pipes, etc.), we tend to neglect the underbody of the sedan car. But, due to the irregular projections, drag is created. To reduce drag, we introduce diffuser which increases fuel economy. In the upcoming decade, electric cars will dominate the automobile industry. Introducing diffuser will serve its purpose irrespective of its type. In this paper, four different ramp angles are taken, and its effect is studied using computational fluid dynamics (CFD). Simulations are done on a sedan type car model (Honda city). By attaching diffuser with increased length from the body and various angles are studied at that length, it is found that the drag force is significantly reduced, and also the vehicle stability is improved.

**Keywords** Drag force · Vehicle stability · Fuel economy · Ramp angles

## 1 Introduction

The necessity to improve fuel efficiency has been increasing as government introduce strict regulations in $CO_2$ emissions. It is also true that the speed of the commercial

S. Ajith Balaa (✉) · S. Aravind · N. Kowshik Santhakumar · S. Saravana Kumar
Department of Mechanical Engineering, PSG College of Technology, Coimbatore, Tamil Nadu, India
e-mail: ajithbalaa03@gmail.com

S. Aravind
e-mail: aravindsakthivel1999@gmail.com

N. Kowshik Santhakumar
e-mail: kowshikchristi@gmail.com

S. Saravana Kumar
e-mail: saravanakumar2622000@gmail.com

© The Editor(s) (if applicable) and The Author(s), under exclusive license to Springer Nature Singapore Pte Ltd. 2021

S. Mohan et al. (eds.), *Materials, Design, and Manufacturing for Sustainable Environment*, Lecture Notes in Mechanical Engineering,
https://doi.org/10.1007/978-981-15-9809-8_49

vehicles keeps increasing as decade pass by. Thus, aerodynamic design plays a major role in improving the fuel efficiency [1]. As the car speed increases, due to which pressure gradient between front and rear of the vehicle increases, drag is created [2]. Also during high-speed cornering, downforce is also required. This can be achieved using undertray diffuser. This idea is not only applicable for fuel-consuming vehicles but also for battery car. The main idea of its working is 'venturi effect', and so it is also knowns as venturi tunnels [3].

Our paper shows how drag force improves fuel efficiency and also improves vehicle stability. Bluffed shapes in the rear of the car creates turbulence, thus creating drag [4]. Whenever there is restriction for passage for air flow, drag force is created pulling the car back [4]. By introducing diffuser, passage for air flow is created thus reducing drag. Due to passage of airflow, velocity of air in the bottom of the car increases [5]. So, the pressure difference between the front and rear of the car decreases, thus reducing the pulling force required for a car [6].

Four different ramp angles are taken, and effect of ramp angle on drag force and lift force is studied. A conclusion of how much percentage of fuel is saved is given, and comparison is made between sedan car with and without diffuser to study the effect of rear undertray diffuser [7].

# 2 Methodology

## 2.1 *Modelling and Design*

CFD analysis of the 3D car model can be done with different turbulence models, but close value is obtained with $k - \varepsilon$ model (SST). 3D car model is first analysed with diffuser and without diffuser body, and the overall drag and lift values are calculated by scaling down the actual model by four times to reduce the computing time and better wind tunnel utilisation by the use of Reynolds number. The vehicle body surfaces were drawn using SolidWorks and imported as Parasolid file in ANSYS environment. For simulation in CFD, geometry creation is required with its grid generation, physical domain integration and choice of computing scheme and turbulence model.

### 2.1.1 3D Car Model

To analyse the characteristics of a three-dimensional car model along with its coefficient of drag and lift calculations, we investigate the flow over and around the car body.

Analysis is done with five different cases

(1) Without diffuser
(2) With diffuser of angle 4°

(3) With diffuser of angle 6°
(4) With diffuser of angle 8°
(5) With diffuser of angle 10°.

By fixing the length (360 mm), we are analysing five different diffuser configurations. All the above cases are tested at a velocity of 64 m/s which is obtained as a result of scaling down the model which is travelling 16.66 m/s. After creating a physical domain around the car using SolidWorks, we are importing it into ANSYS Workbench meshing tool, and triangular elements first cover the surface followed by tetrahedral mesh covering ten layers of inflation extending the entire domain. To understand the effect of pressure and velocity distribution around the car along with coefficient of lift and drag prediction at various location, ANSYS CFX is used as test section for a series of calculation, and the results are presented.

For calculating the velocity of scaled down model, we used Reynolds number. We scaled down a Honda city car to a ratio of 1:4

$$\mathrm{Re} = \frac{\rho V D}{\mu} \tag{1}$$

$\rho$ Density of air (1.2 kg/m$^3$)
$D$ Frontal area of car (2.252 m$^2$)
$V$ Velocity of the car (16.66 m/s)
$\mu$ Dynamic viscosity of fluid ($1.872 \times 10^{-5}$ kg/m s).

$$\mathrm{Re} = \frac{1.2 * 16.66 * 2.252}{1.872 \times 10^{-5}} = 24.07 \times 10^5$$

Since we have reduced the frontal area ($D$) in the ratio of 1:4 scale, then the frontal area becomes $D = 0.56$ m$^2$.

For scaling down the model, we are keeping Reynolds number as constant and finding its velocity.

$$V = \frac{\mathrm{Re} * \mu}{\rho * D} \text{m/s}$$

$$V = \frac{24.07 \times 10^5 * 1.872 \times 10^{-5}}{1.2 * 0.56} = 64\,\text{m/s}$$

The velocity of a scaled down model is 64 m/s (Figs. 1, 2, 3, 4, 5, 6 and 7).

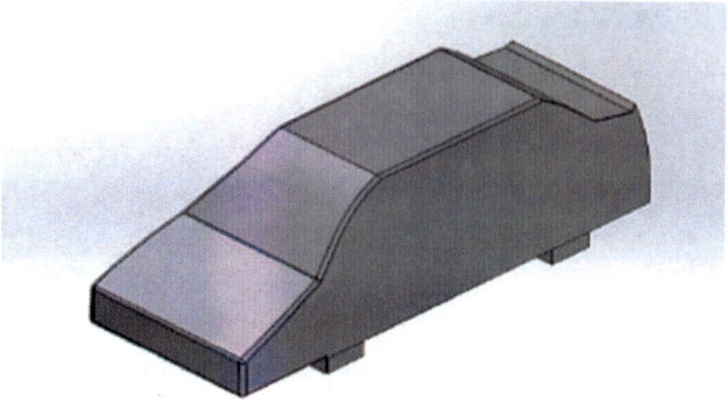

**Fig. 1** Without diffuser

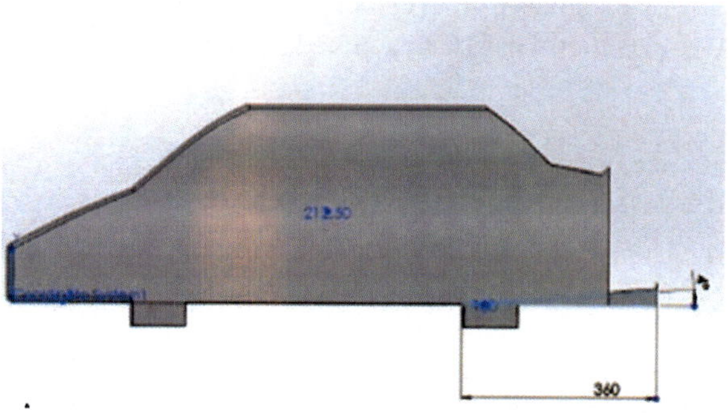

**Fig. 2** Diffuser with ramp angle 4°

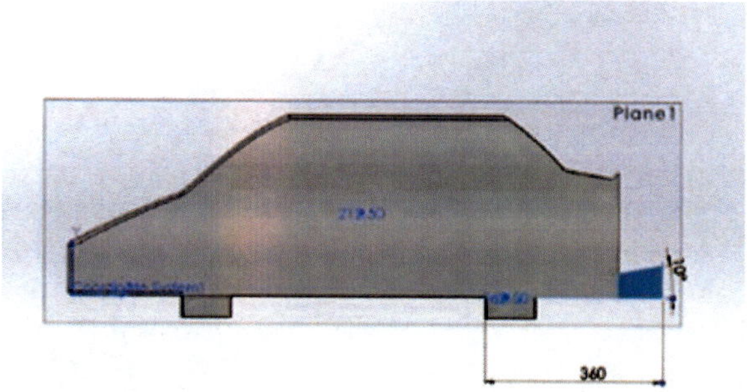


**Fig. 3** Diffuser with ramp angle 10°

## 2.2 *Grid Generation*

The car model is symmetrical. So, only half of the sedan car model is taken to reduce the computational time. For meshing, ANSYS Workbench is used. The result values are varied due to the mesh quality. Mesh is generated by taking a finite volume with unstructured elements [5]. For good quality of mesh, appropriate skewness

Fig. 4 Diffuser with ramp angle 6°

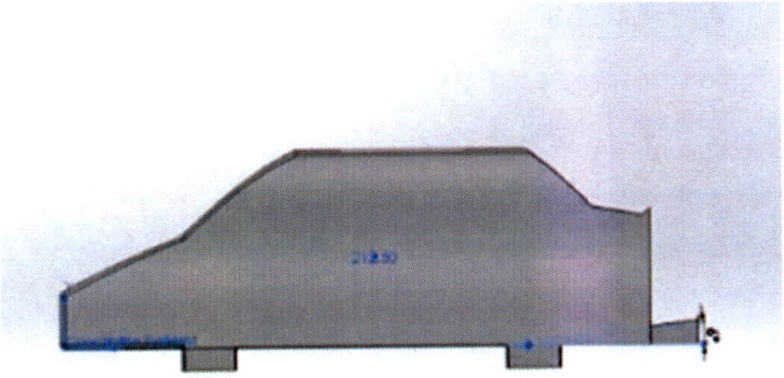

Fig. 5 Diffuser with ramp angle 8°

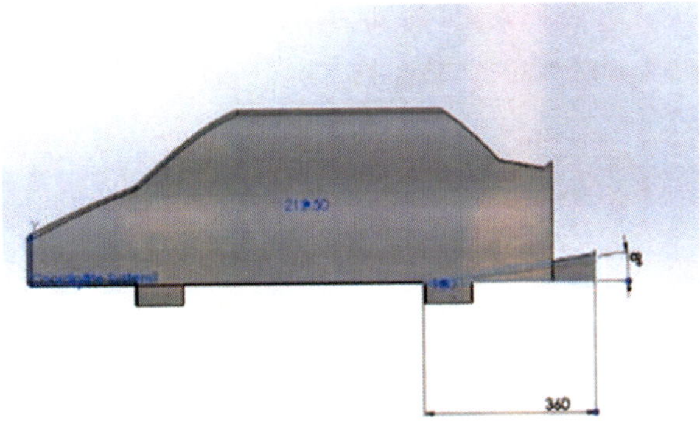


Fig. 6 Enclosure with symmetry

value should be obtained. This can be obtained by varying the scale, shape and size factors. This will be followed by giving desired boundary conditions.

## 2.3 Boundary Conditions

- Inlet velocity: 64 m/s

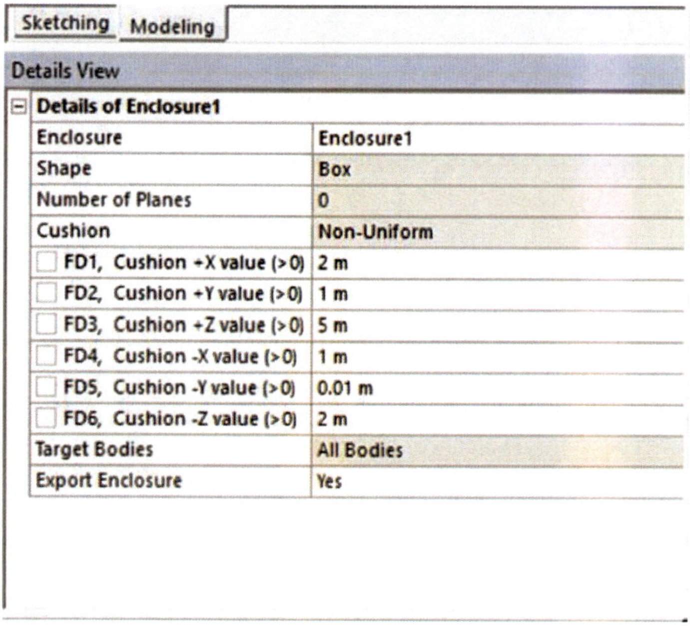

Sketching | Modeling

Details View

| Details of Enclosure1 | |
|---|---|
| Enclosure | Enclosure1 |
| Shape | Box |
| Number of Planes | 0 |
| Cushion | Non-Uniform |
| FD1, Cushion +X value (>0) | 2 m |
| FD2, Cushion +Y value (>0) | 1 m |
| FD3, Cushion +Z value (>0) | 5 m |
| FD4, Cushion -X value (>0) | 1 m |
| FD5, Cushion -Y value (>0) | 0.01 m |
| FD6, Cushion -Z value (>0) | 2 m |
| Target Bodies | All Bodies |
| Export Enclosure | Yes |

**Fig. 7** Dimensions of the enclosure

- Outlet pressure: zero (gauge pressure)
- Wall condition: No slip (Fig. 8).

**Fig. 8** Final meshing

## 3 Results and Discussions

The Fluent and the design procedure with reference to Jory [3] used a simplified geometry of an ideal sedan vehicle (we took Honda city dimensions) with iterative mesh convergence, and the various turbulence models available in Fluent are used. And the value of coefficient of drag for Honda city is taken as a reference value, 0.31.

Unstructured tetrahedral mesh was used for analysis. Ahmed bluff car surface is first covered with triangular elements and is followed by ten layers of inflation with $y+$ value of 0.0001847 m, which is covered with tetrahedral mesh extending the entire domain.

The flow is assumed to be incompressible, and it is also assumed that there is negligible heat transfer (Fig. 9).

Sedan Car Results

In the present analysis, five different slant angles are considered such as 0°, 4°, 6°, 8° and 10°, each having extended length of 360 mm (scaled down) with a velocity of 64 m/s. Drag coefficient values obtained by $k - \varepsilon$ model of these angles are obtained, initial position without any of this diffuser were calculated, and the following results were obtained (Table 1).

$$\text{Length of diffuser} = 360\,\text{mm(from three-fourth position of the car)}$$

**DRAG AND LIFT CALCULATIONS**

CASE 1: WITH DIFFUSER: (ANGLE OF ATTACK = 6°)

1. Calculation of drag coefficient

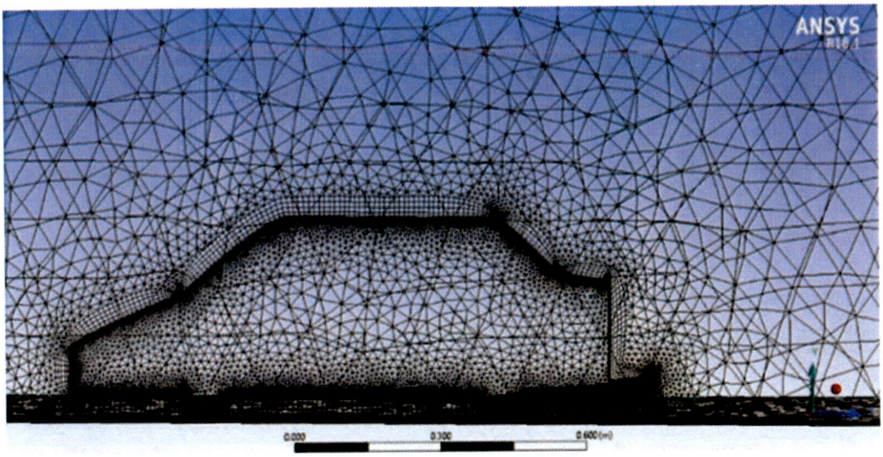


**Fig. 9** Meshed model

**Table 1** Drag and lift values of the car model at different ramp angles

| Type (angle) (°) | Drag (N) | $C_d$ | Lift force (N) | $C_l$ |
|---|---|---|---|---|
| 10 | 120.96 | 0.2895 | −80.58 | 0.1929 |
| 8 | 115.48 | 0.2764 | −57.48 | 0.1376 |
| 6 | 114.36 | 0.2737 | −57.18 | 0.1349 |
| 4 | 118.8 | 0.2843 | −37.94 | 0.0908 |
| 0 | 126 | 0.3015 | 2.37 | 0.0113 |

$$F_D = \frac{1}{2}\rho A V^2 C_d \quad (2)$$

where

$F_D$ Drag force (57.18 N) (half car)
$\rho$ Density of air (1.2 kg/m$^3$)
$A$ Frontal area of the car (0.170 m$^2$)
$V$ Velocity of the model (64 m/s).

From this, we have calculated

$$C_d = \text{Drag coefficient}$$
$$= \frac{57.18 * 2 * 2}{1.2 * 0.17 * 64^2}$$
$$C_d = 0.2737$$

2. Calculation of lift coefficient

$$F_L = \frac{1}{2}\rho A V^2 C_l \quad (3)$$

where

$F_L$ Lift force (−28.59 N) (half car)
$\rho$ Density of air (1.2 kg/m$^3$)
$A$ Frontal area of the car (0.170 m$^2$)
$V$ Velocity of the model (64 m/s).

From this, we have calculated,

$$C_l = \text{Lift coefficient}$$
$$= \frac{28.59 * 2 * 2}{1.2 * 0.17 * 64^2}$$
$$C_l = 0.1349$$

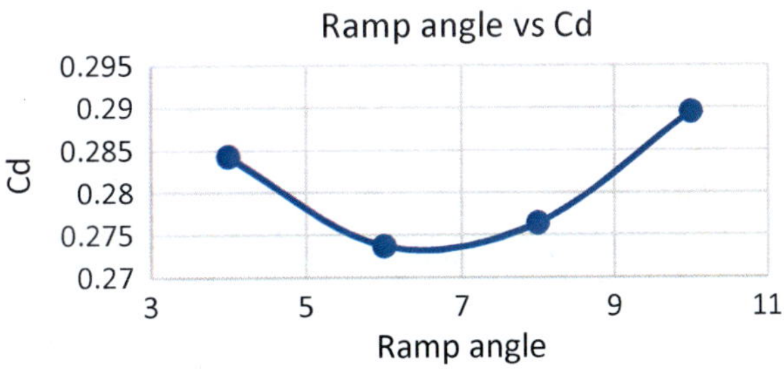


**Fig. 10** Variation of drag value with ramp angles

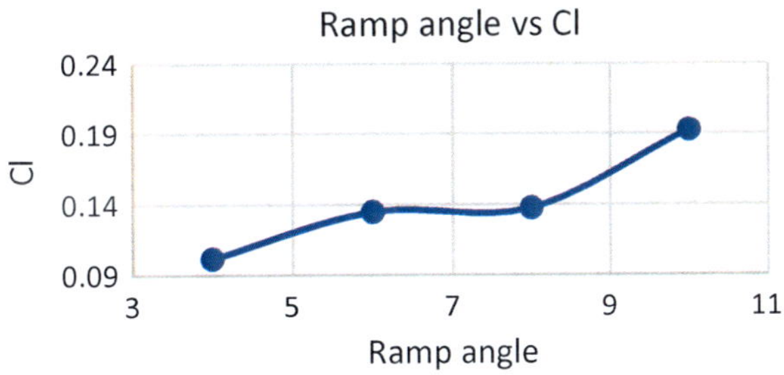


**Fig. 11** Variation of lift with ramp angles

CASE 2: WITHOUT DIFFUSER:

3. Calculation of drag coefficient:
   Using Eq. (2), we calculated coefficient of drag

$$C_{\mathrm{d}} = 0.3015$$

4. Calculation of lift coefficient:
   Using Eq. (3), we calculated coefficient of lift (Figs. 10 and 11)

$$C_{\mathrm{l}} = 0.0113$$

From the above graphs, it is concluded that introducing the diffuser reduces the drag acting on the car considerably, and after 6°, the value got increasing due to the flow separation, and the downforce is linearly increasing with increase in ramp angle (Figs. 12, 13, 14, 15 and 16).
The above pictures show the variation of pressure and velocity with and without diffuser, it is clear that while introducing the diffuser, the wake region behind the car is much reduced, and so the pressure gradient around the car is reduced. It is important to consider the length of the diffuser to obtain minimum value of drag but too lengthy diffuser might be difficult while driving, and the rear section will act as a cantilever.
By this, the fuel efficiency is improved to a considerable extent. The relation between the drag force to fuel efficiency is identified, and percentage of fuel saving is calculated below.

**Fig. 12** Pressure contour (without diffuser)

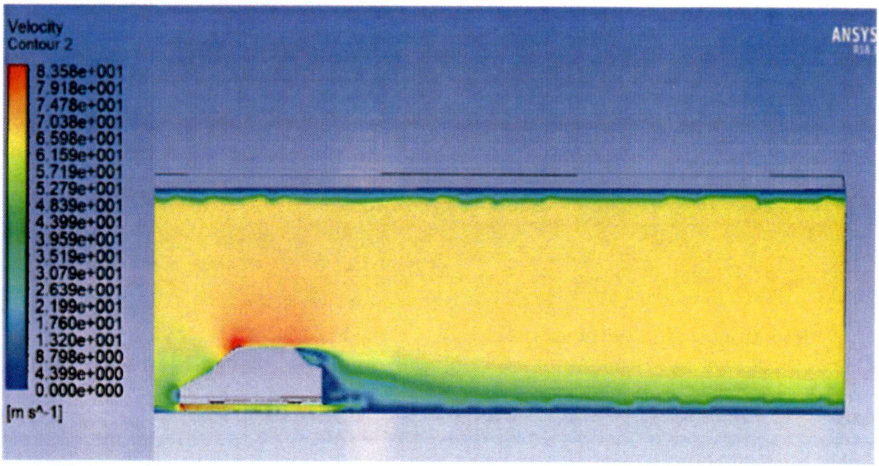


**Fig. 13** Velocity contour (without diffuser)

FUEL EFFICIENCY:

$$\text{Work} = \text{Force} * \text{distance} \tag{4}$$

where
distance = 500 miles (Table 2)

- 1 mile = 1.609 km
- 500 miles = 805 km
- 1 l of diesel produce 9.96 kwh of work, and for 7.7 kwh, it requires 0.77 l.

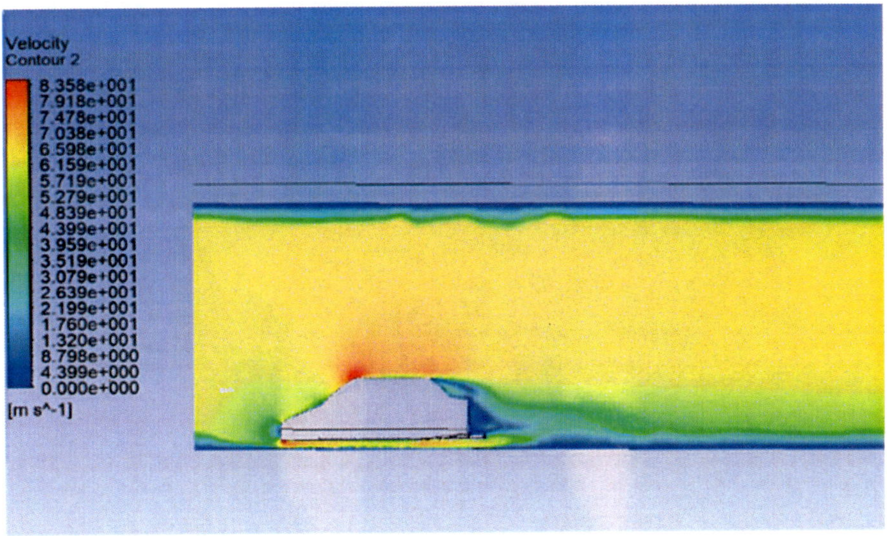


**Fig. 14** Velocity contour with diffuser

**Fig. 15** Pressure contour with diffuser

- For 805 km of distance, 0.77 l of diesel is saved, and for 1 lakh km, 95.65 l of diesel is saved.
- For 1 lakh km of distance, 3906.25 l of diesel is required.
- $$\textbf{Improved efficiency} = \frac{95.65}{3906.25} * 100$$
  $$= \mathbf{2.4}\%.$$

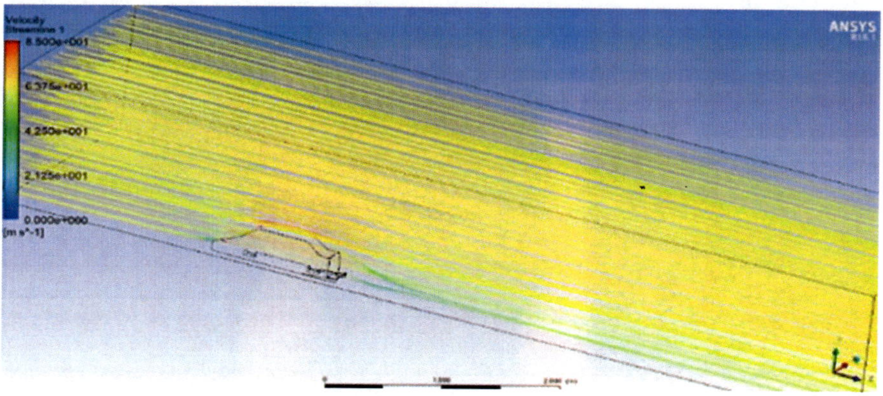


Fig. 16 Velocity streamline with diffuser

Table 2 Calculation of fuel efficiency

| Type | Drag force (N) | Required work (W) (for 500 miles) (kwh) | $\Delta W$ | Fuel power need to be supplied (IC engine) |
|---|---|---|---|---|
| With diffuser (6°) | 114.36 | 25.45 | 2.71 kwh | 7.7 kwh |
| Without diffuser | 126 | 28.16 | | |

Hence, the fuel efficiency is improved by 2.4% (when a car travelling at 60 km/h) as a result of implementing the diffuser.

# 4 Conclusions

Airflow over the sedan car (Honda city) is investigated using the ANSYS CFX, CFD software to understand the flow processes involved in drag production. The configurations included different rear ramp angles of 3D car model. It is found that the diffuser of 3D car model with ramp angle of 6° produces minimum drag force when compared to the 3D car model without diffuser. Later numerical investigation of aerodynamic lift, drag and flow characteristics of the passenger sedan car are carried out. By simulation, it is found that without diffuser, positive lift is created which reduces car stability. For providing car stability and to improve the fuel efficiency by 2.4%, we need to install diffusers. It is also found that drag force is reduced by restricting boundary layer separation and by reducing the form drag by introducing undertray diffuser.

**Acknowledgements** The authors would like to gratefully acknowledge S. O. Kang and Kurian Jory for their outstanding work on diffusers help us to get better ideas for the successful completion of this thesis work.

## References


1. Kang SO (2012) Actively translating a rear diffuser device for the aerodynamic drag reduction of a passenger car. Paper No:583-592-2012
2. Rao et al (2007) External aerodynamic flow for high speed passenger car. SAE Paper No, 2007-26-050
3. Jory K (2009) Computational drag analysis in the underbody for a sedan type car model. SAE Paper No:2009-01-1157
4. Shinsuke et al (2008) Drag force reduction of a bluff-body with an underbody slant and rear flap. SAE paper No. 2008-01-2599
5. Kevin et al (1998) The aerodynamic performance of automotive under-body diffusers. SAE Paper No. 980030
6. Christofer et al (2011) Influence different front and rear wheel designs on aerodynamic drag of a sedan type passenger car. SAE paper No. 2011-36-0165
7. Hu X (2011) Influence of Different Diffuser Angle on Sedan Aerodynamic Characteristics. Paper No:22-29-2011

# Smart Maintenance and Analytics for Indian SMEs

**S. Krishnaraj, M. Gomathi Prabha, and M. Yuvaraja**

**Abstract** Today, in an Industry 4.0 era, every equipment is related to form an integrated community, which requires the employment of advanced predictive gears, in order that records are frequently converted into facts to clarify suspicions and make precise predictive decisions. IoT-based manufacturing and repair improvements are unavoidable trends and demanding situations for today's production. This research delineates the tendencies of production upgradation, further the keenness of smart prognostic technologies for equipments to manage achieving visibility and better productivity. This project is also explained with a case study in which IoT implementation smoothens the flow of production by acquiring the better overall equipment effectiveness of machines in rubber encapsulation industry from 55.09 to 72.91% and 76.07% continually toward the world-class OEE in such a way that every Indian SME could adopt such Kaizen for its steady growth.

**Keywords** Industry 4.0 · Prognostic maintenance · Industrial big data · OEE

## 1 Introduction

Industries are now trying to manage massive records of problems and spontaneous decision-making for productivity enhancement in a fierce corporate culture. Several production systems are not geared up to control massive data because of the dearth of clever analytic tools. Smart industry is based on IoT-enabled production and innovative service. As Internet-enabled systems are incorporated in production, its

S. Krishnaraj (✉) · M. Gomathi Prabha · M. Yuvaraja
Department of Mechanical Engineering, PSG College of Technology, Coimbatore, Tamil Nadu, India
e-mail: conveytokrishnaraj@gmail.com

M. Gomathi Prabha
e-mail: prabhamohan.psg@gmail.com

M. Yuvaraja
e-mail: yuvam81@gmail.com

© The Editor(s) (if applicable) and The Author(s), under exclusive license to Springer Nature Singapore Pte Ltd. 2021
S. Mohan et al. (eds.), *Materials, Design, and Manufacturing for Sustainable Environment*, Lecture Notes in Mechanical Engineering,
https://doi.org/10.1007/978-981-15-9809-8_50

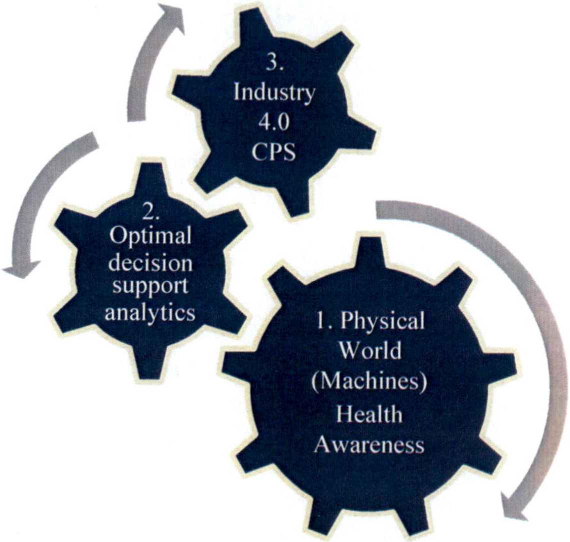


**Fig. 1** Drivers for Industry 4.0 and self-aware machines

predictive technology and facilitation additionally allow electronic aid systems and might tether a fastened intelligence. This technology anticipates machine's overall performance degradation, autonomous manipulation and optimized product needs. The drivers of Industry 4.0 are shown in Fig. 1.

Now, Industry 4.0 aims totally on adaptive mastering with artificial intelligence; thus, better solutions are regularly accomplished through interacting with surrounding structures that have an instantaneous effect to system outcomes. Implementing these tech-based solutions to practice changes every machine into smart and continually develops quality outputs and prognostic control. Though the independent analytics has been implemented effectively in all kind of production, self-aware machines were not in practice in some industries. Upgradation requires further advancement by tackling typical essential hurdles. These hurdles can be divided into few sorts as shown in Fig. 2.

- Manager–Operator Communication: Currently, operator maintains the machines, manager's layout machining schedules by which machines run through assigned obligations. Although those responsibilities are typically optimized by way of senior operators and production managers, a major essential issue is lacking in those decisions like the machine health, i.e., autonomous maintenance which is the prime pillar to achieve total productive maintenance
- Machine Fleet: Comparable or equal machines are being deployed to absolutely specific working situations for numerous responsibilities. As same, maximum feasible methods are designed to help few varieties of equipments and its dealing conditions. Present health management techniques do not seem to be taking gain of combining the identical machines as fleet which is widely known as equipment management

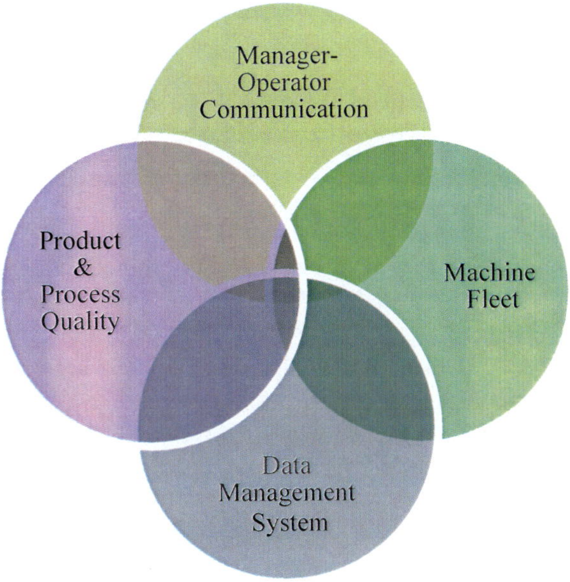


**Fig. 2** Typical issues and hurdles for Industry 4.0

- Process and Product Quality: End results or outcomes of the production and customer satisfactory could offer a better perception on machine performance through backward reasoning. Currently, such system was not found and desires further research
- Data Management System: Data control and sharing in big data surrounding are vital for attaining self-aware machines. The significance of implementing extra flexibilities and accessibilities offered using cloud network is inevitable, but adapting predictive health assessment effectively improves current statistical control methods.

Hence, the goal of the paper is to reveal how modern production firms should evolve by approaching industrial big data environment and sustainable solutions.

## *1.1 Needs and Trends for New Gen Upgrade*

Smart industry concept has booted industry development toward primary adoption of smart analytics, to assist manufacturing and to distinctly automate assembly lines, so that one can be plaint and proactive to contemporary markets' demands and necessities. Under 4.0 concept, remarkable boom within the development of knowledge era and social networks has increasingly motivated consumer belief on innovation, quality, range and robust delivery. These requirements establish the manufacturing abilities of self-awareness, prediction, comparison, reconfiguration and maintenance.

Smart industry enables innovative improvement by receiving more focus in academia and industries through innovative service and data management system.

## 2 Literature Review

Servitization is defined due to the fact enhancing organization's abilities and procedures to shift from product sales, to promoting a new included product and repair imparting which delivers price, i.e., product service system [1]. Mont describes PSS as gadget of services, assisting networks and IT infrastructures which is really designed to be most competitive, assure clients expectations and features than conventional commercial enterprises [2].

Chen et al. [3] introduced a standard smart factory adoption that incorporates physical and cyber technology, and then more complicated and precise integration of previously separate discrete systems. A hierarchical architecture of the digital factory was proposed in this paper; then, the main technologies were evaluated in order to facilitate the machine layer, the network layer and the data layer. Eventually, a candy packaging line was simulated to validate this smart plant, showing that the equipment's OEE has increased comparably. It combines connectivity, networking and process management to fulfill the requirements of the industry [3].

Campos et al. proposed an IoT-based architecture for traceability in the agribusiness context. The International Organization for Standardization's standard 25,000 is used as a metric for determining quality attributes, professional visits to global wine production condition research will be conducted at a winery, and simulations for using the software are tested for validation. They created an architecture in which traceability, scalability, security, reliability and availability support the development of wine tracing applications from production to disposal [4].

### 2.1 *Innovative Maintenance*

Many countries whose financial base is the production planned and achieved remodeling their economic system. They go through bumps from volatile market and consequently the global supply chain. Servitization becomes proposed by means of Vandermerve and Rada[5]. They emphasized the client focus through combining product services, supports and information that are the foremost critical elements. Baines described production servitization with innovation of organizational procedures and skills, from income to included services [6]. Therefore, manufacturing corporations look for production method innovation with specialized induction of service. Thus, the manufacturing and repair industries will prefer production servitization.

### 2.2 *Industrial Big Data—Data Management System*

Recently, information processing becomes a much needed tool to acquire control over any process because human-generated data has been a lift to the social media which is also known as the second web era on account after 2004 [7]. Voluminous present new studies and maximum focus were on industrial mining. This includes forecasts, consumer courting clustering and mining, recommendations, opinions on mining, etc. [8–12]. Hence, these studies focus on human-generated and human-related records as a substitute than gadget-generated facts and industrial records that may include numerous controller, sensor, manufacturing system, etc.

Under 4.0 era, sensible prognostics and cyber physical systems were teaming collectively to appreciate a replacement focusing on the traditional production and manufacturing facilities. Using smart sensor installation, numerous alerts like temperatures, pressures, etc., can be acquired and observed. Additionally, statistical data may be recorded for further record storing. Protocols, like MTConn [13], would assist user the file control indicators. Such aggregated amalgamation is called "industrial big data." The change agent consists of numerous elements such as a collaborative platform, prognostic analysis and visualization tools. The solution deployment platform is preferred based on pace of computation, funding costs, easy practice and update, etc. [14]. Processing of information into needful facts is the prime mover of sustainable upgradation inside a 4.0 factory.

## 3 Self-Aware Machine

Recent development in Internet of things (IoT) framework and also the practice of sensing era have adopted many unified facts that tightly interfaces structures and users together and in addition facilitates a smart informatics inside enterprises. Greater new analysis, the arrival of cloud computation and CPS framework enable self-awareness and clearly ensure ability performance issues. A self-conscious and maintained device with a view to self-check its personal health with its degradation further uses same data from another machine for smart service selections to keep away from capacity issues.

For a system, fitness assessment is carried out through using an information-driven set of rules to investigate statistics/records received from the machine in ambient condition. However, for many business applications, especially for self-aware machines, it remains far from real practice. Current prognostics are normal for selected device and applications, and do not seem to be flexible enough to deal with extra correlated facts. The reasons are summarized as follows:

1. **Lack of closely paired human–machine interaction:**
   An extreme fact for gadget usage and operation is human operation and robust management. Current machines passively hear the operator action and react, even if the mission has no gold standard for its current situation. A better device setup

needs to be enabled to actively suggest corrections and alter process parameters to develop productiveness

2. **Lack of adaptive mastering:**
   In most cases, tracking data from actual machines is done using a predefined procedure. Decision making and breakdown prediction take up a vital role in machine maintenance. This modern era of manufacturing requires smart maintenance system which will be taken care by the operator on his own that gives uninterrupted outcomes as per the schedule.

To fill aforementioned gaps in research, a unified CPS for self-aware machines was developed which extracts useful records with greater effectiveness, and further unravels shrewd decisions.

With the scope of research, physical areas considered were

A machine fleet, with

- Machine condition monitoring
- Control parameters
- Digitized machine records
- Machine configuration and templates
- Utilization charts, duties deployed

Human intervention, with

- Machine maintenance measures
- Man-controlled parameters and utilizations.

In computation area, first, the records and then record formats must be described simply so records from bodily area are easily accessed. Second, cyber area geared up to integrate and gather information on machine health degradation, and then simple understanding is adapted for health evaluation in latest machine. Last, machine health evaluation results fed and returned over the period to the physical area in an order and to take proper action.

## *3.1 Advantages of Self-Aware Machines*

The prime innovation in such framework is self-maintaining machine by using integrated sensor information similarly as a fleet wide record, in order that the data extent is frequently decreased and the identical samples are frequently acquired. This further ensures the statistics left below the big data can be used.

# 4 Case Study: Smart Maintenance System in ABC Polymer Industry

This precise application of IoT was developed for a molding equipment used in rubber encapsulation process. Remote analytics and vision system focus on assessment and prediction of the health of lubrication system.

For this self-aware machine, the developed architecture for data capture and data record consists of managing a data set of parameters and its values from the lubrication tank to the preferred destiny as shown in Fig. 3. The included parameters were fluid flow rate and the viscosity of the oil. Parameters were studied at some process points for the machine, such that at idle performance and at fullest viscosity, where as Raspberry Pi flow sensor helps to determine the cause of the misbehavior of machine and predicts the time of failure of the lubricant.

## 4.1 Data Analysis

Machine-wise OEE parameters were studied monthly and calculated using the below listed formulae as shown in Table 1.

OEE data of the molding machine before implementation is shown in Table 2. Then, its OEE has been calculated and shown in Table 3. Since availability rate falls continually from 82.96 to 82.72%, the OEE falls perpetually from 55.09 to 51.39%.

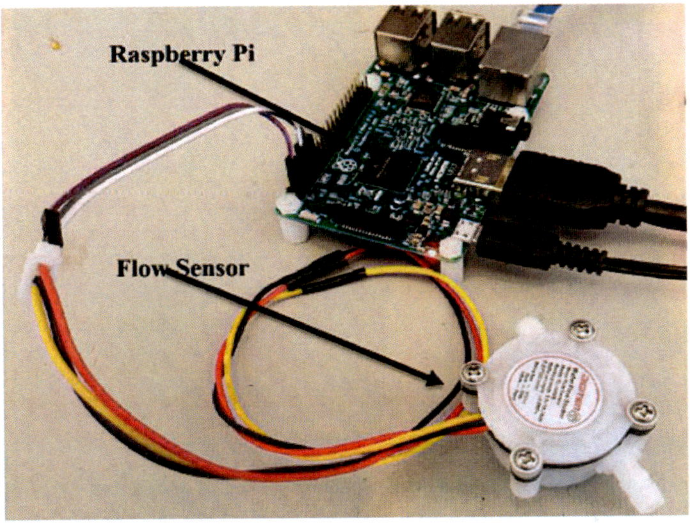


**Fig. 3** Raspberry Pi with flow sensor

**Table 1** OEE calculation formulae

| Net planned production time | Shift time–break time |
|---|---|
| Net operating time | Net planned production time–machine downtime |
| Finished goods | Products produced–products rejected |
| Cycle time per unit | Net operating time/products produced |
| Valuable operating time / unit | Net planned production time/products produced |
| Availability rate (*A*) | Total operating time/total planned production time |
| Performance rate (*P*) | Cycle time/valuable operating time |
| Quality rate (*Q*) | (Products produced-products rejected)/total products produced |
| OEE (%) | Availability * performance rate * quality rate |

**Table 2** OEE data before IoT implementation

| Factors | August 2019 | September 2019 | October 2019 |
|---|---|---|---|
| Total shift time (min) | 31,500 | 31,500 | 31,500 |
| Production break (min) | 4500 | 4500 | 4500 |
| Machine downtime (min) | 4600 | 4700 | 4650 |
| Total components produced (pieces) | 300,000 | 300,000 | 300,000 |
| Total components rejected (pieces) | 60,000 | 65,000 | 75,000 |
| Finished goods (pieces) | 240,000 | 235,000 | 225,000 |
| Total planned production time (min) | 27,000 | 27,000 | 27,000 |
| Total operating time (min) | 22,400 | 22,300 | 22,350 |
| Cycle time/unit (min) | 0.0747 | 0.0743 | 0.0745 |
| Valuable operating time/unit (min) | 0.09 | 0.09 | 0.09 |

**Table 3** OEE before IoT implementation

| | August (%) | September (%) | October (%) |
|---|---|---|---|
| Availability | 82.96 | 82.59 | 82.72 |
| Performance | 83 | 82.55 | 82.78 |
| Quality | 80 | 78.33 | 75 |
| OEE | 55.09 | 53.4 | 51.39 |

## 4.2 Components Used

### 4.2.1 Raspberry Pi

The Raspberry Pi is a low-cost, debit card size board that uses regular keyboard and mouse to attach to a desktop monitor or any other screens. It is a small, adaptable and most capable device that enables people of any sort to explore programming, and learn to program with languages like Python and Scratch, which consists of a 512 MB RAM, ARM V6 Processor, two USB ports and one Ethernet port, HDMI and RCA display ports, audio jack 3.5 mm, dedicated SD card slot, I/O pins for subcomponent general purpose, which runs on 5 v. It has an SD card slot that can read up to 32 GB. The Raspberry Pi's GPIO pins are programmed using the Python language.

### 4.2.2 Flow Sensor

The liquid flow rate is calculated using sensor from flow meter. A flow sensor is a transducer that responds to a magnetic field by varying its output voltage. The flow meter can measure 1–30 LPM which can withstand the fluid pressure less than or equal to 5.0 Mpa, and its working voltage is 5 to 18 V DC with temperature range –25 to + 80 °C and accuracy 90%. It senses liquid flow in the pipe and sends Raspberry Pi the corresponding analog signal.

OEE data of the molding machine before implementation is shown in Table 4. Then, its OEE has been calculated and shown in Table 5. Since availability rate improved continually from 93.52 to 93.7%, the OEE also improved perpetually from 72.91 to 76.07%.

**Table 4** OEE calculation after IoT implementation

| Factors | November 2019 | December 2019 |
|---|---|---|
| Total shift time (min) | 31,500 | 31,500 |
| Production break (min) | 4500 | 4500 |
| Machine downtime (min) | 1750 | 1700 |
| Total components produced (pieces) | 300,000 | 300,000 |
| Total components rejected (pieces) | 50,000 | 40,000 |
| Finished goods (pieces) | 250,000 | 260,000 |
| Total planned production time (min) | 27,000 | 27,000 |
| Total operating time (min) | 25,250 | 25,300 |
| Cycle time/unit (min) | 0.0842 | 0.0843 |
| Valuable operating time/unit (min) | 0.09 | 0.09 |

**Table 5** OEE after IoT implementation

| | November (%) | December (%) |
|---|---|---|
| Availability | 93.52 | 93.7 |
| Performance | 93.56 | 93.67 |
| Quality | 83.33 | 86.67 |
| OEE | 72.91 | 76.07 |

### 4.2.3 IoT Setup Working Principle

The fluid flow has been monitored with the flow sensor which works on Hall effect principle. The fluid flow will be recorded once in a day, the data will be stored, and it is actuated by programming with Python. The discriminative data will be monitored, and when it falls beyond the boundary level, the Message Queuing Telemetry Transport (MQTT) communication protocol encrypted in the Raspberry Pi generates an intimation regarding the lubricant density. Here, the predominant velocity ranges were fed to the program as boundary conditions. Thus technically, the IoT setup checks for number of liters in flow per second and is converted to velocity using the program, where the predefined boundary conditions will be compared and message generated accordingly.

To give the perspective of the problem, assume that a typical press cylinder has a 40-mm bore radius ($A_c$) and 100-mm ($x$) stroke length and it needs to be pressed in 120 s cycle time ($T$) and four press operations per hour. So, the discharge rate ($Q$) is calculated with Eq. (1),

$$\begin{aligned} Q &= \frac{A_c X}{T} \\ &= \pi\,(40)^2(100)\,/\,(4 * 120) \end{aligned} \tag{1}$$

$$\mathrm{Q} = \text{per Second} = 1.047\,\text{LPS}$$

$$\mathrm{Q} = \text{per Hour} = 4.188\,\text{LPH}$$

So, if the flow rate was below 4.1 LPH then it is said that the fluid density has been increased and when it reaches 4.05 LPH then the press machine requires lubrication oil change. Thus when the $Q$ falls below 4.1 LPH, the MQTT program encrypted in the Raspberry Pi generates a warning message to the maintenance personnel. Hence, the maintenance could be completed before the breakdown occurrence which indirectly improves the available time. Finally, since the available time increased the OEE improves gradually.

## 5 Results and Discussion

After implementing the IoT solution in the ABC polymer industry, the OEE is calculated; thus, it shows drastic improvement when compared to the previously calculated OEE. Thus, delay in preforming process is also reduced by lowering the machine downtime. Now, the management system of the industry is balanced and works with high efficiency. The OEE is calculated for next 2 months which shows the result as

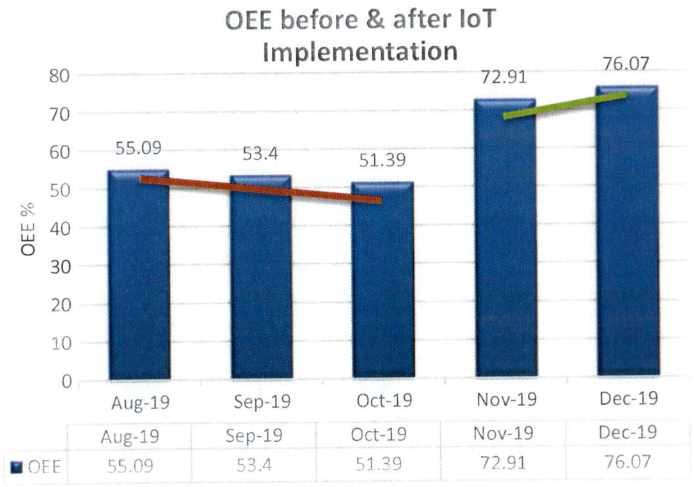


**Fig. 4** OEE graph before and after IoT implementation

72.91 and 76.07% since delay occurrence is reduced a far better achievement than 55.09% OEE as shown in Fig. 4.

Hence, the prognostic monitoring module may be a fashion of the innovative production and big data enterprise. Many regions could be seen before to own an improvement with the emergence of the fourth technological evolution, where four major effect regions emerge:

- Prediction of machine fitness
- Enterprise control levels
- New labor-friendly environment
- Energy saving and optimized protection scheduling.

## 6 Future Scope

Hence, we can clearly conclude that it is not necessary for an Indian SME to have a SAP or ERP system to develop or implement the condition monitoring or machine health assessment or some other technology-based solutions to overcome the maintenance issues. Thus, such simple IoT solution improves the performance and utilization of the machines which perpetually improves the OEE and also the productivity of the machines.

# 7 Conclusion

Thus, Industry 4.0 highlights the need for predictive maintenance for enterprises where the equipments were integrated as unified community, which calls for employment of improved prediction gears, IT tendencies and few other unmet desires. This consists of innovative service that changes manufacturer's price propositions and big facts, also which makes production analysis more critical than the foreseen era. To survive underneath the trends, an innovative framework is proposed to ensure self-conscious machines and maintained machines. It consists of ideas to cyber physical gadget and prognostic decision network. Last, a case study is described in order to show the possibility of the proposed model.

## References

1. Martinez M, Kingston J, Evans S (2010) Challenges in transforming manufacturing organizations into product-service providers. J Manuf Technol Manage 21(4):449–469
2. Mont O (2004) Product-service systems: panacea or myth. Lund University
3. Chen B, Wan J, Shu L, Li P, Mukherjee M, Yin B (2018) Smart factory of Industry 4.0: key technologies, application case, and challenges. IEEE Access 6:6505–6519. https://doi.org/10.1109/access.2017.2783682
4. Campos LB, Cugnasca CE (2015) Towards an IoT-based architecture for wine traceability. In: 2015 international conference on distributed computing in sensor systems. https://doi.org/10.1109/dcoss.2015.31
5. Vandermerwe S, Rada J (1989) Servitization of business: adding value by adding services. Eur Manage J 6(4):314–324
6. Baines TS, Kay JM (2009) The servitization of manufacturing. J Manuf Technol Manage 20(5):547–567
7. Graham P (2005) Web 2.0. Consultado (21/12/2008). www.Nosolousabilidad.Com/articulos/Web20.Htm
8. McAfee A, Brynjolfsson E (2012) Big data: the management revolution. Harvard Business Review
9. Cohen J, Dunlap M, Welton C (2009) MAD skills: NA practices for big data. Proc VLDB Endow 2(2):1481–1492
10. Al-Noukari M, Al-Hussan W (2008, April) Using data mining techniques for predicting future car market demand; DCX case study. In: 2008 3rd International Conference on Information and Communication Technologies: From Theory to Applications, ICTTA 2008. IEEE, pp 1–5
11. Chon SH, Berger J (2006, October) Predicting success from music sales data:a statistical and adaptive approach. In: Proceedings of the 1st ACM workshop on audio and music computing multimedia. ACM, pp 83–88
12. Provost F, Fawcett T (2013) Data science and its relationship to big data and data-driven decision making. Big Data 1(1):51–59
13. Lee BE, Michaloski J, Bengtsson N (2010, January) MTConnect-based kaizen for machine tool processes. In: ASME 2010 international design engineering technical conferences and computers and information in engineering conference. American Society of Mechanical Engineers, pp 1183–1190
14. Lee J, Kao HA (2013) Recent advances and trends in predictive manufacturing systems in big data environment. Manuf Lett 1(1):38–41

# Insight into Smart Fire Detection Systems

**M. Saranya and S. Esakkirajan**

**Abstract** Fire can cause an impact on people, property, and the environment in all countries around the world. A fire can start in an instant and continue to rage until its source is depleted. It can destroy homes and property, causing injuries and taking lives within a matter of seconds. Hence, in building design and management, the fundamental consideration will be fire safety but unfortunately, fire safety is not given prior importance. Today firewalls are more associated with Information Technology (IT) security than with physical safety. In recent days, buildings are more complex but technologies for fire safety are not given enough importance by the constructional engineers and owners of such buildings. This survey paper provides a review report about various technologies used in fire engineering for building smart fire detection systems with different features.

**Keywords** Firewalls · Fire engineering · Smart systems

## 1 Introduction

Today, cities are completely packed with residential and commercial buildings to meet the sudden rise in urbanization and overpopulation in the world. When a fire breaks out in these densely populated buildings, it could seriously threaten many lives of people and causes major economic losses. Hence, the importance of safety has been understood by individuals, Private Organizations & Governments and continues to design a counteract measure to ensure fire safety implementation across all parts

M. Saranya (✉) · S. Esakkirajan
Department of Instrumentation and Control Systems Engineering, PSG College of Technology, Coimbatore, Tamil Nadu, India
e-mail: msa.ice@psgtech.ac.in

S. Esakkirajan
e-mail: ser.ice@psgtech.ac.in

© The Editor(s) (if applicable) and The Author(s), under exclusive license to Springer Nature Singapore Pte Ltd. 2021

S. Mohan et al. (eds.), *Materials, Design, and Manufacturing for Sustainable Environment*, Lecture Notes in Mechanical Engineering,
https://doi.org/10.1007/978-981-15-9809-8_51

of life. To preserve the lives and properties during fire outbreaks, a lot of Fire Detection and Warning Systems (FDWS) are under design consideration. So, Fire Detection and Warning Systems play a vital part in buildings. For installation and annual maintenance of such vital systems, more than billions are spent each year [1]. The maintenance of installed fire detection systems is to be done to ensure the system effectively detects and completely prevents the fire outbreak. Today, Fire Detection and Warning Systems in complex buildings exist with more intelligence and smartness in its arsenals to do more effective detection at any time.

Modern buildings implemented with intelligent and smart fire assistance system should be capable enough to avoid false alarm, assist firefighting, and speed up the evacuation of building when fire outbreak occurs. Firefighters utilize the wireless system to strategize the methodology to fight with an outbreak of fire before their arrival at the incident place. It has been evident that buildings interfaced with Fire Detection and Warning System are safe from catastrophic fire disasters to some extent.

### *1.1 Fire Engineering*

Fire engineering uses various engineering principles ultimately to safeguard the environment, property, and individuals from the damage caused by fire. Safety is achieved by applying the established rules together with in-depth knowledge of the phenomena, effects due to fire accidents, and also the people's reaction & behavior to such an accident. Fire protection engineers identify the risks involved and design safety measures that could prevent and control the fire outbreak [2]. Fire Engineers utilizes the following design considerations to ensure safety during the fire outbreak:

- Prevention

Controls the Ignition, (i.e.) identify & control fuel sources to eradicate the cause of fires to start.

- Communication

At the start of ignition, the information has to be communicated to the occupants, nearby fire stations, and nearby hospitals, so that triggering of active fire systems starts its operation.

- Escape

Ensures the nearby surrounding people and the occupants of buildings are evacuated to safety places.

- Containment

Ignited fire should be contained within the minimum possible area to limit the damage to the property and threats to people's life.

- Extinguishment

Ensuring the extinguishment of fire is done quickly with minimum consequential damage.

### 1.2 Fire Detection and Alarm System

Most of the existing Fire Detection and Warning Systems works on the same principle. On the fire outbreak, immediately triggering of an alarm has to take place to pass the warning sign to the occupant and building managers about fire to plan for evacuation. Fire outbreak can be detected by anyone of the following detectors: heat or flame or smoke or carbon monoxide or multi-sensor. These detectors pass on their identified signals to the processing unit and the corresponding warning signal can be indicated using any of the following devices: bells or sirens or horns or lights or multi-units. Fire Detection and Warning Systems can be categorized into the following types:

- Conventional systems.
- Addressable systems.
- Analogue addressable systems/Intelligent Systems.
- Wireless systems.
- Self-contained systems.

Among various fire detection and alarm systems, Analogue addressable systems, or intelligent systems includes analytical capability along with input detector unit. This allows the accessibility to the local parameters for the determination of the fire existence or a fault or a maintenance requirement. This parameter accessibility completely avoids the raise of false alarm. A pre-alarm warning can also be indicated when the detector is about to approach a trigger condition [3].

This paper gives a detailed review report about various technology used w.r.t. the fire engineering options for building the intelligent fire detection systems with smartness. This review paper is organized as follows: Sect. 2 gives generalized information about the System overview. Sections 3 to 7 provides information about different proposal methods for building a smart fire detection system. Section 8 gives the complete analysis report of the various frameworks in terms of capabilities and limitations. Finally, Sect. 9 gives the conclusion of this paper.

## 2 System Overview

This section gives the complete system overview. Figure 1 shows the overview of smart fire detection system. The system uses sensors to understand the events taking place around the building block and sends the signals continuously to the controller

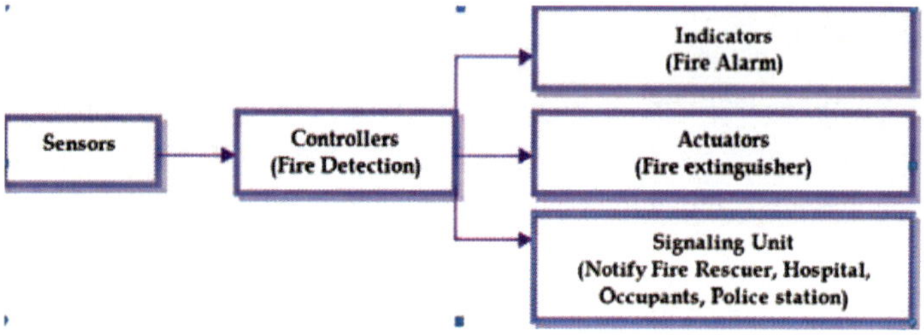


**Fig. 1** System overview

for processing it. The controller then identifies the sensor signal is above a threshold or in case of fire outbreak, it performs the following three actions simultaneously.

- Fire Alarm.
- Fire Extinguisher.
- Fire break and rescue notification to nearby occupant, fire station, hospitals & police station.

## 3 Wireless Fire Protection System [1]

The authors of this paper proposed fire protection system with short message service alerts using a wireless sensor network. The authors propose a system to monitor an individual building of occupants during their sleeping or during their work at the office. The proposed system detects fire with the help of a functional network of smoke detector (i.e.) photoelectric smoke sensors with RF transceivers connected to a microcontroller. Smoke detector helps in sensing the origin and specific location of smoke. The sensor network passes on the sensed signal to the controller interfaced with GSM module to perform the following actions parallel.

- Trigger an alarm (Buzzer/LED).
- Suppress the fire spreading using a sprinkler system ( Water Pump).
- Switching off the building's main electrical switch (Switch/relay).
- Alerts the following members about the outbreak of fire (GSM module).
  - Neighbors and friends.
  - Nearest fire station.
  - Owner of the house.

The authors justify their proposed wireless fire detection system can be implemented in complex building which has numerous floors and rooms. The system is capable to discriminate against the fire and non-fire threats by receiving the distress signal over a specified period of time. Their proposed system was implemented using the following hardware components and the software algorithm. Figure 2 depicts the various blocks involved in this proposed system.

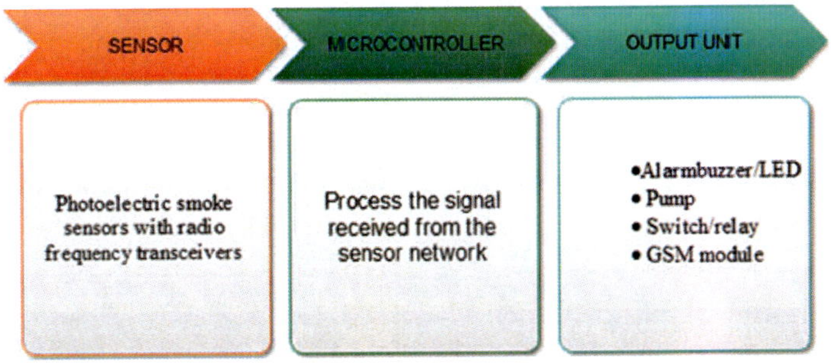


**Fig. 2** Block diagram of the system [1]

The hardware components used for each block in the above figure are given below.

- Smoke detector.
- Buzzer.
- Radio transceiver.
- Relay module.
- GSM modem.
- Atmega328P.

## *3.1 Software Algorithm for Wireless Fire Protection System*

The software algorithm for wireless fire protection system is as follows and their flowcharts are shown in Fig. 3.

- Initialization of the master controller.
- Initialization of the Global System for Mobile communication module.
- Setting Global System for Mobile communication module to Short Message Service Mode (AT + CMGS).
- Initialization of Radio Frequency Module.
- Initiation of transmitter circuitry connected with smoke detector.
- Check for fire detection signal continuously.
- In case of receiving fire detection signals the following actions are performed.
  - Activate alarm.
  - Send SMS indicating fire outbreak to the nearest firefighter station.
  - Send SMS to the owner of house.
  - Activate the water pump.
  - Disable building s main distribution board.
- Else do keep checking for fire signal without doing nothing.

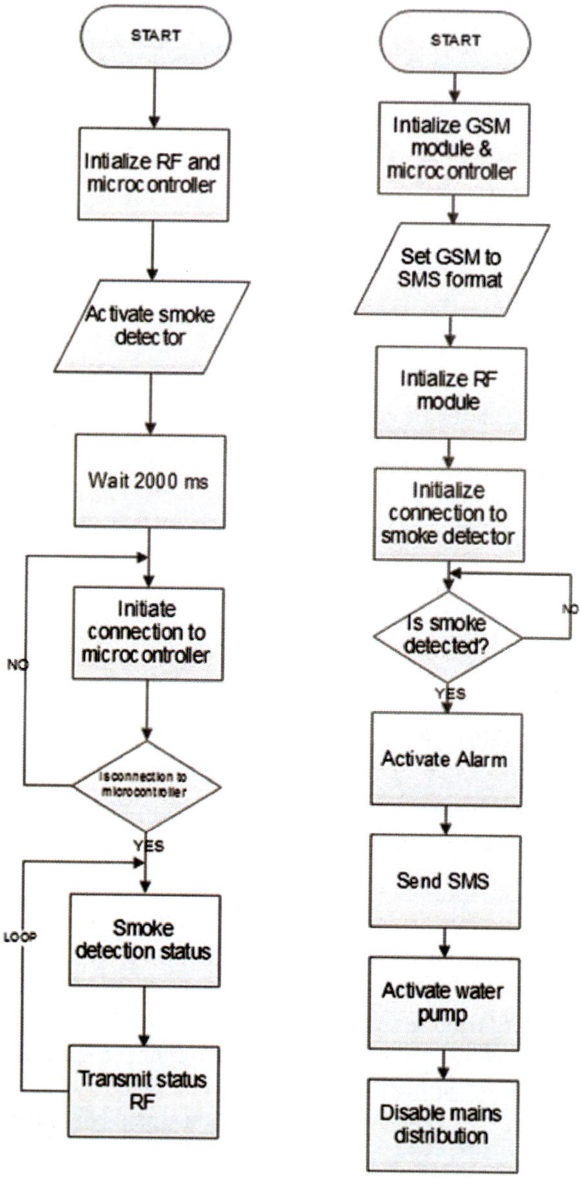


**Fig. 3** Workflow diagram of wireless fire protection system [1]

# 4 Early Notification Fire Detection System [2]

Fire detection system with early notification using ML was proposed by Mohammad et al. Figures 4 and 5 shows the various blocks involved in the proposed system and its network module. Their proposed work intelligently matches the distance gap between the home and fire service departments. They minimized the probability of

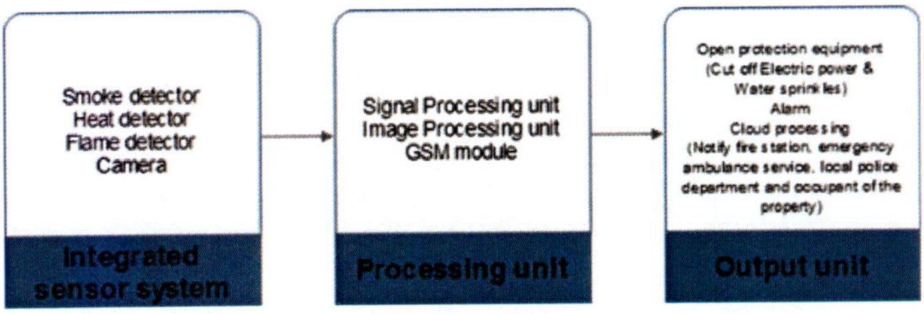


**Fig. 4** Block diagram of early notification fire detection system [2]

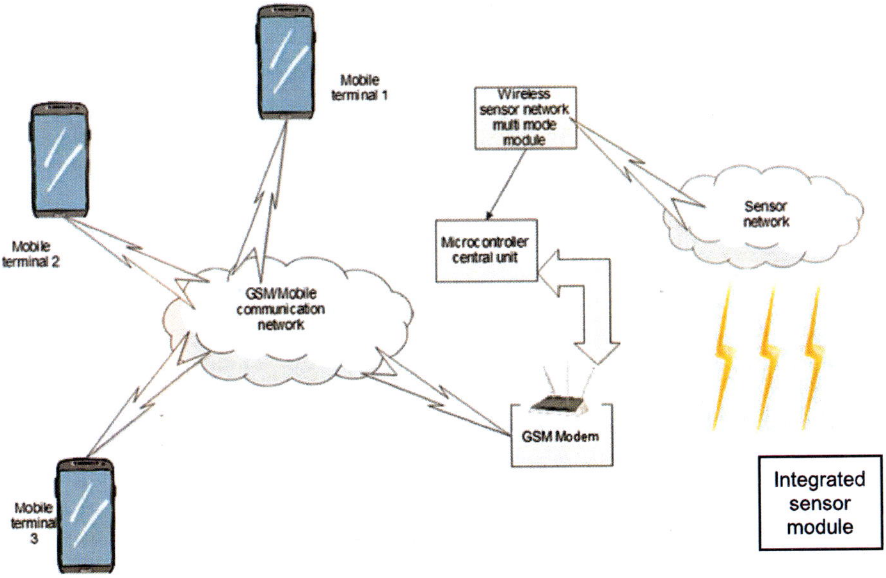


**Fig. 5** Network module [2]

false fire alarm by integrating various sensors that are capable to sense the following parameters.

- Heat.
- Light.
- Smoke.

The authors used individual thresholds for heat, flame, and smoke sensors. This helped them to determine the fire danger (level). The micro-controller determines the fire danger and warns the appropriate subsystems with messages (nearby occupants, fire stations, hospitals, and police stations) with static thresholds to indicate the degree of danger level. Figure 6 depicts the workflow diagram of the system.

The system predicts the output that falls in any one of the following three danger levels

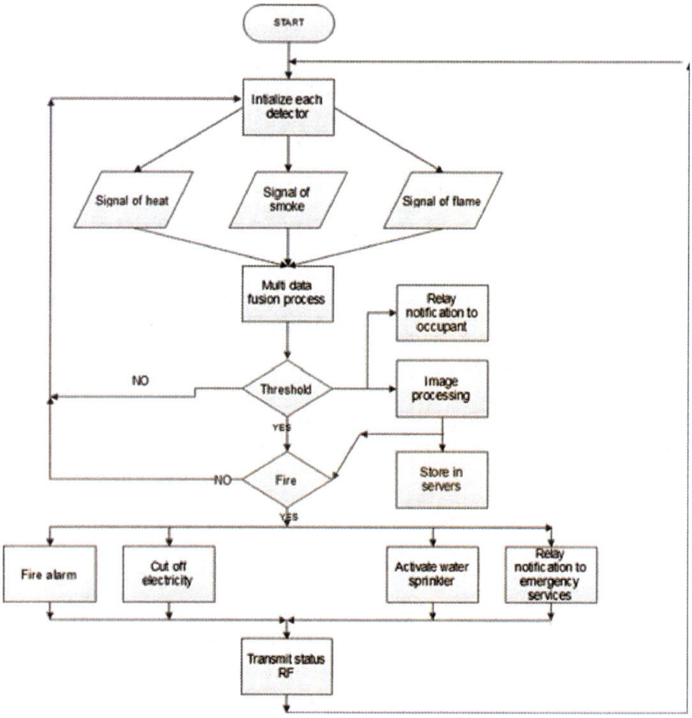


**Fig. 6** Workflow diagram of the system [2]

- Degree of fire smaller than 0.50 (No fire).
- Degree between 0.50 and 0.75 (potential fire).
- Degree greater than 0.75 (Extreme fire).

The authors used Machine learning (data mining classification algorithms) for enhancing the accuracy of their fire detection system. The authors claimed that their system was more reliable to the environment.

## 5 Real-Time Fire Detection System [3]

The author of this paper proposes innovative methods based on multi-feature fusion of flame to address the high rate of false alarm problems existing in traditional fire detection and warning system. The pre-processing stage of the system combines the color and motion detection of the flame. This scheme saves a lot of computational time. The author clearly explains the irregularity of the flame but it has a certain similarity in the image sequences. By using this feature, a novel algorithm based on spatiotemporal relation to check flame centroid stabilization was proposed by the authors. The information contained within the spatiotemporal of the flame was obtained by calculating the temporal information and the centroid of the flame existing in each image frame. Also, they extract the following variability features of the flame to improve the accuracy of the recognition.

- Spatial.
- Shape.
- Area.

Finally, they achieved automatic fire monitoring by using SVM for training and completed their candidate fire image analyses. They showed that their method has reduced false alarm rates with improved accuracy when compared with standard techniques. This method is well suited for domestic and commercial monitoring because of utilizing real-time camera system for monitoring.

## 6 Sensor Fused Fire Detection System (FDS) [4]

Authors of this paper proposed multi-sensor technology and Dempster-Shafer theory based FDS. Figure 7 shows the system overview of the proposed FDS. This system operates in three different stages as follows.

- Sensor Measurement.
- Reception of the fused signal.
- Activation Signal.

The authors relied on using Arduino for understanding the readings measured from different types of sensors. Readings measured from three sensors such as smoke, light, and temperature are fused together for fire detection. The backend Raspberry Pi receives the measured data through wireless media for subsequent processing. The algorithm implemented in the Raspberry Pi uses Dempster-Shafer theory to determine the probability of fire outbreak events. Web server with MySQL database is loaded and executed on the Raspberry Pi for backlog and data analysis purposes. At the end of data analysis, the system provides the following service to notify the outbreak of fire.

- HTTPs.

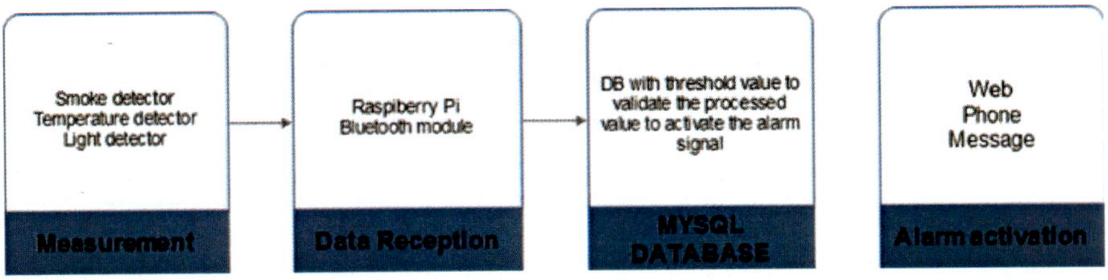


**Fig. 7** Block diagram of the system

- APPs.
- SMS.

The paper also published the statistics collected from the field, test results, and the validation performed on the data. All these tests were conducted in a safe controllable environment by emulating fire events w.r.t. both daytime and nighttime. The experiment emulation undergoes in three phases:

- No fire.
- On-fire.
- Post-fire phases.

Experimental results in this paper projects the following accuracies.

- 98% in both the No-fire and On-fire phases during the daytime.
- 97% during the nighttime under reasonable conditions.
- 97% and 89% accuracy in the daytime and nighttime w.r.t. overall three phases respectively.

## 7 An Intelligent Fire Warning System [5]

The FDWS proposed by the author of this paper is illustrated in Fig. 8. The four different sensors used in this system were smoke, temperature, humidity, and a flame sensor. These sensors collect and then transmit data from the sensor to the MATLAB GUI. The sensor data is transmitted as raw data to fuzzy logic (a linguistic variable) to get trained in ANFIS to detect fire. If the system finds the parameters showing the occurrence of fire in terms of probability as critical, then using GSM message will be sent about the fire condition to the fire rescue station and the building owner. The whole system works in two phases. The first phase is hardware design (the sensor nodes development) and the other phase is a simulation. Figure 8 depicts the architecture of FDWS.

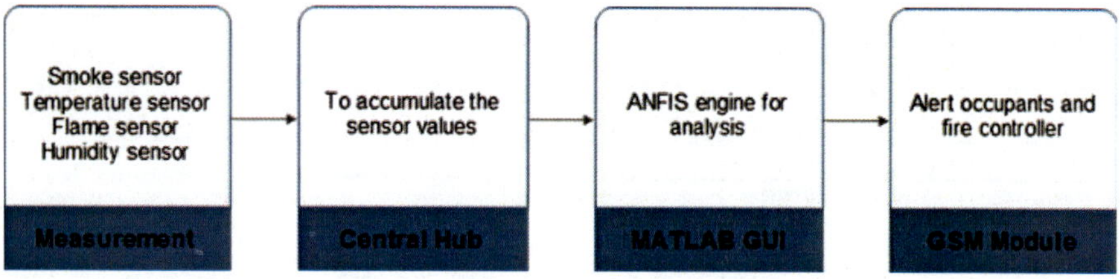


**Fig. 8** Architecture of FDWS [5]

**Table 1** Comparison of various fire detection systems

| Ref.No. | Sensor | Hardware controllers/ simulation toolbox | Communication module | Recipient |
|---|---|---|---|---|
| 1. | Photoelectric smoke sensors | Algorithm implemented in Atmega328P | WiFi, GSM module | Fire station, House owner |
| 2. | Heat, Light, Smoke, Image Camera | Image processing and Machine learning (data mining algorithms) using Matlab | WiFi, GSM module | nearby occupant, fire station, hospitals & police station |
| 3. | Video Camera | Motion and color detection in pre-processing, flame centroid stabilization based on spatiotemporal relation, and SVM for dynamic flame | N/A | N/A |
| 4. | Smoke, Temperature, Light | Algorithm implemented in Raspberry Pi | Bluetooth | Web browsing, SMS, Smartphone APP |
| 5. | Temperature, Humidity, Flame, Smoke | MATLAB ANFIS Simulation | GSM module | Fire station, Nearby occupant |

## 8 Analysis of Various Fire Detection System

Table 1 gives a brief overview of various methodologies to build an FDWS based on the system discussed in Sects. 3 to 7.

## 9 Conclusion

Thus this paper reviews various technologies used for building fire detection and warning system in real-time. With the help of this review, the following conclusion can be done.

- Multi-sensor data fusion network provides the accurate input data.
- Controller implemented with the machine learning and image processing algorithm for processing the input data eliminates the false alarm and increases the overall system performance.
- Motion and color detection, flame centroid stabilization based on spatiotemporal relation increases the speed of raising alarm and notification.

- GSM module is used with every system for sending notification.

Hence, this paper identifies the best technology for building smart fire detection system. So by combining Multi-sensor data fusion input network, Controller with machine learning and image processing algorithm, and GSM module, the overall system performance will increase by reducing the false alarm rate.

## References

1. Okokpujie K et al (2019) Wireless sensor network based fire protection system with sms alerts. Int J Mech Eng Technol (IJMET) 10(02):44–52
2. Mahmud MS, Islam MS, Rahman MA (2017) Smart fire detection system with early notifications using machine learning. Int J Comput Intell Appl 1750009
3. Gong F et al (2019) Real-time fire detection method from video with multifeature fusion. Comput Intell Neurosci. Article ID 1939171
4. Ting YY et al (2018) A data fusion-based fire detection system. IEICE Trans Inf Syst. E101-D(04)
5. Sarwar B et al (2019, July) An intelligent fire warning application using iot and an adaptive neuro-fuzzy inference system. www.mdpi.com/journal/sensors

# Cabbage Discernment Using CNN for Vegebot Application

**M. Thangatamilan, S. J. Suji Prasad, P. Prabhasri, K. Sanjaykumar, S. Sujatha, and S. Sujetha**

**Abstract** Agriculture could provide a unique scope for the headway of the robotic system. Robots are a perfect substitute for human resources to a great extent as they deploy unmanned sensing and machinery system. During the period of plantation and harvesting of crops, the involvement and participation of humans get decreased nowadays. This in turn reduces the cost of labour with improved yield. Agricultural robotics has unique challenges when compared to the robotics in the industrial applications. Cabbage is a cruciferous and most fashionable winter vegetable grown in India. Though India is one of the leading cabbage-producing and cabbage-consuming countries, there is no evidence of automation in cabbage harvesting. The conventional method of using knife for cabbage harvesting is practised till date. Automation in cabbage fields is risky, and its complication level is too high. Complexity starts with identifying the cabbage head size for harvesting. Using a sharp knife for cutting the cabbage head and leaving the stem uneven at some places, even it may damage the cabbage itself. Sometimes, it leads to irregular cabbages, which is futile to meet the market standards. Due to the increased difficulty in harvesting cabbage through labours, automatic cabbage identification and harvesting robot is the only solution. As a part of automation, convolutional neural network is implemented for cabbage localization and classification. In this paper, cabbage localization and classification is performed using CNN, by collecting the different datasets from the cabbage field. The field test results show that the cabbage classification is achieved successfully.

**Keywords** Agriculture · Cabbage · Convolutional neural network · Machine learning · Vegebot

M. Thangatamilan (✉) · S. J. Suji Prasad · P. Prabhasri · K. Sanjaykumar · S. Sujatha · S. Sujetha
Department of Electronics and Instrumentation Engineering, Kongu Engineering College, Perundurai, Erode, Tamil Nadu, India
e-mail: thangatamil.m@gmail.com

S. J. Suji Prasad
e-mail: sjsujiprasad@gmail.com

© The Editor(s) (if applicable) and The Author(s), under exclusive license to Springer Nature Singapore Pte Ltd. 2021
S. Mohan et al. (eds.), *Materials, Design, and Manufacturing for Sustainable Environment*, Lecture Notes in Mechanical Engineering,
https://doi.org/10.1007/978-981-15-9809-8_52

# 1 Introduction

Cabbage (belongs to the species of Brassica family vegetable) is a multi-layered plant grown as an annual vegetable crop due to its thick and dense leafy heads. Leaves are stifled together to form the cabbage head. Full mature cabbage heads weigh from half to three and half kilogram, due to its multi-layered arrangements. It may grow in different shapes and different colours including white, red, purple and green [1, 2]. Cabbage grows only in cool and moist climatic conditions, whereas in hot climatic conditions, cabbage results in poor quality. Best suited soil for cabbage varies from sandy to clay. Soil variation from lighter soil to heavier soil determines the quality of cabbage from early crops to mature crops. Optimum pH range for growing cabbage is 6.0–6.5. Those plants growing in salty soil get easily affected with disease [3]. Harvesting time can be easily identified by looking into the matured heads of cabbage. When the head size reaches its full capacity and is in firm, it indicates the ready condition for harvest. If the demand and prices are high in the market, harvesting may be done with medium head sized cabbages [4]. Sometimes, weather condition sets the situation for harvesting [5]. Cabbage will take around 70 days after planting to reach the desired size for harvest. Cabbage should be harvested immediately once the heads are mature and rigid. Delaying harvest, even a couple of days on the far side maturity, may end up in split heads and inflated incidence of field malady. Harvesting immature heads, however, reduces yield, and therefore the head square measurement too soft to resist handling harm. Immature heads even have a shorter period than mature heads.

The exact manual localization and harvesting of cabbage is shown in Fig. 1.

The head is harvested and curved to one facet and cutting it with a knife. In certain markets, additional leaves are necessary as it reduces the damage for the cabbage. The top of the cabbage head should be carefully handled without any slanting, as it

**Fig. 1** Manual harvesting

harms the top and also stalk length. Broken stalks are more liable to decay. Because the heads are not prepared for harvest at the identical time, thus they are harvested bit by bit based on the maturity of the heads. Harvested cabbages should be kept in the shade before packing. Cabbage head may cut out of the plant for getting more crops, by keeping the roots inside the soil. This root comes up with the new cabbage head. This process is continued for getting more crop heads. Once grown into full mature heads, the whole stem and roots can be taken from the soil. In this work, for detecting suitable cabbage, deep learning approach is used.

Late blind disease affects the volunteer potato. Nieuwenhuizen et al. [6] developed the model for removing this late blind affected potato using two colour-based algorithms and compared for potato plant detection in two sugar beet fields. Results show that both methods matched closely within fields. In one field, 97% and another field only 49% were identified and classified as potato plants. Also, they identified difference is higher when compared between the fields.

Puttemans et al. [7] developed the model for implementing the automatic fruit detector on both strawberries and apples. Their model is compared with the existing model of fruit classification and shows more benefits of harvesting and estimation of crops.

Zhang et al. [8] proposed an algorithm for grabbing the object using a camera and robotic manipulator, based on convolutional neural network and depth first. The camera used for this purpose is RGB-D camera and maintained the six degrees of freedom. Improved PSO is used for calibrating and optimization purpose. The field information is collected using camera and processed using depth algorithm for perfect grabbing of object. Their proposed work proves practicably and effectively.

## 2 Problem Statement and Methodology

Yield of the cabbage mainly depends upon the maturity time of the plant head. Only fully grown cabbage head has the surplus demand in the market. If the mature heads are not properly identified, that is delay in harvesting also leads to head split and is more prone to diseases. Harvesting immature cabbage heads causes damage to them, as heads are half grown and soft to touch. This gives lesser yield to the farmers. Harvesting through labour can also damage the cabbage head by several factors such as improper identification of matured heads, different stalk lengths or broken stalks during the manual cutting of heads from the plant. Proper identification of matured cabbage heads is the major problem in the harvesting.

The system is developed for identifying the mature cabbage heads. The algorithm used for identification is convolutional neural network (CNN). This could be a category of deep learning neural networks. It represents an enormous breakthrough in image recognition. They are most typically accustomed to analyse visual imaging and square head, that are often operating behind the scenes in image classification.

Several images of cabbage are given as training sets to the algorithm. With the use of camera, real-time cabbage is captured and processed and finally localizes as mature cabbage.

## 2.1 Dataset Collection

The development of image classifiers continuously places confidence in data. The dataset contains 2550 cabbage images collected from the field under different lighting conditions (private farm nearby Ooty, Tamil Nadu). All the images are captured through high-definition Logitech C525 portable camera with standardized height and width for the training purpose, at 400 * 400 pixels. The dataset consists of cabbages with different head sizes, starting from smaller to bigger, different coloured cabbages, disease-affected cabbage, cabbages closed with leaves or without leaves, half covered cabbage, cabbage head in different positions [9]. The dataset collected from the field for discernment is shown in Fig. 2.

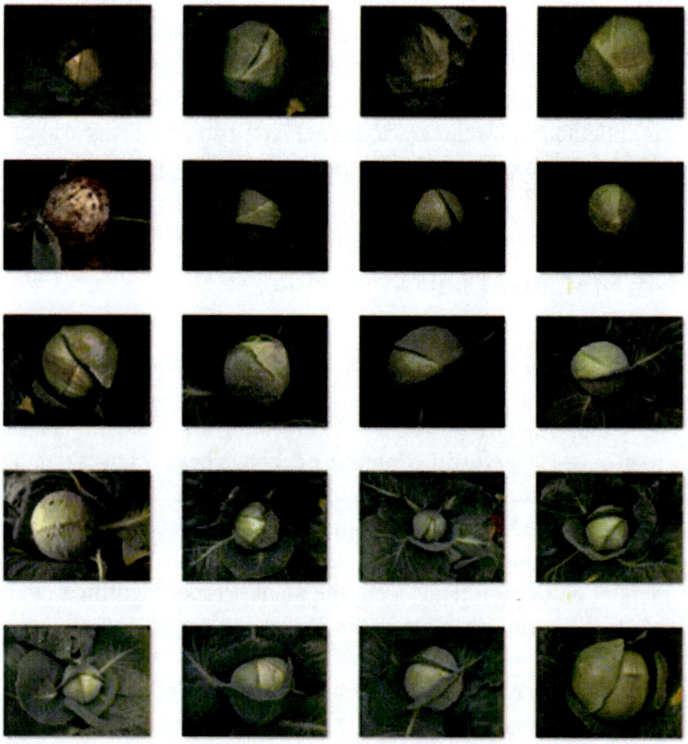

**Fig. 2** Dataset

## 2.2 *Preprocessing*

To train a network and build predictions on new data, the captured cabbage images should match the input size of the network. The dimensions of the captured pictures are to be regulated to match the network, and then only resizing or cropping of captured data is processed to the specified size. The amount of training data can be increased by augmentation to the dataset.

The first step in the preprocessing is to maintain same size and aspect ratio for all the cabbage images. The network model mostly needs the input of square-shaped image. So, cropping can be done to select the squared image. Also, horizontal and vertical rotation of the cabbage image regulates to match the network model. Redundancy of pixels in the input cabbage image should be removed. Preprocessing involves different sections such as cropping, resizing, adding and removing of noise. All the input cabbage images are resized into $400 \times 400 \times 3$, to match the proposed architecture. Noise removal in the input dataset is another preprocessing technique. Normally, noise is always attached to the image taken through camera. Some external disturbances may be added as noise. This can be removed by using median filtering, which removes salt and pepper noise in the input image.

## 2.3 *Building the Structure of CNN Model*

The dataset for cabbage images is collected and processed for building the model. The model for classification of cabbage is based on CNN. CNN consists of convolution, max pooling layers, flatten layer and fully connected layers [10, 11]. Feature extraction is done by convolution and pooling layers, whereas classification is by fully connected layers.

Rectified linear unit (ReLU) is the activation function used in this model for nonlinear operation. Output shows that ***ReLU= max {0, x}*** for ReLU function. As ReLU function is easy and simple to compute, and also performance wise gives good results when compared to other activation functions, many researchers choose this as a default one. The dataset image is convoluted by adding convolutional layers, and then it is send to ReLU function. Input layer should be $400 \times 400$ as they are resized for size. The purpose of pooling layer is for reduction in dimensions without removing important parameters. That is reducing the parameters from the larger image of datasets. Input data is fed into three layers such as convolution and max pooling for image feature extraction. The layers can be added to the model for more complex inputs. The max pooling layer output is given to fully connected layer through flattened layer for obtaining best accuracy.

Figure 3 shows the fully connected CNN architecture.

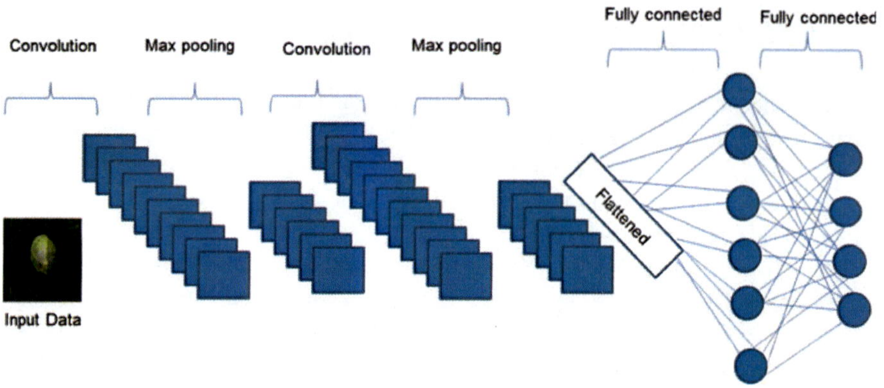


**Fig. 3** CNN architecture

# 3 Results and Discussion

## *3.1 Evaluation*

The dataset obtained for the cabbage classification has undergone into three different sections. The first section was training the model for the particular dataset. Second section validates the dataset, while the third computes the dataset for proposed model. Third section works with resized image of 400 × 400 × 3. Accuracy can be calculated after classification.

$$\text{Accuracy} = \text{tp} + \text{tn}/\text{tp} + \text{tn} + \text{fp} + \text{fn}. \tag{1}$$

where tp is true positive, tn is true negative, fp is false positive and fn is false negative.

Figure 4 shows the cabbage classifier model (Tables 1 and 2).

## *3.2 Obtained Results*

The classified image using the developed model is shown in Fig. 5.

# 4 Conclusions

In this paper, a novel method for classification of cabbages is done with convolution neural network. Datasets were created and processed for classification of cabbage that produces best results using convolutional neural networks. Experimental results show

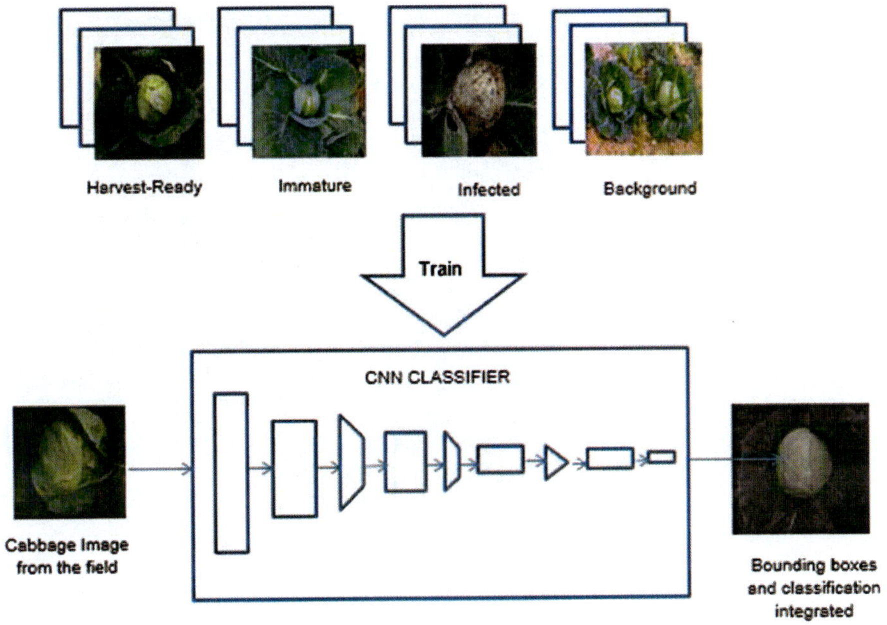


**Fig. 4** CNN cabbage classifier

**Table 1** Result obtained for the proposed method

| Model | Accuracy | Definition |
|---|---|---|
| Proposed approach | 90.86% | Number of cabbage detected/number of real cabbage in field |

**Table 2** Overall performance of proposed method

| Criteria | Result |
|---|---|
| Total real matured cabbages in field | 47 |
| Total matured cabbages detected | 43 |

that the proposed model shows the accuracy result of 90.86% and the cabbage was perfectly classified. Though the accuracy may be high in this method, by fine-tuning some parameters and increasing data, the accuracy rate may be increased.

## 5 Future Work

This research can be further improved with enormous datasets and different machine learning techniques. The cabbage classification and identification can be developed

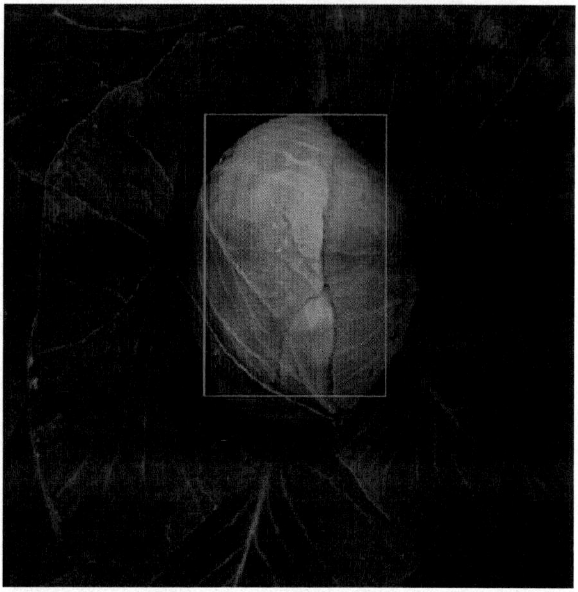

**Fig. 5** CNN cabbage classifier

with automated robotic system which moves in the field over the cabbage plant. Further, the cabbage can be harvested automatically using that moving robot.

**Acknowledgements** We would like to express our gratitude towards Mr. Kanthan, farmer, for giving permission for creating datasets from his cabbage field and sharing the problem of ideology prevailing in his farm.

## References

1. Patra S, Ganguly P, Barik SR, Goon A, Mandal J, Samanta A, Bhattacharyya A (2020) Persistence behaviour and safety risk evaluation of pyridalyl in tomato and cabbage. Food Chem 309:125711
2. Siqi J, Chao W, Shuai W, Xuefang J (2017) (2017) Cabbage: cultivation of unfertilized ovary in vitro and ploidy identification of regenerated plant. Chin Agric Sci Bull 18:9
3. Kumar S, Parkash C, Dhiman MR, Pramanik A, Gautam N, Singh R, Sharma K (2019) Standardization of production technology of cabbage and cauliflower hybrids for off-season cultivation in kullu valley of Himachal Pradesh. IJCS 7(1):869–873
4. Glover MK, Obubuafo J, Agyeman-Duah MO, Doku GD, Glover EK (2017) Constraints associated with the marketing channel of lettuce and cabbage trade in Ghana. J Agric Sustain 10(2)
5. Desai JD, Solanki KD (2013) Extent of adoption of the market intelligence among the summer cabbage growers. Guj J Ext Edu 24:9–13
6. Nieuwenhuizen AT, Tang L, Hofstee JW, Müller J, Van Henten EJ (2007) Colour based detection of volunteer potatoes as weeds in sugar beet fields using machine vision. Precis Agric 8(6):267–278

7. Puttemans S, Vanbrabant Y, Tits L, GoedemT (2016) Automated visual fruit detection for harvest estimation and robotic harvesting. In: International Conference on Image Processing Theory, Tools and Applications (IPTA'16). IEEE, pp 1–6
8. Zhang G, Jia S, Zeng D, Zheng Z (2018) Object detection and grabbing based on machine vision for service robot. In: IEEE 9th Annual Information Technology, Electronics and Mobile Communication Conference (IEMCON), Vancouver, BC, pp 89–94
9. Birrell S, Hughes J, Cai JY, Iida F (2019) A field tested robotic harvesting system for iceberg lettuce. J Field Robot 1–21
10. Voulodimos A, Doulamis N, Doulamis A, Protopapadakis E (2018) Deep learning for computer vision: a brief review. Comput Intell Neurosci
11. Islam, KarimSiddique BMN, Rahman S, Jabid T (2018) Image recognition with deep learning. In: International Conference on Intelligent Informatics and Biomedical Sciences (ICIIBMS), pp 106–110

# Machine Learning Approach for Crop Prediction Based on Climatic Parameters

**S. J. Suji Prasad, R. Suganesh, and M. Thangatamilan**

**Abstract** A machine learning (ML)-based model is proposed to predict the crop which increases the yield of the crop. This will increase the GDP of the country, thereby contributing to the improvement of the economy of the nation. The proposed ML model will recommend a suitable crop for the given agro-climatic parameters like season, rainfall, temperature, humidity, and soil pH. Patternnet, KNN, and SVM algorithms are proposed for the classification and prediction of crop yield. The crop classification accuracy is 90.24% in SVM with a smaller number of samples for each crop. The KNN algorithm, which has high precision of about 92.15% with high specificity and less computing time, is used when there is a large number of classes and samples.

**Keywords** Machine learning · Support vector machines · KNN · Patternnet · Crop prediction

## 1 Introduction

Agriculture is the practice of cultivating crops and raising farm animals which includes the conditioning of the farm and production of plant and animal products for domestic use and also for sales [1]. Agriculture has an essential role in the Indian economy. In recent days, farmers face many problems [2, 3]. The major problems confronting Indian agriculture are those of failure of monsoon, crop pattern, depleted soils, lack of manpower, and lack of modern technology [4, 5]. The crop selection

S. J. Suji Prasad (✉) · R. Suganesh · M. Thangatamilan
Department of Electronics and Instrumentation Engineering, Kongu Engineering College, Erode, Tamil Nadu, India
e-mail: sjsujiprasad@gmail.com

R. Suganesh
e-mail: suganesh15@gmail.com

M. Thangatamilan
e-mail: thangatamil.m@gmail.com

© The Editor(s) (if applicable) and The Author(s), under exclusive license to Springer Nature Singapore Pte Ltd. 2021

S. Mohan et al. (eds.), *Materials, Design, and Manufacturing for Sustainable Environment*, Lecture Notes in Mechanical Engineering,
https://doi.org/10.1007/978-981-15-9809-8_53

problem can be solved by using the ML models to predict the crops based on the data available from sensors and meteorological departments [6, 7]. ML is a division of Artificial Intelligence (AI) that gives the knowledge to the machine. This provides computers the power to learn without being strictly programmed. The main aim is to teach computers to learn automatically without human interference or aid and to make decisions accordingly [8, 9]. The ML models are developed from the learning process. These models are trained from past experiences to develop a particular function. ML has developed along with big data and high-performance computing to produce new chances to untangle, quantitate, and understand data-intensive processes in agricultural environments. In this work, the ML models are developed for predicting the suitable crop based on the prevailing temperature in the area, annual rainfall, relative humidity, and soil pH of the particular agricultural. These models are developed using patternnet, *K*-nearest neighbors, and support vector machines, and the results are compared. These models will help the farmers to maintain the crop pattern and helps them to use the proper fertilizers based on pH value. This will prevent crop failure and helps the farmers to achieve maximum yield.

## 2 Machine Learning Techniques

Machine learning is a subdivision of AI which trains the machine based on past experiences and make predictions based on those experiences. This makes the computers to make data-driven decisions instead of being explicitly programmed. ML aims to give enhancing levels of automation in the engineering process supplacing the time-eating human activity with automation techniques that improve the accuracy and efficiency of the process [10]. In machine learning, the computer is trained by feeding good quality data and then build the models using the above data. The algorithm is chosen based on what type of data available and what kind of task we are trying to automate. ML is the progression of the regular algorithm that makes your programs "smarter," by allowing them to inevitably learn from the provided data [8]. The general block diagram of the machine learning and testing process is shown in Fig. 1.

**Fig. 1** Machine learning process

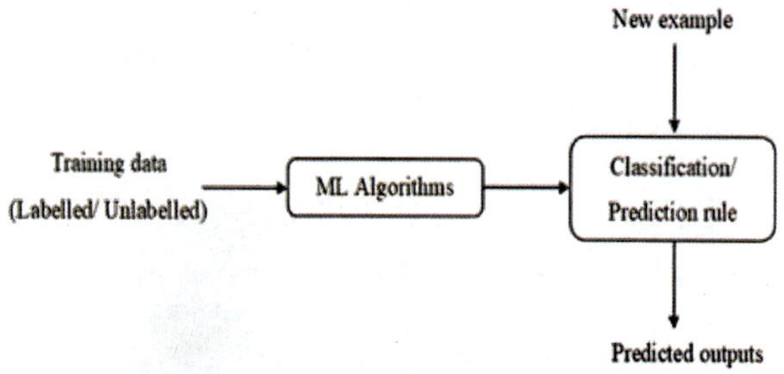

## 2.1 *Learning Process*

The ML model is trained using the training dataset. The model will make predictions based on the input data entered into the algorithm. The ML implementations are classified into supervised learning, unsupervised learning, and reinforced learning [11].

### 2.1.1 Supervised Learning

Supervised learning is also referred to as learning with an instructor. In this method, the teacher is having the knowledge of the environment which is being represented by the set of input and output examples. The dataset will act as a teacher and train the model. The samples of inputs and outputs from the environment are given to the system, and it learns the relationship between the input and target data using the ML techniques. Supervised learning is applicable only when all the data in the inputs are available. It is not possible to predict the output when some of the data are missing [12]. The supervised learning is mostly used to train neural networks and decision trees since the two methods are mainly dependent on the training data. In decision trees, the learning is used to find the attribute which gives the most relevant information which can be used to solve the classification problem. In neural networks, the algorithm is used to find out the error of prediction and then adjust the model to minimize the error [6].

### 2.1.2 Unsupervised Learning

The goal of unsupervised learning is to train the computer something that we do not know how to do. In unsupervised learning, the dataset consists only the input variables without the output variables. In unsupervised learning, a formal framework for unsupervised learning is created based on the opinion that the model's goal is to construct representations of the input that can be used for decision making, predicting future inputs, etc. [13]. Unsupervised learning is a perception finding patterns in the data above and beyond what would be considered pure unstructured noise. Two simple definitive examples of unsupervised learning are clustering and dimensionality reduction learning which is used when the data consist of desired outputs [14].

## 3 Seven Steps of Machine Learning Algorithm

***Data collection:*** The machine learning technique mainly depends on the data. Utmost care should be taken while preparing the data because the accuracy of the model is decided based on the quantity and quality of the data.

***Data preparation:*** In order to employ the collected data to train the model, all the data are put together and processed for determining any relationship between the different variables and any data imbalances. The data collected will be fractioned into training and testing data. The training data will be used for training the model, and testing data will be used for assessing the trained model's performance. Sometimes, the data collected needs to undergo some forms of adjustments like normalization, error correction, and more.

***Choose a model:*** The third step consists of selecting the right model based on the type of data used. Linear regression, logistic regression, decision trees, $K$-means, principal component analysis, SVM, KNN, Naïve Bayes, random forest, and ANN are the commonly used algorithms to build a model.

***Train the model:*** The aim of the training is to make predictions as correct as possible. The objective is to use the data and improve the prediction ability of the model. In each cycle of updating the weights and biases is one training step. In the case of supervised machine learning, the ML model is developed using labelled sample data, but the unsupervised learning tries to draw inferences from non-labelled data (without references to known or labelled outcomes).

***Evaluate the model:*** The developed ML model is tested against the testing dataset which is not used in the training process to check how the model performs. This represents how the model works in the real world, but this does not have to be the case.

***Parameter tuning:*** After evaluating the ML model, it is tested for the originally set parameters to improve the performance of the model. The accuracy of the model can be increased by increasing the number of training cycles.

***Make predictions:*** Prediction or inference is a step where the solution to the problem is found and the value of machine learning is realized. The test data made during the data preprocessing are applied to test the model; a better estimate of how the model will do in the real world [10].

## 4 Performance Evaluation Using Confusion Matrix

The ultimate aim of the ML models is to make accurate predictions to facilitate an organization to create actual values. The training of the developed model is a key step based on which it can be evaluated that how the model extrapolates on unseen data which is an equally important aspect that should be considered in every ML algorithm. It is of the essence to know whether it works and, accordingly, whether we can trust the predictions. The performance of the model is estimated using the confusion matrix.

## *4.1 Confusion Matrix*

A confusion matrix is a summary of results on a classification problem. In the confusion matrix, the number of correct and incorrect predictions are summarized with count values and broken down by each class. The confusion matrix shows in which way the classification model is confused when it makes predictions. It gives an estimation about the errors made by the classifier and also the types of error the classifier made.

### 4.1.1 Classification Rate or Accuracy

The formula for classification rate or accuracy is given by:

$$\text{Accuracy} = \frac{(\text{TP} + \text{TN})}{\text{TP} + \text{TN} + \text{FP} + \text{FN}} \quad (1)$$

There are problems with accuracy as it assumes the same cost for both kinds of errors. The accuracy of 99% can be excellent, good, mediocre, poor, or terrible depending upon the problem.

### 4.1.2 Recall

The recall is defined as the ratio of the total number of correctly classified positive examples to the total number of positive examples. If all the classes are correctly identified, then the recall should be high, i.e., small number of false negatives.

$$\text{Recall} = \frac{\text{TP}}{\text{TP} + \text{FN}} \quad (2)$$

### 4.1.3 Precision

Precision is the ratio of the total number of correctly classified positive examples to the total number of predicted positive examples. If there are more examples labelled as positive, then the precision will be high, i.e., less number of FP

$$\text{Precision} = \frac{\text{TP}}{\text{TP} + \text{FP}} \tag{3}$$

#### 4.1.4 *F*-measure

$F$-measure is the relation between precision and recall. $F$-measure is calculated by using harmonic mean instead of arithmetic mean as it rebukes the extreme values more. The $F$-measure will always be closer to the lesser value of precision or recall.

$$F\text{ - Measure} = \frac{2 * \text{Recall} * \text{Precision}}{\text{Recall} + \text{Precision}} \tag{4}$$

### *4.2 Performance of the ML Models Using a Different Number of Samples*

The ML model developed above is tested with different number of samples for each crop. The model is tested with 1, 10, 100 samples for each crop, and the results are obtained.

#### 4.2.1 The Performance Measure of the Models with One Sample for Each Class

The patternnet, KNN, and SVM algorithms are trained with one sample for all the 41 crops. Then, the confusion matrix is plotted, and the accuracy, sensitivity, specificity, recall, and $F$-measure of each model are calculated using the above formulas.

From Table 1, it is observed that the accuracy and $F$-measure of the SVM classifier is high while using a single sample per class when compared to patternnet and KNN classifier models. The SVM has high precision and low recall which means the SVM classifier recognized more positive examples, and there are more false positives. The KNN has abysmal accuracy than SVM and patternnet classifiers. This shows that SVM is best suited when there is a smaller number of samples for each class in the dataset. This clearly indicates that the precision of all the classifiers is low. In order to improve the precision more of the samples for each class are given in the training data. Also, the accuracy of the KNN is enhanced by increasing the number of neighbors. The SVM also has the computing time near to KNN.

**Table 1** Comparison of patternnet, KNN, and SVM models with one sample for each class

| | Patternnet | KNN | SVM |
|---|---|---|---|
| Correct rate | 0.0488 | 0.2683 | 0.9024 |
| Error rate | 0.9512 | 0.7317 | 0.0976 |
| Sensitivity | 0 | 1 | 1 |
| Specificity | 1 | 0.925 | 1 |
| Positive prediction value | 0 | 0.25 | 1 |
| Negative predictive value | 0.9756 | 1 | 1 |
| Precision | 0 | 0.0323 | 0.2 |
| Recall | 0 | 1 | 1 |
| *F*-measure | 0 | 0.5 | 0.9487 |
| Computing time (s) | 3.33 | 0.93 | 1.05 |

#### 4.2.2 The Performance Measure of the Models with Ten Sample for Each Class

The patternnet, KNN, and SVM algorithms are trained with ten samples for all the 41 crops. Then, the confusion matrix is plotted, and the accuracy, sensitivity, specificity, recall, and *F*-measure of each model are calculated using the above formulas and are given in Table 2.

From Table 2, it is clear that the accuracy of patternnet and KNN increases with an increase in the number of samples. But, the accuracy of the SVM classifier decreases with the increase in samples. The KNN has the highest accuracy of 87.07%, and patternnet has an accuracy of 75.85%. From this table, it is inferred that when there are a large number of classes for classification with more samples for each class, the KNN classifier is preferred. Otherwise, if there is a smaller number of classes with a large number of samples, the patternnet algorithm is used. The computational time

**Table 2** Comparison of patternnet, KNN, and SVM models with ten samples for each class

| | Patternnet | KNN | SVM |
|---|---|---|---|
| Correct rate | 0.7585 | 0.8707 | 0.5 |
| Error rate | 0.2415 | 0.1293 | 0.5 |
| Sensitivity | 0.8 | 1 | 0.6 |
| Specificity | 1 | 1 | 0.9975 |
| Positive prediction value | 1 | 1 | 0.8571 |
| Negative predictive value | 0.995 | 1 | 0.09901 |
| Precision | 0.0762 | 0.1587 | 0.029 |
| Recall | 0.8 | 1 | 0.6 |
| *F*-measure | 0.7785 | 0.9314 | 0.5464 |
| Computing time (s) | 3.781 | 0.88 | 2.194 |

**Table 3** Comparison of patternnet, KNN, and SVM models with 100 samples for each class

| | Patternnet | KNN | SVM |
|---|---|---|---|
| Correct rate | 0.8361 | 0.9215 | 0.419 |
| Error rate | 0.1639 | 0.0785 | 0.581 |
| Sensitivity | 0.83 | 0.89 | 0.47 |
| Specificity | 0.9982 | 1 | 0.999 |
| Positive prediction value | 0.9222 | 1 | 0.9216 |
| Negative predictive value | 0.9958 | 0.9973 | 0.9869 |
| Precision | 0.1125 | 0.2225 | 0.0198 |
| Recall | 0.83 | 0.89 | 0.47 |
| *F*-measure | 0.8331 | 0.906 | 0.4431 |
| Computing time (s) | 15.32 | 4.33 | 26.55 |

of the model increases, if the number of samples is high. Since patternnet is an ANN model, it takes more time for training the model. The training time of the patternnet increases with an increase in the number of hidden layers. But increasing the number of hidden layers improves the accuracy of the network.

### 4.2.3 The Performance Measure of the Models with 100 Sample for Each Class

The above algorithms are again tested with 100 samples per class, and the test results are given in Table 3.

From Table 3, it is observed that increasing the number of samples further improved the accuracy of KNN and patternnet, but the accuracy of SVM is decreased. The KNN has the highest accuracy, precision, and recall compared to the other two classifiers. The KNN has specificity value 1 which shows that the negative classifications are accurate. From comparing all the evaluation parameters, the KNN shows the best result when there is a large number of samples for each class in the training data. While using a large number of data, the computational time of the model is also increasing. The KNN again took advantage in terms of concise computing time when compared to SVM and patternnet. Therefore, the KNN algorithms are mostly preferred for classification problems unless there are only a few samples of data available for each class.

## 5 Conclusion

The proposed ML model recommends a suitable crop, given the agro-climatic parameters like season, rainfall, temperature, humidity, and soil pH. The system was built with three different classifier models using patternnet, KNN, and SVM algorithms and was tested. Properly provided datasets and ML algorithms enable the proposed method to work correctly. The crop classification accuracy is 90.24% in SVM when there are a smaller number of samples for each crop. Also, the computation time is less compared to the patternnet. But, when the number of samples per class increases, the accuracy of the SVM model decreases. In those cases, the KNN and patternnet algorithms come into action. If there are a large number of classes, the recall and precision of patternnet are low. So, there is a possibility of false-positive classification. The computation time of patternnet is high about when compared to KNN and SVM. Therefore, the KNN algorithm which has high precision, specificity, and less computing time is used when there is a large number of classes and samples. Since the crop recommendation system needs high accuracy and precision, the KNN algorithm is recommended to build the ML model. This ML model will help the farmers to choose the right crop and reduces the risk of crop failure and helps them to get maximum yield. The increase in yield of the crop reduces poverty and increases the GDP of the country, thereby contributes to the improvement of the economy of the nation.

## References


1. Supriya J, Vidya Y (2012) The dark side of globalization—in context of India. Int J Eng Manag Sci (IJEMS) 3:29–31
2. Goutam S (2012) Some reflections of globalization on Indian agriculture. Int J Educ Sci Res Rev (IJESRR) 1:157–162
3. Das P (2015) Problems of rural farmer: a case study based on the lowphulabori village under the raha block development area of Nagaon district, Assam. IOSR J Humanit Soc Sci 20:40–43
4. Bhattacharyya R, Ghosh BN, Mishra PK et al (2015) Soil degradation in India: challenges and potential solutions. Sustainability 7:3528–3570
5. Merriott D (2016) Factors associated with the farmer suicide crisis in India. J Epidemiol Glob Health 6:217–227. https://doi.org/10.1016/j.jegh.2016.03.003
6. Vivek MVR, Harsha DVVSSS, Maran PS A survey on crop recommendation using machine learning. Int J Recent Technol Eng (IJRTE) 7:120–125
7. Kamraju M, Vani M, Anuradha T (2017) Crop diversification pattern: a case study of Telangana state. Int J Innovative Sci Res Technol 2:366–371
8. Gramaje A, Thabtah F, Abdelhamid N, Ray SK (2019) Patient discharge classification using machine learning techniques. Annals of Data Science, pp 1–13
9. Cherian S, Murukezhan K (2013) Providing data protection as a service in cloud computing. Int J Sci Res Publ 3:1
10. Liakos KG, Busato P, Moshou D et al (2018) Machine learning in agriculture: a review. Sensors 18:2674
11. Kotsiantis SB, Zaharakis ID, Pintelas PE (2006) Machine learning: a review of classification and combining techniques. Artif Intell Rev 26:159–190
12. Zaffar M, Savita KS, Hashmani MA, Rizvi SSH (2018) A study of feature selection algorithms for predicting students academic performance. Int J Adv Comput Sci Appl 9:541–549

13. Rahman SAZ, Mitra KC, Islam SMM (2018) Soil classification using machine learning methods and crop suggestion based on soil series. In: 2018 21st International Conference of Computer and Information Technology (ICCIT), pp 1–4
14. Shirsath R, Khadke N, More D et al (2017) Agriculture decision support system using data mining. In: 2017 International Conference on Intelligent Computing and Control (I2C2), pp 1–5

# Analysis of Risk Factors in Road Accidents Using Fuzzy ANP Method

**S. Bathrinath, T. Mahendiran, M. Ravikumar, T. Karthi Shesan, R. K. A. Bhalaji, and K. Koppiahraj**

**Abstract** The number of premature death due to road accidents (RA) in recent years has been on vertical horizon. In connection with this, this paper intends to identify, evaluate, and analyze various risks that aggregate the occurrence of RA. For this, through a literature survey and interaction with experts, three main risks, namely, engineering, human, and road-related risks, with a total of twenty-one sub-risks were identified and evaluated in a real-world setting. In this work, an integrated approach comprising both the Delphi method and the fuzzy Analytic Network Process (ANP) is used to evaluate the identified risks. Results reveal that non-uniformity, distracted driving, speeding, tire blowout, and using gadgets are the top five risks triggering the occurrence of RA. This study is not only limited to Indian context, it could be applicable to other developing countries where similar socio-economic condition prevails. Finally, this work proposes managerial implications which could assist the government in providing accident-free road system.

S. Bathrinath (✉) · T. Mahendiran · M. Ravikumar · T. Karthi Shesan · R. K. A. Bhalaji · K. Koppiahraj
Department of Mechanical Engineering, Kalasalingam Academy of Research and Education, Krishnankoil, Tamil Nadu, India
e-mail: bathri@gmail.com

T. Mahendiran
e-mail: mahendirant88@gmail.com

M. Ravikumar
e-mail: ravikumar407a@gmail.com

T. Karthi Shesan
e-mail: shesankarthi@gmail.com

R. K. A. Bhalaji
e-mail: bhalaji28@gmail.com

K. Koppiahraj
e-mail: koppiahraj1993@gmail.com

© The Editor(s) (if applicable) and The Author(s), under exclusive license to Springer Nature Singapore Pte Ltd. 2021
S. Mohan et al. (eds.), *Materials, Design, and Manufacturing for Sustainable Environment*, Lecture Notes in Mechanical Engineering,
https://doi.org/10.1007/978-981-15-9809-8_54

**Keywords** Road accidents · Delphi method · Fuzzy Analytic Network Process (ANP)

## 1 Introduction

The road transportation system plays a significant role in the day-to-day life scenario when compared with the other available inland transportation systems. Road transportation system plays a vital role in the cargo movement and the mass movement of people from one place to another place. Road transportation is preferred over other modes of inland transportation as it has advantages like cheap cost, penetration into deserted places, and offers a wide range of route flexibility. Despite offering many advantages, the road transportation system remains a major concern for the public as it accounts for a considerable number of deaths annually [1]. A road accident may be defined as the failure of the road-vehicle-driver system in performing an operation without any injury or loss [2]. Reasons like defective planning, poor road construction, usage of damaged vehicles, and use of old technologies against rapid population growth aggregate the occurrence of road accidents (RA). Studies carried by global bodies like World Bank, World Health Organization (WHO), and Transport Research Laboratory (TRL) highlights that the incidence of RA is high in low and middle-income countries as compared with high-income countries [3]. Globally, RA is considered the eighth-most leading cause of premature death. As this trend continues, by 2030, RA will become the fifth most cause for premature death. Hence, punitive actions must be taken to prevent the occurrence of RA [4]. Occurrence of RA may result in premature death, permanent disabilities, and hospitalization with major socioeconomic costs all over the world. The United Nations has announced 2011–2020 as the decade of action on road accidents to encourage nations to take measures to make roads safe and ensure the safety of road users.

India, with 5.5 mn Km comprising of national and state highways, and urban and rural roads is the world's second-largest road network. According to the Ministry of Road Transport and Highways ((MoRTH), India, the number of road accidents has increased by 0.5% in 2018 which was decreased by 3.3% in 2017 (https://morth.nic.in/sites/default/files/Road_Accidednt.pdf). Without effective efforts and new initiatives, the total number of road accident deaths in India is likely to cross the mark of 250,000 by 2025. This increasing number of accidents has become a primary concern for both the public and the government. Numerous factors like over speeding, drunk and driving, poor road conditions, multitasking while vehicle driving and animal intervention are cited as the primary causes for RA. Development of a robust framework for accident mitigation in the Indian context often remains a challenge due to traffic heterogeneity, poor quality of accident data, and under-reporting of accidents [5]. Medical Schemes like National Highway Trauma Care Project (NHTCP) and National Highways Accident Relief Services Scheme (NHARSS) have been introduced by the Indian government to curb the deaths due to road accidents. However, in line with the proverb 'prevention is the better cure', the government has to take stringent actions to prevent the occurrence of RA.

From the above information, it is clear that there is an immediate need to identify the risks that are causing RA. In connection with this, this paper aims to identify and evaluate the various risks causing RA.

## 2 Literature Review

Literature review is categorized into two sub-segments namely (a) Risk factors involved in road accidents (b) Research gap.

### 2.1 *Risk Factors Involved in Road Accidents*

Dhruval and Bhagat [6] reviewed the road traffic accidents in India which are inevitable, not preventable, and unpredictable, and also identified some risks in road accidents. From the outcomes, reports stated that injury from road traffic nearly takes the lives of 1.2 million people each year and simultaneously harmed 30–60 million people globally. Research has shown that the involvement of age, alcohol, and risk of road accident that increases with increasing concentration of blood alcohol. Risk was mainly formed by teenage drivers was determined to be maximum than old age drivers [7]. Examine the application of data mining methods related to accidents on road using machine learning algorithms to expect future accident rates and to minimize injuries and crash deaths. Officials can use our outcomes for the control and prevention of road accidents [8]. Elvik [9] reviewed that road accident risks related to the usage of drugs and also meta-analysis has been conducted sixty-six studies recording a total of nearly 260 estimates of the accident risks involvement related to the drug usage while driving. Analyzed the accident in roads as per the condition of the road surface by using $K$-means clustering. From the reports, 70% of accidents are mainly due to human-related risk factors [10]. Singh et al. [7] addressed the challenges and issues of road safety accidents in metropolitan, state, and national city level in India. Without increasing new initiatives and efforts related to road safety, the accident rate will be increasing in the future. There is a necessity to identify the worst scenario in injuries and road deaths and take necessary action. Table 1 shows the risk factors involved in road accidents.

### 2.2 *Research Gap and Focuses*

To achieve zero accidents on the roads, the highways department needs to be more efficient in the process of risk management. Rising nations like India and high way

**Table 1** Risk factors involved in road accidents

| Main Factors | Sub-Factors |
|---|---|
| Engineering related risks (R1) | Speeding ($R_{11}$) |
| | Tire blowout ($R_{12}$) |
| | Vehicle effects ($R_{13}$) |
| | Maintenance error ($R_{14}$) |
| | Monitoring defects ($R_{15}$) |
| | Cargo ($R_{16}$) |
| | Design defects ($R_{17}$) |
| Human related risks (R2) | Non-uniformity ($R_{21}$) |
| | Distracted driving ($R_{22}$) |
| | Using gadgets ($R_{23}$) |
| | Drowsy while driving ($R_{24}$) |
| | Drugged and Drunken ($R_{25}$) |
| | Inexperienced driver ($R_{26}$) |
| | Tailgating ($R_{27}$) |
| Road related risks (R3) | Potholes ($R_{31}$) |
| | Wrong way or improper turns ($R_{32}$) |
| | Animal crossing ($R_{33}$) |
| | Obstacles in the road ($R_{34}$) |
| | Highway construction accident ($R_{35}$) |
| | Dangerous road conditions ($R_{36}$) |
| | Lane splitting ($R_{37}$) |

patrol officers have poor knowledge about road accident risks. Risk in roadsides leads to death and severe injuries to people and they need to be controlled the risks for the upcoming generation. The literature review recognized the risk factors involved in road accidents on various highways. However, the highways department may face various hurdles while undertaking the initiatives of road accident risks. As a result, the impact of a particular risk factor may vary from highway to highway. In the literature, the risk factors are not discussed sufficiently and a ranking of risks is also missing. To fulfill this gap, we have considered this paper which will be useful in finding the most influential risks in road accidents as per the case study. For this purpose, this paper uses MCDM method like Fuzzy ANP for assessing the interrelationship between different risk factors and for ranking of risks in road accidents. This paper contains several focuses which are detailed as follows:

- Recognize risk factors in road accidents from literature review and with the help of highways department officials.
- Proposing a framework to evaluate the ranking of risk in the road accidents.
- Verify the proposed framework using a real case of four national highways.

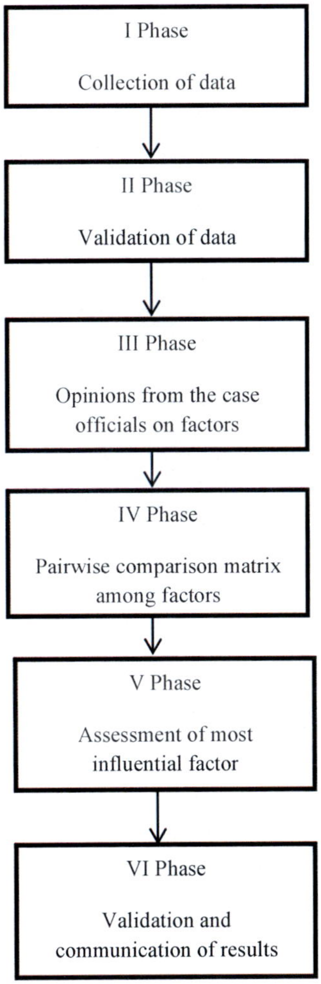


**Fig. 1** Suggested framework to assess the factors involved in road accidents

## 3 Proposed Framework of This Paper

See Fig. 1.

## 4 Methodology

Multi-criteria decision-making method like ANP (Analytic Network Process) which computes the criteria weights and alternatives. The ANP is the lengthy form of AHP, where the key variation between them deliberates the interdependencies between factors. AHP is the structure of hierarchical [11] and ANP is the structure of the network [12], so the network which deliberates the factors interdependencies, outcomes in more clear findings. It is also the model of generalized and it doesn't deliberate any assumptions in the interdependencies factor of the application of decision making. In real-life applications, several researchers recently used MCDM method like ANP because of its significant advantages [13–15]. In previous studies, ANP is used in many applications but additional provision is required to give good outcomes. ANP method can assess the factors, but this paper is purely based on the judgments of human which requires more accuracy and the least ambiguity to diminish potential biases in the outcomes. To overcome this issue, fuzzy is incorporated with ANP to assess our problem. Fuzzy incorporation in decision making enhances the precision of the results [16]. The six phases involved in the Fuzzy ANP method are:

Phase 1: Creation of the construction model

The first phase of ANP is to build a model that defines the influential relationship between the factors shown in Fig. 2. The network model demonstrates both outer and inner dependence between their factors. In this paper, interdependencies among each and every factor are recognized by using the method.

Phase 2: Development of pair-wise comparison in the fuzzy setting

Depends on the interdependencies recognized from the previous phase, the pair-wise comparison matrix is created with the help of decision-makers. The decision-makers deliver their comments about the factor using the form of linguistic is exhibited in Table 2 which is transformed into a triangular fuzzy scale.

Phase 3: Defuzzification

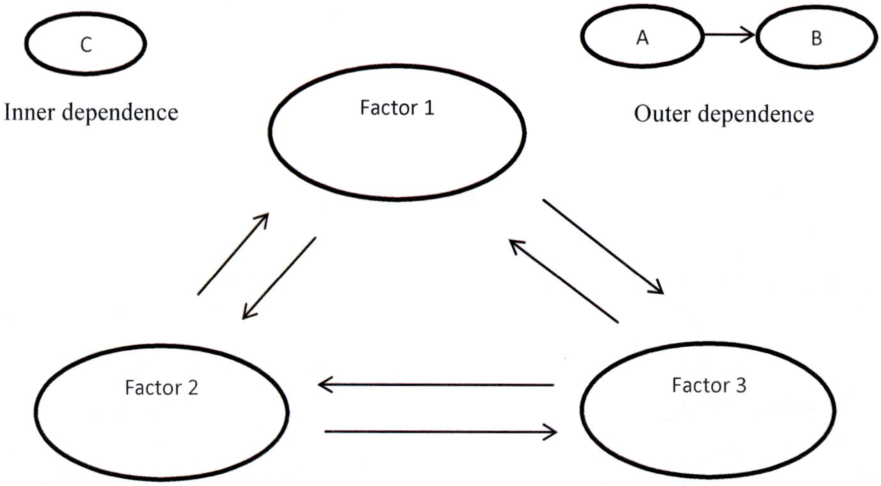


**Fig. 2** Network model with dependencies

**Table 2** Linguistic scales for importance and difficulty

| Linguistic scale for importance | Linguistic scale for difficulty | Triangular fuzzy scale | Triangular fuzzy reciprocal scale |
|---|---|---|---|
| Absolutely more important (AMI) | Absolutely more difficult (AMD) | (5/2,3,7/2) | (2/7,1/3,2/5) |
| Very strongly more important (VSMI) | Very strongly more difficult (VSMD) | (2,5/2,3) | (1/3,2/5,1/2) |
| Strongly more important (SMI) | Strongly more difficult (SMD) | (3/2,2,5/2) | (2/5,1/2,2/3) |
| Weakly more important (WMI) | Weakly more difficult (WMD) | (1,3/2,2) | (1/2,2/3,1) |
| Equally important (EI) | Equally difficult (ED) | (1/2,1,3/2) | (2/3,1,2) |
| Just equal | Just equal | (1,1,1) | (1,1,1) |

Using the method of defuzzification, the next phase is to transform the TFNs (Triangular fuzzy numbers) into crisp no's.

Phase 4: Finding the relative weights

Once the TFNs are defuzzified, then we find the relative weights for the factor along with their interdependencies.

Phase 5: Development of supermatrix

By using the weights from the previous phase, the supermatrix is developed. The weights are positioned in the specific column where each factor influences the other factor. The common shape of a super-matrix is demonstrated in Eq. (1).

$$A = \begin{bmatrix} A_{11} & 0 & 0 & A_{14} \\ A_{21} & A_{22} & 0 & A_{24} \\ A_{31} & 0 & A_{33} & A_{34} \\ A_{41} & A_{42} & A_{43} & 0 \end{bmatrix} \quad (1)$$

Phase 6: Limit matrix

The final weights of the factors are attained by using the notion of limit matrix by increasing the supermatrix to limiting powers using Eq. (2) (Tables 3, 4, 5, 6, 7, 8, 9 and 10).

**Table 3** Pairwise comparison matrix and local weights of the main factors

| Factors | R1 | R2 | R3 | Local weights |
|---|---|---|---|---|
| R1 | (1,1,1) | (2,5/2,3) | (1,3/2,2) | 0.4966 |
| R2 | (1/3,2/5,1/2) | (1,1,1) | (3/2,2,5/2) | 0.4368 |
| R3 | (1/2,2/3,1) | (2/5,1/2,2/3) | (1,1,1) | 0.2317 |

**Table 4** Pairwise comparison matrix and local weights for engineering related sub-factors

| R1 | $R_{11}$ | $R_{12}$ | $R_{13}$ | $R_{14}$ | $R_{15}$ | $R_{16}$ | $R_{17}$ | Local weights |
|---|---|---|---|---|---|---|---|---|
| $R_{11}$ | (1,1,1) | (1,3/2,2) | (1/2,1,3/2) | (3/2,2,5/2) | (2,5/2,3) | (1,3/2,2) | (1/2,1,3/2) | 0.1812 |
| $R_{12}$ | (1/2,2/3,1) | (1,1,1) | (1/2,1,3/2) | (1,3/2,2) | (2,5/2,3) | (3/2,2,5/2) | (1/2,1,3/2) | 0.1627 |
| $R_{13}$ | (2/3,1,2) | (2/3,1,2) | (1,1,1) | (1/2,1,3/2) | (3/2,2,5/2) | (1/2,1,3/2) | (1,1,1) | 0.1482 |
| $R_{14}$ | (2/5,1/2,2/3) | (1/2,2/3,1) | (2/3,1,2) | (1,1,1) | (1,3/2,2) | (1/2,1,3/2) | (1,3/2,2) | 0.1280 |
| $R_{15}$ | (1/3,2/5,1/2) | (1/3,2/5,1/2) | (2/5,1/2,2/3) | (1/2,2/3,1) | (1,1,1) | (3/2,2,5/2) | (3/2,2,5/2) | 0.1192 |
| $R_{16}$ | (1/2,2/3,1) | (2/5,1/2,2/3) | (2/3,1,2) | (2/3,1,2) | (2/5,1/2,2/3) | (1,1,1) | (1/2,1,3/2) | 0.1251 |
| $R_{17}$ | (1/2,2/3,1) | (1/2,2/3,1) | (1/2,2/3,1) | (1/2,2/3,1) | (1/2,2/3,1) | (1/2,2/3,1) | (1,1,1) | 0.1352 |

**Table 5** Pairwise comparison matrix and local weights for human-related sub-factors

| R2 | $R_{21}$ | $R_{22}$ | $R_{23}$ | $R_{24}$ | $R_{25}$ | $R_{26}$ | $R_{27}$ | Local weights |
|---|---|---|---|---|---|---|---|---|
| $R_{21}$ | (1,1,1) | (1/2,1,3/2) | (3/2,2,5/2) | (1,3/2,2) | 2,5/2,3 | (3/2,2,5/2) | (1/2,1,3/2) | 0.1930 |
| $R_{22}$ | (2/3,1,2) | (1,1,1) | (2,5/2,3) | (1,3/2,2) | (1/2,1,3/2) | (1,3/2,2) | (1,3/2,2) | 0.1833 |
| $R_{23}$ | (2/5,1/2,2/3) | (1/3,2/5,1/2) | (1,1,1) | (2,5/2,3) | (2,5/2,3) | (1/2,1,3/2) | (1,3/2,2) | 0.1582 |
| $R_{24}$ | (1/2,2/3,1) | (1/2,2/3,1) | (1/3,2/5,1/2) | (1,1,1) | (1/2,1,3/2) | (3/2,2,5/2) | (1,3/2,2) | 0.1239 |
| $R_{25}$ | 1/3,2/5,1/2 | (2/3,1,2) | (1/3,2/5,1/2) | (2/3,1,2) | (1,1,1) | (2,5/2,3) | (3/2,2,5/2) | 0.1449 |
| $R_{26}$ | (2/5,1/2,2/3) | (1/2,2/3,1) | (2/5,1/2,2/3) | (2/5,1/2,2/3) | (1/3,2/5,1/2) | (1,1,1) | (1,3/2,2) | 0.0913 |
| $R_{27}$ | (2/3,1,2) | (1/2,2/3,1) | (1/2,2/3,1) | (1/2,2/3,1) | (2/5,1/2,2/3) | (1/2,2/3,1) | (1,1,1) | 0.1051 |

**Table 6** Pairwise comparison matrix and local weights for road-related sub-factors

| R3 | $R_{31}$ | $R_{32}$ | $R_{33}$ | $R_{34}$ | $R_{35}$ | $R_{36}$ | $R_{37}$ | Local weights |
|---|---|---|---|---|---|---|---|---|
| $R_{31}$ | (1,1,1) | (1,3/2,2) | (1/2,1,3/2) | 3/2,2,5/2 | (1/2,1,3/2) | (1,1,1) | (1,3/2,2) | 0.1686 |
| $R_{32}$ | (1/2,2/3,1) | (1,1,1) | (1,3/2,2) | (1/2,1,3/2) | (1,3/2,2) | (3/2,2,5/2) | (2,5/2,3) | 0.1797 |
| $R_{33}$ | (2/3,1,2) | (1/2,2/3,1) | (1,1,1) | (1,3/2,2) | (3/2,2,5/2) | (1/2,1,3/2) | (1,3/2,2) | 0.1611 |
| $R_{34}$ | 2/5,1/2,2/3 | (2/3,1,2) | (1/2,2/3,1) | (1,1,1) | (1,3/2,2) | (3/2,2,5/2) | (1/2,1,3/2) | 0.1443 |
| $R_{35}$ | (2/3,1,2) | (1/2,,2/3,1) | 2/5,1/2,2/3 | (1/2,2/3,1) | (1,1,1) | (1/2,1,3/2) | (3/2,2,5/2) | 0.1252 |
| $R_{36}$ | (1,1,1) | (2/5,1/2,2/3) | (2/3,1,2) | (2/5,1/2,2/3) | (2/3,1,2) | (1,1,1) | (2,5/2,3) | 0.1377 |
| $R_{37}$ | (1/2,2/3,1 | (1/3,2/5,1/2) | (1/2,2/3,1) | (1/2,2/3,1) | (2/5,1/2,2/3) | (1/3,2/5,1/2) | (1,1,1) | 0.0830 |

**Table 7** The inner reliance matrix of the factors regarding "Engineering Factors"

| R1 | R2 | R3 | Relative importance weights |
|---|---|---|---|
| R2 | (1,1,1) | (1,3/2,2) | 0.58 |
| R3 | (1,3/2,2) | (1,1,1) | 0.41 |

**Table 8** The inner reliance matrix of the factors regarding "Human Factors"

| R2 | R1 | R3 | Relative importance weights |
|---|---|---|---|
| R1 | (1,1,1) | (3/2,2,5/2) | 0.66 |
| R3 | (2/5,1/2,2/3) | (1,1,1) | 0.33 |

**Table 9** The inner reliance matrix of the factors regarding "Road Factors"

| R3 | R1 | R2 | Relative importance weights |
|---|---|---|---|
| R1 | (1,1,1) | (2,5/2,3) | 0.71 |
| R2 | (1/3,2/5,1/2) | (1,1,1) | 0.28 |

**Table 10** Calculated global weights for sub-factors

| Factors and local weights | Sub-factors | Local weights | Global weights |
|---|---|---|---|
| Engineering-related risks (R1) (0.4322) | $R_{11}$ | 0.1813 | 0.0784 |
| | $R_{12}$ | 0.1628 | 0.0704 |
| | $R_{13}$ | 0.1482 | 0.0641 |
| | $R_{14}$ | 0.1281 | 0.0554 |
| | $R_{15}$ | 0.1193 | 0.0516 |
| | $R_{16}$ | 0.1251 | 0.0541 |
| | $R_{17}$ | 0.1352 | 0.0585 |
| Human-related risks (R2) (0.4428) | $R_{21}$ | 0.1930 | 0.0855 |
| | $R_{22}$ | 0.1834 | 0.0812 |
| | $R_{23}$ | 0.1583 | 0.0701 |
| | $R_{24}$ | 0.1240 | 0.0549 |
| | $R_{25}$ | 0.1449 | 0.0642 |
| | $R_{26}$ | 0.0913 | 0.0554 |
| | $R_{27}$ | 0.1051 | 0.0599 |
| Road-related risks (R3) (0.2899) | $R_{31}$ | 0.1687 | 0.0489 |
| | $R_{32}$ | 0.1797 | 0.0521 |
| | $R_{33}$ | 0.1612 | 0.0467 |
| | $R_{34}$ | 0.1444 | 0.0419 |
| | $R_{35}$ | 0.1253 | 0.0363 |
| | $R_{36}$ | 0.1578 | 0.0457 |
| | $R_{37}$ | 0.0830 | 0.0241 |

$$\lim_{n \to \infty} (W)^n \tag{2}$$

## 5 Case Study

The suggested framework is validated using data from four national highways located in the southern part of India. Each and every day more than 1000 vehicles are crossing the roads. These highways are more useful to many organizations for easy transportation and smooth in their business. As per the recent reports from the highways department, many accidents are happened due to some risks and it leads to deaths and injuries to peoples. Highway patrol officers are difficult to manage the risks. In this scenario, risk assessment is needed to control the risks. Therefore they require to recognize the risk factor and ranking of risks for achieving good transportation. This paper suggests the multi-criteria decision-making method like Fuzzy ANP to assess the most influential risk factors and their rankings. Principal Secretary, project director, director-general, chief engineer, and highway patrol officials are in the team of decision-makers and they are denoted as ($DM_1, DM_2, DM_3, DM_4, DM_5$), and they are chosen depends on the experience and knowledge in decision making. Delphi method is used to gather risk factors from the experts. A team consists of five decision-makers who given their ratings to risk factors for obtaining the pair-wise comparison matrix. For implementing the highway roads and mitigating the risks effectively, highway officials can use our outcomes.

## 6 Result and Discussions

As per the current scenario, there are various MCDM methods obtainable for solving the problem and to make decisions for the proposed targets. In this paper, Fuzzy ANP is used to recognize the most influential risks in road accidents. In routine life, high way patrol officials are difficult to manage road accidents. This paper identifies three main risk factors and 21 sub-factors related to road accidents are recognized and their interrelationships with each other are analyzed by using Fuzzy ANP method. Figure 3 shows the ranking of the main factors.

Human-related risk (R2) is the most influential one in road accidents based on Fig. 3. According to national road safety, human error contributes to 57% of road accidents. Human errors are caused due to a lack of concentration and crossing off the limits of road safety rules and off the regulation. Humans made mistakes by their behavior and the situation around off. Human error is made by a human without proper knowledge of driving on the road. In this paper, human error has to contain maximum weight of the problem. For reducing the accident caused by human error, we need to improve the methods of testing the drivers and to improve the methods of

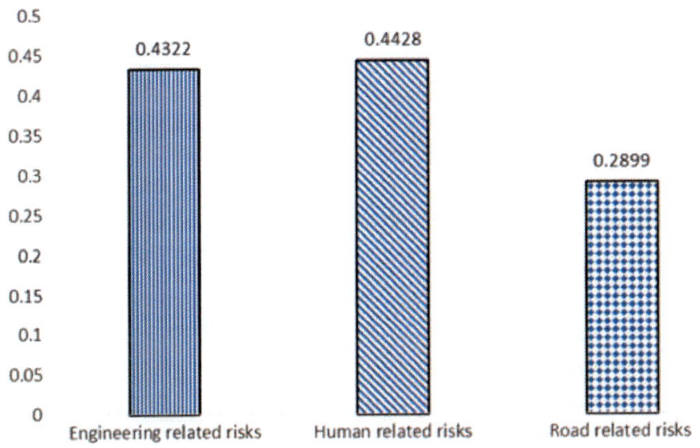


**Fig. 3** Ranking of main factors

teaching to the learner. Lack of attention needed by the government road sectors for reducing this kind of activity. Distraction of driver on the road causes a serious case of trouble, it needs to be avoided. Second, engineering-related risks is another thread of road accidents. The term engineering risk (R1) is involved in the activities of road controlling units like monitoring capabilities, maintenance error, speeding, etc. This risk involves a huge amount of damage to the government and private sectors. Vehicles are needed to manufacture as per the law and regulations of country for safe and secured ride. This kind of accident is taken place by the carelessness of the vehicle owner or the driver. To prevent the engineering risks from the accident, we need more control and take care of vehicles for their good and neat performance for the owner long-lasting. Both the human error risk and the engineering risk are separated by their activities involved in the form of accidents. Each and every accident is made by failure of modes and reckless.

There are some sub-factors that will make a huge amount of accidents on the road. In our paper, non-uniformity ($R_{21}$) is a big threat for road accidents based on Fig. 4. This risk is mainly due to not following the road regulation of the national road sectors. It means disobeying the rules and regulations of road safety control. For example, on highways if a vehicle needs to overtake the vehicle in-front off, the driver needs to go on the right side of that vehicle crossing of it. If he/she overtakes the vehicle by the left side of the vehicle, it leads to serious cause of the accident and the damage of the vehicle. Because on left side of overtaking, the crossing vehicle may turn or slow down for any prepossess. So this kind of non-violating rules need be not penetrated. Secondly, distracted driving ($R_{22}$) is the second most influential sub-factor. Nowadays distraction plays a vital role in road accidents because of doing unnecessary things in the vehicle while driving. For example, if you are driving on the highway ride and you are in rush of traffic. At that time you are using mobile phones or any other gadgets, it leads to an accident to the other drivers or yourself by did not have the concentration on the road. This kind of accident is prevented by not involving in other activities while driving. Thirdly, speeding ($R_{11}$) is the most

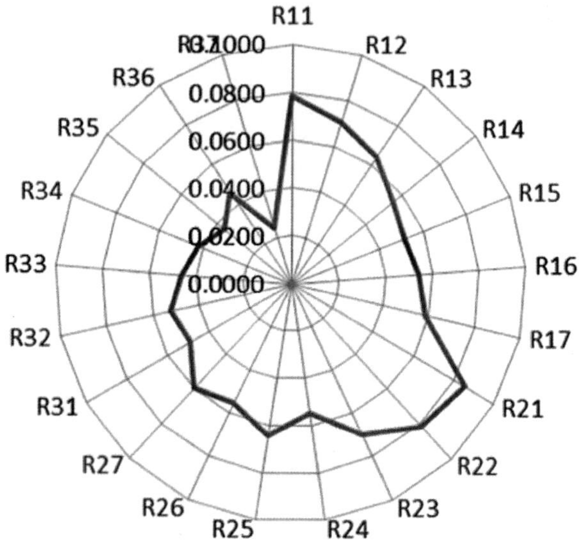


**Fig. 4** Ranking of sub-factors

common and wide cause of an accident that happened from the start age of using vehicles. Speeding takes place in the event of youngsters driving, they all very eager to drive fast one better than others. In speeding cases, the drivers are maximum leads to death or serious cause of injury. Speeding on the roads like national highways is normal, but they speed up in the roads like municipal, state roads may cause serious accidents that lead to death. Youngsters are the first threat to the speeding event of an accident. Try to control the speed of the vehicles in case of emergencies. Vehicle speeding meter needs to be inserted in some of the vehicles like college, school, staff buses, etc.

Nowadays highways accidents are gaining more no number of accidents because of roadside construction like road construction, tollgate building, excavator for land drilling, bridge road construction, etc. The second last one is highway road construction accidents ($R_{35}$) are needed to be more prevention by the state or national government for reducing the accident in this domain. The road materials like black rocks (Gravel stone) are stored or dropped on the roadside for road construction and it may reduce the size of the road it may cause traffic and it leads to accident of the drivers. Lane splitting ($R_{37}$) is another cause of accidents for the road user and this accident is mainly due to non-educator and also it is the least sub-factor. It is the risk mainly associated without the knowledge of crossing the road as well as lane crossing and splitting of the road. A divider is used to split the different road lanes or the ways of road to the highways. Lane splitting is commonly is used for the route turning, bus stand, etc.

## 7 Conclusion

In this paper, we carried out the important risk factors that need to be concentrated on controlling the road accidents. To analyze the road accident risks are carried out by using Fuzzy ANP method for a better outlook. Road safety rules are regulated towards the safety of the driver and passengers. This method carried out the significance of the factors and the priority among the road accident and also it is used to calculate the weightage of the factor. This paper has been taken for reducing road accident on the state or national highways. Human error and engineering risk are the two key factors in road accidents. Our outcomes will be helpful for the road safety services like Indian Roads Construction Corporation (IRCC), National Highways and Infrastructure Development Corporation Limited, National highway authority of India (NHAI), and Indian Academy of Highway Engineers (IAHE) for controlling the risks and they need to maintain the rules and regulations of road safety. Rules and regulations are created nor generated for the avoiding of accident, damage, death, etc.

This paper has some limitations. Firstly, this work has been carried out only in the south part of India. Future research work will be carried out in other parts of India and other countries. Secondly, risk factors are analyzed by using Fuzzy ANP method. Other ranking methods like DEMATEL, ISM, and AHP will be used in the future perspective.

## 8 Managerial Implications

From the obtained results, the following managerial implications are proposed.

- Installation of speed governance in all vehicles may reduce RA due to vehicle speeding.
- Provision of the separate lane for non-motorized vehicles and designated truck/bus.
- Install and upgrade the median barriers and edge barriers at the road edges.
- Improve the narrow bridges and roads into a wider lane network.
- Provision of adequate pedestrian facilities like crossings, safety zones, and footways.

## References


1. Arora YK, Kumar S (2020) Statistical approach to predict road accidents in India. In: Computing in engineering and technology. Springer, Singapore, pp 189–196
2. Maqbool Y, Sethi A, Singh J (2019) Road safety and road accidents: an insight. Int J Inf Comput Sci 6:93–105

3. Panda P, Boyanagari VK, Ayyanar R (2019) Burden, pattern, and causes of road traffic accidents in South India: estimate of years of life lost. CHRISMED J Health Res 6(1):52
4. Dindi K, Bachani D, Singhal M, Singh AR (2019) Road traffic injuries: epidemiology, challenges and initiatives in India. Nat Med J India 32(2):113
5. Park J, Yun D (2017) Analysis of road accident according to road surface condition
6. Dhruval P, Bhagat S (2019) A review study on road accident in India
7. Singh G, Pal M, Yadav Y, Singla T (2020) Deep neural network-based predictive modeling of road accidents. Neural Comput Appl 1–10
8. Mayhew DR, Simpson HM, Kaye S, Meier G (1985) Alcohol, age, and risk of road accident involvement. In: Proceedings of the 9th international conference on alcohol, drugs and traffic safety, pp 937–947
9. Elvik R (2013) Risk of road accident associated with the use of drugs: a systematic review and meta-analysis of evidence from epidemiological studies. Accid Anal Prev 60:254–267
10. Makkar A, Gill HS (2017) A radical approach to forecast the road accident using data mining techniques
11. Singh SK (2017) Road traffic accidents in India: issues and challenges. Transp Res Procedia 25:4708–4719
12. Zhang W (2019) Analytic hierarchy process (AHP): Matlab computation software. Netw Biol 9(1):10
13. Bonyani A, Alimohammadlou M (2019) A novel approach to solve the problems with network structure. Oper Res, 1–19
14. Chen Z, Ming X, Zhang X, Yin D, Sun Z (2019) A rough-fuzzy DEMATEL-ANP method for evaluating sustainable value requirement of product service system. J Clean Prod 228:485–508
15. Zhou X, Wang L, Qin J, Chai J, Muñoz CQG (2019) Emergency rescue planning under probabilistic linguistic information: an integrated FTA-ANP method. Int J Disaster Risk Reduct 37:101170
16. Mahdiyar A, Tabatabaee S, Durdyev S, Ismail S, Abdullah A, Rani WNMWM (2019) A prototype decision support system for green roof type selection: a cybernetic fuzzy ANP method. Sustain Cities Soc 48:101532

# Selection of the Best Crop for Farming Using Machine Learning

S. J. Suji Prasad, M. Thangatamilan, V. Aravindan, A. Harish, S. Janani, and S. D. Kausika

**Abstract** The Indian economy largely depends on agriculture and agro-industry products. The factors affecting the yield of agriculture are unpredictable rain, diverse soil parameters and changing weather conditions. The crop yield prediction based on the above said parameters is preserved in this project. Machine learning (ML) is used for predicting the crop yield and the parameters considered as nitrogen, phosphorus, potassium and pH. The proposed system is developed for predicting the yield of four crops, namely soybean, sugarcane, groundnut and sorghum. The proposed system will assist the farmers in suitable farming and increasing their yield.

**Keywords** Crop yield · Farming · Machine learning · Nitrogen · Phosphorus · Potassium · Virtual instrumentation

S. J. Suji Prasad (✉) · M. Thangatamilan · V. Aravindan · A. Harish · S. Janani · S. D. Kausika
Department of Electronics and Instrumentation Engineering, Kongu Engineering College, Erode, Tamil Nadu, India
e-mail: sjsujiprasad@gmail.com

M. Thangatamilan
e-mail: thangatamil.m@gmail.com

V. Aravindan
e-mail: aravinderode1999@gmail.com

A. Harish
e-mail: harishaanbu@gmail.com

S. Janani
e-mail: jananisamiappan99@gmail.com

S. D. Kausika
e-mail: sdkausika@gmail.com

© The Editor(s) (if applicable) and The Author(s), under exclusive license to Springer Nature Singapore Pte Ltd. 2021
S. Mohan et al. (eds.), *Materials, Design, and Manufacturing for Sustainable Environment*, Lecture Notes in Mechanical Engineering,
https://doi.org/10.1007/978-981-15-9809-8_55

## 1 Introduction

Agriculture is the backbone of India's economy, and it largely depends on agriculture yield and agro-industry products. Unpredictable rainwater influences the agriculture sector in India. Agricultural growth also depends on various soil parameters, i.e., nitrogen, phosphorous, potassium, crop rotation, soil moisture and surface temperature. Thus, technology will prove to be advantageous to agronomy, which will increase crop productivity resulting in better yields to the farmer.

The crop selection approaches are statistical, crop simulation and machine learning. Out of these, the statistical and crop simulation models are quite popular. The ML grants several approaches to define rules and patterns in large data sets related to crop yields and have well known to predict capability. Also, it can self-extemporize the predictive model. Regrettably, so far, the ML has not been applied on a large scale in the country, mainly because we are used to the statistical and crop simulation methods. Several methods of predicting and modeling crop yields which have been developed are empirical, and the success rate is due to the consideration of fewer parameters. This effort offers to help farmers check the soil quality depending on the investigation done based on soft computing approach. Thus, the system focuses on testing the soil quality to predict the crop suitable for cultivation according to their soil type and maximize the crop yield by recommending appropriate fertilizer.

A case study on the problems faced by rural farmers was conducted by Das [1] through a survey based on the questionnaire and the interview from the farmers in Assam. From the collected data, the significant problems faced by the farmers are lack of mechanization, poor yield, variety of seeds, flood and drought, lack of knowledge about demandable crops and inadequate marketing facilities. These can be overcome by introducing new technologies and by introducing new government schemes to provide the required farming equipment. The brief study was made by Kamraju et al. [2] on the various crop patterns followed by farmers in the Telangana state. Crop diversification implies the cultivation of a variety of crops in a region. Greater the number of crops in combination, greater will be the degree of diversification. It gives a more extensive choice to produce a variety of crops to lessen the risk in the areas of drought or with distinct soil problems. Crop substitution and shift are also taking place in regions with different soil problems. Crop diversification changes mainly due to the availability of irrigational facilities particularly lift irrigation facilities, fertile soil, adoption of fertilizers and mechanization of agriculture, etc. In the future, crop pattern changes in low diversified areas would lead to serious environmental consequences like groundwater depletion, soil fertility, waterlogging and salinity, which can reduce the crop productivity and growth potential of agriculture.

The classification rules were generated by Deepa et al. [3] for five agriculture crops, specifically paddy, groundnut, sugarcane, clubman rage supported the soft call system victimization soft set approach. Aquino et al. [4] proposed a methodology for predicting yield for the grape and wine industry. The calibrated model predicts the yield based on the number of detected berries in images with a root-mean-square error. An adjective neuro-fuzzy logical thinking system (ANFIS) was used by Naderloo et al. [5] which was accustomed to predict the gain yield of irrigated wheat in Abyek city of Ghazvin Province, Iran.

Mishra et al. [6] explored the application of ML techniques in agricultural crop production. The accurate and timely forecast is necessary for agricultural crop production because of policy decisions. The ML techniques like SVM, KNN, random forest, RBF, Bayesian network, ANN and *k*-means clustering are applied in the agricultural domain. A large number of data presently available from various resources can be analyzed to obtain hidden knowledge. The prediction model built using these techniques will have high accuracy over the traditional forecasting methods. Paul et al. [7] gave a summary of the works, highlighting different seeds, crops and fruits with the paper also try to deliver an analysis, which can help scholars to look at some related problems in the context of India.

A contactless, stress-free method of swine live weight estimation by machine vision technology was developed by Okinda et al. [8]. Adaptive neuro-fuzzy inference system (ANFIS) is used for modeling where feature extraction is done through image processing. The quality analysis of meat and fish, fruits, vegetables and bread were carried out by computer vision techniques. In the fruits and vegetables, the quality inspection can be done with the application of computer vision technology and image processing techniques [9]. The approach for the disease classification of the apple using color, texture and shape was introduced by Dubey et al. [10]. *K*-means clustering method has been used for infected fruit part detection, and support vector machine is used for the classification of healthy and unhealthy apples. Jaiswal et al. [11] used digital image processing techniques for finding the average rate of sedimentation in the Kharkhara reservoir. Linear discriminant analysis (LDA) was used by Radhika et al. [12] for the soil texture classification and developed a machine learning model to find out the physical and chemical properties of the soil.

ANN was used to predict the suitable crop by sensing various parameters of soil and also parameter related to the atmosphere [13]. Parameters like the type of soil, PH, nitrogen, phosphate, potassium, organic carbon, calcium, magnesium, sulfur, manganese, copper, iron, depth, temperature, rainfall and humidity were taken for analysis. Garg et al. [14] used the data of wheat yield of the University of Agriculture and Technology for forecasting. The study focuses on predicting data values on a broad spectrum of fuzzy logic computations based on second- and third-degree relationships. For the efficient defuzzification operation, the regression analysis model had been enforced.

## 2 Materials Used

The existing methods are statistical, crop simulation and machine learning. The most used way is data mining. It determines large data sets and defines rules and patterns. In data mining, two techniques are used. They are multiple linear regressions (MLRs) and the density-based clustering technique. Machine learning methods are not applied on a large scale. A new approach for predicting crop yield using a neural network is presented in this work.

## 2.1 Overview of Proposed System

The proposed system is based on the prediction of crop yield using soft computing method. In this proposed system, soil testing is done with the help of the Agrinex soil testing component. Agrinex soil testing consists of potassium, nitrogen and phosphorous capsule. Those capsules are mixed with water and made into a solution. The solution is analyzed by an image processing method. Using LabVIEW, the NPK values are detected, and using MATLAB, the values are classified with the help of the neural network.

## 2.2 Block Diagram of Prediction of Crop Yield

This section includes a brief overview of all the components used in the system. Figure 1 describes the system in brief in the form of block diagram.

From Fig. 1, the block diagram of prediction of crop yield system consists of a webcam, laptop, National Instrumentation LabVIEW 2015 version, Agrinex soil testing kid and MATLAB (neural network). The webcam is connected to the laptop for image processing analysis. The output from MATLAB is given to neural network. Depending upon the range selected, which crop is suitable for a particular land is displayed.

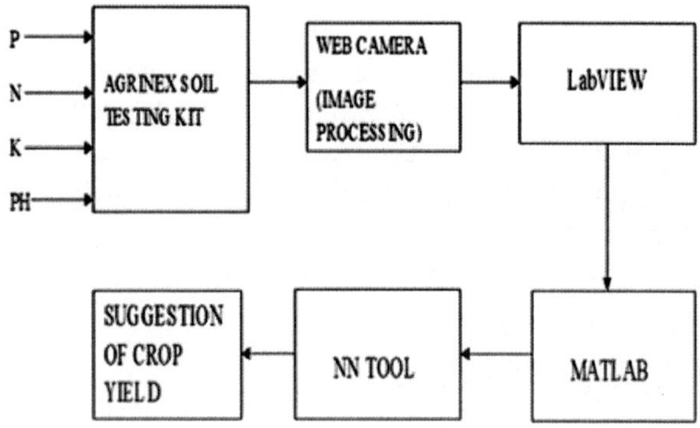


**Fig. 1** Block diagram of prediction of crop yield

### 2.3 *Block Diagram for Image Detection*

Using LabVIEW, the numeric value of the solution contain Agrinex doctor capsule is detected. The block diagram for camera detection is shown in Fig. 2. Using Agrinex capsules, the test solution is made, and through image processing, the numeric values of the NPK are displayed in the LabVIEW.

## 3 Methodology

The three primary nutrients that are important to growing plants effectively are nitrogen, phosphorous and potassium (N–P–K). Nitrogen (N) analyzed by the colorimetric method results in pink color, and it depicts the availability of nitrate in the soil sample. Phosphorous (P) appears blue, and it represents the availability of phosphorus in soil sample while potassium ($k$)-colorimetric color, turbid in appearance in the presence of potassium in the test sample, and pH appear different color as per their $H^+$ ion concentration. The standard pH scale for soils typically ranges between 4 and 8.5. Soil acidity or pH scale influences plants to take up nutrients from the soil.

The sample soil is mixed with water (rainwater or distilled water) with a 1:2 ratio in volume and complete blending is done. After ten minutes, the soil sinks in the bottom and the water is separated. The chemical within the tube is carefully added and transfer extract with the pipette. After blending, the compound gets dissolved. The new color gets developed after twenty minutes, and it is compared with the color of the N–P–K of the sphere.

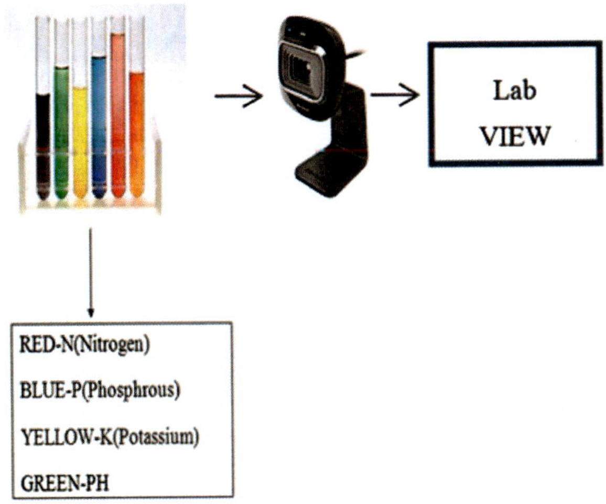


**Fig. 2** Block diagram for image detection

**Table 1** Nutrient content of crops

| Crops | Nitrogen content (kg/ha) | Phosphorus content (kg/ha) | Potassium content (kg/ha) |
|---|---|---|---|
| Sugarcane | 225 | 62.5 | 112.5 |
| Groundnut | 25 | 50 | 75 |
| Sorghum | 90 | 45 | 45 |
| Soybean | 20 | 40 | 20 |

### *3.1 Nutrient Content of Crops*

The N–P–K content in the selected four crops TNAU [15, 16] is shown in Table 1.

## 4 Virtual Instrumentation-Based Image Detection and Analysis

Virtual instrumentation-based image detection and analysis: Using LabVIEW, the numeric value of the solution which contains Agrinex doctor capsule is detected. LabVIEW (block diagram) program stores identified the numeric values in the array of arrays. After that, the average of all the array values is taken. The block diagram and front panel for image detection are shown in Figs. 3 and 4.

### *4.1 Analysis of the N–P–K Values*

The analysis data is used for the understanding of the N–P–K values. It helps to visualize the capabilities of the data which allows for effective prediction. The data is analyzed by the comparison between standard values and solution values. The LabVIEW data table is arrived based on Fig. 5.

#### 4.1.1 Nitrogen Content Analysis

Nitrogen is the most important component in the prediction of crop yield. Nitrogen has major compounds of amino acids, chlorophyll (which plants use for photosynthesis). The nitrogen content for the soil taken for analysis is shown in Table 2.

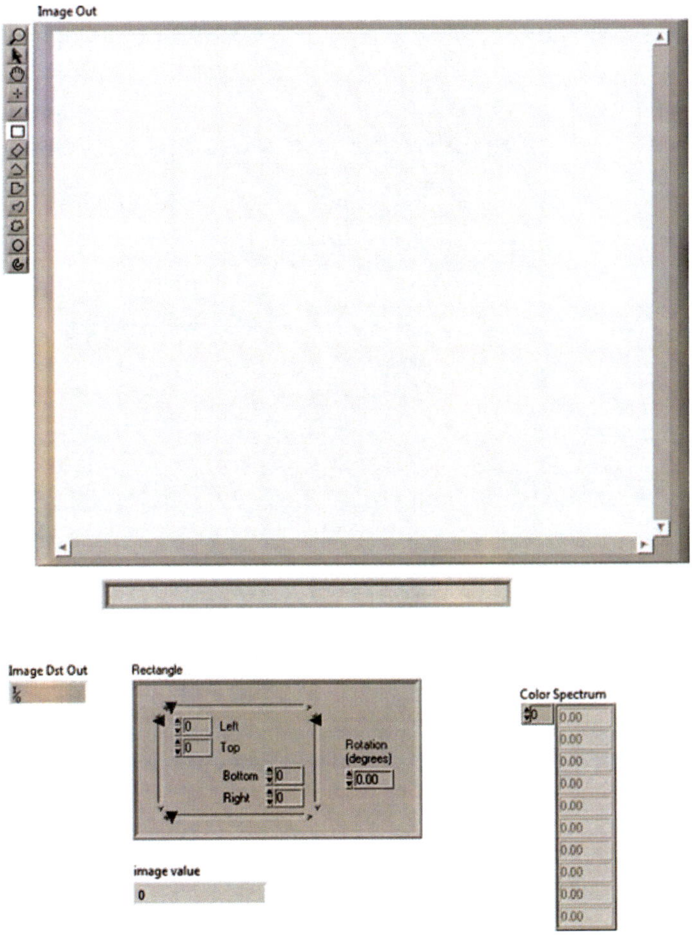


**Fig. 3** Front panel window for image detection

### 4.1.2 Phosphorous Content Analysis

Higher grain production, improved crop quality, greater stalk strength, increased root growth and crop maturity are determined by adequate phosphorous. The phosphorus content for the soil taken for analysis is shown in Table 3.

### 4.1.3 Potassium Content Analysis

Increases in the amount of potassium increase root growth, drought tolerance and cellulose. Thus, potassium is also a significant component for crop yields. It helps to translate sugars and starches and also gains rich in starch. The potassium content for the soil taken for analysis is shown in Table 4.

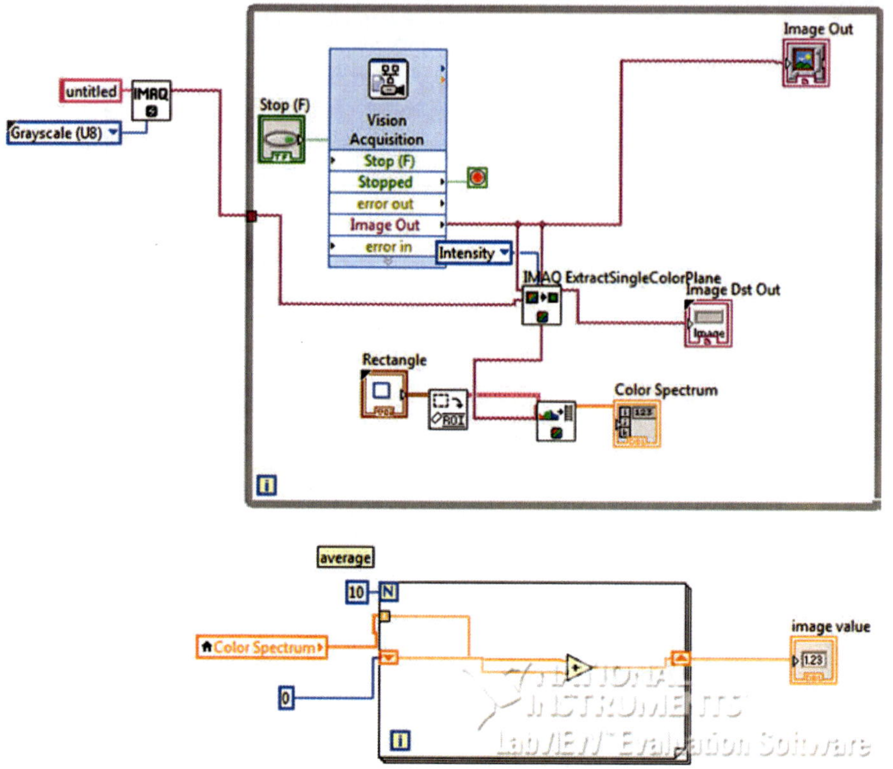


**Fig. 4** Block diagram window for camera detection

## 5 Results and Discussion

The set of rules were generated for the cultivation of suitable crop based on the NPK values. Sorghum, groundnut, sugarcane and soybean are selected for analysis, and programming is done for the selected crops. According to this algorithm, the user can manually feed the soil into the test kit, and based on nitrogen, phosphorus, potassium present in the tested soil, the best-suited soil will be displayed. The best-predicted crop for the selected soils tested is shown in Table 5.

From the above table, the soils collected from different five villages are tested, and two samples are suitable for the cultivation of sugarcane, and one sample is suitable for groundnut. The other two samples are not under the selected four crops.

## 6 Conclusion

The experimental setup developed predicts the growth of a crop in all geographical areas. By implementing this smart setup, farmers will be able to improve their yield

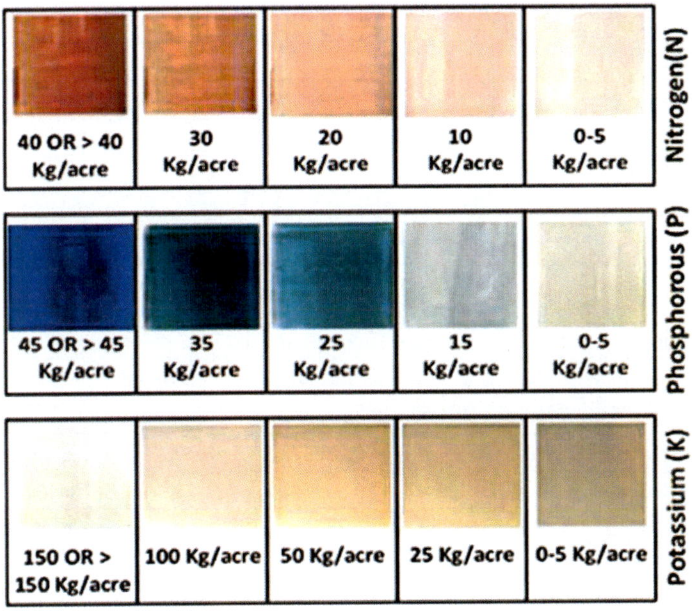


**Fig. 5** N–P–K color image data set [17]

**Table 2** Nitrogen content analysis

| Nitrogen | Numeric | LabVIEW data obtained |
|---|---|---|
| ≥40 | 0.7229 | |
| 30 | 0.82136 | 0.8389 |
| 20 | 0.933 | |
| 10 | 0.9361 | |
| 0–5 | 0.7166 | |

**Table 3** Phosphorous content analysis

| Phosphorous | Numeric | LabVIEW data obtained |
|---|---|---|
| ≥45 | 0.998 | |
| 35 | 0.545 | |
| 25 | 0.9677 | 0.8721 |
| 15 | 0.588 | |
| 0–5 | 0.7227 | |

and their income. This device is more compatible with the farmers by providing a handheld device for predicting the crop yield. A real-time model that can control and monitor the complete status of crop parameters using smart sensors was developed. The developed system works well for selected four crops, namely sorghum,

**Table 4** Potassium content analysis

| Potassium | Numeric | LabVIEW data obtained |
|---|---|---|
| 150 or 150 above | 0.4498 | 0.2721 |
| 100 | 0.86293 | |
| 50 | 0.93699 | |
| 25 | 0.9417 | |
| 0–5 | 0.7841 | |

**Table 5** Results of suitable crop for farming from the soil test

| S. No. | Soil sample | Nitrogen | Phosphorus | Potassium | Suitable crop for farming |
|---|---|---|---|---|---|
| | | LabVIEW data obtained | | | |
| 1 | Village-1 | 0.5696 | 0.7723 | 0.6982 | NO DATA |
| 2 | Village-2 | 0.8389 | 0.8721 | 0.2721 | Sugarcane |
| 3 | Village-3 | 0.9298 | 0.9541 | 0.9263 | Groundnut |
| 4 | Village-4 | 0.7786 | 0.4658 | 0.3696 | NO DATA |
| 5 | Village-5 | 0.7691 | 0.8486 | 0.9398 | Sugarcane |

groundnut, sugarcane and soybean; it can be extended for the other crops in the future.

## References

1. Das P (2015) Problems of rural farmer: a case study based on the lowphulabori village under the raha block development area of Nagaon district, Assam. IOSR J Humanit Soc Sci 20:40–43
2. Kamraju M, Vani M, Anuradha T (2017) Crop diversification pattern: a case study of Telangana state. Int J Innov Sci Res Technol 2:366–371
3. Deepa N, Ganesan K (2018) Multi-class classification using hybrid soft decision model for agriculture crop selection. Neural Comput Appl 30:1025–1038. https://doi.org/10.1007/s00521-016-2749-y
4. Aquino A, Millan B, Diago M-P, Tardaguila J (2018) Automated early yield prediction in vineyards from on-the-go image acquisition. Comput Electron Agric 144:26–36
5. Naderloo L, Alimardani R, Omid M et al (2012) Application of ANFIS to predict crop yield based on different energy inputs. Measurement 45:1406–1413
6. Mishra S, Mishra D, Santra GH (2016) Applications of machine learning techniques in agricultural crop production: a review paper. Indian J Sci Technol 9:1–14
7. Paul A, Ghosh S, Das AK et al (2020) A review on agricultural advancement based on computer vision and machine learning. In: Emerging technology in modelling and graphics. Springer, pp 567–581
8. Okinda C, Liu L, Zhang G, Shen M (2018) Swine live weight estimation by adaptive neuro-fuzzy inference system. Indian J Anim Res 52:923–928
9. Chopde S, Patil M, Shaikh A et al (2017) Developments in computer vision system, focusing on its applications in quality inspection of fruits and vegetables—a review. Agric Rev 38:94–102

10. Dubey SR, Jalal AS (2016) Apple disease classification using color, texture and shape features from images. SIViP 10:819–826
11. Rai AK, Jaiswal RK, Galkate R, Nayak TR (2019) Assessment of sedimentation in Kharkhara and Paralkot Reservoir using digital image processing techniques. Adv Innov Res 6:162
12. Radhika K, Latha DM (2019) Machine learning model for automation of soil texture classification. Indian J Agric Res 53:78–82
13. Dahikar SS, Rode SV (2014) Agricultural crop yield prediction using artificial neural network approach. Int J Innov Res Electr Electron Instrum Control Eng 2:6
14. Garg B, Aggarwal S, Sokhal J (2018) Crop yield forecasting using fuzzy logic and regression model. Comput Electr Eng 67:383–403
15. Tamilnadu Agricultural University (2016) Soil rating chart. https://agritech.tnau.ac.in/agriculture/agri_soil_soilratingchart.html
16. TNAU (2016) Nutrient management for field crops. htttp://agritech.tnau.ac.in/agriculture/agri_nutrientmgt_nutrientmgtforfieldcrops.html
17. Agrinex soil testing kid. https://agrinex-corporation.business.site/

# Numerical Analysis on Flow Characteristics of Air Through Human Respiratory Airway Using OpenFOAM

**Borra Mohan Krishna and Vikas Rajan**

**Abstract** The pulmonary system in the human body plays a major role in respiration of air into lungs by inhalation and exhalation which helps in exchange of oxygen and carbon dioxide in human lungs. The airflow through the respiratory system, starting from oral cavity to alveoli, undergoes significant changes in its behaviour and its thermophysical properties. A deep understanding is required to clearly predict the flow behaviour of air through respiratory passage. This understanding helps us to diagnose or to prevent some of the pulmonary diseases or disorders like asthma, chronic obstructive pulmonary disease (COPD), etc. In this present study, a numerical study is carried out on a small part of respiratory airway bifurcation. The study is carried out in open-source CFD package OpenFOAM-5 using SimpleFOAM solver. And the results are obtained for the velocity profiles of the air inside respiratory airway for different mass flow rates of air 12, 30, 60 lit/min corresponding to rest, normal activity and exercise breathing conditions.

**Keywords** CFD · COPD · SIMPLE · CT · MRI

## 1 Introduction

Human respiratory airways are bifurcated into 23 bifurcations, and each bifurcation is called as generations; from trachea, each bifurcation is reduced in its diameter, and these bifurcations finally end up with fine sac-like alveoli structures where the oxygen diffuses into blood [1]. The airflow through respiratory airway varies with respect to age and person to person, the velocity distribution, wall shear stress and the pressure contours which are different, and the wall shear stresses are higher in case of infants

B. M. Krishna · V. Rajan (✉)
Department of Mechanical Engineering, Amrita Vishwa Vidyapeetam, Amritapuri, Kollam, Kerala, India
e-mail: vikasrajan@am.amrita.edu

B. M. Krishna
e-mail: krishnaborra95@gmail.com

© The Editor(s) (if applicable) and The Author(s), under exclusive license to Springer Nature Singapore Pte Ltd. 2021

S. Mohan et al. (eds.), *Materials, Design, and Manufacturing for Sustainable Environment*, Lecture Notes in Mechanical Engineering,
https://doi.org/10.1007/978-981-15-9809-8_56

when compared to adults [2]. Hence, the flow of air is completely dependent upon age and health condition of the person. The mechanism of the airflow distribution inside lung bronchioles has to be studied in order to understand or to predict the behaviour of the flow in case of any abnormalities like bronchitis or cystic fibrosis to cure those abnormalities or to administer drugs into particular disease affected sites. The experimental method or subjective study to examine the flow behaviour costs more a lot of time-consuming and effort consuming, because constructing experimental models for different subjects with different health conditions is a difficult task.

Numerical studies using CFD have been proven best way of analysing the flow behaviour virtually, and the flow patterns and the flow characteristics obtained in the CFD studies are matching with each other [3]; hence, the CFD has been a best tool to investigate flow patterns. The models are developed from CT scan or MRI scan files form the initial diagnosis, geometries are extracted, and a CFD study is made over the extracted model to virtually study the abnormalities in the health condition or flow of fluid over the tumours, and even growth of the tumour causes a lot of behavioural changes in the flow characteristics [4–6]. The effective diagnosis of a disease inside respiratory airway has been increased using CFD rather than the conventional diagnosis, because using CFD the severity of the disease and also the magnitude and direction of the flow in case of disease affected parts in the respiratory airway can be easily detected [7, 8], which would give a deep understanding of the flow characteristics of air inside the respiratory airway, whereas in case of conventional method, the magnitude and the velocity for the flow of fluid inside airway cannot be determined [9, 10]. A lot of studies have been made on the realistic models extracted from the CT and MRI scans, and also many idealistic models have been analysed, and the obtained results show good agreement with those of the realistic models with a negligible error percentage [11, 12].

In this present study, the CFD analysis is made over a respiratory airway bifurcations from 3rd generation to 6th generation in healthy condition in order to clearly understand the flow property variations in that particular three bifurcations using an open-source CFD tool OpenFOAM-5 using in-build solver code.

## 2 Methodology Adopted

An idealistic model is constructed using SolidWorks, and the meshing for the model is done in Ansys-18.2 (Fig. 1).

The mesh developed is an unstructured mesh with tetrahedral cells with different cell count ranging at 3.1, 4.3 and 6.2 lakhs for the grid-independent study. The gird is then imported into OpenFOAM to solve flow governing equations.

The case is solved using existing solvers in OpenFOAM CFD package. Using SimpleFOAM solver, the case is solved for fluid flow governing equations. As the flow involves only air with low temperature, energy equation is neglected and the case is only solved for mass and momentum equations. Even there is no turbulence in this case because the mass flow rates in the geometric volume are less and the flow

**Fig. 1** Geometry of airway

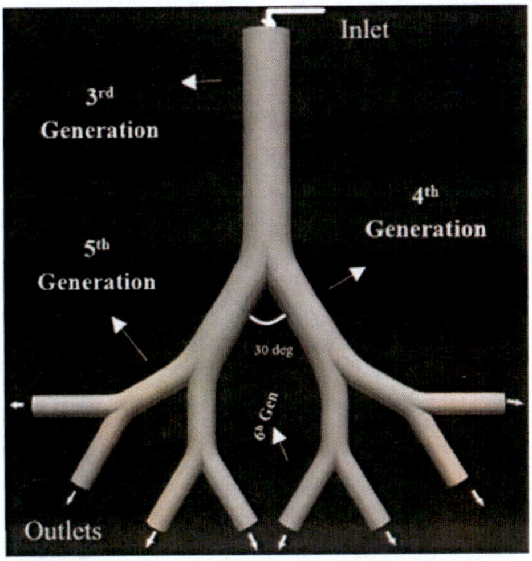


does not lead to any disturbances and chaos. Hence, the laminar model is sufficient to solve the flow phenomena.

The flow governing equations are as follows:

The continuity equation is as follows:

$$\nabla.\bar{v} = 0 \tag{1}$$

The momentum equation corresponding to $x$-, $y$- and $z$-directions in the fluid flow is as follows:

$$\frac{\partial u}{\partial t} + u\frac{\partial u}{\partial x} + v\frac{\partial u}{\partial y} + w\frac{\partial u}{\partial z} = -\frac{1}{\rho}\frac{\partial p}{\partial x} + \nu\left(\frac{\partial^2 \mathrm{u}}{\partial x^2} + \frac{\partial^2 v}{\partial y^2} + \frac{\partial^2 w}{\partial z^2}\right) \tag{2}$$

$$\frac{\partial v}{\partial t} + u\frac{\partial v}{\partial x} + v\frac{\partial v}{\partial y} + w\frac{\partial v}{\partial z} = -\frac{1}{\rho}\frac{\partial p}{\partial y} + \nu\left(\frac{\partial^2 \mathrm{v}}{\partial x^2} + \frac{\partial^2 v}{\partial y^2} + \frac{\partial^2 v}{\partial z^2}\right) \tag{3}$$

$$\frac{\partial w}{\partial t} + u\frac{\partial w}{\partial x} + v\frac{\partial w}{\partial y} + w\frac{\partial w}{\partial z} = -\frac{1}{\rho}\frac{\partial p}{\partial y} + \nu\left(\frac{\partial^2 \mathrm{w}}{\partial x^2} + \frac{\partial^2 w}{\partial y^2} + \frac{\partial^2 w}{\partial z^2}\right) \tag{4}$$

where

| | |
|---|---|
| $\rho$ | is the density of air 1.225 kg/m$^3$. |
| $p$ and $v$ | are pressure and velocity. |
| $\nu$ | is the dynamic viscosity. |

## 3 Numerical Method

OpenFOAM consists of built-in solvers to solve the flow governing equations depending upon the type of flow problem. In this case, SimpleFOAM solver is used. As the case is steady-state, Semi-Implicit Method for Pressure Linked Equations (SIMPLE) algorithm is used to solve the coupling between the pressure and velocity fields. The solution residuals are set to $1e^{-5}$ to achieve better convergence in the solution. As the mesh used is a tetrahedral mesh, a non-orthogonal correction is given in order to get better results. The time step size for the case is taken as 0.02 s. The grid-independent study is also been made in order to obtain the best computational and optimal mesh, so as to reduce the time computation with the best accuracy. The solutions obtained at 4.1 lakh cells have been best when validated, so the grid independency is achieved with 4.1 lakh tetrahedral cells.

## 4 Results and Discussion

The case is run with flow of air within the respiratory airway, and the results are obtained for the fluid flow characteristics and validated with Justus Kavita Mutuku et al. The velocity contours are obtained for different mass flow rates of the air (Fig. 2).

The results obtained are compared with the Justus Kavita Mutuku et al., and the results have shown good agreement with their model. Hence, the open-source code is verified and proved to be correct for this present geometry (Fig. 3).

The velocity contours are obtained for different mass flow rates, and the velocity contour changes are been observed to follow the same trend at all the mass flow rates but with different magnitudes. For low mass flowrates as the flow of fluid is less in the whole domain. As a result of this, the velocity of the fluid is also very less. When

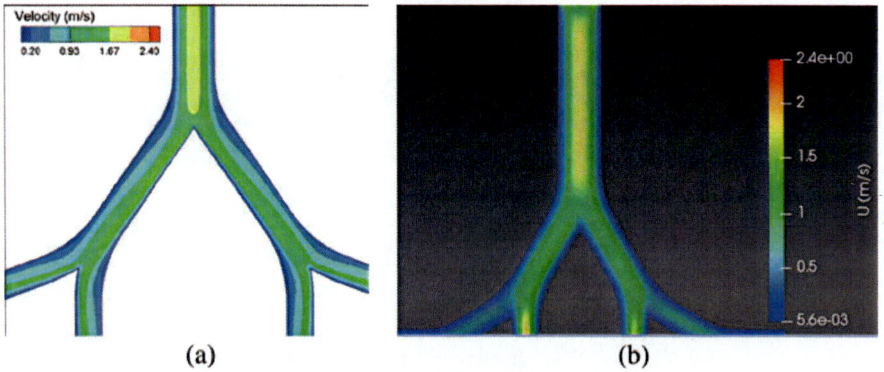


**Fig. 2** Comparison of present results with **a** Justus Kavita Mutuku et al. **b** present study

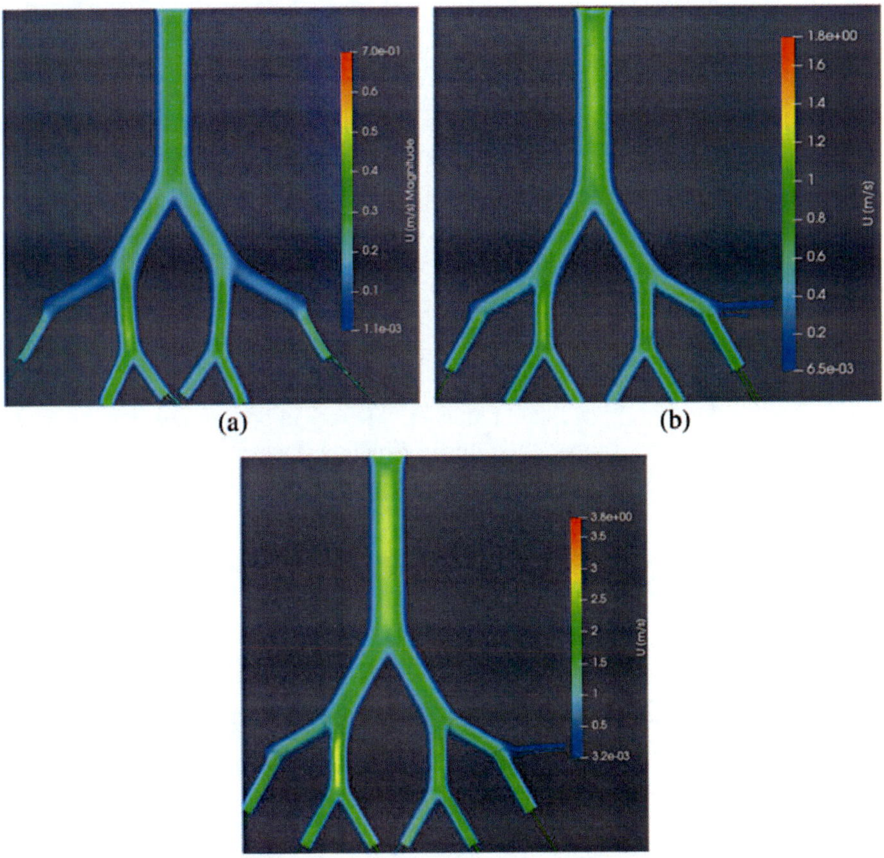


(c)

**Fig. 3** Velocity streamlines at different mass flow rates at 12, 15, 30 lit/min

the mass flow rate is increased, the velocity magnitude also seems to be increased in the domain.

As the domain is a bifurcated geometry with different bifurcation angles, the flow is deviated at each bifurcation, and there are significant changes observed in the flow phenomena. Flow after entering the 3rd generation the velocity magnitude is high as the diameter is more. As the diameter is decreased, the flow velocity is also been decreased and maximum velocity at all the mass flow rates is observed at centreline of the domain in all the generations. At all the bifurcations as the flow is approaches the bifurcation there is deceleration in the flow and that deceleration at the bifurcation wall gives rise in static pressure of fluid and in that case there is a boundary layer formed near the wall and here will be a backflow and a recirculation regions are developed at those bifurcations, and hence the pressure inside recirculation regions is also high and there is flow separation observed at these bifurcations with decrease in velocity. When flow is at the bifurcation walls due to retardation of flow, the velocity magnitude is drastically dropped such that the flow at each bifurcation is approaching rest condition.

The pressure inside the whole fluid domain did not give significant changes in all over the domain because the domain is very small and that too the operating conditions of study are in atmospheric conditions; hence, the pressure changes inside the domain are not that significant (Fig. 4).

The flow at near bifurcation walls has a very low such that there is a static pressure build-up, and also, there are some recirculation regions developed at the bifurcation angles, so inside the recirculation regions, the pressure is high and that increase in pressure which eventually reduces the velocity. The flow at bottommost final bifurcations is observed to be very low, because the final bifurcation is highly deviated from the centre axis, and even the gravity also pulls the fluid downwards, and the flow is more in the straight deviated branches, whereas in bottommost left and right bronchioles are deviated towards the upper direction, mass of fluid entering those airway tubes is very less, and also the velocity in those bronchioles is very low such

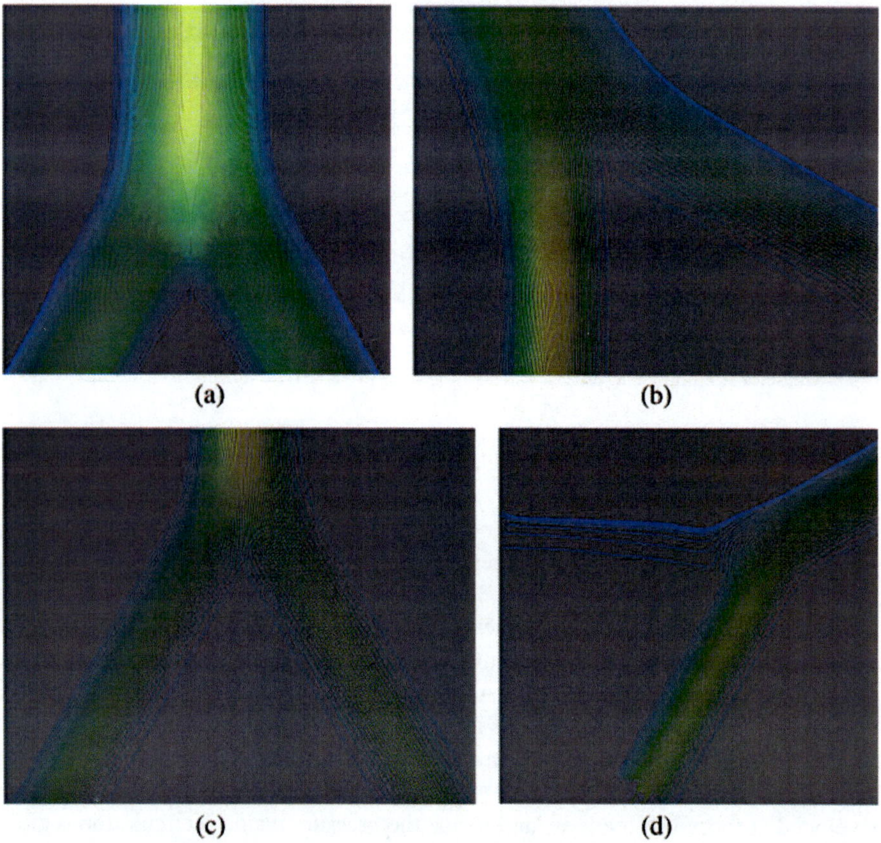

**Fig. 4** Velocity streamlines at each and every bifurcation of this physical domain **a** 1st bifurcation **b** 2nd bifurcation **c** 3rd bifurcation **d** extreme left bronchiole

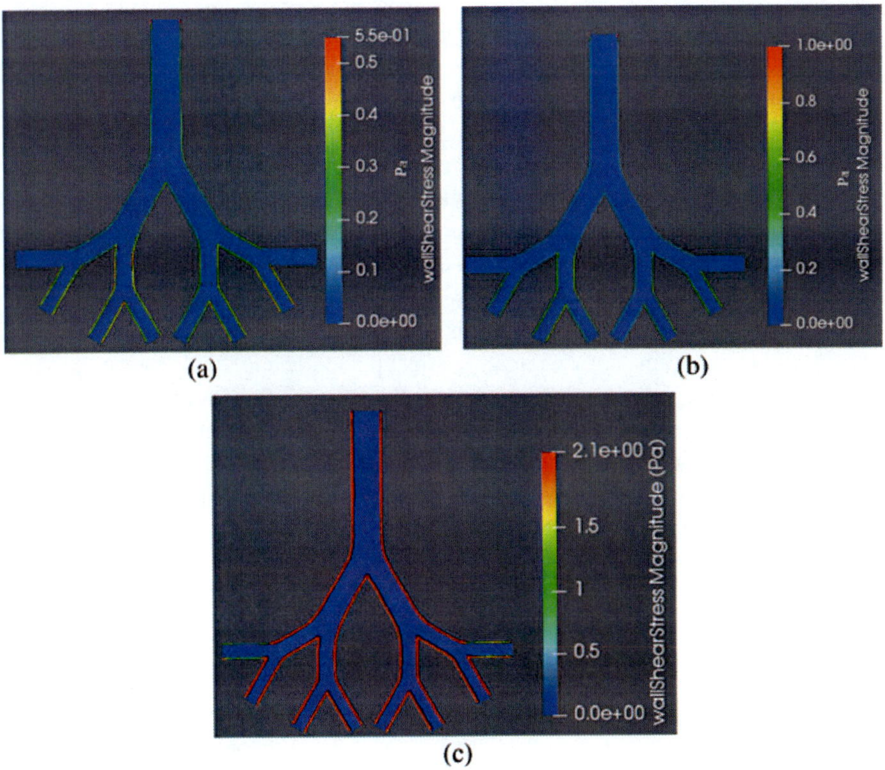


**Fig. 5** Figure showing wall shear stresses at different mass flow rates **a** rest condition **b** moderate breathing **c** exercise breathing

that the flow is almost to approach static condition with slight increase in pressure. At the walls, the flow is still and static, and there the wall shear stresses are also observed to be more. The highest peak value in the wall shear stresses is observed in maximum flow rate condition (Fig. 5).

The wall shear stress contours are obtained for the three mass flow rates of the fluid. The wall shear stresses in the rest breathing condition are very less or even negligible on the passage walls. But gradually when the mass flow rate and the velocity of the fluid have been increased, the wall shear stresses are also being increased. The significant stresses are obtained in the normal inhalation, and for the exercise breathing case, the wall shear stresses are changed significantly, and a peak value is attained in exercise breathing condition.

But for all the cases of the wall shear stresses, the flow of air or the wall shear stresses are minimum in outermost bifurcated bronchioles towards extreme left and right side of model. The wall shear stresses obtained are very low and negligible in those branches, and comparatively higher values are obtained in the centre bronchioles which are inclined downwards directly without any deviation.

## 5 Conclusion and Future Scope

The numerical analysis is made on a triple bifurcation of respiratory airway geometry with different mass flow rates of air. And the results show that the less mass flow rate in the domain in rest breathing condition has a very less velocity, and also there is a little increase in pressure value in that case, slightly when the mass flow rate is increased in case of the normal breathing and exercise breathing condition, the velocity contours are been observed to be more in these two cases, and also there are some recirculation zones observed at large bifurcations of the upper part. And also due to the basic structure of the airway, there is influence on the fluid flow, where the bottommost bifurcations are deviated towards upper side and the flow through those bronchioles is less. This may lead to abnormalities in the flow in case of diseases.

The study can be further extended to effectively study action of drugs administered in case of any diseases so that the action of drug particles inside respiratory airway can be studied effectively in order to increase the bioavailability of the drugs administered, and also disease diagnosis can be done virtually.

## References

1. Weibel ER, Gomez DM (1962) Architecture of the human lung. Sci 137(3530)
2. Tsega EG (2018) Computational fluid dynamics modeling of respiratory airflow in tracheobronchial airways of infant, child, and adult. Hindawi Comput Math Meth Med 9603451:9
3. Cheng YS, Zhou Y, Chen TB (1999) Particle deposition in a cast of human oral airways. Aerosol Sci Technol 31(4):286–300
4. Feng Y, Xu Z, Haghnegahdar A Computational fluid-particle dynamics modeling for unconventional inhaled aerosols in human respiratory systems. In: Aerosols—science and case studies
5. Nowak N, Kakade PP, Annapragada AV (2003) Computational fluid dynamics simulation of airflow and aerosol deposition in human lungs. Ann Biomed Eng 31:374–390
6. Qi S, Li Z, Yue Y, van Triest HJ, Kang Y (2014) Computational fluid dynamics simulation of airflow in the trachea and main bronchi for the subjects with left pulmonary artery sling. Biomed Eng Online
7. Augusto LLX, Lopes GC, Gonçalves JAS A CFD study of deposition of pharmaceutical aerosols under different respiratory conditions. Braz J Chem Eng 33(3):549–558
8. Koullapis PG, Kassinos SC, Lin C-L (2015) Computational fluid dynamics (CFD) simulations of aerosol deposition in the lungs. In: International symposium on turbulence and shear flow phenomena, Melbourne, Australia
9. Kolanjiyil AV, Kleinstreuer C (2016) Computationally efficient analysis of particle transport and deposition in a human whole-lung-airway model. Part I: theory and model validation. Comp Biol Med 79
10. Xi J, Yang T, Talaat K, Wen T, Zhang Y, Klozik S, Peters S Visualizationoflocal deposition of nebulized aerosols in a human upper respiratory tract model. J Vis 21(2):225–237
11. Zhang B, Qi S, Yue Y, Shen J, Li C, Qian W, Wu J Particle disposition in the realistic airway tree models of subjects with tracheal bronchus and COPD. Hindawi Biomed Res Int 2018(7428609):15
12. Xi J, Si XA (2018) Review of feature extraction from exhaled aerosol fingerprints to diagnose lung structural remolding. Biomed J Sci Tech Res 11(3). BJSTR.MS.ID.002097

# Investigation of Green Manufacturing in Motor and Pump Industries Through a System Model 'GREEN-6S'

**R. Gnanaguru, Rama Thirumurugan, and I. Rajendran**

**Abstract** This article reports the adoption of green environment through 5S and safety principles in motor and pump industries at Coimbatore. Even though the importance of green manufacturing is increasing day by day among the industrial and research communities, a definite implementation system model does not exist in practice for industries. It is not easy to convert conventional manufacturing to eco-friendly manufacturing without the support of quality and process improvement tools. The appreciation of impact and implementation of green manufacturing a different level of manufacturing is also very difficult. To overcome these issues, 6S principles are introduced with green manufacturing to develop a new process paradigm. This work describes a system model called as 'GREEN-6S' to transform less green into greener in motor and pump industries.

**Keywords** Sustainability · Green manufacturing · 5-S · 6-S · Kaizen · A3 reports · GREEN-6S

## 1 Introduction

In recent decades, Coimbatore district has witnessed tremendous progress in its industry, economy and living standards of people. However, all these progress obtained at the expense of environment damage. Hence, it is the need of the hour to understand how motor and pump industries can be eco-friendly without affecting their profit. Green manufacturing activity in motor and pump industries is categorist

R. Gnanaguru (✉) · R. Thirumurugan · I. Rajendran
Department of Mechanical Engineering, Dr. Mahalingam College of Engineering and Technology, Pollachi, Tamilnadu, India
e-mail: drguru@drmcet.ac.in

R. Thirumurugan
e-mail: rtm@drmcet.ac.in

I. Rajendran
e-mail: irus_rajendran@yahoo.co.in

© The Editor(s) (if applicable) and The Author(s), under exclusive license to Springer Nature Singapore Pte Ltd. 2021
S. Mohan et al. (eds.), *Materials, Design, and Manufacturing for Sustainable Environment*, Lecture Notes in Mechanical Engineering,
https://doi.org/10.1007/978-981-15-9809-8_57

as two aspects: one is introducing green energy through wind and solar another is applying green in the manufacturing process by eliminating the waste and reducing the activities against the sustainability.

However, a long-term commitment towards the adaptation of green in motor and pump industry benefits not only through cost saving but also enriches the brand name with customers. The implementation of green has differed across the industry to industry. Some of the industries come under compulsion of regulatory and others towards the establishing brand with customers. Still, there is a gap in awareness on green adaptation and integration with business.

Green manufacturing is a system which combining manufacturing problems with the process and product design to identify and assess the flow of environmental waste to minimize the environmental impact and maximize resource efficiency. The objective of 'GREEN-6S' is to design and develop the products which reduce the negative impact on the environment by production and disposal.

The motor and pump industries in Coimbatore have environmental issues which cannot be solved by the environmental-related techniques. Hence, a system model is required to prepare the plan for implementation green and also control mechanism. After the implementation, the assessment is another critical area to show the areas where the system to be improved. At the same time, the support of any one of the improvement technique helps to improve the process against environmental damage.

The first 'S' is representing sorting, which is focused on the identification, separation and reduction of environmental damage material in the process. The second 'S' simplify helps to arrange the material in a proper environment based on the ecosystem. Shine is third 'S', which keeps the working environment is clean and green. Fourth 'S' is standardization which is focused on the procedure to maintain the green environment. Sustain is a fifth 'S' to maintain green and sixth 'S' focuses on safety.

Hence, this article addresses a system model of 'GREEN-6S' to implement the green in the motor and pump industries at Coimbatore and also minimize the barriers to implementing green. The proposed 'GREEN-6S' system model is a qualitative method to improve green in motor and pump industries as well as a roadmap for further research work on quantitative evaluation.

After the literature review on 6S (5S + Safety), green manufacturing results that 6S principles are underutilized during the implementation of green manufacturing due to non-availability of proper system model. In order to overcome this issue, the work being reported in this article was carried out by the methodology shown in Fig. 1.

## 2 Literature Review

Before designing a model for implementing green manufacturing through the application of fundamental lean tools, the literature arena was surveyed to locate such models contributed by the researchers. This search resulted in the identification

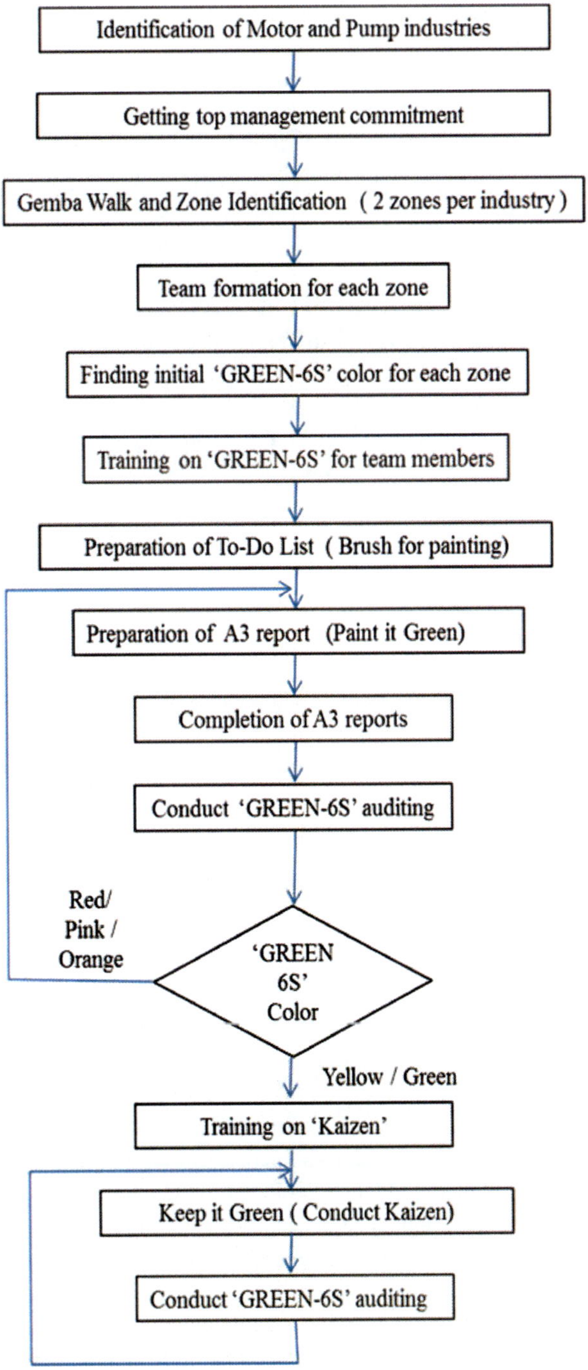


**Fig. 1** Research methodology

of eight such models. The characteristics and features of these models are briefly described in this section.

Sawhney et al. [15] have presented a model which is named as En-lean (short form of environmentally lean) which has been created to establish linkage between lean and green manufacturing principles. It has been pointed out that certain activities complement both lean and green manufacturing. On the other hand, certain lean and green manufacturing principles contradict with each other. Hence, several aspects of lean and green manufacturing paradigms are required to be studied and models for integrating both lean and green manufacturing paradigms are required to be developed. In order to fulfil this need, En-lean framework has been designed. En-lean has been designed to implement lean and green manufacturing principles in integrated manner in metal cutting industry.

Duarte and Cruz-Machado [4] have laid a strong theoretical foundation for developing a model that would facilitate the implementation of lean and green manufacturing paradigms in an integrated manner. To begin with, the contents and stipulations of 12 models which have been under implementation in the world over the past three decades to achieve continuous improvement have been briefly described. These models are generally addressed under three titles namely awards, standards and frameworks.

Pampanelli et al. [13] have contributed a model named as lean and green model which is applicable in production cells. Before explaining about this model, the commonalties between lean and green manufacturing paradigms have been appraised. It has been claimed that the waste elimination concept of lean manufacturing facilitates to achieve the goals of green manufacturing to a significant extent. It has been claimed that in order to achieve the goals of green manufacturing, the lean manufacturing is required to be implemented to a basic level. After describing these fundamental principles, the lean and green model has been presented. This model stipulates the achievement of the goals of green manufacturing through the application of the fundamental lean tool called kaizen in five steps.

White and James [18] have mentioned that the relationship between lean and green manufacturing paradigms has been well established. Yet, the effective integrated implementation of lean and green manufacturing paradigms is seldom reported in literature arena. In order to fill this research gap, a method of using process mapping (P mapping) technique to apply lean and green manufacturing paradigms in an integrated manner is proposed. It is claimed that like value stream mapping, P mapping is also a fundamental lean manufacturing tool which has found wide applications.

Kurdve et al. [11] have reported a study conducted among leading Swedish manufacturing companies in which the integration of manufacturing management and system models with environmental management system is being carried out. The manufacturing management and system model considered in the study was incorporated with lean elements which have been tailor-made to suit specific companies. These models have been termed as 'company-specific production system models' (XPS). The integration of XPS with environmental management system was studied in five leading companies situated in Sweden from six perspectives under the titles 'values and vision', 'principles', 'tools, methods and techniques', 'key performance indicators', 'organization' and 'auditing system'. After studying from these perspectives, the integration of EMS with XPS of these companies was analysed.

Verrier et al. [17] have thoroughly studied the research papers reporting researches on lean and green manufacturing integration. By making use of the findings reported in these papers, theoretical formulations have been made to create foundation on lean and green manufacturing integration. Subsequently, the tools that facilitate the integration of lean and green manufacturing paradigms have been listed. Some of them include Gemba walk, value stream mapping (VSM) and standard work. These tools have been identified as most powerful fundamental tools that can be used to achieve lean and green manufacturing integration. The authors referred Capability Maturity Model Integration (CMMI) to develop a model called lean and green maturity model. This model suggests the integration of lean and green manufacturing paradigms in organizations in five steps.

Fu et al. [6] have reported a research in which a model for implementing lean and green manufacturing paradigms in an integrated manner was developed. This model was implemented in a chemical manufacturing company situated in China. These authors have mentioned that quite a number of researchers have reported researches on the integration of lean and green manufacturing paradigms. Yet some deficiencies are persisting in this direction of research. One of the deficiencies is that most of the researches reported are conceptual and theoretical in nature.

Another deficiency is that the case studies reported in literature arena on implementing the integrated models of lean and green manufacturing paradigms have largely been conducted in developed countries [16]. These case studies were conducted in companies in which lean manufacturing paradigm was being implemented effectively. However, this kind of effective lean manufacturing implementation is not practically observed in developing countries like China and India [5, 9].

Developing models for integrated implementation of lean and green manufacturing paradigms has been emerging as an effective research domain in the recent years. This statement is made here on encountering the appearance of Abreu et al. [1] in which an exclusive literature review on lean–green models have been presented. Like the way it is stated in Fu et al. [6], Abreu et al. [1], Chiarini [3] and King and Lenox [10] have also mentioned that a significant number of researchers have been working on developing lean–green models. Yet these models have not been effective in achieving full-fledged integration of lean and green manufacturing paradigms.

The models described above have indicated that necessary ingredients and components are to be included when green manufacturing is implemented through the application of lean manufacturing principles [2, 8, 14]. Particularly, the information and knowledge presented in the above papers have indicated that the fundamental lean tools like 5S and Kaizen are eco-triggering lean tools [7, 13, 12]. Although the lean–green integrated models have been implemented in few industries, a model to implement green manufacturing through the application of fundamental lean tools in motor and pump industry is yet to be developed by the researchers. In order to fill this research oriented practice gap, while pursuing the research being reported here, a model named as 'GREEN-6S' was designed.

## 3 System Model for 'GREEN-6S'

The objective of the proposed 'GREEN-6S' framework is to understand the following:

- Identifying the different activities involved to assess the present 6-S and green level of the motor and pump industries.
- Developing the transformation plan needed in 'GREEN-6S' implementation.
- Explaining how to maintain the obtained green level.

The structure of the 'GREEN-6S' framework (Fig. 2) consists of two modules, the first one describes the planning stages and another model describes the tools required for each level to control the process. The tools in each stage are based on performance measurement that shows the strategic objectives and constraints specify by the decision-makers.

### *3.1 Find 'GREEN-6S' Colour*

The initial stage of any improvement process is assessing the current situation. In this stage, the 6-S and level of green manufacturing are assessed with the help of 'GREEN-6S' audit sheet. The colour shows how or towards the green transformation in terms of colour. The assessment sheet is and each heading having five evaluation criteria to identify the level of the system. It is divided into six headings, G6S-1 to

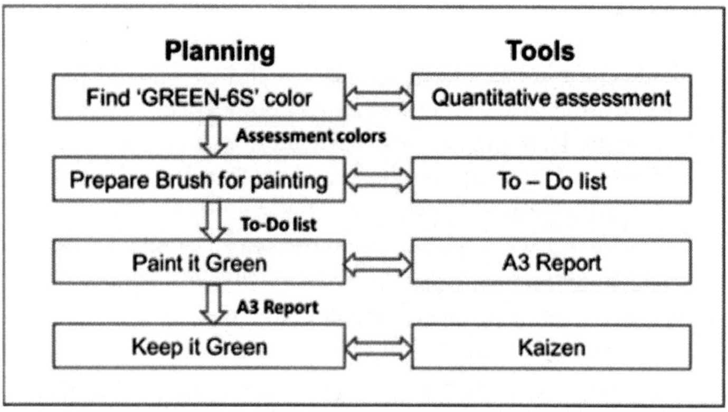


**Fig. 2** 'GREEN-6S' system model

**Table 1** GREEN-6S rubrics—sort

| GREEN 6S Rubrics | | | | | | |
|---|---|---|---|---|---|---|
| Criteria | | 1.1 Unwanted Materials | 1.2 Machines, equipment, furniture | 1.3 Hazardous | 1.4 Defective items | 1.5 Excess storage |
| Sort G6S-1 | Red / 0 | Unwanted materials are found on the floor | Unusable machines or machine parts are not disposed | Hazardous items are not identified | Defective items are neither disposed nor repositioned as per standards | Excess of materials (including packing) are kept in the shop floor |
| | Pink / 1 | Routine Removal of unwanted materials from work floor | Unusable machines or machine parts are identified | Standards are made for disposal of Hazardous materials | 25% of defective items are either disposed or repositioned as per standards | 25% of excess of materials are relocated |
| | Orange / 2 | Regular use of Red Tag zone | 50% of unusable machines or machine parts are disposed | In very few areas practicing disposal is happening as per standards | 50% of defective items are either disposed or repositioned as per standards | 50% of excess of materials are relocated |
| | Yellow / 3 | Proper standard for the disposal of items | All unusable machines or machine parts are disposed | Habitually, systematically disposal is happening | All defective items are either disposed or repositioned as per standards | Only required materials are kept in the shop |
| | Green / 4 | Identification of new techniques to prevent waste materials | Preventing machines becoming repentant | Implementation of new ideas for Disposal of Hazardous materials | Visual management initiated | Practice of Just In Time &Kanban. |

G6S-6. The colours are identified with the help of behaviour mentioned below in the Rubrics Tables 1, 2, 3, 4, 5 and 6.

**Table 2** GREEN-6S rubrics—simplify

| Criteria | | 2.1 Dividing lines | 2.2 Layout and identification | 2.3 Cluster of materials equipment & instruments | 2.4 Shelves and Cupboards | 2.5 Order lines |
|---|---|---|---|---|---|---|
| Simplify G6S - 2 | Red / 0 | Dividing lines are not exiting on the floor and items are kept haphazardly | Item locations are not clearly marked with Sings. | Scrap, rework & repair items are mixed up and instruments have no identification. | All items are not arranged properly in Shelves Cupboards and without identification | Overall appearance is shabby and without Visuals. |
| | Pink / 1 | Only Gangways are marked | 25% of items are clearly marked with large signs and labels. | Scrap, rework & repair items are separated without identifications. | Items are arranged properly in Shelves and Cupboards without identification | Overall appearance is good but without Charts ,Graphs, Visuals, etc. |
| | Orange / 2 | Gangways and all machine locations are marked | 50% of items are clearly marked with large signs. Retrieval time more than 30 seconds. | Scrap, rework & repair items are separated and identified clearly with labels. | Items are arranged properly in Shelves and Cupboards without colour identification | Overall appearance is good with Charts Graphs, Visuals, etc. |
| | Yellow / 3 | Place for everything and everything in its place | All items are clearly marked with large signs. | Scrap, rework & repair items are separated and identified clearly with colour labels. | Only required items are arranged properly in Shelves and Cupboards with identification | Effective usage of colours, signs, charts and to avoid searching. |
| | Green / 4 | Implementation of new marking systems for Ergonomics | Visual management | Hazardous materials are separated and identified properly | Kitting and Shadow boards Techniques | Show Room / Show case concept |

**Table 3** GREEN-6S rubrics—shine

| GREEN 6S Rubrics | | | | | | |
|---|---|---|---|---|---|---|
| **Criteria** | | **3.1 Floors** | **3.2 Shelves and lockers** | **3.3 Walls, Entrance and Windows** | **3.4 Spills** | **3.5 Machines** |
| Shine G6S - 3 | Red / 0 | Floors are dirty and damage, everywhere litter | Shelves and lockers are not clean inside, top, under and behind | Walls , Entrance and Windows are dirt and cobwebs | Oil and chemical spills are everywhere and accumulated on machines and floors | All machines are full of dirt, grime & rust. |
| | Pink / 1 | Floors are clear but damage, everywhere litter | Shelves and lockers are clean inside but not clean in top, under and behind | Walls are clean but Windows and Entrance are dirt and cobwebs | Oil and chemical spills are not exits on floor but accumulated on machines | Few machines are without dirt, grime and rust. . |
| | Orange / 2 | Floors are clear and without damage, everywhere litter | Shelves and lockers are clean inside and top but not clean in under and behind | Walls and Windows are neat and clean but Entrance is dirt and cobwebs | Oil and chemical spills are not exits on machines and floors without accumulation | All machines are neat and clean. |
| | Yellow / 3 | Floors are clean and without damage, without litter | Shelves and lockers are clean inside, top, under and behind | Walls , Entrance and Windows are neat and clean. | Mechanisms is provide to avoid spills of oil and chemical | All machines are neat and clean but having aesthetic look. |
| | Green / 4 | With new ideas preventing dirt and dust | Prevention of dirt and dust | Ownership | New ideas to avoid spillage | Total Productive Maintenance |

**Table 4** GREEN-6S rubrics—standardize

| | | | | | | |
|---|---|---|---|---|---|---|
| Standardise G6S - 4 | Red / 0 | No CheckSheet for housekeeping | No visuals regarding schedule, responsibility. | No standard procedures available with proper documents | Machines does not have cautions, label for controls and schedule. | No mechanism is available to reduce emission of hazardous to environment |
| | Pink / 1 | Checksheets for House keeping are in place | Visuals regarding schedule without responsibilities | Few standard procedures available without proper documents | Machines have cautions but no label for controls and schedule. | Few visual aids are exits reduce emission of hazardous to environment |
| | Orange / 2 | Check sheet for the Machine Maintainence schedule | Visuals placed regarding schedule and responsibility. | All standard procedures available without proper documents | Machines have cautions and label for controls but no schedule. | Some visual aids are exits reduce emission of hazardous to environment |
| | Yellow / 3 | Use of other Standard tools | Visuals placed regarding controls and functions | All standard procedures available with proper documents | Machines have cautions, label for controls and schedule. | Sufficient visual aids are exits reduce emission of hazardous to environment |
| | Green / 4 | Finding new tools for the process improvement in done regularly | Visual placed to create awareness on Green culture | Layout to be changed to reduce man and material movement. | Andon to be provided to save power. | New mechanism to be placed to eliminate hazardous to environment |

**Table 5** GREEN-6S rubrics—sustain

| GREEN 6S Rubrics | | | | | | |
|---|---|---|---|---|---|---|
| Criteria | | 5.1 Daily routine | 5.2 Auditing and Review | 5.3 Training | 5.4 Competition and Campaign | 5.5 Willingness |
| Sustain G6S - 5 | Red / 0 | No one takes part in daily stand up meeting and 5s oath and follow standards / work instructions. | Regular 'Green-5S' auditing and review not in practice. | Proper 'Green-5S' training not in practice. | 'Green-5S' Competition and 'Green-5S' Campaign are not in practice | Overall there is no willingness and no desire to make it a habit to have excellent 'Green-5S' |
| | Pink / 1 | Very few takes part in daily stand up meeting and 5s oath and follow standards / work instructions. | Regular 'Green-5S' auditing in practice but review meeting not exits. | Very few supervisors under gone 'Green-5S' training | 'Green-5S' Competition in practice but Campaign not in practice. | Some willingness and desire to make it a habit to have excellent 'Green-5S' |
| | Orange / 2 | Most of then takes part in daily stand up meeting and 5s oath and follow standards / work instructions. | Regular 'Green-5S' auditing and review in practice. | Some supervisors and senior operators under gone 'Green-5S' training | 'Green-5S' Competition and 'Green-5S' Campaign are in practice | Most of the employees have willingness to make it a habit to have excellent 'Green-5S' |
| | Yellow / 3 | All of them takes part in daily stand up meeting and 5s oath and follow standards / work instructions. | Top management participating review meeting. | Every on under gone Proper 'Green-5S' training in regular intervals. | 'Green-5S' New letter in process | Overall there is willingness and desire to make it a habit to have excellent 'Green-5S' |
| | Green / 4 | New ideas for habitual practice | Innovative auditing system | Training department with experts | Suggestion scheme in practice | Motivate with rewards and taking them to world class companies of 'Green-5S' |

**Table 6** GREEN-6S rubrics—safety

| Criteria | | 6.1 PPE | 6.2 Fire Safety | 6.3 Work Place Safety | 6.4 OSHA posters | 6.5 Environmental safety |
|---|---|---|---|---|---|---|
| Safety G6S - 6 | Red / 0 | No Proper PPE standards | No proper Fire Fighting Equipments | Sharp Edges, Electrical wirings on floor, Poor lit area, Congested gangways | No awareness of OSHA at workplace | Air Polluting process, improper sewage disposal, Improper garbage disposal, |
| | Pink / 1 | Only Basic PPE (footwear) are used | Fire Fighters are used | Medical and First aid assistance are in place | No awareness of OSHA to workers, NO poster on OSHA information | Proper Emission Control, STP and Segregation of garbage and disposal |
| | Orange / 2 | PPE Standards are not properly followed | Fire Fighters are marked and check regularly | Well lit workplace, Organised electrical wirings, Adequate work space | Posters are not up to date / not relevant to process | Identification /reduction of environmental damage causing wastes in the process materials |
| | Yellow / 3 | PPE Standards are properly followed and monitored | Fire Exits are marked and Visible, Safe Assembly Zone are identified and marked | Identification of Safety hazard materials/ Process | All required information are posted and up to date | Identification/ Usage of Alternate energy Source |
| | Green / 4 | Wearing PPE is followed as a culture | Fire Detection/ Prevention Methods and technologies are identified and used | Methods and Awareness for workplace injury Prevention | Safety procedures are identified and posted at Hazardous area or injury prone area | Green manufacturing/ Zero Emission |

| | | |
|---|---|---|
| | Adoption of GREEN - 6S | MCET |
| | Green - 6S Audit Sheet | GREEN 6S |
| | KRIPA PUMPS | 01 |

Name of the Auditor: Dr. I. Rajendran
Name of the Zone : 01
Date of Audit: 31/7/18

| Evaluation | 1 | 2 | 3 | 4 | 5 | 'S' Total out of 20 | Colour |
|---|---|---|---|---|---|---|---|
| G6S-1 | 0 | 1 | 0 | 1 | 0 | 2 | R |
| G6S-2 | 1 | 0 | 1 | 0 | 0 | 2 | R |
| G6S-3 | 1 | 0 | 0 | 0 | 0 | 1 | R |
| G6S-4 | 0 | 0 | 0 | 0 | 0 | 0 | R |
| G6S-5 | 0 | 0 | 0 | 0 | 0 | 0 | R |
| G6S-6 | 1 | 0 | 0 | 0 | 0 | 1 | R |
| Auditor Signature | | | | | Total | 6 | R |
| | | | | | Weighted Average | 1 | |

Colour Key:

| Colour | Red - R | Pink - P | Orange - O | Yellow - Y | Green - G |
|---|---|---|---|---|---|
| Score | 0-4 | 5-9 | 10-14 | 15-19 | 20 |
| Total Score | 0-23 | 24-47 | 48-71 | 72-95 | 96-120 |

**Fig. 3** Sample 'GREEN-6S' audit sheet

The major benefit of this assessment system is less time consuming and it is clearly indicating the current state in the form of visual. For assessing the overall level of 6S and green manufacturing of a particular zone, the weighted average has to be calculated. The final assessment of 'GREEN-6S' will be reported as shown in Fig. 3.

### *3.2 Prepare Brush for Painting*

After the completion of 'GREEN-6S' audit, the 'To-Do List' (Fig. 4) is introduced to prepare a 'Brush' to change the colour of the current situation. All the issues raised during the assessment is considered as opportunity and entered in 'To-Do List' with raised date and raised by. Then, the implementation team should suggest a suitable solution to overcome the issues. The 'To-Do List' to be completed in this stage with the information of a responsible person's name and the due date for completion.

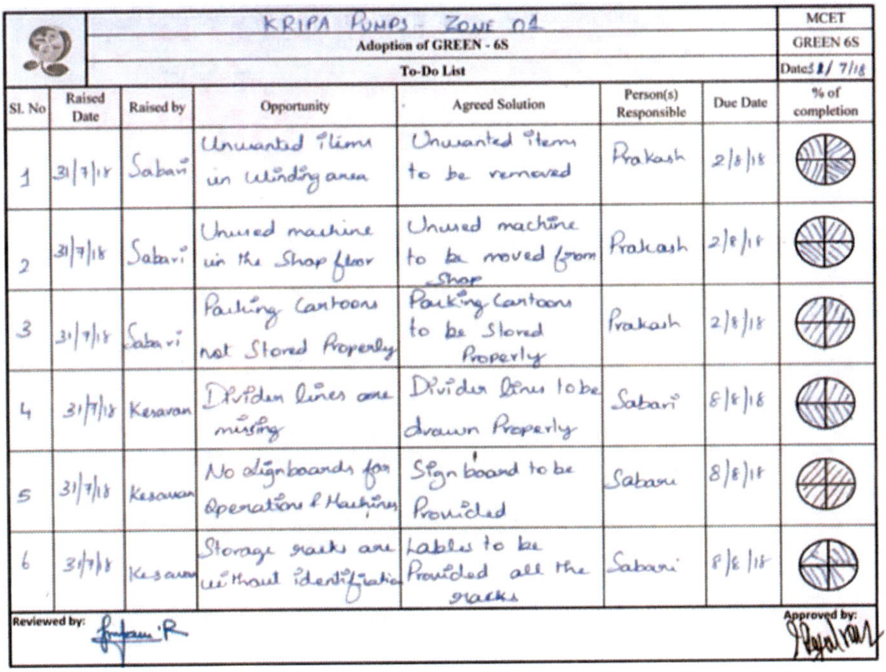

KRIPA PUMPS - ZONE 01
Adoption of GREEN - 6S
To-Do List

MCET
GREEN 6S
Date: 31/7/18

| Sl. No | Raised Date | Raised by | Opportunity | Agreed Solution | Person(s) Responsible | Due Date | % of completion |
|---|---|---|---|---|---|---|---|
| 1 | 31/7/18 | Sabari | Unwanted items in winding area | Unwanted items to be removed | Prakash | 2/8/18 | |
| 2 | 31/7/18 | Sabari | Unused machine in the shop floor | Unused machine to be moved from shop | Prakash | 2/8/18 | |
| 3 | 31/7/18 | Sabari | Packing cartoons not stored properly | Packing cartoons to be stored properly | Prakash | 2/8/18 | |
| 4 | 31/7/18 | Kesavan | Divider lines are missing | Divider lines to be drawn properly | Sabari | 8/8/18 | |
| 5 | 31/7/18 | Kesavan | No sign boards for operation & machinery | Sign board to be provided | Sabari | 8/8/18 | |
| 6 | 31/7/18 | Kesavan | Storage racks are without identification | Lables to be provided all the racks | Sabari | 8/8/18 | |

Reviewed by:
Approved by:

**Fig. 4** Sample 'To-Do List'

## 3.3 Paint it Green

After the development of an implementation plan in the form of 'To-Do List', Toyota 'A3 report' has to be prepared for implementation actions. Each opportunity is implemented either individually or concurrently with other opportunities. A structured methodology should be developed in the form of 'A3 report' to understand the current situation and implement the solution to change the colour towards green (Fig. 5).

## 3.4 Keep it Green

The challenges of this stage are how to sustain the achieved green transformation and after the attainment of green how to obtain the green improvement which is called as greenness. The success of 'GREEN-6S' improvement is measured by the degree of its sustainability. The sustainability of 6S and green manufacturing is obtained through Kaizen events which are focused on continuous improvement.

The Kaizen team should be part of the regular planning activities faculties. The Kaizen Sheet is shown in Fig. 6. Most of the quantitative assessment is ending with the final score after auditing and it cannot provide the score after reaching the maximum value. Hence, the system is stopped after reaching the maximum score and it does not

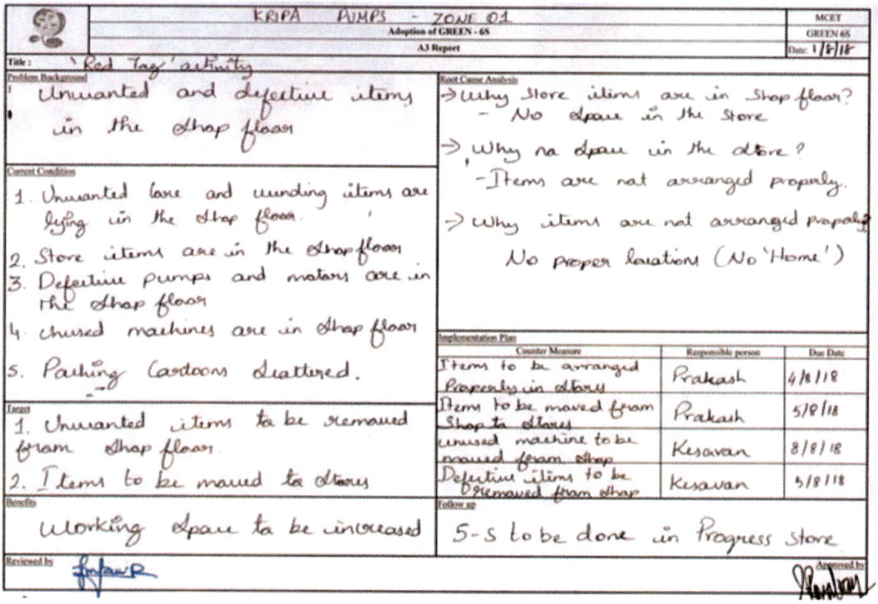

KRIPA PUMPS - ZONE 01

Adoption of GREEN - 6S

A3 Report

MCET | GREEN 6S | Date: 1/8/18

Title: 'Red Tag' activity

Problem Background: Unwanted and defective items in the shop floor

Current Condition:
1. Unwanted bore and winding items are lying in the shop floor.
2. Store items are in the shop floor
3. Defective pumps and motors are in the shop floor
4. Unused machines are in shop floor
5. Packing Cartons scattered.

Root Cause Analysis:
→ Why store items are in shop floor? – No space in the store
→ Why no space in the store? – Items are not arranged properly.
→ Why items are not arranged properly. No proper locations (No 'Home')

Implementation Plan

| Counter Measure | Responsible person | Due Date |
| --- | --- | --- |
| Items to be arranged properly in stores | Prakash | 4/8/18 |
| Items to be moved from shop to stores | Prakash | 5/8/18 |
| Unused machine to be moved from shop | Kesavan | 8/8/18 |
| Defective items to be removed from shop | Kesavan | 5/8/18 |

Target:
1. Unwanted items to be removed from shop floor.
2. Items to be moved to stores

Benefits: Working space to be increased

Follow up: 5-S to be done in Progress store

Reviewed by | Approved by

Fig. 5 Sample 'A3 report'

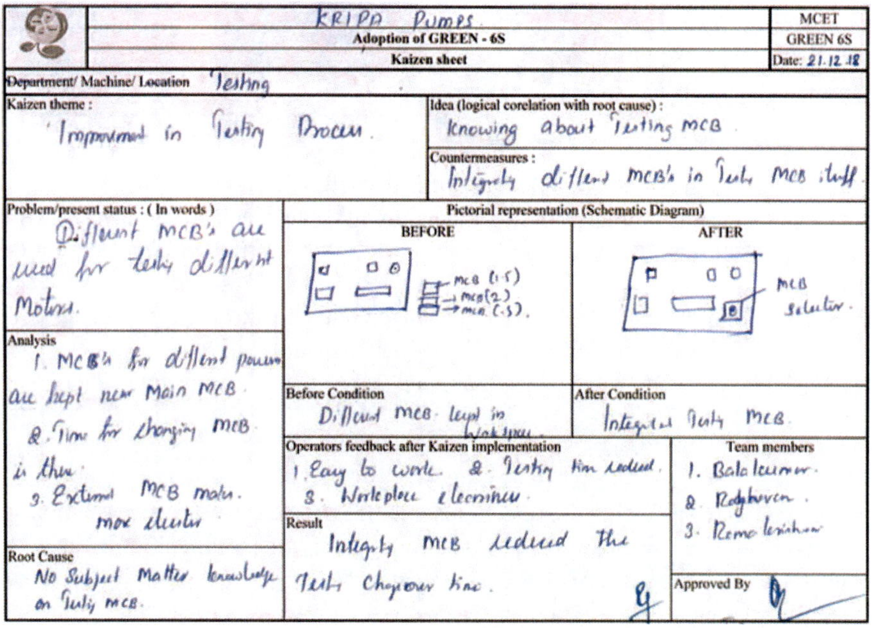

KRIPA PUMPS

Adoption of GREEN - 6S

Kaizen sheet

MCET | GREEN 6S | Date: 21.12.18

Department/ Machine/ Location: Testing

Kaizen theme: 'Improvement in Testing Process.

Idea (logical corelation with root cause): Knowing about Testing MCB.

Countermeasures: Integrity different MCB's in Testing MCB itself.

Problem/present status : ( In words ): Different MCB's are used for testing different Motors.

Analysis:
1. MCB's for different powers are kept near Main MCB.
2. Time for changing MCB is there.
3. Extend MCB makes more electric

Root Cause: No Subject Matter knowledge on Testing MCB.

Pictorial representation (Schematic Diagram)

BEFORE: MCB (1.5), MCB (2), MCB (.5).

AFTER: MCB selector.

Before Condition: Different MCB kept in work place.

After Condition: Integrated Testing MCB.

Operators feedback after Kaizen implementation: 1. Easy to work. 2. Testing time reduced. 3. Workplace cleanliness.

Team members: 1. Balakumar. 2. Raghavan. 3. Ramakrishnan

Result: Integrity MCB reduced the Testing Changeover time.

Approved By

Fig. 6 Sample 'Kaizen sheet'

provide the room for further process improvement and creative implantation. But, the 'GREEN-6S' system model provides the mechanism to keep the green is stable and green become greener.

## 4 Investigation of 'Green-6S' in Motor and Pump Industries

The following five pump and motor industries are identified in Coimbatore and a MSME cluster is formed to adopt 'GREEN-6S' system model.

1. KRIUPA PUMPS, 165/3, Palaniandavarkoilthottam, Ganapathy.
2. KERA PUMP INDUSTRIES, 16, P.N. Palayam road, Ganapathy.
3. AQUVA Pumps, 63/1C, Elangonagar, Avarampalayam.
4. TEXS Pumps, 15, Jayasimmapuram, P.N. palayam.
5. AGRO Pumps, 148/2, Pappaiyammalgaren, Avarampalayam.

### *4.1 Green 6S Adoption Procedure*

After the Gemba walk, two zones are identified for 'GREEN-6S' adoption at different companies and named according to each company, then the initial audit is done on each zone. The current status of the companies is measured by performance indicators such as cycle time, searching time and walking time with the help of standard worksheet and combination sheet. The training on 'GREEN-6S' was given to the technicians. The team members prepared 'To-Do list' and three 'A3 reports' for each zone. After that, the second and third audits are carried to measure the improvement. After the completion of the third audit, the training on 'KAIZEN' was given to team members. Two events are identified for Kaizen, the first event is a preparation of shadow boards for fitting tools and another was resizing the painting booth for proper utilization. The fourth audit is also conducted during the period of kaizen activities to identify the current colour. After the fourth audit, the performance indicator is measured to check the improvement in the process.

After the adoption of 'GREEN-6S', searching, man and part movement, down-time are reduced along with the eco-friendly working environment. The change in colour shows the success of 'GREEN-6S' system model and also the involvement of employees. The A3 reports help to improve the audit score and it also helps to complete the To-Do list as per schedule.

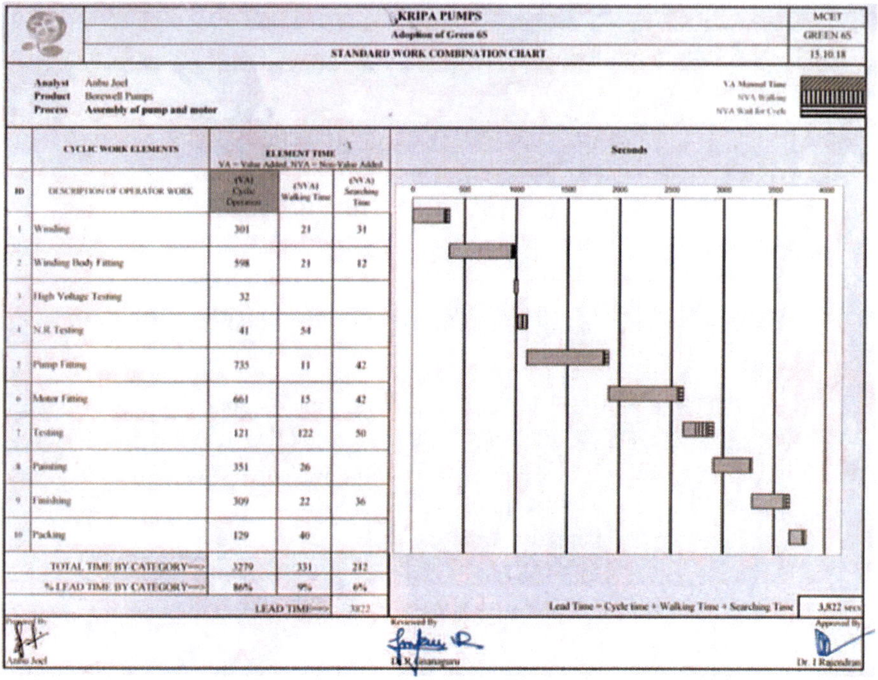


**Fig. 7** Sample 'standard work sheet'

## 4.2 Standard Work Sheet

The standard work sheet (Fig. 7) is a tool which provides a graphical view of the workstation, the path of the operator, to understand the workflow and make the workflow smoothly. This sheet visually shows the layout and work sequence of the process. This form goes hand-in-hand with the standard work combination sheet.

## 4.3 Standard Work Combination Sheet

The standard work combination sheet (Fig. 8) is used to visually represent the time taken to do the various activities in the process. It also represents the various time involves in the process such as cycle time searching time and walking time.

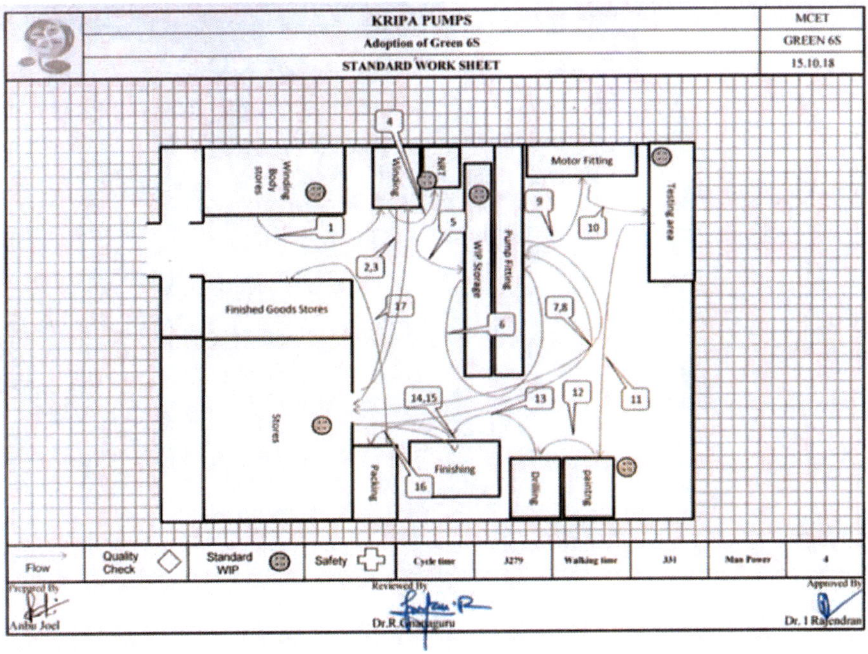


**Fig. 8** Sample 'standard work combination sheet'

## *4.4 Document Reality*

After a first audit of 'GREEN-6S', the following opportunities are identified as common:

1. Unwanted items and unused machines are lying in shop floor.
2. Divider lines are missing.
3. Identification is missing all the places (Sign boars/Stickers).
4. Floors, walls and windows are covered with dirt.
5. Dust bins not in use.
6. No standard operation procedure (SOP).
7. First aid box and fire extinguishers not in use.
8. No visuals in the shop.
9. Shop floor having paint odour.
10. Personnel productive equipment (PPE) not in practice.

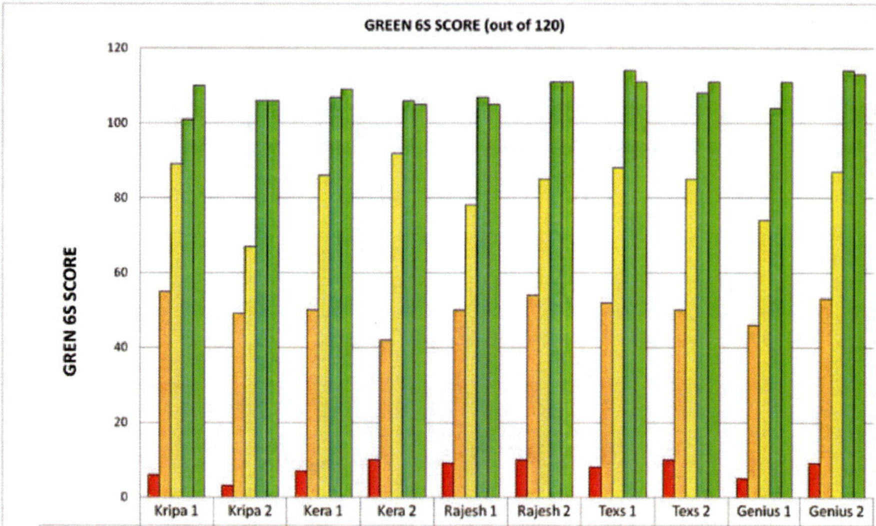


| | Kripa 1 | Kripa 2 | Kera 1 | Kera 2 | Rajesh 1 | Rajesh 2 | Texs 1 | Texs 2 | Genius 1 | Genius 2 |
|---|---|---|---|---|---|---|---|---|---|---|
| Assesment 1 | 6 | 3 | 7 | 10 | 9 | 10 | 8 | 10 | 5 | 9 |
| Assesment 2 | 55 | 49 | 50 | 42 | 50 | 54 | 52 | 50 | 46 | 53 |
| Assesment 3 | 89 | 67 | 86 | 92 | 78 | 85 | 88 | 85 | 74 | 87 |
| Assesment 4 | 101 | 106 | 107 | 106 | 107 | 111 | 114 | 108 | 104 | 114 |
| Assesment 5 | 110 | 106 | 109 | 105 | 105 | 111 | 111 | 111 | 111 | 113 |

**Fig. 9** Green-6S audit score of 10 zones

# 5 Results

## *5.1 GREEN-6S Audit Scores*

GREEN-6S scores of each zone are consolidated below for four assessments (Fig. 9)

## *5.2 Performance Indicators*

After the implementation of GREEN-6S, not only the environment affecting factors are reduced, the improvement of performance is measured by the performance indicator which shows a reduction in the cycle time and a most significant reduction in non-value adding activity like searching and walking. The cycle time (Fig. 10), searching time (Fig. 11) and walking time (Fig. 12) comparison is given below for each industry. The cycle time, walking time and searching time are measured in seconds.

The above comparison shows that the implementation of GREEN-6S improves the workplace productivity also. The above comparison concludes that on implementing GREEN-6S at the workplace will reduce cycle time by 10%, walking time by 25% and

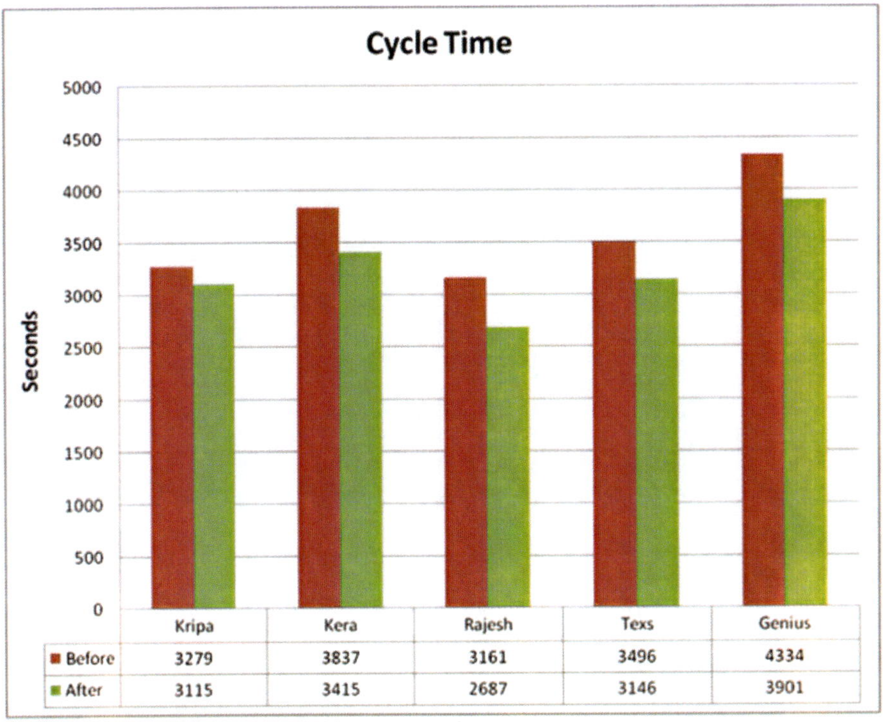


**Fig. 10** Cycle time comparison

searching time 42% on average. Thus, the GREEN-6S has an effect on productivity also likewise it has in green manufacturing.

The results of five industries show that the worth implementing 'GREEN-6S' in the motor and pump industries. Implementation of 6S principles is complementary to environmental performance and greater source reduction. The following benefits are absorbed towards the improvement of productivity in terms of lead time (Fig. 13).

## *5.3 Feel Good Factor*

Feel Good Factor is derived from Herzberg's motivation theory model. Herzberg's motivation theory or two-factor theory argues that there are two factors that an organization can adjust to influence motivation in the workplace.

These factors are:

Motivators are encouraging employees to work harder. It increases the job satisfaction.

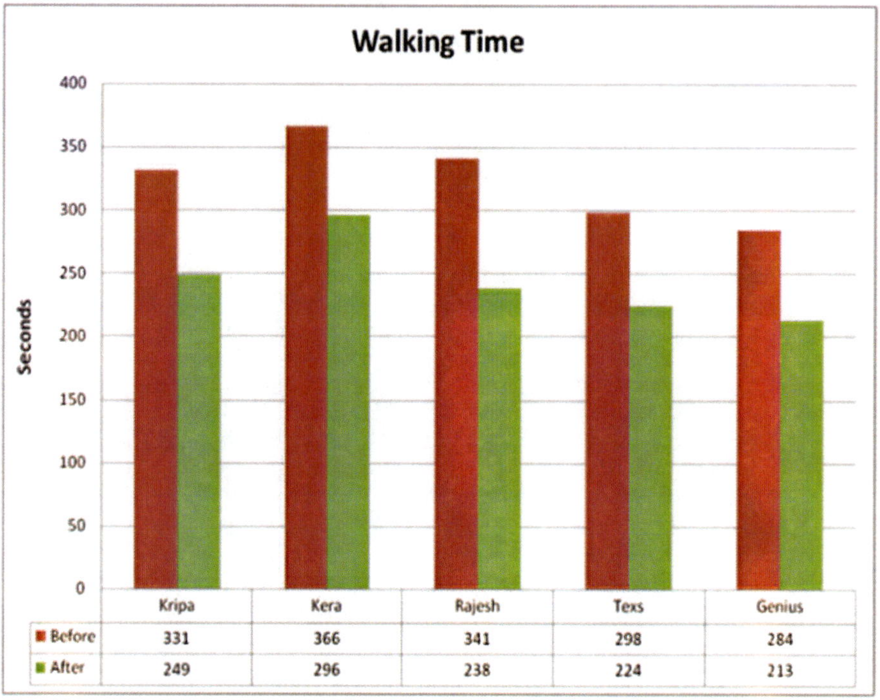


**Fig. 11** Walking time comparison

**Fig. 12** Searching time comparison

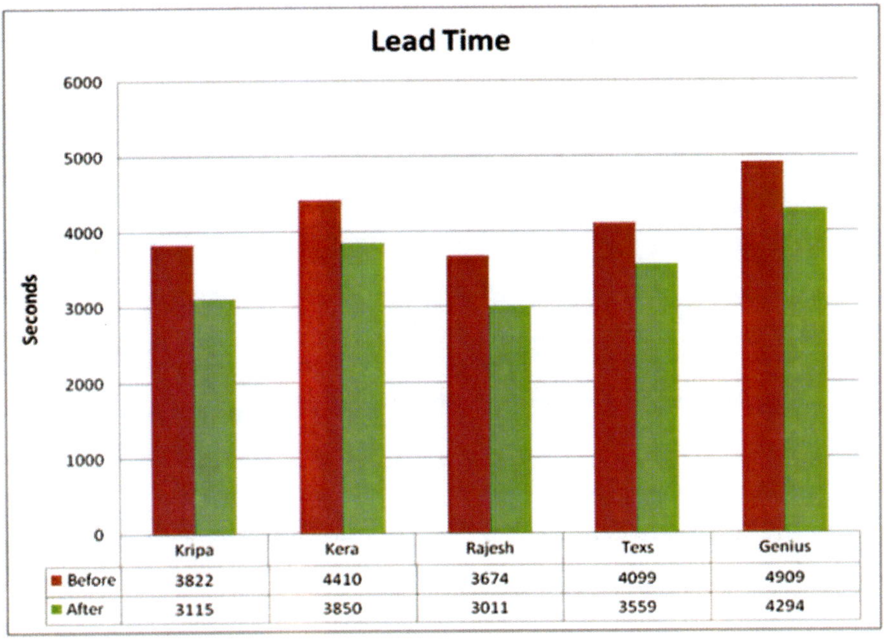


**Fig. 13** Lead time comparison

Hygiene factors that would not encourage employees to work harder but they will cause them to become unmotivated if they are not present. This decreases the job satisfaction.

Motivational factors such as achievement, recognition and responsibility are achieved by implementing 'GREEN-6S'. The hygiene factors such as supervision, relationship and work condition are improved. Thus, implementing the 'GREEN-6S' will increase the job satisfaction. Thus, the Feel Good Factor of the employees is measured by attendance of the employees. The absenteeism is indirectly proportional to the Feel Good Factor.

$$\text{Feel Good Factor } \alpha \ \frac{1}{\text{Absenteeism}}$$

The Feel Good Factor can be measured by the absenteeism of employees. As per the above analysis, Fig. 14 gives the absenteeism data from January 2018 to March 2019.

Figure 14 shows the decrease in the percentage of absenteeism after the implementation of the 'GREEN-6S'. Hence, it clearly indicates that the Feel Good Factor of the employees is improved.

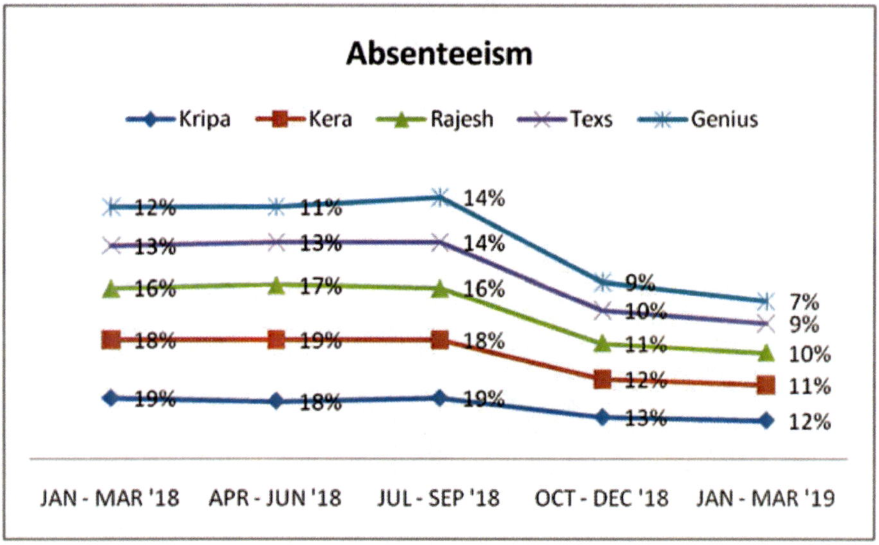


**Fig. 14** Absenteeism comparison

# 6 Conclusion

This research study addresses the development of the green environment through 6S (5S + Safety) principles in motor and pump industries of Coimbatore. These industries are thought to have a collective impact on environmental issues, they lack behind in larger area in terms of environmental activeness and performance. To meet with the challenges of the business environment, the company strive to rearrange their production by implementing green environment through a system model called 'GREEN-6S'.

The 'GREEN-6S' model captures various planning activities to migrate from a less green into a greener and more eco-efficient manufacturing. The various planning stages are accompanied by the required control metrics as well as various lean tools in an open mixed architecture. The proposed model is a comprehensive qualitative approach to improve green in manufacturing as well as a roadmap for future quantitative research to better evaluate this new paradigm.

The implementation of 'GREEN-6S' in motor and pump industries reducing cycle time by 10%, walking time by 25% and searching time 42% in average which helps to improve productivity by 10% in average. Hence, the research study concludes that 'GREEN-6S' model have impact on productivity and also employee Feel Good Factor.

**Acknowledgements** The authors wish to acknowledge the Entrepreneurship Development and Innovation Institute (EDII), Chennai, for their support to conduct this research study with Rs. 500,000 during the period of July 2018 to June 2019.

## References

1. Abreu MF, Alves AC, Moreira F (2017) Lean-green models for eco-efficient and sustainable production. Energy 137:846–853
2. Ahuja R, Sawhney A, Arif M (2017) Driving lean and green project outcomes using BIM: a qualitative comparative analysis. Int J Sustain Built Environ 6(1):69–80
3. Chiarini A (2014) Sustainable manufacturing-greening processes using specific lean production tools: an empirical observation from European motorcycle component manufacturers. J Clean Prod 85:226–233
4. Duarte S, Cruz-Machado V (2013) Modelling lean and green: a review from business models. Int J Lean Six Sigma 4(3):228–250
5. Dües CM, Tan KH, Lim M (2013) Green as the new lean: how to use lean practices as a catalyst to greening your supply chain. J Clean Prod 40:93–100
6. Fu X, Guo M, Zhanwen N (2017) Applying the green embedded lean production model in developing countries: A case study of china. Environ Devel 24:22–35
7. Gnanaguru R, Puvaneswari K, Mallick J, Jegadheesan C, Mohan Sivakumar V, Devadasan SR (2011) Toyota's A3 reports for improving 6-S activities: an aeronautical industry case study. Int J Serv Oper Manage 10(2):239–254
8. Ho S, MohdHashim AGB, MohdIdris MA (2015) Applicability of SIRIM Green 5-S model for productivity & business growth in Malaysia. TQM J 27(2):185–196
9. Jabbour CJC, de Sousa Jabbour ABL, Govindan K, Teixeira AA, de Souza Freitas WR (2013) Environmental management and operational performance in automotive companies in Brazil: the role of human resource management and lean manufacturing. J Clean Prod 47:129–140
10. King AA, Lenox MJ (2001) Lean and green? An empirical examination of the relationship between lean production and environmental performance. Prod Oper Manage 10(3):244–256
11. Kurdve M, Zackrisson M, Wiktorsson M, Harlin U (2014) Lean and green integration into production system models–experiences from Swedish industry. J Clean Prod 85:180–190
12. O'hEocha M (2000) A study of the influence of company culture, communications and employee attitudes on the use of 5Ss for environmental management at Cooke Brothers Ltd. TQM Mag 12(5):321–330
13. Pampanelli AB, Found P, Bernardes AM (2014) A lean & green model for a production cell. J Clean Prod 85:19–30
14. Rehman MAA, Shrivastava RR, Shrivastava RL (2013) Validating green manufacturing (GM) framework for sustainable development in an Indian steel industry. Univ J Mech Eng 1(2):49–61
15. Sawhney R, Teparakul P, Bagchi A, Li X (2007) En-lean: a framework to align lean and green manufacturing in the metal cutting supply chain. Int J Enterp Netw Manage 1(3):238–260
16. Thanki S, Govindan K, Thakkar J (2016) An investigation on lean-green implementation practices in Indian SMEs using analytical hierarchy process (AHP) approach. J Clean Prod 135:284–298
17. Verrier B, Rose B, Caillaud E (2016) Lean and green strategy: the lean and green house and maturity deployment model. J Clean Prod 116:150–156
18. White GRT, James P (2014) Extension of process mapping to identify "green waste". Benchmarking: An Int J 21(5):835–850

# Concrete Bridge Crack Detection Using Convolutional Neural Network

**R. Vignesh, B. Narenthiran, S. Manivannan, R. Arul Murugan, and V. RajKumar**

**Abstract** The conventional image processing algorithms are not found to perform well in detecting crack problems. Generously the crack classification performance is also not clear with traditional deep learning neural network. To mitigate these issues, a convolutional neural network-based detection of bridge crack is presented in this work. The arous space pyramid pool (ASPP)-based feature extraction with depthwise separable convolution is modeled. With ASPP, the multiscale information of image features can be obtained, and the proposed convolution model provides large reception field so as to effectively fuse huge amount of contextual data on feature maps. Hence, the computational complexity of the model is greatly reduced. The results are verified in simulation which shows that the proposed method has achieved a highest detection accuracy of 96.68% which is higher than the conventional deep learning model.

**Keywords** Deep learning · Convolutional neural networks · Arous space pyramid pool · Bridge crack detection

R. Vignesh · R. Arul Murugan
Department of Mechanical Engineering, Karpagam College of Engineering, Coimbatore, Tamil Nadu, India
e-mail: vigneshbe1211@gmail.com

R. Arul Murugan
e-mail: arulmuruganraju@gmail.com

B. Narenthiran · S. Manivannan
Department of Mechanical Engineering, Karpagam Academy of Higher Education, Coimbatore, Tamil Nadu, India
e-mail: narenthirannani@gmail.com

S. Manivannan
e-mail: manivannan.s@kahedu.edu.in

V. RajKumar (✉)
Department of Mechanical Engineering, PSG Institute of Technology and Applied Research, Coimbatore, Tamil Nadu, India
e-mail: rajkumarv@psgitech.ac.in

© The Editor(s) (if applicable) and The Author(s), under exclusive license to Springer Nature Singapore Pte Ltd. 2021



S. Mohan et al. (eds.), *Materials, Design, and Manufacturing for Sustainable Environment*, Lecture Notes in Mechanical Engineering,
https://doi.org/10.1007/978-981-15-9809-8_58

# 1 Introduction

Bridges assume a significant job in everyday life as a sort of transportation framework. As per the statistics, 90% of the bridge harm happens in every year because of cracks. Thus, recognizing the state of cracks is essential for the sheltered utilization of scaffolds. The conventional discovery techniques fundamentally depend on manual visual assessment, which has less precision, in addition it is also hard to ensure the nature of assessment. As of late, AI and PC vision innovation are effectively utilized in the field of programmed split discovery [1–5] and have pulled in an ever increasing number of engineers' consideration.

The conventional crack identification strategies are depending on advanced processing system. Simi et al. [6] utilize four edge recognition strategies to discover solid crack. Chowdhury et al. [7] presented a strategy for extricating splits by setting up a multiscale 2D wavelet transformation. Kilic et al. [8] developed a FoSA calculation by utilizing F* seed development to associate break focuses as seed focuses to detect crack. Yamaguchi et al. [4] investigated a strategy which combines seed identification and tensor vote to acquire higher discovery precision. Garrett et al. [9] developed the ITV break calculation to distinguish splits by utilizing repeated tensor vote depending on Zou's strategy. As input pictures have enormous commotion, these picture preparing strategies cannot precisely distinguish the breaks.

The idea of profound learning was initiated by Al Qadi et al. [10]. The two perspectives have multiple layers; ANN model has solid component learning capacity, and the significant information got by the profound learning model has progressive portrayal of the first information, which will significantly encourage grouping and representation. For profound NNs, it is hard to accomplish ideal performance during preparation, and this issue is tackled by layer-by-layer preparation. Doebling et al. [11] proposed CNN to utilize picture space data for decreasing the quantity of preparation parameters, in this manner an incredible improvement is achieved with this model.

Profound learning gain proficiency with a profound nonlinear system describes input information and executes complex estimation. It has an incredible capacity to be familiar with the basic highlights of informational collections from a little pattern set. Various PC vision assignments show the adequacy of profound highlights extricated by profound neural systems. The convolutional neural network of profound learning has made extraordinary progress in picture processing, video acknowledgment, discourse acknowledgment, target discovery, and so forth [12–15]. The profound learning has not been applied to the field of break discovery since 2016, and Zhang et al. [2] utilized CNN to detect split. Nonetheless, since the technique did not utilize exact criteria for positive and negative preparation patterns, the precision was not high.

A solid break discovery classifier is constructed in this paper depending on CNN. The proposed strategy has the accompanying commitments:

- The proposed classifier is minimum influenced by clamors like light, water stains, and foundation concealing, so it has great flexibility.

- A deformity identification model has been built up to identify different kinds of structure harm like delamination, voiding, spalling, and consumption and has wide flexibility.
- A novel feature extraction technique depends on ASPP module and profundity detachable convolution, which can be joined with the system to enhance execution and extricate better picture highlights.

## 2 Proposed Method

Bridge crack detection is essentially a classification task, which aims to judge whether a given picture contains cracks. Characterizing the image is the main research content of the task. To solve this problem, a CNN is used to train the pictures with given ground truth tag. This section describes the implementation of the algorithm and the overall structure of the network.

### *2.1 Data Preparation*

There is no amalgamate bridge fracture database in academia. Thus, in this work, the bridge fracture dataset is combined and preprocessed to satisfy the needs of CNN training. The total quantity of original images is 2068, which was acquired by the CMOS area array camera that comes with the DJI Phantom 4 pro. The size of each original image is 1024 × 1024, and the final dataset is generated from the original image by the following steps:

1. Since the original images all contain cracks, it is not conducive to our classification task. Therefore, each original image is cropped to a size of 512 × 512, and after the blurred image is removed, 6069 images are finally obtained as our dataset. It contains 4058 crack pictures, 2011 background pictures, and the rate of crack pictures to background pictures is about 2:1. In the end, we used 4856 pictures as our preparation set and 1213 pictures as test data.
2. These pictures are further random cropped to 224 × 224 to satisfy the input needs of the Resnet50 network. At the same time, a network trained on a relatively small picture can scan any image larger than the design size.

The resulting data set comprises a wide spectrum of image changes for generating a powerful crack classification, as depicted in Fig. 1.

As shown in Fig. 2, our dataset also contains cropped images of crack locations at the four edges of the image space. In these pictures, the crack area only accounts for a small part of the picture. After passing through the CNN, the picture size is further reduced, so the edge crack is more difficult to be recognized by the network than the center crack. The use of such images can increase the generalization of the

**Fig. 1** Examples of pictures utilized in preparation stage. **a** fine picture, **b** fine picture, **c** Back ground shaded picture, **d** Back ground shaded picture, **e** Robust light illuminated image

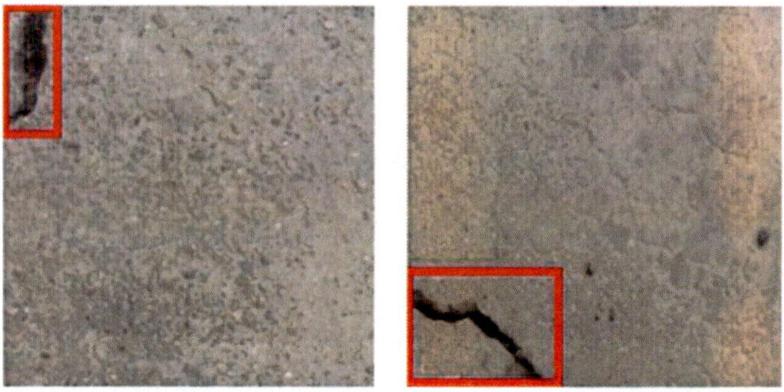

**Fig. 2** Pictures of crack at four edges

network we train, because the proportion of cracked areas in the actual detection is usually small.

## 2.2 *Gabor Filter*

A joint domain analysis was proposed by Gabor in 1946 by combining time and frequency domain functions to eliminate the loss of time domain data of conventional Fourier transformation. The Gabor function resolves the best analytical representation in the joint domain by utilizing the lower bound. The Gabor kernel is a BPF and functions like multiple channel filer in the human visual structures.

In 1985, Daugman modified the single-dimensional Gabor filter design to two-dimensional filter. The 2D Gabor function has 8 degrees of options in: $x_0$ and $y_0$ to determine the location of the portion in 2D spatial space, $u_0$ and $v_0$ are regulation directions which indicate the location of the channel in 2D recurrence field, $\theta$ is required direction, $\omega$ is spatial recurrence, $\psi$ is the stage counterbalanced of the balance factor, which chooses the evenness or antisymmetry of the channel and the width and the length of the circular Gaussian in 2D envelope and the point between direction of sinusoidal wave vector and the 2D Gaussian tomahawks. The general expression of the Gabor filter is

$$g(x, y) = s(x, y) * W(x, y) \tag{1}$$

where the Gaussian envelope is defined as

$$s(x, y) = \frac{1}{2\pi\sigma_x\sigma_y} e^{\frac{1}{2}\left[\left(\frac{x'}{\sigma_{x'}}\right)^2 + \left(\frac{y'}{\sigma_{y'}}\right)^2\right]} \tag{2}$$

where $x' = x\cos\theta + y\sin\theta$ and $y' = -x\sin\theta + y\cos\theta$. Here, $\sigma_x$ and $\sigma_y$ are the scaled factors of the neighborhood.

A complex sinusoidal with a phase offset $\Psi$ is given by

$$w(x, y) = e^{j(2\pi\omega_0 x' + \Psi)} = \cos\left(2\pi\omega_0 x' + \Psi\right) + j\sin\left(2\pi\omega_0 x' + \Psi\right) \tag{3}$$

The Gabor consists of real and imaginary components. The real component is

$$g(x, y) = \frac{1}{2\pi\sigma_x\sigma_y} e^{\frac{1}{2}\left[\left(\frac{x'}{\sigma_{x'}}\right)^2 + \left(\frac{y'}{\sigma_{y'}}\right)^2\right]\{\cos[2\pi\omega_0(x\cos\theta + y\sin\theta + \Psi)]\}} \tag{4}$$

The imaginary part is

$$g(x, y) = \frac{1}{2\pi\sigma_x\sigma_y} e^{\frac{1}{2}\left[\left(\frac{x'}{\sigma_{x'}}\right)^2 + \left(\frac{y'}{\sigma_{y'}}\right)^2\right]\{\sin[2\pi\omega_0(x\cos\theta + y\sin\theta + \Psi)]\}} \tag{5}$$

Equations (4) and (5) represent the ellipticity (aspect ratio) and wavelength of the Gabor filter.

$$g(x, y) = \frac{Y}{2\pi\sigma^2} e^{-\frac{1}{2}\left[\left(\frac{x'^2 + y^2 y'^2}{\sigma^2}\right)\right]} e^{\left(j\frac{2\pi x'}{\lambda} + \Psi\right)} \tag{6}$$

where $Y = \frac{\sigma_x}{\sigma_y}$.

A Gabor filter with wavelength 20, variance 20, and zero offset is given below (Figs. 3 and 4).

## 2.3 *Enhanced Adaptive Threshold-Based Segmentation*

Segmentation is a process of splitting the image based on the consistency and the inconsistency of the regions. The performance of the conventional threshold segmentation algorithm (Otsu method) is enhanced using the proposed approach. Initially crack marks from the images are to be removed. The image histogram contains two tops. The conventional Otsu threshold segmentation algorithm performs excellently

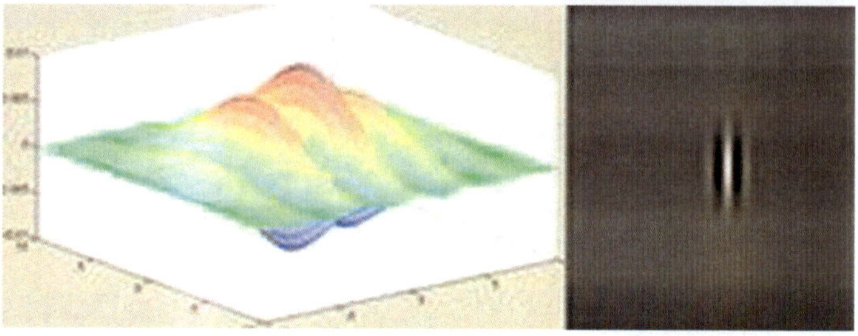

**Fig. 3** Zero degree rotation of Gabor filter

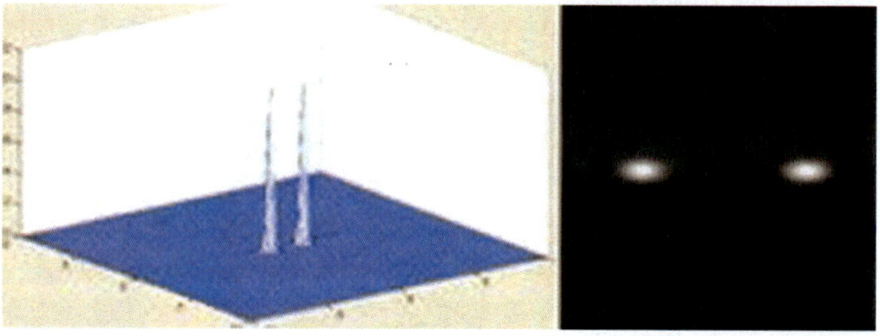

**Fig. 4** Frequency response of Gabor filter

for the images whose histogram has two tops. Using threshold-based segmentation algorithm, the segmentation threshold $T$ of the given image can be obtained. Based on the threshold esteem $T$, the segmentation algorithm replaces the pixel esteems by 0 for those that are higher than $T$ of the original image and utilizes 255 to replace the pixel esteems that are lower than $T$ in the original image, by which the isolation of target and background is achieved. In this method, the pixel esteem that is higher than $T$ is converted into 0, and remaining are not changed. This method can remove the edges from the gray image.

Then, the image segmentation using our enhanced adaptive threshold algorithm is performed. After the first step, the image histogram is modified to have a single top. The conventional segmentation algorithm is not able to satisfy the needs of the segmentation.

The steps to implement the proposed method are given below:

(1) The highest gray value $t_1$ and lowest gray value $t_k$ of the image are acquired. The median $T_0$ is taken as the initial threshold.
(2) According to the threshold $T_0$, the image is split into two components as the foreground and the background image. The average gray values $\mu_1$ and $\mu_2$ of the two components are obtained. The average values are calculated as below

$$\mu_1 = \frac{\sum_{i=0}^{T_i} i n_i}{\sum_{i=0}^{T_i} n_i}, \quad \mu_2 = \frac{\sum_{i=T_i}^{L-1} i n_i}{\sum_{i=T_i}^{L-1} n_i} \tag{7}$$

(3) Measure the new threshold: $T_{k+1} = 1/2(\mu_1 + \mu_2)$
(4) If $T_k = T_{k+1}$, end : or $k + 1 \to k$, repeat the step (2) to begin the process.

The best threshold $T_k$ can be obtained by the above steps. The updating formula used to obtain the crack image is

$$d(x, y) = \begin{cases} 0, & g(x, y) \geq T_k \text{ or } g(x, y) = 0 \\ 255, & g(x, y) < T_k \end{cases} \tag{8}$$

## 2.4 *Hyperparameters*

The classifier is prepared using the momentum optimization algorithm. As low preparing rate is suggested, the network applies a scientifically minimized preparation rate. There are 32 patterns per batch during training, the momentum is fixed at 0.9, the starting preparation rate is set to 0.005, and the weight decay is 0.2.

## 2.5 *Overall Network Structure*

The network structure we designed is shown in Fig. 15, and the architecture of the feature extraction block is given in Fig. 5. The network uses Resnet50 as the backbone. In this work, a CNN module is developed in which ASPP was applied to capture multidimensional context data. At the same time, a depthwise separate convolution is applied after each 3 × 3 convolutional layer of the ASPP module to reduce the computational loss and the parameters quantity without sparing the accuracy of the results. Our module can be inserted anywhere in the convolutional neural network model, and it can reuse the underlying image feature information.

### 2.5.1 Atrous Convolution Model

Advanced picture classifiers coordinate multidimensional setting data via persistent pooling layer and sub-examining layer, bringing about loss of data regarding article edges and debasement of picture goals. Since the crack detection task focuses on the crack edge information in the image, we introduce an atrous convolution in the network structure. Compared to traditional convolution, atrous convolution (or dilated convolution) can provide a larger receptive field with a comparable amount of computation, thus enabling the extraction of more denses feature maps.

Let $F : Z^2 \to R$ denote a discrete operator. Assume $\Omega_r = [-r, r^2]\ \Omega Z^2$ and $k : \Omega_r \to R$ denote size of discrete filter. Then, the convolution operator * is defined as

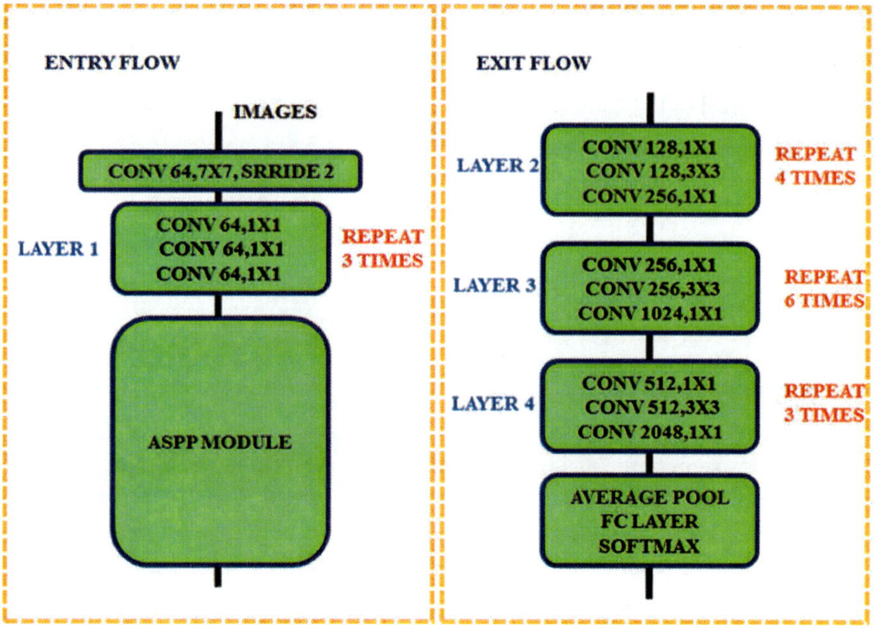


**Fig. 5** Illustration of the architecture of the proposed ConvNet

$$(K * k)(p) = \sum_{s+t-p} F(s)k(t) \tag{9}$$

Then, this operator is normalized. Let 1 be a dilation factor and let $*l$ be defined as

$$(K * lk)(p) = \sum_{s+lt-p} F(s)k(t) \tag{10}$$

Here $*l$ is an atrous convolution or l-atrous convolution. The familiar convolution is atrous convolution.

Atrous convolution empowers exponential extension of the open field without loss of resolution and coverage [24]. Let $F_0, F_1, \ldots, F_{n-1} : Z^2 \rightarrow R$ be a discrete operator and let $F_0, F_1, \ldots, F_{n-2} : \Omega^2 \rightarrow R$ be a discrete $3 \times 3$ filter. Applying a cavity channel with an exponential increase:

$$F_{i+1} = F_i * 2ik_i, \quad i = 0, 1, \ldots, n-2 \tag{11}$$

The open field of the elemental pin $F_{i+1}$ is called as a element set in $F_0$ which modifies the esteem of $F_{i+1}$. Let the size of the open field $p$ in $F_{i+1}$ is the number of these elements. Then, the open field size of every element in $F_{i+1}$ is $(2i + 2-1) \times (2i + 2-1)$. The receptive field is an exponentially increasing square, as given in Fig. 6.

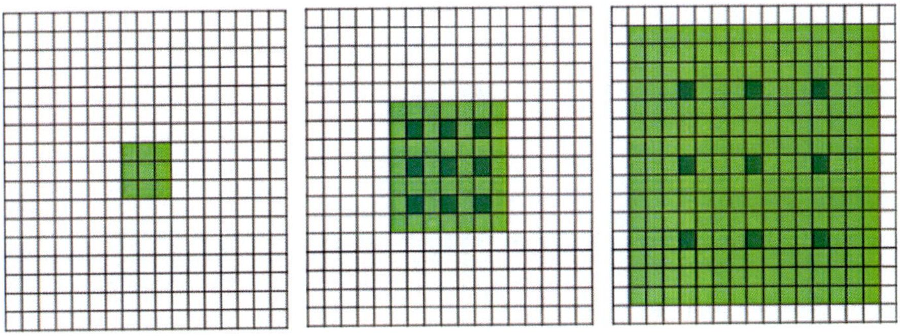

**Fig. 6** Diagrammatic representation of atrous convolution

### 2.5.2 ASPP Module

The bridge crack image usually occupies a small portion of the overall image. So as to detect the crack in the image, it is necessary to accurately extract the feature information of the crack, so in this work, the ASPP module is utilized to obtain multidimensional crack characteristic information.

ASPP is part of the DeepLabv2 system developed by the Google. Aspired by spatial pyramid pooling, a parallel sampling is done with various sampling rates of atrous convolution of input, which is proportional to catching the setting of the picture in numerous scales, in this manner acquiring multiscale picture highlight data. It has several notable features for deep convolutional neural networks: (1) ASPP uses multilevel spatial sampling and has proven multilayer pooling can effectively cope with object deformation and (2) due to the flexibility of the input rate, ASPP can collect features extracted in variable scale. These factors can improve the recognition accuracy of deep neural networks.

Three convolutional layers with different atrous rates were used, with atrous rates of 2, 4, and 8, respectively. In order to generate the result, the features extracted at each sampling rate are processed by separate branches, and the feature maps are bilinearly interpolated from the parallel network branches to the original images, and they are fused to obtain the maximum response of the location at different scale. Multiscale processing significantly enhances performance, but requires computation of the feature response on all depth convolutional neural network layers for multiple dimensions of the input image (Fig. 7).

### 2.5.3 Depthwise Separated Convolution Structure

The depthwise separated convolution was developed and utilized in Inception models to reduce the computational complexity of the first few layers. By adopting the depthwise separated convolution method, the effect of reducing the number of parameters and increase in the operation speed is achieved. The Google team's used a depthwise separated convolution structure with atrous convolution and demonstrated that

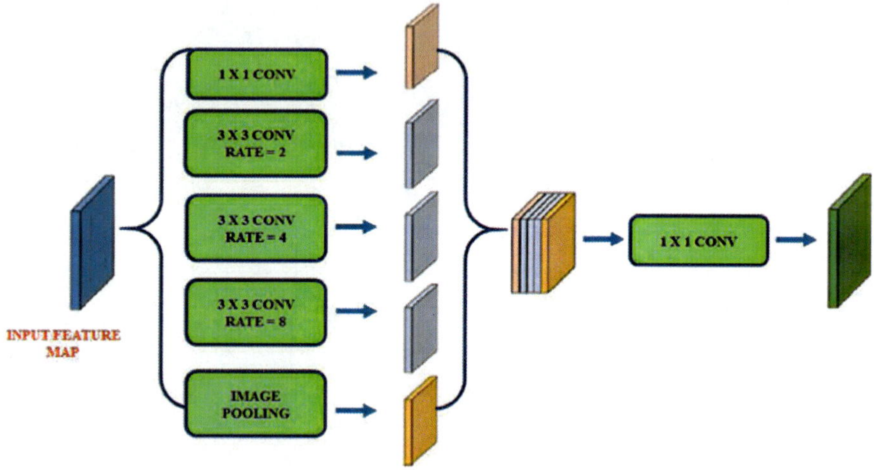


**Fig. 7** Architecture of the ASPP module

the atrous separable convolution significantly reduced the computational complexity while maintaining good performance.

The depthwise separated convolution converts the traditional convolution into deep convolution and a $1 \times 1$ convolution called point-by-point convolution, and its calculation process is shown in Fig. 8. Deep convolution utilizes a solitary channel for each information channel. The point-by-point convolution is applied on a $1 \times 1$ convolution to merge the yields of the profound convolution. The general convolution channels the contribution to one stage and then joins the contributions to another arrangement of yields. The depthwise distinct convolution isolates it into two layers,

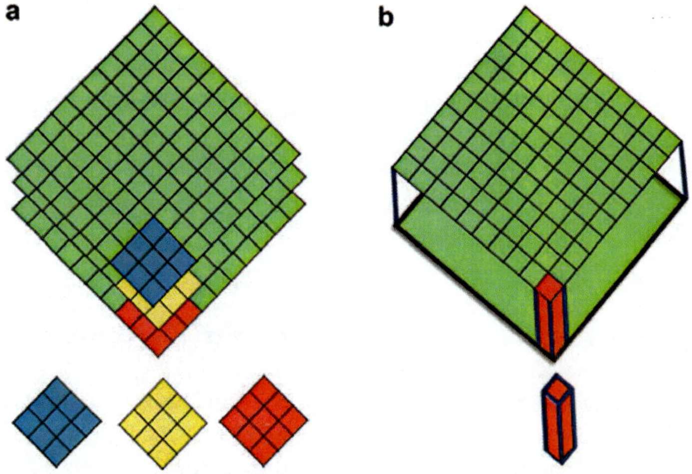


**Fig. 8** Separable convolution. **a** depthwise convolution **b** pointwise convolution

one for sifting and one for blending. This disintegration significantly lessens the measure of figuring and the intricacy of the model.

For the convolutional neural network in this work, the addition of deep distinct convolution can improve the computational efficiency of the network on a limited GPU, and on the other hand, it also saves computation time, thereby improving the effectiveness of the network in practical applications.

## 3 Results and Discussion

The proposed CNN-based crack recognition model has been trained and tested in MATLAB simulation environment. A sample of three crack images with different backgrounds was shown below has a size of 256 × 256 pixels (Figs. 9, 10, 11 and 12).

The preprocessing step of crack image processing has three modules such as grayscale conversion, filtration, and image enhancement. The frequency content of the crack portions is filtered using Gabor filter. The proposed enhanced adaptive threshold-based segmentation adopted in this paper has efficiently detected the crack edges. This two-stage segmentation extracts almost most the crack portions with high accuracy (Figs. 13 and 14).

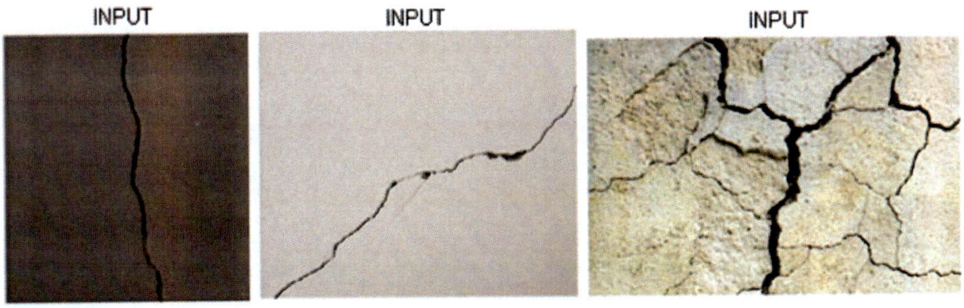


**Fig. 9** Input images

**Fig. 10** Grayscale image (preprocessed image)

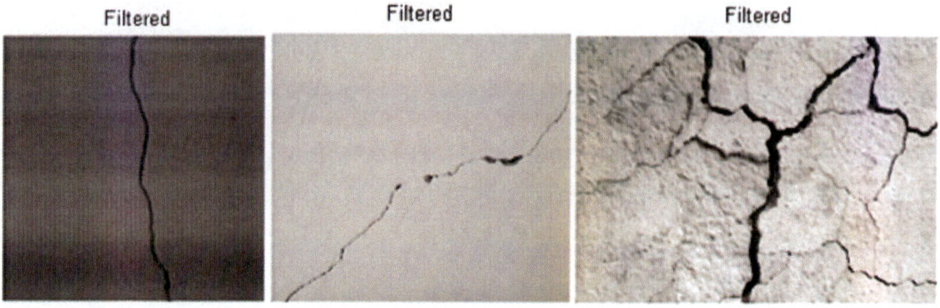


**Fig. 11** Filtered image (preprocessing output)

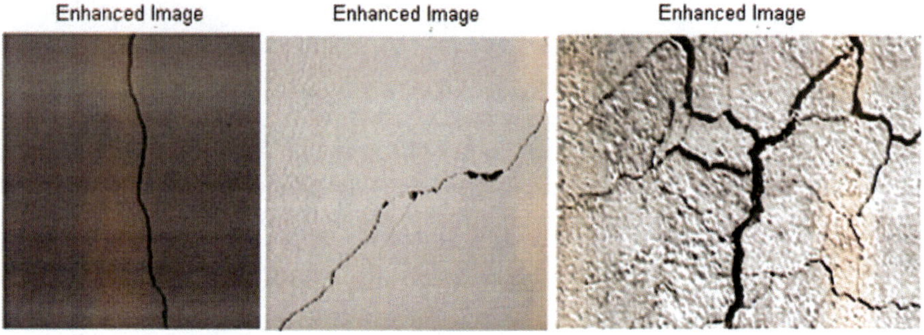


**Fig. 12** Enhanced image (preprocessing output)

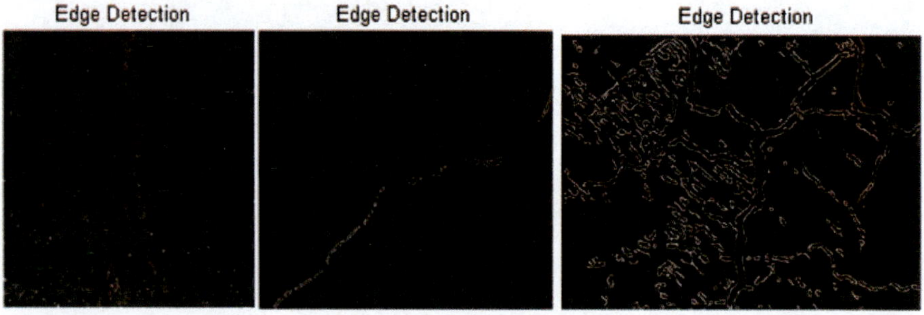


**Fig. 13** Edge detected image

A satisfactory performance has been achieved with the proposed method, and the results also confirmed the viability. This approach has the potential to classify the bridge cracks from the back ground, and hence, they can be used as an appraisal in estimating the reliability of the concretes

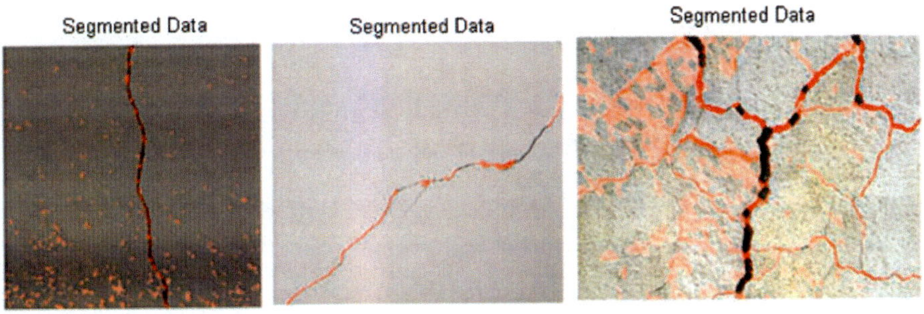


**Fig. 14** Segmented images

## *3.1 Model Evaluation*

### 3.1.1 Evaluation Indicators

We use four evaluation indicators to evaluate the model proposed in this paper, namely the recall rate, the missed alarm rate, the accuracy rate, and the false alarm rate. We test 1213 images in the test set, calculate the four indicators based on the corresponding ground truth, and give the value of these indicators.

When calculating these indicators, we count the number of images instead of the pixel count. This is because when we do ground truth, we only mark which images contain cracks, which images are backgrounds, and are not accurate to pixels. Therefore, when evaluating, it is not accurate to the pixel level and can only be counted and evaluated by whether the current picture contains cracks.

Several concepts commonly used in detecting problems are listed below:

(1) True positive (TP): refers to the total quantity of crack pictures classified as cracks.
(2) True negative (TN): indicates the total quantity of background pictures classified as background.
(3) False positive (FP): indicates the total quantity of background pictures classified as cracks.
(4) False negative (FN): indicates the total quantity of crack pictures as background.

The recall rate $R$ indicates how many true crack pictures (reference ground truth) in the sample are correctly classified as cracks, and its estimation method is given in Eq. (5):

$$R = \frac{\mathrm{TP}}{\mathrm{TP} + \mathrm{FN}} \tag{12}$$

The missed alarm rate MA indicates how many true crack pictures (reference ground truth) in the sample are misclassified as background, and its calculation method is as shown in Eq. (6):

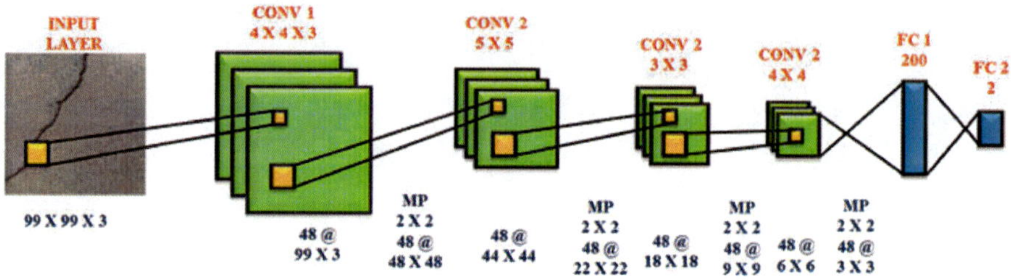


**Fig. 15** CNN structure proposed by L. Zhang

$$\mathrm{MA} = \frac{\mathrm{FN}}{\mathrm{TP} + \mathrm{FN}} \tag{13}$$

The accuracy rate $P$ indicates how many pictures that are predicted to be cracks (reference output) are true cracks, and its estimation method is given in Eq. (14):

$$P = \frac{\mathrm{TP}}{\mathrm{TP} + \mathrm{FP}} \tag{14}$$

The false alarm rate FA indicates how many true background pictures (reference ground truth) in the sample are misclassified as crack, and it is calculated as in Eq. (15):

$$\mathrm{FA} = \frac{\mathrm{FP}}{\mathrm{TP} + \mathrm{FP}} \tag{15}$$

### 3.1.2 Other Detection Methods

In order to evaluate the CNN-based crack detection as proposed, we also introduce a method for comparison, which is crack detection method based on the CNN model developed by L. Zhang in the first section.

In the article, the author used the CNN network structure shown in Fig. 15 to train and detect the unprocessed original image. The system comprises of four convolutional layers and two totally associated layers.

### 3.1.3 Comparison of Test Results

We evaluate the model proposed in this paper according to the evaluation indicators mentioned in Sect. 3.1.1. The test results of proposed method are compared and the evaluation results of the original Resnet50 and L. Zhang proposed structure is given in Table 1.

**Table 1** Comparison of different models

| Methods | *R* (%) | MA (%) | *P* (%) | FA (%) |
|---|---|---|---|---|
| Resnet50 | 99.03 | 96.70 | 93.66 | 6.34 |
| CNN (Zhang 2016 ICIP) | 99.61 | 94.31 | 86.96 | – |
| Proposed method | 99.55 | 94.64 | 96.69 | 3.31 |

It can be seen from Table 1 that the proposed crack detection model is superior to the original Resnet50 and the one proposed by L. Zhang in all aspects. Compared with L. Zhang's CNN-based crack detection model, the advantages of the model structure of this paper are mainly reflected in:

(1) When designing the CNN network, considering that the first layer of CNN is usually used to extract image edge information, we use a $3 \times 3$ convolution kernel because the odd-numbered convolution kernel can extract edge features. L. Zhang designed the first layer of CNN as a $4 \times 4$ convolution kernel, which violated the principle of extracting edge features of the first layer of CNN.
(2) The number of convolution kernels of each layer of the Resnet50 used in this paper is increasing, which are 64, 128, 256, and 512, respectively. This incremental relationship ensures that each layer can obtain complete and effective feature expression. Compared with the model structure of 48 convolution kernels in each layer of L. Zhang, this paper has more advantages in extracting features.
(3) We add ASPP module to Resnet50, which is parallel with diverse inspecting rate at given information, which is identical to catch the setting of pictures in numerous scales, in this way getting multiscale picture highlights data, improving the recognition accuracy of the network. In parallel, the use of atrous convolution can provide large receptive fields with a comparable amount of computation, thereby enabling the extraction of denser feature maps.
(4) We add a depth wise separable convolution to the model, which greatly reduces the computational complexity and model complexity.

According to Table 1, the crack detection model of this paper has a great advantage in accuracy rate, which is 9.93% higher than that of L. Zhang. This is mainly because the proposed ConvNets can extract crack features more effectively and distinguish cracks from non-cracks. At the same time, compared to the original Resnet50, proposed method has a lower false alarm rate. This shows that the situation of mistakenly identifying the background as a crack occurs even less in our method.

## 4 Conclusions

This paper proposed a bridge crack detection algorithm based on profound learning CNN for accurately detecting the cracks in solid bridges. This paper is designed using Gabor filter to remove noises, enhanced adaptive threshold-based segmentation to identify the edges in the crack image, and a feature extraction module based on atrous spatial pyramid pooling and depthwise distinct convolution module. Our classification model can extract image feature information in a better way and improves bridge crack identification accuracy with a maximum value of 96.69%.

## References

1. Prasanna P, Dana KJ, Gucunski N, Basily BB, La HM, Lim RS, Parvardeh H (2016) Automated crack detection on concrete bridges. IEEE Trans Autom Sci Eng 13(2):591–599
2. Zhang L, Zhou G, Han Y, Lin H, Wu Y (2018) Application of internet of things technology and convolutional neural network model in bridge crack detection. IEEE Acc 6:39442–39451
3. Lim RS, La HM, Shan Z, Sheng W (2017) Developing a crack inspection robot for bridge maintenance. IEEE J Comput Eng 6288–6293
4. Yamaguchi T, Mizutani T, Tarumi M, Su D (2019) Sensitive damage detection of reinforced concrete bridge slab by time-variant deconvolution of SHF-band radar signal. IEEE Trans Geosci Remote Sens 57(3):1478–1488
5. Yamaguchi T, Mizutani T, Nakamura N, Tarumi M, Ando Y, Hara I (2017) Bridge slab damage detection by signal processing of UHF-band ground penetrating radar data. IEEE J Comput Eng 12(3):415–421
6. Simi A, Benedetto A, Manacorda G, Tosti F (2012) Novel perspectives in bridges inspection using GPR. IEEE Trans Geosci Remote Sens 27(3):239–251
7. Chowdhury M, Khan SM, Atamturktur S, Rahman M (2016) Integration of structural health monitoring and intelligent transportation systems for bridge condition assessment: current status and future direction. IEEE Int Image Proc 17:2107–2122
8. Kilic G, Alani AM, Aboutalebi M (2013) Applications of ground penetrating radar (GPR) in bridge deck monitoring and assessment. IEEE Trans Pattern Anal Mach Intell 97:45–54
9. Garrett JH, Chen S, Cerda F, Rizzo P, Bielak J, Kovacevic J (2014) Semi-supervised multi resolution classification using adaptive graph filtering with application to indirect bridge structural health monitoring. IEEE Trans Sig Proc 62(11):2879–2893
10. Al-Qadi IL, Zhao S (2015) Development of an analytic approach utilizing the extended common midpoint method to estimate asphalt pavement thickness with 3D ground-penetrating radar. IEEE Trans Autom Sci Eng 78:29–36
11. Doebling SW, Farrar CR, Nix DA (2015) Vibration–based structural damage identification. IEEE Trans Autom Sci Eng 359(19):131–149
12. Nix DA, Zhao S, Al-Qadi IL (2017) Development of regularization methods on simulated ground-penetrating radar signals to predict thin asphalt overlay thickness. IEEE Sig Process 132:261–271
13. Wu MC, Tung PC, Hwang YR (2015) The development of a mobile manipulator imaging system for bridge crack inspection. IEEE Trans Autom Sci Eng 11:717–729
14. Mehrani E, Ayoub A, Ayoub A (2016) Evaluation of fiber optic sensors for remote health monitoring of bridge structures. IEEE Trans Autom Sci Eng 42(2):183–199
15. Sheng W, Lim RS, La HM (2014) A robotic crack inspection and mapping system for bridge deck maintenance. IEEE Trans Autom Sci Eng 11(2):367–378

# Experimental Investigation of Duplex Stainless Steel Using RSM and Multi-objective Genetic Algorithm (MOGA)

**Mahesh Gopal**

**Abstract** Machining of hard stainless steel requires more tremendous effort to ensure high quality of the material with minimum machining costs. To machine a particular material, the cutting tool generates high compressive deformation and high temperature impact on the tool, also leading to phase transformation. So, it is very essential to select cutting parameters and cutting conditions in order to get perfect output. The model is developed to study the cause of cutting parameters. Experiments are conducted to predict surface roughness and machining time on duplex 2205 ASTM A276 round bar material using carbide tip tool material of hard turning on CNC lathe. Considering the machining parameters, the second-order mathematical model is developed by CCD of RSM. The performance characteristic study is done by ANOVA method. The design software Design-Expert V12 is used to examine the effect. Cutting speed is the significant influencing parameter compared to other parameters. For the best possible solutions, multi-objective genetic algorithm (MOGA) is trained and tested. MOGA recommends the most excellent lowest predicted value. The confirmatory analysis results and predicted values are found to be in commendable agreement with experiential values.

**Keywords** Duplex stainless steel · Carbide tool · ANOVA · RSM · MOGA · Surface roughness · Machining time

## 1 Introduction

In recent decades, enormous research has been taken place to improve efficiency, productivity, quality of output product while machining. The selection of suitable processes for the manufacture of a particular part is based upon a matching of the required characteristic of the work material and various machining parameters.

M. Gopal (✉)
Department of Mechanical Engineering, College of Engineering and Technology, Wollega University, Post Box No: 395, Nekemte, Ethiopia
e-mail: doctorgmahesh@gmail.com

© The Editor(s) (if applicable) and The Author(s), under exclusive license to Springer Nature Singapore Pte Ltd. 2021

S. Mohan et al. (eds.), *Materials, Design, and Manufacturing for Sustainable Environment*, Lecture Notes in Mechanical Engineering,
https://doi.org/10.1007/978-981-15-9809-8_59

Machinability of workpiece is defined as the combination of physical factors, cutting force, speed, rate of feed, cutting depth, spindle speed, etc., to obtain superior surface finish and less tool wear rate and complete the work at specified time [1]. Nowadays, manufacturing industries adopt high-speed machining techniques to decrease production costs and to increase the quality of the machined parts. Engineering components such as valves, flanges, fittings and other pressure-containing parts made of steel need higher accuracy and better surface finish. Many numbers of studies have been conducted in the field of machinability.

The experiment is conducted to measure the surface roughness of hardened AISI H11 (X38CrMoV5-1) using CBN7020 tools. The analysis technique: RSM and ANOVA is employed to determine the parameter affect of the output responses, results rate of feed and workpiece hardness is the most influencing variable [2]. The Taguchi method is used to examine the cutting variables while turning of S45C steel workpiece by aid of tungsten carbide cutting tools. This is simple, effective and efficient methodology to optimize design [3]. While turning of AISI austenitic stainless steel, when cutting speed increases the flank wear rate and roughness of material decrease [4]. Several studies have been conducted in machining steel material: X38CrMoV5-1 steel [5], mild steel [6, 7], AISI 202 austenitic stainless steel [8], AISI 52100 steel [9, 10], AISI H13 steel [11], martensitic stainless tool steel [12], AISI 202 austenitic stainless steel [13], bearing steel (AISI 52100) [14], E0300 alloy steel [15], AISI 1040 steel [16].

## *1.1 Duplex Stainless Steel (DSS)*

Duplex stainless steel has the property of corrosion resistance and other characteristics. There are three primary classifications of stainless steel. They are austenitic steel, ferritic steel and duplex (ferritic–austenitic) steel. Many researches have been done for past 3 decades in the field of steel, but there is a limited research done in duplex (ferritic–austenitic) steel. Duplex stainless steels have an extensive use in the areas of oil and gas industries, petrochemical industries, and pulp and paper industries. Duplex stainless steels have superior strength and less ductility compared to other austenitic grades [17].

Nitrogen and alloy content plays a significant role in modern duplex stainless steel, it is very difficult to machine and tends machinability decreases rapidly [18]. The modern duplex grades [19] possess the twice corrosion resistance and higher strength compared to other austenitic grades, enabling gauge and also by weight and cost reductions. The researchers [20] experimentally investigated using duplex stainless steel as workpiece and PVD-coated inserts as cutting tool. Test was performed, using cutting speed and cooling conditions at low and fluid pressure as high, results the tool withstand for a long period without wear, better quality surface roughness. The author [21] suggests that to improve the surface finish, cutting parameters must be in medium cutting speed and lower rate of feed, and to reduce tool wear rate the rate of feed should be very low. The experimental work has been conducted by a different speed, rate of feed and stable cutting depth of cast duplex stainless steel to measure surface unevenness in dry machining operation; when the speed increases, the roughness decreases, and when rate of feed decreases the roughness also decreases [22]. The rate of feed is the influencing variable while turning duplex stainless steel. A mathematical model is developed using RSM methodology to predict roughness of the surface [23]. Two different nitrogen alloy duplex stainless steels are employed in cutting operation. The optimization is carried out to predict surface roughness using Taguchi method. The author [24] suggests that the cutting speed affects the tool and rate of feed affects the surface roughness. Experiment is conducted to measure active wear rate of TiN-coated cemented carbide tool during drilling of DSS-ASTM A8190 [25]. An optimization is carried out using multi-regression model to predict the wear rate of tool while cutting PH-hardened duplex stainless steels [26]. The author [27] compared the performance of cemented carbide tool by conducting an experiment in machining of DSS and high alloyed austenitic steel. The experiment is performed on duplex alloy (SAF 2205 and SAF 2507) to compare and to analyze the roughness value, wear rate of tool, force. The author [28] suggests that the SAF 2205 and SAF 2507 have poor machinability characteristics.

## 2 Surface Roughness and Machining Time

RSM is the best tool to investigate the outcome attained from factorial experimentation. It analyzes all engineering-related problems. RSM delivers extra information with less quantity of experiments. The experiments are conducted to develop the model. The model shows the relation between the response variable and the independent variables. To predict surface roughness theoretically the equations are

$$\text{Surface Roughness (Ra)} = cV^{k_1}F^{k_2}D^{k_3}N^{k_4} \tag{1}$$

where

| | |
|---|---|
| Ra | predicted surface roughness (μm), |
| $V$ | cutting speed (m/min), |
| $F$ | rate of feed (mm/rev), |
| $D$ | cutting depth (mm), |
| $N$ | nose radius (mm), |
| $k_1, k_2, k_3, k_4$ | model parameters, |
| $c$ | response error. |

The optimization is carried out using speed, rate of feed and cutting depth as input parameters in order to maximize rate of production and production cost minimization [29]. The equation of the machining time is [30]

$$\text{Tm} = \frac{\pi DL}{10VF} \tag{2}$$

where

| | |
|---|---|
| Tm | machining time/piece (min/pc), |
| $D$ | diameter of the specimen (mm), |
| $L$ | length of the specimen (mm), |
| $V$ | cutting speed (m/min), |
| $F$ | rate of feed (mm/rev). |

## 3 Procedure for Experimentation

The cutting operation is performed on Kirloskar Turnmaster-35 Lathe. The carbide-coated insert (Tungaloy—SNMG 120 408 MT AH925) is selected for machining operation under no lubrication, considering the input parameters like cutting speed, rate of feed, cutting depth, nose radius. The workpiece of duplex 2205 ASTM A276 round bar of length is 60 mm considered for conducting tests. The specimen is machined 20 mm to measure surface roughness and machining time. The design matrix V12 software is used to conduct the experiment. The second-order mathematical quadratic model consists of 30 sets of experiments allowed to estimate linear, quadratic and interaction effect of the process parameters. Analysis variance technique is used to authenticate the satisfactoriness of the model. In order to optimize surface roughness and machining time, MATLAB software with multi-objective genetic algorithm (MOGA) is used. Surface roughness is measured using Mitutoyo SJ201 surface roughness tester and machining time using stopwatch.

### 3.1 *Identification of Design Matrix*

Experimental design is prepared by CCD of RSM, and each factor is located at one of five uniformly spaced values, frequently coded as −2, −1, 0, +1, +2 [31]. The four factors, five levels CCD consisting of 30 experimental sets of actual state are used for experimentation purpose (Table 1). The ranges of cutting conditions are selected as suggested by the Practical Guidelines Handbook [32]. The chemical composition

**Table 1** Process parameters and their levels

| Cutting parameter | Units | Levels | | | | |
|---|---|---|---|---|---|---|
| | | −2 | −1 | 0 | 1 | 2 |
| Cutting speed ($V$) | m/min | 80 | 100 | 120 | 140 | 160 |
| Rate of feed ($F$) | mm/rev | 0.2 | 0.3 | 0.4 | 0.5 | 0.6 |
| Cutting depth ($D$) | mm | 2.5 | 3.0 | 3.5 | 4.0 | 4.5 |
| Nose radius ($N$) | mm | 0.8 | 1.2 | 1.6 | 2.4 | 2.8 |

**Table 2** Typical chemical composition of duplex 2205 stainless steel

| Elements | % of composition |
|---|---|
| Carbon | 0.03 max |
| Manganese | 2 |
| Silicon | 1 |
| Phosphorous | 0.03 |
| Sulfur | 0.02 |
| Chromium | 21–23 |
| Molybdenum | 2.5–3.5 |
| Nickel | 4.5–6.5 |
| Nitrogen | 0.08–0.2 |
| Copper | Nil |

of duplex 2205 stainless steel grade is shown in Table 2. The trail consists of 30 runs as shown in Table 3.

## 4 Results and Discussion

The experiments have been conducted as per design matrix and are given in Table 3. ANOVA is done and given in Table 4, for prediction of Ra. The $F$-value of 462.78 implies model is significant. In the table, 0.01% chance of an $F$-value could occur due to noise. $P < 0.0500$ indicates model terms are significant. In this case, $V$, $F$, $D$, VF, VN, DN, $V^2$, $F^2$, $N^2$ are significant model terms. Values greater than 0.1000 indicate the model terms are not significant. The lack of fit $F$-value of 0.29 implies the lack of fit is not significant relative to the pure error. There is a 95.60% chance that a lack of fit $F$-value this large could occur due to noise. Nonsignificant lack of fit is good. Table 5 shows the results of ANOVA for prediction of machining time (Tm). The model $F$-value of 2.96 implies the model is significant. There is only a 2.26% chance that an $F$-value this large could occur due to noise. $P$-values less than 0.0500 indicate model terms are significant. In this case, DN, $V^2$, $D^2$ are significant model terms. Values greater than 0.1000 indicate the model terms are not significant. The lack of fit $F$-value of 0.28 implies the lack of fit is not significant relative to the pure error. There is a 95.94% chance that a lack of fit $F$-value this large could occur due to noise. Nonsignificant lack of fit is good.

**Regression equations**

The regression equation of actual machining parameters

$$\text{Surface Roughness (Ra)} = +4.07990 - 0.029924 * V + 0.795276 * F + 0.700579 * D - 0.338271 * N$$

**Table 3** Experimentation and observation

| Trail No. | Cutting speed ($V$) (m/min) | Rate of feed ($F$) (mm/rev) | Cutting depth ($D$) (mm) | Nose radius ($N$) (mm) | Surface roughness (Ra—Exp) (μm) | Machining time (Tm—Exp) (s) | Surface roughness (Ra—Pred) (μm) | Machining time (Tm—Pred) (s) |
|---|---|---|---|---|---|---|---|---|
| 1. | 120 | 0.4 | 3.5 | 1.6 | 2.47 | 1.54 | 2.49 | 1.42 |
| 2. | 100 | 0.5 | 3.0 | 1.2 | 3.01 | 1.46 | 2.99 | 1.49 |
| 3. | 140 | 0.5 | 3.0 | 1.2 | 2.01 | 1.56 | 2.00 | 1.50 |
| 4. | 100 | 0.3 | 4.0 | 2.4 | 2.86 | 1.61 | 2.87 | 1.59 |
| 5. | 120 | 0.4 | 4.5 | 1.6 | 2.74 | 1.52 | 2.73 | 1.57 |
| 6. | 100 | 0.3 | 3.0 | 2.4 | 2.91 | 1.48 | 2.91 | 1.50 |
| 7. | 140 | 0.5 | 4.0 | 1.2 | 2.29 | 1.53 | 2.30 | 1.51 |
| 8. | 140 | 0.3 | 4.0 | 2.4 | 2.43 | 1.60 | 2.44 | 1.59 |
| 9. | 120 | 0.4 | 3.5 | 1.6 | 2.53 | 1.51 | 2.49 | 1.42 |
| 10. | 120 | 0.4 | 3.5 | 1.6 | 2.52 | 1.47 | 2.49 | 1.42 |
| 11. | 140 | 0.5 | 4.0 | 2.4 | 2.21 | 1.54 | 2.21 | 1.50 |
| 12. | 120 | 0.4 | 3.5 | 1.6 | 2.45 | 1.36 | 2.49 | 1.42 |
| 13. | 80 | 0.4 | 3.5 | 1.6 | 3.39 | 1.58 | 3.40 | 1.59 |
| 14. | 120 | 0.4 | 3.5 | 1.6 | 2.52 | 1.37 | 2.49 | 1.42 |
| 15. | 100 | 0.5 | 4.0 | 1.2 | 3.34 | 1.51 | 3.35 | 1.47 |
| 16. | 120 | 0.4 | 3.5 | 1.6 | 2.47 | 1.28 | 2.49 | 1.42 |
| 17. | 140 | 0.3 | 3.0 | 1.2 | 2.17 | 1.61 | 2.17 | 1.60 |
| 18. | 120 | 0.6 | 3.5 | 1.6 | 2.69 | 1.35 | 2.68 | 1.39 |
| 19. | 140 | 0.3 | 4.0 | 1.2 | 2.51 | 1.52 | 2.50 | 1.50 |
| 20. | 100 | 0.3 | 3.0 | 1.2 | 2.86 | 1.53 | 2.85 | 1.51 |
| 21. | 140 | 0.5 | 3.0 | 2.4 | 2.31 | 1.25 | 2.32 | 1.28 |
| 22. | 100 | 0.5 | 4.0 | 2.4 | 2.97 | 1.57 | 2.97 | 1.58 |
| 23. | 120 | 0.4 | 2.5 | 1.6 | 2.31 | 1.53 | 2.32 | 1.55 |
| 24. | 100 | 0.5 | 3.0 | 2.4 | 3.01 | 1.44 | 3.02 | 1.38 |
| 25. | 120 | 0.2 | 3.5 | 1.6 | 2.74 | 1.45 | 2.75 | 1.47 |
| 26. | 100 | 0.3 | 4.0 | 1.2 | 3.23 | 1.41 | 3.22 | 1.39 |
| 27. | 160 | 0.4 | 3.5 | 1.6 | 1.88 | 1.59 | 1.88 | 1.64 |
| 28. | 120 | 0.4 | 3.5 | 2.8 | 2.72 | 1.32 | 2.70 | 1.37 |
| 29. | 140 | 0.3 | 3.0 | 2.4 | 2.54 | 1.51 | 2.53 | 1.48 |
| 30. | 120 | 0.4 | 3.5 | 0.8 | 2.70 | 1.34 | 2.72 | 1.38 |

$$
\begin{aligned}
& - 0.040625 * V * F - 0.001250 * V * D \\
& + 0.006205 * V * N - 0.100000 * F * D \\
& - 0.138393 * F * N - 0.341071 * D * N \\
& + 0.000090 * V^2 + 5.59742 * F^2
\end{aligned}
$$

**Table 4** ANOVA—prediction of Ra

| Source identified | Sum of square value | df | Mean square value | *F*-value | *P*-value | |
|---|---|---|---|---|---|---|
| **Model** | 3.90 | 14 | 0.2784 | 462.78 | <0.0001 | Significant |
| *V*-cutting speed | 2.99 | 1 | 2.99 | 4964.79 | <0.0001 | |
| *F*-feed | 0.0095 | 1 | 0.0095 | 15.82 | 0.0012 | |
| *D*-cutting depth | 0.1057 | 1 | 0.1057 | 175.77 | <0.0001 | |
| *N*-nose radius | 0.0010 | 1 | 0.0010 | 1.60 | 0.2257 | |
| VF | 0.1056 | 1 | 0.1056 | 175.59 | <0.0001 | |
| VD | 0.0025 | 1 | 0.0025 | 4.16 | 0.0595 | |
| VN | 0.0920 | 1 | 0.0920 | 152.95 | <0.0001 | |
| FD | 0.0004 | 1 | 0.0004 | 0.6650 | 0.4276 | |
| FN | 0.0011 | 1 | 0.0011 | 1.90 | 0.1881 | |
| DN | 0.1737 | 1 | 0.1737 | 288.79 | <0.0001 | |
| $V^2$ | 0.0359 | 1 | 0.0359 | 59.65 | <0.0001 | |
| $F^2$ | 0.0869 | 1 | 0.0869 | 144.42 | <0.0001 | |
| $D^2$ | 0.0020 | 1 | 0.0020 | 3.31 | 0.0889 | |
| $N^2$ | 0.0935 | 1 | 0.0935 | 155.43 | <0.0001 | |
| **Residual** | 0.0090 | 15 | 0.0006 | | | |
| Lack of fit | 0.0033 | 10 | 0.0003 | 0.2869 | 0.9560 | Not significant |
| Pure error | 0.0057 | 5 | 0.0011 | | | |
| **Cor total** | 3.91 | 29 | | | | |

$$+ 0.033897 * D^2 + 0.231020 * N^2 \tag{3}$$

The machining time (Tm) of the factors is given below

$$\begin{aligned} \text{Machining Time (Tm)} = &+ 5.58776 - 0.023294 * V - 0.510281 * F \\ &- 1.53034 * D - 0.023342 * N \\ &- 0.010000 * V * F + 0.000500 * V * D \\ &- 0.002467 * V * N + 0.525000 * F * D \\ &- 0.404018 * F * N + 0.179911 * D * N \\ &+ 0.000125 * V^2 + 0.393431 * F^2 \\ &+ 0.140737 * D^2 - 0.043329 * N^2 \end{aligned} \tag{4}$$

**Table 5** ANOVA—prediction of Tm

| Source identified | Sum of square value | df | Mean square value | *F*-value | *P*-value | |
|---|---|---|---|---|---|---|
| **Model** | 0.2183 | 14 | 0.0156 | 2.96 | 0.0226 | Significant |
| *V*-cutting speed | 0.0002 | 1 | 0.0002 | 0.0290 | 0.8670 | |
| *F*-feed | 0.0192 | 1 | 0.0192 | 3.65 | 0.0755 | |
| *D*-cutting depth | 0.0140 | 1 | 0.0140 | 2.65 | 0.1242 | |
| *N*-nose radius | 0.0004 | 1 | 0.0004 | 0.0779 | 0.7840 | |
| VF | 0.0064 | 1 | 0.0064 | 1.22 | 0.2875 | |
| VD | 0.0004 | 1 | 0.0004 | 0.0760 | 0.7865 | |
| VN | 0.0145 | 1 | 0.0145 | 2.76 | 0.1173 | |
| FD | 0.0110 | 1 | 0.0110 | 2.10 | 0.1683 | |
| FN | 0.0098 | 1 | 0.0098 | 1.85 | 0.1935 | |
| DN | 0.0483 | 1 | 0.0483 | 9.19 | 0.0084 | |
| $V^2$ | 0.0698 | 1 | 0.0698 | 13.27 | 0.0024 | |
| $F^2$ | 0.0004 | 1 | 0.0004 | 0.0816 | 0.7791 | |
| $D^2$ | 0.0343 | 1 | 0.0343 | 6.52 | 0.0220 | |
| $N^2$ | 0.0033 | 1 | 0.0033 | 0.6250 | 0.4415 | |
| **Residual** | 0.0789 | 15 | 0.0053 | | | |
| Lack of fit | 0.0282 | 10 | 0.0028 | 0.2787 | 0.9594 | Not significant |
| Pure error | 0.0507 | 5 | 0.0101 | | | |
| **Cor total** | 0.2972 | 29 | | | | |

## *4.1 Interaction Effect of Variables*

### 4.1.1 Direct Effect—Surface Roughness

In this study, the direct result of cutting speed, rate of feed, cutting depth and nose radius over surface roughness are examined. Figure 1 illustrates the small changes in cutting speed producing large changes in surface roughness. Figure 2 illustrates that the rate of feed between 0.3 mm/rev and 0.5 mm/rev results in better surface roughness, so rate of feed can be controlled to achieve the better results. In Fig. 3 at low cutting depth, the surface roughness is very low. But in Fig. 4, it is evident that the nose radius between 1.2 and 2.4 mm results in better surface roughness.

### 4.1.2 Direct Effect—Machining Time

In this direct effect diagram, Fig. 5 illustrates the machining time is very low between 100 and 140 m/min of cutting speed. So, cutting speed is an influencing factor and foremost role in machining time. In Fig. 6, there is an increase in the rate of feed the

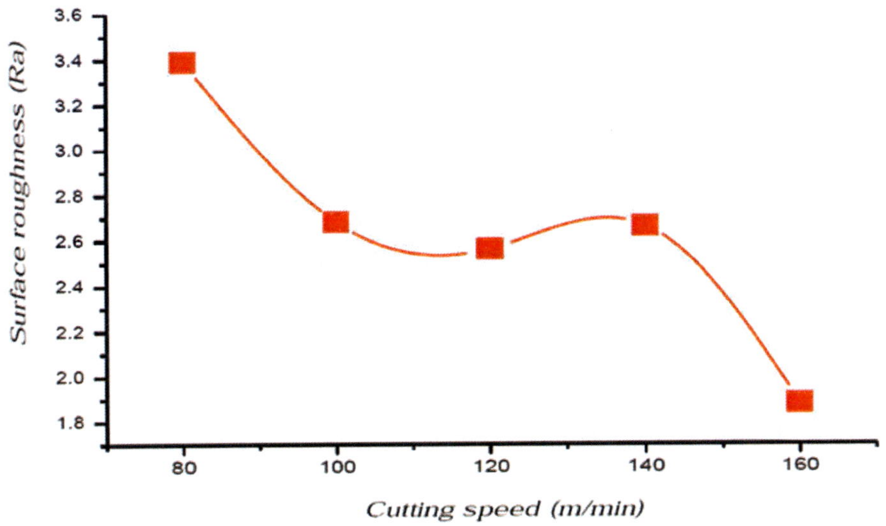


**Fig. 1** Direct effect of cutting speed

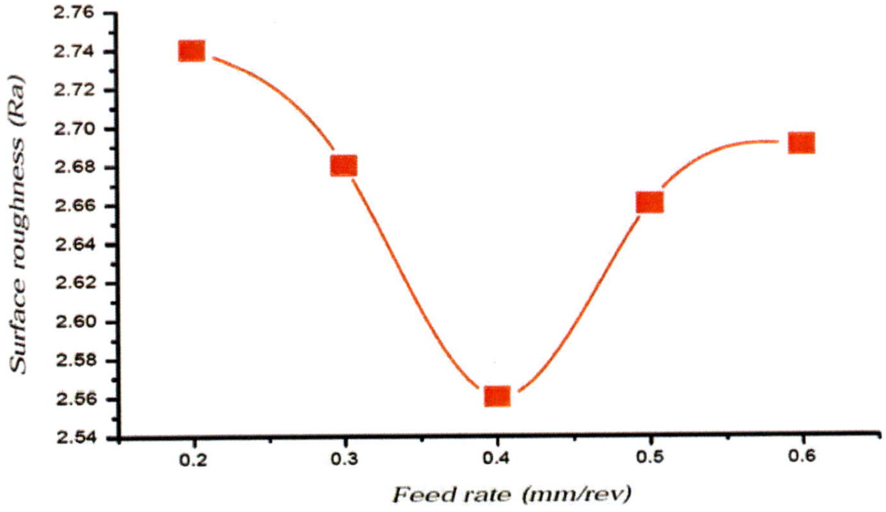


**Fig. 2** Direct effect of rate of feed

machining time is reduced. In Fig. 7, the machining time is very low between 3.00 and 4.00 mm. In Fig. 8, the nose radius should be in between 1.2 and 2.4 mm to get low machining time.

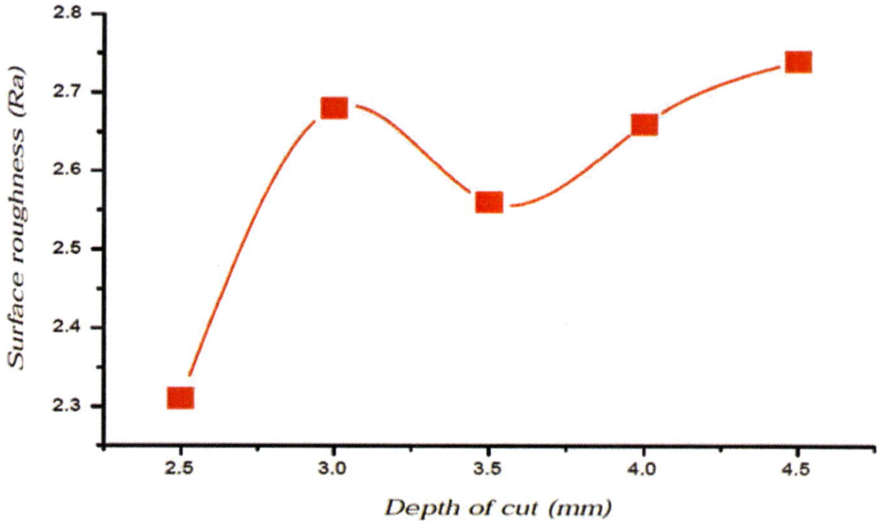


**Fig. 3** Direct effect of cutting depth

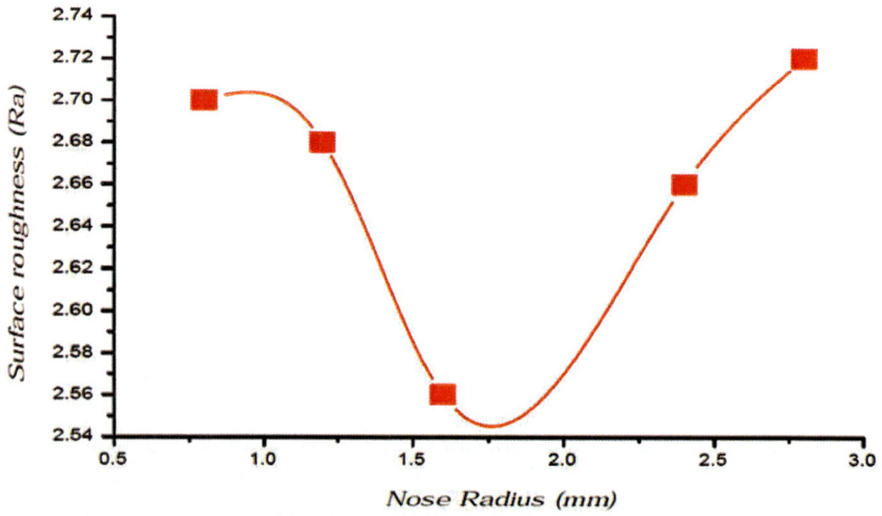


**Fig. 4** Direct effect of nose radius

## *4.2 Interaction Effect on Surface Roughness*

Figures 9, 10 and 11 depict the interaction effect of surface roughness (Ra). Figure 9 confirms that the cutting speed and the rate of feed have a considerable cause on the surface roughness. In the diagram if the cutting speed increases, roughness value decreases. At the same time, the lower rate of feed, the Ra, is very low, and it is

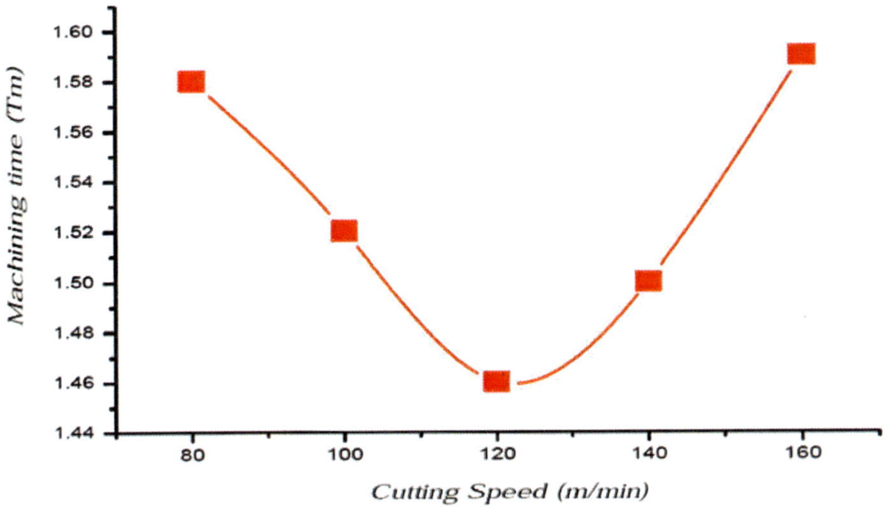


**Fig. 5** Direct effect of cutting speed

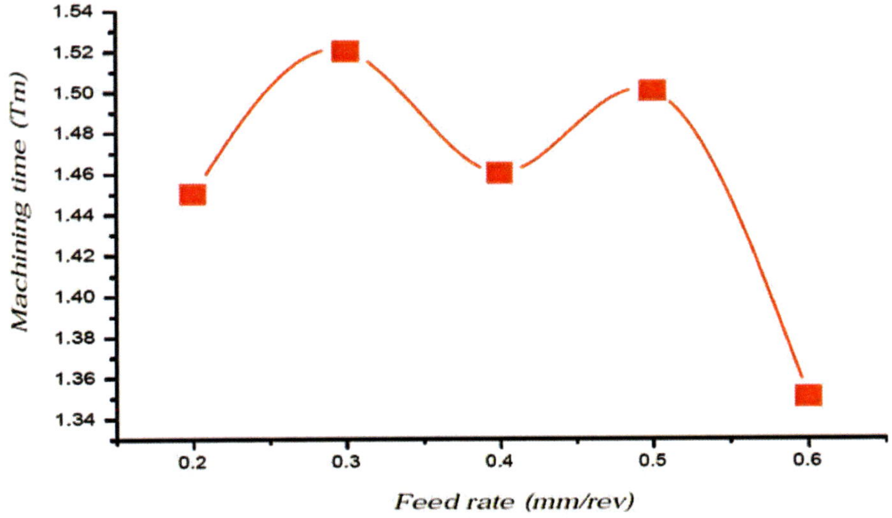


**Fig. 6** Direct effect of rate of feed

evidenced at lower cutting speed and higher rate of feed the surface roughness value increased. The results are confirmed from the analysis of variance in Table 4. From Fig. 10, surface roughness increases when there is an increase in cutting depth. So, high cutting speed, low rate of feed and low cutting depth have a considerable effect, and careful consideration must be taken while selecting the speed. From Fig. 11, the Ra is low between 1.0 and 2.5 mm of nose radius. From the result, it is evidenced for

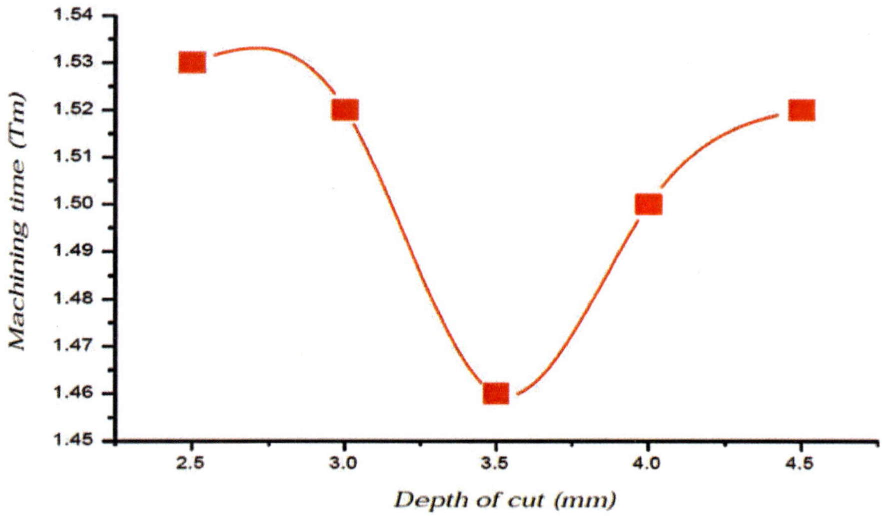


**Fig. 7** Direct effect of cutting depth

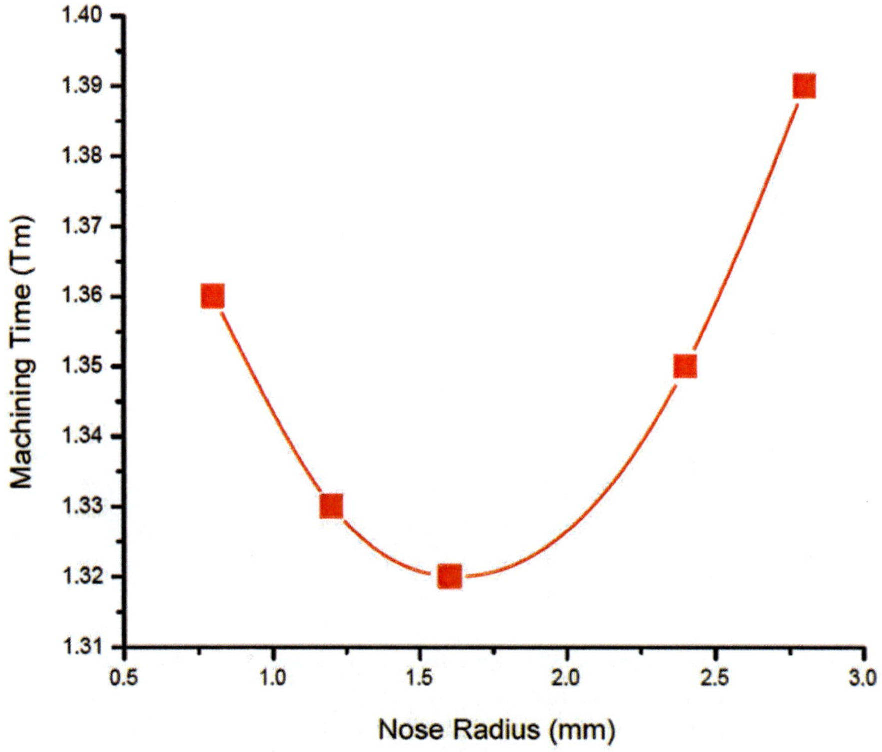


**Fig. 8** Direct effect of nose radius and machining time

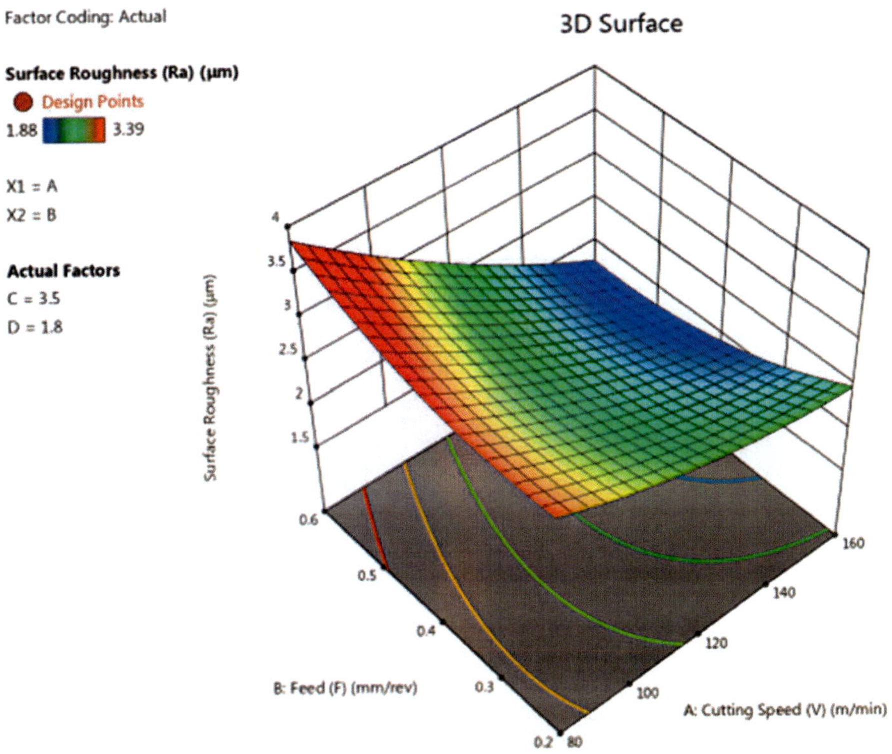


**Fig. 9** Interaction plot of cutting speed and rate of feed over surface roughness

turning operation of duplex steel in direct effect diagram also; the nose radius angle should be in between 1.00 and 2.5 mm for better surface roughness.

## 4.3 Interaction Effect for Machining Time

Figures 12, 13 and 14 depict the interaction effect of process variable of duplex stainless steel on machining time (Tm). Figure 12 illustrates the cutting speed and rate of feed over machining time. To achieve low Tm, the cutting speed should be in between 100 and 140 m/min. When the rate of feed increases subsequently machining time decreases. In Fig. 13, cutting speed and depth occur over machining time, and the machining time is low between 2.5 and 4.0 mm of cutting depth. It is evidenced that in Figs. 12 and 13, the cutting speed must be in between 100 and 140 m/min. From Fig. 14, the nose radius should be in between 1.2 and 2.4 mm to achieve minimum machining time.

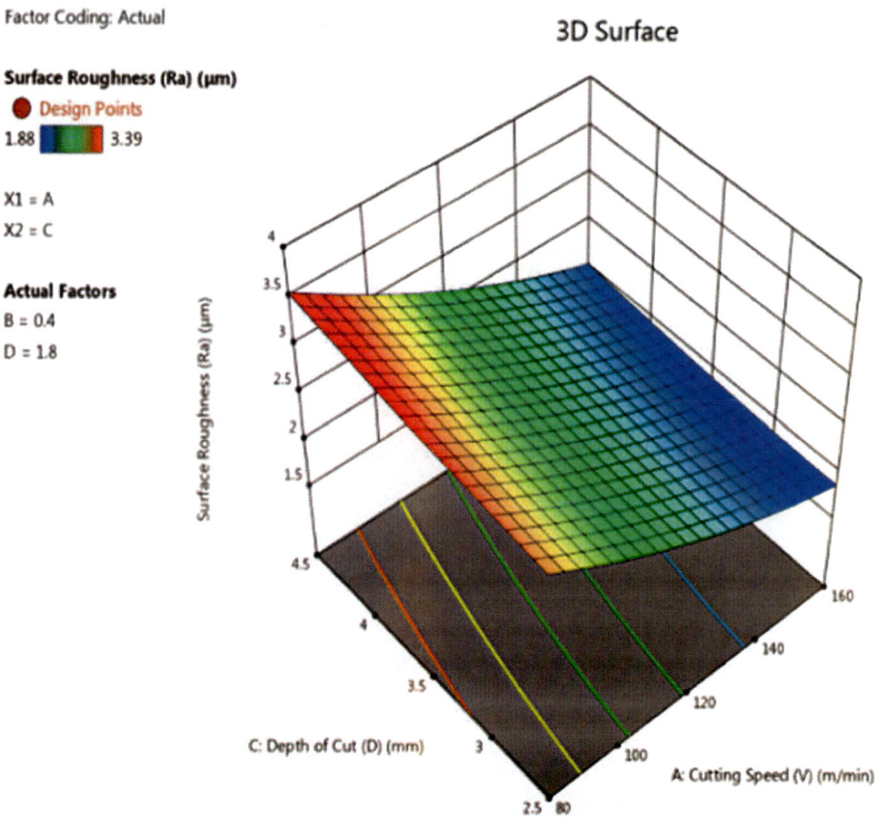


**Fig. 10** Interaction plot of cutting speed and cutting depth over surface roughness

## 5 Multi-objective Genetic Algorithm (MOGA)

In this investigation, multi-objective genetic algorithm is employed with controlled limits. The developed regression model is employed in genetic algorithm to predict the optimal correlation between the parameters and responses.

The objective function of the input using MOGA

$$\text{To minimize: Surface Roughness: Ra}(V, F, D, N) \tag{5}$$

$$\text{To minimize: Machining Time: Tm}(V, F, D, N) \tag{6}$$

Subject to constraints:

Ra $\leq$ Ra limit, Tm $\leq$ Tm limit

where Ra limit and Tm limit indicate the lower and upper limits that necessitate to be satisfied for responses.

Within parameter ranges:

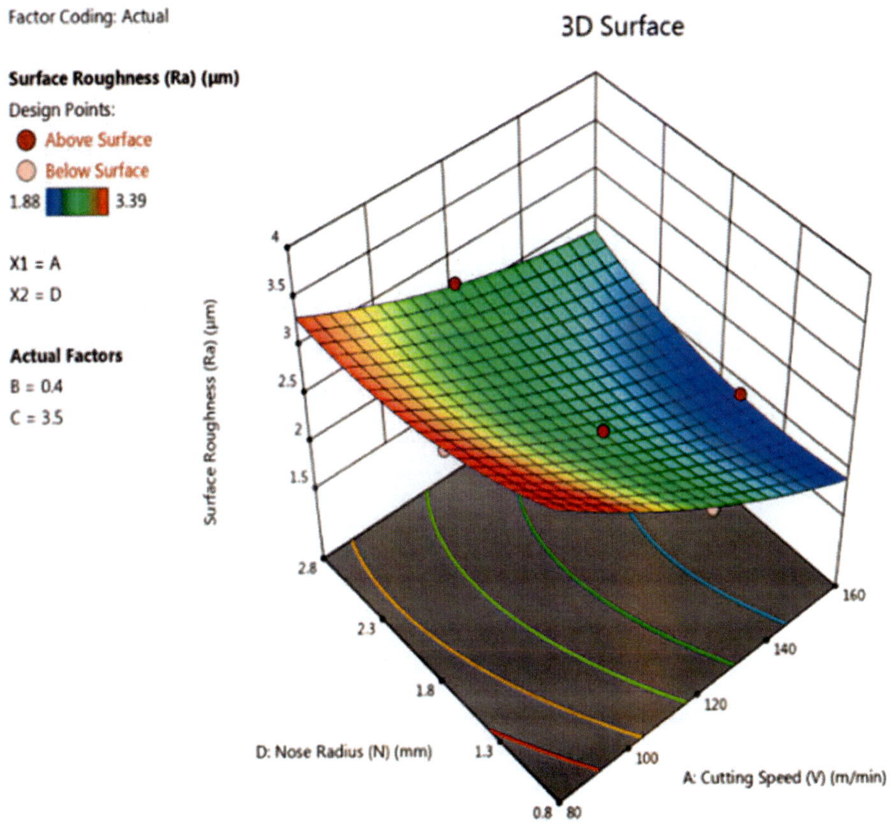


**Fig. 11** Surface interaction plot of cutting speed and nose radius over surface roughness

$$80\,\text{m/min} \leq V \leq 160\,\text{m/min} \tag{7}$$

$$0.2\,\text{m/rev} \leq F \leq 0.6\,\text{m/rev} \tag{8}$$

$$2.5\,\text{mm} \leq D \leq 4.5\,\text{mm} \tag{9}$$

$$0.8\,\text{mm} < N < 2.8\,\text{mm} \tag{10}$$

To obtain the best possible result, population size, selection function, crossover rate and mutation rate are shown in Table 6.

Figure 15 shows the genetic algorithm fitness value. Figure 16 illustrates the Pareto frontier distributed points of MOGA for their responses. From the investigation using MOGA, it is observed that the input parameter grouping of twenty-one sets of non-dominated Pareto optimal solutions is given in Table 7. In Table 7, it has been observed that all the solutions created from MOGA are equally excellent agreement. All the

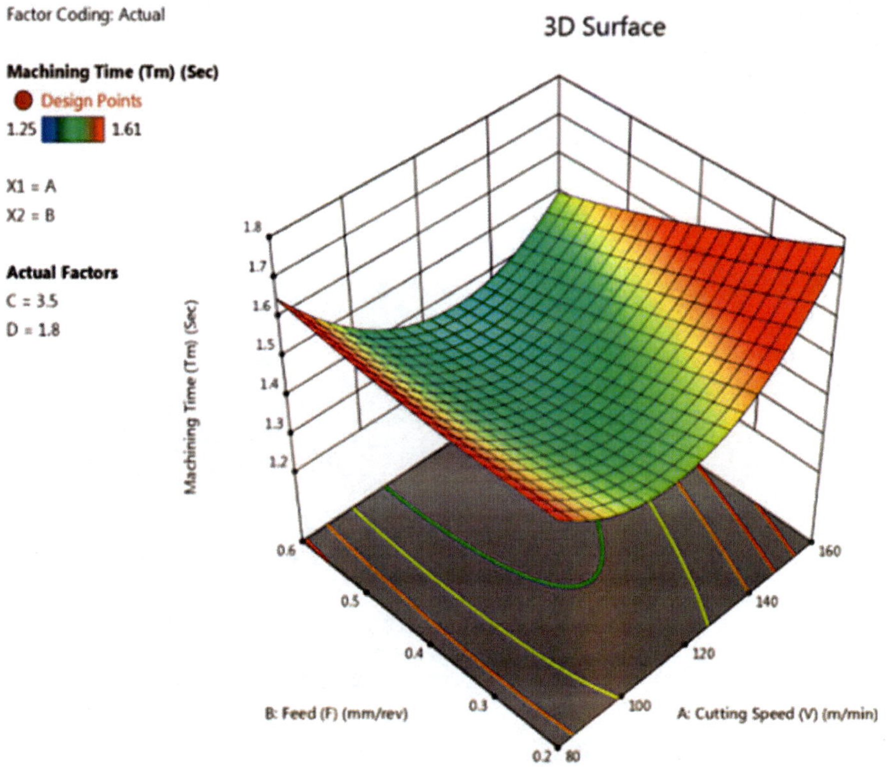


**Fig. 12** Surface interaction plot of cutting speed and rate of feed over machining time

achieved results of turning operation of duplex stainless steel are further validated by physical dimensions. The percentage of error is found to be within ±2%, and error is considerable.

## 6 Conclusion

The experiment is performed on DSS using carbide insert tool in CNC lathe. CCD of RSM is used to predict surface roughness (Ra) and machining time (Tm). The effects of the cutting parameter are analyzed. The multi-objective genetic algorithm (MOGA) is employed to optimize the machining variables and their responses. In this research, the subsequent conclusion has been attained.

- Cutting speed is a foremost important variable to minimize the surface roughness (Ra) and machining time (Tm); if the cutting speed increases the surface roughness (Ra), machining time (Tm) reduces. Surface roughness and machining time are very low when the rate of feed is in between 0.3 and 0.5 mm/rev. The nose radius

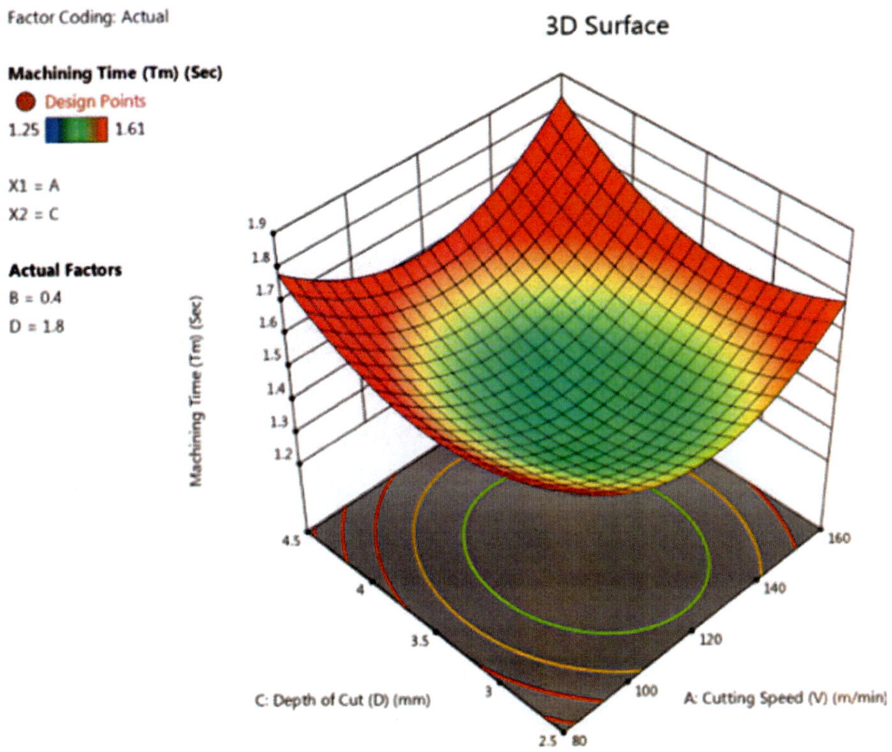


**Fig. 13** Surface interaction plot of cutting speed and cutting depth over machining time

should be in between 1 and 2.5 mm to achieve better surface roughness and minimum machining time.

- Based on multi-objective optimization using MOGA, the 21 non-dominated Pareto optimal solutions are acquired, the confirmatory test is conducted and the percentage of error is in good agreement with observed values.

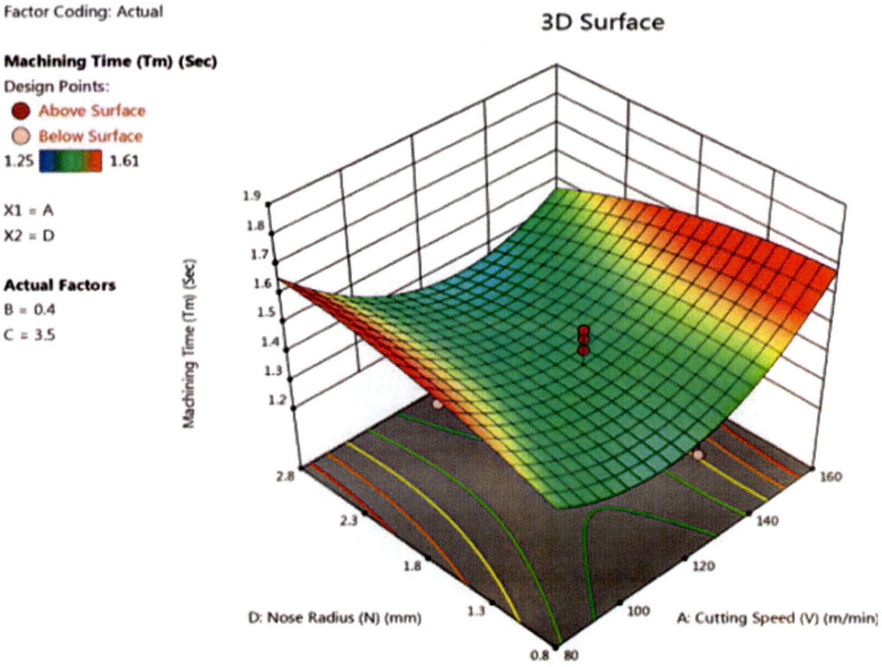


**Fig. 14** Surface interaction plot of cutting speed and nose radius over machining time

**Table 6** Genetic algorithm parameter

| Parameter | Setting value |
|---|---|
| Size of population | 100 |
| Scaling function rank | Function, stochastic uniform |
| Mutation function | Gaussian |
| Mutation rate | 0.1 |
| Crossover function | Scattered |
| Crossover rate | 1.0 |
| Generations | 1000 |

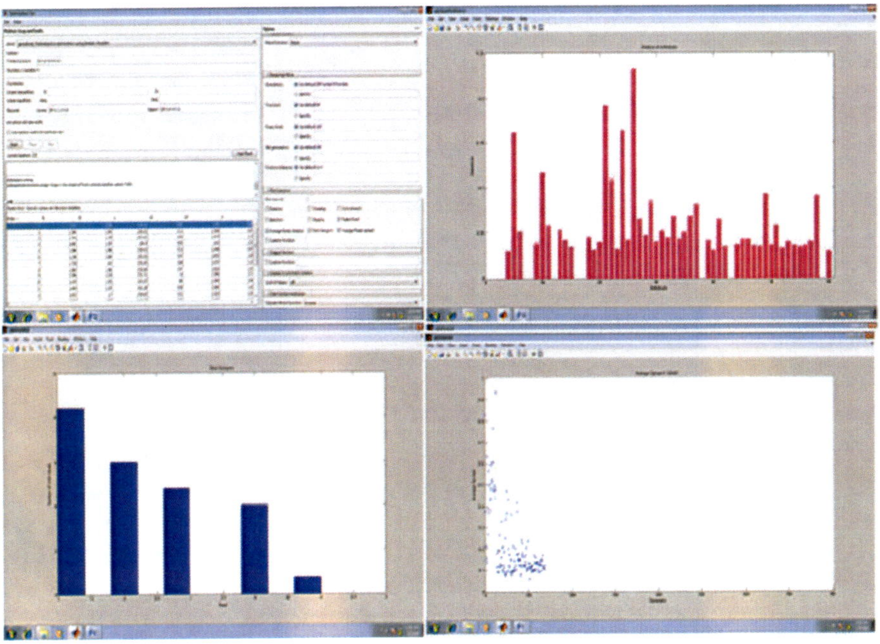

Fig. 15 Performance of fitness value generation and the best individual performances of variables in coded form

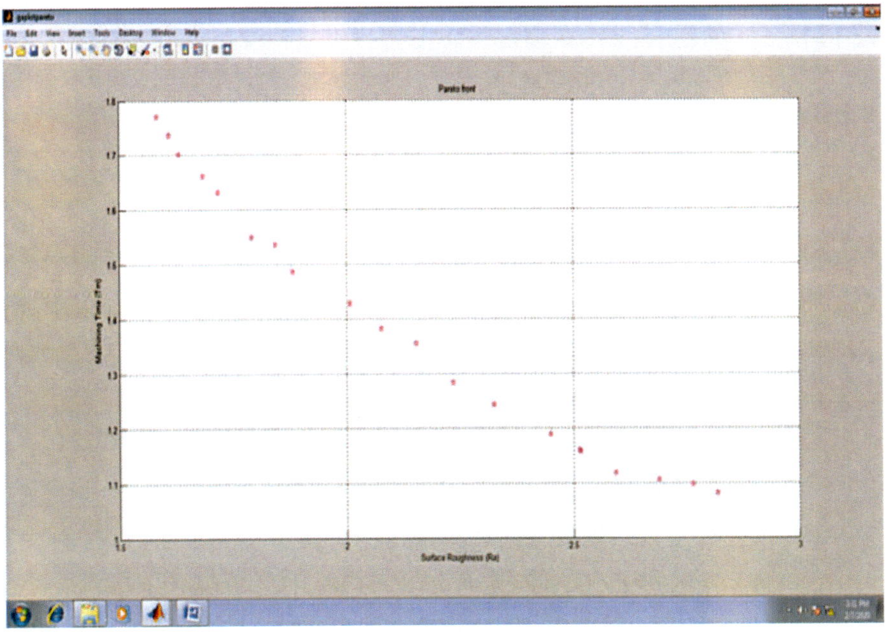


Fig. 16 Pareto optimal frontier chart

**Table 7** MOGA—pareto optimal solutions

| Trail No. | Cutting speed (m/min) | Rate of feed (mm/rev) | Cutting depth (mm) | Nose radius (mm) | Ra—predicted (MOGA) (μm) | Tm—predicted (MOGA) (s) |
|---|---|---|---|---|---|---|
| 1. | 153.14 | 0.60 | 2.55 | 2.49 | 2.51 | 1.16 |
| 2. | 155.02 | 0.51 | 2.51 | 0.88 | 1.58 | 1.77 |
| 3. | 154.64 | 0.58 | 2.58 | 1.44 | 1.79 | 1.55 |
| 4. | 144.97 | 0.59 | 2.62 | 2.57 | 2.69 | 1.11 |
| 5. | 155.02 | 0.51 | 2.51 | 0.88 | 1.58 | 1.77 |
| 6. | 152.52 | 0.60 | 2.65 | 2.14 | 2.24 | 1.28 |
| 7. | 153.48 | 0.55 | 2.54 | 1.23 | 1.72 | 1.63 |
| 8. | 153.64 | 0.60 | 2.61 | 1.55 | 1.88 | 1.49 |
| 9. | 152.72 | 0.60 | 2.59 | 2.57 | 2.59 | 1.12 |
| 10. | 152.07 | 0.56 | 2.58 | 2.05 | 2.15 | 1.36 |
| 11. | 152.29 | 0.56 | 2.58 | 1.48 | 1.85 | 1.53 |
| 12. | 154.64 | 0.55 | 2.52 | 1.01 | 1.63 | 1.70 |
| 13. | 140.09 | 0.60 | 2.65 | 2.57 | 2.77 | 1.10 |
| 14. | 150.85 | 0.58 | 2.60 | 1.73 | 2.01 | 1.43 |
| 15. | 154.39 | 0.53 | 2.52 | 0.92 | 1.61 | 1.74 |
| 16. | 154.80 | 0.53 | 2.52 | 1.24 | 1.68 | 1.66 |
| 17. | 153.12 | 0.60 | 2.56 | 2.49 | 2.52 | 1.16 |
| 18. | 152.83 | 0.60 | 2.61 | 2.26 | 2.33 | 1.24 |
| 19. | 152.51 | 0.58 | 2.60 | 1.92 | 2.08 | 1.38 |
| 20. | 138.89 | 0.60 | 2.65 | 2.61 | 2.82 | 1.08 |
| 21. | 151.84 | 0.59 | 2.60 | 2.40 | 2.45 | 1.19 |

# References


1. Montgomery DC (1990) Introduction to statistical quality control, 2nd edn. Wiley, New York
2. Benlahmidi S, Aouici H, Boutaghane F, Khellaf A, Fnides B, Yallese MA (2016) Design optimization of cutting parameters when turning hardened AISI H11 steel (50 HRC) with CBN7020 tools. Int J Adv Manufact Technol
3. Yang WH, Tarng YS (1998) Design optimization of cutting parameters for turning operations based on the taguchi method. J Mater Process Technol 84:122–129
4. Korkut I, Kasap M, Ciftci I, Seker U (2004) Determination of optimum cutting parameters during machining of AISI 304 austenitic stainless steel. Mater Des 25:303–305
5. Fnides B, Aouici H, Yallese MA (2008) Cutting forces and surface roughness in hard turning of hot work steel X38CrMoV5-1 using mixed ceramic. Mechanika 2:70
6. Kohli A, Dixit US (2005) A neural-network-based methodology for the prediction of surface roughness in a turning process. Int J Adv Manuf Technol 25:118–129
7. Pal SK, Chakraborty D (2005) Surface roughness prediction in turning using artificial neural network. Neural Comput Appl 14:319–324

8. Kaladhar M, Subbaiah KV, Rao CN, Rao KN (2011) Application of taguchi approach and utility concept in solving the multi-objective problem when turning AISI 202 austenitic stainless steel. J Eng Sci Technol Rev 4:55–61
9. Thiele JD, Melkote SN (1999) Effect of cutting edge geometry and workpiece hardness on surface generation in the finish hard turning of AISI 52100 steel. J Mater Process Technol 94:216–226
10. Thiele JD, Melkote SN, Peascoe RA, Watkins TR (2000) Effect of cutting-edge geometry and workpiece hardness on surface residual stresses in finish hard turning of AISI 52100 steel. Trans ASME 122:642–649
11. Özel T, Hsu T-K, Zeren E (2005) Effects of cutting edge geometry, workpiece hardness, rate of feed and cutting speed on surface roughness and forces in finish turning of hardened AISI H13 steel. Int J Adv Manuf Technol 25:262–269
12. Noordin MY, Venkatesh VC, Sharif S (2007) Dry turning of tempered martensitic stainless tool steel using coated cermet and coated carbide tools. J Mater Process Technol 185:83–90
13. Kaladhar M, Subbaiah KV, Rao CS, Rao KN (2010) Optimization of process parameters in turning of AISI202 austenitic stainless steel. ARPN J Eng Appl Sci 5:9
14. Singh D, Rao PV (2007) A surface roughness prediction model for hard turning process. Int J Adv Manuf Technol 32:1115–1124
15. Manna A, Salodkar S (2008) Optimization of machining conditions for effective turning of E0300 alloy steel. J Mater Process Technol 2:147–153
16. Asiltürk I, Çunkas M (2011) Modeling and prediction of surface roughness in turning operations using artificial neural network and multiple regression method. Expert Syst Appl 38:5826–5832
17. Armas IA, Moreuil SD (2009) Duplex stainless steels. ISTE Ltd./ John Wiley & Sons Inc, UK/USA
18. Gunn R (1997) Duplex stainless steels. Abington Publishing, Cambridge, England
19. Olsson J, Snis M (1997) Duplex—a new generation of stainless steels for desalination plants. Desalination 205:104–113
20. de Oliveira Junior CA, Diniz AE, Bertazzoli R (2014) Correlating tool wear, surface roughness and corrosion resistance in the turning process of super duplex stainless steel. J Braz Soc Mech Sci Eng 36:775–785
21. Selvaraj DP, Chandramohan P, Mohanraj M, Rajesh PK (2013) Experimental investigations on surface roughness, cutting force and tool wear of duplex stainless steel in end milling using taguchi method. Inter Rev Mech Eng 7:1133–1141
22. Selvaraj DP, Chandramohan, P (2010) Influence of cutting speed, rate of feed and bulk texture on the surface finish of nitrogen alloyed duplex stainless steels during dry turning. Engineering 2:453–460
23. Krolczyk G, Legutko S, Gajek M (2013) Predicting the surface roughness in the dry machining of duplex stainless steel (dss) for different cutting conditions. Metalurgija 52:259–262
24. Philip SD, Chandramohan P, Mohanraj M (2014) Optimization of surface roughness, cutting force and tool wear of nitrogen alloyed duplex stainless steel in a dry turning process using taguchi method. Measurement 49:205–215
25. Paro J, Hanninen H, Kauppinen V (2001) Tool wear and machinability of HIPed P/M and conventional cast duplex stainless steels. Wear 249:279–284
26. Ahmadi E, Homami RM (2012) Experimental investigation and mathematical modeling of composite ceramic cutting tools with alumina base in the machining process of PH-hardened austenitic-ferritic (duplex) stainless steel. Int J Adv Des Manufact Technol 5
27. Carlborg C (1991) Machinability of duplex stainless steel. Proc duplex stainless steel 1:683–696
28. Nomani J, Pramanik A, Hilditch T, Littlefair G (2013) Machinability study of first generation duplex (2205), second generation duplex (2507) and austenite stainless steel during drilling process. Wear 20:304
29. Gopalakrishnan B, Fraiz AK (1991) Machining parameter selection for turning with constraints: an analytical approach based on geometric programming. Int J Prod Res 29(9):1897–1908
30. Okushima K, Hitomi K (1964) A study of economical machining: an analysis of the maximum profit cutting speed. Int J Prod Res 3(1):73–78

31. Box GEP, Draper NR (1987) Empirical model-building and response surfaces. John Wiley, New York
32. Practical guidelines for the fabrication of duplex stainless steel, 2nd Edn. (2009) International Molybdenum Associations(IMOA), London, UK

# CFD Analysis of Combined Thermal Radiation and Conjugate Heat Transfer in a 3D FFS

**G. S. Krishnendu and J. S. Jayakumar**

**Abstract** Mathematical investigation of conjugate heat transfer in a 3-d rectangular duct having a vertical forward-facing step with a participating medium is carried out using OpenFOAM. The stepped interface splits the rectangular domain into two regions (viz., solid and fluid). The fluid considered is carbon-dioxide, which is a participating medium. The ability of the code has been verified by validating conjugate heat transfer and radiative heat transfer. The fluid flows at a Reynolds number 250 through a rectangular domain with a contraction ratio of 0.5. The case study mainly focused on the effect of optical thickness and linear scattering albedo on thermal characteristics of the flow. The changes in average Nusselt number with optical thickness and linear scattering albedo are analyzed. Also, the sense of including radiation effects into a thermal analysis are discussed.

**Keywords** Radiation heat transfer · Participating media · Conjugate heat transfer · Forward-facing step · fvDOM · OpenFOAM

## Abbreviation

| | |
|---|---|
| BFS | Backward-Facing Step |
| DOM | Discrete Ordinate Method |
| FFS | Forward-Facing Step |
| FVM | Finite Volume Method |
| RTE | Radiative Transfer Equation |

G. S. Krishnendu · J. S. Jayakumar (✉)
Department of Mechanical Engineering, Amrita Vishwa Vidyapeetham, Amritapuri, Clappana, Kerala, India
e-mail: jsjayan@gmail.com

G. S. Krishnendu
e-mail: krishnendujyothi@gmail.com

© The Editor(s) (if applicable) and The Author(s), under exclusive license to Springer Nature Singapore Pte Ltd. 2021

S. Mohan et al. (eds.), *Materials, Design, and Manufacturing for Sustainable Environment*, Lecture Notes in Mechanical Engineering,
https://doi.org/10.1007/978-981-15-9809-8_60

## Nomenclature

| | |
|---|---|
| CR | Contraction ratio (h/H) |
| $H$ | Height of the inlet section (m) |
| $h$ | Height of the outlet section (m) |
| $I$ | Radiation intensity ($Wsr^{-1}$) |
| $L$ | Characteristics length (m) |
| $n$ | Refractive index |
| Nu | Nusselt number |
| Pr | Prandtl number |
| $q$ | Heat Flux ($Wm^{-2}$) |
| Re | Reynolds number |
| $S$ | Step Height (m) |
| $T$ | Temperature (K) |
| $\theta$ | Dimensionless temperature |

## Subscript

| | |
|---|---|
| b | Bulk temperature |
| c | Convection |
| f | Fluid |
| h | Hot wall |
| $i, j$ | 1, 2, 3 corresponds to $x$, $y$, $z$ directions respectively |
| in | Inlet |
| r | Radiation |
| s | Solid |
| w | Interface wall |

## Greek Symbol

| | |
|---|---|
| $\alpha$ | Thermal diffusivity ($\alpha = \kappa/\rho c_p$ $m^2s^{-1}$) |
| $\beta$ | Extinction coefficient ($m^{-1}$) |
| $\kappa$ | Thermal conductivity ($Wm^{-1}K^{-1}$) |
| $\mu$ | Dynamic viscosity (Pas) |
| $\omega$ | Linear scattering albedo |
| $\Phi$ | Scattering phase function |
| $\rho$ | Density (kg $m^{-3}$) |
| $\sigma$ | Stefan–Boltzmann-Constant $5.67 \times 10^{-8}$ $Wm^{-2}K^{-4}$ |
| $\sigma_a$ | Absorption coefficient ($m^{-1}$) |
| $\sigma_s$ | Scattering coefficient ($m^{-1}$) |

$\tau$ Optical thickness

## 1 Introduction

Mostly analysis is done by considering only the surface to surface radiation between the solid walls by assuming the fluid is fully transparent media. Thermal radiation is the "heat transfer caused by electromagnetic waves" [1]. By varying the molecular energy level, everything around us continuously emits and absorbs electromagnetic waves which imply that the fluid medium can be transparent or semi-transparent. The combustion process of IC engines, gas turbines, nuclear explosions, etc. are the main applications of thermal radiation. In all these cases, the gas phase is most probably a participating medium. So it is worthwhile to analyze in an absorbing/emitting medium, namely the Participating medium. In a participating media, absorption, emission, and scattering happen at each location within the medium. So it is somewhat difficult to model thermal radiation in such a media. The radiative transfer equation (RTE), an equation having a history of more than a century describes the thermal radiation mathematically [2].

The main challenging thing regarding RTE is the selection of a suitable method of solution. Many methods are available to find the solution to thermal radiation problems in participating medium. Among them, the discrete ordinate method (DOM) and the finite volume method (FVM) methods are most used in computational fluid dynamics. In recent years the CFD analysis of thermal radiation witnessed several discoveries. Kim and Huh [3] found a new scheme of FVM and studied the advantages of the new scheme over basic FVM and DOM schemes in a 3-d rectangular enclosure. Some researchers developed code in OpenFOAM for modeling thermal radiation. In this category, recent works were done by Wang et al. [4] and Harijishnu and Jayakumar [5], in which they developed new solvers for solving RTE in OpenFOAM. Samson et al. [6] modeled thermal radiation and convective heat transfer in an electronic system to study the heat dissipation associated with electronic devices.

Talukdar and Simonson [7] investigated the coupled convection-radiation heat transfer through parallel-plate channels with heated and cooled walls. Young Kim and Baek [8] studied all the heat transfer modes in a 2-d expanding geometry. Kuznetsov et al. [9] investigated the interaction of conduction, natural convection, and radiation in a cavity with a heater. Horikiri et al. [10] modeled the conjugate heat transfer flow in a 3-d and 2-d model with ventilation and without ventilation, and fluid and thermal changes studied. Favre et al. [11] numerically simulated combined conjugate convection and surface thermal radiation. Bahrami and Nassab [12] done a study on the forward-facing step (FFS) to find the entropy generation in the fluid flow. Abu-Mulaweh [13] researched laminar flow over the backward-facing step (BFS) and FFS. The author has given a summary of the effect of Re, Pr, etc. on thermal and fluid characteristics. Cintolesi et al. [14] conducted a 2-dimensional analysis of conjugate heat transfer and surface radiative heat transfer. They found a new thermodynamic solver named *coupledRadiationFoam.*

The sudden contraction and expansion in the geometry always have a challenging part in the design of engineering equipment. Because of such geometric transformations, flow separation and subsequent reattachment occur. Such spatial variation in the velocity field causes turbulence production which in turn increases the size of the separation bubble by interacting with the mean flow. In the case of BFS geometry, there occurs only one recirculation region after the step. But in the FFS geometry, there are more than one recirculation region adjacent to the step. i.e., one region on the upstream near to the bottom corner of the step, and another one downstream near to the step edge. This separation and following reattachment make the simple geometry of FFS into a complicated one. Due to these complications related to separation and reattachment, the FFS geometry had only got less attention compared to that of the BFS. Most of the practical applications of these geometries have multi regions because sometimes fluids flow around a solid or flows between the sidewalls. That is heat transfer in both regions is combined. Also, radiation can occur in both. But, from the literature study, it is evident that most of the available literature related to FFS is limited to combined forced convective and radiative heat transfer studies. Nevertheless, there is no previous works are available regarding the study of combined thermal radiation and conjugate heat transfer in a 3-d FFS. Hence the main objective of the present study is to find out how the combined thermal radiation and conjugate heat transfer influences the overall heat transfer in a 3-d FFS with a participating medium.

## 2 Mathematical Modelling

### *2.1 Physical Domain and Boundary Conditions*

The physical domain considered for the current simulation is given (Fig. 1). The geometry is provided with a vertical FFS of height $S$, with a contraction ratio (CR) of 0.5. Total length, height, and width of the duct are $6H$, $1.5H$, and $0.6H$. To get a fully developed flow at the outlet the downstream length of the step is considered to be $5H$. The fluid enters the domain with a velocity $u$ in which can be calculated from Re, and temperature $T_{\text{in}}$. All the walls except the bottom wall of the solid, are insulated. The bottom wall of the solid is heated at a constant temperature of ($T_{\text{h}}$) which is greater than the $T_{\text{in}}$. Inlet and outlet section have a radiation boundary condition of $\in = 1$, and emissivity of the adiabatic walls is fixed to 0.8.

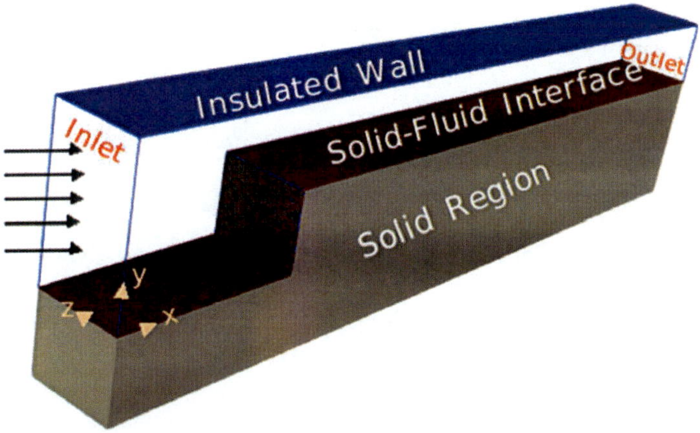


**Fig. 1** Geometry for the problem

## 2.2 *Governing Equations*

The major assumptions that had been taken into consideration for the mathematical modeling are: (1) Laminar flow, (2) fluid is a perfect gas, (3) the thermophysical properties of the solid are constant, and (4) the emissivity of the medium does not vary with wavelength. With these assumptions, the basic equations of the fluid flow and heat transfer are modified.

$$\frac{\partial \rho}{\partial t} + \frac{\partial \rho u_j}{\partial x_j} = 0 \quad (1)$$

$$\frac{\partial(\rho u_i)}{\partial t} + \frac{\partial}{\partial x_j}\left(\rho u_{rj} u_i\right) + \rho \epsilon_{ijk}\omega_i u_j = -\frac{\partial p_{rgh}}{\partial x_i} - \frac{\partial \rho g_j x_j}{\partial x_i} + \frac{\partial}{\partial x_j}\left(\tau_{ij} + \tau_{tij}\right) \quad (2)$$

where $u$ represent the velocity, $u_r$ the relative velocity, $g$ the gravitational acceleration, and $\tau_{ij}$ and $\tau_{tij}$ are the viscous and turbulent stresses.

$$\frac{\partial(\rho h)}{\partial t} + \frac{\partial}{\partial x_j}\left(\rho u_j h\right) + \frac{\partial(\rho k)}{\partial t} + \frac{\partial}{\partial x_j}\left(\rho u_j k\right) = -\frac{\partial(q_i + q_{ti})}{\partial x_i} + \rho r + \text{Rad} + \frac{\partial p}{\partial t} - \rho g_j u_j + \frac{\partial}{\partial x_j}\left(\tau_{ij} u_i\right) \quad (3)$$

where $h$ is specific enthalpy, $k$ is kinetic energy, $q_i + q_{ti}$ is heat transferred to fluid element by diffusion and turbulence, $r$ is the heat source term, and Rad is the heat source by radiation.

$$\frac{\partial(\rho h)}{\partial t} = \frac{\partial}{\partial x_j}\left[\alpha \frac{\partial h}{\partial x_j}\right] \quad (4)$$

The thermal radiation is modeled using RTE. The RTE for a participating media is written as [1]:

$$\frac{\mathrm{d}I(\vec{r},\vec{s})}{\mathrm{d}s} = -\sigma_\mathrm{a} I(\vec{r},\vec{s}) - \sigma_\mathrm{s} I(\vec{r},\vec{s}) + an^2\frac{\sigma T^4}{\pi} + \frac{\sigma_\mathrm{s}}{4\pi}\int_0^{4\pi}\left(I(\vec{r},\vec{s})\Phi(\vec{s},\vec{s})\mathrm{d}\Omega'\right) \quad (5)$$

To model thermal radiation, Finite Volume Discrete Ordinate Method (FVDOM) has been chosen for the present analysis. It is an FVM implementation of the DOM. In DOM, the angular domain of 4Ω is discretizing into a finite number of discrete directions and integrate over these directions. By implementing FVM on DOM, the physical domain is subdivided into several control volumes, i.e. the RTE is again discretized by FVM. This peculiarity of this model gives a more accurate solution. Hence in FVDOM, the RTE is integrating over control volume d$V$ and control angle dΩ. And the unique feature of this model is, the accuracy can be increased by using a finer discretization.

## 2.3 Solution Procedure

The complex governing equations are replaced by a system of algebraic equations, which results in matrices with millions of coefficients. The OpenFOAM-solver *chtMultiRegionFoam*, which is made for conjugate heat transfer study, is the base solver for the present analysis. To incorporate *fvDOM* radiation model added *IDefault* and *G* files in the 0 folders, made changes in the *radiationProperties* located in the *constant* folder, added boundary conditions for *G* and *IDefault* in *changeDictionaryDict* file located in the *system* folder, and finally modified the *fvSolutions* and *fvSchemes* to get the accurate results. The *Gauss linear Upwind grad*(*Ii_h*) divSchemes is used for *div*(*Ji, Ii_h*). The Geometric agglomerated algebraic multigrid solver, *GAMG* is used for discretizing RTE in each direction. Each octant of the angular space $4\pi$ is discretized in a finite number of solid angles, $N_\theta$ and $N_\Phi$. Due to this discretization, the total number of solid angles become $4 * N_\theta * N_\Phi$.

## 2.4 Validation of the Code

To check the ability of the code to model both phenomena, both conjugate heat transfer and radiative heat transfer are validated separately. To validate the code for conjugate heat transfer, laminar flow over the FFS was carried out and the results are compared with the results of Jayakumar et al. [15]. They considered a laminar flow over an FFS with conjugate heat transfer in a rectangular domain and studied the effect of Re on temperature distribution in the solid material. The simulation was done for a fixed Pr of 5.6, Re of 250, and $k_\mathrm{s}/k_\mathrm{f}$ ratio of 10. Temperature profile

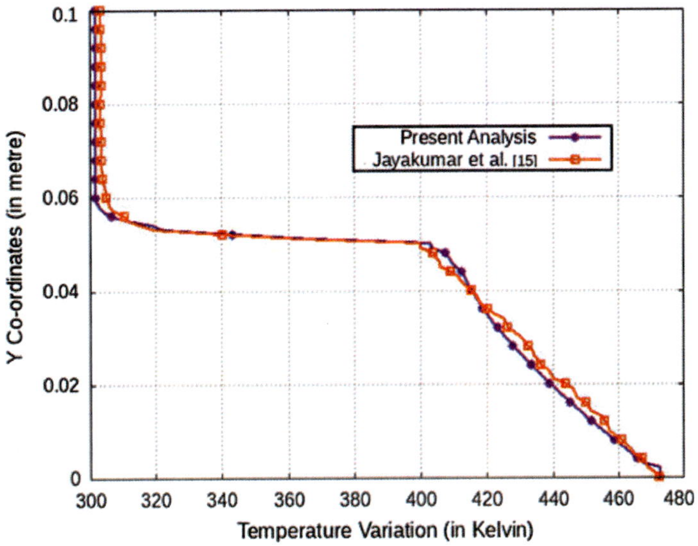


**Fig. 2** Temperature profile at $x = 1.6$ (Re = 250)

variation (Fig. 2) at a distance of 1.6 m from the leading edge, obtained closely matches (less than 3% error) with the results of Jayakumar et al. [15]. Hence the ability of the code has been tested and confirmed that this code is good enough to model the conjugate heat transfer phenomenon.

Radiative heat transfer was validated by reproducing the work of Rayeni and Nassab [16] in which combined radiation and convection heat transfer in a 3-d FFS was considered. And compared the variation in Nu on the upstream wall of the step with Rayeni and Nassab [16] for different $\tau$ value. The simulation was done for a Re of 250, $N$ of 100, and $\omega$ of 0.5, and plotted the convective Nu distribution over the upstream of the step (Fig. 3). The obtained results match with the results of Rayeni and Nassab [16]. Hence the ability of the code to model radiative heat transfer has been tested and confirmed.

## 2.5 *Grid Independency Study*

Accuracy of the present study is confirmed by conducting a grid independency study, to find a grid-independent solution. A rectangular domain with CR = 0.5, laminar flow of carbon dioxide with Re= 250, optical thickness $\tau = 0.2$, and scattering albedo $\omega = 0$ is considered for the grid independency study. To study the grid independency, plotted, and compared (Fig. 4) the $q$ value on the upstream of the step from (0.02, 0.084, 0.05) to (0.2, 0.084, 0.05). In Fig. 4, wall heat flux values at $5.04 \times 10^5$, $4.4 \times$

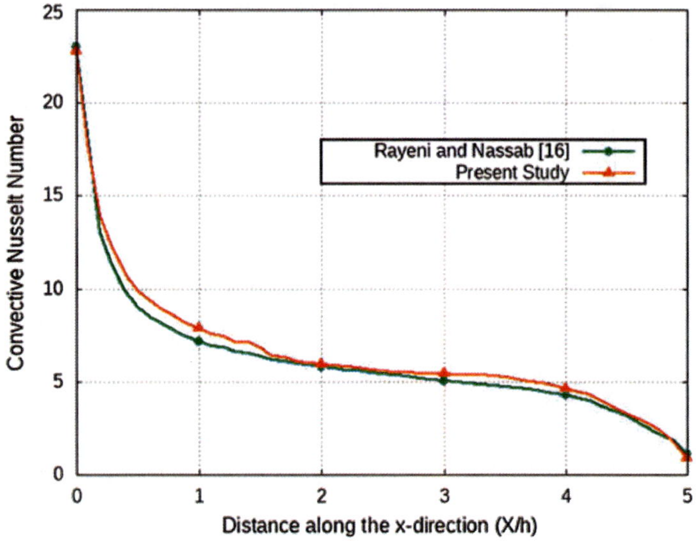


**Fig. 3** Convective Nu distribution on the bottom wall for $\tau = 0.05$

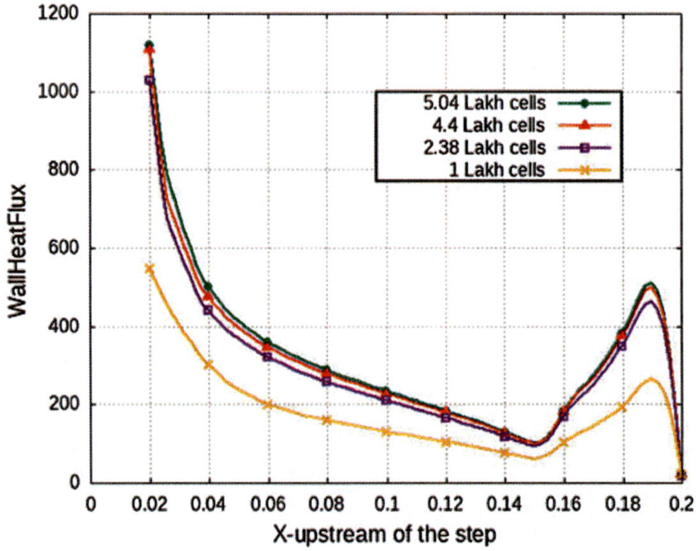


**Fig. 4** Grid independency study

$10^5$, and $2.38 \times 10^5$ are matching. But at some points values are different for $2.38 \times 10^5$. So for the analysis, $4.4 \times 10^5$ cells are considering for further analysis. Because a further increase in the number of cells will consume more time to produce the same result.

## 3 Results and Discussion

Numerical investigation of conjugate heat transfer in a 3-d rectangular duct having vertical FFS with a participating medium was done in OpenFOAM. The effect of radiative properties on convective Nu at the solid-fluid interface of the domain is considered for the analysis. Since the fluid flow can carry energy to a large distance, the usually used method to achieve a high heat transfer rate is forced convection. But by efficiently combining the heat transfer in fluids and solids we can increase the heat transfer rate more. Hence in the present analysis, carbon dioxide is used to flow through a rectangular domain with a constant CR at Re= 250 and wall emissivity $\epsilon = 0.8$ to find out how the combined thermal radiation and conjugate heat transfer influences the overall heat transfer in a 3-d FFS with a participating medium. Here we are mainly focused on the effect of $\tau$ and $\omega$ on thermal characteristics of the flow, and the effect of these parameters is analyzed by discussing the change in average Nu.

### 3.1 *Importance of Radiative Heat Transfer*

Most of the thermal analysis is carried out by taking only the conduction and convection heat transfer. But it will affect drastically the overall heat transfer, which will lead to inaccurate results. Hence it necessary to study the effect of radiation in thermal analysis. A considerable change in total heat transfer can be observable in the downstream region of those figures. The 3-d view of FFS (Fig. 5) is a good graphical representation of the effect of radiation on total Nu ($Nu_c + Nu_r$) distribution. In all the figures, unlike the other two regions, the region of the step doesn't have much change in total heat transfer with radiation. But the contraction of the geometry caused a much more increased heat transfer in the downstream region.

Since the geometry is 3-d, the 3-d effect of the vortices causes a radical change in the $z$-direction (Fig. 5), which leads to the needfulness of averaging the value of Nu in that direction. Hence the distribution of average convective Nu for different regions (upstream, step, and downstream) are plotted (Figs. 6 and 7) without considering the radiation and without considering the radiation. Compared to the downstream region, the vortices formed at the step corner of the upstream are less effective in heat transfer. That's why there is a significant variation in Nu value downstream. In Fig. 6, due to the different data ranges of upstream, step, and downstream regions, the radiation effect not visible. So to interpret the variation in various regions, the exact variation at each region is plotted separately in Fig. 7 ($X$ is only a notation to represent the co-ordinate points, it is not the $x$-axis). The importance of including radiation effects into a thermal analysis is well observed in Fig. 7. Averaged the local $Nu_c$ value in $z$-direction for each region to plot average $Nu_c$. Except for regions of step, the variation in parameters is plotted by taking the co-ordinate points in the $x$-direction.

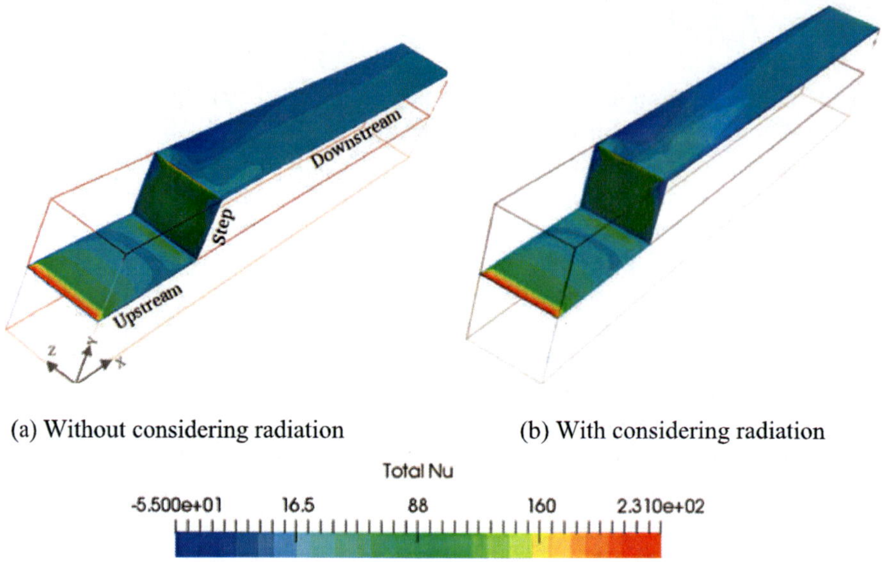


**Fig. 5** 3D view of total Nu distribution over the fluid-solid interface

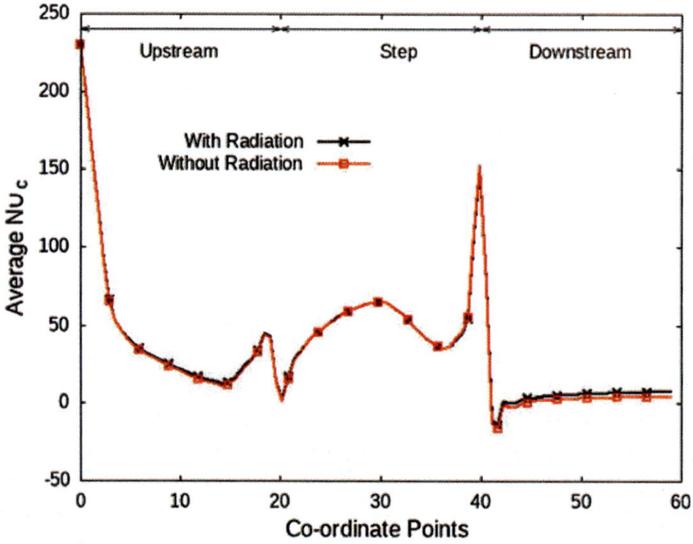


**Fig. 6** Distribution of average $Nu_c$ in different regions of the FFS with and without radiation

The thermal energy is transferred to the fluid layer close to the surface by random molecular motion (due to no-slip condition). The maximum value of $Nu_c$ (Fig. 6) is encountered in the inlet section where the interface temperature is at the lowest with a minimum change in temperature. Since the surface and fluid temperature are differ at the fluid–surface interface, a region of the fluid is developed through which the

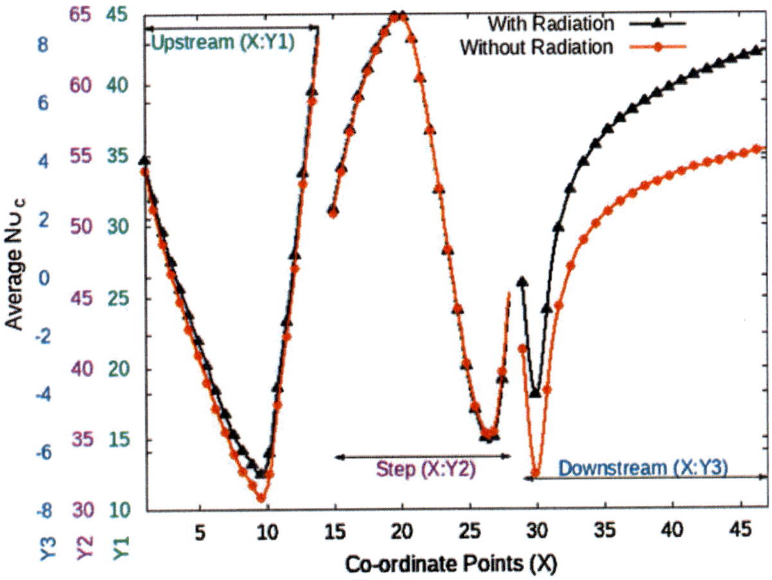


**Fig. 7** Zoomed view of variation of average $Nu_c$ in different regions of the FFS with and without radiation

temperature varies, this region is the thermal boundary-layer. Inside this layer, the change in temperature is the same at each point. Also with the flow progression in the $x$-direction, boundary layer thickness increases, and temperature gradient ($\nabla T$) goes on decreasing. This will decrease the wall heat flux and corresponding Nu value (Fig. 6) in the region where the flow is developing.

In the upstream, the least value of Nu is observed in the bottom corner of the step where the flow hits the step and slows down to zero velocity (Fig. 8a). This causes a rotating flow near the step, which leads to a sudden increase and decrease in Nu (Fig. 6) in the recirculation region. The variation of Average $Nu_c$ along the $y$-direction is plotted in the step region (Figs. 6 and 7). Along the step region, a substantial decrease in temperature is due to relatively cooler fluid in the downstream region which leads to a higher Nu value at the step edge. It's the area where more active convection occurs. After the recirculation region (Fig. 8), the Nu settles to a constant value, which means the flow becomes developed in that region. Large Nu values (Fig. 6) at the downstream region show that more active convection occurring in this region. After the peak point, Nu decreases and settles to a constant value. The constant value region (Fig. 6) corresponds to a thermally fully developed region, i.e. the dimensionless form of temperature $\theta(x, y)$ becomes independent of $x$ (does not vary along the length in flow direction), and produces constant flux.

The pattern in the distribution of Nu doesn't change with any parameters, because it fully depends on the geometry of the FFS. As observed, a higher increase in heat transfer with radiation in the downstream region is due to an increase of $q$. The

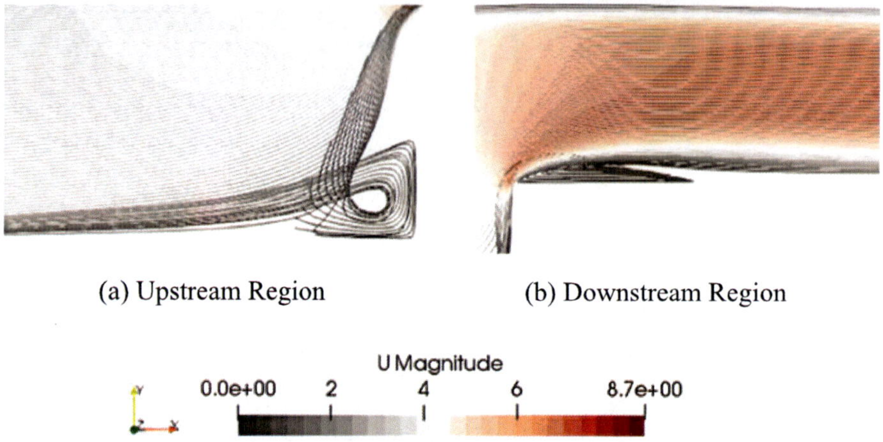


**Fig. 8** Recirculation region in upstream and downstream region of the step

addition of radiation into the analysis increased the overall heat transfer. This study envokes the requirement of including radiative heat transfer in thermal analysis.

## 3.2 *Effect of* $\tau$

$\tau$ is an important parameter that governs the thermal radiation in a participating medium. $\tau$ is the product of the extinction coefficient and characteristics length. Extinction happens in a volume when the radiation coming in and coming out are different. In the case of air at room temperature, there is no extinction (because of the presence of diatomic molecules such as $O_2$ and $N_2$, and comparatively less amount of water vapor) i.e. there is no scattering or absorption occurs. But in the case of $CO_2$, there will be an extinction because it is a highly absorbing gas. So for the analysis different optical thickness ($\tau$ = 0.1, 0.4, 1.0, 5.0, 10, 25 and 50) are considered by varying $\beta$. The $\kappa$ of solid and fluid, and $\omega$ (= 0.5) are kept constant to analyze the effect of optical thickness. The variation of average convective Nu with $\tau$ for upstream, step, and downstream region are plotted in Fig. 9.

A detailed explanation of the variation of average $Nu_c$ at different regions of FFS with distance is given in Sect. 3.1. Increased the $\tau$ value by increasing the extinction coefficient. The extinction coefficient consists of two processes, absorption, and scattering. Since for studying optical thickness variation, we had taken $\omega$ as 0.5, both $\sigma_a$ and $\sigma_s$ are considered equal. i.e., an increase in $\tau$ increases $\sigma_a$ and $\sigma_s$ equally. Usually increased $\tau$ of the medium increases the absorption/emission properties of the participating medium. This increased absorption/emission increases the internal energy stored in the solid-fluid interface, which is converted into heat. i.e. increased $\tau$ keeping $\omega = 0.5$ increases the average Nu. From Fig. 9 we can observe that for $\tau \leq$ 5, the usual trend is showing. i.e. increase in $\tau$ increased the convective heat transfer

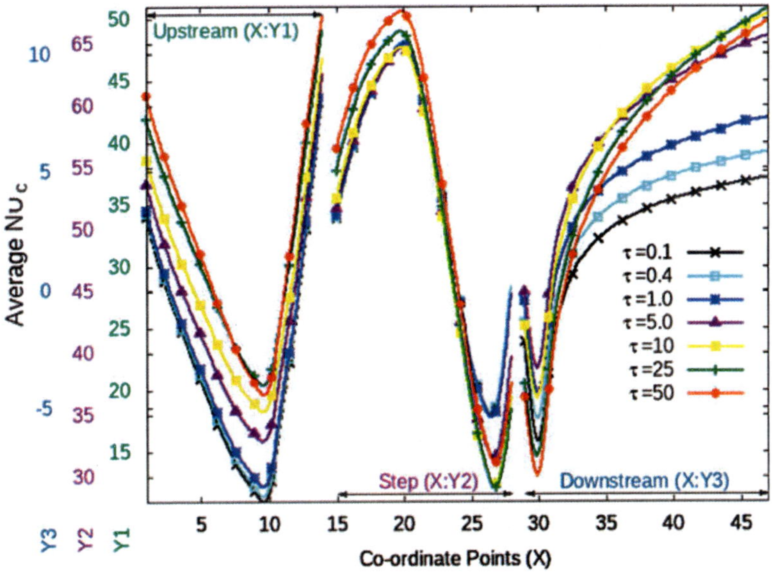


**Fig. 9** Zoomed view of variation of average $Nu_c$ with τ in different regions of the FFS

by increasing the absorption/emission. But for $\tau > 5$, there is a variation in the usual trend. The variations are clearly visible downstream (Fig. 9) due to the contracted area. Because for $\tau > 5$, higher $\tau$ increases absorption/emission as well as scattering. So the increase in scattering radiates the energy in all directions and reduces the effect of radiation. Also, the increased absorption and reemission increase the diffusion, which may lead to an increase in conductive heat transfer and a decrease in average Nu. i.e. for very higher values of $\tau$ with equal importance to absorption/emission and scattering, decreases the average Nu.

## 3.3 *Effect of* ω

Scattering is not a surface phenomenon, it is all related to the domain volume, also it need not take place equally in all direction. But isotropic scattering is considered equal in all directions. The intensity of the scattering in different directions depends on the molecules of the participating gas and wavelength. The term Linear Scattering Albedo, $\omega = \sigma_s/\beta$ is used to compare the effect of scattering and absorption/emission in a participating media. So for the analysis different $\omega$ (=0, 0.25, 0.5, 0.75 and 1), varied the value of $\sigma_s$. The $\kappa$ of solid and fluid, $\beta$, and $\tau$ (=0.2) are kept constant to analyze the effect of $\omega$. The variation of average convective Nu with $\omega$ for upstream, step, and downstream region are plotted in Fig. 10.

Linear scattering albedo is varied from non-scattering condition ($\omega = 0$) to a pure scattering condition ($\omega = 1$). In another way, we can also say that Linear scattering

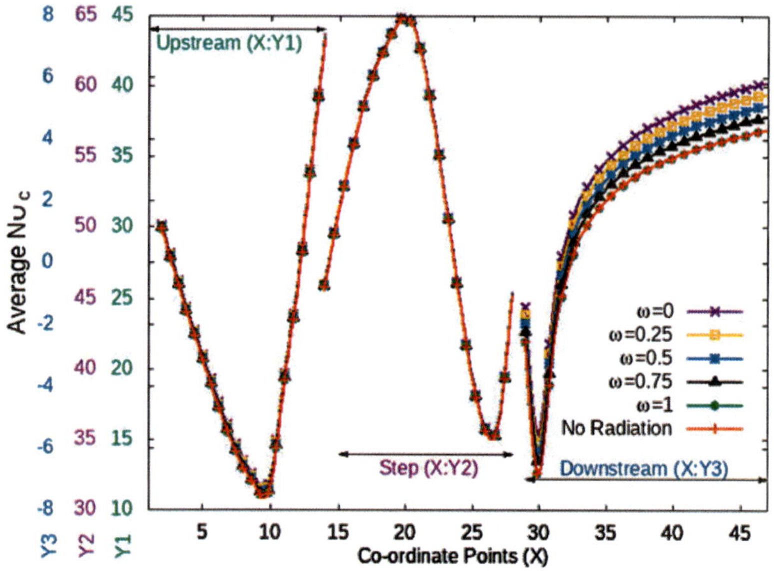


**Fig. 10** Zoomed view of variation of average $Nu_c$ with τ in different regions of the FFS

albedo varied from pure absorbing condition ($\omega = 0$) to a non-absorbing condition. The condition is just a reversed form of Sect. 3.2, but the difference is in Sect. 3.2 both $\sigma_a$ and $\sigma_s$ are varied equally. Here both are changing inversely, but the similarity is with the decreasing pattern of overall heat transfer with a decrease in absorption (Refer Figs. 9 and 10) for low values of $\tau$. Increased scattering effect, reduces the effect of radiation by scattering the radiation energy in all directions. This may lead to an increase in temperature gradient at the interface, which results in a decrease in overall heat transfer. For the analysis, $\tau$ is considered as 0.2, i.e. an optically thin medium. From Fig. 9 for optically thin medium, decrease in absorption decreased the average $Nu_c$. Here also the same is true and proved that for optically thin medium change in scattering doesn't have as much effect as absorption/emission. And in Fig. 9 we can see that variation in Nu of pure scattering medium and medium without considering radiation are merged. i.e., a pure scattering medium or a non-absorbing medium has the same behavior as a medium without considering radiation.

## 4 Conclusion

Numerical analysis of combined conjugate heat transfer and thermal radiation on a 3-dimensional rectangular domain with FFS was carried out in OpenFOAM-6. A laminar flow of $CO_2$ at a Re of 250 is considered as the fluid medium which can emit, absorb, or scatter radiation. The analysis was carried out for different $\tau$ and

$\omega$. The necessary computation time for convergence of each case is about two and a half days in 8 processors. Majors findings of the present analysis are pointed below:

- Without including radiation into the analysis decreased the overall heat transfer, which provokes the importance of radiation in thermal analysis.
- Higher values of $\tau$ with equal importance to absorption/emission and scattering decreases the average Nu.
- For $\tau \leq 5$, change in scattering doesn't have as much effect as absorption/emission.
- Pure scattering medium or non-absorbing medium results in the same heat transfer of a medium without considering radiation.

## References


1. Modest MF (2013) Radiative heat transfer. Academic press
2. Mishchenko MI (2013) 125 years of radiative transfer: enduring triumphs and persisting misconceptions. In: American institute of physics conference proceedings, vol. 1531, no. 1, pp. 11–18)
3. Kim SH, Huh KY (2000) A new angular discretization scheme of the finite volume method for 3-d radiative heat transfer in absorbing, emitting and anisotropically scattering media. Int J Heat Mass Transf 43(7):1233–1242
4. Wang CA, Shen TR, Gao JP, Tan JY (2019) Development of RTE solver for radiative transfer in absorbing-emitting medium using finite volume based CFD library OpenFOAM. Int J Therm Sci 140:36–42
5. Harijishnu R, Jayakumar JS (2017) Thermal radiation heat transfer in participating media by finite volume discretization using collimated beam incidence. IOP Conf Ser Mater Sci Eng 243(1):012009
6. Samson A, Torzewicz T, Raszkowski T, Janicki M, Zubert M, Napieralski A (2016) Modelling of average radiation and convection heat transfer coefficient value in electronic systems. In: MIXDES-23rd international conference mixed design of integrated circuits and systems, IEEE, pp 271–275
7. Talukdar P, Simonson CJ (2007) Effect of axial radiation on heat transfer in a thermally and hydrodynamically developing flow between parallel plates. Numer Heat Transfer Part A: Appl 52(10):911–934
8. Young Kim M, Baek SW (1996) Numerical analysis of conduction. Convection, and radiation in a gradually expanding channel. Numer Heat Transfer Part A: Appl 29(7):725–740
9. Kuznetsov GV, Kurilenko NI, Nee AE (2018) Mathematical modelling of conjugate heat transfer and fluid flow inside a domain with a radiant heating system. Int J Therm Sci 131:27–39
10. Horikiri K, Yao Y, Yao J (2014) Modelling conjugate flow and heat transfer in a ventilated room for indoor thermal comfort assessment. Build Environ 77:135–147
11. Favre F, Colomer G, Lehmkuhl O, Oliva A (2016) Numerical simulations of conjugate convection combined with surface thermal radiation using an immersed boundary method. J Phys Conf Ser 745(3):032017
12. Bahrami A, Nassab SG (2010) Study of entropy generation in laminar forced convection flow over a forward-facing step in a duct. Int Rev Mech Eng 4(4):399–404
13. Abu-Mulaweh HI (2003) A review of research on laminar mixed convection flow over backward-and forward-facing steps. Int J Therm Sci 42(9):897–909
14. Cintolesi C (2016) Large-eddy simulations of conjugate heat transfer with evaporation-condensation and thermal radiation, Ph.D. Thesis, School of environmental and industrial fluid mechanics, University of Trieste

15. Jayakumar JS, Sharma A, Rakshit D, Eswaran V (2007) Conjugate heat transfer study of laminar flow over a forward facing step. In: Proceedings of Asian computational fluid dynamics (CFD), pp 232–248
16. Rayeni AD, Nassab SG (2016) Analysis of combined radiation and forced convection heat transfer in 3D laminar flow over an inclined forward facing step. J Electron Cooling Ther Control 6(1):1–18

# An Extensive Review on Befitting Batteries and Drives for Electric Vehicles

**S. Baskaran, Ramkumar Venkatasamy, D. Venkatesa Prabu, T. Ramesh Kumar, and B. Meenakshi Priya**

**Abstract** Sooner or later inevitably the world is poised towards witnessing a huge technology transfer in the Automobile sector. Huge investments are happening in the Electric vehicle segment. Hence, the prime intention of this paper is to present an overview of the various potential battery and drive technologies commonly preferred for Electric vehicles since almost all-electric vehicles are classified mainly by the type of battery and drive installed in it, this paper attempts to impart the basic knowledge in that direction alongside its performance analysis, advantages and limitations.

**Keywords** Technology transfer · Electric vehicle · Batteries · Drives

## 1 Introduction

This Review paper attempts to solve ambiguities surrounding the batteries and drive motors for electric vehicles based on extensive compiling of useful information on batteries together with drives. Energy requirements of Electric vehicle drives are satisfied with various battery technologies. The choice of battery type solely depends upon the type of drive, cost, charging time, lifespan, power rating, and mileage. A huge variety of batteries are being used in contemporary Ev's viz lead-acid, Lithium polymer, Lithium-ion, Nickel-cadmium, Nickel-metal hydride and sodium Sulphur. Apart from that, there are also some of the potential future battery technologies which

S. Baskaran (✉) · R. Venkatasamy · D. Venkatesa Prabu
Department of Mechatronics Engineering, Dr. Mahalingam College of Engineering and Technology, Coimbatore, Tamil Nadu, India
e-mail: baskaranbannari@hotmail.com

T. Ramesh Kumar
Department of Mechanical Engineering, Bannari Amman Institute of Technology, Coimbatore, Tamil Nadu, India

B. Meenakshi Priya
Department of Mechatronics Engineering, Kongu Engineering College, Erode, Tamil Nadu, India

© The Editor(s) (if applicable) and The Author(s), under exclusive license to Springer Nature Singapore Pte Ltd. 2021

S. Mohan et al. (eds.), *Materials, Design, and Manufacturing for Sustainable Environment*, Lecture Notes in Mechanical Engineering,
https://doi.org/10.1007/978-981-15-9809-8_61

keep evolving rapidly into a new energy source like sand battery, Gold nanowire battery, ultracapacitors, $Al^+$ air, $Na^+$ ion, $Zn^+$ air, and $C^+$ ion.

Governments of countries have roadmap to meet the environmental targets which includes providing incentives for BEVs, but current BEVs need to evolve rapidly to be competitive than fuel-based vehicles for acceptance [1].

Diverse electric motor drives used in electric vehicles like SRM, BLDC, IM with its specifications, and its power converters are compared for appropriate selection of drives to suitable applications [2].

This paper envisions the cradle to grave characteristics of battery units in electric vehicles, it mainly focuses on negatives aspects of battery technologies in electric vehicles throughout the European Union. On the negative side, electric battery technology contributes to greenhouse emissions in such a way more amount of combustion takes place in producing the electricity especially in countries which don't have renewable energy sources, toxic components inside the batteries raise concerns about effective waste management techniques for disposal of batteries [3].

This paper suggests for selecting a suitable motor for EVs numerous parameters has to be taken into account, most of the times EVs manufactures are conservative about introducing new drive technologies due to lack of reliable data the authors recommend standardized drive cycle plots be used for comparing different drives [4].

A critical overview of electric storage unit of EVs apart from batteries such as Ultracapacitor, Flywheel energy storage, Hydrogen energy as well as Energy generation units like Fuel cell, Photovoltaic cell, Automotive thermoelectric generator, and Regenerative braking are discussed [5].

This manuscript emphasizes competent details of EV design to be specific battery energy sources, electrical machines, charging techniques, impacts, trends, and possible directions of future developments thus it gives an overall picture of the future of electric vehicles [6].

With regard to air pollution concerns, electric vehicles are a good choice for the environment but to replace the IC engine thermal problems associated with battery management of electric vehicles viz heat generation and heat dissipation due to intrinsic temperature gradients in cell module have to be solved [7].

Current stature of infrastructure for EVs viz battery charging points, servicing its merits and demerits are compared [8].

## 2 Contemporary Battery Technologies

The following subsection gives a picture of some of the commonly used batteries in EVs in the present situation.

### 2.1 Lead Acid Batteries

Lead-acid batteries are considered to be a viable option in terms of economy and power factor but some of the shortcomings are more weight, short lifespan, and susceptible to malfunctioning at low temperatures.

### 2.2 Lithium-Ion Batteries

It is a kind of battery commonly used in mobiles, laptops, and many other electronic gadgets. It possesses the advantage of high power density, high energy/mass ratio, and the ability to work under different temperature gradients. On the negative side, Li-ion batteries have a short life span and high cost.

### 2.3 Nickel-Metal Hydride (NiMH) Batteries

Electronic appliances like digital cameras, trimmers, etc. which are used for a short duration of time use NiMH batteries. NiMH batteries have the ability to store a high amount of energy by virtue of its specific power capacity.

## 3 Impending Prospects of Battery Technologies

### 3.1 Sand Battery

Currently, available lithium-ion battery uses graphite in anodes. In sand battery graphite is replaced with silicon in anodes that will result in 3 times higher performance than graphite counterpart alongside lifespan of a battery is also increased.

### 3.2 Gold Nanowire Batteries

When compared to human hair gold nanowire are thinner by 100 times. It is focused on nullifying the degradation of cells at the time of recharge. Although complete degradation of the battery cannot be avoided, the research is focused on controlling it to a significant level. So far these kinds of batteries are recharged for over 200,000 times but no degradation is noticed, notably, it uses gel-type electrode.

### 3.3 Graphene Batteries

Graphene batteries could offer a mileage limit of about 500+ miles in electric cars for a single charge. It has the ability to get charged to its full capacity within a few minutes, it recharges 5X better than Li-ion batteries.

### 3.4 Aluminium-Air Battery

Aluminium-air battery exploits oxygen content in the air to fill its cathode, within a single charge a car can drive 1100 miles as the recent test results.

### 3.5 Sodium-Ion Batteries

The main constituent of lithium iron battery is scarcely available lithium, obviously, it adds up to the cost of the battery. A great pace of R & D is underway to replace lithium with sodium since it is available in abundance. Usage of sodium ion will result in 7 times more energy-efficient batteries. Also, research on zinc-air battery is underway.

### 3.6 Liquid Flow Batteries

Molecules in PH neutral water have the ability to store energy that is the underlying principle behind flow battery. Flow batteries sustain exceptionally better than lithium-ion batteries. It may not be suitable for electronic gadgets but it is more relevant to tap the high-rated energies produced by renewable energy sources like wind and solar installations.

## 4 Assortment of Battery Specifications

### 4.1 Cell and Modules

Cell is the smallest element in a battery. A Battery consists of 1 or more cells, for instance, if a cell produces 2 V output for producing 12 V output 6 cells are connected in series. Every cell is encompassed through a vault called a module.

## 4.2 C and E Rates

Rate at which battery gets discharged in regard to its maximum capacity is given by C rate. A 1 C rate for a battery means that the current discharged will drain the whole battery in 1 h. E rate affiliates with discharge power 1 E rate is discharge power to release the total battery in 1 h.

## 4.3 Primary and Secondary Cells

Primary cells cannot be recharged, to the contrary the one that can be recharged is called a secondary cell.

## 4.4 mA Hours

mA stands for milliAmpere, very useful in measuring the amount of time a battery can supply a current. Time for which the current can be supplied by a battery in hours. $T$ is calculated using

$$T = C_p / I_k$$

$C_p$ mA hours of the battery
$I$ Current flow
$K$ Peukert's constant (1.1–1.6, build upon age and life of the battery).

## 4.5 AA and AAA

Specifies the size of alkaline batteries AAA is smaller than AA.

## 4.6 kWh

kWh stands for kilowatt-hour. Capacity of energy storage of a battery is measured in kWh. More kWh means more power it can store as well as more devices can be connected to the battery.

## 4.7 Specific Energy (Wh/Kg)

It is the stated battery energy per unit mass, it is also called gravimetric energy density.

## 4.8 Specific Power (W/Kg)

It is the maximum available power per unit mass.

## 4.9 Energy Density (Wh/L)

It is the stated battery energy per unit volume.

## 4.10 State of Charge (SOC)

Battery capacity as a percentage of maximal holding ability is expressed in SOC.

$$\mathrm{SOC}(\%) = 100\left[1 - \frac{1}{Q}\int_0^t i(t)\mathrm{d}t\right]$$

## 4.11 Depth of Discharge (DOD)

DOD is an expression of percentage of battery capacity that has been discharged as percentage of maximum capacity.

$$\mathrm{DOD}(\%) = 100\left[\frac{1}{Q}\int_0^t i(t)\mathrm{d}t\right]$$

DOD describes how deeply the battery is discharged. DOD describes how much load current $i(t)$ in certain time divided by maximum battery capacity (Q).

**Table 1** Open circuit voltage of batteries

| Battery type | Open circuit voltage (V) |
|---|---|
| Lead-acid | 2 |
| Ni–MH | 1.7 |
| Li-ion | 3.5 |

### *4.12 Terminal Voltage*

It is the voltage or potential difference between battery terminal subsequently it is coupled with load.

### *4.13 Open-Circuit Voltage*

Voltage between battery terminals without load. Under no-load condition lead-acid battery has an open circuit voltage of 2 V likewise Ni–MH has 1.7 V and Li-ion has 3.5 V as depicted in Table 1.

### *4.14 Internal Resistance*

Resistance inherent with the battery is termed as internal resistance. Efficiency of battery is inversely proportional to resistance. Moreover, it also reduces the thermic stableness of the unit as more energy during charge is converted into heat.

## 5 Prominent Battery Charging Methods

### *5.1 Constant Voltage*

As the name suggests in this method keeping the voltage as a constant current is varied continuously. DC power supply is used throughout the process. Lithium-ion and lead-acid batteries apply this type of charging with combined circuitry to prevent the battery from damage.

## 5.2 *Constant Current*

It is reverse of constant voltage constant current chargers alter the voltage applied to the battery to provide a constant current flow. This kind of circuitry is normally well suited for Ni–Cd and nickel-metal hydride batteries.

## 5.3 *IUI Charging*

It consists of three modes i.e. I-U-I more suited for fast charging standard. Battery is charged at constant current ($I$) rate initially until cell voltage attains a set point. Again the charger switches to constant voltage ($U$) phase. In the meantime current through the battery comes down gradually up to reference level. In the last phase charger again switches to constant current mode to enable the voltage rise to a higher pre-set limit.

## 5.4 *Percentage Capacity Discharged Versus Cell Voltage of Different Batteries*

Energy cells or batteries are materialized for an ample amount of end products using a variety of scientific knowledge, that results in a wide variety of performance components. The plots depict a few vital aspects an engineer must consider when specifying a battery to satisfy the needs of the end product. Figure 1 depicts the discharge curve of Li-ion cell voltage is slowly reduced with regards to capacity discharge.

Figure 2 depicts the discharge curve of Lead-acid cell, compared to Li-ion cell voltages of lead-acid batteries gets decreased at a quicker rate.

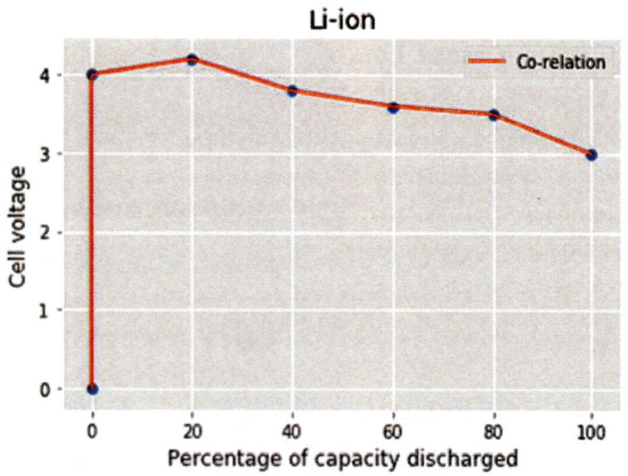


**Fig. 1** Li-ion Percentage discharge versus cell voltage curve

Fig. 2 Lead-acid Percentage discharge versus cell voltage curve

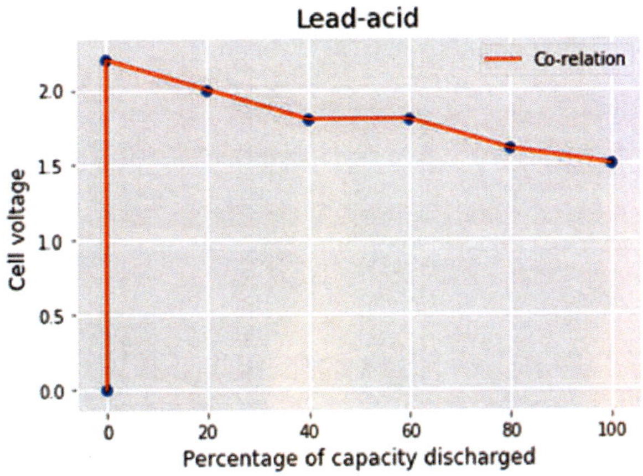


Fig. 3 Ni–MH Percentage discharge versus cell voltage curve

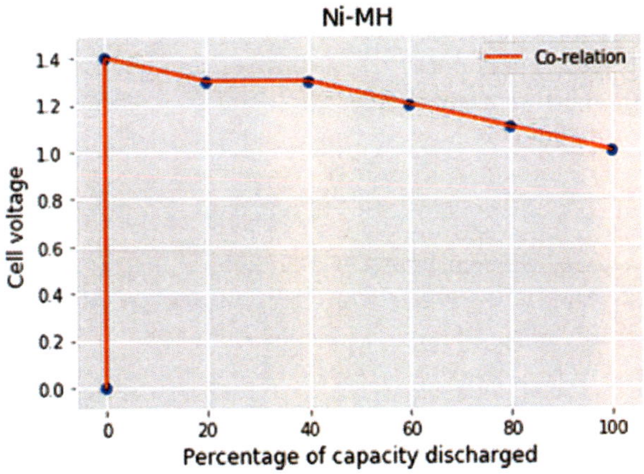


Figure 3 depicts the discharge curve of Ni–MH, compared to Li-ion and Lead-acid cell voltages of Ni–MH gets reduced at a very slow rate, thus it has to be used for some unique end product.

## 6 Electric Storage Capacity of Different Types of Batteries

See Fig. 4.

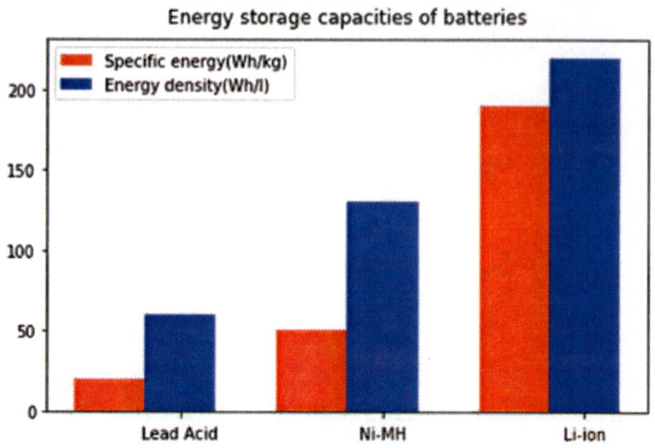


**Fig. 4** Electric storage capacity of batteries

## 7 Electric Motor Drive Requirements of EVs

Selecting the appropriate electric motor drive is a very essential step that determines the performance of EVs. Motor of EVs needs to offer high torque at low speed for starting and acceleration, high power at high speed for cruising. The reference characteristic of Ev's drive requirement is shown in Fig. 5. The basic requirements of EVs electric motor are summed up as follows

- High Spontaneous power and power density
- Ability to produce more torque at less speed
- Broad speed range with constant power
- Weightless construction with low moment of inertia
- Less cost and less electromagnetic interference.

Desired output characteristics of EVs drive motor is shown in Fig. 6.

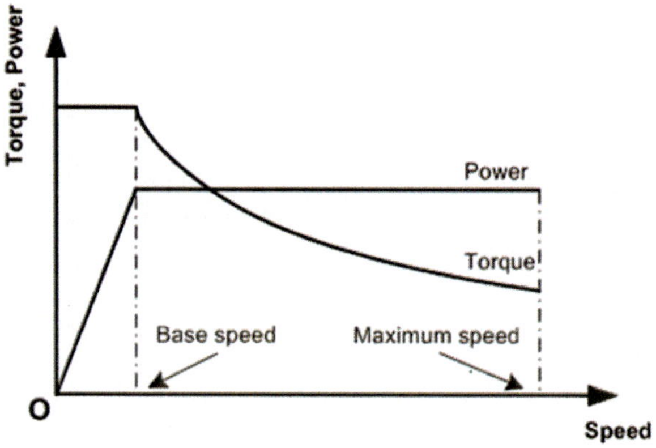


**Fig. 5** Desired output characteristics of electric motor drives in EVs

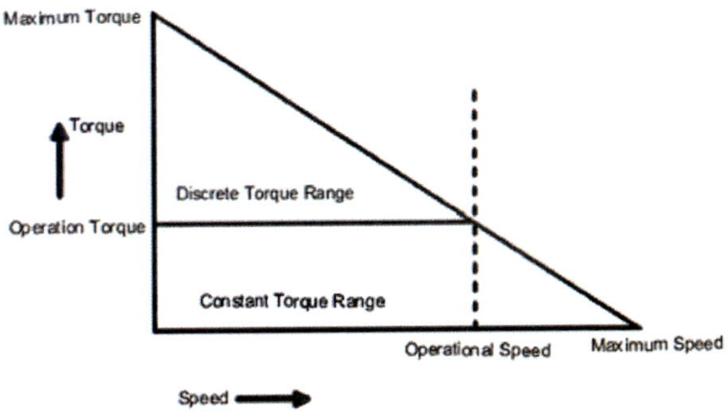


**Fig. 6** BLDC-brushless DC motor

# 8 Characteristics of 4 Types of Motor Drives Suitable for Evs

## *8.1 Brushed DC Motor Drives*

**Pros**:

- BDC drives are widely popular due to its high torque-low speed combination as it is well suited for traction application.
- Desired Motor speed can be set by varying the voltage, that aids in accelerating and controlling the Evs.
- BDC drives satisfy traction and adjustable speed requirements, this in turn augments Evs performance.

**Cons**:

- Heavily built construction.
- Low Efficiency and reliability.
- Increased weight due to commutator and brushes, results in high maintenance.

## *8.2 Induction Motor Drives*

**Pros**:

- IMDs have simple assembly with robust construction, Low maintenance, and initial cost.
- Speed variation is possible by altering the frequency voltage.
- Suitable for all types of environments.

**Cons**:

- Due to breakdown torque motor will stall during operation.

- Although the motor cost is less, to attain speed variation it requires costly controllers.
- IMs have low starting torque, at light loading conditions it operates under low power factor.

### 8.3 Permanent Magnet Brushless DC Motor Drives

**Pros**:

- PM BLDC produces higher efficiency with high power density.
- By using the conduction angle control method up to four-speed variations is possible with reference to its base speed.
- Pulse width modulation technique can also be used to attain infinite variations in speed without harmonics.
- Size to speed ratio is high, also it produces less heat.

**Cons**:

- For controlling the PM BLDC additional software's and hardware's are required.
- Initial cost of PM BLDC motor is high.
- Magnetic field produced inside the PM BLDC is sensitive to ambient conditions, it results in inconsistent speed.

### 8.4 Switched Reluctance Motor Drives

**Pros**:

- Exceptionally good torque-speed, constant power characteristics in alignment with EVs load requirement.
- Structure of rotor is very simple as it is free from coil windings, magnets, and brushes.
- It lacks magnetic sources that make it insensitive to external magnetic field as well as temperature and offer fault-tolerant design.

**Cons**:

- Periodic fluctuations in torque results in torque ripple effect.
- Operation of motor produces acoustic noise.
- External controlling of motor parameter is possible but more complex than IMs and PM BLDC.

# 9 Comparison of Important Performance Metrics of Electric Drives

Decision matrix analysis method is used to arrive at the right choice of drive for EVs among the 4 motor drives. Values are concluded based on rating scale with a maximum value of 10 desired aspects are assigned a maximum value, weights are in the range of 0–1 with 1 as maximum and 0 as least. Table 2 illustrates the decision matrix for choosing the right motor for the end application.

The results suggest that BLDC motors are more suitable for EVs based on the 5 metrics essential for Electric vehicle drives, the next choice is SRM motors. But it is seemingly the decision of Electric vehicle manufacturer to harness the relative advantage of drive. Tesla prefers to use AC induction motor in its car models produced.

Table 3 shows the different drives used by prominent Electric car manufactures, although BLDC drive is installed in Evs in sizable amount due to good traction and maintenance-free operation, few manufactures use AC induction motor for its

**Table 2** Decision matrix for drive selection

| Factors | Efficiency | Weight | Motor cost | Power to weight ratio | Power factor |
|---|---|---|---|---|---|
| Weights | – | – | – | – | – |
| BDC | 4 | 4 | 6 | 5 | 7 |
| IM | 8 | 8 | 4 | 7 | 6 |
| PM BLDC | 10 | 9 | 9 | 9 | 9 |
| SRM | 9 | 10 | 6 | 7 | 4 |

**Table 3** Batteries and drives used in Modern EVs

| Company | EV modal | EV motor | Battery |
|---|---|---|---|
| Tesla | X and S | AC Induction motor (variable voltage variable frequency) | 70 kWh and 90 kWh, 18650 NCA cells of 3.4 Ah |
| Nissan | Leaf | AC synchronous electric motor | 30 kWh; Li-manganese, 192 cells |
| BMW | I3 | Synchronous Reluctance Motor | 42 kWh, LMO/NMC, large 60 A prismatic cells |
| Toyota | Prius | Brushless DC motor | 4.4 kWh Li-ion |
| Renault | Fluence Z.E | AC synchronous electric motor | 22 kWh lithium-ion battery |
| Mahindra | E20 | 3 phase AC induction motor | 16 kWh Li-ion |
| Honda | EV plus | PM brushless motor | 288 V NiMH |
| General motors | GM EV1 | AC induction motor | 18.7 kWh lead-acid |

smooth operation nevertheless AC-IM drives are bulky due to sophisticated circuit required for varying the frequency to control speed and direction.

## 10 Torque—Speed Curves for Commonly Used Drives

Torque-speed curve is the important criterion that decides the purchase of motor, a motor can be mentioned to have some rated torque but if it delivers the rated torque at little speed it may not be suitable for the desired application. Hence torque-speed curve helps in deciding the appropriate choice of drive-in context with the application. Figures 6 and 7 depicts the torque versus speed curve for BLDC motor and induction motor.

Figures 8 and 9 depicts the torque-speed curve for BLDC and SRM drives. Plots can be inferred by comparing with desired torque-speed curves for Electric vehicles shown in Fig. 6.

### Conclusions

- In this study 4 types of drives suitable for EVs with its expected performance output characteristics are discussed, based on the weight matrix permanent magnet

**Fig. 7** IM-induction motor

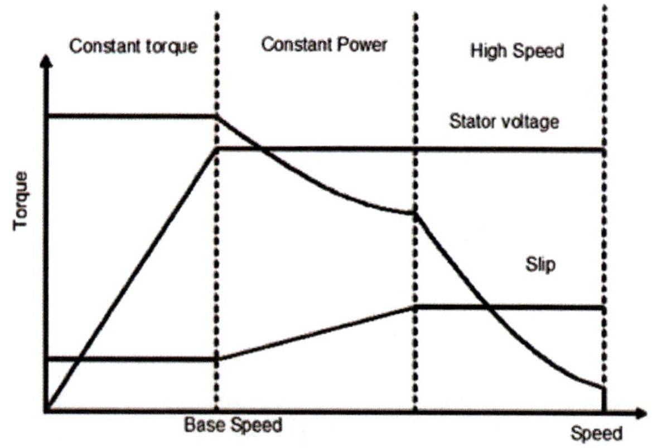


**Fig. 8** BLDC-brushless DC Motor

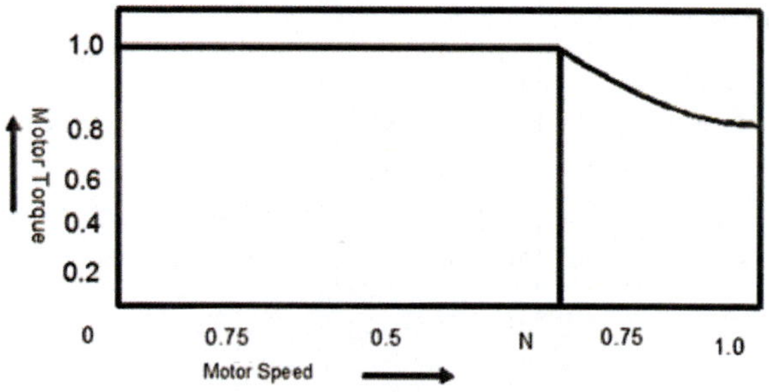

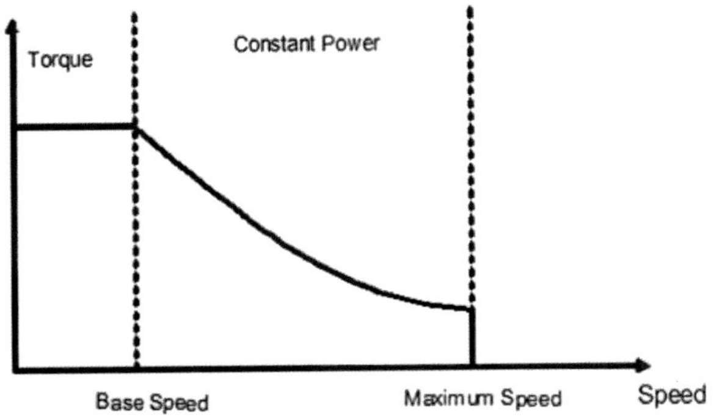


**Fig. 9** SRM-switched reluctance motor

BLDC is more suitable for electric vehicles, it also provides high power to weight ratio.

- State of art research progress in battery technologies, common specifications of battery, and charging methods are also examined.
- Due to climate and energy security concerns, commercial vehicle makers have started to focus on battery-based transportation, but still the future of battery vehicles are unclear an attempt to made in this literature to give crisp information about the fundamental building blocks of Electrical vehicle technology.

## References

1. Andwari AM, Pesiridis A, Rajoo S, Martinez-Botas R (2017) A review of battery electric vehicle technology and readiness levels. Renew Sustain Energy Rev 78(2):414–430
2. Lulhe AM (2015) A technology review paper for drives used in electrical vehicle (EV) and hybrid electrical vehicles (HEV). IEEE Trans 10(2):632–636
3. Manzetti S, Mariasiu F (2015) Electric vehicle battery technologies : from present state to future systems. Renew Sustain Energy Rev 51:10041012
4. De Santiago J et al (2012) Electrical motor drivelines in commercial all-electric vehicles: a review. IEEE Trans Veh Technol 61(2):475–484
5. Tie SF, Tan CW (2013) A review of energy sources and energy management system in electric vehicles. Renew Sustain Energy Rev 20:82–102
6. Un-noor F et al (2017) Future direction of development. Energies 1217(10):2–84
7. Xia G, Cao L, Bi G (2017) A review on battery thermal management in electric vehicle application. J Power Sources 367:90–105
8. Yilmaz M, Krein PT (2013) Review of battery charger topologies, charging power levels, and infrastructure for plug-in electric and hybrid vehicles. IEEE Trans Power Electron 28(5):2151–2169

# Classification of Rice Grains Based on Quality Using Probabilistic Neural Network

**S. Mohanraj, B. Narenthiran, S. Manivannan, R. Arul Murugan, and V. Raj Kumar**

**Abstract** Rice is the most significant cultivated harvests everywhere throughout the world, specifically in Asian nations. Nowadays the evaluation of rice quality has a great impact on the market due to adulteration by plastic rice and stones. The assessment of rice is prepared physically by experienced a rancher which is a tedious and monotonous errand and in particular it is a dangerous strategy where the rice might be pulverized by growth pollution. In this paper, a quick, programmed and non-destructive assessment practice is endeavored to measure the nature of rice based on deep learning neural system model. The pre-processing is started by the Median channel to expel noises from the input pictures. By utilizing Fuzzy c-means clustering, the edges of the rice pictures are appropriately depicted. The features like contour, edge, and region are chosen with the assistance of a genetic algorithm. A probabilistic neural system is created to characterize the portioned rice picture. The presentation of PNN model is introduced to show its viability with regards to exactness, accuracy, f-score, and review, and the outcomes are compared with the existing SVM method.

S. Mohanraj · R. Arul Murugan
Department of Mechanical Engineering, Karpagam College of Engineering, Coimbatore, Tamil Nadu, India
e-mail: sbmohan03@gmail.com

R. Arul Murugan
e-mail: arulmuruganraju@gmail.com

B. Narenthiran · S. Manivannan (✉)
Department of Mechanical Engineering, Karpagam Academy of Higher Education, Coimbatore, Tamil Nadu, India
e-mail: manivannan.s@kahedu.edu.in

B. Narenthiran
e-mail: narenthirannani@gmail.com

V. Raj Kumar
Department of Mechanical Engineering, PSG Institute of Technology and Applied Research, Coimbatore, Tamil Nadu, India
e-mail: rajkumarv@psgitech.ac.in

© The Editor(s) (if applicable) and The Author(s), under exclusive license to Springer Nature Singapore Pte Ltd. 2021

S. Mohan et al. (eds.), *Materials, Design, and Manufacturing for Sustainable Environment*, Lecture Notes in Mechanical Engineering,
https://doi.org/10.1007/978-981-15-9809-8_62

**Keywords** Median filter · Segmentation · Genetic algorithm · SVM · PNN

## 1 Introduction

Agriculture is one of the broad occupations on the planet in which rice is a significant rural item in Asian nations. It is fundamentally ordered by its grain shape, shading, and so on. This paper displays the utilization of the machine vision framework for the grain arrangement. Machine vision has been utilized in the most utilization of grain in order to separate rice assortments dependent on uncommon highlights, for example, shape, length, pastiness, shading, and inside harm of rice. As the innovation is developing, ventures and people groups are versatile to new advances as opposed to utilizing old methods. More individuals devour the rice as nourishment, so the request is expanding.

The primary objective of computer vision is to decipher the picture and to extricate its physical, geometrical, or topological properties. The area of computer vision includes the picture securing framework, pixel preparing as a piece of the execution of mechanical tasks. The applications of computer vision intend to examine an enormous number of items so as to accomplish necessary quality controls. Assessment of farming items by utilizing image processing has become a significant research enthusiasm in recent years. In contrast to the majority of the other modern items, the shape, size, shading, and surface of the farming items that are not administered by a special numerical capacity. It has become a difficult undertaking for the computer framework to perceive and characterize normally shifting the appearance of natural elements like rice, grain, etc. The systematic framework has gotten an option in contrast to the manual assessment of rice tests for piece trademark properties and the measure of contaminated materials in that. Data about rice types and quality are required at a few phases during rice dealing activities. The existing rice dealing arrangements are obvious, but they require fast access to rice types and quality through visual review. But, the procedures included are dreary and tedious. The administrator loses fixation following a couple of long stretches of working because of weakness, vision, and inappropriate lighting. Consequently, a programmed classifier can forestall human blunders in the quality assessment process, which has gotten an option in contrast to manual investigation of rice tests.

Researchers have been conducted to characterize rice dependent grain in a well arrangements and phenomenal imaging condition by utilizing morphological highlights like region, edge, significant pivot, minor hub, lengthening, minimization, etc., in [1–3]. Portrayal model depending on shading to separate the shading qualities from the information shading picture and surface to extricate data with respect to the course of action of pixels in a picture was additionally done in [4–6]. The greater part of the distributed work depending on distinguishing proof of rice types by putting rice pieces in a non-contacting style. An additional study has been done to order rice pieces by consolidating all the previously mentioned highlights and to improve arrangement exactness [5, 6]. Such this distinguishing proof procedure requires lumbering arrangement and is hard to actualize. Also, the framework requires a gadget to exhibit a bit in a non-contacting design. The pre-handling activities like, picture division, background evacuation, and item extraction are essential and tedious activities. A considerable picture pre-handling steps are redundant if recognizable proof of rice is completed utilizing sample pictures [7, 8]. In addition, a sample picture doesn't contain singular items in it. Subsequently, there is no compelling reason to pre-process the picture for background expulsion and article extraction. Characterization of rice types by utilizing rice picture tests with shading and surface was done in [9]. Later distributed works are the order of shading pictures by utilizing SVM and other factual classifiers [10]. Grouping of rice by diminishing the component vectors and utilizing step circle and highlight's mistake was done in . This paper portrays the arrangement of rice picture tests by utilizing a probabilistic neural system (PNN) with a minimum number of highlights without trading off the order exactness. The grouping forms include four phases—first-stage manages getting of the input picture, the subsequent stage manages division, the third stage centres around highlight extraction and the last stage involves characterization by utilizing the proposed PNN and contrasting their outcomes and other existing systems.

## 2 Proposed Method

We approach the evaluation of rice quality by using PNN classification. The probabilistic neural system (PNN) is a feed-forward neural architecture (ANN). In the ordering of rice quality assessment with the proposed PNN design, the essential advances (Reprocessing, preparing, Activation work choices, Regularizations, and Ensemble the numerous strategies) are utilized to accomplish the most extreme precision from the pictures. The block diagram of the proposed strategy is provided beneath (Fig. 1).

At first, the procedure begins by considering rice pictures as an input image. In the wake of stacking the picture, it will go for pre-processing steps. The highlights can be extricated from rice pictures (Fig. 2).

It is trailed by arrangement part where it will order, whether the picture is proliferative or non-proliferative picture or wait picture. The system is implemented in MATLAB. The thoughts of Digital Imaging are canvassed in the computerized picture, picture pre-processing, picture examination, and arrangement.

### *2.1 Pre-processing by Wiener Filter*

The probabilistic neural framework tackles spatial data of the rice picture. A fundamental advance related to the pre-processing is the resizing of images. Before supporting the designing for gathering, convert the pictures into a diminished scale. Then, a while later, converts it into $L$ illustrates. It is a monochrome picture that is

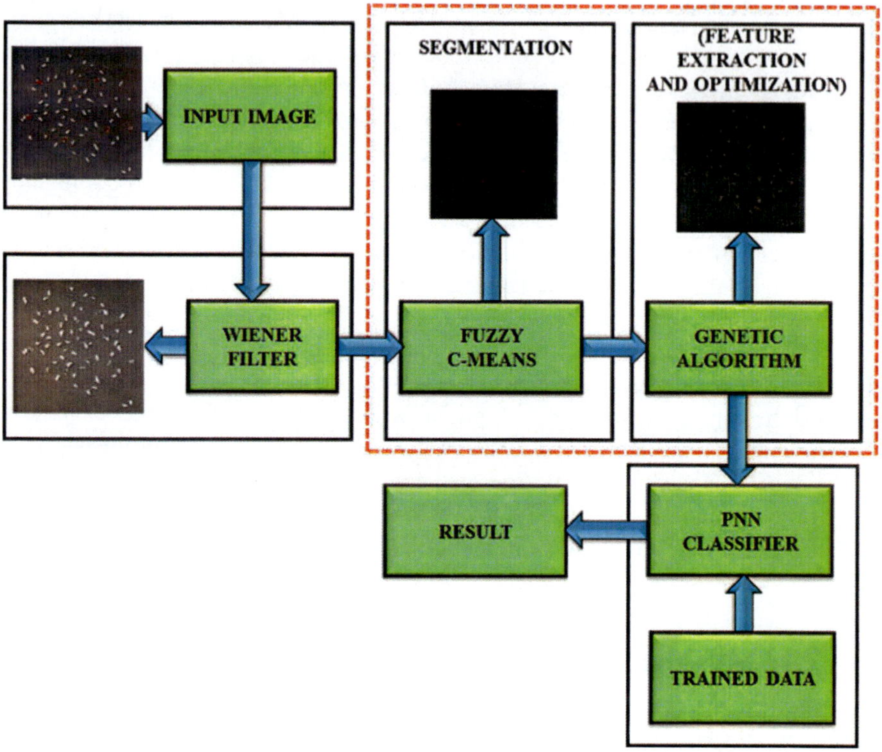


**Fig. 1** Block diagram

used to highlight the rice seed [13]. Additionally, fix the photos in a single dimension for processing.

#### 2.1.1 Weiner Filter

The backward sifting is a rebuilding procedure of deconvolution, i.e., when the picture is obscured by a low pass channel, it is conceivable to recoup the picture by reverse separating or summed up converse separating. But, the opposite separating is exceptionally touchy to additive clamor. The Wiener filter executes an ideal tradeoff between converse sifting and clamor smoothing. It evacuates the added substance clamor and rearranges the obscuring all the while. The Wiener filter measures the mean square mistake. At the end of the day, it limits the general mean square blunder during the time spent opposite separating and commotion smoothing. The Wiener filter is a direct factual estimation of the picture. It is also called as least mean square mistake filter and is established on thinking about pictures and commotion as irregular factors (Fig. 3).

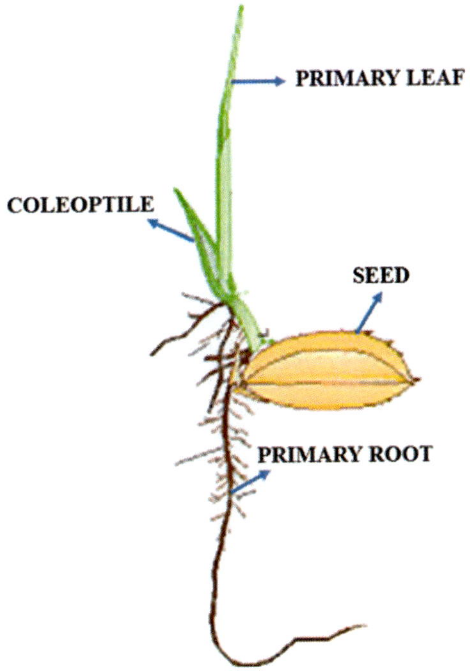


**Fig. 2** Components of germinated rice seed [11, 12]

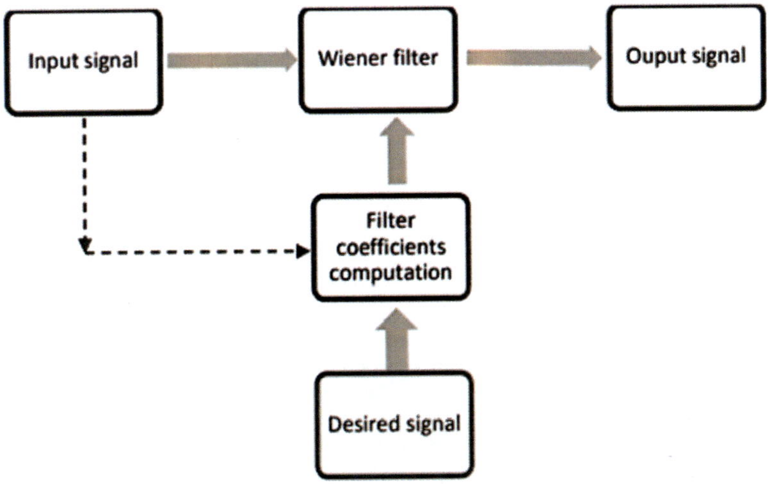


**Fig. 3** Block diagram of Wiener filter

These channels are intended for an ideal recurrence reaction. One is accepted to know about the unearthly properties of the first sign and the commotion, and one looks for the LTI channel whose yield would come as near the first sign as could reasonably be expected. Wiener channels are described by the accompanying:

(a) Presumption: sign and (added substance) commotion are stationary straight arbitrary procedures with known ghostly qualities.
(b) Necessity: the channel must be physically feasible, for example, causal (this necessity can be dropped, bringing about a non-causal arrangement).
(c) Execution criteria: least mean-square mistake.

In the recurrence area, the corrupted picture $x(n, m)$ is spoken to as $X(u, v)$. The first picture is gotten by

$$\hat{S}(u, v) = G(u, v)X(u, v) \tag{1}$$

The Wiener filter is

$$H(u, v) = \frac{G^*(u, v)P_s(u, v)}{|G(u, v)|^2 P_s(u, v) + P_n(u, v)} \tag{2}$$

where

$G(u, v)$ is degradation function
$P_s(u, v)$ is the signal power spectrum
$P_n(u, v)$ is the noise power spectrum.

The output of wiener filter is,

$$Y(u, v) = H(u, v) * X(u, v) \tag{3}$$

Wiener channels give the straight least mean square measure of the input picture from the perceptions and have been utilized broadly for the reclamation of boisterous and obscured pictures. The fundamental thought behind the Wiener channel is to utilize the data contained in the current picture just as in the imaging framework utilized.

### 2.1.2 Segmentation Using Fuzzy C-Means Algorithm

Clustering is the method of separating the data into homogenous units by considering the relationship of objects. The clustering method is the allocation of the feature vectors into $N$ clusters. Every $n$th cluster has $C_n$ as its center. Among various kinds of clustering methods, Fuzzy C-Mean clustering (FCM) is the extensively utilized one. FCM utilizes reciprocal distance to determine fuzzy weights. The input of this process is a pre-known number of clusters, $N$. The mean position of every the member of a cluster is identified. The output is the segregating of $N$ clusters on a class of objects. The goal of the FCM cluster is to reduce the total weighted mean square error, (MSE). The FCM consents each feature vector to match with several clusters of different fuzzy membership values. The final division is dependent on the optimum weight of the feature vector over all clusters.

The fluffy c-means clustering is one of the basic methods that is used to divide the picture by partitioning the space of the picture into different bunch locales with comparative picture's pixels esteems. For medicinal pictures division, the appropriate grouping type is fluffy bunching. The Fuzzy c-means (FCM) can be viewed as the fuzzified form of the k-means method. It is a grouping calculation which empowers the information to have a level of having a place with each bunch of level of participation. It is created by Dunn [14] and modified by Bezdek [15]. The calculation is an iterative bunching strategy that delivers an ideal $c$ segment by limiting the weight inside the gathering entirety of squared mistake target work [15] is generally utilized in picture division and example acknowledgment.

The calculation is an iterative bunching strategy that creates an ideal $c$ parcel by limiting the weighted inside gathering entirety of squared mistake target work $\mathrm{J_{FCM}}$

$$\mathrm{J_{FCM}} = \sum_{k=1}^{n}\sum_{i=1}^{c} u_{ik}^{q} d^2(x_k, v_i) \tag{4}$$

where $X = \{x_1, x_2, \ldots, x_n\} \subseteq R^p$ is the dataset in the p-D vector spaces, $n$ is the quantity of information, $c$ is the cluster number with $2 \le c \le n$, $u_{ik}$ is the membership degree of $x_k$ in the $i$th cluster, $q$ is a mass exponent of every fuzzy member, $v_i$ is the centre of the bunch $i$, $d^2(x_k, v_i)$ is a space between $x_k$ and $v_i$. The answer to the objective function $\mathrm{J_{FCM}}$ is obtained in a repetitive process, which is performed as:

1. Assign the esteems of $c$, $q$, and $\epsilon$.
2. Assume the fuzzy separation array $U = [u_{ik}]$.
3. Fix the loop count as b−0.
4. Measure the $c$ cluster centres $\{v_i^{(b)}\}$ with $U^{(b)}$

$$v_i^{(b)} = \frac{\sum_{k=1}^{n} u_{ik}^{(b)q} x_k}{\sum_{k=1}^{n} u_{ik}^{(b)q}} \tag{5}$$

5. Estimate the membership $U^{(b+1)}$. For $k = 1$ to, measure the following: $I_k = \{i | 1 \le i \le c, d_{ik} = \|x_k - v_i\| = 0\}$, $/I$; for the $k$th column of the array, calculate the membership esteems:

   (a) If $I_k = \varphi$ then

$$u_{ik}^{(b)} = \frac{1}{\sum_{j=1}^{c} \left(\frac{d_{ik}}{d_{jk}}\right)^{\frac{2}{(q-1)}}} \tag{6}$$

   (b) Else $= 0$ for all $i \not\subset I$ and $\sum_{i \in I_k} u_{ik}^{(b+1)} = 1$; next $k$.

6. $\left\|U^{(b)} - U^{(b+1)}\right\| < \epsilon$ stop; otherwise, set $b = b + 1$ and go to step 4.

### 2.1.3 Genetic Algorithm Based Feature Extraction

Essentially the feature extraction is utilized to decrease the measure of pixels expected to portray an immense arrangement of information. It is a general structure for creating a mixture of factors with adequate memory necessity, low force, and fulfilling exactness. The extraction of printed highlights assumes a significant job in design acknowledgment and picture handling.

Space for the GA to work is the binary search space. This is because the chromosomes are strings of bits. Similar to human evolution, the GA calculates the limited digital population. Initially, the first population is arbitrarily generated and verified by a fitness function. A gene value of '1' shows that $s$ certain highlights indexed by the location of '1' is chosen, for a binary chromosome employed. If it is '0', the feature will not be chosen for the chromosome to be evaluated. Their grading is performed by the position index of the highlighted index as '1's. On the basis of these grades, the most $n$ fittest kid (Elitism of size $n$) is chosen to be carried over to the subsequent generation.

After automatically pushing the elite offspring to the subsequent generation, the remainder of offspring in the current population goes through genetic function hybridization and alteration for respectively forming hybridization and alteration offspring. The offspring which comprise the elite, hybridization, and alteration, formulate the novel population or new generation. Hybridized offspring are generated using a genetic functional crossover process for combining two individuals (chromosomes). Whereas, by means of flipping of bits that depend on its probability, mutation operator is utilized for the genetic disturbance of genes in every chromosome. For this work, the initial population considered is a matrix of dimensions of population size × length of the chromosome. Population size is how many chromosomes are contained in the population. The quantity of bits or genes in each chromosome is known as the chromosome length. The population size should be at least equal to the value of the chromosome length, to enable the individuals in the population to span the whole search space (Fig. 4).

*Energy*

The picture homogeneity is estimated by the whole square of second request minute. At the point when pixels are same or when pictures are homogeneous, the vitality or second request minute is likewise high.

$$\text{Energy} = \sum_{i=0}^{N_g-1} \sum_{j=0}^{N_{g-1}} P_{ij}^2 \tag{7}$$

where $N_g$ is grey tone.

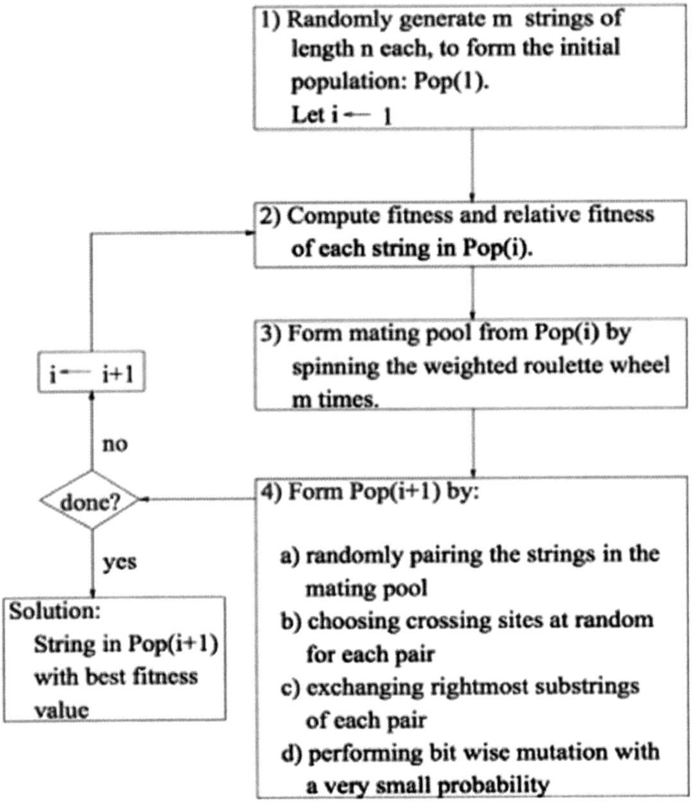


**Fig. 4** Genetic algorithm

*Inertia Moment*

It indicates the nearby homogeneity and will be high for the uniform dim level and high reverse GLCM. Its weight esteem is the converse of the heaviness of complexity.

$$\text{IM} = \frac{\sum_{i=0}^{N_g-1} \sum_{j=0}^{N_{g-1}} P_{ij}}{1 + (i - j)^2} \tag{8}$$

*Entropy*

It is the measure of data in the packed picture. It gauges the loss of data in the transmitted data.

$$\text{Entropy} = \sum_{i=0}^{N_g-1} \sum_{j=0}^{N_{g-1}} P_{ij} * \log(P_{ij}) \tag{9}$$

*Correlation*

The direct conditions of the dim degree of neighbor pixels are estimated by connection.

$$\text{Correlation} = \frac{\sum_{i=0}^{N_g-1}\sum_{j=0}^{N_{g-1}}(ij)P_{ij} - \mu_x\mu_y}{\sigma_x\sigma_y} \quad (10)$$

### 2.1.4 PNN Classification

A PNN is a feed-forward neural system, which was gotten from Bayesian system and a factual calculation called Kernel Fisher discriminant investigation. It was presented by D. F. Specht.

PNN has made three presumptions: on the off chance that every characterization likelihood thickness work has a similar kind, at that point it is the Gaussian appropriation and it is likewise the typical conveyance. Every characterization in which Gaussian dissemination likelihood thickness capacity of the covariance lattice is corner to corner grid and every worth is the equivalent [10].

A PNN comprises of $d$ input units, $n$ design units, and $c$ class units. Each example unit shapes the internal result of its weight vector and the standardized example vector $x$ to frame $z = w^t x$, and afterward produces $\exp\left[(z-1)/\sigma^2\right]$ where $\sigma$ is a parameter set by the client and is equivalent to 1.414 occasions the width of the successful Gaussian window. Every class unit totals such commitments from the example unit associated with it. This safeguards the action in every one of the classification units speaks to the Parzen window thickness gauge utilizing a circularly symmetric Gaussian window of covariance $\sigma^2 I$, where $I$ is the $d \times d$ identity matrix. The structure of PNN is shown in Fig. 1. In the event that we indicate the parts of the $j$-th design as $x_{jk}$ and the loads to the $j$-th pattern unit $w_{jk}$, for $j = 1, 2, \ldots, n$ and $k = 1, 2, \ldots, d$, then

Given below is the PNN training algorithm:

(1) Start assume $j = 0$, $n =$ number of patterns
(2) Let $j = j + 1$
(3) Assign $x_{jk} \leftarrow x_{jk} \Big/ \left(\sum_i^d x_{ji}^2\right)^{0.5}$
(4) Train: $w_{jk} \leftarrow x_{jk}$
(5) If $x \in w_i$ then $a_{ic} \leftarrow 1$
(6) Till $j = n$
(7) Finish.

For the classification of pattern, $x$ is as follows: Each pattern is kept at the input unit. Every pattern unit estimates their inner product:

$$z_k = w_k^t x$$

which radiates a nonlinear capacity $z_k$. Every example unit adds to its related classification unit, a sign equivalent to the likelihood the test point was created by a Gaussian focused on the related preparing point. The whole of these neighborhood gauges (figured at the related classification unit) gives the discriminant work $g_i(x)$—the Parzen window gauge of the hidden appropriation. The maximum $g_i(x)$ activity gives the ideal classification for the test point.

PNN Classification Algorithm

(1) Start assume $k = 0$, $x =$ test sample
(2) Do $k \leftarrow k + 1$
(3) $z_k = w_k^t x$
(4) If $a_{kc} = 1$ Then $g_c \leftarrow g_c + \exp\left[\frac{z-1}{\sigma^2}\right]$
(5) Till $k = n$
(6) Return class ←argmax $g_i(x)$
(7) Finish.

There are two significant features of PNN; since the learning rule is easy and needs just a single pass via the training information, the speed of learning is high. Another advantage is the fresh training samples can be added quite easily with an already trained classifier (Fig. 5).

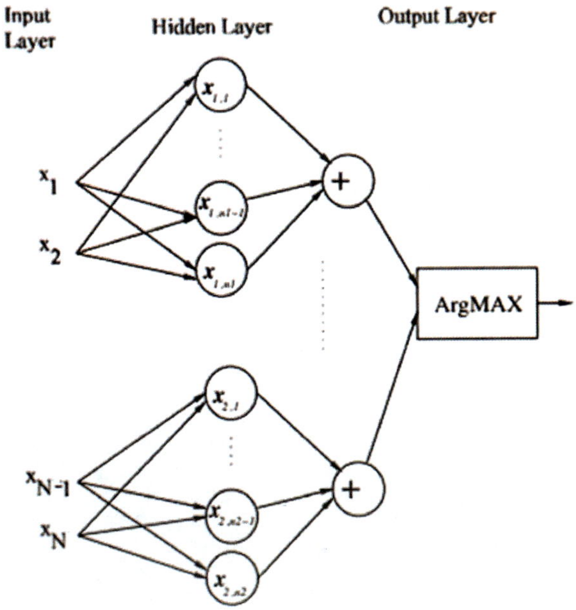


**Fig. 5** Schematic of PNN

## 3 Results and Discussion

In the detection of rice quality, two samples of rice grains are dispersed on a plate. So as to remove the shadow effects, a black background is chosen. The detestable classes of rice grains are can be efficiently separated using the proposed approach. The proposed method is tested and verified in MATLAB (Fig. 6).

It is essential to remove the noises from the input image sets before proceeding with the algorithm. Preprocessing operation is performed as the initial stage of our algorithm. The linear motion of the rice grains causes blur and poor sampling. The Weiner is a suitable choice to reduce the amalgamation in intensities. The input color image is converted into grayscale to find the noise coefficients in the input image. The filtered image thus generated is shown below in Figs. 7 and 8.

The filtered images are converted into binary scale to enhance all the features of the rice grains. To make the features of digital image stand out, contrast enhancement is performed which changes the range values of the image pixels. The result of this process is given in Figs. 9 and 10.

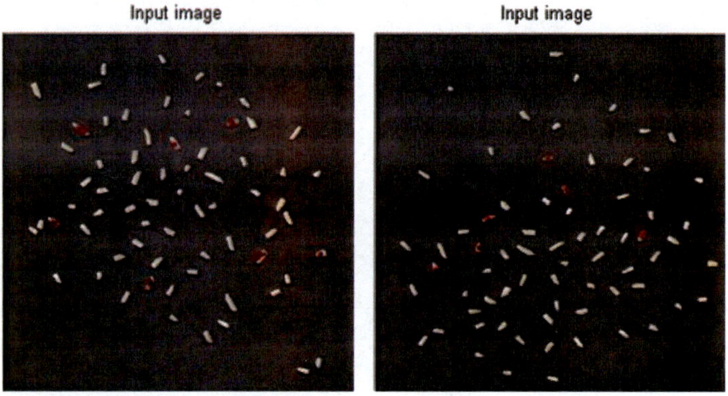


**Fig. 6** Input rice image set 1 and set 2

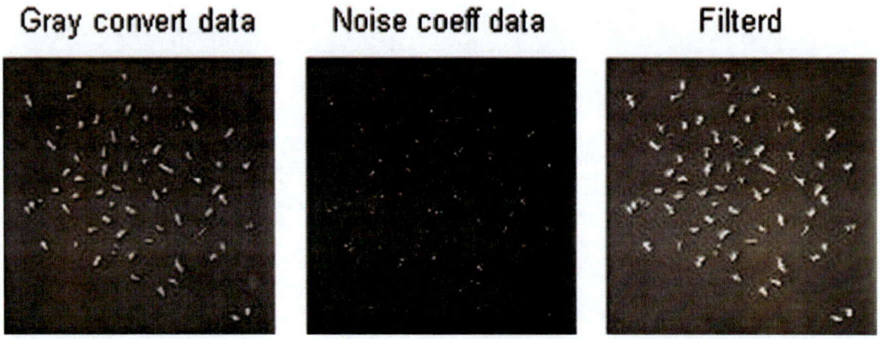


**Fig. 7** Pre processed output of input image set 1

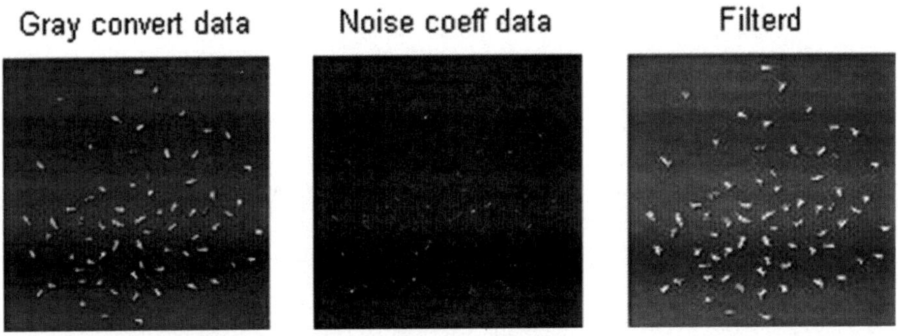


**Fig. 8** Pre processed output of input image set 2

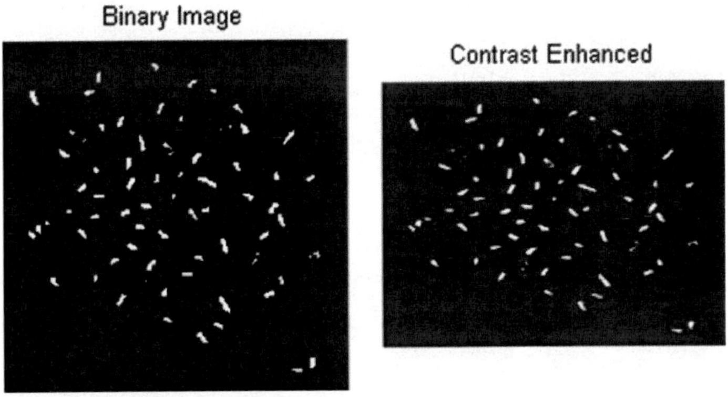


**Fig. 9** Binarization of input image set 1

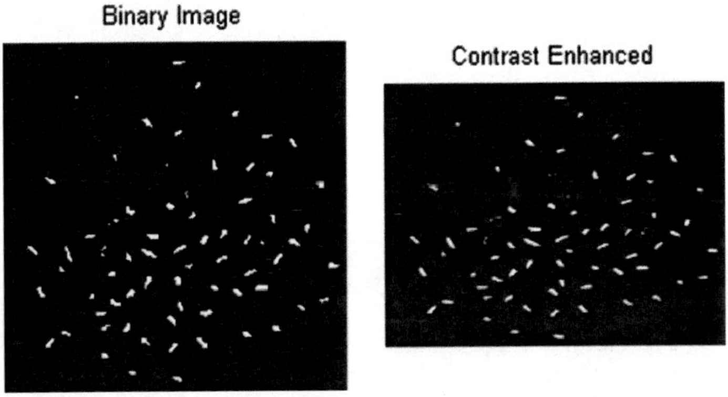


**Fig. 10** Binarization of input image set 2

Now the digital image is processed to get the morphological features of the rice grains. This nonlinear convolution procedure extracts the original shape information and creates an output image of the same size. The sensitivity of the proposed method is also improved using a morphological operation. The output after a morphological operation is shown in Figs. 11 and 12.

The rice grains with similar traits and data points are grouped into separate clusters so as to detect the defected rice grains. The degree to which the data point belongs to every cluster is graded by the membership function of Fuzzy C-means clustering. The classes of rice grains can be segmented once clustering is finished. Figures 13 and 14 depicts the results of the clustered image for image set 1 and image set 2.

The final output of detestable rice grains is shown below (Figs. 15 and 16).

The features like energy, entropy, contrast, correlation, homogeneity, and variance are used as distinguishing parameters of the proposed rice grain images. Using these measures, the dimensions of the image sets are reduced to the compatible range without compromising the content of most discriminating information. These feature values are compared with the features' values of the conventional neural network to measure the performance of the proposed method [16, 17]. The comparison chart

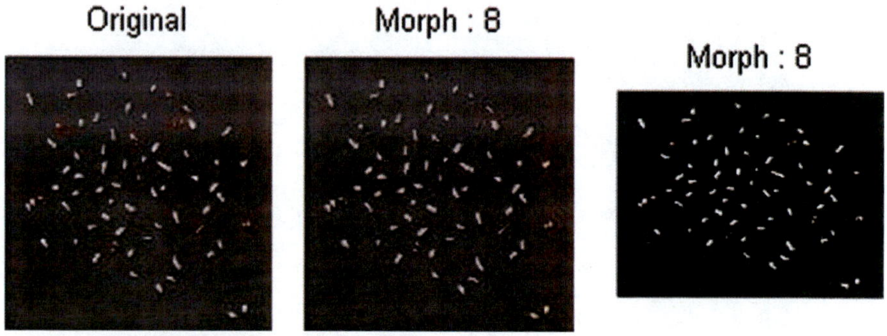


**Fig. 11** Morphological features image set 1

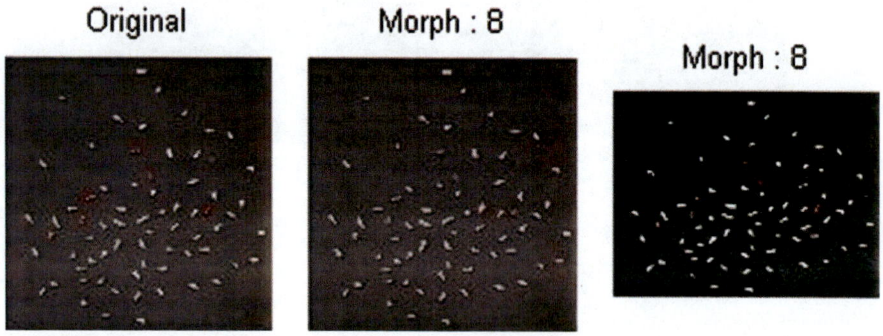


**Fig. 12** Morphological features image set 2

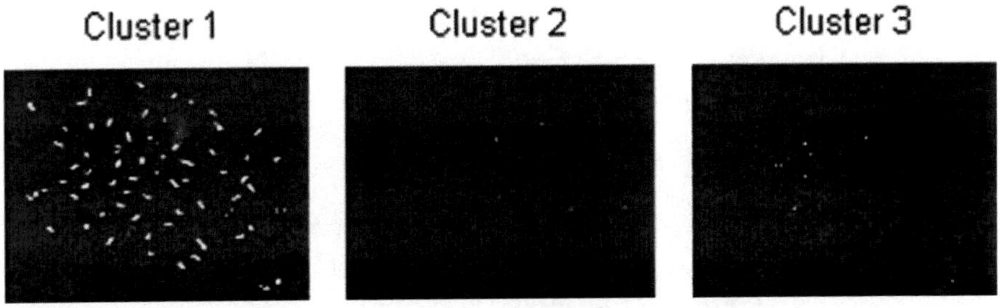


**Fig. 13** Clustering of image set 1

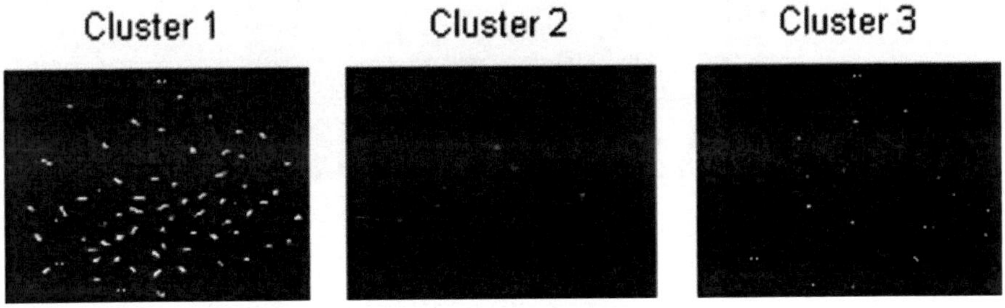


**Fig. 14** Clustering of image set 2

**Fig. 15** Segmented output of image set 1

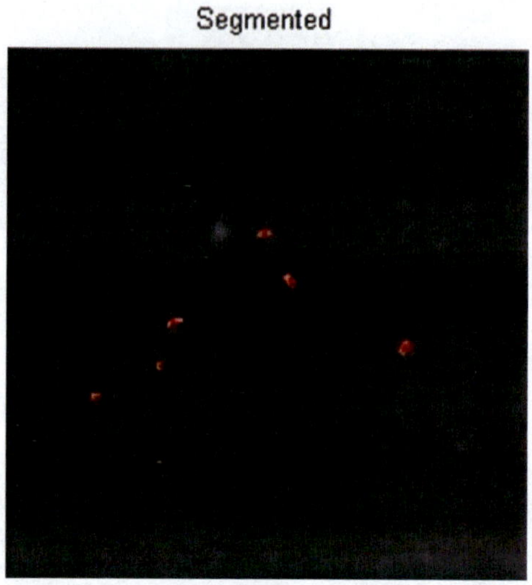


**Fig. 16** Segmented output of image set 2

clear depicts that the proposed method has noticeable values more than the existing methods (Fig. 17).

**Performance comparison proposed methods**
Also, the performance of our method is measured in terms of accuracy, sensitivity, and specificity. For various ranges of image sets, the accuracy is not compromised

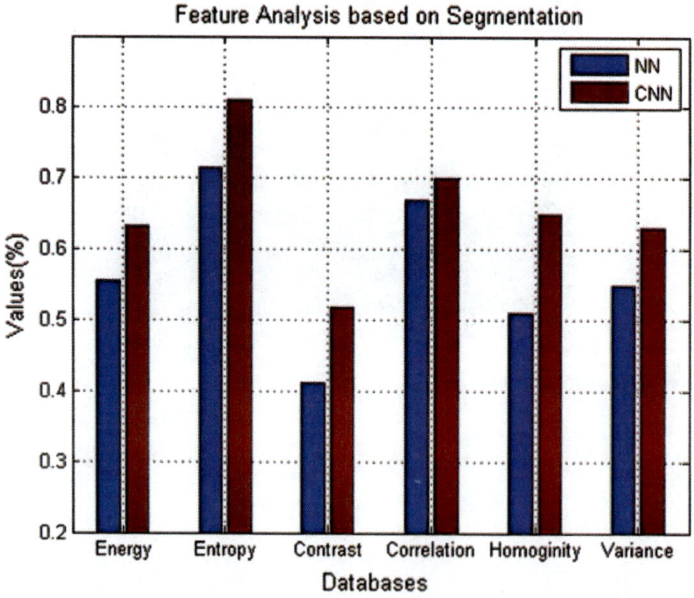


**Fig. 17** Feature values of proposed method

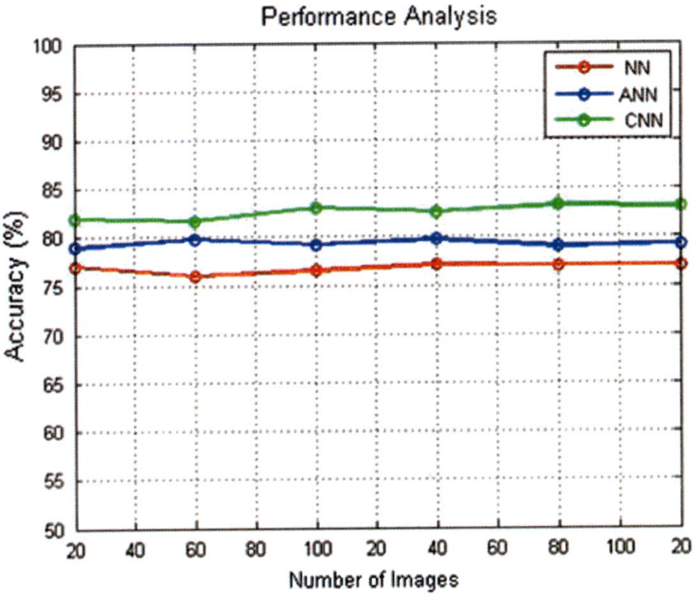


**Fig. 18** Accuracy comparison of proposed method with existing methods

and shown to be outperformed with a maximum value of 83%. For the same image sets the experiment is conducted with ANN and conventional neural networks [18], in which they show constant performance for various image sets (Fig. 18).

The comparison of the sensitivity is tabulated below. The sensitivity performance is increased by up to 85% till 40 image sets. When the number of image sets reaches 100, the sensitivity value is decreased to 80% and then again increases steadily up to 84% whereas the comparable performance was given by considered methods with a maximum value of 82% for ANN and 79.5% for the conventional neural network (Fig. 19).

The true negative rate is an important measuring performance that designates the rice grains that do not satisfy quality level as detestable [19]. A high specificity reduces false-positive rate. The performance is consistent for various image sets with values more than the existing methods (Fig. 20; Table 1).

## 4 Conclusions

A new approach for evaluation of rice quality is proposed using a convolutional neural network in this paper. The result of the proposed CNN technique is compared with ANN and conventional neural network classifiers. The maximum accuracy of this method is 84%. It is found that with CNN, the system gives consistently better performance for various image sets when compared with other classifiers.

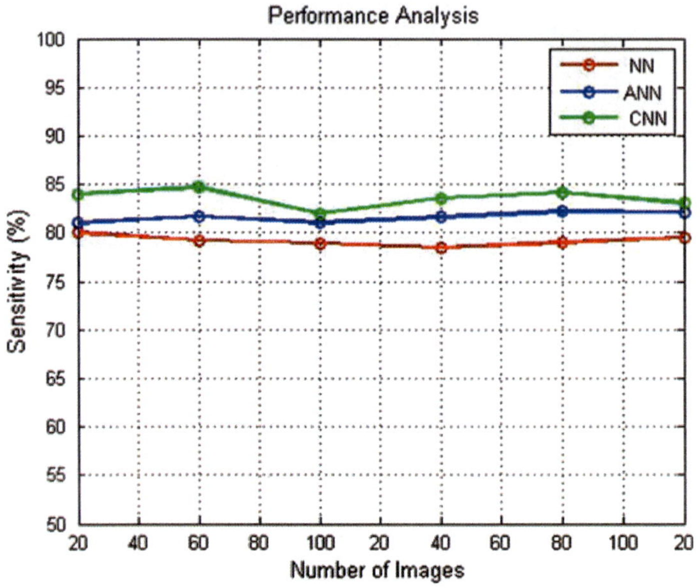


**Fig. 19** Sensitivity comparison of proposed method with existing methods

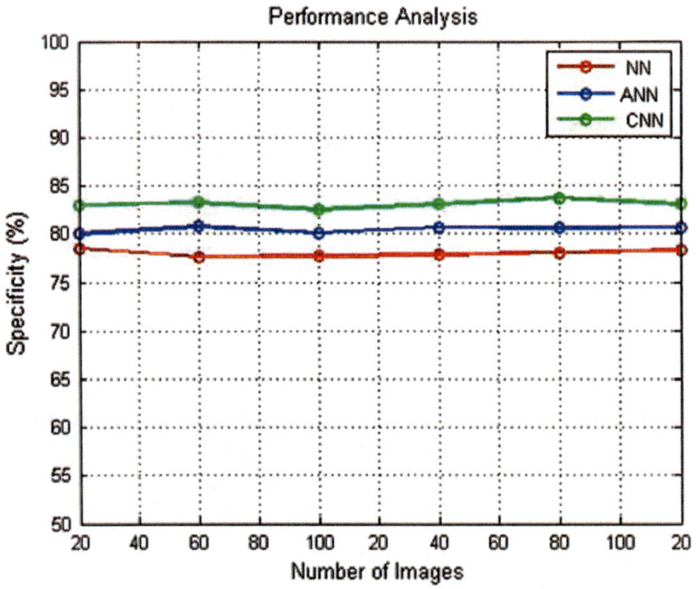


**Fig. 20** Specificity comparison of proposed method with existing methods

**Table 1** Performance comparison

| Method used | Accuracy (%) | Sensitivity (%) | Specificity (%) |
|---|---|---|---|
| Proposed CNN classifier | 83.5 | 85 | 84 |
| ANN classifier | 80 | 82 | 80.5 |
| Convolutional neural network | 77 | 80 | 78 |

The proposed classifier can classify rice grain quality without any degradation in accuracy, sensitivity, and specificity. It is visualized from the result that CNN is more suitable in evaluating the quality of rice grain with the maximum accuracy, sensitivity, and specificity of 83.5%, 85%, and 84%, respectively.

## References


1. Douik A (2010) Cereal grain classification by optimum features and intelligent classifiers. Int J Comput Commun Control 5:506–516
2. Choudhary R, Paliwal J, Jayas DS (2008) Classification of cereal grain using wavelet, morphological, color and texture features of non-touching kernel. Biosyst Eng 99:330–337
3. Visen NS, Paliwal J, Jayas DS (2004) Image analysis of bulk grain samples using neural network. Can Biosys Eng 46:7.11–7.15
4. Majundar S, Jayas DS (2000) Classification of cereal grain using machine vision: III texture mode. Trans ASAE 43(6):1681–1687
5. Paliwal J, Visen NS, Jayas DS (2003) Cereal grain and dockage identification using machine vision. Biosys Eng 85(1):51–57
6. Majundar S, Jayas DS (2000) Classification of cereal grain using machine vision: IV combined morphology, color and texture model. Trans ASAE 6:1689–1694
7. Vanitha V (2019) Rice disease detection using deep learning. Int J Recent Technol Eng 7(5):534–542
8. Singh AK, Rubiya A, Raja BS (2015) Classification of rice disease using digital image processing and SVM classifier. Int J Electr Electron Eng 7(1):294–299
9. Gullu A, Ozan AKI, Ucar E (2015) Classification of rice grain using image processing and machine learning techniques. In: International scientific conference, pp. 352–354
10. S-J Chuang, Y-L Chou (2011) Neural networks for the recognition of traditional chinese handwriting. Presented at the 14th IEEE international conference on computational science and engineering
11. Sivakumar D, SuriyaKrishnaan K, Akshaya P, Anuja GV, Devadharshini GT (2019) Computerized growth analysis of seeds using deep learning method. Int J Recent Technol Eng 7(6):257–268
12. Lurstwut Benjamapon (2016) Rice seed germination analysis. Int J Comput Appl Technol Res 5(4):176–182
13. Eke BO, Egbono F (2016) Evaluation of seeding growth rate using image processing. Int J Comput Appl Technol Res 5(4):183–190
14. Dunn JC (1973) A fuzzy relative of the isodata process and its use in detecting compact well-separated clusters. Cybernetics 3(3):32–57
15. Bezdek JC (1981) Pattern recognition with fuzzy objective function algorithms. Plenum, New York
16. Khanchaitit P, Boonkoom T (2019) Fuzzy-filtered neural network for rice disease diagnosis using image analysis. Int J Inno Technol Exploring Eng 8(8):437–446

17. Chatnuntawech I, Tantisantisom K, Khanchaitit P, Boonkoom T, Bilgic B, Chuangsuwanich E (2018) Rice classification using hyperspectral imaging and deep convolutional neural network. Int J Comput Appl Technol Res 5(4):176–182
18. Mitchell M (1998) An introduction to genetic algorithms. MIT Press, Cambridge, MA
19. Pazoki AR, Farokhi F (2014) Classification of rice grain varieties using two artificial neural networks (MLP and neuro-fuzzy. J Anim Plant Sci 24(1):336–343

# Modeling, Optimization and Corrosion Analysis of FS Welded LM25-SiC MMCs

**P. Venugopal and N. Natarajan**

**Abstract** LM25-SiC metal matrix composites were manufactured by stir casting technique. The composites were welded by friction stir (FS) welding process. Experimental studies were carried out for specimens treated at different temperatures for FS welded composites, and pitting corrosion properties were evaluated. Mathematical models were developed to correlate welding process parameters to pitting potential. Excel solver software was used to optimize the pitting potential for FS welded MMCs. Composites reinforced with 10% SiC show maximum pitting corrosion resistance when treated at cryogenic temperature. FS welded MMCs show higher resistance to corrosion when processed at high temperature compared to MMCs of as-welded composites processed at cryogenic atmosphere.

**Keywords** FS welding · Elevated temperature · Cryogenic temperature · Pitting corrosion

## 1 Introduction

Superior properties such as stiffness, specific strength, good wear resistance, lesser weight and thermal resistance were achieved when metal matrix composites (MMCs) were produced using aluminum alloys as matrix [1]. MMCs reinforced with particulates are found to be popular because of less production costs and ease in secondary processes.

Different fabrication techniques are employed for manufacturing MMCs. Stir casting [2–5] was found to be the efficient way of manufacturing MMC. The

P. Venugopal (✉)
Department of Mechanical Engineering, Muthayammal College of Engineering, Rasipuram, Tamil Nadu 637408, India
e-mail: venunaveen73@gmail.com

N. Natarajan
Department of Mechanical Engineering, Muthayammal Engineering College, Rasipuram, Tamil Nadu 637408, India

© The Editor(s) (if applicable) and The Author(s), under exclusive license to Springer Nature Singapore Pte Ltd. 2021

S. Mohan et al. (eds.), *Materials, Design, and Manufacturing for Sustainable Environment*, Lecture Notes in Mechanical Engineering,
https://doi.org/10.1007/978-981-15-9809-8_63

**Table 1** Chemical content of LM 25 aluminum alloy

| Elements | Si | Cu | Fe | Mn | Cr | Mg | Ti | Pb | V | Co | Al |
|---|---|---|---|---|---|---|---|---|---|---|---|
| % by weight | 7.475 | 0.084 | 0.462 | 0.024 | 0.020 | 0.432 | 0.013 | 0.014 | 0.004 | 0.011 | Remainder |

most successful MMCs produced till date were SiC reinforced aluminum MMCs. Aluminum-based MMCs found widespread application in aerospace structure, IC engines, electronic packaging and in recreational products [6].

Welding plays an important role in all these applications. Maximum studies [7] on welding of aluminum MMCs have dealt with tungsten inert gas welding, metal inert gas welding and other fusion welding processes. Little attention has been paid for FS welding of aluminum matrix composites reinforced with SiC.

Friction stir welding (FSW) [8, 9] is a process involving joining of metals without fusion or filler material. FSW can be used for joining particulate reinforced aluminum metal matrix composites. Therefore, LM25 was chosen as matrix reinforced with different volume % of SiC for the experiments. Little investigations were done in joining of LM25—SiC MMCs. Table 1 shows the chemical content of LM 25.

Aluminum alloys are principal materials that have found usage in many industries as a result of its mechanical properties, structural virtues, fabrication, high resistance to corrosion, weldability, etc. [10]. Pitting corrosion commonly occurs in aluminum. Pitting leads to corrosion fatigue and stress corrosion cracking. Therefore, corrosion properties of LM25-SiC are essential to be analyzed. In this investigation, FS welded composites were analyzed for pitting corrosion properties. Excel solver software is used to optimize the process parameters of pitting potential of FS welded composites.

## 2 Experimental Procedure

### *2.1 Stir Casting*

Figure 1 shows the commonly used setup of stir casting process to produce aluminum composites. A cylindrical ingot of size of diameter 80 mm and height 150 mm was cast. Aluminum alloy was stirred above liquidus temperature in an electrical resistance furnace using three-bladed mechanical impeller [11]. After degassing with nitrogen, the reinforcement particle size of 25 μm is added slowly. A bimetallic thermocouple was used to control and maintain constant temperature. 5% SiC, 10% SiC and 15% of SiC were added to LM 25, and specimens were prepared with different specifications. The stirring was continued at 750 °C for 15 min till solidification of slurry. 3 K/s rate of cooling was maintained for solidification of cast composite. This makes the reinforcement particles to incorporate in matrix alloy. Specimens with thickness of 4 mm were prepared for welding from the cylindrical cast.

**Fig. 1** Manufacturing arrangement for stir casting technique

**Table 2** Levels of process parameters for FSW

| S. No. | Parameter | Units | Notation | Levels | | |
|---|---|---|---|---|---|---|
| | | | | −1 | 0 | 1 |
| 1 | % of SiC | – | *A* | 5 | 10 | 15 |
| 2 | Post weld heat treatment temperature | °C | *B* | −140 | 27 | 194 |
| 3 | Tool rotational speed (FSW) | rpm | *C* | 1000 | 1200 | 1400 |

## 2.2 *Friction Stir Welding*

Table 2 shows levels of process parameters for joining the composite plates by FSW process. The tool made of C40 carbon steel has 18 mm shoulder diameter and the pin height 2.9 mm with a truncated cone having head 5.5 mm diameter and 6.5 mm base. Table 3 shows design matrix for carrying out friction stir welding process. A constant axial load of 0.5 kN is applied with a feed of 0.67 mm/s. Better resistance to corrosion is achieved when the rotational speed is 800 rpm [12]. The FS welded composites are shown in Fig. 2.

## 2.3 *Post Weld Heat Treatment*

As per design matrix, the FS as-welded composites were treated in a muffle furnace to an elevated temperature of 300 °C for an hour and cooled. The FS as-welded composites were treated in CTS H-82 CHEST Model cryostat chamber at cryogenic temperature of −140 °C. A control accuracy of ±5 °C could be varied from ambient to −190 °C. Flow can be controlled using solenoid valve and pressure regulator. Cryogenic treatment was carried out for 96 h.

**Table 3** Design matrix measured properties of FS welded MMCs

| Trial run | Design matrix | | | Measured properties of FS welded MMCs | |
|---|---|---|---|---|---|
| | (% of SiC), *A* | Post weld heat treatment temperature (°C), *B* | Tool rotational speed (rpm), *C* | Corrosion potential mV, $E_{Corr}$ | Pitting potential mV, $E_p$ |
| 1 | 5 | −140 | 1200 | −1095 | −735 |
| 2 | 5 | 194 | 1200 | −1057 | −730 |
| 3 | 15 | −140 | 1200 | −1098 | −730 |
| 4 | 15 | 194 | 1200 | −1050 | −720 |
| 5 | 5 | 27 | 1000 | −1100 | −760 |
| 6 | 5 | 27 | 1400 | −1075 | −758 |
| 7 | 15 | 27 | 1000 | −1080 | −745 |
| 8 | 15 | 27 | 1400 | −1064 | −738 |
| 9 | 10 | −140 | 1000 | −1115 | −750 |
| 10 | 10 | −140 | 1400 | −1068 | −740 |
| 11 | 10 | 194 | 1000 | −1024 | −722 |
| 12 | 10 | 194 | 1400 | −1030 | −723 |
| 13 | 10 | 27 | 1200 | −1055 | −735 |
| 14 | 10 | 27 | 1200 | −1047 | −740 |
| 15 | 10 | 27 | 1200 | −1061 | −730 |

**Fig. 2** FS welded composites

## 2.4 *Microstructure*

Optical microscope was used to examine the different zones on weld metal. Photographs were taken to analyze the areas of welded zone. Keller's reagent was

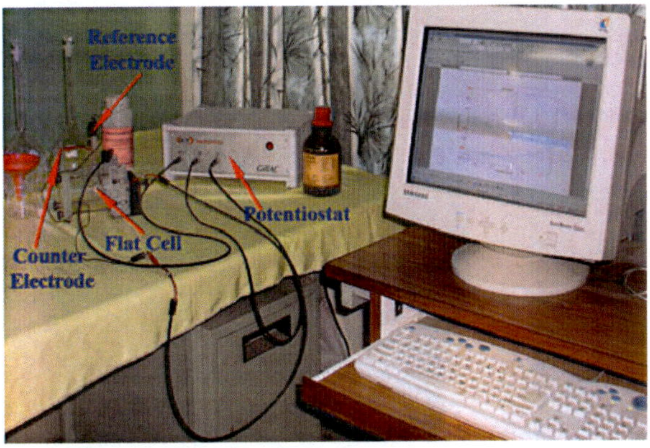


**Fig. 3** Corrosion test experimental setup

used as etchant to expose the microstructures. Optical microstructures of sample specimens are offered.

### 2.5 Corrosion Test

Single surface of the composites was exposed to the aggressive solution when embedded in epoxy resin and copper wire was connected electrically in a glass tube. Emery papers of 1000 grit size were used to polish the composites. The composites are washed using distilled water which was used to wash the composites, and smooth polished surface was obtained using degreaser. A flat cell was used to conduct electrochemical test using aqueous solution with 3.5 wt% NaCl (6.9 pH).

Tafel curves were drawn to analyze the resistance toward pitting. Stabilization of potential is obtained after immersing the specimen in flat cell for 50 min. A complete hardware AC Gill 5500 Potentiostat/Galvanostat was used to obtain the data on reaction of electrochemical system to dc excitation. Figure 3 shows a personal computer, a potentiostat and electrochemical cell. Potential range varied from −1000 to +400 mV versus SCE was investigated with rate of scan 0.5 mV/s. FS welded samples were tested on three zones (weld metal, LM25 alloy and TMAZ). Figure 4 shows a sample current versus potential Tafel curve.

## 3 Modeling

### 3.1 Formulation of Mathematical Models

$y = f\ (A, B, C)$ express the response function representing the pitting potential. $A$ represents % of SiC, $B$ represents post weld heat treatment, and $C$ represents tool

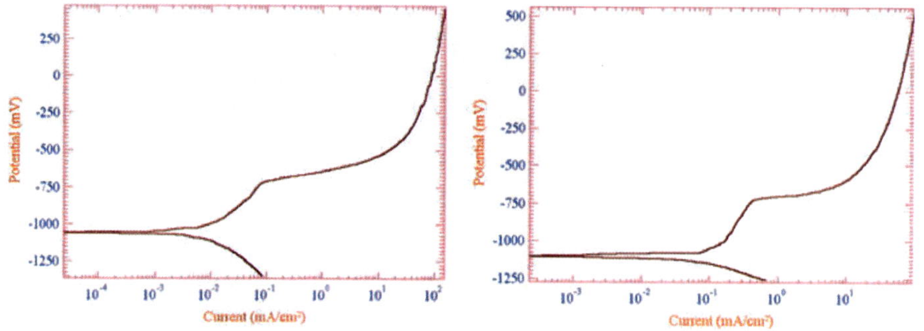


a. Base Material b. FS Welding

**Fig. 4** Current versus potential: **a** Base material, **b** friction stir welding

rotational speed for FSW process. Equation (1) depicts the regression polynomial equation to characterize the response surface.

$$Y = b_o + \sum_{i=1}^{k} b_i x_i + \sum_{\substack{i=1 \\ i \neq j}}^{k} b_{ii} x_{ii} + \sum_{i,j=1}^{k} b_{ij} x_{ij} \quad (1)$$

The selected polynomial for three factors is expressed as

$$Y = b_0 + b_1\,A + b_2\,B + b_3\,C + b_{11}\,A^2 + b_{22}\,B^2 + b_{33}\,C^2 + b_{12}\,AB + b_{13}\,AC + b_{23}\,BC$$

Minitab software was used to carry out the mathematical models, and Eqs. (3) and (4) depict the model for pitting and corrosion potential.

Pitting potential

$$\text{(FSW)}, \mathrm{E_{PF}} = -735 + 1.375A + 8.250B - 0.375C - 5.625A^2 + 11.625B^2 - 12.125C^2 - 1.000AB + 3.750AC - 2.000BC \quad (3)$$

Corrosion potential

$$\text{(FSW)}, \mathrm{E_{FCorr}} = -1049.00 + 7.25A + 17.63B + 3.88C - 23.00A^2 + 5.25B^2 - 5.25C^2 - 0.75AB - 4.75AC - 3.00BC \quad (4)$$

**Table 4** ANOVA and adequacy results for FS welded MMCs

| Details of responses | Source | Sum of squares | D. F. | Mean squares | *F*-ratio | Square multiple *R* (%) | Adjusted square multiple *R* (%) | Model adequacy |
|---|---|---|---|---|---|---|---|---|
| Pitting potential | Regression | 1882.18 | 9 | 209.13 | 2.41 | 89.08 | 76.54 | Adequate |
| | Residuals | 230.75 | 5 | 46.15 | | | | |
| | Lack of fit | 180.75 | 3 | 60.25 | | | | |
| | Pure error | 50.00 | 2 | 25.00 | | | | |
| Corrosion potential | Regression | 5349.68 | 9 | 594.41 | 4.11 | 93.02 | 80.46 | Adequate |
| | Residuals | 401.25 | 5 | 80.25 | | | | |
| | Lack of fit | 345.25 | 3 | 115.08 | | | | |
| | Pure error | 56.00 | 2 | 28.00 | | | | |

*F*-ratio = Mean square of lack of fit/mean square of pure error
S.S. = Sum of squares
D.F. = Degrees of freedom
Mean square = Sum of squares/degrees of freedom

## 3.2 *Testing the Adequacy of Models*

ANOVA is used to test the adequacy of models, and Table 4 confirms the adequacy of all models.

## 3.3 *Model Validation*

Same experimental setup used for conformity tests to analyze the accuracy of mathematical models. Intermediate values were assigned between upper and lower limits of process parameters, and the responses were measured. Comparison was done between observed and predicted values. 99% mean accuracy confirms the results for $E_P$ and $E_{Corr}$. Table 5 depicts conformity test results.

$$\%\text{Error} = \frac{\text{Observed Value} - \text{Predicted Value}}{\text{Predicted Value}} \times 100$$

# 4 Optimization

Equation (3) was considered as objective function, and Excel solver software was used to optimize the pitting potential for FS welded composites. Equation (5) depicts the objective function for the pitting potential of welded MMCs.

**Table 5** Confirmation of corrosion property tests for FS welded MMCs

| S. No. | Process parameters | | | Pitting potential (mV) | | | Corrosion potential (mV) | | |
|---|---|---|---|---|---|---|---|---|---|
| | % SiC, *A* | PWHT temperature (°C), *B* | Tool rotational speed (rpm), *D* | Predicted | Observed | % error | Predicted | Observed | % error |
| 1. | 5 | 119 | 1132 | −733.02 | −749 | 2.18 | −1070.53 | −1054 | −1.54 |
| 2. | 10 | 154 | 1328 | −728.19 | −743 | 2.03 | −1033.7 | −1019 | −1.42 |
| 3. | 15 | −45 | 1136 | −742.79 | −731 | −1.59 | −1071.84 | −1052 | −1.85 |
| Mean error | | | | | | 0.87 | | | −1.60 |

**Table 6** Optimized solutions for FS welded MMCs

| Optimal process parameters |
|---|
| $A = 15\%$ SiC |
| $B = -11$ °C |
| $D = 1400$ rpm |
| Optimized pitting potential = −748.6 mV |

Maximize pitting potential

$$\text{(FSW)}E_{\text{PF}} = -735.000 + 1.375A + 8.250B - 0.375C - 5.625A^2 + 11.625B^2 - 12.125C^2 - 1.000AB + 3.750AC - 2.000BC \quad (5)$$

The following constraints were assumed to optimize the pitting potential of FS welded composites

$$-1 < A < 1; -1 < B < 1; -1 < C < 1$$

After iterations, the optimized solutions obtained are presented in Table 6.

Conformity test was conducted to verify the results of optimization. Optimum parameter was set for performing weld trials. FS welded composite were subjected to corrosion tests as discussed in 2.5, and corrosion properties were measured. Table 7 shows the recorded average values.

## 5 Results and Discussions

Figure 5 shows the optical micrographs of 5, 10 and 15 vol.% of SiC composites. Uniform distribution of SiC is obtained if volume fraction of SiC is increased. Asthana [13] reports that the particle distribution is uniform when volume fraction of particles increases.

Optical microstructures [14] of the FS welded joint for different zones of composite containing 15% SiC are shown in Fig. 6. The typical microstructure of FS welded composite containing 15% SiC reveals that plastic deformation and thermal variation were found in the TMAZ.

**Table 7** Conformity test results for optimization of corrosion properties of FS welded MMCs

| S. No. | Process parameters | | | Pitting potential | | | Corrosion potential | | |
|---|---|---|---|---|---|---|---|---|---|
| | % SiC, *A* | PWHT temperature (°C), *B* | Tool rotational speed (rpm), *D* | Predicted | Observed | % error | Predicted | Observed | % error |
| 1. | 15 | −11 | 1400 | −748.6 | −759 | 1.38 | −1073.73 | −1097 | 2.17 |

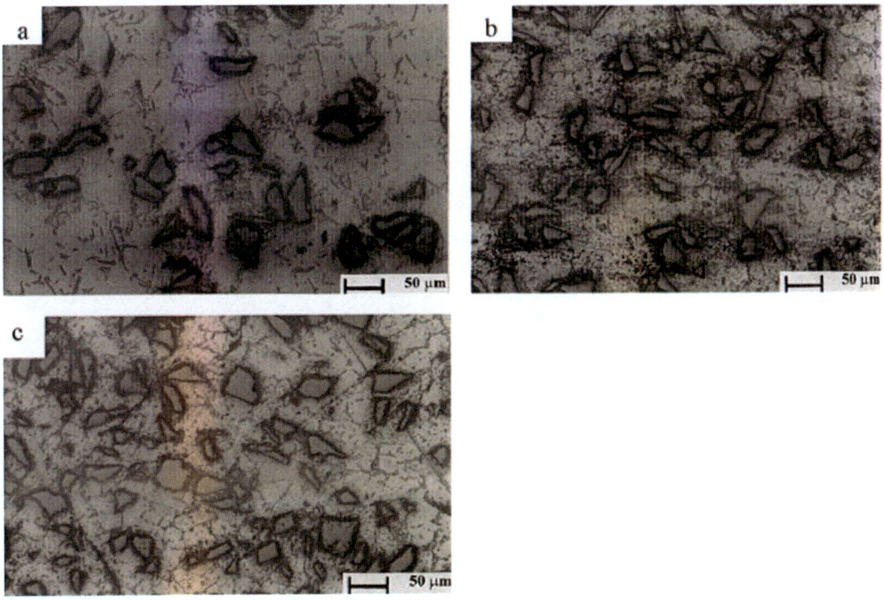


**Fig. 5** Optical microstructure of MMCs having: **a** 5% SiC, **b** 10% SiC and **c** 15% SiC

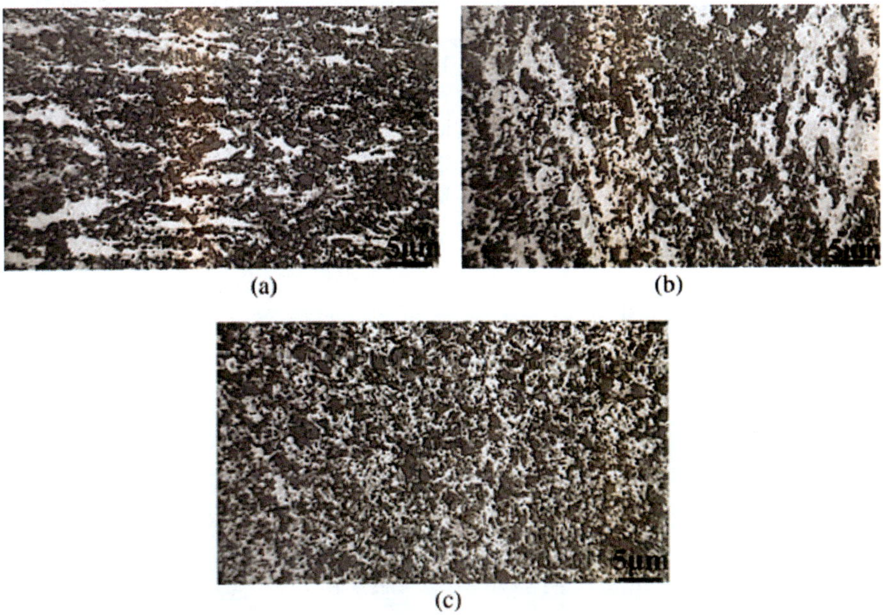


**Fig. 6** Optical microstructures of the FS welding of 15% SiC composite (specimen 8): **a** LM25, **b** TMAZ, and **c** weld nugget

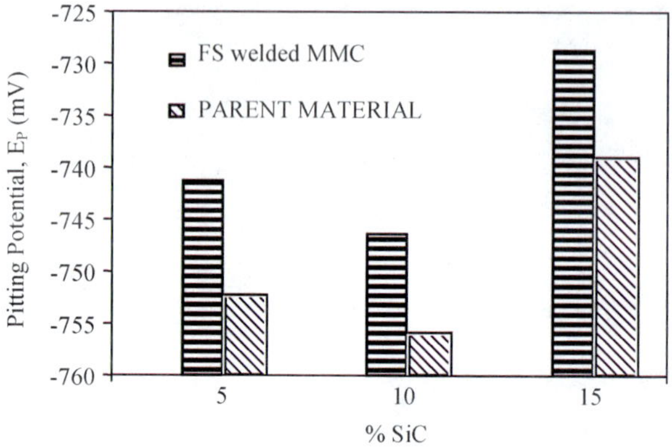


**Fig. 7** Comparison of pitting potential of FS welded MMCs and parent material for different % composition of SiC

More uniform distribution of SiC particles which was identified in the welded zone suggests that during FS welding, particles rearrange when stirred and deformed. Microstructural examination of FS welded composites using optical microscope shows no typical defects such as porosity or reinforcement segregation which were observed in the welded zone of composites.

The outcome of % SiC versus pitting potential is shown in Fig. 7. The pitting potential elevates in MMCs containing 15% SiC joined FS welding compared to MMCs containing 5 and 10% SiC. The reason may be due to occurrence of large amount of SiC particles uniformly distributed in matrix alloy.

The outcome of % SiC versus corrosion potential is shown in Fig. 8. The corrosion potential is high in MMCs containing 15% SiC compared to MMCs containing 5% and 10% SiC.

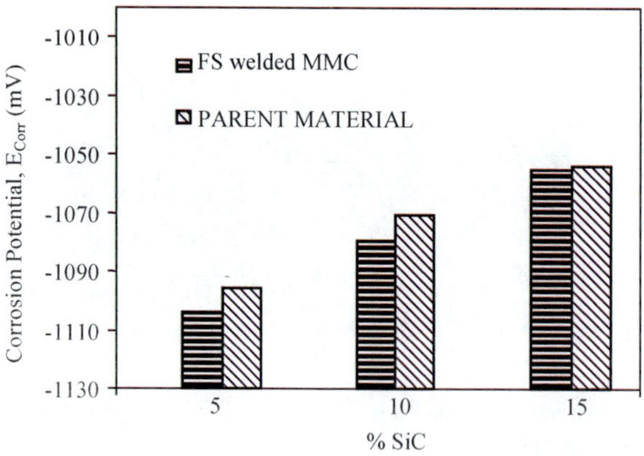


**Fig. 8** Comparison of corrosion potential of FS welded MMCs and parent material for different % composition of SiC

Figure 9 depicts the outcome of post weld heat treatment temperature on pitting potential for FS welded and parent composites. The figure confirms that pitting potential is high in specimens treated at elevated temperature. The pitting potential is less in parent material compared to FS welded composites.

Figure 10 depicts the outcome of post weld heat treatment temperature on corrosion potential. It is evident that corrosion potential is high in specimens treated at high temperature compared to specimens treated in cryogenic temperature and as-welded composites. The corrosion potential is less in FS welded composites compared to parent material.

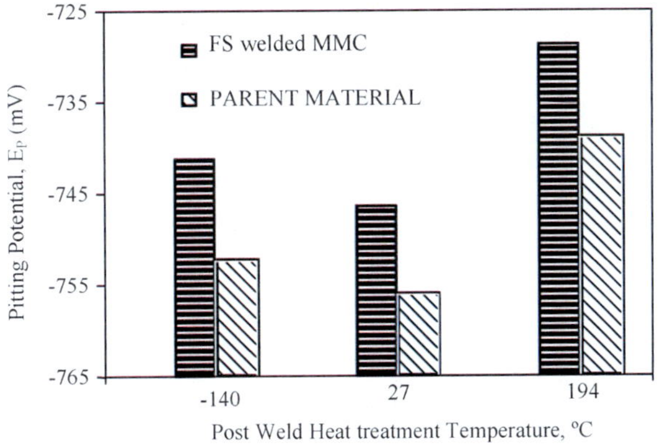


**Fig. 9** Comparison of pitting potential of FS welded MMCs and parent material heat treated after welding at different temperatures

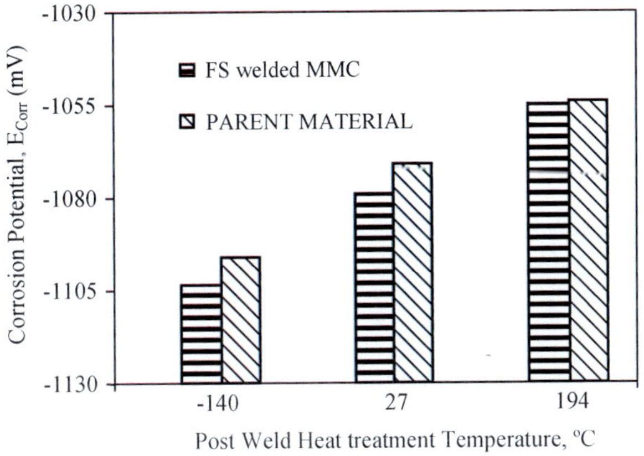


**Fig. 10** Comparison of corrosion potential of FS welded MMCs and parent material heat treated after welding at different temperatures

## 6 Conclusions

The following conclusions are drawn based on the experimental work carried out on FS welded LM25-SiC metal matrix composites:

- SiC particles are uniformly distributed when stir casting process is used for manufacturing MMCs. Uniform of distribution of SiC was due to increase in volume of SiC particles.
- Uniform distribution in welded zone suggests particle rearrangement occurred when highly stirred and deformed during FSW process.
- FS welded MMCs with 10% SiC and processed at cryogenic atmosphere has maximum pitting resistance.
- FS welded MMCs processed at high temperature reveals high pitting potential.
- Pitting potential was high in MMCs processed at high temperature when analyzed to as welded and processed at cryogenic atmosphere.

## References

1. McDanels DL (1985) Analysis of stress-strain, fracture and ductility behavior of aluminum matrix composites containing discontinuous silicon carbide reinforcement. Metal Trans A 16:1105
2. Ulhas K, Annigeri Veeresh Kumar GB (2017) Method of stir casting of aluminum metal matrix composites: a review. Mater Today: Proc 4(2):1140–1146
3. Zhang D, Zhao LG, Roy A (2019) Mechanical behavior of silicon carbide under static and dynamic compression. J Eng Mater and Technol, ASME, 141(1):011007. (10 pages)
4. Skolianos S (1996) Mechanical behavior of cast SiCp–reinforced Al–4.5%Cu–1.5%Mg alloy. Mater Sci Eng, A 210:76–82
5. Kang CG, Yoon JH, Seo YH (1997) The upsetting behavior of semi-solid aluminum material fabricated by a mechanical stirring process. J Mat Proc Tech 66:30
6. Hanumanth GS, Irons GA (1993) Particle incorporation by melt stirring for the production of metal-matrix composites. J Mat Sci 28:245
7. Luo ZP (2006) Crystallography of SiC/$MgAl_2O_4$/Al interfaces in a pre-oxidized SiC reinforced SiC/Al composite. Act Mat 54:47–58
8. Uzun Huseyin (2007) Friction stir welding of SiC particulate reinforced AA2124 aluminium alloy matrix composite. Mater Des 28:1440–1446
9. Liu G, Murr LE, Niou CS, McClure JC, Vega FR (1997) Microstructural aspects of the friction-stir welding of 6061-T6 aluminum. Scr Mat 37:355
10. Ouyang Jiahu, Yarrapareddy Eswar, Kovacevic Radovan (2006) Microstructural evolution in the friction stir welded 6061 aluminum alloy (T6-temper condition) to copper. J Mat Proc Tech 172:110–122
11. Venugopal P, Murugan N (2012) Corrosion analysis of GTA welded metal matrix composites. J Eng, Des Technol 10(1)
12. Elatharasan G, Kumar VS (2014) Corrosion analysis of friction stir welded AA 7075 aluminum alloy. J Mech Eng 60(1):29–34

13. Asthana R (1977) Solidification processing of reinforced metals. Transtech Publishers, Switzerland
14. Shamsipur Ali, Anvari Amir, Kayvani Ahmad (2018) Improvement of microstructure and corrosion properties of friction stir welded AA5754 by adding Zn layer. Int J miner Metall Mater 25:967–973

# Experimental Investigation on Tool Wear Using Sub-zero Treated Insert in Turning of EN 24 Steel

**C. Ramesh Kannan, S. Manivannan, and V. Raj Kumar**

**Abstract** In this paper, experimental investigations on tool wear have been made to facilitate increasing the tool life by sub-zero treatment of carbide insert tool. The effect of computer numerical control (CNC) inserts during orthogonal machining of AISI4340/EN 24 steel, cutting parameters are originated, and the tool life is predictable and compared with untreated and sub-zero treated insert. Generally, the inserts are treated under sub-zero temperature treatment as there is an increase in tool life. Turning parameters such as depth of cut, feed and speed are considered to analyze tool wear. Micro-hardness measurements have been carried out to identify the hardness of treated and untreated insert. Equations are developed using Buckingham's $\pi$ theorem, and the results have been validated against experiments. For further implementation, manufacturing with respect to the tool life optimization is discussed.

**Keywords** Sub-zero treatment · Buckingham's $\pi$ theorem · Carbide insert · Steel

C. Ramesh Kannan · S. Manivannan
Department of Mechanical Engineering, Karpagam Academy of Higher Education, Coimbatore, Tamil Nadu, India
e-mail: crkannanmech@gmail.com

S. Manivannan
e-mail: manivannan.s@kahedu.edu.in

V. Raj Kumar (✉)
Department of Mechanical Engineering, PSG Institute of Technology and Applied Research, Coimbatore, Tamil Nadu, India
e-mail: rajkumarv@psgitech.ac.in

© The Editor(s) (if applicable) and The Author(s), under exclusive license to Springer Nature Singapore Pte Ltd. 2021

S. Mohan et al. (eds.), *Materials, Design, and Manufacturing for Sustainable Environment*, Lecture Notes in Mechanical Engineering,
https://doi.org/10.1007/978-981-15-9809-8_64

## 1 Introduction

A lathe is a machine tool used principally for shaping pieces of metal, wood or other materials by causing the workpiece to be held and rotated by the lathe while a tool bit is advanced into the work causing the cutting action. Tool life is an important parameter in evaluating the performance of the cutting tools. Tool wear affects dimensions and surface quality of the workpiece, and it is also one of the important criteria in determining tool life. Sub-zero treatment (SZT) is sometimes conducted within the outlined temperature and holding time at this temperature from one to two hours. Prolonged holding at temperature when hardening can stabilize the primary solid solution of the many grades of steels and scale back the impact if sub-zero treatment directly following the hardening operation. A particular quantity of maintained primary solid solution might continuously be found in hardened steel. Maintaining primary solid solution reduces hardness, wear resistance and thermal conduction of steel and makes its dimensionally unstable.

The effect of sub-zero treatments alters the mechanical and microstructure properties. Retained austenite gets transformed to Martensite. The micro-structural modification studies concluded on enhancing mechanical properties of tool steel [1]. Wear behavior of steel specimens has been studied on varying the sub-zero treatments against conventionally heat treated ones. Wear behavior of the specimens has been explained on the basis of their characterization [2]. This work deals with effect of cryogenic treatment at sub-zero level in high speed tool steel which is widely used in automobile industries. Wear resistance was experimented by using sliding wear abrasion technique, and its hardness was observed using Vickers microhardness tester (MMTX7) [3]. This paper deals with the operative modes and mechanisms of wear. The above depends on the temperature of sub-zero treatment and the sliding velocity of the wear test [4]. Investigations have dealt with built-up edge formation, and layers of the tool rake face during turning process [5]. The sub-zero treatments alter the precipitation behavior of secondary carbides and decrease the temperature of sub-zero treatment where bulk hardness will increase marginally; however, actual hardness of the matrix is increased significantly through deep refrigerant treatment.

Fracture toughness was measured by bend tests exploiting miniature chevron-notched spherical bar specimens [6]. Multiple tempering at moderately higher temperature and for longer duration is done to minimize the quantity of roughness content in tool steels [7]. The C45 steel shows the force required for cutting the steel was lower for treated condition than untreated condition (carbide tool) [8]. The machinability of hardened steel was analyzed by measuring cutting forces, tool wear and surface finish of the workpiece [9]. Studies vividly show the influence of the hardness on the friction coefficient, while the wear rates have great understanding of the sliding speed and hardness [10, 11]. Savings in machining costs employing multilayer TiN-coated insert are compared to uncoated carbide inserts in hard machining considering flank wear. This advocates economical possibility of utilizing multilayer TiN-coated carbide inserts in hard turning applications [12–15].

# 2 Details of the Experiment

## 2.1 *Work Specimen and Tool Material*

EN 24 steel is used as workpiece material. EN 24 is a high tensile, medium carbon alloy steel. It combines high tensile strength, shock resistance and high ductility. Chemical compositions of EN 24 steel in weight percentages are carbon 0.38%, silicon 0.35%, manganese 0.70%, sculpture 1.8% and molybdenum 0.35%, respectively.

Tungsten carbide is the material used which has elevated strength, and it is highly compression resistant than any melted/casted or forged metals and alloys. Tungsten carbide retains toughness and impact power in high cryogenic temperature ranges (−453 °F.). The compositions show simple dry coefficient of friction.

## 2.2 *Sub-zero Treatment*

Sub-zero treatments were done in which the tool is soaked in liquid nitrogen and placed under liquid nitrogen bath. Initially, the temperature is maintained under liquid nitrogen vacuum atmosphere. The tool is maintained at −72 °C for 24 h. Then the tools are kept to be under room temperature. Micro-hardness of the tool is checked in best treatments. The micro-hardness is taken at a magnification of 400×. Figure 1 shows the micro-hardness of the tool before the treatment process is carried out.

The micro-hardness is taken off at a magnification of 400×. The following figure shows the micro-hardness of the tool after the treatment process is conceded. The treatment is carried out at sub-zero level in which at off −72 °C. The tool is treated for the same at about 24 h which shows improvement in the hardness level.

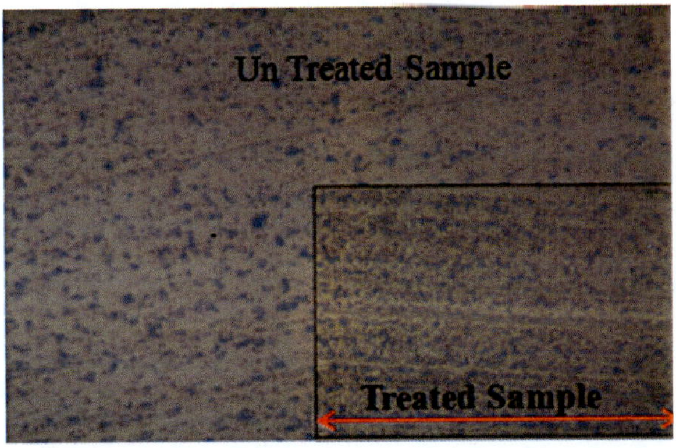


**Fig. 1** Micro-hardness image of cutting tool

### 2.3 Experimental Procedure for Turning

The objective of the turning experiment is to study the trends on wear and tool life of sub-zero treated and untreated samples in a continuous cutting instance. Determination of wear resistance and tool life is based on the magnitude of flank wear on the insert during a turning operation. For the experiments performed with continuous cuts, a new cutting edge was used for each length of time that the insert was to perform a cut. For experiments performed with repeated cuts, the same cutting edge was used for a specified length of time throughout the whole experiment, and there was a 5-min break in between each successive cut. Coolant was not used in any of the turning experiments.

## 3 Result and Discussion

In the process of conducting experiments using the lathe, certain phenomena were observed that greatly affected the outcome of the tests. It took quite some time before some of these phenomena were discovered, and it required whole sets of tests to be repeated. However, by doing so, the reliability of the results was very much increased and gave confidence to conclusions made in the later chapters of this report.

$$\text{Material removal rate} = \pi \times D_{\text{avg}} \times f \times d \times N$$

where

$N$ Rotational speed of workpiece, rpm
$f$ Feed rate, mm/rev
$D_{\text{avg}}$ $(D_{\text{a}} + D_{\text{b}})/2$
$D_{\text{a}}$ Workpiece diameter after machining
$D_{\text{b}}$ Workpiece diameter before machining
$d$ Depth of cut, mm.

The optimization of tool life is carried out by using Buckingham's pie theorem. Buckingham pie theorem deals with the dimension variables, and this study derives dimensionless correlations to describe the life of treated and untreated samples.

The procedure adopted to find out the dimensionless parameters, $\pi$

- Roll of corresponding parameters;
- Apply the MLT system;
- Details of dimensional parameters; and
- Let $f$, $V$, $d$, $t$, @, $M$, $O$ and m indicate the repeating variables.

Let the variables be feed ($f$), depth of cut ($d$), speed ($s$), time ($t$), material removal rate ($m$), wear ($w$), density ($\rho$).

Let feed, density and speed be a constant factor for a set of readings.

Depth of cut, material removal rate, wear and time as variables

$$\prod_1 = f^{a1}\,\rho^{b1}\,s^{c1}\,d$$

$$\prod_2 = f^{a2}\,\rho^{b2}\,s^{c2}\,m$$

$$\prod_3 = f^{a3}\,\rho^{b3}\,s^{c3}\,w$$

$$\prod_4 = f^{a4}\,\rho^{b4}\,s^{c4}\,t$$

$$F\left(\prod_1, \prod_2, \prod_3, \prod_4\right) = 0$$

$$\prod_1 = (L)_1^a\left(ML^{-3}\right)^{b1}\left(T^{-1}\right)^{c1}L$$

$$\prod_2 = (L)_2^a\left(ML^{-3}\right)^{b2}\left(T^{-1}\right)^{c2}MT^{-1}$$

$$\prod_3 = (L)_3^a\left(ML^{-3}\right)^{b3}\left(T^{-1}\right)^{c3}L$$

$$\prod_4 = (L)_4^a\left(ML^{-3}\right)^{b4}\left(T^{-1}\right)^{c4}T$$

By solving the above equations, we get

$$a_1 = -1\ b_1 = 0\ c_1 = 0$$

$$a_2 = -3\ b_2 = -1\ c_2 = -1$$

$$a_3 = -1\ b_3 = 0\ c_3 = 0$$

$$a_4 = 0\ b_4 = 0\ c_4 = 1$$

By applying the solution, we get

$$\text{Tool life} = f\left(d/f,\ m/\rho\ sf^3\right)$$

Substituting these values, we get

$$\prod_{1} = d/f$$

$$\prod_{2} = m/(\rho s f^3)$$

$$\prod_{3} = w/f$$

$$\prod_{4} = t/s$$

By applying the solution obtained for tool life is a function of below-mentioned parameters:

$$\text{Tool life} = f\left(d/f, m/\rho s f^3, w/f, t/s\right)$$

Taking the values of speed as 700 rpm and feed as 0.2 mm/rev and depth of cut as 0.1, 0.15, 0.2 mm, the pie values are solved and analytical and theoretical calculations were made by integrating the function values. The closure of values occurs when the limit is about 2–5. When the limit is of 0–1, the error is higher, and thus, the normal limit value is taken and plotted (Figs. 2, 3, 4 and 5; Table 1).

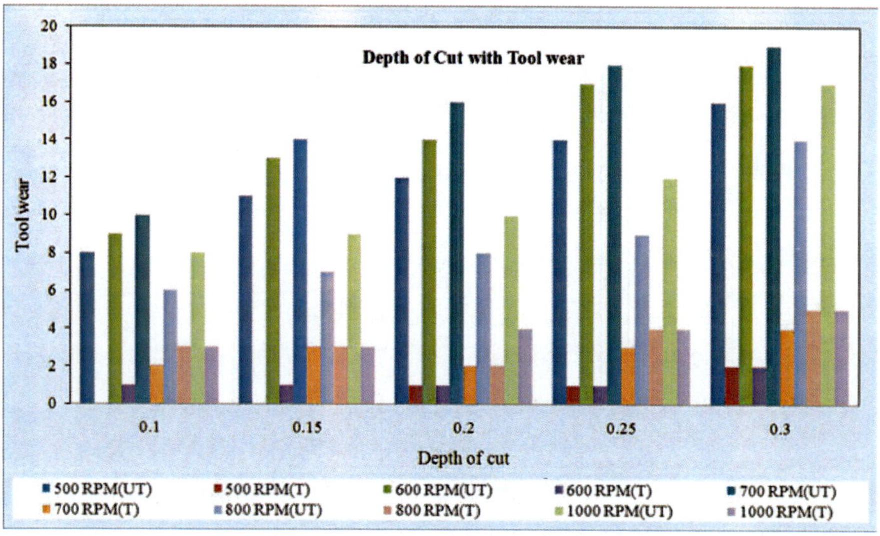


**Fig. 2** Comparison of untreated and treated tool wear with depth of cut

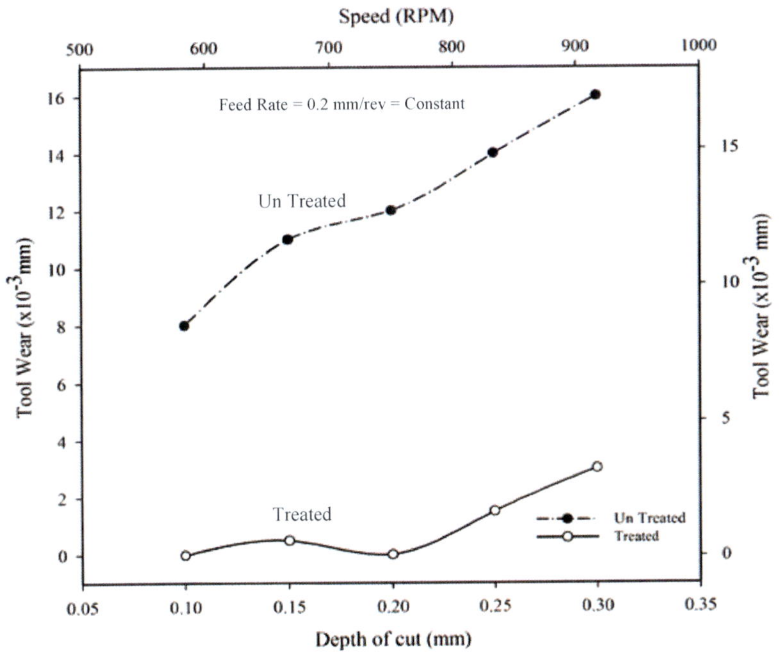


**Fig. 3** Effect of depth of cut and speed on tool wear

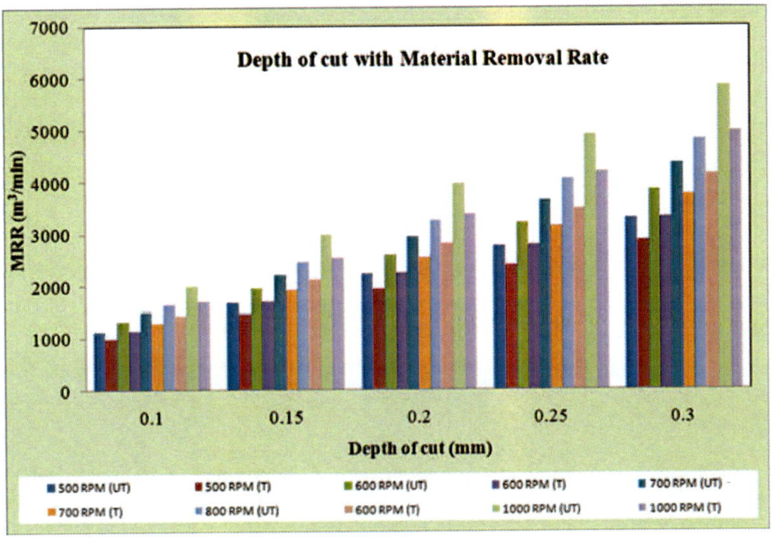


**Fig. 4** Comparison of untreated and MRR with depth of cut

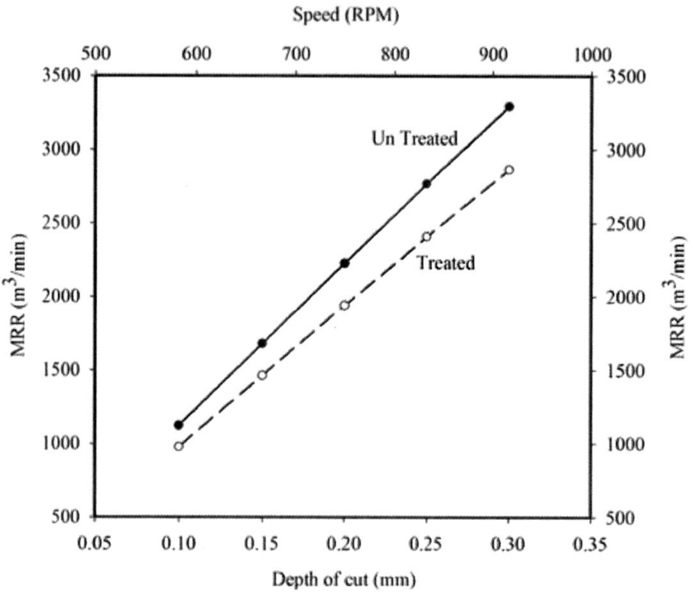


**Fig. 5** Effect of speed and depth of cut with MRR

## 4 Conclusions

EN 24 steel was (60 mm length × 35.8 mm diameter) used as work specimen, and turning operation was carried out. Flank wear and crater wear measured using Tool Maker's Microscope in various trials. Weight difference of tool inserts observed before the trial and after the trial using precision balance. The wear and the material removal rate is calculated after and before sub-zero treatment optimum tool life which is estimated in terms of time and total material to be consumed per material removed to find out. Theoretical and analytical results were compared to 1 which the theoretical results exploit over analytical results. The treated tool has fewer tool wear than the untreated tool. The hardness increase is shown in the microstructure. As a result, the tool wear is reduced and optimum cutting speed is found to be 700 rpm, feed rate being 0.2 mm/rev and the depth of cut of about 0.2 mm.

**Table 1** Tool wear and MRR for treated and untreated tool

| Sl. No. | Feed (mm/rev) | Speed (rpm) | DoC (mm) | Untreated tool | | Treated tool | |
|---|---|---|---|---|---|---|---|
| | | | | Wear rate ($\times 10^{-3}$ mm) | Material removal rate ($mm^3$/min) | Wear ($\times 10^{-3}$ mm) | Material removal rate ($mm^3$/min) |
| 1. | 0.2 | 500 | 0.1 | 8 | 1123.37 | 1 | 0979.36 |
| 2. | 0.2 | 500 | 0.15 | 16 | 1679.16 | 0 | 1463.16 |
| 3. | 0.2 | 500 | 0.2 | 12 | 2227.89 | 0 | 1939.89 |
| 4. | 0.2 | 500 | 0.25 | 11 | 2767.20 | 2 | 2407.20 |
| 5. | 0.2 | 500 | 0.3 | 16 | 3294.74 | 3 | 2862.74 |
| 6. | 0.2 | 600 | 0.1 | 10 | 1310.36 | 1 | 1137.56 |
| 7. | 0.2 | 600 | 0.15 | 19 | 1958.47 | 1 | 1699.27 |
| 8. | 0.2 | 600 | 0.2 | 13 | 2598.11 | 1 | 2252.51 |
| 9. | 0.2 | 600 | 0.25 | 9 | 3226.44 | 0 | 2794.44 |
| 10. | 0.2 | 600 | 0.3 | 14 | 3840.65 | 2 | 3322.25 |
| 11. | 0.2 | 700 | 0.1 | 6 | 1484.22 | 2 | 1283.19 |
| 12. | 0.2 | 700 | 0.15 | 5 | 2217.23 | 3 | 1916.55 |
| 13. | 0.2 | 700 | 0.2 | 8 | 2940.92 | 2 | 2540.01 |
| 14. | 0.2 | 700 | 0.25 | 7 | 3651.43 | 3 | 3150.28 |
| 15. | 0.2 | 700 | 0.3 | 5 | 4345.45 | 4 | 3744.07 |
| 16. | 0.2 | 800 | 0.1 | 6 | 1645.36 | 3 | 1416.27 |
| 17. | 0.2 | 800 | 0.15 | 7 | 2458.62 | 3 | 2114.98 |
| 18. | 0.2 | 800 | 0.2 | 7 | 3260.58 | 2 | 2802.39 |
| 19. | 0.2 | 800 | 0.25 | 9 | 4047.46 | 4 | 3474.72 |
| 20. | 0.2 | 800 | 0.3 | 10 | 4815.50 | 5 | 4128.22 |
| 21. | 0.2 | 1000 | 0.1 | 8 | 1993.90 | 3 | 1707.53 |
| 22. | 0.2 | 1000 | 0.15 | 8 | 2979.08 | 3 | 2549.52 |
| 23. | 0.2 | 1000 | 0.2 | 10 | 3950.12 | 4 | 3377.38 |
| 24. | 0.2 | 1000 | 0.25 | 12 | 4902.33 | 4 | 4186.41 |
| 25. | 0.2 | 1000 | 0.3 | 16 | 5830.98 | 5 | 4971.88 |

## References


1. Koneshlou M, Asl KM, Khomamizadeh F (2011) Effect of cryogenic treatment on microstructure, mechanical and wear behaviors of AISI H13 hot work tool steel. Cryogenic 51:55–61
2. Das D, Dutta AK, Ray KK (2011) Sub-zero treatments of AISI D2 steel: Part II. Wear behavior. Cryogenic 527:2194–2206
3. Da Silva FJ, Franco SD, Machado AR, Ezugwu EO, Souza AM (2006) Performance of cryogenically treated HSS tools. Wear 261:674–685
4. Das D, Ray KK, Dutta AK (2008) Influence of temperature of sub-zero treatments on the wear behavior of die steel. Wear 267:1361–1370

5. List G, Nouari M, Gehin D, Gomez S, Manaud JP, Le Petitcorps Y, Girot F (2005) Wear behavior of cemented carbide tools in dry machining of aluminium alloy. Wear 259:1177–1189
6. Das D, Sarkar R, Dutta AK, Ray KK (2010) Influence of sub-zero treatments on fracture toughness of AISI D2 steel. Mater Sci Eng 528:589–603
7. Kalyan Kumar KVBS, Choudhury SK (2008) Investigation of tool wear and cutting force in cryogenic machining using design of experiments. J Mater Process Technol 203:95–101
8. Sreerama Reddy TV, Sorna Kumar T, Venkatarama Reddy M, Venkat Ram R (2008) Machining performance of low temperature treated P-30 tungsten carbide cutting tool inserts. Cryogenics 48:458–461
9. Senthil Kumar A, Rajadurai A, Sornakumar T (2003) Machinability of hardened steel using alumina based ceramic cutting tools. Int J Refract Metal Hard Mater 21:109–117
10. Agrawal S, Chakrabarti AK, Chattopadhyay AB (1995) A study of the machining of cast austenitic stainless steel with carbide tools. J Mater Process Technol 52:610–620
11. Dhar NR, Paul S, Chattopadhyay AB (2002) Machining of AISI 4140 steel under cryogenic cooling–tool wear, surface roughness and dimensional deviation. J Mater Process Technol 123:483–489
12. Sahoo AK, Sahoo B (2012) Experimental investigations on machinability aspects in finish hard turning of AISI 4340 steel using uncoated and multilayer coated carbide inserts. Measurement 45:2153–2165
13. Das D, Dutta AK, Ray KK (2010) Sub-zero treatments of AISI D2 steel: Part I. Microstructure and hardness. Mater Sci Eng 527:2182–2193
14. Talib RJ, Zaharah AM, Selamat MA, Mahaidin AA, Fazira MF (2013) Friction and wear characteristics of WC and TiCN-coated insert in turning carbon steel workpiece. Procedia Eng 68:716–722
15. Kagnaya T, Boher C, Lambert L, Lazard M, Cutard T (2014) Microstructural analysis of wear micro mechanisms of WC–6Co cutting tools during high speed dry machining. Int J Refract Metal Hard Mater 42:151–162

# Design Fabrication and Control of Soft Robotic Gripper for Material Handling

D. Venkatesa Prabu, B. Meenaskhi Priya, and E. B. Priyanka

**Abstract** In recent years, soft robotics attracts the attention of robotic gripper research community in the wider range. Currently, many research analyses have been conducted on the area of material handling by utilizing soft robotic gripper technology for experimenting the various grasping techniques such as bioinspired, biomimetic design, silicon soft structure analysis, deformation and reinforcement sensing, various fabrication techniques, soft machine learning and soft composite. The presented research results enumerate the detailed fabrication process and control scheme for the soft robotic finger mainly toward the role of manipulator. The necessity of developing the soft robotic grippers specifically to perform various grasping action with standard size and shapes of objects. The proposed soft finger contains three axially aligned chambers, which are actuated independently to provide the imitator contribution of the human finger. The developed three-dimensional model of the soft finger is established using SolidWorks software and analyzed in ABAQUS and finite element analysis software. The analysis results reveal that the proposed model can withstand larger pressure and achieve different deformation in each chamber to perform material handling with excellent grasping techniques.

**Keywords** Robotic control · Soft gripper · Three chambers · Grasping · Reinforcement · Material handling

D. Venkatesa Prabu (✉) · B. Meenaskhi Priya · E. B. Priyanka
Department of Mechatronics Engineering, Kongu Engineering College, Erode, India
e-mail: venkatprabu13@gmail.com

B. Meenaskhi Priya
e-mail: b.meenakshipriya@gmail.com

E. B. Priyanka
e-mail: priyankabhaskaran1993@gmail.com

© The Editor(s) (if applicable) and The Author(s), under exclusive license to Springer Nature Singapore Pte Ltd. 2021

S. Mohan et al. (eds.), *Materials, Design, and Manufacturing for Sustainable Environment*, Lecture Notes in Mechanical Engineering,
https://doi.org/10.1007/978-981-15-9809-8_65

# 1 Introduction

In the industrial automation, using robotics technology is paving the large span of research analysis especially in the sector of material handling. Rather than the conventional robots, in addition its convoluting parts play more attention while handling the objects with trajectory planning [1–3]. By analyzing this, a soft robotics gripper fabrication and control scheme varies with respect to the load capacity and industrial consideration by pointing toward softness and deformable rate. Soft robotics is the emerging experimental category of soft grippers fabricating using silicone materials and living exploits similar to exhibit the living characterization. Robotic soft gripper goes with the way of self-exploiting and understanding the environmental needs based on the application and objects specifications [4–7]. By comparing with the robots developed using rigid natured substances, soft robots offer enhanced elasticity, adaptability and reinforcement in accomplishing tasks with more human comfort toward safety when it is under working environments. Soft robotics end-effector is shaving three-motion pitch, yaw and roll for grabbing and the external material [8–10].

Section 2 describes the different patterns of soft finger and designing of the gripper part. Section 3 describes the analysis that has been taken for finding the best pattern from the previous section. Section 4 concludes the result from the previous sections and describes where this model can be used for getting better results.

# 2 Design of Softgripper

The finger is designed with single chamber to mimic the animal finger [11]. The finger shown in Fig. 2 provides maximum bending for the given load.

## *2.1 3D Design for Pattern 1*

In this finger design, they have three parts, and they are actuator, top layer and bottom layer. These parts are helpful to produce soft finger as said in [12]. Additionally, the finger has stretcher part. The stretcher is provided with 2 mm at both ends of the finger (Fig. 1).

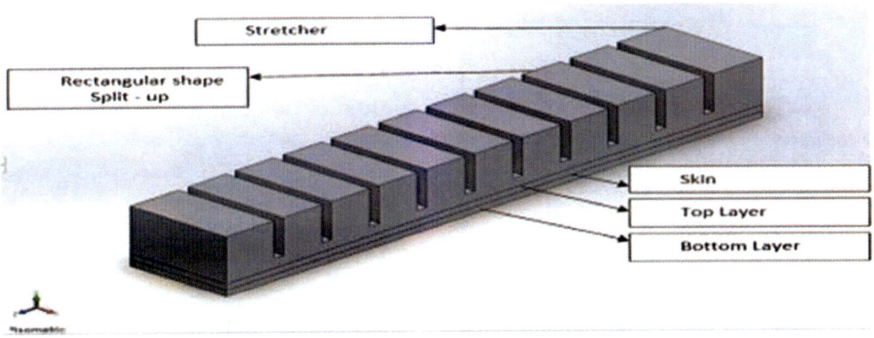


**Fig. 1** Isometric view of Pattern 1

## 2.2 2D Design for Pattern 1

Before going to the 3D design, make sure with the 2D—design for exact dimension to extrude the finger. Each split-up has 8 mm, and total length of the finger is 112 mm as shown in Fig. 2.

In between two split-ups, the length is 2 mm and height of the finger is 19 mm. Top layer and bottom layer have the 2 mm dimension. Breadth of the soft finger is 20 mm as shown in Figs. 3 and 4.

**Fig. 2** Front view of Pattern 1

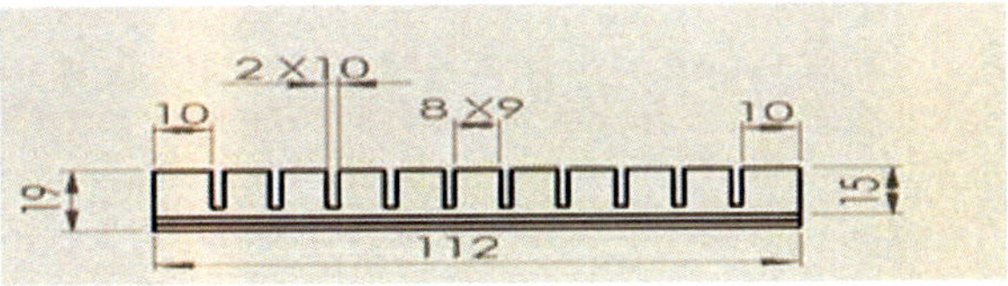


**Fig. 3** Bottom view 1

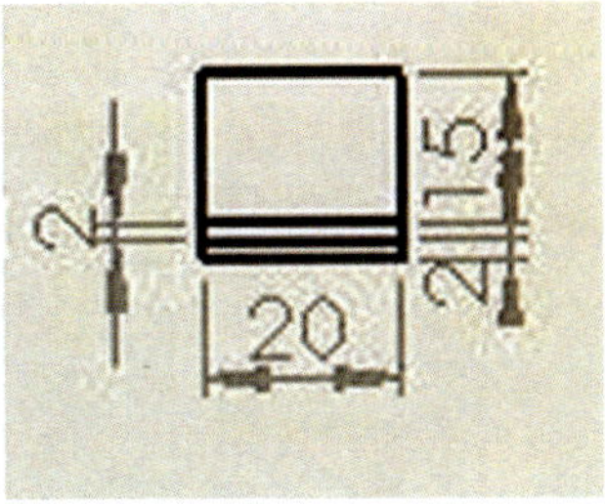

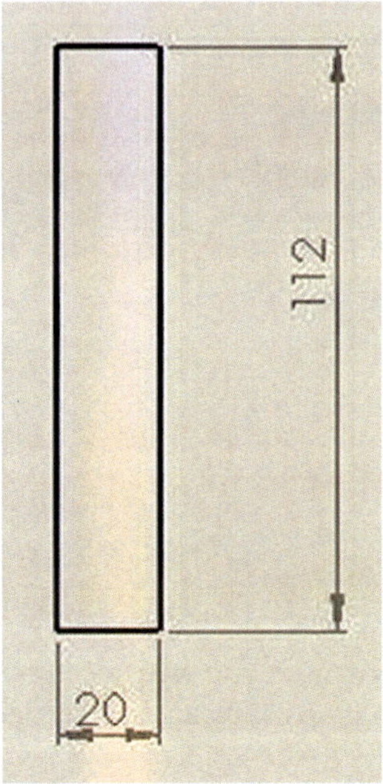


**Fig. 4** Side view 1

## 2.3 3D Design for Pattern 2

The finger shown in Fig. 5 is used for providing maximum bending that can be produced by this model for the given loads. In this finger design, they have three parts as same as in previous pattern. These parts are helpful to produce soft finger. Additionally, the finger has stretcher similar to previous one.

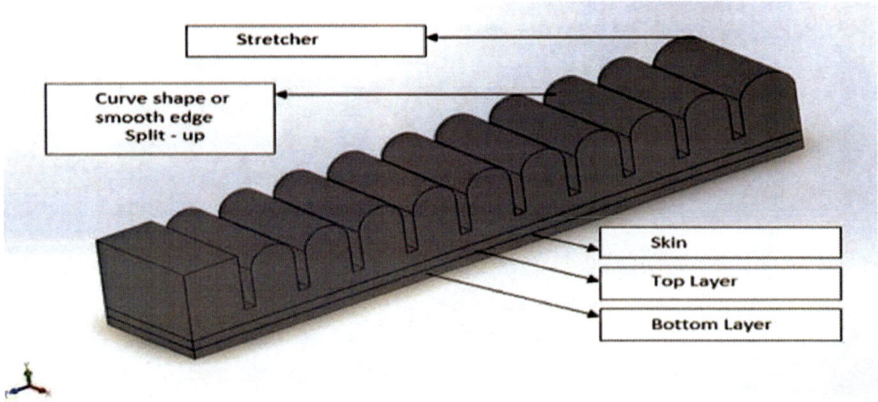


**Fig. 5** Isometric view of Pattern 2

**Fig. 6** Front view of Pattern 2

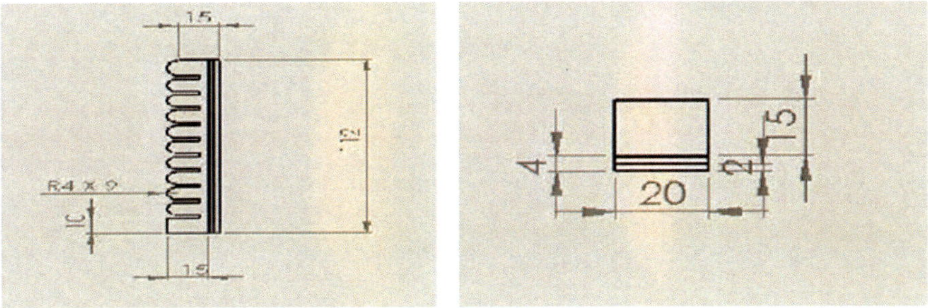


**Fig. 7** Right view 2 and side view 2

## *2.4 2D Design for Pattern 2*

Before going to the 3D design, ensure that the 2D design for exact dimension is to extrude the finger. Each split-up has 8 mm, and total length of the finger is 112 mm as shown in Fig. 6. The length of each split-up between the fingers is 2 mm, and height of the finger is 19 mm. Top layer and bottom layer have 2 mm. Breadth of the soft finger is 20 mm, and radius of the split-up is 4 mm as shown in Fig. 7.

## *2.5 3D Design for Pattern 3*

The finger is designed with single chamber to mimic the animal behavior. The finger shown in Fig. 8 is used for providing maximum bending that can be produced by this model for the given load. The top layer and bottom layer are designed as same as in the previous models except that there are some changes at the edge of the split-up areas. These parts are helpful to produce soft finger. Additionally, the finger has stretcher which has the 2 mm dimension.

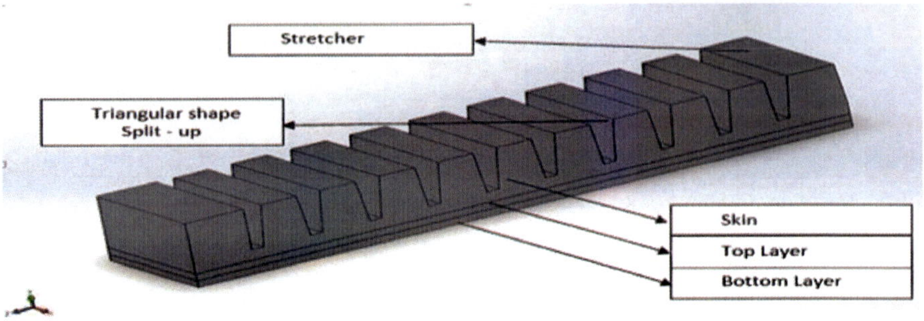


**Fig. 8** Isometric view of Pattern 3

**Fig. 9** Holder dimension

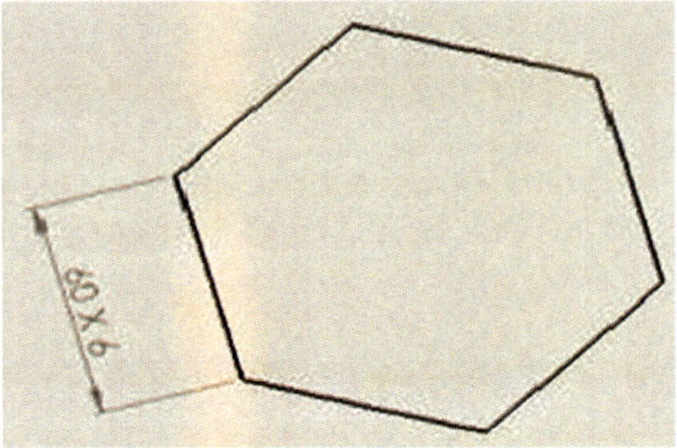


## 2.6 *Holder Design*

Hexagon structure is initially chosen for connecting six fingers. In future, more number of fingers can be added by altering the structure of the model as said in [13]. Each surface of the hexagon is of 60 mm dimension as shown in Fig. 9. To extrude the hexagon structure, a 30 mm dimension is selected in both directions as shown in Fig. 10.

The palm of the finger is made up of gray cast iron since the holder has to be fixed which the manipulator which is made of a hard material gray cast iron as shown in Fig. 10. So particularly, the same material is chosen for holder to fix the multiple fingers.

## 2.7 *Gripper Design*

The 3D model of soft gripper undertaken for the analysis is visualized in Fig. 11. Mainly silicone rubber is considered as the fabricating material for the soft gripper due to its flexibility nature and the hard material such as gray cast iron which is rigid one [14]. When the pressure is applied, the flexible part bends according to grasp the object.

Fig. 10 View of holder

Fig. 11 3D view of soft gripper

# 3 Analysis of Single Finger by Applying Air Pressure

These three types of pattern results discussed with finite element analysis are done by ABAQUS 6.14 software [15].

## 3.1 Pattern 1

The soft gripper bends for the applied pressure. In the following sections, analysis is taken by applying the positive pressures on the finger. It can obtain 360-degree deformation by using ABAQUS 6.14 simulation software as shown in Fig. 12. In that, 360-degree deformation is achieved in 55 kPa pressure for Pattern 1. Top layer

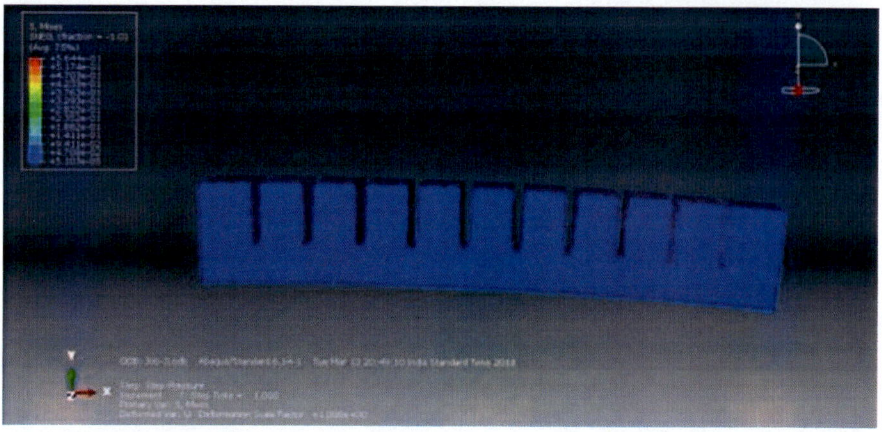

**Fig. 12** 5 kPa pressure applied

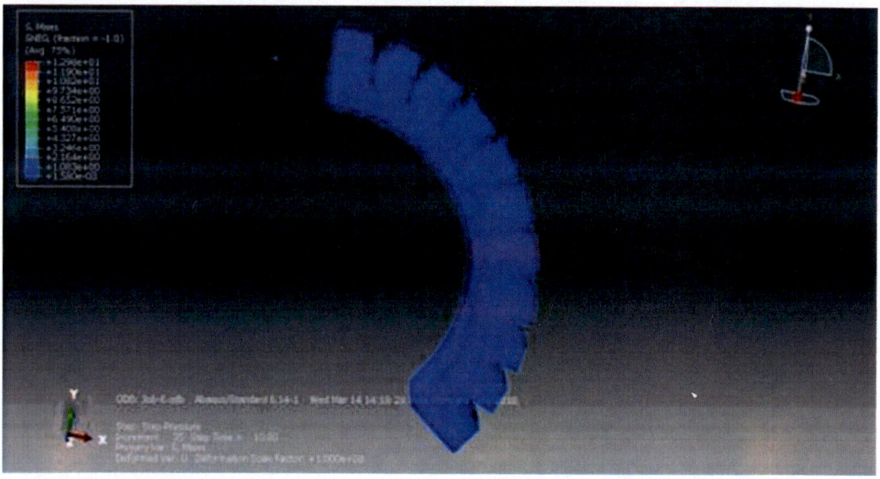

**Fig. 13** 30 kPa pressure applied

has split up for flexible rotation in that split-up has rectangular shape. Deformation of finger is shown in Figs. 13 and 14.

## *3.2 Deformation Graph*

In this graph, we have shown displacement (m) versus time (s). At 45 and 55 kPa, the finger gets deformation, and it can measure by displacement respected by time [16] as shown in Fig. 15, respectively.

**Fig. 14** 55 kPa pressure applied

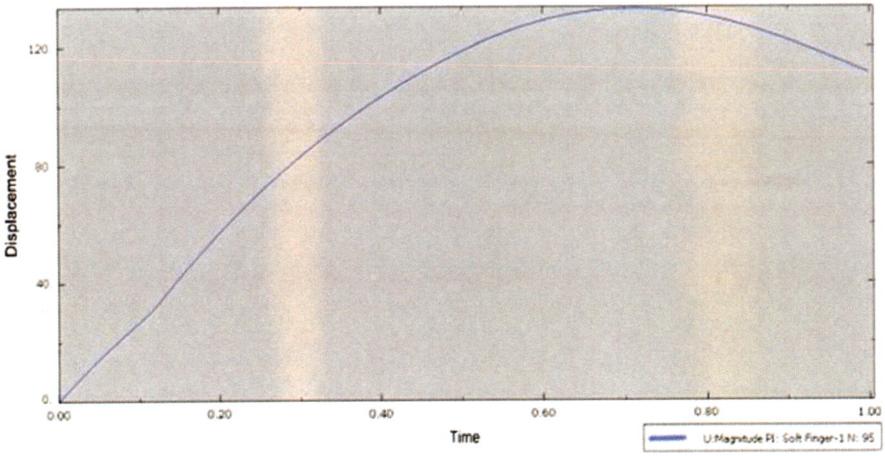


**Fig. 15** 55 kPa pressure graph

## 3.3 *Pattern 2*

The soft gripper is bending due to applied finite element analysis. In the following sections, analysis is taken the positive pressure which is considered for analysis Dong et al. [17]. It can obtain 360-degree deformation by ABAQUS 6.14 simulation software as shown in Fig. 16. In that, 360-degree deformation is achieved by 65 kPa pressure for this Pattern 2. This is done by ABAQUS 6.14 software [18]. Top layer has split up for flexible rotation in that split-up has curved shape or smooth edges. Deformation of finger is shown in Figs. 16, 17 and 18.

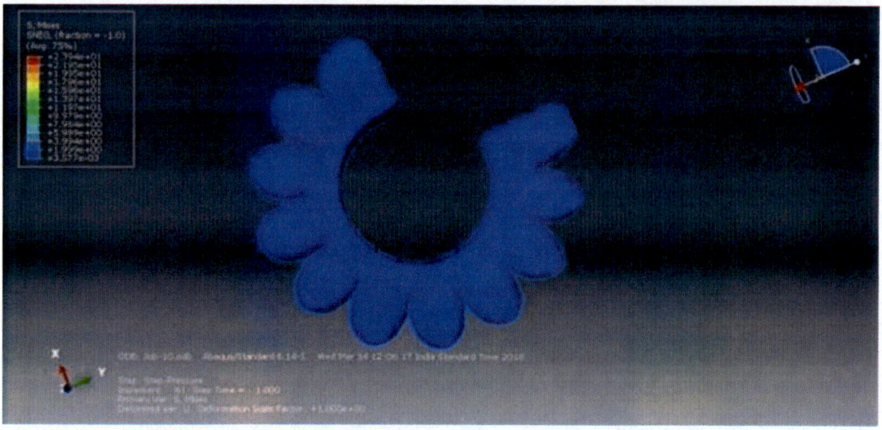

**Fig. 16** 60 kPa pressure

**Fig. 17** 65 kPa pressure

## *3.4 Gripper Analysis*

The analysis of gripper has been done using SolidWorks platform for experimentation. The analysis has taken by applying different load pressures throughout the chambers as revealed in Fig. 19. On comparing the various load pressures, the responses related to stress, strain and displacements [19, 20] are considered and exposed in Table 1.

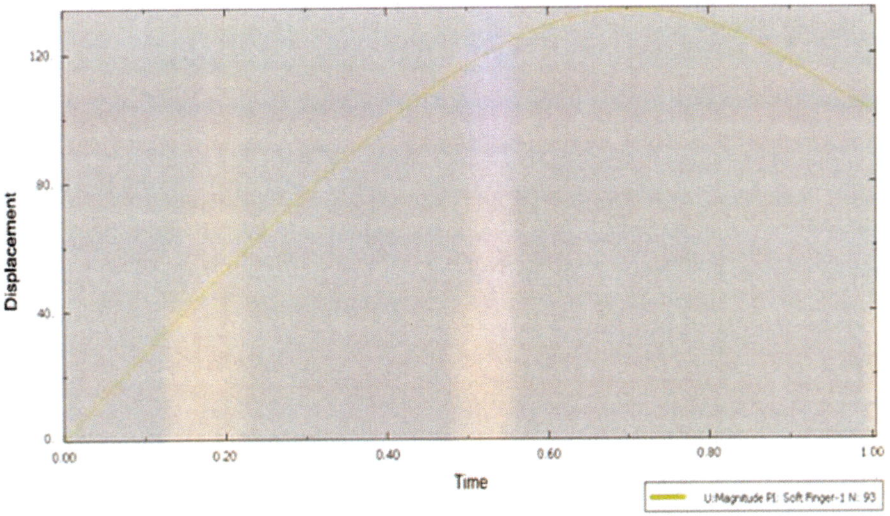


**Fig. 18** Response of 65 kPa pressure

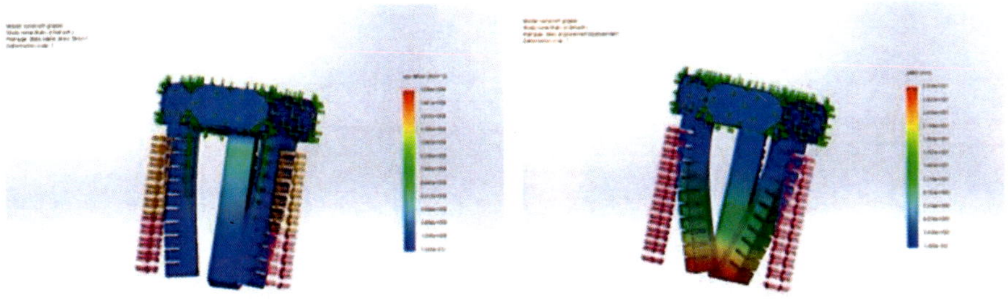

**Fig. 19** 4–6 kPa stress–strain–displacement

**Table 1** Gripper analysis

| S. No. | Load (kPa) | Stress | Strain | Displacement () |
|---|---|---|---|---|
| 1. | 1 | $5.313e^{+005}$ | $5.490e^{-003}$ | $4.728e^{+000}$ |
| 2. | 2 | $1.058e^{+006}$ | $1.095^{-002}$ | $9.504e^{+000}$ |
| 3. | 3 | $1.594e^{+006}$ | $1.653e^{-002}$ | $1.439e^{+001}$ |
| 4. | 4 | $2.314e^{+006}$ | $2.218e^{-002}$ | $1.933e^{+001}$ |
| 5. | 5 | $2.626e^{+006}$ | $2.786e^{-002}$ | $2.430e^{+001}$ |
| 6. | 6 | $3.212e^{+006}$ | $3.352e^{-002}$ | $2.926e^{+001}$ |

## 4 Conclusion

The gripper was designed in SolidWorks 2016 and analyzed in ABAQUS 6.14 software. It can reduce the time in food industry and helps in soft material handling factories. It can improve the efficiency of work and quantity of materials. From the above analysis, it is found that the pressure should be at maximum extent to be followed in the designed gripper which is around 55–70 kPa. If more pressure is applied, then it will lead to tare the material. In future, by tuning the control circuit, the grasp force can be analyzed.

## 5 Future Scope

The manufacturing technique other than FDM technology will produce better grasping performance in soft finger. The hollow structured supporting upper half will improve performance of gripper. The construction of two materials combined structure will improve grasping performance of gripper.

## References

1. Deimel Raphael, Brock Oliver (2013) A novel type of compliant and underactuated robotic hand for dexterous grasping. Int J Robot Res 35(1–3):161–185
2. Heidari H et al (2017) Design and fabrication of robotic gripper for grasping in minimizing contact force. Adv Space Res
3. Petković D et al (2016) Adaptive control algorithm of flexible robotic gripper by extreme learning machine. Robot Comput-Integr Manuf 37:170–178
4. Rodrigue H et al (2017) Curved shape memory alloy-based soft actuators and application to soft gripper. Compos Struct 176:398–406
5. Takizawa T et al (2018) Grasping force estimation in robotic forceps using a soft pneumatic actuator with a built-in sensor. Sens Actuators A: Phys
6. Tan N, Gu X, Ren H (2018) Design, characterization and applications of a novel soft actuator driven by flexible shafts. Mech Mach Theory 122:197–218
7. Wang Z, Torigoe Y, Hirai S (2017) A prestressed soft gripper: design, modeling, fabrication, and tests for food handling. IEEE Robot Autom Lett 2(4):1909–1916
8. Wang W et al (2016) Soft composite hinge actuator and application to compliant robotic gripper. Compos Part B: Eng 98:397–405
9. Welcome DE et al (2016) Tool-specific performance of vibration- reducing gloves for attenuating fingers-transmitted vibration. Occup Ergon 13(1):23–44
10. Yang H et al (2017) A novel pneumatic soft sensor for measuring contact force and curvature of a soft gripper. Sens Actuators A: Phys 266:318327
11. Hirose S, Umetani Y (1978) The development of soft gripper for the versatile robot hand. Mech Mach Theory 13(3):351–359
12. Hirai S et al (2014) Design and control of a bio-inspired soft wearable robotic device for ankle–foot rehabilitation. Bioinspir Biomim 9(1):016007
13. Rosati G, Minto S, Oscari F (2017) Design and construction of a variable-aperture gripper for flexible automated assembly. Robot Comput-Integr Manuf 48:157–166

14. Kim G-S (2007) Development of a three-axis gripper force sensor and the intelligent gripper using it. Sens Actuators A: Phys 137(2):213–222
15. Walsh CJ, Galloway KC, Polygerinos P, Wood RJ (2013) Mechanically programmable bend radius for fiber-reinforced soft actuators. IEEE
16. Jain RK et al (2015) Design and manufacturing of mobile micro manipulation system with a compliant piezoelectric actuator based micro gripper. J Manuf Syst 35:76–91
17. Dong H et al (2018) Geometric design optimization of an under- actuated tendon-driven robotic gripper. Robot Comput-Integr Manuf 50:80–89
18. Garces-Schroder M et al (2015) Micro-gripper: a new concept for a Monolithic single-cell manipulation device. Sens Actuators A: Phys 236:130–139
19. Trimmer B, Laschi C, Kim S (2013) A bioinsipred revolution in robotics. Trends Biotechnol 31(5):287–294
20. Bemfica JR et al (2013) Mechatronic design of a three-fingered gripper for underwater applications. IFAC Proc 46(5):307–312